ELECTROCHEMICAL METHODS for NEUROSCIENCE

ELECTROCHEMICAL METHODS for NEUROSCIENCE

Edited by

Adrian C. Michael
University of Pittsburgh
Pennsylvania

Laura M. Borland
University of Pittsburgh
Pennsylvania

CRC Press
Taylor & Francis Group
Boca Raton London New York

CRC Press is an imprint of the
Taylor & Francis Group, an informa business

CRC Press
Taylor & Francis Group
6000 Broken Sound Parkway NW, Suite 300
Boca Raton, FL 33487-2742

International Standard Book Number-13: 978-0-8493-4075-8 (Hardcover)

Library of Congress Cataloging-in-Publication Data

Electrochemical methods for neuroscience / [edited by] Adrian C. Michael & Laura M. Borland.
p. ; cm. -- (Frontiers in neuroengineering series ; 1)
Includes bibliographical references and index.
ISBN-13: 978-0-8493-4075-8 (hardcover : alk. paper)
ISBN-10: 0-8493-4075-6 (hardcover : alk. paper)
1. Neurochemistry. 2. Electrochemical analysis. I. Michael, Adrian C. II. Borland, Laura M. III. Title. IV. Series.
[DNLM: 1. Electrochemistry--methods. 2. Neurochemistry--methods. 3. Biosensing Techniques--methods. 4. Neurotransmitter Agents--metabolism. QD 115 E379 2007]

QP356.3.E44 2007
612.8'042--dc22 2006031628

Visit the Taylor & Francis Web site at
http://www.taylorandfrancis.com

and the CRC Press Web site at
http://www.crcpress.com

Series Preface

The Frontiers in Neuroengineering series presents the insights of experts on emerging experimental techniques and theoretical concepts that are or will be at the vanguard of neuroscience. Books in the series cover topics ranging from methods to investigate apoptosis to modern techniques for neural ensemble recordings in behaving animals. The series also covers new and exciting multidisciplinary areas of brain research, such as computational neuroscience and neuroengineering, and describes breakthroughs in fields like insect sensory neuroscience, primate audition, and biomedical engineering. The goal is for this series to be the reference that every neuroscientist uses in order to get acquainted with new methodologies in brain research. These books can be given to graduate students and postdoctoral fellows when they are looking for guidance to start a new line of research.

Each book is edited by an expert and consists of chapters written by the leaders in a particular field. Books are richly illustrated and contain comprehensive bibliographies. Chapters provide substantial background material relevant to the particular subject. Hence, they are not the usual type of method books. They contain detailed "tricks of the trade" and information as to where these methods can be safely applied. In addition, they include information about where to buy equipment and web sites helpful in solving both practical and theoretical problems. Finally, they present detailed discussions of the present knowledge of the field and where it should go.

We hope that, as the volumes become available, the effort put in by us, the publisher, the book editors, and the individual authors will contribute to the further development of brain research. The extent to which we achieve this goal will be determined by the utility of these books.

Sidney A. Simon, Ph.D.
Miguel A.L. Nicolelis, M.D., Ph.D.
Series Editors

Preface

Electrochemical methods of analysis are more frequently playing a central role in achieving one of the prime objectives of neuroscience research, the understanding of the role of specific neurotransmitters in normal brain function and the pathophysiology of central nervous system disorders. About thirty five years ago, Ralph N. Adams at the University of Kansas took the first step of implanting a carbon microelectrode into the brain of a rat in an effort to measure the extracellular concentration of catecholamine neurotransmitters. Subsequent years were spent sorting out the meanings of electrochemical responses observed under various experimental conditions and reliable identification of what was being detected was accomplished by the time that Jay Justice summarized the field in his 1987 volume *Voltammetry in the Neurosciences*. Technical progress continued through the 1990s to this day, leading to advances in the contribution of electrochemical methods to the understanding of brain function. The efforts of numerous research groups, many represented in the chapters of this book, have enhanced the sensitivity, selectivity and spatio-temporal resolution of electrochemical techniques to the point that they now provide information that is not accessible by any other means. The objective of this book is to provide the interested reader with an updated summary of the present, albeit still evolving, state of the art and to establish the basis upon which to reach sound decisions about incorporating electrochemical techniques into ongoing or newly emerging research programs in the neuroscience disciplines.

We are indebted to many individuals without whom this volume would not have reached fruition. We are certainly grateful to the series editors, Sid Simon and Miguel Nicolelis, for encouraging us to embark upon this project. Sid's enthusiasm is unrelenting, to say the least! We are also grateful to Barbara Norwitz and Jill Jurgensen of Taylor & Francis for their assistance, encouragement, and patience. Finally, it goes without saying that this volume would not have been completed were it not for the willing efforts of the contributing authors. We apologize to the numerous groups making invaluable contributions to the field whom we were unable to invite to contribute a chapter. It was not our objective to cover the field exhaustively, but rather to provide a sampling that would introduce and orient the newcomer to the fundamental aspects and applications of electrochemistry in the neurosciences.

Adrian C. Michael
Laura M. Borland

Editors

Adrian C. Michael began his career in electrochemistry as a graduate student in the laboratory of Joseph B. Justice, Jr., in the Department of Chemistry at Emory University in Atlanta, Georgia. He held postdoctoral positions with R. Mark Wightman in the Department of Chemistry at Indiana University in Bloomington, Indiana, and with Adam Heller in the Department of Chemical and Petroleum Engineering at the University of Texas at Austin, Texas. He has been a member of the Department of Chemistry and the Center for Neuroscience at the University of Pittsburgh since 1990. His research group applies voltammetry to in vivo measurements of dopamine and has developed enzyme-modified microelectrodes for the in vivo amperometric detection of glutamate. He is an active member of the Pittsburgh Conference on Analytical Chemistry and Applied Spectroscopy, which organizes the annual Pittcon Conference and Exposition. He is also an active member of the Society for Analytical Chemists of Pittsburgh.

Laura M. Borland obtained her Ph.D. in the Department of Chemistry at the University of Pittsburgh under the direction of Adrian Michael. At the time this biography was prepared, she was a postdoctoral fellow in the laboratory of Nancy Allbritton at the University of California at Irvine, California.

Contributors

Andy Alexander
Department of Chemistry
Indiana University
Bloomington, Indiana

Guillermo Alvarez de Toledo
Department of Physiology and
Biophysics School of Medicine
University of Seville
Seville, Spain

Christian Amatore
Département de Chimie
UMR Centre National de la Recherche
Scientifique and ENS-UPMC 8640
Ecole Normale Supérieure
Paris, France

Anne Milasincic Andrews
Department of Veterinary and
Biomedical Sciences
Penn State Neuroscience Institute
Huck Institutes of the Life Sciences
The Pennsylvania State University
University Park, Pennsylvania

Stéphane Arbault
Département de Chimie
UMR Centre National de la Recherche
Scientifique and ENS-UPMC 8640
Ecole Normale Supérieure
Paris, France

Marat V. Avshalumov
Departments of Neurosurgery and
Physiology & Neuroscience
New York University School of Medicine
New York, New York

Marianne Benoit-Marand
Centre National de la Recherche
Scientifique and UMR 5541
Université Victor Segalen Bordeaux 2
Bordeaux, France

Laura M. Borland
Department of Physiology and Biophysics
University of California, Irvine
Irvine, California

Amanda J. Bressler
Department of Chemistry
The Pennsylvania State University
University Park, Pennsylvania

Jason J. Burmeister
Morris K. Udall Parkinson's Disease
Research Center of Excellence
Center for Sensor Technology
University of Kentucky College of Medicine
Lexington, Kentucky

Ann-Sofie Cans
Department of Chemistry
The Pennsylvania State University
University Park, Pennsylvania

Joseph M. Casto
Cellular and Integrative Physiology Section
Department of Biological Sciences
Illinois State University
Normal, Illinois
and
Central Illinois Neuroscience Foundation
Bloomington, Illinois

Stephanie J. Cragg
University Department of Pharmacology
Oxford University
Oxford, United Kingdom

Lynette C. Daws
Department of Physiology
The University of Texas Health
Sciences Center
San Antonio, Texas

B. Keith Day
Department of Anatomy and Neurobiology
Morris K. Udall Parkinson's Disease
Research Center of Excellence
Center for Sensor Technology
University of Kentucky College of Medicine
Lexington, Kentucky

Gregor Dernick
F. Hoffmann-La Roche Ltd.
Basel, Switzerland

Robert Ensman
Ensman Instrumentation and
Department of Chemistry
Indiana University
Bloomington, Indiana

Daniel Eves
Department of Chemistry
The Pennsylvania State University
University Park, Pennsylvania

Andrew G. Ewing
Department of Chemistry, Integrated
Biosciences Program, and
Department of Neural and
Behavioral Sciences
The Pennsylvania State University
University Park, Pennsylvania

Paul A. Garris
Cellular and Integrative Physiology Section
Department of Biological Sciences
Illinois State University
Normal, Illinois

Greg A. Gerhardt
Department of Anatomy and Neurobiology
Morris K. Udall Parkinson's Disease
Research Center of Excellence
Center for Sensor Technology
University of Kentucky College of Medicine
Lexington, Kentucky

François Gonon
Centre National de la Recherche
Scientifique and UMR 5541
Université Victor Segalen Bordeaux 2
Bordeaux, France

Phillip G. Greco
Cellular and Integrative Physiology Section
Department of Biological Sciences
Illinois State University
Normal, Illinois

Joshua M. Gulley
Department of Psychology and Neuroscience
Program
University of Illinois at Urbana-Champaign
Champaign, Illinois

Kevin N. Hascup
Department of Anatomy and Neurobiology
Morris K. Udall Parkinson's Disease
Research Center of Excellence
Center for Sensor Technology
University of Kentucky College of Medicine
Lexington, Kentucky

Byron A. Heidenreich
Department of Psychology
Illinois State University
Normal, Illinois

Greg Howes
Cellular and Integrative Physiology Section
Department of Biological Sciences
Illinois State University
Normal, Illinois
and
Central Illinois Neuroscience Foundation
Bloomington, Illinois

Sabina Hrabětová
Department of Physiology and Neuroscience
New York University School of Medicine
New York, New York

Peter Huettl
Department of Anatomy and Neurobiology
Morris K. Udall Parkinson's Disease
Research Center of Excellence
Center for Sensor Technology
University of Kentucky College of Medicine
Lexington, Kentucky

Carrie E. John
Department of Physiology and Pharmacology
Wake Forest University School of Medicine
Winston-Salem, North Carolina

David A. Johnson
Pinnacle Technology, Inc.
Lawrence, Kansas

Sara R. Jones
Department of Physiology and Pharmacology
Wake Forest University School of Medicine
Winston-Salem, North Carolina

Gaynor A. Larson
Department of Pharmacology
University of Colorado at Denver and
Health Sciences Center at Fitzsimons
Aurora, Colorado

Manfred Lindau
School of Applied and Engineering Physics
Cornell University
Ithaca, New York

Marc Maxson
Integrated Biosciences Program
The Pennsylvania State University
University Park, Pennsylvania

Mark A. Messerli
BioCurrents Research Center
Program in Molecular Physiology
Marine Biology Laboratory
Woods Hole, Massachusetts

Adrian C. Michael
Department of Chemistry
University of Pittsburgh
Pittsburgh, Pennsylvania

Charles Nicholson
Department of Physiology and Neuroscience
New York University School of Medicine
New York, New York

Justin R. Nickell
Department of Anatomy and Neurobiology
Morris K. Udall Parkinson's Disease
Research Center of Excellence
Center for Sensor Technology
University of Kentucky College of Medicine
Lexington, Kentucky

Jyoti C. Patel
Departments of Neurosurgery and
Physiology & Neuroscience
New York University School of Medicine
New York, New York

Xiomara A. Perez
Department of Chemistry
The Pennsylvania State University
University Park, Pennsylvania

John Poehlman
Department of Chemistry
Indiana University
Bloomington, Indiana

Francois Pomerleau
Morris K. Udall Parkinson's Disease
Research Center of Excellence
Center for Sensor Technology
University of Kentucky College of Medicine
Lexington, Kentucky

Sirinun Pongmaytegul
Cellular and Integrative Physiology Section
Department of Biological Sciences
Illinois State University
Normal, Illinois

Jorge E. Quintero
Department of Anatomy and Neurobiology
Morris K. Udall Parkinson's Disease
Research Center of Excellence
Center for Sensor Technology
University of Kentucky College of Medicine
Lexington, Kentucky

Stephen Rayport
Department of Psychiatry
Columbia University
New York, New York

George V. Rebec
Program in Neuroscience
Department of Psychological and
Brain Sciences
Indiana University
Bloomington, Indiana

Margaret E. Rice
Departments of Neurosurgery and Physiology & Neuroscience
New York University School of Medicine
New York, New York

Donita L. Robinson
Bowles Center for Alcohol Studies and Department of Psychiatry
University of North Carolina
Chapel Hill, North Carolina

Erin C. Rutherford
Department of Anatomy and Neurobiology
Morris K. Udall Parkinson's Disease Research Center of Excellence
Center for Sensor Technology
University of Kentucky College of Medicine
Lexington, Kentucky

Stefan G. Sandberg
Cellular and Integrative Physiology Section
Department of Biological Sciences
Illinois State University
Normal, Illinois

Richard H. Sanger
BioCurrents Research Center
Program in Molecular Physiology
Marine Biology Laboratory
Woods Hole, Massachusetts

Wolfgang Schuhmann
Analytical Chemistry–Elektroanalytik & Sensorik
Ruhr-Universität Bochum
Bochum, Germany

Albert Schulte
Department of Chemistry, Faculty of Science and Agriculture
The University of the West Indies
St. Augustine, Trinidad and Tobago

Michael J. Serpe
Department of Chemistry
World Precision Instruments, Inc.
Sarasota, Florida

Peter J. S. Smith
BioCurrents Research Center
Program in Molecular Physiology
Marine Biology Laboratory
Woods Hole, Massachusetts

Roland G. W. Staal
Department of Neurology
Columbia University
New York, New York

Marie-Francoise Suaud-Chagny
INSERM
Université Claude Bernard
Laboratoire de Neuropharmacologie et Neurochimie
Lyon, France

David Sulzer
Departments of Neurology, Psychiatry, and Pharmacology
Columbia University
New York, New York

Sarah Threlfell
University Department of Pharmacology
Oxford University
Oxford, United Kingdom

Glenn M. Toney
Department of Physiology
The University of Texas Health Sciences Center
San Antonio, Texas

R. Mark Wightman
Department of Chemistry
University of North Carolina
Chapel Hill, North Carolina

George S. Wilson
Department of Chemistry
University of Kansas
Lawrence, Kansas

Nathan Wittenberg
Department of Chemistry
The Pennsylvania State University
University Park, Pennsylvania

Hua Yang
Waters Corporation
Milford, Massachusetts

Nancy R. Zahniser
Department of Pharmacology and
Program in Neuroscience
University of Colorado at Denver and
Health Sciences Center at Fitzsimons
Aurora, Colorado

Xueji Zhang
Department of Chemistry
World Precision Instruments, Inc.
Sarasota, Florida

Table of Contents

1 An Introduction to Electrochemical Methods in Neuroscience

Laura M. Borland and Adrian C. Michael

CONTENTS

INTRODUCTION

Electrochemical methods comprise a collection of extremely useful measurement tools for neuroscience. A central feature of these methods is an electrode that provides a surface or interface where some form of a charge-transfer process occurs. This charge-transfer process gives rise to potentials and/or currents that can be measured and related either by theory or by calibration to the concentration of substances in the solution that bathes the electrode. Broadly speaking, these methods can be divided into two groups: measurements that do not involve current, known generally as potentiometric methods, and measurements that involve current flow at an electrode under potential control, known as amperometry, voltammetry, or polarography, depending upon the details of the experimental design. Amperometric and voltammetric methods are the main, but not exclusive, focus of this book. It is completely natural that neuroscientists should find electrochemical methods of value because the nervous system contains numerous electrochemically detectable targets including simple inorganic ions, catecholamine and indolamine neurotransmitters and their metabolites, glutamate, nitric oxide (NO), glucose, lactate, ascorbate, urate, hydrogen peroxide (H_2O_2), oxygen (O_2), and pH. Moreover, the high ionic strength of biological environments creates a perfect milieu for electrochemistry, which requires conductive media.

This chapter briefly introduces the history of amperometry and voltammetry in the neurosciences and briefly surveys their applications. Although not intended as a thorough explanation of the methods themselves, the chapter provides a tutorial and primer on their basic operating principles and points the interested reader to sources of more detailed information.

A Bit of Electrochemical History: Early Lessons in Selectivity

The foundations of the amperometric and voltammetric methods employed in neuroscience today find their roots in the work of Jaroslav Heyrovsky, who received the 1959 Nobel Prize in Chemistry for the discovery and development of polarographic methods of analysis. Heyrovsky discovered that it was possible to cause the reduction of organic molecules at a mercury electrode by the application of an appropriate voltage. The reduction gave rise to a current quantitatively related to the concentration of the substance being reduced. Furthermore, the position and shape of the polarographic wave, i.e., a plot of the observed current as a function of the applied potential, was useful for identifying the substance. These are the key principles underlying the use of amperometric and voltammetric methods in neuroscience today.

Ralph Adams, formerly of the University of Kansas, is generally credited with being the first to implant a carbon microelectrode into the brain of a rat with the objective of measuring the in vivo concentration of catecholamine neurotransmitters and their metabolites in the extracellular fluid (Kissinger, Hart, and Adams 1973; Adams 1976; Wightman et al. 1976). These early studies provided an early lesson in the importance of selectivity and the need to identify and confirm the chemical source of oxidation currents recorded in the brain since it was quickly discovered that catecholamines and their metabolites are not the only substances in the brain that will oxidize at carbon electrodes. Early in the 1980s, Mark Wightman and his colleagues devised what are now colloquially known as the "Five Golden Rules" of in vivo electrochemistry. These rules, which are still cited, are a list of criteria that must be met for claiming the certain identification of the substance detected during in vivo electrochemical measurements. These criteria are discussed fully in Justice's previous book on voltammetry in the neurosciences (Justice 1987) and elsewhere (Phillips and Wightman 2003).

The invention of the carbon fiber microelectrode by Gonon and colleagues was a milestone in the development of significant electrochemical methods in neuroscience (Ponchon et al. 1979; Gonon et al. 1980). These electrodes, which are still popular, could be electrochemically pretreated to allow resolution of the oxidation currents recorded in the brain into distinct peaks at different positions along the potential axis (Gonon et al. 1981; Gonon, Navarre, and Buda 1984). These studies demonstrated the first of numerous approaches to the development of a variety of selective electrochemical recording techniques that are currently used in the laboratories of the contributors to this volume. Another major milestone in the development of selective recording techniques was the introduction of carbon electrodes coated with the permselective polymer material called Nafion (Gerhardt et al. 1984), which is a sulfonated derivative of Teflon. The sulfonate groups ionize in wet Nafion to produce fixed anionic sites in the membrane. These fixed anionic sites limit the ability of anionic substances in the brain from reaching the electrode surface, which has proven very useful as a way to eliminate signal contributions from anionic substances such as ascorbate, urate, and the acidic metabolites of the monoamine neurotransmitters. Hence, Nafion increases the selectivity of the electrochemical techniques for cationic substances such as the monoamine neurotransmitters themselves. In addition, the Teflon-like nature of Nafion imparts the electrode surface with the nonstick characteristics that make Teflon itself of interest in cookware. Nafion seems to improve the biocompatibility of the electrode surface and diminish fouling of the electrodes during in vivo experiments (Turner, Harrison, and Rajotte 1991).

IN VIVO ELECTROCHEMISTRY: THE BENEFITS OF SIZE AND SPEED

The challenges associated with establishing the selectivity of in vivo electrochemical recordings encouraged many neuroscience laboratories to focus their efforts on the technique of microdialysis. Microdialysis is a sampling technique whereby small molecules are recovered from the extracellular fluid of the brain by passing a perfusion fluid through hollow fiber of dialysis membrane. Samples of the dialysate are analyzed by methods that involve chromatography or, more recently, capillary electrophoresis with a variety of detection strategies based on electrochemistry, fluorescence spectroscopy, or mass spectrometry, to name a few. The selectivity of these analytical tools is tremendous: no one has ever had reason to doubt the identification of a peak in the chromatogram or electropherogram of a dialysate sample. Given the availability of this alternative, superbly selective approach to in vivo neurochemical analysis, it is certainly appropriate to question the merits of the electrochemical techniques.

As the subsequent chapters in this volume will demonstrate, the merits of the electrochemical techniques lie in their ability to perform measurements that provide answers that are not presently accessible by microdialysis or any other measurement technique. Although microdialysis offers superior selectivity, electrochemical recording techniques offer high spatial and temporal resolution. Carbon fibers are available for the construction of electrodes with single-digit micrometer dimensions, which are more then twentyfold smaller than typical microdialysis probes. Due to their small dimensions, microelectrodes inflict minimal damage when they are implanted into living brain tissue (Peters et al. 2004), meaning they can be placed within micrometer distances of neuronal terminals (Venton et al. 2003; Peters et al. 2004). For this reason, electrochemical recordings suffer less diffusional distortion in the recording of dynamic events (Wightman et al. 1988) such as those associated with electrical stimulation procedures. High-speed, highly localized recordings have permitted, for example, the evaluation of the kinetics of a variety of transporters such as the dopamine transporter (Wightman and Zimmerman 1990; Garris et al. 1994; Wu et al. 2001, 2002; Garris et al. 2003; Michael et al. 2005) crucial in terminating the actions of dopamine on presynaptic and postsynaptic dopamine receptors in a number of brain regions. Other transporters have been the targets of electrochemical assays in recent years (Daws et al. 1998; Frazer and Daws 1998; Perez and Andrews 2005).

The skills honed in the refinement of techniques for monitoring electrically evoked dopamine release in the striatum and accumbens of the rat have evolved very recently into the ability to detect spontaneous, i.e., non-evoked, dopamine transients presumably associated with the naturally occurring electrical activity of dopaminergic fibers (Robinson, Heien, and Wightman 2002; Robinson et al. 2003; Cheer et al. 2004; Robinson and Wightman 2004; Roitman et al. 2004). These transients are virtually invisible to microdialysis measurements because of the temporal blurring associated with diffusion across the dialysis membrane. Hence, it is becoming increasingly apparent that the high spatiotemporal resolution of the electrochemical recordings provides unique information regarding the operational characteristics of intact neuronal systems in the living brain.

The high spatiotemporal resolution of electrochemical recording techniques are also proving powerful in a variety of measurement contexts outside the living brain. Neurochemical studies in tissue slices and synaptosomal suspensions generally lack temporal resolution due to the superfusion methods employed in these preparations; however, direct insertion of a microelectrode into a slice or suspension restores the ability to record dynamic events otherwise lost in conventional measurements (Schenk et al. 1983; Jones, Garris, and Wightman 1995; Jones et al. 1996; Avshalumov and Rice 2003; Avshalumov et al. 2003). Microelectrodes with dimensions in the single micron range are compatible with measurements at the cellular and subcellular domains. Microelectrodes and microsensors have been used in conjunction with numerous electrochemical techniques to monitor the release of substances from single cells (Boudko et al. 2001; Kumar et al. 2001; Smith and Trimarchi 2001). Quantal size and vesicular volume have also been investigated using microelectrodes (Kozminski et al. 1998; Pothos et al. 1998, 2000;

Edwards and Sulzer 2000; Colliver et al. 2000). High-speed recordings of adrenaline provided the first direct observation of exocytosis (Leszczyszyn et al. 1991; Mosharov et al. 2003). The value of the high spatiotemporal resolution offered by these electrochemical techniques is unquestionable.

New electrochemical methods and designs continue to be developed, producing better, smaller, and faster electrodes, improving this already versatile technique, while simultaneously discovering new features and functions of the brain and other neurochemical environments. Nearly 30 years after Adams and Gonon first began their in vivo experiments, thousands of papers demonstrate the versatility and functionality of electrochemistry for in vivo detection, with the field continuing to grow, expand, and explore new neurochemical arenas.

THE SCOPE OF ELECTROCHEMISTRY IN THE NEUROSCIENCES

Many neurochemical species have been studied using electrochemistry. The electrochemically active catecholamines, including dopamine, norepinephrine, and their metabolites, are perhaps the most-studied neurochemicals. In physiological media (pH 7.4) or in vivo, the catechol moiety of the catecholamines is oxidized to an *ortho*-quinone, as shown in Figure 1.1. Serotonin and its tryptamine-derived metabolites are also electro-active and have been studied in vivo with

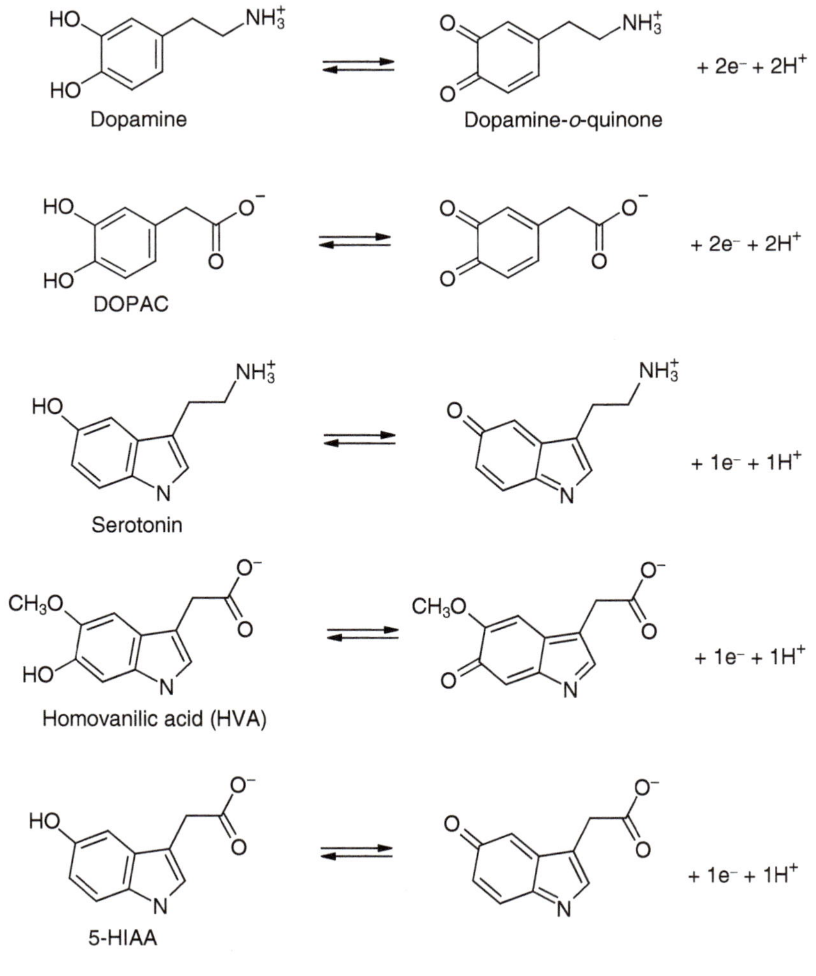

FIGURE 1.1 Electroactive neurotransmitters and their metabolites.

voltammetry and microelectrodes (Daws et al. 1998). The phenollic portion of serotonin and its metabolites is also oxidized, to form a ketone, as shown in Figure 1.1. Numerous other electrochemically detectable substances encountered in brain tissues, including ascorbic acid, NO, O_2, and H_2O_2, are easily measured with microelectrodes, with their electrochemical schemes represented in Figure 1.2.

The ascorbate anion is a particularly interesting neurochemical. Levels of ascorbate in the brain range from 200 to 400 μ*M*, making it one of the most prevalent molecules in the brain. Ascorbate is oxidized at potentials close to that of dopamine and norepinephrine, so avoiding the detection of ascorbate was one of the early selectivity challenges mentioned earlier. The interferant effects of ascorbate on the electrochemical detection of dopamine can be controlled by increasing the scan rate of the applied potential. Interferant signals from the oxidation of ascorbate are decreased by increasing the scan rate because of the slow rate of charge transfer of ascorbate at the electrode surface. Ascorbate can also interfere with the reduction of dopamine-*o*-quinone back to dopamine, resulting in an increase in the dopamine signal at the electrode, but this effect can also be diminished by using a fast scan rate. Despite its reputation among electrochemists as an interferent in the in vivo detection of monoamines, electrochemical measurements have illuminated ascorbate's role as a neuromodulator in brain function (Christensen, Wang, and Rebec 2000; Rebec and Wang 2001).

The electrochemical methods used in the neurosciences are not exclusive to the detection of neurotransmitters and their metabolites but include the detection and measurement of small molecules and biologically significant ions as well. Ion-selective electrodes, such as pH electrodes, are often used in vitro but can be miniaturized for use in various in vivo neurochemical applications (Boudko et al. 2001). The Clark O_2 sensor is a small-molecule, rather than an ion-selective, electrode that has applicability in vivo. O_2 is reduced to produce a measurable current at the electrode surface. The O_2 sensor is constructed simply of a membrane-coated platinum wire electrode; however carbon-fiber cylinder electrodes have also been used, in place of platinum, to detect O_2 (Kennedy, Jones, and Wightman 1992). NO is another small molecule with relevance in vivo. NO sensors, which strongly resemble O_2 sensors, have been constructed to study and

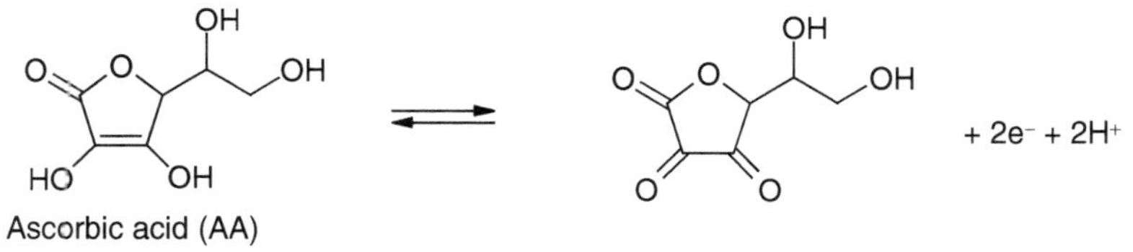

$$NO \longrightarrow NO^+ + 1e^-$$

Nitric oxide

$$O_2 + 4H^+ + 4e^- \longrightarrow 2H_2O$$

Oxygen

$$H_2O_2 + 2H^+ + 2e^- \longrightarrow 2H_2O$$

Hydrogen peroxide

FIGURE 1.2 Additional electroactive species found in the brain—ascorbic acid, nitric oxide, oxygen, and hydrogen peroxide.

measure NO in a number of different environments (Zhang et al. 2000, 2002; Kumar et al. 2001). H_2O_2 is another small molecule of interest to neuroscientists. This molecule is easily detected at carbon-fiber or platinum electrodes; however, it is often oxidized at the electrode in the same potential window as many other electroactive species. Consequently, a more selective electrode or sensor is required to detect and measure H_2O_2 in vivo. Luckily, naturally occurring enzymes selective for peroxide are available and can be coupled with electrodes to make a H_2O_2 sensor (Kulagina and Michael 2003).

Electrode selectivity for electroactive compounds like peroxide and ascorbic acid can also be improved with the addition of electrode coatings and additives. Electrode coatings include electro-deposited polymers (1, 2 diaminobenzene), electronically conducting polymers (polypyrrole), and polyelectrolytes (Nafion). Polymers and polyelectrolyte films, along with a number of different conducting and nonconducting electropolymerizable coatings, can be used to capture or immobilize enzymes on an electrode surface, creating a species-selective sensor (Kulagina and Michael 2003). Enzyme-based electrochemical microsensors are also used to increase selectivity for electroactive analytes, like peroxide and AA, and to create novel detection schemes for non-electroactive compounds, like glucose, glutamate, choline, and lactate (Gregg and Heller 1990; Heller 1992; Barlett and Cooper 1993; Kulagina, Shankar, and Michael 1999; Cui, Kulagina, and Michael 2001; Kulagina and Michael 2003; Wilson and Gifford 2005; Day et al. 2006; McMahon et al. 2006).

Electrochemical measurements of non-electroactive species, such as glucose, glutamate, lactate, and choline, are conducted using electrodes constructed with specialized coatings and additives, which selectively generate electrochemical signals in response to a particular analyte. Oxidase enzymes are the most commonly encapsulated or immobilized enzymes for use in electrochemical sensors, particularly in the detection of glucose, glutamate, choline, and lactate. These enzymes catalyze the conversion of a particular substrate between the reduced or oxidized state, which can then be detected at the electrode surface (Bartlett and Cooper 1993). These *redox* enzymes typically undergo the following reaction scheme to generate an electrochemically sensitive species, which is monitored at the electrode surface:

$$\text{substrate(RED)} + \text{enzyme(OX)} \rightarrow \text{product(OX)} + \text{enzyme(RED)},$$
$$\text{enzyme(RED)} + \text{co-substrate(OX)} \rightarrow \text{enzyme(OX)} + \text{co-product(RED)}.$$

Oxidase enzymes use O_2 as the primary cosubstrate, generating H_2O_2 as the final co-product, which can be electrochemically oxidized at the electrode surface. Ideally, the rate of H_2O_2 production is directly related to the concentration of the analyte detected by the enzymes at the electrode. For example, the oxidation of glutamate and subsequent oxidation of H_2O_2 is as follows:

$$\text{glutamate} + \text{GluOx(OX)} \rightarrow \alpha\text{-ketoglutarate} + NH_3 + \text{GluOx(RED)},$$
$$\text{GluOx(RED)} + O_2 \rightarrow \text{GluOx(OX)} + H_2O_2,$$
$$\text{at electrode surface}:\ H_2O_2 \rightarrow 2H^+ + O_2 + 2e^-.$$

In this type of electrochemical sensor, the redox enzymes are immobilized in thin films to decrease the distance that H_2O_2 has to travel to the electrode surface to be oxidized. Thin films assist in limiting peroxide diffusion or superfluous reaction of H_2O_2, which would result in low sensor sensitivity. The response time of the electrode also decreases with decreasing film thickness, as the diffusion distance of the substrate into the film to react with the enzyme decreases (Day et al. 2006). Thin films of encapsulated or immobilized redox enzymes are often-used electrochemical methods of detection that could be easily and readily extended for use in the neurosciences.

Another type of electrochemical sensor utilizes enzymes for the detection of different species but does not require immobilization of the enzyme on the electrode surface. Instead, this type of sensor "wires" enzymes to the electrode using a cross-linked polymeric redox macromolecule composed of a poly(vinylpyridine) complexed to Os(bipyridine)$_2$Cl (Gregg and Heller 1990; Heller 1992; Kulagina, Shankar, and Michael 1999; Kulagina and Michael 2003). The redox polymer forms a three-dimensional gel that incorporates, but does not immobilize, enzymes, while at the same time allowing for rapid in-and-out diffusion of substrates and products. Electrons transferred during the reduction of the enzyme are rapidly wired through the redox centers of the polymer to the electrode surface and are detected as changes in the current response. Electrochemical sensors of this nature can be constructed to detect and measure different non-electroactive species simply by changing the functional enzyme from glutamate oxidase to choline oxidase. One type of enzyme-wired electrode for the detection of glutamate utilizes a cascade consisting of a glutamate oxidase and horseradish peroxidase (HRP). HRP is used in this sensor to reduce the H_2O_2 produced by the glutamate oxidase reaction with glutamate. The HRP subsequently oxidizes the osmium redox polymer, generating a current change in response to glutamate detection (Kulagina, Shankar, and Michael 1999).

It should be mentioned that electrochemistry is used in the neurosciences in many capacities, frequently as an online detector for sampling and separation devices like microdialysis and high performance liquid chromatography (HPLC); however, this introduction will focus on the utility of this technique as an in vivo analytical device. The benefits of using microelectrodes coupled to voltammetry in vivo include real-time, microsecond recording and improved spatial resolution. The small size of microelectrodes also limits implantation-induced damage to the brain tissue, while still allowing for extremely fast recording of analytes in the nanomolar concentration range. The numerous features that make microelectrodes excellent tools for in vivo electrochemical applications in neuroscience will be described in more detail later in this introduction. This introduction will concentrate on the applications of electrochemical methods in vivo, but with the awareness that microelectrodes and electrochemical detection are also widely used in other capacities in the neurosciences.

ELECTROCHEMISTRY FUNDAMENTALS

This section provides a very brief introduction to the principles of the electrochemical methods of analysis by way of introduction and orientation to the subsequent chapters in this volume. The fundamental aspects of the numerous electrochemical methods of analysis are fully explained in a number of excellent resources, which the interested reader is encouraged to review for a deeper understanding. It should be emphasized that these methods are not mysterious. Instead, they have evolved steadily from the pioneering and Nobel Prize-winning work of Heyrovsky, and they find innumerable applications in many fields of science and technology. In fact, the measurement of pH, which is most often performed electrochemically, is reputed to be the most frequently performed analytical measurement, outnumbering all others combined. At present, sales of home glucose testers generate more revenue than the sales of all other devices for analytical determinations combined. The home glucose tester is a gold microelectrode that detects H_2O_2 generated by glucose oxidase immobilized in a thin polymer film attached to the electrode.

The best sources of more detailed information regarding these techniques are the textbooks used in courses on analytical chemistry. Virtually all of these textbooks (Skoog, Holler, and Nieman 1998, Harris 2003; Christian 2004) have numerous chapters devoted to electroanalytical methods. One particularly noteworthy textbook is entirely devoted to electrochemical methods and is more advanced, intended for use in semester-long, graduate-level chemistry courses (Bard and Faulkner 2001). The remainder of this chapter introduces the principles of amperometry and voltammetry, i.e., methods that involve the flow of current at electrodes under potential control. Most of the subsequent chapters in this book discuss the fundamental aspects of the particular electrochemical technique used in the authors' hands.

CHRONOAMPEROMETRY

During an electrochemical measurement, the current that provides information about the composition of the solution bathing the electrode is derived from the oxidation and reduction of solution components at the surface of the electrode. When a molecule undergoes oxidation at the electrode surface, electrons flow from the molecule, into the electrode material, and up a contact wire into the terminal of an ammeter, a device that measures and reports the magnitude of the current. When a molecule undergoes reduction, the electrons flow in the opposite direction, i.e., from the electrode material to the molecule. Thus, it is completely appropriate to view the electrode as analogous to either an oxidizing agent or a reducing agent, depending on the direction of current flow. The deciding factor in whether the electrode behaves as an oxidant or reductant at any point in time is the electrode potential, which is under the control of the experimenter. When the electrode potential is adjusted towards negative values, electrons accumulate on the surface of the electrode where they experience coulombic repulsion. One way to escape this repulsion is to leave the electrode, i.e., to enter the adjacent solution. Thus, negative potentials make the electrode a stronger reducing agent. If the potential is adjusted to more positive values, then the electrode begins to attract electrons from its surroundings so that it acts as an oxidizing agent.

How positive does the electrode potential have to be in order to carry out an oxidation reaction? This depends on the nature of the substance undergoing the reaction. Each substance has its own formal potential, the potential at which the substance would be 50% oxidized and 50% reduced if sufficient time is allowed for the reaction to reach equilibrium. Electrochemical methods do not bother to wait for equilibrium to be reached, and instead, measure the current that flows as the reaction moves toward equilibrium; Therefore, faster reactions produce more current than slower ones. Differences between the formal potential values and the kinetics of different reactions give rise to unique electrochemical responses from different substances that turn out to be very useful for confirming the identity of detected substances.

Diffusion of molecules in the solution adjacent to the electrode plays a central role in determining the magnitude of the current observed during electrochemical measurements. In order to undergo electrochemical oxidation or reduction reactions, molecules in the bathing solution must be in molecular-scale contact with the electrode material. Free electrons have virtually no existence in aqueous solutions, so transfer of electrons between a molecule and electrode requires physical contact between them. Hence, the speed at which molecules diffuse to the electrode is also an important determinant in the rate of electrochemical oxidation and reduction reactions. Diffusion is a mode of mass transport that occurs as a result of the tendency of molecules in fluids to undergo continuous random motion, roughly equivalent to the Brownian motion of small particles. In the presence of a concentration gradient, random motion carries more molecules from regions of high concentration to low concentration than vice versa, resulting in net-mass transport down the concentration gradient. In electrochemical systems, the concentration gradient of interest is that caused by oxidation and reduction reactions at the electrode surface. The reaction lowers the concentration of the substance at the surface of the electrode, which initiates diffusion from the surroundings towards the electrode. Ultimately, the rate of diffusion and the rate of the electrochemical reaction are mutually interdependent.

Diffusion is described by Fick's laws. Fick's first law relates the diffusion flux, J, to the steepness of the concentration gradient:

$$J = -D\frac{\partial C}{\partial x}, \tag{1.1}$$

where D is the diffusion coefficient, C is the concentration, and x is distance. In Equation 1.1, the units of J are moles $cm^{-2}\,s^{-1}$. Converting this quantity to the corresponding current, coulombs per second, requires multiplication by the area of the electrode and a factor for the number of coulombs per mole of reactant:

$$i = nFAD\frac{\partial C}{\partial x}\bigg|_{x=0}, \tag{1.2}$$

where i is the current, n is the number of electrons transferred per molecule of reactant, F is Faraday's constant, and A is the surface area of the electrode. In Equation 1.2, the subscript $x = 0$ specifies that it is the concentration gradient at the surface of the electrode that determines the current.

Perhaps the best way to appreciate Equation 1.2 is to consider its application to an example of an electrochemical measurement. Chronoamperometry is one of the simplest methods to describe, and do. In this method, the electrode is initially held at some rest potential that does not cause oxidation or reduction of molecules in the bathing solution. To perform the measurement, the potential is suddenly stepped to a new potential to initiate the electrochemical reaction (Figure 1.3a). Because of the step-wise change in the electrode potential, chronoamperometry is a member of the so called potential pulse methods. In some cases, it is possible to step the potential to a value that causes the immediate and complete reaction of any molecule of the target substance in contact with the electrode. Hence, the concentration of the substance at the electrode surface goes immediately to zero, creating a concentration gradient between the electrode surface and the bathing solution that initiates diffusion of the substance towards the electrode (Figure 1.3b).

The concentration gradient evolves over time according to Fick's second law of diffusion:

$$\frac{\partial C}{\partial t} = \frac{\partial J}{\partial x} = D\frac{\partial^2 C}{\partial x^2}. \tag{1.3}$$

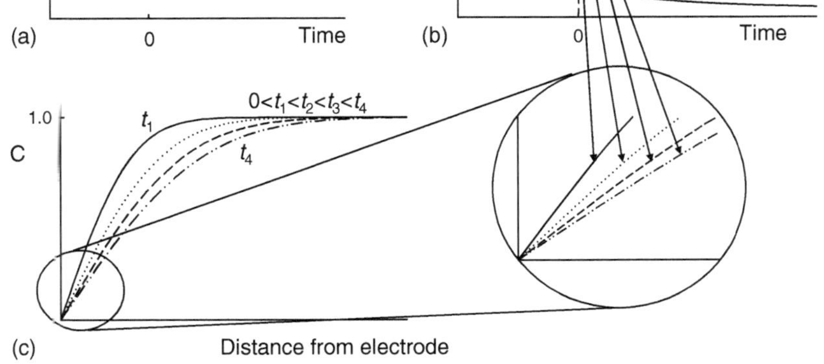

FIGURE 1.3 (a) Diagram of the step potential used in chronoamperometry. (b) Current vs. time response generated in a chronoamperometric experiment. (c) Concentration profile for times after the start of the chronoamperometric experiment (Circled Inset). The decrease in the current response during the experiment corresponds to a decrease in the slope of the concentration, as described by the Cottrell equation (Equation 1.4).

As differential equations go, Equation 1.3 is quite simple to solve, and solutions under many conditions are known. In the case of the chronoamperometry experiment under discussion, the solution is very well known (Figure 1.3c). As the reaction at the electrode continues, the solution next to the electrode becomes further depleted, meaning that with the passage of time, the concentration gradient at the electrode surface becomes less and less steep. Hence, the current decreases, as it turns out, with the reciprocal of the square root of time. The current-time relationship for this experiment is known as the Cottrell equation:

$$i(t) = \frac{nFA\sqrt{D}C}{\sqrt{\pi t}}, \tag{1.4}$$

where all the variables have the same meaning as before.

It is necessary to bear in mind that there might be more than one contribution to the total current observed during an electrochemical measurement. Equation 1.4 describes only that contribution to the current associated with the oxidation or reduction reaction, sometimes called the Faradaic current in deference to Michael Faraday, who first related electrical charge to mass. However, when the potential is suddenly stepped to a new value, there is a second source of current associated with the process of actually changing the potential of the electrode surface. This component is called the charging current and usually decays exponentially after a potential step. A quick glance at Figure 1.3c might lead one to suspect that measuring the current at short times after the potential step would provide the greatest sensitivity. However, this is not so because at very short times the Faradaic current is overwhelmed by the charging current, so typically one waits for the charging current to decay before making the measurement. With the very small electrodes used in most neuroscience applications, the charging current decays within a few milliseconds of the potential step, which is one of the reasons that fast electrochemical recordings are possible.

A Comment about Charging Currents

It might seem odd to emphasize the charging current contribution so early in the explanation of electrochemical methods, but, it turns out that the charging current has a very important consequence for the design of electrochemical measurements for neuroscience applications. In part because of the charging current, all electrochemical measurements have a nonzero background signal. This makes it almost impossible to use electrochemical methods to measure static concentrations of target substances because it is usually impossible to know what fraction of the current being measured is due to the charging current and what fraction is due to the presence of the target substance. It is for this reason that electrochemical methods have come to be regarded as best suited to monitoring dynamic events such as the change in extracellular dopamine concentration during electrical stimulation of a dopaminergic axons or the change in adrenaline associate with the fusion of vesicle with the surface of a chromaffin cell. Thus, electrochemical methods are not well suited to measuring the resting or basal concentration of substances in any preparation.

Other Potential Step Methods

An extension of the chronoamperometric technique is normal pulse voltammetry (NPV), which involves the application of a series of potential pulses each to sequentially different values (Figure 1.4). As in chronoamperometry, the current produced during each step is proportional to the analyte concentration, but it also reflects how the applied potential during each pulse controls the electrochemical reaction so that the voltammogram, i.e., a plot of the current vs. the potential applied during each pulse, is diagnostic of the substance being detected. The voltammogram associated with NPV is sigmoidal in shape, with the Faradaic current starting out at a small value and increasing to a plateau once the potential pulse reaches a value that causes complete

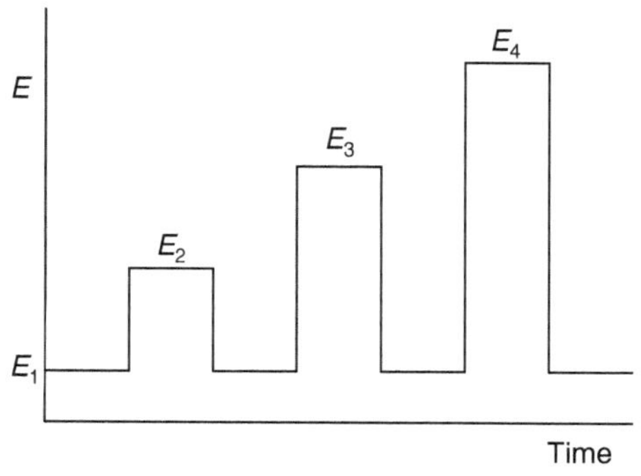

FIGURE 1.4 Diagram of the potential waveform used in NPV.

oxidation or reduction of the target substance. The current reaches its half-maximal value at the formal potential.

Differential pulse voltammetry (DPV) is another technique that provides increased sensitivity and more efficient differentiation and resolution of different species. The pulse waveform is different from that used in NPV in that the potential between each pulse is not returned all the way back to the same rest potential (Figure 1.5a). This has the effect of diminishing the charging current between each pulse and allowing for slightly quicker recording of the voltammogram. The current is sampled immediately before each potential pulse is applied (i_a) and immediately before each pulse ends (i_b), and it is the difference between these current responses ($i_b - i_a$), that is used to produce the voltammogram (Figure 1.5b), hence, the term differential in the name of the technique. DPV produces a peaked voltammogram, roughly corresponding to the derivative of the NPV voltammogram; this peaked voltammogram can improve both the qualitative and quantitative aspects of the technique.

Cyclic Voltammetry

Chronoamperometry, NPV, and DPV are potential pulse techniques. Potential sweep methods are also available. Cyclic voltammetry (CV) is widely used in neuroscience applications. As the name implies, the potential is swept from an initial potential to a final potential and then returned to the initial potential, usually at the same sweep rate (Figure 1.6a). The current measured continuously during the sweep is reported against the applied potential (Figure 1.6b). Hence, the technique provides voltammetric information (current vs. potential) about the substance being detected. This is useful for the purposes of qualitative identification as the voltammogram of a substance

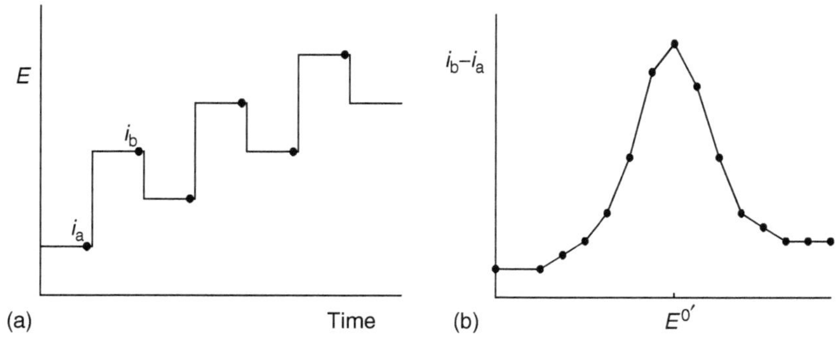

FIGURE 1.5 (a) Diagram of the potential waveform used in DPV. (b) Differential pulse voltammogram.

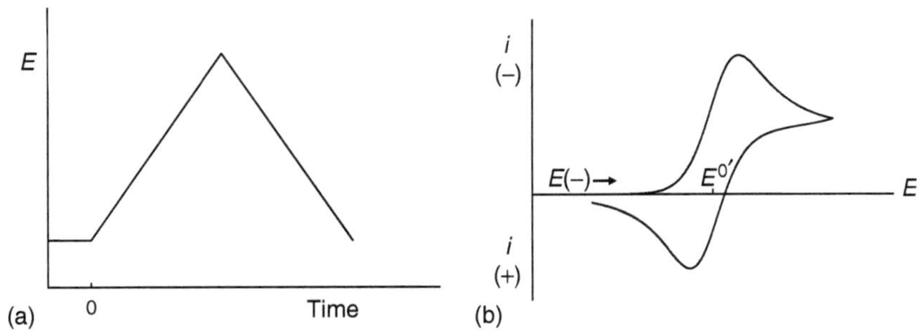

FIGURE 1.6 (a) Diagram of a potential waveform used in linear sweep (cyclic) voltammetry. (b) A cyclic voltammogram.

is generally unique in its position along the potential axis and its shape. For example, the cyclic voltammogram of dopamine is easily distinguished from the voltammogram of ascorbate. With a small electrode that rapidly charges to the new applied potential, sweep rates of several hundred volts per second are quite feasible, so the entire CV can be recorded in a matter of milliseconds. This is interesting because the entire CV can be completed in about the same amount of time that it takes to record a single chronoamperometric pulse. Thus, CV is the fastest voltammetric recording technique available, attracting quite a bit of attention since temporal resolution is one of the performance attributes of electrochemistry compared to microdialysis.

Amperometry

In many cases, the potential is held constant throughout an experimental procedure. This technique has come to be known as amperometry. It must be used with caution because no information about the voltammetric properties (current vs. potential) of the detected substance is obtained, so amperometry is best suited to situations where there is reason to be confident of the identity of the substance to be detected. Such cases are not uncommon. For example, it is by now very well established that electrical stimulation of the medial forebrain bundle releases dopamine in the striatum and that bovine chromaffin cells release adrenaline during exocytosis. The benefit of amperometry is that it permits the fast gathering of data: the time scale of the experiment is limited only by the speed of the electronics used to record the current; there are no timing limitations associated with the pulse or sweep waveforms described earlier.

Amperometry is also used in conjunction with enzyme-modified electrodes such as those being developed in a number of laboratories for the detection of glutamate. Although there are different schemes for this process, glutamate oxidase is immobilized in a polymer film placed over the surface of an electrode. The electrode actually detects the H_2O_2 generated by the immobilized enzyme. To confirm that the signal is derived from glutamate, a sentinel electrode is used. The sentinel electrode is made as identically as practicable to the glutamate electrode but without glutamate oxidase, so in this case, it is the design of the electrode modification layer that provides selectivity information rather than the voltammetric properties. Therefore, amperometry is a suitable technique.

CONCLUSION

Electrochemical methods are versatile tools for detecting, monitoring, and measuring various neurochemical species. Microelectrodes, in conjunction with the electrochemical techniques presented within this chapter, have been used to explore environments too small or too fragile to be examined in other ways. Electrochemical techniques have also been used to examine chemical events on a subsecond time scale, providing new kinetic information and real-time information about exocytosis and spontaneous neurotransmitter release. Overall, the ease of use and flexibility in application make electrochemical methods ideal for studies requiring superior temporal and spatial resolution of neurochemicals in physiological environments.

REFERENCES

Adams, R. N., Probing brain chemistry with electroanalytical techniques, *Anal. Chem.*, 48, 1128A, 1976.

Avshalumov, M. V. and Rice, M. E., Activation of ATP-sensitive K^+ (K_{ATP}) channels by H_2O_2 underlies glutamate-dependent inhibition of striatal dopamine release, *Proc. NatlAcad. Sci.*, 100, 11729, 2003.

Avshalumov, M. V., Chen, B. T., Marshall, S. P., Peña, D. M., and Rice, M. E., Glutamate-dependent inhibition of dopamine release in striatum is mediated by a new diffusible messenger, H_2O_2, *J. Neurosci.*, 23, 2744, 2003.

Bard, A. J. and Faulkner, L. R., *Electrochemical Methods: Fundamentals and Applications*, Wiley, New York, 2001.

Barlett, P. N. and Cooper, J. M., A review of the immobilization of enzymes in electropolymerized films, *J. Electroanal. Chem.*, 362, 1, 1993.

Boudko, D. Y., Moroz, L. L., Linser, P. J., Trimarchi, J. R., Smith, P. J. S., and Harvey, W. R., In situ analysis of pH gradients in mosquito larvae using noninvasive, self-referencing, pH-sensitive microelectrodes, *J. Exp. Biol.*, 204, 691, 2001.

Cheer, J. F., Wassum, K. M., Heien, M. L. A. V., Phillips, P. E. M., and Wightman, R. M., Cannabinoids enhance subsecond dopamine release in the nucleus accumbens of awake rats, *J. Neurosci.*, 24, 4393, 2004.

Christensen, J. C., Wang, Z., and Rebec, G. V., γ-Aminobutyric acid infusion in substantia nigra pars reticulata in rats inhibits ascorbate release in ipsilateral striatum, *Neurosci. Lett.*, 280, 191, 2000.

Christian, G. D., *Analytical Chemistry*, 6th ed., Wiley, New York, 2004.

Colliver, T. L., Pyott, S. J., Achalabun, M., and Ewing, A. G., VMAT-mediated changes in quantal size and vesicular volume, *J. Neurosci.*, 20, 5276, 2000.

Cui, J., Kulagina, N. V., and Michael, A. C., Pharmacological evidence for the selectivity of in vivo signals obtained with enzyme-based electrochemical sensors, *J. Neurosci. Meth.*, 104, 183, 2001.

Daws, L. C., Toney, G. M., Gerhardt, G. A., and Frazer, A., In vivo chronoamperometric measures of extracellular serotonin clearance in rat dorsal hippocampus: contribution of serotonin and norepinephrine transporters, *J. Pharmacol. Exp. Ther.*, 286, 967, 1998.

Day, B. K., Pomerleau, F., Burmeister, J. J., Huettl, P., and Gerhardt, G. A., Microelectrode array studies of basal and potassium-evoked release of L-glutamate in the anesthetized rat brain, *J. Neurochem.*, 96, 1626–1635, 2006.

Edwards, R. H. and Sulzer, D., Synaptic vesicle transporter expression regulates vesicle phenotype and quantal size, *J. Neurosci.*, 20, 7297, 2000.

Frazer, A. and Daws, L. C., Serotonin transporter function in vivo: assessment by chronoamperometry, In *Advances in serotonin receptor research: molecular biology, signal transduction and therapeutics*, Annals New York Academy of Sciences, 861, 217–229, 1998.

Garris, P. A., Ciolkowski, E. L., Pastore, P., and Wightman, R. M., Efflux of dopamine from the synaptic cleft in the nucleus accumbens of the rat brain, *J. Neurosci.*, 14, 6084, 1994.

Garris, P. A., Budygin, E. A., Phillips, P. E. M., Venton, B. J., Robinson, D. L., Bergstrom, B. P., Rebec, G. V., and Wightman, R. M., A role for presynaptic mechanisms in the actions of nomifensine and haloperidol, *Neuroscience*, 118, 819, 2003.

Gerhardt, G. A., Oke, A. F., Moghaddam, B., and Adams, R. N., Nafion-coated electrodes with high selectivity for CNS electrochemistry, *Brain Res.*, 290, 390, 1984.

Gonon, F., Buda, M., Cespuglio, R., Jouvet, M., and Pujol, J. F., In vivo electrochemical detection of catechols in the neostriatum of anaesthetized rats: dopamine or DOPAC? *Nature*, 286, 902, 1980.

Gonon, F. G., Fombarlet, C. M., Buda, M. J., and Pujol, J. F., Electrochemical treatment of pyrolytic carbon fiber electrodes, *Anal. Chem.*, 53, 1386, 1981.

Gonon, F. G., Navarre, F., and Buda, M. J., In vivo monitoring of dopamine release in the rat brain with differential pulse voltammetry, *Anal. Chem.*, 56, 573, 1984.

Gregg, B. A. and Heller, A., Crosslinked redox gels containing glucose oxidase for amperometric biosensor applications, *Anal. Chem.*, 62, 258, 1990.

Harris, D. C., *Quantitative Chemical Analysis*, 6th ed., W. H. Freeman, New York, 2003.

Heller, A., Electrical connection of enzyme redox centers to electrodes, *J. Phys. Chem.*, 96, 3579, 1992.

Jones, S. R., Garris, P. A., and Wightman, R. M., Different effects of cocaine and nomifensine on dopamine uptake in the caudate-putamen and nucleus accumbens, *J. Pharm. Exp. Ther.*, 274, 396, 1995.

Jones, S. R., O'Dell, S. J., Marshall, J. F., and Wightman, R. M., Functional and anatomical evidence for different dopamine dynamics in the core and shell of the nucleus accumbens in slices of rat brain, *Synapse*, 23, 224, 1996.

Justice, J. B. Jr., *Voltammetry in the Neurosciences: Principles, Methods, and Applications*, Humana Press, Clifton, NJ, 1987.

Kennedy, R. T., Jones, S. R., and Wightman, R. M., Simultaneous measurement of oxygen and dopamine—coupling of oxygen consumption and neurotransmission, *Neuroscience*, 51, 55, 1992.

Kissinger, P. T., Hart, J. B., and Adams, R. N., Voltammetry in brain tissue—a new neurophysiological measurement, *Brain Res.*, 55, 209, 1973.

Kozminski, K. D., Gutman, D. A., Davila, V., Sulzer, D., and Ewing, A. G., Voltammetric and pharmacological characterization of dopamine release from single exocytotic events at rat pheochromocytoma (PC12) cells, *Anal. Chem.*, 70, 3123, 1998.

Kulagina, N. V. and Michael, A. C., Monitoring hydrogen peroxide in the extracellular space of the brain with amperometric microsensors, *Anal. Chem.*, 75, 4875, 2003.

Kulagina, N. V., Shankar, L., and Michael, A. C., Monitoring glutamate and ascorbate in the extracellular space of brain tissue with electrochemical microsensors, *Anal. Chem.*, 71, 5093, 1999.

Kumar, S. M., Porterfield, D. M., Muller, K. J., Smith, P. J. S., and Sahley, C. L., Nerve injury induces a rapid efflux of nitric oxide (NO) detected with a novel NO microsensor, *J. Neurosci.*, 21, 215, 2001.

Leszczyszyn, D. J., Jankowski, J. A., Viveros, O. H., Diliberto, E. J., Near, J. A., and Wightman, R. M., Secretion of catecholamines from individual adrenal medullary chromaffin cells, *J. Neurochem.*, 56, 1855, 1991.

McMahon, C. P., Rocchitta, G., Serra, P. A., Kirwan, S. M., Lowry, J. P., and O'Neill, R. D., Control of the oxygen dependence of an implantable polymer/enzyme composite biosensor for glutamate, *Anal. Chem.*, 78, 2352–2359, 2006.

Michael, A. C., Borland, L. M., Mitala, J. J. Jr., Willoughby, B. M., and Motzko, C. M., Theory for the impact of basal turnover on dopamine clearance kinetics in the rat striatum after medial forebrain bundle stimulation and pressure ejection, *J. Neurochem.*, 94, 1202, 2005.

Mosharov, E. V., Gong, L. W., Khanna, B., Sulzer, D., and Lindau, M., Intracellular patch electrochemistry: regulation of cytosolic catecholamines in chromaffin cells, *J. Neurosci.*, 23, 5835, 2003.

Perez, X. A. and Andrews, A. M., Chronoamperometry detects differential changes in synaptosomal uptake in serotonin transporter knockout mice, *Anal. Chem.*, 77, 818, 2005.

Peters, J. L., Miner, L. H., Michael, A. C., and Sesack, S. R., Ultrastructure of carbon fiber microelectrode implantation sites after acute voltammetric measurements in the striatum of anesthetized rats, *J. Neurosci. Methods.*, 137, 9, 2004.

Phillips, P. E. M. and Wightman, R. M., Critical guidelines for validation of the selectivity of in vivo chemical microsensors, TrACs, *Trends Anal. Chem.*, 22, 509, 2003.

Ponchon, J. L., Cespuglio, R., Gonon, F., Jouvet, M., and Pujol, J. F., Normal pulse polarography with carbon fiber electrodes for in vitro and in vivo determination of catecholamines, *Anal. Chem.*, 51, 1483, 1979.

Pothos, E. N., Przedborski, S., Davila, V., Schmitz, Y., and Sulzer, D., D_2-like dopamine autoreceptor activation reduces quantal size in PC12 cells, *J. Neurosci.*, 18, 5575, 1998.

Pothos, E. N., Larsen, K. E., Krantz, D. E., Liu, Y., Haycock, J. W., Setlik, W., Gershon, M. D., Edwards, R. H., and Sulzer, D., Synaptic vesicle transporter expression regulates vesicle phenotype and quantal size, *J. Neurosci.*, 20, 7297, 2000.

Rebec, G. V. and Wang, Z., Behavioral activation in rats requires endogenous ascorbate release in striatum, *J. Neurosci.*, 21, 668, 2001.

Robinson, D. L. and Wightman, R. M., Nomifensine amplifies subsecond dopamine signals in the ventral striatum of freely-moving rats, *J. Neurochem.*, 90, 894, 2004.

Robinson, D. L., Heien, M. L. A. V., and Wightman, R. M., Frequency of dopamine concentration transients increases in dorsal and ventral striatum of male rats during introduction of conspecifics, *J. Neurosci.*, 22, 10477, 2002.

Robinson, D. L., Venton, B. J., Heien, L. A. V., and Wightman, R. M., Detecting subsecond dopamine release with fast-scan cyclic voltammetry in vivo, *Clin. Chem.*, 49, 1763, 2003.

Roitman, M. F., Stuber, G. D., Phillips, P. E. M., Wightman, R. M., and Carelli, R. M., Dopamine operates as a subsecond modulator of food seeking, *J. Neurosci.*, 24, 1265, 2004.

Schenk, J. O., Miller, E., Rice, M. E., and Adams, R. N., Chronoamperometry in brain slices: quantitative evaluations of in vivo electrochemistry, *Brain Res.*, 277, 1, 1983.

Skoog, D. A., Holler, F. J., and Nieman, T. A., *Principles of Instrumental Analysis*, 5th ed., Saunders College Publishing, Philadelphia, PA, 1998.

Smith, P. J. S. and Trimarchi, J. R., Noninvasive measurement of hydrogen and potassium ion influx from cells and epithelial structures, *Am. J. Physiol. Cell Physiol.*, 280, 1, 2001.

Turner, R. F. B., Harrison, D. J., and Rajotte, R. V., Preliminary in vivo biocompatibility studies on perfluorosulfonic acid polymer membranes for biosensor applications, *Biomaterials*, 12, 361, 1991.

Venton, B. J., Zhang, H., Garris, P. A., Phillips, P. E. M., Sulzer, D., and Wightman, R. M., Real-time decoding of dopamine neurotransmission in the caudate-putamen during tonic and phasic firing, *J. Neurochem.*, 87, 1284, 2003.

Wightman, R. M. and Zimmerman, J. B., Control of dopamine extracellular concentration in rat striatum by impulse flow and uptake, *Brain Res. Rev.*, 15, 135, 1990.

Wightman, R. M., Strope, E., Plotsky, P. M., and Adams, R. N., Monitoring of transmitter metabolites by voltammetry in cerebrospinal fluid following neural pathway stimulation, *Nature*, 262, 145976, 1976.

Wightman, R. M., Amatore, C., Engstrom, R. C., Hale, P. D., Kristensen, E. W., Kuhr, W. G., and May, L. J., Real-time characterization of dopamine overflow and uptake in the rat striatum, *Neuroscience*, 25, 513, 1988.

Wilson, G. S. and Gifford, R., Biosensors for real-time in-vivo measurements, *Biosens. Bioelect.*, 20, 2388–2403, 2005.

Wu, Q., Reith, M. E. A., Kuhar, M. J., Carroll, F. I., and Garris, P. A., Preferential increases in nucleus accumbens dopamine after systemic cocaine administration are caused by unique characteristics of dopamine neurotransmission, *J. Neurosci.*, 21, 6338, 2001.

Wu, Q., Reith, M. E. A., Walker, Q. D., Kuhn, C. M., Carroll, F. I., and Garris, P. A., Concurrent autoreceptor-mediated control of dopamine release and uptake during neurotransmission: an in vivo voltammetric study, *J. Neurosci.*, 22, 6272, 2002.

Zhang, X., Cardosa, L., Broderick, M., Fein, H., and Lin, J., An integrated nitric oxides sensor based on carbon fiber coated with selective membranes, *Electroanalysis*, 12, 1113, 2000.

Zhang, X., Kislyak, Y., Lin, J., Dickson, A., Cardosa, L., Broderick, M., and Fein, H., Nanaometer size electrode for nitric oxide and S-nitrosothiols measurement, *Electrochem. Commun.*, 4, 11, 2002.

2 Rapid Dopamine Release in Freely Moving Rats

Donita L. Robinson and R. Mark Wightman

CONTENTS

THE PHENOMENON OF DOPAMINE TRANSIENTS

Dopamine concentrations in the striatum of rats fluctuate during significant behavioral and pharmacological events, such as copulation and administration of addictive drugs. These changes have traditionally been measured in awake animals using microdialysis with offline electrochemical analysis (Westerink 1995). Microdialysis integrates the dopamine signal over space and time. Typically, the dialysis probe is 1–2 mm long and dialysate samples are collected over 5–15 min.

However, many of the behavioral or environmental events at which dopamine release is expected occur on a much faster timescale than microdialysate sampling. Fast-scan cyclic voltammetry (FSCV) measures dopamine concentrations in much smaller spatial and temporal dimensions: typically at approximately 6 μm × 100 μm cylindrical electrodes and sampling every 100 ms. Thirteen years after FSCV was used to monitor electrically-evoked dopamine release in anesthetized rats (Stamford et al. 1984; Millar et al. 1985), the method was used to detect discrete, naturally occurring dopamine signals in the nucleus accumbens (NAc) shell of awake rats during behavior (Rebec et al. 1997). In parallel, detection limits for in vivo voltammetric methods have improved from 5 μM (Ewing, Bigelow, and Wightman 1983) to less than 20 nM (Cheer et al. 2004; Stuber et al. 2005). As a result, several papers have now documented dopamine concentration transients in response to a variety of behavioral and pharmacological events.

Physiology of Dopamine Transients

Dopamine transients are subsecond fluctuations in dopamine concentrations, lasting from 0.2 to several seconds and ranging from 10 nM to more than 1 μM. For example, Figure 2.1 shows a spontaneous dopamine transient preceding an electrically stimulated dopamine signal recorded in the NAc core. Transients occur at baseline frequencies in the dorsal and ventral striatum of rats (Robinson, Heien, and Wightman 2002), and increase in frequency during novelty (Rebec et al. 1997; Robinson and Wightman 2004), social interaction (Robinson et al. 2001; Robinson, Heien, and Wightman 2002), pharmacological challenge (Cheer et al. 2004; Robinson and Wightman

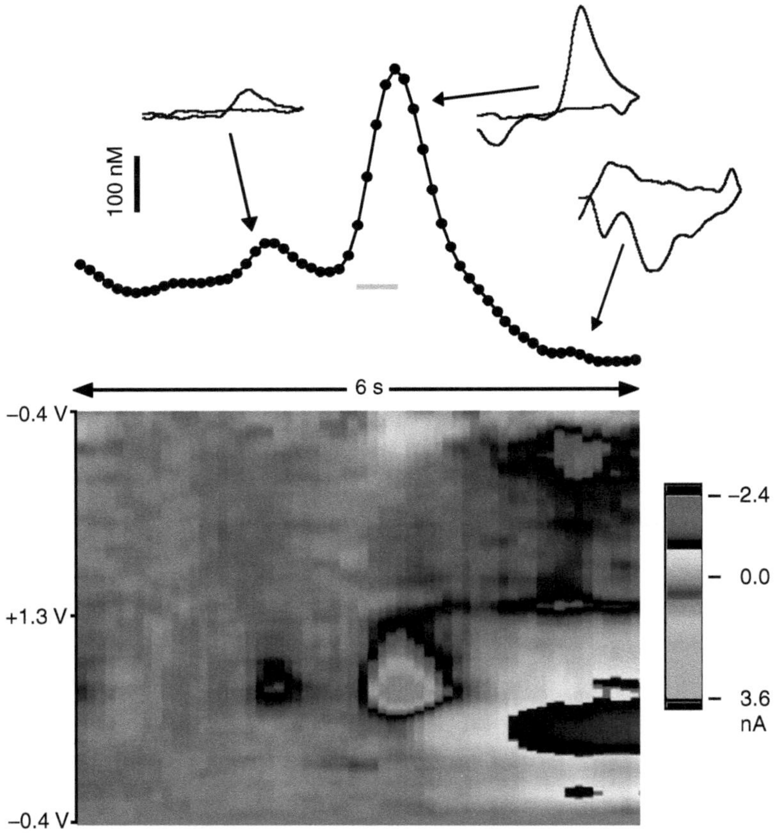

FIGURE 2.1 **(See color insert following page 272.)** Spontaneous and electrically evoked dopamine transients from the NAc of a rat. The color plot (bottom, Michael et al. 1998) shows changes in current (color) by applied potential (*y*-axis) over 6 s (*x*-axis). The unfiltered data have been smoothed three times with a two-dimensional smooth. Fluctuations in current at the oxidation potential of dopamine, approximately 600 mV (versus Ag/AgCl reference), are shown in the trace above the color plot. A small increase in current occurs at approximately 2.5 s, and is confirmed to be due to dopamine oxidation by the cyclic voltammogram (above); this is a spontaneous dopamine transient. One second later, an electrical stimulation (indicated by the green bar; 20 Hz, 12 p, biphasic, 2 ms/phase, 125 μA) was delivered to the ventral tegmental area, producing a robust increase in dopamine release, as indicated by the cyclic voltammogram (above).

2004; Stuber et al. 2005) and self-administration of reinforcers (Phillips et al. 2003b; Roitman et al. 2004; Stuber, Wightman, and Carelli 2005, Stuber et al. 2005).

Dopamine transients measured in terminal regions are presumed to arise from burst firing of dopamine neurons (Garris and Rebec 2002). In awake rats, burst firing of dopamine neurons occurs spontaneously (Freeman, Meltzer, and Bunney 1985) and can be triggered by operant responding and reinforcement (Hyland et al. 2002), as well as by drug administration (Freeman and Bunney 1987). Moreover, approximately 25% of dopamine neurons in awake rats appear to fire the majority of spikes in near-coincidence with another neuron (Freeman, Meltzer, and Bunney 1985; Freeman and Bunney 1987; Hyland et al. 2002), suggesting electrical coupling or shared afferent input. Although electrical coupling does not appear to synchronize tonic firing of dopamine neurons (Vandecasteele, Glowinski, and Venance 2005), it may promote synchrony of NMDA-driven bursts (Komendantov and Canavier 2002). Together, these electrophysiological data predict that dopamine concentrations might exhibit local, subsecond fluctuations, such as the dopamine transients observed via FSCV.

Indeed, dopamine transients can be measured after electrical stimulation trains delivered to dopamine fibers or cell bodies. While the majority of FSCV studies have used stimulation trains (e.g., 24 pulses at 60 Hz) that are faster or longer than measured bursts of spikes in vivo, reproducible changes in dopamine concentrations have been observed time-locked to more physiological patterns, such as 12 pulses at 20 Hz (Figure 2.1). It is noteworthy, however, that the occurrence of detected dopamine transients in freely moving rats is much less frequent than the occurrence of firing bursts (>3 spikes) by dopamine neurons: 0.01–0.02 Hz (approximated from data in Table 2.1) versus 0.1 Hz (Hyland et al. 2002). This discrepancy suggests that only the largest bursts or simultaneous bursts from multiple neurons produce changes in dopamine concentrations detectable at the carbon-fiber electrode.

Advances in Recording Technique

The ability to observe dopamine transients in vivo has followed technical advances in FSCV. Fast dopamine measurements were first made in freely moving rats following electrical stimulation of

TABLE 2.1
Basal Frequency, Amplitude, and Duration of Dopamine Transients in the Dorsal and Ventral Striatum of Freely Moving Rats

Waveform	Transients, min^{-1} Average	Transients, min^{-1} Range	$[DA]_{max}$, nM	Duration, s
−0.4 to +1.0 V, 300 V/s (Robinson, Heien, and Wightman 2002)	CP: 0.04±0.01 NAc: 0.24±0.05 OT: 0.15±0.07	CP: 0.0–0.10 NAc: 0.05–0.43 OT: 0.0–0.29	CP: 210±20 NAc: 250±40 OT: 240±20	CP: 4.4±0.5 NAc: 4.3±0.2 OT: 4.5±0.5
−0.4 to +1.0 V, 300 V/s (Robinson and Wightman 2004)	NAc: 0.5±0.1 OT: 0.7±0.3	NAc: 0.11–1.22 OT: 0.06–1.89	NAc: 140±20 OT: 120±30	NAc: 4.8±0.4 OT: 4.6±0.7
−0.4 to +1.3 V, 400 V/s (Cheer et al. 2004)	NAc: 1.4±0.4	NAc: 1.1–5.8	NAc: 51±9	n/a
−0.6 to +1.4 V, 400 V/s (Stuber et al. 2005)[a]	NAc: 1.0±0.3	n/a	NAc: 40±7	n/a

[a] Measured during saline infusion.

Source: Robinson, D. L. and Wightman, R. M., *J. Neurochem.*, 90, 894–903, 2004; Cheer, J. F., Wassum, K. M., Heien, M. L., Phillips, P. E , and Wightman, R. M., *J. Neurosci.*, 24, 4393–4400, 2004; Stuber, G. D., Wightman, R. M., and Carelli, R. M., *Neuron*, 46, 661–669, 2005.

dopamine fibers (Garris et al. 1997). These early experiments used disk electrodes with an electroactive surface of approximately 0.5 mm^2. While the disk electrodes yielded low noise the spatial resolution made it difficult to find adequate dopamine release sites because they are about 4 μm apart in the dorsal striatum and approximately 6 μm apart in the ventral striatum of the rat. Switching to a cylindrical electrode with a surface area of approximately 2 mm^2 allowed much easier placement and more sensitivity, as more release sites were sampled (Cahill et al. 1996).

Other methodological changes that resulted in improved signal-to-noise ratios were using a glass seal instead of epoxy at the carbon–glass interface, pretreating the electrodes with purified isopropyl alcohol (Bath et al. 2000), securing the electrodes in sturdy micromanipulators (Garris et al. 1997; Phillips et al. 2003a), upgrading the instrumentation, refining the headstage construction, optimizing the applied potential (Heien et al. 2003), and mathematically subtracting contributions of interfering species (such as pH) from the dopamine signal (Runnels et al. 1999; Heien et al. 2005). These issues will be discussed in greater detail in the following sections. These advances have allowed dopamine detection limits to drop to less than 20 nM, and research reports have progressed from measuring no spontaneous dopamine transients, to detecting only the largest transients, to observing basal frequencies of greater than 1 transient per minute.

MEASUREMENT OF DOPAMINE TRANSIENTS: FAST-SCAN CYCLIC VOLTAMMETRY

In voltammetric recording, a potential is applied to an electrode and the resulting current changes due to oxidation and reduction of chemicals at the electrode surface are measured. The basic difference between voltammetric techniques such as FSCV, constant-potential amperometry, chronoamperometry, or differential normal-pulse voltammetry, is the pattern of the applied potential. FSCV employs a triangular waveform that typically ramps from a negative potential to a positive potential and back, e.g., −0.4 to +1.0 V and back at 300 V/s, repeated at 10 Hz, and holding at −0.4 V between scans. Thus, current changes due to pH shifts and redox reactions are measured at the range of potentials in the triangular waveform. The charging of the double layer around the electrode results in a large, relatively stable background current that can be subtracted to reveal smaller variations in current due to oxidation of dopamine. This subtraction of the background current defines FSCV as a differential technique that measures changes in current as opposed to absolute currents.

Changes in current due to dopamine or other compounds are determined by the background-subtracted cyclic voltammogram, which is a plot of the current versus applied potential. When measured versus a Ag/AgCl electrode, dopamine has an oxidation potential of approximately 0.6 V and the *ortho*-quinone reduces back to dopamine at approximately −0.2 V. This profile is different from other compounds that might rapidly change in vivo, including serotonin, pH, and nitric oxide (Heien et al. 2003). However, the cyclic voltammogram of norepinephrine has the same profile as dopamine, requiring that anatomical recording sites be considered when identifying a signal as "dopamine" or simply "catecholamine."

Many of the methodological details of FSCV in freely moving rats are described elsewhere in great detail (Phillips et al. 2003a). The main points are reviewed here, along with with some updates and explanations.

ELECTRODE PREPARATION

For freely-moving rat experiments, carbon-fiber microelectrodes that are encased in glass and secured in micromanipulators were used (more details in Phillips et al. 2003a; also see (Millar and Pelling 2001) for instructions on carbon-fiber electrode construction). T-650 carbon fibers (Thornel, Greenville, SC) that are formed by pyrolysis of polyacrylonitrile (PAN) were used. The fiber is loaded into a glass capillary (0.6 mm O.D., 0.4 mm I.D.; A-M Systems, Carlsborg, WA) and pulled in an electrode puller (Narishige, Tokyo, Japan). The seal can be made with epoxy or by adjusting the parameters of the electrode puller such that a glass seal of at least 100 μm is formed. The fiber extending from the seal is trimmed under the microscope to the preferred length, typically 50–100 μm (although earlier studies used electrodes up to 250 μm length). The maximum length of the electrode is generally determined by the noise limitations of the amplifiers in the associated instrumentation. The higher the noise of the amplifiers, the larger the electrode that must be used.

The electrode is then secured in a micromanipulator that allows stability and precise placement into brain tissue. While a commercial model is available for electrophysiological measurements (Biela microdrives, Crist Instrument Co., Hagerstown, MD), we have used a manipulator made in-house at the University of North Carolina Department of Chemistry instrumentation shop. This fits into the locking guide cannula from Bioanalytical Systems (West Lafayette, IN) and allows for at least 75 μm precision. A silver wire is inserted and sealed in the open glass end of the electrode to connect to the headstage. Before use, the electrodes are pretreated by soaking in purified isopropyl alcohol for at least 10 min to increase sensitivity (Bath et al. 2000).

Electrode Calibration

The electrodes are calibrated in vitro to establish the relationship between current and dopamine concentration. This can be carried out either pre- or post-experiment; the more conservative method is to calibrate post-experiment (Logman et al. 2000) as this accounts for changes at the electrode surface associated with brain tissue, such as protein adhesion. Immediately after the experiment, the electrode is removed from the rat and the tip is rinsed with a strong, focused stream of water. Next, it is placed in purified isopropyl alcohol until calibration. A minimal calibration solution is used that consists of HEPES buffer (20 mM HEPES, 150 mM NaCl, 1.2 mM $CaCl_2$, pH = 7.4) or tris buffer (12 mM tris–HCl, 140 mM NaCl, 3.25 mM KCl, 1.2 mM $CaCl_2$, 1.25 mM NaH_2PO_4, 1.2 mM $MgCl_2$, 2 mM $NaSO_4$, pH = 7.4); tris buffer is required for applied waveforms of greater than 1 V, as the HEPES will oxidize at those potentials. The electrode is placed in a flow cell in a constant stream of buffer, and periodically the fluid is switched to buffer with known quantities of dopamine, typically 0.5–2 μM. The voltammetric signal resulting from the pulse of dopamine at the electrode is used to estimate the current/concentration ratio for naturally occurring dopamine transients. Calibrations are typically conducted at room temperature, as diffusion is minimally different at room temperature versus 37°C (Bard and Faulkner 2001).

Surgical Preparation

The goal of the surgical procedure is to implant the guide cannula (for later electrode insertion), the reference electrode, and the stimulating electrode to electrically stimulate dopamine transients. Although only the reference and the guide cannula (with the carbon-fiber electrode) are needed to record dopamine transients, the ability to electrically evoke dopamine release via the stimulating electrode is crucial for the positive identification of dopamine transients (see "Targeting Dopamine Transients" in this chapter).

The surgery is completed in two stages (more details in Phillips et al. 2003a). The first stage consists of securing the guide cannula (locking guide with stylette, Bioanalytical Systems, West Lafayette, IN) and reference electrode (Ag/AgCl, ~5 mm length) with cranioplastic cement and screw anchors. The second stage is placing the stimulating electrode (stainless steel, 0.2 mm, bipolar; Plastics One, Roanoke, VA). To do this, a carbon-fiber electrode is inserted into the brain, typically the caudate–putamen (CP), via the guide cannula and connected to the FSCV instrumentation. Next, the stimulating electrode is placed above a dopamine fiber area, typically the medial forebrain bundle or the substantia nigra/ventral tegmental area, and incrementally lowered. Current (125 μA, 24–60 p, 60 Hz) is delivered through the stimulating electrode at each increment until a dopamine signal is observed at the carbon-fiber electrode. After the dopamine signal is optimized, the stimulating electrode is cemented in place.

Electrochemical Measurements

Instrumentation and Software

High-sensitivity dopamine measurements require instrumentation that is low-noise and capable of measuring in the subnanoampere regime with a broad frequency bandpass (Cahill et al. 1996). The requirements are similar to those in patch-clamp recordings; indeed, patch-clamp amplifiers have been used for recordings with carbon-fiber electrodes. The first commercial instrument specifically designed for FSCV with carbon-fiber electrodes was introduced by Ensman Instrumentation in the 1980s. This instrument was well suited for use with anesthetized animals or brain slices and is still available from Cypress Systems, Lawrence, Kansas. Another commercially available system for FSCV is the Millar Voltammeter, P.D. Systems, West Moseley, U.K. However, to make measurements in awake animals, a new instrument, termed the *universal electrochemical instrument* (UEI), has been designed by the instrument shop in the Department of Chemistry, University of North Carolina at Chapel Hill. The UEI is modular so that different components can be developed and used within a single chassis that contains the power supply. The instrument is designed to be used with computer-controlled interface boards (National Instruments, Austin, TX) controlled by locally written software (*LabVIEW*, National Instruments). An early version of the software has been described in detail (Michael et al. 1999), and a description of recent modifications is available (Heien et al. 2003) and is discussed below.

In its present form, the UEI consists of two modules and a headstage amplifier unit that is mounted on the head of the preparation (Figure 2.2). One of the modules serves as a

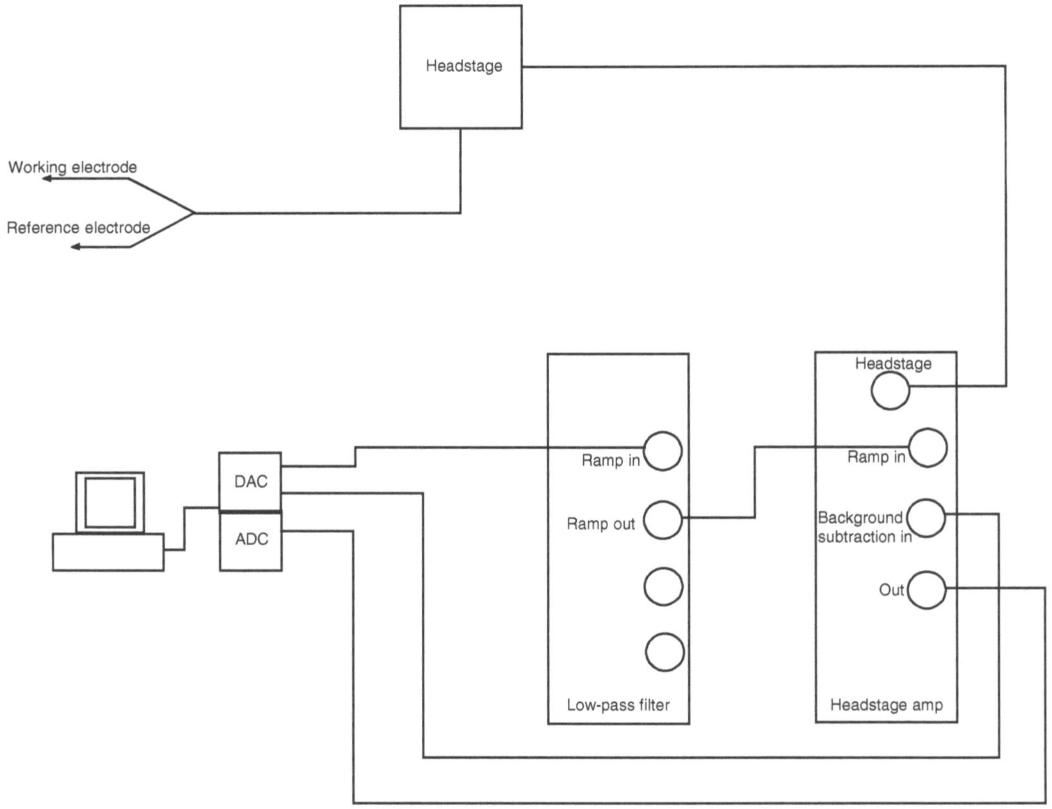

FIGURE 2.2 Block diagram of the components used in the instrument for cyclic voltammetric measurement of dopamine. (Figure courtesy of Colin McKinney, University of North Carolina. With permission.)

signal-conditioning unit for the applied, computer-generated waveform. The module filters and attenuates the applied waveform to remove digital noise. These steps are necessary because any noise on the applied waveform is amplified in the voltammetric recordings (Howell et al. 1986). The output of the first unit goes to the second module that combines all of the lines into a single cable so that they can be routed through a commutator to the headstage. The lines in this cable consist of ± 15 V and common (to drive the operational amplifier in the headstage), the applied waveform, and an output voltage that is proportional to the voltammetric current. The output voltage can be further amplified with the second module and it is then sent to the analog-to-digital converter of the computer interface board.

The critical unit of the instrument is the headstage. It contains surface-mount amplifiers and components so that it is lightweight. The applied waveform contained in the cable from the commutator is connected to the noninverting input of the current transducer, and the inverting input is connected directly to the carbon-fiber electrode to convert the voltammetric current to an output voltage. The headstage circuitry is designed such that the applied waveform is removed from the output voltage. The reference electrode is connected to system common. This design was adopted so that future applications could use multiple carbon-fiber electrodes or could be adapted for use with electrophysiological probes, with all voltages referenced to the reference electrode held at system common.

The software used to interface the instrumentation, *TarHeel CV*, is being commercialized by ESA Biosciences, Inc., Chelmsford, MA. This software controls the timing of the experiment and generates the applied triangular waveform and outputs it from a digital-to-analog converter that is a component on the interface board. The output voltages from the experiment are digitized and stored in computer memory and subsequently recorded on the computer disk. The applied potential can be customized: waveform shape, number of points, voltages, scan rate, and repetition rate are all variable. Aspects of the electrical stimulation to the dopamine fibers can be specified, such as the polarity, current, frequency, and pulses of the electrical stimulation, as well as the timing between voltammetric scans and stimulations to reduce artifacts. Electrical stimulations can also be externally triggered by events such as lever presses, or programmed as in intracranial self-stimulation play-back patterns. A digital oscilloscope is also included in *TarHeel CV*, showing the applied waveform and the nonbackground-subtracted cyclic voltammogram. Other data collection parameters that can be set are the gain of the signal and data collection parameters (single or sequential files). The oxidation current at a specified potential as well as the background current of the electrode are shown in real-time during data collection.

TarHeel CV has an analysis component that can be used within the data collection program or separately as a stand-alone program. One file at a time can be analyzed, with the user setting the background (typically 10 consecutive scans, or 1 s) and the cyclic voltammograms (typically 3–10 consecutive scans). The instrumentation does not filter the electrochemical signal as it is collected. Instead, the data can be smoothed or filtered as appropriate in the analysis component, including filtering the color plot or current versus time data or performing a three-dimensional smoothing of the data. Data are presented as a color plot, current versus time trace, and cyclic voltammogram. Data from these graphs can be written to text files for use in other programs. The signal to noise, based on the user-specified background and cyclic voltammograms, is automatically calculated. This screen is the link to further analyses targeting dopamine transients (see "Targeting Dopamine Transients" in this chapter).

Experimental Protocol

When designing an experimental protocol, technical aspects of electrical stimulation of dopamine neurons and the waveform parameters as well as practical aspects of timing must be considered.

Electrically Stimulated Dopamine Release

Dopamine release can be reproducibly evoked by stimulation of the dopamine cell bodies or fibers. This has been a valuable technique for analyzing drug effects on dopamine release and uptake, as the signal can be modeled to estimate the amount of dopamine release per stimulation pulse, as well as the Michaelis–Menten parameters, V_{max} and K_m, for dopamine uptake (Garris et al. 2003).

Electrically stimulated dopamine release is also necessary in experiments measuring spontaneous dopamine transients. First, it ensures that the electrode is positioned in a site with dopamine terminals. Second, the comparison of electrically evoked dopamine release before and after the experiment can be used to evaluate signal-to-noise and the integrity of the carbon fiber. Finally, it provides the template by which dopamine transients are identified (see "Targeting Dopamine Transients" in this chapter).

Electrical stimulations used in our studies range from 4 to 24 pulses (biphasic, 2 ms/phase), applied at 10–60 Hz, at a current of 80–200 μA. The larger stimulation trains produce larger and more reproducible signals, whereas the smaller trains are more physiological (e.g., <4 pulses at 10–20 Hz). To evoke a reproducible signal, the stimulations must be spaced in time to allow for replenishment of the readily releasable pool. Electrically stimulated dopamine release has been modeled to reveal facilitation and depression mechanisms that are dependent on previous stimulations (Montague et al. 2004). Therefore, the larger the stimulation, the more time must elapse before applying a subsequent stimulation. For example, 2 min is an adequate interval between stimulations of 24 pulses at 60 Hz and 125 μA.

Waveform Selection

Traditionally, we have used a triangle waveform with applied potentials of -0.4 to $+1.0$ V and back (i.e., the "traditional waveform") to measure electrically evoked dopamine release, and some studies of dopamine transients have also reported use of these applied potentials. However, we have also used waveforms with extended anodic and/or cathodic potentials to enhance sensitivity to dopamine (i.e., "extended waveforms"). With the traditional waveform, the electrode potential is held at -0.4 V between scans; this promotes the adsorption of dopamine to the carbon surface to enhance sensitivity (Baur et al. 1988). When this holding potential is increased to more negative values, sensitivity increases; however, the response time of the electrode to changes in dopamine slows (Heien et al. 2003). Extending the anodic potential from $+1.0$ to $+1.4$ V also increases sensitivity to dopamine, but at the expense of selectivity to potential interfering compounds (Heien et al. 2003). Heien and colleagues directly compared the traditional waveform to an extended waveform of -0.6 to $+1.4$ V and back: although there was a ninefold increase in sensitivity to dopamine in vivo, there were slower response kinetics and less selectivity between dopamine and possible interfering species such as dihydroxyphenylacetic acid (DOPAC), serotonin, 5-hydroxy indoleacetic acid (5-HIAA), ascorbic acid, and uric acid. Some of these compounds can be excluded as interfering species by targeting fast current changes that would be observed for vesicular neurotransmitter release but not for metabolic products. Nevertheless, strict criteria are needed to separate naturally occurring dopamine transients from other sources of current change in vivo.

Because the extended waveforms exhibit decreased time resolution compared to the traditional waveform, they are less suitable for making kinetic measurements of dopamine transients such as $[DA]_{max}$, half-width, and area under the peak. While extended waveforms are useful for comparing relative kinetics of dopamine transients (e.g., a percent increase in $[DA]_{max}$ associated with cocaine delivery), they will not yield the absolute kinetics required for modeling dopamine release and uptake parameters (Wu et al. 2001).

The FSCV scan rate determines the time required for a single voltage excursion. Typically, a scan rate of 400 V/s allows a cyclic voltammogram to be acquired in 10 ms. The scan rate also affects the amplitude of the current waveform. The current arising from charging the double-layer capacitance of the electrode, along with any currents associated with species confined to the electrode surface, increases linearly with the scan rate. Thus, doubling the scan rate from 400 to 800 V/s causes the background current to double. Because the majority of the signal amplitude from dopamine oxidation arises from dopamine that is adsorbed to the electrode (Bath et al. 2000; Heien et al. 2003), the dopamine signal will double as well. Serotonin and norepinephrine also adsorb and exhibit similar behavior. However, molecules that are not cations at physiological pH, such as DOPAC and ascorbate, do not adsorb and exhibit different behaviors. For those species, the current during the cyclic voltammogram is controlled by their diffusion to the electrode. Diffusion-controlled electrochemical processes increase in amplitude with the square root of the scan rate (Bard and Faulkner 2001), i.e., the faster the scan rate, the lower the contribution from these interfering species.

The time between cyclic voltammograms (the repetition rate) also affects the amplitude of species that adsorb to the electrode. The greater the time between scans, the more the adsorption processes can approach equilibrium (Bath et al. 2000). The net result is a greater signal for dopamine. We typically use 100 ms, which is a compromise between allowing time for dopamine to accumulate on the electrode and maintaining sufficient time resolution to follow dopamine's concentration fluctuations.

In summary, when choosing a waveform, the researcher must balance temporal precision and selectivity with sensitivity and know the limitations of each waveform. Early reports of dopamine transients used the traditional waveform (Rebec et al. 1997; Robinson, Heien, and Wightman 2002; Robinson and Wightman 2004), but an extended waveform of −0.6 to +1.4 V and back was used to monitor small changes in dopamine associated with operant responding for cocaine (Phillips et al. 2003b; Stuber, Wightman, and Carelli 2005, Stuber et al. 2005). Currently, our lab uses an extended waveform of −0.4 to +1.3 V and back, which yields almost as much sensitivity as the −0.6 to +1.4 V waveform but with better time resolution and selectivity (Cheer et al. 2004; Heien et al. 2005).

Note that the extended waveforms require longer for the electrode to stabilize after initial insertion into brain tissue. Although the reason for this is unknown, it is likely due to adaptations at the carbon surface. After the electrode is inserted into the brain, we typically apply the waveform at 60 Hz for 10–15 m and then apply it at 10 Hz (the typical frequency of data collection) for another 5 m before optimizing the electrode placement for recording.

Other Considerations

Files for behavioral experiments are typically collected in sequence and are limited to 90 s or less to minimize electrode drift during dopamine targeting (Heien et al. 2005). The length of the total experiment is kept as short as possible (usually about 60 min) to guard against electrode drift, tissue damage at the electrode site due to the rat hitting the manipulator or excessive movement, or changes in sensitivity over time. The experimenter can check these possibilities by monitoring the signal-to-noise ratios of electrically stimulated release before and after the experiment.

Data Analysis

Targeting Dopamine Transients

With thousands of electrochemical scans per experiment, dopamine transients are targeted using objective, high-throughput algorithms. One method is the "CV match" algorithm written into the *Tar Heel CV* software (Robinson et al. 2003). This automatically determines the correlation of each background-subtracted voltammogram with a dopamine cyclic-voltammogram template. The template voltammogram is taken from an electrically stimulated dopamine signal in the same rat, preferably at the same recording site. The background to be subtracted is specified as a certain number of scans (typically 10 scans or 1 s) and distance from the cyclic voltammogram under analysis; this is a moving background, as it is always relative to the target voltammogram. The CV match is typically run with 3–4 backgrounds (e.g., +25, +15, +10, and −5 scans from the target voltammogram). Alternatively, the program can be modified to use a single background for subtraction from the whole data file. The correlations are then displayed in graphic form, with scan number on the x-axis and r or r^2 on the y-axis. The program can generate text files of the r or r^2 values and the corresponding scans that exceed a threshold specified by the user. We generally use a threshold of $r^2 = 0.75$ or $r = 0.866$; this cutoff was determined after calculating the correlation coefficient of dopamine with interfering species, such as DOPAC, serotonin, and 5-HIAA (Heien et al. 2003). Earlier versions of the software also calculated the inverse mean squared error, which yielded a larger number as the variation between the target and the template decreased. The CV match algorithm is a very conservative method, as only clear dopamine cyclic voltammograms with no overlapping pH signal or other interfering species will be targeted.

A second, more thorough method to identify dopamine transients is the "principal component regression" algorithm written into the *Tar Heel CV* software (explained in more detail in Heien, Johnson, and Wightman 2004). In brief, this algorithm uses principal component clustering to separate changes in current due to dopamine from noise or pH shifts. For this calculation, a training set is generated with cyclic voltammograms of dopamine, pH changes, and any other identifiable contributor, such as electrode drift. Typically, five voltammograms of each compound or factor are used, each illustrating different concentrations or amplitudes. The dopamine and pH signals are taken from electrically stimulated dopamine release and subsequent pH shifts, and the dopamine signals are converted to concentration using the in vitro electrode calibration. The training set is used to calculate eigenvalues that are employed to analyze experimental recordings by a principal component regression. The result is a concentration versus time plot for dopamine, as well as for pH shifts and any other factor used in the training set. To test the goodness of fit, the data can be reproduced with the regression values, then compared to the original data. The residual (Q) is calculated as the least-squared difference between the two data sets. The peaks in the dopamine concentration versus time plot that surpass the noise threshold can then be analyzed for quantity, amplitude, and duration by hand or using commercially available software such as *MiniAnalysis 6.0.3.* (Synaptosoft, Inc.).

The advantage of the principal component analysis is that it uses more information to identify dopamine and can separate overlapping contributions of dopamine and pH, resulting in greater signal-to-noise ratios. An additional advantage of the principal component analysis is that slower (<90 s) changes in dopamine concentrations can be resolved, as well as dopamine transients. These slower changes are often masked by pH changes (Heien et al. 2005) and would not be detected using the CV match algorithm. However, the analysis is only as good as the training set, so several low-noise, electrically stimulated dopamine signals and pH shifts of varying amplitudes are required. Another considering factor is electrical noise or the presence of a compound not in the training set. This possibility is monitored with the residual Q: when Q exceeds the 95% confidence interval of the data, the principal component analysis is invalid at those time points.

With either analysis method, it is important to remember that dopamine cannot be electrochemically differentiated from norepinephrine. Thus, transients measured in brain regions with substantial norepinephrine input, such as the prefrontal cortex, should be termed *catecholamine transients*, even when electrically stimulated dopamine is used in the training set.

Measuring Amplitude and Duration

Once dopamine transients are identified, the attributes of amplitude and duration can be measured. In early studies, researchers measured these aspects manually by selecting an appropriate background for each transient and then examining scan-by-scan to determine the peak amplitude ($[DA]_{max}$) and duration. The background could be selected from the CV match algorithm, e.g., when multiple backgrounds were run, the background yielding the best r^2 value can be used, then the amplitude of the peak current could be measured. However, as sensitivity improved and the number of transients detected in an experiment increased, this approach became very time-consuming. When principal component analysis (PCA) is used, the resulting concentration versus time plot can be analyzed with a peak-analysis program such as *MiniAnalysis* to yield $[DA]_{max}$, half-width, and area under the peak.

Shifts in pH

Electrically stimulated dopamine signals are often followed by an alkaline change in pH (Runnels et al. 1999). These pH shifts have been verified using ion-selective electrodes and shown to be time-locked to changes in oxygen concentration (Venton, Michael, and Wightman 2003). Moreover, they can be modulated in concert with oxygen by compounds that are vasoactive, indicating that they reflect blood-flow changes (Venton, Michael, and Wightman 2003). Indeed, pH shifts are not observed at every electrode placement, but wax and wane as the electrode is lowered, consistent with the concept that they are dependent upon proximity to a blood vessel. The shape of the pH voltammogram varies according to the applied waveform (e.g., compare Runnels et al. 1999 with Heien et al. 2003), but in each case the changes in current can overlap with those of dopamine oxidation. The pH contribution at the oxidation potential of electrically evoked dopamine can be mathematically subtracted as described by Venton, Michael, and Wightman (2003), but is much more readily handled by principal component regression.

Spontaneous changes in pH shifts have also been observed in freely moving animals following dopamine transients (Robinson D. L. and Wightman R. M., unpublished observations). A subset of recordings ($n = 5$, -0.4 to $+1.0$ V applied waveform) from a published paper (Robinson and Wightman 2004) contained robust pH shifts after electrical stimulations in the ventral striatum. Templates of the pH signals were used to find spontaneous pH fluctuations using the CV match algorithm. In these recording sites, 53% of the spontaneous dopamine transients were followed within 5 s by a basic shift in pH. Moreover, using principal component analysis, Heien et al. (2005) revealed acidic pH shifts accompanying tonic increases in dopamine following acute cocaine infusion. Inasmuch as pH changes reflect blood flow and neuronal activity (Venton, Michael, and Wightman 2003), the study of pH signals might provide insight into postsynaptic activity and functional effects of dopamine transients.

Statistical Analysis

Many aspects of spontaneous dopamine transient data are nonlinear and have inhomogeneous variance, making a traditional ANOVA inappropriate. Instead, we have often used analyses that take into account nonhomogenous population variances and nonlinear association. For example, the frequency of dopamine transients is count data (the number of transients measured per unit time) and is bounded by zero (i.e., no transients detected during a sample), so analyses for data with Poisson distributions are used. Likewise, because the amplitudes of dopamine transients are bounded by and skewed toward the detection limit of the instrumentation, we use analyses for data with gamma distributions.

We construct statistical models of the FSCV data using the generalized linear model, or "genmod" command, in *SAS* software (SAS Institute Inc., Cary, NC). Unlike traditional ANOVA models that use least squares regression, this analysis uses maximum likelihood. The main advantage of the *SAS* generalized linear model for voltammetric data is that the regression can be tailored to various distributions, by using a linear, logistic, Poisson, or gamma regression. Added advantages of the *SAS* generalized linear model are that it can calculate a repeated-measures analysis without having an equal n in each group and it allows you to use the complete data for subjects when a datum is missing (i.e., unit of observation is subject-time). Post-hoc comparisons following the *SAS* generalized linear model are made with Wald-type statistics that are analogous to *t*-tests and *f*-tests for regression comparisons.

Verification of Recording Site

As the diameter of the carbon fiber is less than 10 μm, it is impossible to detect the recording site with light microscopy. However, two approaches have been used to determine the recording site. The first is to estimate the position by examination of damage left by the guide cannula and dorsal electrode tract, where the diameter of the glass capillary was larger (Robinson, Heien, and Wightman 2002). The depth of the electrode is known by documenting the number of turns made on the micromanipulator; each turn equals 300 μm. The recording site is then estimated using the known depth at the trajectory indicated by the observable damage. A second method is to lesion the site while the rat is anesthetized (Phillips et al. 2003a). A stainless steel electrode is inserted via the guide cannula to the same depth as the carbon-fiber electrode. An electrolytic lesion is made at the electrode tip by delivering a current (100 μA, 5 s) that is visualized in the tissue section with light microscopy. Note that lesions can be made directly with the carbon fiber (Rebec et al. 1997), but doing so renders the electrode unusable for post-experiment calibration.

NATURALLY OCCURRING DOPAMINE TRANSIENTS

Several papers in the past few years have documented naturally occurring dopamine transients in freely-moving rats in a variety of behavioral and pharmacological experiments. These data will be discussed below.

Environmental Stimuli

The first measurements of dopamine concentration transients were at the presentation of salient stimuli to the rat, using the traditional waveform of -0.4 to $+1.0$ V. Rebec et al. (1997) recorded dopamine transients in the NAc shell of rats at entry into a novel environment. The transients were sharp increases in current at the oxidation potential of dopamine, with amplitudes of 2.2 ± 0.7 nA and durations of 8 ± 1 s. The signals were not detected in the NAc core or the CP, although blunted signals (1.4 ± 0.4 nA amplitude, 24 ± 2 s duration) were observed in the core-shell boundary of the NAc. Moreover, no subsequent transients were observed during the exploratory phase following entry to the novel environment. The concentration of dopamine was unknown, as the electrodes could not be post-calibrated to convert the electrochemical current into concentration. Nevertheless, these data were important not only as the first report of spontaneous dopamine transients, but also as the first report of regional variation in phasic dopamine signaling.

Subsequently, dopamine transients were measured in the NAc core of male rats at the presentation of a sexually receptive female and during subsequent interaction (Robinson et al. 2001). Dopamine transients were observed at the presentation of the female, as well as during subsequent appetitive sexual behavior, such as approach and sniffing. The transients were fast (0.2–0.9 s) and reached concentrations of 200–500 nM dopamine. Nomifensine amplified the amplitude and duration of the dopamine transients and increased the frequency of detection, thereby providing pharmacological evidence that the signals were indeed dopaminergic.

Subsequent studies extended these findings in several important ways. First, transients occurred in the CP, NAc shell and OT, as well as the NAc core both at a baseline rate and at the presentation of stimuli (Robinson, Heien, and Wightman 2002; Robinson and Wightman 2004). Second, the presentation of nonreceptive female and male rats were also effective at triggering dopamine transients (Robinson, Heien, and Wightman 2002), as were novel odors, unexpected noises, and food treats (Robinson and Wightman 2004). Finally, the effectiveness of these stimuli to trigger dopamine transients waned with repeated presentation, indicating that the neurochemical response habituated (Robinson, Heien, and Wightman 2002).

Basal Conditions

While the initial studies used environmental stimuli to trigger rapid dopamine release (Rebec et al. 1997; Robinson et al. 2001), later studies included basal measurements of transients (Table 2.1). Using an applied waveform of −0.4 to +1.0 V, Robinson, Heien, and Wightman (2002) described average basal frequencies of approximately 0.2 per min in the NAc and OT, which were significantly higher than the frequency of 0.04 per min detected in the CP. In a subsequent study, Robinson and Wightman (2004) reported frequencies of 0.5 per min in the NAc and 0.7 per min in the OT, which were not statistically different. Moreover, the amplitude and duration of transients did not vary among brain regions in either study (Table 2.1). However, the detection limits of these studies were 60 nM (Robinson, Heien, and Wightman 2002) and 40 nM (Robinson and Wightman 2004); these regional measurements should be replicated now that detection limits have improved.

Later studies used waveforms that were extended in the anodic and/or cathodic direction and detected more small dopamine transients (<50 nM), resulting in higher basal frequencies. Cheer et al. (2004) measured approximately 1.5 transients per min in the NAc during basal conditions using an applied waveform of −0.4 to +1.3 V, whereas Stuber et al. (2005) measured 1 transient per min in the NAc during saline infusions using the −0.6 to +1.4 V applied waveform. The detection limits in these studies were approximately 13 nM (Cheer et al. 2004) and approximately 18 nM (Stuber et al. 2005), and the mean amplitude of dopamine transients was approximately 50 nM, suggesting that many transients were undetected in earlier studies using the −0.4 to +1.0 V waveform.

All of these studies found a large range of transient frequency—from recording sites where no transients were detected to sites where more than 5 transients per min were measured (Table 2.1). The differential propensity to measure spontaneous dopamine transients was not due simply to a lack of dopamine release sites near the carbon-fiber electrode because dopamine release was evoked at each site with electrical stimulation of the dopamine cell bodies or fibers. Rather, the phenomenon of "hot" and "cold" sites for dopamine transients may reflect terminal regions of dopamine neurons that have higher versus lower rates of firing in bursts or in synchrony with other neurons (Freeman, Meltzer, and Bunney 1985; Freeman and Bunney 1987; Hyland et al. 2002).

Pharmacology

Many studies have used FSCV to measure the presynaptic effects of drugs on dopamine transmission—specifically, effects on dopamine release and uptake after electrical stimulation of the dopamine neurons. For example, dopamine transporter antagonists increase the magnitude of electrically evoked dopamine signals by slowing dopamine uptake as measured in slices, anesthetized animals, and awake animals (Garris and Wightman 1995; Giros et al. 1996; Garris et al. 2003). To determine whether nomifensine has similar results on the kinetics of spontaneous dopamine transients, Robinson and Wightman (2004) examined the effects of nomifensine (7 mg/kg, i.p.) in the ventral striatum of awake rats. Nomifensine amplified $[DA]_{max}$ of dopamine transients by 10% and duration by 50%, consistent with the previous work on electrically stimulated dopamine signals. Likewise, Stuber et al. (2005) compared the amplitude of dopamine transients during saline versus cocaine infusion (4 i.v. infusions of 0.33 mg, 5 min apart) in drug-naïve rats and found approximately a twofold increase in $[DA]_{max}$ after cocaine infusion. Interestingly, cocaine did not affect transient amplitude in rats with a history of cocaine self-administration.

Administration of cocaine or nomifensine also increased the frequency of detected dopamine transients. Overall, dopamine transients were approximately eight times more frequent during cocaine infusions as compared to saline infusions in drug-naïve rats, and three times more frequent in rats with a history of cocaine self-administration; similar increases were seen in rats self-administering cocaine (Stuber et al. 2005). When analyzed minute-by-minute, the rate of dopamine transients in the drug-naïve rats was well correlated with theoretical brain concentrations of cocaine (Pan, Menacherry, and Justice 1991). Nomifensine also increased the frequency of detected dopamine transients, by approximately sevenfold in the NAc and approximately twofold in the CT (Robinson and Wightman 2004). It is unclear whether the increase in transients reflects an amplification of transients previously below the detection limit or a change in firing rate of the dopamine neurons. In either case, the practical effect is an increase in the number of dopamine signals reaching extrasynaptic dopamine receptors.

The effects of acute intravenous cocaine administration on dopamine release and pH changes have recently been examined in more detail by Heien et al. (2005). Using principal component analysis, they distinguished three components to the electrochemical signal in the 90 s following cocaine infusion: dopamine transients, a tonic increase in dopamine, and an acidic pH shift. Transients were evident immediately after the cocaine infusion, then increased in amplitude and duration as cocaine was distributed to the brain. However, these transients were on top of a tonic increase in dopamine concentration that reached approximately 700 nM after 3 mg/kg cocaine, reaching half-maximal concentration at approximately 30 s. The amplitude of the tonic signal was dose-dependent, but the time course was similar at all doses tested (0.3–3 mg/kg, i.v.). An acidic shift in pH followed the same time course as the tonic dopamine signal, likely due to the vasoconstrictive effects of cocaine as it was distributed to brain tissue. These data emphasize the fact that multiple compounds can contribute to the electrochemical signal, so it is crucial to use techniques that allow differentiation. Moreover, these data provide evidence that the tonic dopamine signal is not simply a composite of dopamine transients over time, but a separate component of dopamine transmission.

Cheer et al. (2004) investigated the effects of WIN55,212-2, a cannabinoid CB_1 receptor agonist, on rapid dopamine signaling. After administration of WIN55,212-2 (125 μg/kg, i.v.), the rate of dopamine transients increased almost threefold, while $[DA]_{max}$ of the transients increased twofold, as compared to the pre-drug values. Moreover, the effect of WIN55,212-2 on transient frequency was dose-dependent and reversed by administration of the CB_1 receptor antagonist SR141716A. These findings strengthen the hypothesis that transients arise from burst firing of dopamine neurons, as WIN55,212-2 increases both the number of spikes in a burst and the frequency of burst firing of dopamine neurons (Cheer et al. 2003).

Self-Administration of Reinforcers

In addition to the pharmacological effects of cocaine, dopamine transients can be triggered by the cues associated with cocaine self-administration and may contribute to the maintenance of self-administration behavior. FSCV, using the -0.6 to $+1.4$ V extended waveform, was employed to measure phasic dopamine fluctuations in rats during operant responding for cocaine (Phillips et al. 2003b). In this behavioral paradigm, each lever-press triggered an infusion pump that delivered cocaine (0.33 mg, i.v.) and simultaneously triggered a 20 s light/tone cue that signaled the infusion. Dopamine transients occurred immediately following the lever-press at the initiation of the light/tone cue. In fact, the unexpected presentation of the light/tone cue in the absence of a lever-press also elicited a dopamine transient in trained animals, but not naïve animals, indicating that the learned association between the cue and the cocaine infusion was an important component in the dopaminergic response.

In addition, dopamine transients were observed in the seconds preceding the lever-press, although the exact timing varied from trial to trial. These dopamine signals accompanied breaks in the stereotypical movements associated with cocaine's psychostimulant effects and approach to the lever. Indeed, inducing a dopamine transient with electrical stimulation of the dopamine fibers often triggered a break in stereotypy and a lever-press, suggesting that dopamine transients contribute to the maintenance of drug-taking behavior. Similarly, electrical stimulation of the medial forebrain bundle/lateral hypothalamus has been shown to trigger behavior dependent on the environmental context (Valenstein 1969). These data also support the hypothesis that phasic dopamine signals facilitate behavioral switching (Redgrave, Prescott, and Gurney 1999) and approach responses (Ikemoto and Panksepp 1999).

Subsequent studies used within-session extinction and reinstatement to provide further evidence that the transients preceding the lever-press were associated with lever approach, whereas the transients immediately following the lever-press were associated with learning about the cues (Stuber, Wightman, and Carelli 2005). After 30 min of responding for cocaine, saline was substituted so that the light/tone cue no longer predicted cocaine administration; rats pressed the lever approximately 20 times before the behavior was extinguished. When responding stopped for at least 30 min, cocaine replaced the saline and the rats were given a noncontingent infusion of cocaine to reinstate responding. Analysis of dopamine signals during the extinction phase revealed that the pre-response transients remained stable during extinction, but post-response dopamine transients decreased in amplitude even during the first ten presses, and were often absent during the last five presses. When cocaine pressing was reinstated, however, the post-response dopamine signals returned to pre-extinction amplitudes.

Similar dopamine fluctuations were observed during operant responding for sucrose (Roitman et al. 2004). In this study, a cue onset predicted availability of a sucrose-reinforced lever, and triggered a dopamine transient in the NAc within 1 s. The same cue elicited no dopaminergic response in another group of rats, for which the cue did not predict availability of the lever. The rats pressed the lever at the peak of the dopamine signal, with a mean of 5.5 s after the onset of the cue. In contrast, when the sucrose was delivered (0.3 M, 200 μL, intraoral administration) and consumed, no further dopamine transients were recorded. These data link phasic dopamine fluctuations with associative and appetitive aspects of sucrose reinforcement, rather than consummatory aspects.

Some of the first measurements of electrically stimulated dopamine signals in awake rats were made while the rats self-administered the electrical stimulations to the ventral tegmental area and substantia nigra, or intracranial self-stimulation (Garris et al. 1999; Kilpatrick et al. 2000). When the electrical stimulations (60 Hz, 24 p, biphasic, 80–250 μA) were delivered singly to the rats by the experimenter, reproducible dopamine transients resulted (i.e., electrically evoked release). However, when the rats self-administered trains of these stimulations during continuous reinforcement (FR1 schedule), dopamine release was measured at the beginning of the session then quickly dropped to undetectable levels. These results were replicated using chronoamperometry during ICS in mice at an FR1 schedule (Yavich and Tiihonen 2000). In contrast, when Yavich and Tiihonen (2000) increased the response requirement to FR8, which spaced the stimulations out about 5 s apart, they measured robust electrochemical signals. Indeed, previous impulse flow can determine the amount of dopamine release from subsequent stimulations; this can be modeled with facilitation and depression factors (Montague et al. 2004) and may be largely dependent on readily releasable pools of dopamine (Yavich and MacDonald 2000). It is yet unknown whether pre-response dopamine signals, such as those seen during cocaine and sucrose self-administration, would be observed during intracranial self-stimulation.

FUNCTION OF DOPAMINE TRANSIENTS

Current theory of dopamine function is that it is a neuromodulator positioned to influence afferent excitatory input onto medium spiny neurons (Nicola, Surmeier, and Malenka 2000). Tonic dopamine appears to have a gating function, enabling various behaviors. Phasic dopamine, on the other hand, may have a distinct function, such as attention and behavior switching (Ikemoto and Panksepp 1999; Redgrave, Prescott, and Gurney 1999) or reward-associated learning (Schultz 2002). Considering the high concentrations that can be achieved via dopamine transients and their precise timing, phasic dopamine signals are positioned to modulate highly specific inputs to the target cells. Furthermore, although dopamine transients in response to salient, unexpected stimuli appear to be simultaneous across the dorsal and ventral striatum (Robinson, Heien, and Wightman 2002; Robinson and Wightman 2004), the baseline rates of dopamine transients varies widely, both across the striatum and within specific nuclei (Table 2.1).

To address these functionality issues, methods are under development via collaboration between Mark Wightman and Regina Carelli to combine electrochemical and electrophysiological measurements (Williams and Millar 1990a, 1990b) in freely moving rats (Cheer et al. 2005). This combined technique will allow the simultaneous measurements of dopamine transients and cell firing at the carbon-fiber electrode, providing real-time associations between dopamine release and postsynaptic firing rates. The addition of iontophoresis of pharmacological agents will allow testing of causal relationships between the two components.

REFERENCES

Bard, A. J. and Faulkner, L. R., *Electrochemical Methods, Fundamentals and Applications*, 2nd ed., Wiley, New York, 2001.

Bath, B. D., Michael, D. J., Trafton, B. J., Joseph, J. D., Runnels, P. L., and Wightman, R. M., Subsecond adsorption and desorption of dopamine at carbon-fiber microelectrodes, *Anal. Chem.*, 72, 5994–6002, 2000.

Baur, J. E., Kristensen, E. W., May, L. J., Wiedemann, D. J., and Wightman, R. M., Fast-scan voltammetry of biogenic amines, *Anal. Chem.*, 60, 1268–1272, 1988.

Cahill, P. S., Walker, Q. D., Finnegan, J. M., Mickelson, G. E., Travis, E. R., and Wightman, R. M., Microelectrodes for the measurement of catecholamines in biological systems, *Anal. Chem.*, 68, 3180–3186, 1996.

Cheer, J. F., Heien, M. L., Garris, P. A., Carelli, R. M., and Wightman, R. M., Simultaneous dopamine and single-unit recordings reveal accumbens GABAergic responses: implications for intracranial self-stimulation, *Proc. Natl. Acad. Sci. USA*, 102, 19150–19155, 2005.

Cheer, J. F., Kendall, D. A., Mason, R., and Marsden, C. A., Differential cannabinoid-induced electrophysiological effects in rat ventral tegmentum, *Neuropharmacology*, 44, 633–641, 2003.

Cheer, J. F., Wassum, K. M., Heien, M. L., Phillips, P. E., and Wightman, R. M., Cannabinoids enhance subsecond dopamine release in the nucleus accumbens of awake rats, *J. Neurosci.*, 24, 4393–4400, 2004.

Ewing, A. G., Bigelow, J. C., and Wightman, R. M., Direct in vivo monitoring of dopamine released from two striatal compartments in the rat, *Science*, 221, 169–171, 1983.

Freeman, A. S. and Bunney, B. S., Activity of A9 and A10 dopaminergic neurons in unrestrained rats: further characterization and effects of apomorphine and cholecystokinin, *Brain Res.*, 405, 46–55, 1987.

Freeman, A. S., Meltzer, L. T., and Bunney, B. S., Firing properties of substantia nigra dopaminergic neurons in freely moving rats, *Life Sci.*, 36, 1983–1994, 1985.

Garris, P. A. and Rebec, G. V., Modeling fast dopamine neurotransmission in the nucleus accumbens during behavior, *Behav. Brain Res.*, 137, 47–63, 2002.

Garris, P. A. and Wightman, R. M., Distinct pharmacological regulation of evoked dopamine efflux in the amygdala and striatum of the rat in vivo, *Synapse*, 20, 269–279, 1995.

Garris, P. A., Christensen, J. R. C., Rebec, G. V., and Wightman, R. M., Real-time measurement of electrically evoked extracellular dopamine in the striatum of freely moving rats, *J. Neurochem.*, 68, 152–161, 1997.

Garris, P. A., Kilpatrick, M., Bunin, M. A., Michael, D., Walker, Q. D., and Wightman, R. M., Dissociation of dopamine release in the nucleus accumbens from intracranial self-stimulation, *Nature*, 398, 67–69, 1999.

Garris, P. A., Budygin, E. A., Phillips, P. E., Venton, B. J., Robinson, D. L., Bergstrom, B. P., Rebec, G. V. et al., A role for presynaptic mechanisms in the actions of nomifensine and haloperidol, *Neuroscience*, 118, 819–829, 2003.

Giros, B., Jaber, M., Jones, S. R., Wightman, R. M., and Caron, M. G., Hyperlocomotion and indifference to cocaine and amphetamine in mice lacking the dopamine transporter, *Nature*, 379, 606–612, 1996.

Heien, M. L., Johnson, M. A., and Wightman, R. M., Resolving neurotransmitters detected by fast-scan cyclic voltammetry, *Anal. Chem.*, 76, 5697–5704, 2004.

Heien, M. L., Phillips, P. E., Stuber, G. D., Seipel, A. T., and Wightman, R. M., Overoxidation of carbon-fiber microelectrodes enhances dopamine adsorption and increases sensitivity, *Analyst*, 128, 1413–1419, 2003.

Heien, M. L., Khan, A. S., Ariansen, J. L., Cheer, J. F., Phillips, P. E., Wassum, K. M., and Wightman, R. M., Real-time measurement of dopamine fluctuations after cocaine in the brain of behaving rats, *Proc. Natl Acad. Sci. USA*, 102, 10023–10028, 2005.

Howell, J. O., Kuhr, W. G., Ensman, R. E., and Wightman, R. M., Background subtraction for rapid scan voltammetry, *J. Electroanal. Chem.*, 209, 77–90, 1986.

Hyland, B. I., Reynolds, J. N., Hay, J., Perk, C. G., and Miller, R., Firing modes of midbrain dopamine cells in the freely moving rat, *Neuroscience*, 114, 475–492, 2002.

Ikemoto, S. and Panksepp, J., The role of nucleus accumbens dopamine in motivated behavior: a unifying interpretation with special reference to reward-seeking, *Brain Res. Brain Res. Rev.*, 31, 6–41, 1999.

Kilpatrick, M. R., Rooney, M. B., Michael, D. J., and Wightman, R. M., Extracellular dopamine dynamics in rat caudate–putamen during experimenter-delivered and intracranial self-stimulation, *Neuroscience*, 96, 697–706, 2000.

Komendantov, A. O. and Canavier, C. C., Electrical coupling between model midbrain dopamine neurons: effects on firing pattern and synchrony, *J. Neurophysiol.*, 87, 1526–1541, 2002.

Logman, M. J., Budygin, E. A., Gainetdinov, R. R., and Wightman, R. M., Quantitation of in vivo measurements with carbon fiber microelectrodes, *J. Neurosci. Methods*, 95, 95–102, 2000.

Michael, D., Travis, E. R., and Wightman, R. M., Color images for fast-scan CV measurements in biological systems, *Anal. Chem.*, 70, 586A–592A, 1998.

Michael, D. J., Joseph, J. D., Kilpatrick, M. R., Travis, E. R., and Wightman, R. M., Improving data acquisition for fast-scan cyclic voltammetry, *Anal. Chem.*, 71, 3941–3947, 1999.

Millar, J. and Pelling, C. W., Improved methods for construction of carbon fibre electrodes for extracellular spike recording, *J. Neurosci. Methods*, 110, 1–8, 2001.

Millar, J., Stamford, J. A., Kruk, Z. L., and Wightman, R. M., Electrochemical, pharmacological and electrophysiological evidence of rapid dopamine release and removal in the rat caudate nucleus following electrical stimulation of the median forebrain bundle, *Eur. J. Pharmacol.*, 109, 341–348, 1985.

Montague, P. R., McClure, S. M., Baldwin, P. R., Phillips, P. E., Budygin, E. A., Stuber, G. D., Kilpatrick, M. R. et al., Dynamic gain control of dopamine delivery in freely moving animals, *J. Neurosci.*, 24, 1754–1759, 2004.

Nicola, S. M., Surmeier, J., and Malenka, R. C., Dopaminergic modulation of neuronal excitability in the striatum and nucleus accumbens, *Annu. Rev. Neurosci.*, 23, 185–215, 2000.

Pan, H. T., Menacherry, S., and Justice, J. B., Differences in the pharmacokinetics of cocaine in naive and cocaine-experienced rats, *J. Neurochem.*, 56, 1299–1306, 1991.

Phillips, P. E., Robinson, D. L., Stuber, G. D., Carelli, R. M., and Wightman, R. M., Real-time measurements of phasic changes in extracellular dopamine concentration in freely moving rats by fast-scan cyclic voltammetry, *Methods Mol. Med.*, 79, 443–464, 2003a.

Phillips, P. E., Stuber, G. D., Heien, M. L., Wightman, R. M., and Carelli, R. M., Subsecond dopamine release promotes cocaine seeking, *Nature*, 422, 614–618, 2003b.

Rebec, G. V., Christensen, J. R., Guerra, C., and Bardo, M. T., Regional and temporal differences in real-time dopamine efflux in the nucleus accumbens during free-choice novelty, *Brain Res.*, 776, 61–67, 1997.

Redgrave, P., Prescott, T. J., and Gurney, K., Is the short-latency dopamine response too short to signal reward error? *Trends Neurosci.*, 22, 146–151, 1999.

Robinson, D. L. and Wightman, R. M., Nomifensine amplifies subsecond dopamine signals in the ventral striatum of freely-moving rats, *J. Neurochem.*, 90, 894–903, 2004.

Robinson, D. L., Heien, M. L., and Wightman, R. M., Frequency of dopamine concentration transients increases in dorsal and ventral striatum of male rats during introduction of conspecifics, *J. Neurosci.*, 22, 10477–10486, 2002.

Robinson, D. L., Phillips, P. E., Budygin, E. A., Trafton, B. J., Garris, P. A., and Wightman, R. M., Sub-second changes in accumbal dopamine during sexual behavior in male rats, *Neuroreport*, 12, 2549–2552, 2001.

Robinson, D. L., Venton, B. J., Heien, M. L., and Wightman, R. M., Detecting subsecond dopamine release with fast-scan cyclic voltammetry in vivo, *Clin. Chem.*, 49, 1763–1773, 2003.

Roitman, M. F., Stuber, G. D., Phillips, P. E., Wightman, R. M., and Carelli, R. M., Dopamine operates as a subsecond modulator of food seeking, *J. Neurosci.*, 24, 1265–1271, 2004.

Runnels, P. L., Joseph, J. D., Logman, M. J., and Wightman, R. M., Effect of pH and surface functionalities on the cyclic voltammetric responses of carbon-fiber microelectrodes, *Anal. Chem.*, 71, 2782–2789, 1999.

Schultz, W., Getting formal with dopamine and reward, *Neuron*, 36, 241–263, 2002.

Stamford, J. A., Kruk, Z. L., Millar, J., and Wightman, R. M., Striatal dopamine uptake in the rat: in vivo analysis by fast cyclic voltammetry, *Neurosci. Lett.*, 51, 133–138, 1984.

Stuber, G. D., Wightman, R. M., and Carelli, R. M., Extinction of cocaine self-administration reveals functionally and temporally distinct dopaminergic signals in the nucleus accumbens, *Neuron*, 46, 661–669, 2005.

Stuber, G. D., Roitman, M. F., Phillips, P. E., Carelli, R. M., and Wightman, R. M., Rapid dopamine signaling in the nucleus accumbens during contingent and noncontingent cocaine administration, *Neuropsychopharmacology*, 30, 853–863, 2005.

Valenstein, E. S., Behavior elicited by hypothalamic stimulation: a prepotency hypothesis, *Brain Behav. Evol.*, 2, 295–316, 1969.

Vandecasteele, M., Glowinski, J., and Venance, L., Electrical synapses between dopaminergic neurons of the substantia nigra pars compacta, *J. Neurosci.*, 25, 291–298, 2005.

Venton, B. J., Michael, D. J., and Wightman, R. M., Correlation of local changes in extracellular oxygen and pH that accompany dopaminergic terminal activity in the rat caudate-putamen, *J. Neurochem.*, 84, 373–381, 2003.

Westerink, B. H., Brain microdialysis and its application for the study of animal behaviour, *Behav. Brain Res.*, 70, 103–124, 1995.

Williams, G. V. and Millar, J., Differential actions of endogenous and iontophoretic dopamine in rat striatum, *Eur. J. Neurosci.*, 2, 658–661, 1990a.

Williams, G. V. and Millar, J., Concentration-dependent actions of stimulated dopamine release on neuronal activity in rat striatum, *Neuroscience*, 39, 1–16, 1990b.

Wu, Q., Reith, M. E., Wightman, R. M., Kawagoe, K. T., and Garris, P. A., Determination of release and uptake parameters from electrically evoked dopamine dynamics measured by real-time voltammetry, *J. Neurosci. Methods*, 112, 119–133, 2001.

Yavich, L. and MacDonald, E., Dopamine release from pharmacologically distinct storage pools in rat striatum following stimulation at frequency of neuronal bursting, *Brain Res.*, 870, 73–79, 2000.

Yavich, L. and Tiihonen, J., Patterns of dopamine overflow in mouse nucleus accumbens during intracranial self-stimulation, *Neurosci. Lett.*, 293, 41–44, 2000.

3 Presynaptic Regulation of Extracellular Dopamine as Studied by Continuous Amperometry in Anesthetized Animals

Marianne Benoit-Marand, Marie-Francoise Suaud-Chagny, and François Gonon

CONTENTS

ELECTRICALLY EVOKED DOPAMINE RELEASE MONITORED BY CONTINUOUS AMPEROMETRY

INTRODUCTION

The Quest for Speed

In the peripheral and central nervous systems, neurotransmission is achieved by the release of neurotransmitters from nerve terminals into the extracellular fluid bathing postsynaptic receptors. Therefore, locally monitoring neurotransmitters in the extracellular fluid is an important technical goal. Because neurotransmission in the brain is a rapid process, with the transfer of chemical messages between neurons occurring in milliseconds, monitoring techniques that provide high temporal resolution are especially significant. Dopamine (DA) is an important neurotransmitter in several structures of the central nervous system such as the striatum, which contains the highest density of dopaminergic terminals. Since the 1970s, two main approaches to monitoring DA in the extracellular fluid have been developed: in vivo microdialysis and various in vivo electrochemical techniques. Electrochemical techniques can be used to monitor DA because the DA molecule is easily oxidized at the surface of carbon electrodes. Among the electrochemical techniques, continuous amperometry provides the fastest available temporal response with a low selectivity for DA over other oxidizable molecules. This chapter explains how to take advantage of the high temporal resolution of continuous amperometry while simultaneously avoiding the potential pitfalls associated with its low selectivity.

Principle of Continuous Amperometry

Continuous amperometry involves the application of a constant potential difference between a working electrode and a reference electrode. This potential is fixed at a level sufficient to oxidize the molecule of interest. In the case of DA, the carbon electrode is held at $+0.4$ V versus an Ag/AgCl reference electrode. Molecules that come into contact with the electrode surface are oxidized, with each DA molecule transferring two electrons to the electrode, giving rise to a measurable oxidation current. Molecules are delivered to the electrode surface by diffusion, which occurs at a rate proportional to the concentration in the solution surrounding the electrode. Hence, the DA oxidation current is also proportional to the DA concentration in the solution. However, any substance present in the solution that oxidizes at or below the potential applied to the electrode may contribute to the total oxidation current. For this reason, the total oxidation current might include contributions from other oxidizable substances, so continuous amperometry is not selective and must be used accordingly.

Methods: Carbon Fiber Electrodes and Potentiostats

With all electrochemical techniques, the aim is to oxidize molecules in the solution; therefore, the material selected for the working electrode must be conductive and highly resistant to being oxidized itself. This latter requirement especially applies to continuous amperometry because the working electrode is continuously held at an oxidative potential; most other electrochemical techniques involve alternating between oxidative and reductive potentials. Carbon, which is both conductive and more resistant to oxidation than metals, is the best choice of electrode material for electrochemical techniques used in the oxidation range. Among the various forms of carbon, carbon fibers are the most suitable because they are made of pyrolitic graphite, a material akin to pure graphite crystal. Gonon and coworkers designed the first carbon fiber microelectrode in 1978.[1]

The flow of oxidation current between the working and reference electrode might alter the applied potential by ohmic drop or electrode polarization. If not avoided, this would lead to irreproducible results. Some workers use a three-electrode potentiostat design in which the current through the reference electrode is maintained at zero by means of an auxiliary electrode. However, the active surface area of a carbon fiber microelectrode is very small indeed, so the current actually recorded during continuous amperometry is likewise very small, i.e., 1 nA or less. Thus, the ohmic drops are too small to be of any concern, and polarization effects are eliminated by using a reference electrode with sufficient surface area to assure that the current density at the reference electrode is small. Both potentiostats designed for continuous amperometry (AMU 130 by Radiometer Analytical, Lyon, France and Micro-C by WPI, Sarasota, FL) are two-electrode potentiostats.

Validity and Field of Application

Monitoring Dopamine in the Presence of Other Oxidizable Compounds

The use of continuous amperometry is completely valid when the compound of interest represents the main oxidizable compound released in the vicinity of the carbon fiber electrode. This condition exists during the monitoring of adrenaline or noradrenaline release by single chromaffin cells[2] and of DA release by single PC12 cells.[3] Likewise, in vitro preparations of sympathetic nerves release only noradrenaline, which can be unequivocally monitored by continuous amperometry.[4] The technique can also be used for monitoring electrically evoked DA release in striatal slices because, in this preparation, oxidizable molecules in the extracellular space of the slice remain below the detection limit of continuous amperometry unless DA release is evoked either by electrical stimulation or by amphetamine.[5]

In contrast, the in vivo monitoring of DA release, either in anesthetized or in freely moving animals, is a much more challenging task because the extracellular DA level remains below 100 n*M*, even after amphetamine stimulation, whereas the main metabolite of DA (i.e., DOPAC) is present in the extracellular space at a concentration of 20 μ*M*, and ascorbic acid is present at a concentration of 200 μM.[6,7] Both DOPAC and ascorbic acid are easily oxidizable substances that contribute to the oxidation current when the carbon fiber electrode is continuously held at +0.4 V versus Ag/AgCl. Several voltammetric techniques have been developed to monitor released DA without interference from DOPAC or ascorbic acid.[8–12] These studies were performed in anesthetized rats and showed that brief electrical stimulations of dopaminergic neurons, at a distance from the striatum, evoked a detectable increase in extracellular DA level but did not alter the extracellular levels of DOPAC and ascorbic acid. Brief electrical stimulations, lasting for 40 s or less, evoked an immediate increase in the exocytotic release of DA, which did not overlap with a modest and delayed increase in extracellular DOPAC[12,13] because the half-life of DOPAC is about 10 min[6] and because DOPAC is not sequestered in synaptic vesicles of dopaminergic neurons.[14]

In anesthetized animals, electrical stimulation of dopaminergic neurons does not enhance the extracellular ascorbic acid level because dopaminergic terminals do not release ascorbic acid. Indeed, in the striatum of rats rendered hemi-parkinsonian by a unilateral 6-hydroxydopamine lesion, the striatal extracellular ascorbic acid level is not altered in the lesioned striatum compared to the intact one.[7] Moreover, the contribution of ascorbic acid to the secretion spikes, corresponding to exocytotic release from single chromaffin cells, has been proved to be negligible compared to the contribution of catecholamines[15]; the same is likely true regarding exocytotic release from dopaminergic fibers *in vivo*.[14] Nevertheless, the extracellular ascorbic acid level is not always stable. For example, it is enhanced by amphetamine or dopaminergic agonists and depressed by D2 dopaminergic antagonists and anesthetics. These pharmacological effects are explained by the link between ascorbic acid release and the activity of the glutamate transporters.[16] These variations in the extracellular ascorbic acid level affect the basal oxidation current recorded with continuous amperometry; however, they do not contribute to the transient oxidation signal evoked by brief electrical stimulations because pharmacologically induced ascorbic acid variations are much slower.[7]

Amperometric Monitoring of Dopamine Release Evoked by Brief Electrical Stimulation

From the considerations developed above, it follows that, in anesthetized animals, DA is the only oxidizable compound released to the striatal extracellular fluid by brief electrical stimulation of dopaminergic neurons. Therefore, although continuous amperometry cannot separate DA from other oxidizable compounds, it is suitable for monitoring evoked DA overflow in this particular condition. Indeed, the brief changes in oxidation current, evoked by electrical stimulation and monitored with continuous amperometry in the striatum of anesthetized rats, exhibited all the expected characteristics of an impulse flow-dependent DA release process: (1) The evoked responses closely fit the anatomy of the dopaminergic neurons. Indeed, electrical stimulation of sites just above the medial forebrain bundle (MFB) did not evoke any response.[17] Moreover, the amplitude of the evoked responses was closely related to the density of dopaminergic terminals.[18] (2) When the carbon fiber electrode was held at 0 V instead of +0.4 V, i.e., at a potential that does not cause DA oxidization, no change in the oxidation current was recorded except transient electrical artifacts associated with the stimulation pulses.[17,19] (3) A blockade of impulse flow-dependent DA release by 6-hydroxydopamine injection in the MFB completely suppressed the evoked responses.[20] Moreover, partial degeneration of striatal dopaminergic terminals induced a corresponding decrease in the amplitude of the evoked responses.[21] (4) The evoked responses were abolished by the local application of tetrodotoxin and strongly diminished by cadmium, a blocker of calcium channels.[17] (5) The evoked responses were closely time-correlated with the onset and duration of the stimulation.[17,22] (6) Inhibition of DOPAC synthesis by pargyline did not decrease the evoked responses.[17] (7) Inhibition of DA uptake strongly slowed down the decay phase of the evoked responses.[17,23] Moreover, in the striatum of mice lacking the DA transporter, this decay phase was 100 times slower than in normal mice.[18] (8) Potentiation of the evoked DA release by D2 antagonists enhanced the amplitude of the evoked responses,[17,22] but this effect was not observed in mice lacking D2 receptors.[24]

Field of Application

Continuous amperometry can be used only in those restricted experimental conditions where only one oxidizable compound is responsible for the variations in the oxidation current recorded by the carbon fiber electrode. In contrast, continuous amperometry cannot be used to monitor DA release in freely moving animals in behavioral situations or to monitor variations of the basal extracellular DA level induced by pharmacological treatments such as amphetamine or dopaminergic agonists. Continuous amperometry can be used in anesthetized animals to monitor DA release evoked by brief electrical stimulations. It is sensitive enough to record the DA overflow evoked by single pulse stimulations in brain regions densely innervated by dopaminergic terminals.[17,18,20,25] In brain regions exhibiting a lower density of dopaminergic innervation, such as the basolateral nucleus of the amygdala, high-frequency stimulus trains (24 pulses, 60 Hz) must be used to evoke detectable DA overflow.[26] Finally, continuous amperometry can also be used in brain regions densely innervated by noradrenergic terminals to monitor the noradrenaline overflow evoked by brief electrical stimulations.[27]

Comparison with Other Approaches

Comparison with In Vivo Microdialysis

Microdialysis has proved to be suitable for monitoring the basal extracellular DA level in freely moving animals. However, the time resolution of microdialysis is in the range of one to several minutes even if faster sampling rates are used. Indeed, this low time resolution is inherent in the fact that variations in the DA content of collected samples reflect variations in the extracellular DA level in the intact tissue with a time distortion caused by: (1) DA diffusion from intact release sites to the microdialysis probe through tissues lesioned by the implantation of the probe; (2) DA diffusion through the dialysis membrane; and (3) DA dilution in the perfusion medium filling the probe and the tubing connecting the probe to the fraction collector.[28] Therefore, pronounced DA overflow evoked by brief electrical stimulation cannot be monitored with microdialysis unless animals are treated with an uptake inhibitor. Indeed, uptake inhibition promotes DA diffusion from intact release sites to the probe, as elegantly shown by studies combining microdialysis with fast scan cyclic voltammetry (FSCV).[29,30]

Comparison with Fast Scan Cyclic Voltammetry

Compared to continuous amperometry, FSCV is much more selective for DA over other oxidizable compounds. Therefore, FSCV can be used to monitor DA transients in freely moving rats in behavioral situations (see Chapter 2 by Robinson and Wightman). FSCV has also been used extensively in anesthetized animals to monitor the DA overflow evoked by electrical stimulation. Observations obtained with FSCV and with continuous amperometry are generally in excellent agreement.[26,31,32] Minor discrepancies between some of the earliest studies have since been resolved. For example, the uptake blocker GBR 12909 was found to potentiate DA release rather than affecting DA clearance according to an early FSCV study.[33] In contrast, continuous amperometry clearly showed that GBR 12909 slows down DA clearance[23]; this observation was confirmed by a subsequent FSCV study from Wightman's group.[34]

The main advantage of continuous amperometry over FSCV is that the former exhibits a faster temporal response.[26,31,32] First, the sampling rate usually used with FSCV is 100 ms; the amperometric response to a rectangular pulse of catecholamine is more rectangular than the FSCV response because of the temporal lag due to adsorption and desorption of DA to and from the electrode.[32] With continuous amperometry, the time resolution is that of the data acquisition system and might be also limited by the filter of the current amplifier. A sampling interval of 20 μs is used to monitor amperometric events corresponding to the exocytotic release of DA.[14] When the electrically evoked DA overflow is recorded with continuous amperometry, the sampling interval is usually 1 ms.[25] This is sufficient because the kinetics of evoked DA overflow is constrained by DA reuptake. Maximal DA uptake kinetics occur in the in the dorsal striatum, where the apparent DA half-life is in the range of 60–80 ms.[18,20,23] Models of DA diffusion from release sites to carbon fiber electrodes suggest that the real half-life of DA in the intact striatum is in the range of 25–40 ms.[25] Therefore, the decay phase of the evoked DA overflow reflects DA elimination by reuptake with a much better accuracy when it is monitored with continuous amperometry than with FSCV.

Comparison with Other Electrochemical Techniques

Apart from FSCV and continuous amperometry, two other electrochemical techniques have been used to monitor electrically evoked DA overflow: differential normal pulse voltammetry

(DNPV) and differential pulse amperometry (DPA). Both techniques were mainly used in combination with electrochemically treated carbon fiber electrodes. With DNPV it is possible to record, every 30 s, two distinct oxidation peaks corresponding to ascorbic acid and catechols. In untreated rats, the main compound contributing to the catechol peak is DOPAC.[6] When rats are treated with pargyline to inhibit DOPAC synthesis, the remaining catechol peak is entirely due to extracellular DA.[10,11] With this approach, it was possible to show that electrical stimulation of the MFB evoked the release of DA but did not alter the extracellular ascorbic acid level.[11] Therefore, this approach provided definitive information to characterize the in vivo monitoring of the evoked DA overflow. Currently, the interest of DNPV is low because of its slow time resolution.

DPA represents an improvement of DNPV with similar selectivity and sensitivity and a sampling rate of 500 ms. When DPA is combined with electrochemically treated carbon fiber electrodes, it is sensitive enough to detect the evoked DA overflow in the prefrontal cortex.[37] However, the real-time resolution is limited by the response time of electrochemically treated carbon fiber electrodes (10–30 s, depending on the treatment).[12,13,38] Comparison between DPA and continuous amperometry is more extensively discussed in a previous review.[39]

DPA has also been used with untreated carbon fiber electrodes.[40] With this approach, the time resolution is given by the sampling rate of DPA, which is 500 ms. The selectivity of DPA for DA over ascorbic acid[40] is similar to that of continuous amperometry (Dugast, Suaud-Chagny, and Gonon 1994). The only advantage of DPA is that the carbon fiber electrode is held at a negative potential (−240 mV) between measurement pulses (+200 mV versus Ag/AgCl). Therefore, with DPA, DA molecules, which are oxidized by the carbon fiber to the corresponding quinone, can be reduced back by the electrode to DA. In contrast, because with continuous amperometry the electrode is maintained at an oxidizing potential, the electrode may contribute, by oxidative degradation, to eliminate the released DA. Comparison between continuous amperometry and DPA, combined with the same untreated carbon fiber electrode in the same experimental conditions, made it possible to resolve this issue. In this experiment, the rate of elimination of the released DA was monitored in the striatum of mice lacking the DA transporter. With both techniques, the measured DA half-life was exactly the same.[18] This comparison demonstrates that, even in the absence of DA uptake, oxidation of DA by the electrode in the continuous amperometric mode does not contribute to DA elimination because, in the extracellular fluid, ascorbic acid is present at a much higher concentration than DA. Thus, the quinones formed by oxidation of DA at electrodes are immediately reduced back to DA by ascorbic acid, which is a potent reducing agent.[18,26] From these considerations, it follows that in vitro tests for measuring the sensitivity to DA in the continuous amperometric mode must be performed with solutions containing high levels ($\geq 100\ \mu M$) of ascorbic acid.[17,26,32]

CHARACTERISTICS OF DOPAMINERGIC TRANSMISSION

Dopaminergic Transmission Is Extrasynaptic but Local

Whether dopaminergic transmission is confined to the synaptic cleft or occurs through stimulation of postsynaptic receptors far away from release sites has been a matter of debate for decades. However, experimental evidence in favor of extrasynaptic transmission is growing.[36] First, most dopaminergic receptors are not located inside dopaminergic synapses.[41–44] Second, dopaminergic transporters are also extrasynaptic.[45] Third, most of the released DA diffuses out of the synaptic cleft and can be detected with carbon fiber electrodes.[20,35] Diffusion of DA in the extrasynaptic extracellular space is limited by DA uptake. Therefore, in the striatum, where uptake is very active, half of the released DA is cleared at a diffusion distance of 7–12 μM.[20,35,36] Thus, dopaminergic transmission is extrasynaptic but also highly localized. This conclusion points out the value of carbon fiber electrodes over microdialysis probes. The size of the former (5–8 μM in diameter) is in the same range as the average diffusion distance of DA, whereas that of the microdialysis probes (200 μM in diameter) is much larger.

RELEASE AND ELIMINATION OF DOPAMINE FROM THE EXTRACELLULAR SPACE

Release and Diffusion of Dopamine in the Extracellular Fluid to the Electrode

The exocytotic release of DA from dopaminergic terminals is extremely fast: it occurs within less than 1 ms, as reported from the amperometric monitoring of quantal release in cultured dopaminergic neurons.[14] Unfortunately, this fast quantal release cannot be recorded in vivo or in striatal slices because it is not possible to position a carbon fiber electrode in direct contact with one or several intact release sites. The electrode penetration creates a dead zone of lesioned tissue between the intact release sites and the active surface of the electrode. The thickness of this dead zone has been estimated to be between 6 and 9 μM.[5,26,36] These estimates seem reasonable given the fact that the diameter of the carbon fibers range from 5 to 8 μM.

Elimination of Released Dopamine by Reuptake

Many pharmacological studies have shown that reuptake by the DA transporter plays a major role in the clearance of released DA although other pathways of elimination, such as extracellular degradation, have been considered. Recent studies using mice lacking the DA transporter show that reuptake is the only mechanism responsible for the clearance of released DA. In the striatum and nucleus accumbens of these mice, the half-life of released DA is slowed down by two orders of magnitude.[18,46] The half-life is not affected by inhibitors of catechol-O-methyl transferase or noradrenaline or serotonin transporters and is only slightly slowed by inhibition of monoamine oxidase.[18,46] Likewise, in the basal nucleus of the amygdala, DA uptake seems to be the only mechanism responsible for the elimination of released DA.[47] In the prefrontal cortex it appears likely that the noradrenaline transporter also contributes to the clearance of DA.[48] DA reuptake has been modeled with Michaellis–Menten kinetics. It is a slow phenomenon compared to exocytotic release. Indeed, the half-life of DA has been estimated to range from 25 to 40 ms (see above).

Monitoring the Evoked Dopamine Overflow with Carbon Fiber Electrodes

The extracellular DA concentration results from a dynamic equilibrium between release and reuptake. When DA overflow is evoked by a brief electrical stimulation, the rising phase of the oxidation signal reflects release minus reuptake and the decay phase only reuptake. Because release is much faster than reuptake, the rising phase should be much sharper than the decay phase. However, diffusion between intact dopaminergic terminals and the electrode surface through the dead zone equally slows down both phases. This explains why the transient oxidation signal evoked by brief electrical stimulation appears only slightly asymmetrical when recorded in the striatum (Figure 3.1). The asymmetry increases in the nucleus accumbens and olfactory tubercle because reuptake is less efficient in these limbic areas than in the striatum[23] and is even more pronounced in the basal nucleus of the amygdala where reuptake is much slower.[26] Nevertheless, when compared under the same experimental conditions with the same electrode, transient oxidation signals appear more asymmetrical when recorded with continuous amperometry than with FSCV.[5] Therefore, continuous amperometry is more accurate than FSCV for the independent estimate of release and reuptake parameters from evoked oxidation responses.

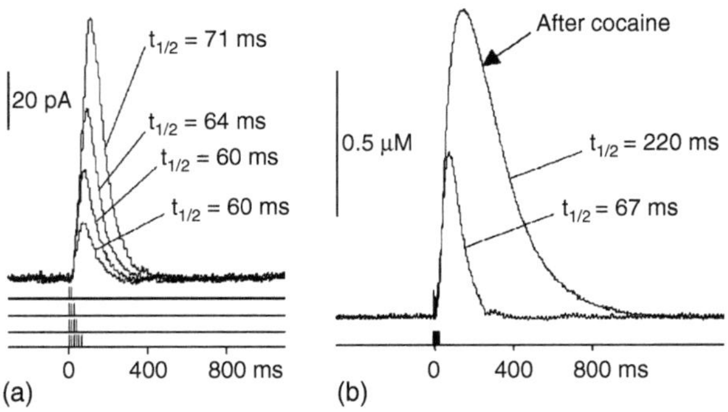

FIGURE 3.1 Effects of train length and of cocaine on evoked DA overflow. The transient changes in oxidation current were monitored with a carbon fiber electrode held at +0.4 V by continuous amperometry in the striatum of a C57 mouse as described.[18] MFB electrical stimulations were applied every 15 s, and the traces represent the average of ten responses to ten successive stimulations. Stimulation artifacts were removed by subtracting recordings obtained when the electrode was held at 0 V.[17] The carbon fiber electrode was calibrated for DA, after in vivo recordings, using a flow injection system.[17] For each averaged trace, the half-life ($t_{1/2}$), i.e., the time for 50% decrease from the maximum, is indicated. (a) Responses evoked by stimulations consisting in 2, 3, 4, and 6 pulses at 100 Hz. Notice that the maximal amplitude of the response is linearly related to the number of pulses and that the half-life is only moderately enhanced. (b) Responses evoked by MFB stimulations (4 pulses at 100 Hz) either before or 20 min after cocaine (20 mg/kg subcutaneous). The DA uptake inhibitor cocaine strongly enhanced half-life and also the maximal amplitude of the response.

TECHNICAL RECOMMENDATIONS

EXPERIMENTAL PROTOCOLS

Carbon Fiber Electrodes

Two shapes of carbon fiber electrodes have been developed: disk electrodes, in which the active surface is the section of one carbon fiber, and cylindrical carbon fiber electrodes, in which the active surface is that of a single carbon fiber protruding for 25–250 μM out of an insulating glass pipette. The first advantage of cylindrical electrodes over disk electrodes is that their active area is much larger. Because the oxidation current is proportional to the active area, this shape facilitates the detection of low DA concentrations. The second advantage derives from the fact that the cylindrical surface represents the main active area and parallels the penetration axis. During implantation into brain tissues, the tip of the electrode is covered with tissue fragments, but most of the cylindrical surface remains uncovered (see Figure 1 in Gonon[49]). Early on, FSCV was performed with disk-shaped carbon fiber electrodes, but cylindrical electrodes are increasingly used in combination with FSCV in recent in vivo studies (see Chapter 2). Continuous amperometry has always been combined with cylindrical electrodes.[17,26]

Several procedures have been developed to produce cylindrical carbon fiber electrodes. The main difficulty is that one must use a resin to ensure a good seal between the carbon fiber and the glass pipette, but the resin must not cover the active electrode surface. We developed a procedure in which a polyester resin is pushed into the glass pipette from its back, unpulled end.[50] The resin reaches the end of the pulled tip of the pipette by capillary action and, normally, does not cover the protruding part of the carbon fiber. However, this procedure is quite delicate. With other procedures, an epoxy resin fills the pipette tip by capillary action from the pulled end of the pipette, and the electrode is immediately washed in acetone to remove residual epoxy from the active part of the carbon fiber.[26]

Several types of carbon fibers have been used to produce electrodes. Because carbon fibers are usually produced to strengthen composite products, producers treat them to improve the adhesion of resin to carbon. This treatment is called *sizing* and alters the electrochemical properties of carbon fibers. Therefore, electrodes must be produced with unsized carbon fibers. In our team, we compared three types of unsized carbon fibers for their efficacy in combination with continuous amperometry: (1) AGT 10000, produced by SEROFIM, France, with a diameter of 8 μM (these fibers are no longer produced); (2) T300J, produced by SOFICAR, France, under the TORAYCA license, with a diameter of 7 μM; and (3) unsized carbon fiber supplied by Goodfellow Corporation, Devon, PA (reference product C 005722), with a diameter of 7 μM. The third type of fibers was found to be the most efficient and was used in our most recent study.[51]

Implantation of Carbon Fiber Electrodes In Vivo

The implantation procedure must minimize the covering of the electrode with tissue fragments. The surgical approach must be performed under a microscope to avoid damaging the cortical surface. In rats, the cortical surface is protected by the pia mater, a sticky and transparent layer adhering to the cortical surface, and by the dura mater, a very strong membrane separated from the pia mater by cerebrospinal fluid. The dura mater must be surgically removed to give carbon fiber electrodes access to brain tissue. Direct implantation of carbon fiber electrodes through the pia mater is mechanically possible but results in fouling the active tip of the electrode with pia matter. To minimize fouling, we puncture the pia mater with a previously used carbon fiber electrode that has been stored for a few months in the open air. The resulting hole is marked with a thin, black ring of dust at the cortical surface. Then, a new carbon fiber electrode is implanted through this ring. In mice, the dura mater is thinner and adheres to the cortical surface. Therefore, it cannot be surgically removed. New carbon fiber electrodes are implanted through a hole formed in the dura mater as described above.

After an initial in vivo experiment, carbon fiber electrodes may be reused after washing off any tissue fragments with a water solution containing 5% triton X100. However, the electrode performance tends to decrease with successive in vivo experiments. The maximal number of experiments performed in our lab with the same electrode is three in rats and five in mice.

Reference Electrode

As in almost all electrochemical studies applied to neuroscience, our reference electrode is of the Ag/AgCl type. The active part is a silver wire 0.5 mm in diameter and 10 mm long (reference product 7890, supplied by A-M Systems, INC, Everet, WA). It is electrochemically coated with silver chloride by anodizing it in hydrochloric acid (1 *N*) at +0.65 V for 1 min. During in vivo recording, the active portion of the reference electrode is put between two sponges (2 mm thick) moistened by physiological saline solution (NaCl 9 g/L) and spread out on the skull. Reference electrodes must be renewed every five to ten in vivo experiments.

In Vitro Calibration of Carbon Fiber Electrodes

After in vivo experiments, the sensitivity of the carbon fiber electrodes to DA is tested in a flow injection system, originally developed for FSCV[52] and used for continuous amperometry.[17,32] For reasons given above (section entitled Comparison with Fast Scan Cyclic Voltammetry), the perfusion fluid must contain at least 100 μM of ascorbic acid. Because the comparison of in vitro calibrations performed before and after in vivo implantation shows a substantial decrease in sensitivity induced by implantation, reliable in vitro calibration must be performed after in vivo implantation. However, the use of in vitro calibration to estimate the absolute amplitude of an evoked overflow in terms of DA concentration must be considered with caution. Indeed, electrochemical detection is affected by several parameters that cannot be easily controlled under in vivo conditions: diffusion coefficient, pH, and the presence of compounds that are not electroactive by themselves but which might interfere with DA detection.

Analysis of the Data

Direct Estimate of Dopamine Release and of Dopamine Uptake

With high frequency stimulation (100 Hz), the maximal amplitude of the oxidation responses is linearly related to the number of pulses between one and four pulses, either in untreated animals (Figure 3.1; see also [18]), or when DA uptake is inhibited by nomifensine.[19] Indeed, during these brief stimulations, DA uptake is negligible over DA release. Therefore, the maximal amplitude of the responses evoked by these brief stimulations is a good index of the amplitude of the DA release per pulse. For example, this index was used to show that DA release was reduced by 65% in the striatum of mice treated with MPTP[21] and that STOP null mice exhibit a higher DA release than control mice.[53]

As illustrated in Figure 3.1, the half-life, measured from the decay phase of the evoked responses, is a good index of the efficiency of DA uptake. Indeed, half-life is increased by up to one order of magnitude after pharmacological inhibition[23,51] and by two orders of magnitude in mice lacking the DA transporter.[18] However, uptake inhibition also enhances the maximal amplitude of the evoked responses, even if the stimulation parameters (high frequency and number of pulses $\leq$4) are chosen to minimize uptake influence (Figure 3.1).[51] Indeed, uptake inhibition favors diffusion of DA from release sites to the electrode surface.[26] Therefore, variations in the maximal amplitude of the responses evoked by brief stimulations reflect changes in the DA release per pulse provided that they are not associated with alterations of DA uptake. For example, Brun et al.[53] observed that the response evoked by single pulse stimulation was larger in STOP null mice compared to wild type mice. They attributed this increase to enhanced DA release because they showed that, in a wide range of comparable overflow amplitudes, $t_{1/2}$ was similar in both genotypes.

DA reuptake is often modeled with Michaellis–Menten kinetics, in which two parameters are defined: the affinity parameter (K_M) and the maximal velocity (V_{max}). The half-life ($t_{1/2}$ in Figure 3.1) depends on both K_M and V_{max}. Estimation of K_M from the evoked DA overflow requires a mathematical model (see below). However, a rough estimate of V_{max} can be directly obtained from the evoked DA overflow. Indeed, V_{max} roughly corresponds to the maximal slope of the decay phase after stimuli of sufficient intensity (e.g., 20–50 pulses at 50–100 Hz) to cause DA to reach levels sufficient to saturate the DAT.[53]

Mathematical Models

As discussed above, the evoked DA overflow results from DA release minus DA uptake. However, Wightman et al. already reported in 1988 that models based only on these two processes poorly fit FSCV recordings. Better fit was obtained by taking into account the diffusion of DA from intact release sites to the electrode.[54] Three models have been recently developed to simulate recordings obtained with continuous amperometry.[5,26,55] All of them take into account the diffusion of DA from intact release sites to the electrode. The model proposed by Venton et al.[26] represents an improvement over the former ones because it takes into account the fact that ascorbic acid reduces DA molecules oxidized by the electrode. The fit of this model with experimental recordings performed under highly diverse experimental conditions is indeed impressive.

CONCLUSIONS

Continuous amperometry has been used in brain slices to monitor the electrically evoked DA overflow and compared to FSCV in the same experimental conditions. In slices, continuous amperometry provides the same pieces of information as FSCV but with a better time resolution.[5]

The use of continuous amperometry, both in striatal slices and in the striatum of anesthetized rodents, might help to resolve a puzzling question: the characteristics of the electrically evoked DA release strongly differ when studied in vitro or in vivo. Up to now, attempts to resolve this issue have been unsuccessful.[56]

Continuous amperometry, combined with untreated cylindrical carbon fiber electrodes, is a valid technique to monitor the DA overflow evoked by brief electrical stimulations in anesthetized rats and mice. In these restricted experimental conditions, continuous amperometry is superior to FSCV because it provides the same pieces of information but with better time resolution.[26] Moreover, continuous amperometry is an easier technique, with minimal requirement in terms of electrochemical equipment. Direct measurement of the maximal amplitude of the evoked signals, and of the time for 50% decay from the maximum, provides fairly independent measurements of the two mechanisms that control extracellular DA levels: DA release per pulse and DA uptake. These direct estimates are sufficient in most experimental conditions for studying the parameters that regulate release and uptake. In more demanding experimental conditions, which associate prominent changes in both release and uptake, mathematical models can be used for a more accurate and independent measurement of release and uptake.[26] Thus, FSCV, and, more recently, continuous amperometry, have greatly contributed to the knowledge of the presynaptic mechanisms regulating dopaminergic transmission (also see the review by Schmitz et al.)[57] In the future, the use of continuous amperometry to monitor electrically evoked DA release, either in slices or in anesthetized animals, will undoubtedly help to tackle unresolved issues such as short-term plasticity of dopaminergic transmission.[57]

REFERENCES

1. Gonon, F. et al., Mesure électrochimique continue de la libération de dopamine réalisée in vivo dans le néostriatum du rat, *CR Acad. Sci. Paris*, 286, 1203, 1978.
2. Wightman, R. M. et al., Temporally resolved catecholamine spikes correspond to single vesicle release from individual chromaffin cells, *Proc. Natl. Acad. Sci.*, 88, 10754, 1991.
3. Pothos, E. N., Desmond, M., and Sulzer, D., L-3,4-dihydroxyphenylalanine increases the quantal size of exocytotic dopamine release in vitro, *J. Neurochem.*, 66, 629, 1996.
4. Gonon, F., Msghina, M., and Stjärne, L., Kinetics of noradrenaline released by sympathetic nerves, *Neuroscience*, 56, 535, 1993.
5. Schmitz, Y. et al., Amphetamine distorts stimulation-dependent dopamine overflow: effects on D2 autoreceptors, transporters, and synaptic vesicle stores, *J. Neurosci.*, 21, 5916, 2001.
6. Gonon, F. et al., In vivo electrochemical detection of catechols in the neostriatum of anaesthetized rats: dopamine or DOPAC? *Nature*, 286, 902, 1980.
7. Gonon, F. et al., Voltammetry in the striatum of chronic freely moving rats: detection of catechols and ascorbic acid, *Brain Res.*, 223, 69, 1981.
8. Ewing, A. G., Bigelow, J. C., and Wightman, R. M., Direct in vivo monitoring of dopamine release from two striatal compartments in the rat, *Science*, 221, 169, 1983.
9. Kuhr, W. G. et al., Monitoring the stimulated release of dopamine with in vivo voltammetry. I: characterization of the response observed in the caudate nucleus of the rat, *J. Neurochem.*, 43, 560, 1984.
10. Gonon, F., Navarre, F., and Buda, M., In vivo monitoring of dopamine release in the rat brain with differential normal pulse voltammetry, *Anal. Chem.*, 56, 573, 1984.
11. Gonon, F. and Buda, M., Regulation of dopamine release by impulse flow and by autoreceptors as studied by in vivo voltammetry in the rat striatum, *Neuroscience*, 14, 765, 1985.
12. Gonon, F., Nonlinear relationship between impulse flow and dopamine release by rat midbrain dopaminergic neurons as studied by in vivo electrochemistry, *Neuroscience*, 24, 19, 1988.
13. Suaud-Chagny, M. F., Buda, M., and Gonon, F. G., Pharmacology of evoked dopamine release studied in the olfactory tubercle by in vivo electrochemistry, *Eur. J. Pharmacol.*, 164, 273, 1989.
14. Pothos, E. N., Davila, V., and Sulzer, D., Presynaptic recording of quanta from midbrain dopamine neurons and modulation of the quantal size, *J. Neurosci.*, 18, 4106, 1998.

15. Leszczyszyn, D. J. et al., Secretion of catecholamines from individual adrenal medullary chromaffin cells, *J. Neurochem.*, 56, 1991.
16. Rebec, G. V. and Pierce, R. C., A vitamin as neuromodulator: ascorbate release into the extracellular fluid of the brain regulates dopaminergic and glutamatergic transmission, *Prog. Neurobiol.*, 43, 537, 1994.
17. Dugast, C., Suaud-Chagny, M. F., and Gonon, F., Continuous in vivo monitoring of evoked dopamine release in the rat nucleus accumbens by amperometry, *Neuroscience*, 62, 647, 1994.
18. Benoit-Marand, M., Jaber, M., and Gonon, F., Release and elimination of dopamine in vivo in mice lacking the dopamine transporter: functional consequences, *Eur. J. Neurosci.*, 12, 2985, 2000.
19. Chergui, K., Suaud-Chagny, M. F., and Gonon, F., Nonlinear relationship between impulse flow, dopamine release and dopamine elimination in the rat brain in vivo, *Neuroscience*, 62, 641, 1994.
20. Gonon, F., Prolonged and extrasynaptic excitatory action of dopamine mediated by D1 receptors in the rat striatum in vivo, *J. Neurosci.*, 17, 5972, 1997.
21. Bezard, E. et al., Adaptative changes in nigrostriatal pathway in response to increased 1-methyl-4-phenyl-1,2,3,6-tetrahydropyridine-induced neurodegeneration in the mouse, *Eur. J. Neurosci.*, 12, 2892, 2000.
22. Benoit-Marand, M., Borrelli, E., and Gonon, F., Inhibition of dopamine release via presynaptic D2 receptors: time course and functional characteristics in vivo, *J. Neurosci.*, 21, 9134, 2001.
23. Suaud-Chagny, M. F. et al., Uptake of dopamine released by impulse flow in the rat mesolimbic and striatal systems in vivo, *J. Neurochem.*, 65, 2603, 1995.
24. Rougé-Pont, F. et al., Changes in extraces lular dopamine induced by morphine and cocaine: crucial control by D2 receptors, *J. Neurosci.*, 22, 3293, 2002.
25. Dugast, C. et al., On the involvement of a tonic dopamine D2-autoinhibition in the regulation of pulse-to-pulse-evoked dopamine release in the rat striatum in vivo, *Naunyn Schmiedebergs Arch. Pharmacol.*, 355, 716, 1997.
26. Venton, B. J. et al., Real-time decoding of dopamine concentration changes in the caudate-putamen during tonic and phasic firing, *J. Neurochem.*, 87, 1284, 2003.
27. Dugast, C., Cespuglio, R., and Suaud-Chagny, M. F., In vivo monitoring of evoked noradrenaline release in the rat anteroventral thalamic nucleus by continuous amperometry, *J. Neurochem.*, 82, 529, 2002.
28. Westerink, B. H., Analysis of biogenic amines in microdialysates of the brain, *J. Chromatogr. B Biomed. Sci. Appl.*, 747, 21, 2000.
29. Lu, Y., Peters, J. L., and Michael, A. C., Direct comparison of the response of voltammetry and microdialysis to electrically evoked release of striatal dopamine, *J. Neurochem.*, 70, 584, 1998.
30. Yang, H., Peters, J. L., and Michael, A. C., Coupled effects of mass transfer and uptake kinetics on in vivo microdialysis of dopamine, *J. Neurochem.*, 71, 684, 1998.
31. Michael, D. J. and Wightman, R. M., Electrochemical monitoring of biogenic amine neurotransmission in real time, *J. Pharm. Biomed. Anal.*, 19, 33, 1999.
32. Venton, B. J., Troyer, K. P., and Wightman, R. M., Response times of carbon fiber microelectrodes to dynamic changes in catecholamine concentration, *Anal. Chem.*, 74, 539, 2002.
33. May, L. J., Kuhr, W. G., and Wightman, R. M., Differentiation of dopamine overflow and uptake processes in the extracellular fluid of the rat caudate nucleus with fast-scan in vivo voltammetry, *J. Neurochem.*, 51, 1060, 1988.
34. Budygin, E. A. et al., Effect of tolcapone, a catechol-O-methyltransferase inhibitor, on striatal dopaminergic transmission during blockade of dopamine uptake, *Eur. J. Pharmacol.*, 370, 125, 1999.
35. Garris, P. A. et al., Efflux of dopamine from the synaptic cleft in the nucleus accumbens of the rat brain, *J. Neurosci.*, 14, 6084, 1994.
36. Gonon, F. et al., Geometry and kinetics of dopaminergic transmission in the rat striatum and in mice lacking the dopamine transporter, *Prog. Brain Res.*, 25, 291, 2000.
37. Claustre, Y. et al., SSR181507, a dopamine D(2) receptor antagonist and 5-HT(1A) receptor agonist. I: neurochemical and electrophysiological profile, *Neuropsychopharmacology*, 28, 2064, 2003.
38. Marcenac, F. and Gonon, F., Fast in vivo monitoring of dopamine release in the rat brain with differential pulse amperometry, *Anal. Chem.*, 57, 1778, 1985.
39. Suaud-Chagny, M. F., In vivo monitoring of dopamine overflow in the central nervous system by amperometric techniques combined with carbon fibre electrodes, *Methods*, 33, 322, 2004.

40. Suaud-Chagny, M. F. et al., Fast in vivo monitoring of electrically evoked dopamine release by differential pulse amperometry with untreated carbon fibre electrodes, *J. Neurosci. Methods*, 45, 183, 1992.
41. Levey, A. I. et al., Localization of D1 and D2 dopamine receptors in brain with subtype-specific antibodies, *Proc. Natl. Acad. Sci.*, 90, 8861, 1993.
42. Sesack, S. R., Aoki, C., and Pickel, V. M., Ultrastructural localization of D-2 receptor-like immunoreactivity in midbrain dopamine neurons and their striatal targets, *J. Neurosci.*, 14, 88, 1994.
43. Yung, K. K. et al., Immunocytochemical localization of D1 and D2 dopamine receptors in the basal ganglia of the rat: light and electron microscopy, *Neuroscience*, 65, 709, 1995.
44. Caillé, I., Dumartin, B., and Bloch, B., Ultrastructural localization of D1 dopamine receptor immunoreactivity in the rat striatonigral neurons and its relation with dopaminergic innervation, *Brain Res.*, 730, 17, 1996.
45. Nirenberg, M. J. et al., The dopamine transporter: comparative ultrastructure of dopaminergic axons in limbic and motor compartments of the nucleus accumbens, *J. Neurosci.*, 17, 6899, 1997.
46. Giros, B. et al., Hyperlocomotion and indifference to cocaine and amphetamine in mice lacking the dopamine transporter, *Nature*, 379, 606, 1996.
47. Garris, P. A. and Wightman, R. M., Distinct pharmacological regulation of evoked dopamine efflux in the amygdala and striatum of the rat in vivo, *Synapse*, 20, 269, 1995.
48. Valentini, V., Frau, R., and Di Chiara, G., Noradrenaline transporter blockers raise extracellular dopamine in medial prefrontal but not parietal and occipital cortex: differences with mianserin and clozapine, *J. Neurochem.*, 88, 917, 2004.
49. Gonon, F., Monitoring dopamine and noradrenaline release in central and peripheral nervous systems with treated and untreated carbon-fiber electrodes, In *Voltammetric Methods in Brain Systems*, Boulton, A., Baker, G., and Adams, R. N., Eds., Humana Press, Clifton, NJ, 1995, (chap. 5).
50. Gonon, F., Buda, M., and Pujol, J. F., Treated carbon fiber electrodes for measuring catechols and ascorbic acid, In *Measurement of Neurotransmitter Release In Vivo*, Marsden, C. A., Ed., Wiley, Chichester, U.K., 1984, chapter 7.
51. Samaha, A. N. et al., The rate of cocaine administration alters gene regulation and behavioral plasticity: implications for addiction, *J. Neurosci.*, 24, 6362, 2004.
52. Kristensen, E. W., Wilson, R. L., and Wightman, R. M., Dispersion in flow injection analysis measured with microvoltammetric electrodes, *Anal. Chem.*, 58, 986, 1986.
53. Brun, P. et al., Dopaminergic transmission in STOP null mice, *J. Neurochem.*, 94, 63, 2005.
54. Wightman, R. M. et al., Real-time characterization of dopamine overflow and uptake in the rat striatum, *Neuroscience*, 25, 513, 1988.
55. Schonfuss, D. et al., Modelling constant potential amperometry for investigations of dopaminergic neurotransmission kinetics in vivo, *J. Neurosci. Methods*, 112, 163, 2001.
56. Benoit-Marand, M., Schmitz, Y., Gonon, F., Zhang, H., and Sulzer, D., Monitoring dopamine release under in vivo and in vitro conditions, In *Monitoring Molecules in Neorosciences, Proceedings of the 9th International Conference on In Vivo Methods*, National University of Ireland, Dublin, p. 227, 2001.
57. Schmitz, Y. et al., Presynaptic regulation of dopaminergic neurotransmission, *J. Neurochem.*, 87, 273, 2003.

4 Fast Scan Cyclic Voltammetry of Dopamine and Serotonin in Mouse Brain Slices

Carrie E. John and Sara R. Jones

CONTENTS

INTRODUCTION

Fast scan cyclic voltammetry (FSCV) is an electrochemical technique that can be used to monitor release and uptake dynamics of endogenous monoamine levels both in vitro [1–3] and in vivo.[4–6] Fast scan cyclic voltammetry is able to measure the three major monoamine neurotransmitters, dopamine (DA), serotonin (5-HT), and norepinephrine (NE), because these substances can be oxidized at low voltages, providing selective electrochemical detection based on voltage-dependent oxidation and reduction processes.

Briefly, a voltage that is sufficient to oxidize monoamines is applied to a carbon fiber microelectrode. The oxidation process results in current flow at the electrode surface; the amount of current is converted into concentration of monoamine in the vicinity of the electrode tip by means of calibration in a flow injection system. The current is measured rapidly, every 100 ms, and when plotted against time, can provide information as to how, for example, stimulation parameters and pharmacological agents can alter the dynamics of the monoamine released near the surface of the electrode. Ultimately, these measurements make it possible to distinguish kinetic parameters for release and uptake processes, a distinguishing feature of electrochemical techniques like FSCV. This chapter is designed to describe the use of FSCV to measure DA and 5-HT in mouse brain slices.

ADVANTAGES OF THE USE OF MOUSE BRAIN SLICES

Guinea pigs and rats have been most commonly used to explore monoamine dynamics, particularly in brain slices.[1,2] It was nearly 10 years ago with the advent of the DA transporter (DAT) knockout (DAT-KO) mouse[7] that FSCV was first routinely used in mouse brain slices. Recent increases in the use of genetically defined mouse lines have made mice an important animal model for monoamine research. Thus far, FSCV research in mice has focused almost exclusively on DA dynamics in striatal brain slices. While species generality does seem to apply to striatal DA detection, it is important to note that species variability in midbrain monoamine recording has been shown between guinea pig and rat,[8] and we have recently documented further variations in the mouse midbrain.[9] Therefore, the reasoned choice of animal models can aid in obtaining desired information about specific monoamines in specific brain areas. This fact highlights the importance of characterizing mouse neurochemistry in its own right and suggests that direct translation of knowledge gained from one species is not necessarily applicable to another species.

Brain slices were chosen as the in vitro assay in which to assess monoamine dynamics because several slices can be made from each region of interest to provide the opportunity to test one pharmacological agent on several slices to reduce statistical variability or to test several pharmacological agents in a single mouse brain. Additionally, one animal can be used to test slices from multiple brain regions. These facts have made brain slice voltammetry an exemplary tool to study DA dynamics in ex vivo studies; we have had the unique opportunity to study changes in the DA system (release, uptake, autoreceptor control, and drug effects) in hard-to-get brain tissue from rats and monkeys that had long histories of drug or alcohol self-administration.[10,11] Brain slices are also convenient because anatomical structures within the slice are visible and electrode placement can be made vs. anatomical landmarks, rather than by stereotaxic coordinates, as is the case of in vivo experiments. Due to the small size of the carbon fiber microelectrode (7 μm diameter, 50–200 μm in length), it is possible to differentiate structures that are spatially closely associated, even in the mouse brain, for example: nucleus accumbens (NAc) core vs. NAc shell or ventral tegmental area (VTA) vs. substantia nigra (SN). The use of brain slices also allows greater control of the physiological environment (e.g., temperature), and experimentally manipulated stimulation parameters can greatly reduce interference from such contaminants as pH changes, which are not possible to control in vivo. Furthermore, brain slice work does not suffer from the presence of anesthetic drugs that are necessary in an anesthetized in vivo preparation. Finally, the slice perfusion system makes drug application simple and reproducible, and cumulative drug concentration–effect curves can be collected within a single slice. We have determined that cumulative concentrations of pharmacological agents (most notably DA uptake inhibitors) applied to brain slices to not differ in their ability to alter uptake or release when compared to the same pharmacological agent applied at a single concentration. Furthermore, an advantage of brain slices is that the pharmacological action of drugs can be localized to the brain region of interest, as there are no contributions of pharmacological actions in the periphery or at other central locations. These features of brain slice voltammetry allow for an efficient way to assess monoamine dynamics in a variety of brain areas and carry out several pharmacological manipulations in each region, while reducing the number of animals necessary to collect large data sets.

METHODOLOGY FOR FAST SCAN CYCLIC VOLTAMMETRY IN MOUSE BRAIN SLICES

Electrode Fabrication and Calibration

A triad of reference, auxiliary and working electrodes are typically used in brain slice FSCV recordings. We use a silver/silver chloride reference electrode made from silver wire (A-M Systems, Inc., Carlsborg, WA) that is coated with a thin layer of silver chloride by electroplating in a solution of 1 *M* HCl. A 22 AWG tinned copper wire (Belden, Parkville, Victoria, Australia) is used for the auxiliary electrode.

To prepare the cylindrical carbon fiber microelectrodes,[12] a single carbon fiber (7 μm radius, type P-450, Amoco, Greenville, SC) is aspirated into a glass capillary with a microfilament (1.6 mm inside diameter, 1.2 mm outside diameter, A-M Systems, Inc., Carlsborg, WA). Each capillary is pulled into two electrodes using a micropipette puller (Model PE-2, Narashige, Tokyo, Japan or Model P-97, Sutter Instrument Co., Novato, CA) so that carbon fiber is exposed at the tapered (pulled) end. The exposed carbon fiber is trimmed to a length of approximately 50–200 μm with the aid of a non-inverting microscope at 40× magnification (Model BH-2, Olympus Corp., Lake Success, NY). Before use, the electrodes are backfilled with 150 m*M* potassium chloride using a spinal needle (25 gauge, 3.5 in., Becton–Dickinson and Co., Franklin Lakes, NJ), and a 4.5 in Kynar UL 1422 28/1 g wire (Squires Electronics, Inc., Cornelius, OR) is inserted in the blunt end to make an electrical connection with the carbon fiber.

When the carbon-fiber aspirated capillaries are pulled, it is important that the taper of the glass make a tight seal around the carbon fiber. If a good seal between the glass and the carbon is not obtained, the electrode can be sealed with epoxy; however, we have moved away from this practice because it is not necessary for the majority of electrodes produced. Traditionally, carbon fiber microelectrodes have been fabricated on vertical pullers, for example, Model PE-2 from Narashige Inc. On this style of puller, two settings, magnet and heat, are adjusted to create the desired type of glass seal on the carbon fiber. Recently, we have begun to fabricate electrodes on a horizontal puller, the Sutter Instruments Model P-97, similar to those used to fabricate electrophysiology electrodes. This style of puller requires four settings—heat, pull, velocity, and time—to be adjusted. In this case, the electrodes are heated and pulled in a manner controlled by a program entered into the puller's computerized controller. This allows a greater amount of flexibility and control over working electrode fabrication. On either style puller, it is initially necessary to adjust the puller settings to obtain the best seal of glass to carbon fiber and desired shape. However, very few adjustments are necessary after this initial determination of optimal settings.

A flow injection system[13] is used to calibrate electrode responses to DA and 5-HT. The calibration buffer consists of (in m*M*) NaCl (126), KCl (2.5), NaH_2PO_4 (1.2), $CaCl_2$ (2.4), $MgCl_2$ (1.2), $NaHCO_3$ (25), and HEPES acid (20), pH 7.4 at 23°C (room temperature). It is ideal to calibrate electrodes in a solution that is identical to that in which brain slice recording are obtained, i.e. artificial cerebrospinal fluid (aCSF), however, we have determined that the inclusion of ascorbate and glucose in aCSF used in electrode calibration did not change the sensitivity of our electrodes to DA or 5-HT.[9] It is important to note that calibration of electrodes must occur with concentrations of monoamines that are relevant to the biological sample, as decreased sensitivity of electrodes is evident with high monoamine concentrations.[9] In our hands, monoamine release in brain slices is generally ≤ 1 μ*M*, therefore we use calibration factors obtained with concentrations of DA or 5-HT that are between 0.1 and 1 μ*M*.

FSCV Data Acquisition

The commercially available EI-400 (Cypress Systems, Inc., Lawrence, KS) is both a waveform generator and a biopotentiostat which measures and amplifies the current at the working electrode relative to the reference and sends that information to the National Instruments (Austin, TX) interface boards in the computer for analog-to-digital conversion and subsequent analysis with locally written data analysis programs. For FSCV, an EI-400 is used in B-channel and two- or three-electrode mode (three-electrode mode if an auxiliary reference electrode is utilized). In this configuration, the EI-400 uses the reference electrode as ground, and alters the potential of the carbon fiber microelectrode relative to that ground. For DA recording, the potential of the carbon fiber microelectrode is linearly scanned from −400 to 1200 mV and back to −400 mV vs. Ag/AgCl (Figure 4.1a, triangle waveform); for 5-HT recording, the potential is scanned from 0 to 1200 to −600 mV and back to 0 vs. Ag/AgCl (Figure 4.1b, N waveform). The difference in applied waveforms is due to the difference in the adsorption properties of DA and 5-HT. Both DA and 5-HT are positively charged molecules and are attracted to the carbon surface when sitting at negative potentials. Dopamine and 5-HT are both adsorbed onto the electrode surface when a negative resting potential is applied between scans. However, adsorption occurs to a much greater extent with 5-HT phenolic and hydroxyindole oxidation products, blocking portions of the electrode surface, thereby greatly reducing electrode sensitivity and also causing slow electrode response times. To combat this problem, 5-HT recording uses an applied waveform that is maintained at 0 mV vs. Ag/AgCl between scans, preventing 5-HT accumulation around the surface of the electrode; the electrode is scanned positively to oxidize 5-HT, then negatively to reduce 5-HT, and returned to a 0 mV potential vs. Ag/AgCl. For both DA and 5-HT waveforms, each is applied at 300 V/s, lasts 9–10 ms and occurs every 100 ms. While it is not possible to directly distinguish the catecholamines DA and NE with FSCV, catecholamines can be distinguished from the indolamines, including 5-HT. Catecholamines and indolamines produce distinct current vs. voltage curves (voltammograms), and as seen in Figure 4.2, DA and 5-HT can be distinguished based on their reduction current peaks. Typically, DA and 5-HT oxidation peaks are both at approximately 500–700 mV vs. Ag/AgCl, while the single DA reduction peak is around −200 mV and the double 5-HT reduction peaks are around 0 and −500 vs. Ag/AgCl.

We utilize background-subtraction techniques in order to create voltammograms. This is because the fast scan rate generates a large background current which can obscure the voltammetric current due to monoamines. Fortunately, the background current is fairly stable. Therefore, to obtain voltammograms, the background current obtained before electrical stimulation of the slice

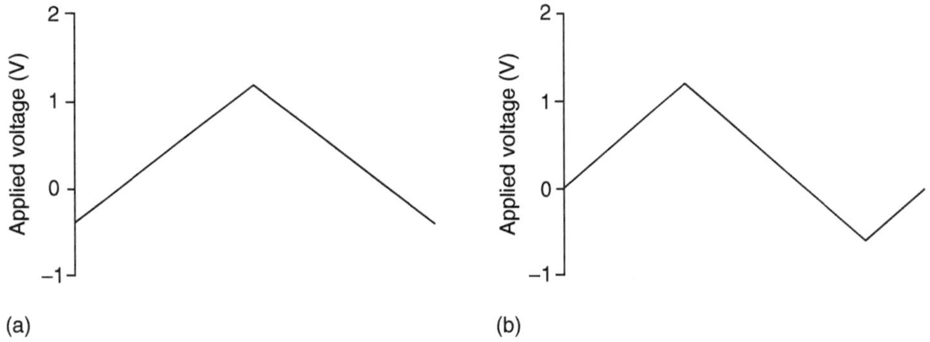

FIGURE 4.1 Voltages applied to carbon fiber electrodes during voltammetric recordings of dopamine (DA) and serotonin (5-HT). (a) The triangle waveform, −400 to 1200 and back to −400 mV vs. Ag/AgCl, is used for DA recording. (b) The N waveform, 0–1200 to −600 and back to 0 mV vs. Ag/AgCl, is used for 5-HT recording.

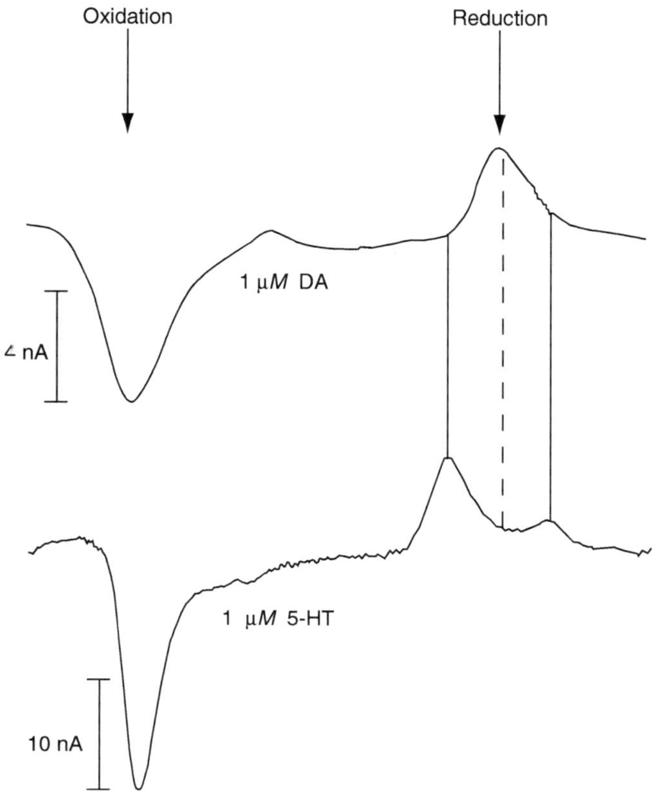

FIGURE 4.2 Dopamine and 5-HT voltammograms. Dopamine (1 μ*M*) voltammograms (current vs. voltage curves) show one oxidation and one reduction peak. Serotonin (1μ*M*) voltammograms show one oxidation and two reduction peaks. Voltammograms were obtained using the N waveform in Figure 4.1b. Note that the reduction peaks of DA and 5-HT occur at different potentials.

is digitally subtracted from that of the voltammograms of interest (generally the peak of monoamine release). In this work, five scans are averaged together to obtain both the background and stimulation-induced signals. To obtain current vs. time plots, the current at the oxidation potential (typically 500–700 mV vs. Ag/AgCl for both DA and 5-HT) is converted to concentration based on post-calibration of the electrode in 1 μ*M* DA or 5-HT.

Brain Slice Preparation

Mice between 2 and 4 months of age are asphyxiated with CO_2. Once the animal has stopped breathing, it is removed and quickly decapitated. To remove the brain, the skin and fur on the top of the head are removed with large dissecting scissors, cuting from the nape of the neck to the nose. Then, any remaining spinal column and membranous tissue from the posterior part of the skull over the cerebellum is removed with large dissecting scissors. The skull is removed by cutting the lateral aspects of both the occipital and interparietal bones to the lambdoid suture and then removing the occipital and interparietal bones. Then, the remaining skull removed by cutting along the line of the lateral sutures of the parietal and frontal bones, to the nasal septum. Then, a flat spatula is inserted at the base of the skull along the rostral–caudal axis. The brain is lifted up slightly, any remaining muscles or nerves connecting the brain to the skull are cut, and then the brain is fully removed and placed in approximately 100 mL of ice cold oxygenated (95% O_2/5% CO_2) aCSF. The aCSF consists of (in m*M*) NaCl (126), KCl (2.5), NaH_2PO_4 (1.2), $CaCl_2$ (2.4), $MgCl_2$ (1.2), $NaHCO_3$ (25), glucose (11), HEPES (20), L-ascorbic acid (0.4), pH 7.4.

After the brain chills for 1 min in ice-cold aCSF, it is placed on a Petri dish containing frozen aCSF that is covered with a paper towel soaked with ice-cold aCSF. There, a single-edged carbon steel blade (Electron Microscopy Sciences, Hatfield, PA) is used to remove the olfactory bulbs and cerebellum; the brain is then cut in half along the coronal midline with a single-edged blade. The cut end of the brain is glued to the cutting stage of the specimen chamber (which is pre-chilled in a −20°C freezer) of the vibratome (Leica VT1000S, Vashaw Scientific Inc., Norcross, GA) with cyanoacrylate glue (Instant Krazy Glue, Advanced Formula, Columbus, OH). The specimen chamber is then mounted within the vibratome, a small cube of frozen aCSF is added to the specimen chamber to keep the buffer cold, and the chamber is then filled with ice-cold oxygenated aCSF. The area surrounding the chamber is then filled with crushed ice and ice-cold water, with attention paid to avoiding getting ice or water within the chamber. One half of a double-edged stainless steel razor blade (Electron Microscopy Sciences, Hatfield, PA) is placed in the blade holder of the vibratome to cut slices 400 μm thick. All slices are made within 20 min of brain removal. Using a spatula and small horse-hair paint brush, slices are transferred to a 100 mL beaker containing room temperature aCSF that is constantly oxygenated where they incubate for at least 1 h.

Electrically Stimulated Monoamine Release

Electrical Stimulation

Slices used for FSCV are placed in a locally constructed slice perfusion chamber, weighted down with 2–3 mm sections of platinum wire (0.5 mm diameter, Alfa Aesar, Ward Hill, MA) and superfused with oxygenated aCSF (either 23 or 32°C) at 1 mL/min. We have carried out brain slice voltammetry experiments in both room temperature baths and those closer to physiological temperature. While evoked monoamine release and subsequent uptake can be monitored at room temperature, monoamine transporters act in a temperature-dependent manner[14] and uptake kinetics are slower at room temperature, compared to physiological temperature.

The electrical stimulation is generated by the computer (Dell Optiplex GX280, Dell Inc., Round Rock, TX) and sent to digital to analog and input–output interface boards (National Instruments, Austin, TX), which sends the stimulus to optical isolators (Neuro Log NL-800, Medical Systems, Greenvale, NY). Optical isolators are used to separate the electrical stimulation from the electrochemical measurements. After optical isolation, the stimulus is sent to the bipolar stimulating electrodes (MS303/3, Plastics One, Roanoke, VA). Monophasic stimulation pulses, 1 or 4 ms wide, 350 μA, are generated by locally constructed software. The stimulating electrode is mounted on a micromanipulator (Model MM3, Narishige International U.S.A. represented by Pacer Scientific, Los Angeles, CA) and positioned so that it rests on the surface of the slice. The carbon-fiber microelectrode, mounted on a separate but identical micromanipulator, is positioned 75 μm into the slice, 100 μm from the center of the tips of the bipolar stimulating electrode, creating an equilateral triangle. Electrode placements are made with the aid of a stereo-microscope (Sterozoom 6, Leica, Vashaw Scientific, Norcross, GA). Stimulation parameters differ between each brain region of interest. Generally, a 1-pulse stimulation (4 ms wide) is used to elicit DA from the caudate-putamen (CPu), NAc core and NAc shell. 30-pulse, 30 Hz stimulations (1 ms wide) are used to elicit monoamine release from VTA, SN pars compacta (SNc), SN pars reticulata (SNr) and dorsal raphe nuclei (DRN). Occasionally, stimulus artifacts (noise) are generated by the stimulation. Reducing the duration (width) or amplitude of the stimulus pulse(s) is usually effective in reducing or eliminating the artifacts.

Advantages of the Use of Electrical Stimulation

Electrical stimulation is used to elicit the release of endogenous monoamines from the brain slice and then uptake kinetics can be monitored. This offers a great advantage over exogenously applied monoamine studies. While transporter uptake rates can be reasonably estimated from low concentrations of exogenously applied monoamines, the high concentrations of often used exogenously applied monoamines can complicate the interpretation of uptake data. For example, high concentrations of DAT substrates, such as DA or amphetamine, can cause substrate transport both into (forward direction) and out of (reverse direction) the cytoplasm. The kinetics of transporters that are dually functioning to take up substrate, as well as release substrate, are complicated and transporter kinetics under these conditions cannot be attributed to uptake alone.

Experimental Design

In a typical brain slice FSCV experiment, evoked overflow is first determined by obtaining responses to electrical stimulation in aCSF. Interstimulus intervals vary in each brain region and we generally stimulate every 5 min in the CPu, NAc core, and NAc shell and every 10 min in the VTA, SNc, SNr, and DRN. It was noted in midbrain regions that interstimulus intervals of less than 10 min caused depletion of the releasable pool of monoamine. This is presumably due to the slower rates of uptake and lower synthesis enzyme levels of these regions, compared to the striatum. Responses usually stabilize within 1 h and are required to be stable for at least 30 min before pharmacological agents are applied. Pharmacological agents are applied to the slices via the superfusate, and responses are recorded throughout the drug application. It is important to determine the time for the maximal effect of acute superfusion of a pharmacological agent for each individual drug, as factors such as lipophilicty are important determinants of the ability of the drug to act at the recording site.

Data Analysis

Uptake Rate Determination

The extracellular concentration of monoamines primarily reflects a balance between release and uptake; pharmacological agents can change one or both of these processes and these drug effects are of neurobiological interest. Generally, a Michaelis–Menten based kinetic model is used to evaluate release and uptake kinetics.[15]

$$\frac{d[DA]}{dt} = f[DA]p - \frac{V_{max}}{(K_m/[DA]) + 1} \tag{4.1}$$

[DA] is the instantaneous extracellular concentration of DA, f is the pulse frequency, [DA]p is the concentration of DA elicited per stimulus pulse, and V_{max} and K_m are Michaelis–Menten uptake parameters. For this type of modeling, three assumptions are made: (1) each stimulus pulse releases a fixed quantity of monoamine into the extracellular space, [DA]p or [5-HT]p, (2) uptake is a saturable process, and (3) uptake is the primary mechanism-clearing, released monoamine. Furthermore, this modeling assumes a control K_m value (inversely related to the affinity of the transporter for its monoamine) of 0.16 μM for DA at DAT and 0.17 μM for 5-HT and the 5-HT transporter (SERT). An additional kinetic parameter, V_{max}, which is related to the number and turnover rate of the monoamine transporters, is also determined. These assumptions are best suited for evaluating release and uptake in striatal regions that use 1-pulse stimulations and have rapid uptake, but can also be used with multiple pulse stimulations in areas with low uptake rates, such as the midbrain.[9]

Michaelis–Menten kinetic determination of uptake parameters would be inaccurate to describe DA clearance in DAT-KO mice, with genetically deleted DAT, because, as stated above, there must be the assumption that uptake is the primary mechanism clearing released monoamine. This is because Michaelis–Menten kinetics assumes two processes: (1) monoamine binding to the transporter (with the affinity described by K_m), and (2) subsequent "catalysis" of DA, i.e. transport into the intracellular space, which is maximal (described by V_{max}) under saturating conditions. Because DAT-KO mice are genetically lacking DAT, the first step of this process is impossible. Previously in DAT-KO mice, DA clearance was estimated using a Michaelis–Menten kinetic modeling program with the K_m value set at a high concentration. Then a pseudo first-order rate constant, k, was calculated by dividing V_{max} by K_m.[16] Jones et al.[16] reported that the k value for DA clearance in CPu slices from DAT-KO mice was 0.05 s^{-1}. When using the Michaelis–Menten kinetic model in monoamine transporter knockout mice, one must assume a K_m of greater than 10 μM. This makes intuitive sense, given that the absence of the transporter implies no affinity of the transporter for its monoamine. Figure 4.3 illustrates an example of k values obtained using various presumed K_m values in the Michaelis–Menten kinetic model in a single CPu slice from a DAT-KO mouse. As is apparent, the K_m value assumed in wild type animals (approximately 0.2 μM) would greatly overestimate the k value, which should be approximately 0.05 s^{-1}, reflective of the first-order elimination process of diffusion.

Due to the differences in stimulation parameters and uptake kinetics between regions of the striatum and regions of the midbrain, and the fact that there is no active uptake in monoamine transporter knockout animals, a different method of modeling data has proven useful for the midbrain as well as knockout mice. In areas where release is in the low micromolar range, uptake can be simplified to fit first order kinetics, as described by Cragg et al.[17] In our analyses, a one-phase exponential decay fit well. Figure 4.4 shows examples of VTA uptake modeled with both Michaelis–Menten kinetics and one-phase exponential decay, both of which determine fits of the data with $r^2 \geq 0.90$. Additionally, this method can also be used to determine uptake rates in monoamine transporter knockout mice. We have independently determined the uptake rate constant k using slices from midbrain and from animals with a genetic deletion of monoamine transporters by fitting the clearance portion of the current vs. time curves to a one-phase exponential decay. In the VTA of DAT-KO mice, we have found k values of 0.06 ± 0.02[18] and 0.07 ± 0.01[9] s^{-1} utilizing

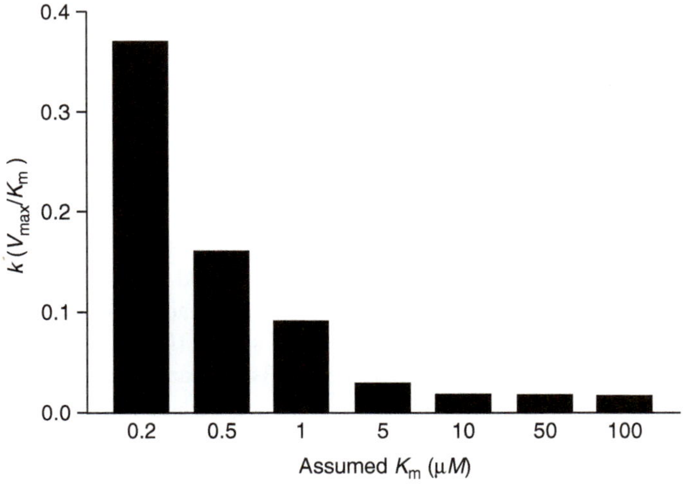

FIGURE 4.3 Pseudo first-order rate constant k values, using the equation $k = V_{max}/K_m$, obtained using various assumed K_m values in the Michaelis–Menten kinetic model (obtained from modeling a single DA transporter knockout (DAT-KO) CPu slice). It is necessary to use an assumed K_m value of 10 μM or greater to accurately describe uptake kinetics in transporter knockout mice.

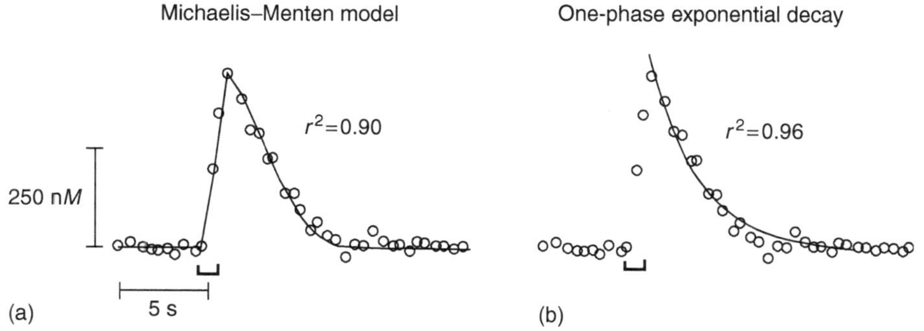

FIGURE 4.4 Representative efflux curves measured by fast scan cyclic voltammetry (FSCV) in response to 30-pulse, 30-Hz (350 μA, 1 ms, indicated by the line under the efflux curves) stimulation in an individual wild-type mouse ventral tegmental area (VTA) slice. Open circles are raw data (plotted every 5th point for visual clarity). (a) Michaelis–Menten kinetic modeling fit is overlaid (solid line). (b) One-phase exponential decay fit is overlaid on the falling phase of the curve. The r^2 values for fits of the model to the raw data are indicated.

Michaelis–Menten modeling ($k = V_{max}/K_m$) and one-phase exponential decay, respectively (Figure 4.5), which are both in agreement with the 0.05 s^{-1} published for DAT-KO CPu.[16]

Regional Variation in Uptake Rates

As suggested above, the uptake rate of a given transmitter is brain-region specific. Striatal DA uptake rates are faster in the CPu than in the NAc, which are in turn faster than in amygdala, prefrontal cortex or somatodendritic regions like the VTA and SN.[2,19–21] This suggests that DA signaling is regionally distinct, and has led to the evolution of the term "volume transmission" to describe extracellular diffusion to reach sites of action distant from the synapse.[17,22–24] It is important to understand that active uptake and volume transmission are not mutually exclusive concepts. Diffusion to sites of action and clearance by active uptake both occur during DA volume transmission.[25] In fact, it is well established that DATs are located outside the synapse,[26,27] as are the majority of DA receptors,[28,29] making extrasynaptic diffusion necessary and traditional synaptic neurotransmission unlikely. Furthermore, quantitative evaluation of diffusion and uptake in midbrain suggests that diffusion is the most important determinant of the lifetime of DA.[17]

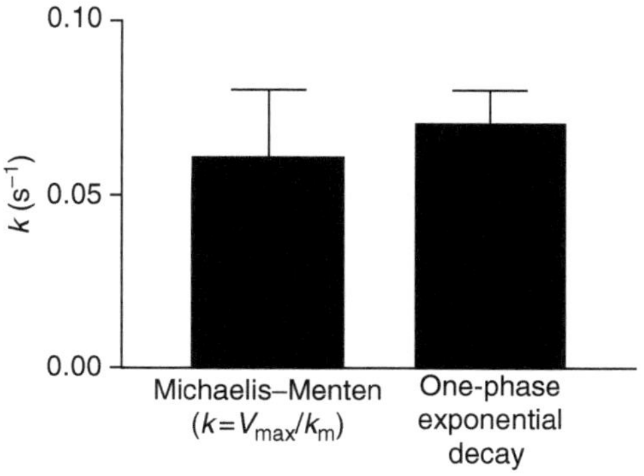

FIGURE 4.5 Two methods to determine rate constant *k* values in averaged data from VTA slices from DAT-KO mice. Michaelis–Menten modeling (left bar, from Mateo, Y. et al., *Proc. Natl Acad. Sci. U.S.A.*, 101, 372, 2004) and one-phase exponential decay modeling (right bar, from John, C. E. et al., *J. Neurochem.*, 2005, in press) yield similar results.

Even in the dorsal striatum, where DA uptake is very rapid and predominately determines DA time course, diffusion is an important component of DA transmission.[30] Even so, the relative contribution of uptake mechanisms is a factor in determining the role of diffusion,[25] which can vary greatly between brain regions, and the role of uptake can contribute to envisioning transmission in a particular brain area as synaptic-like or paracrine-like.

MONOAMINE UPTAKE INHIBITORS AND RELEASERS

MECHANISMS OF ACTION

Psychostimulants, which interact with monoamine transporters to increase extracellular monoamine levels, come in two varieties: those that are pure uptake inhibitors and those that function as releasers. Uptake inhibitors bind to transporter proteins and inhibit transporter function, slowing the uptake of monoamines from the extrasynaptic space and thereby increasing the extracellular concentration of monoamines.[31–34] Releasers, like amphetamine, increase extracellular monoamine levels primarily by acting as false substrates, competitively inhibiting neurotransmitter reuptake and promoting reverse transport.[31,35–39] In addition, releasers displace monoamines from synaptic vesicles[38,40] and act as monoamine oxidase inhibitors;[41,42] both of these effects aid to increase the cytoplasmic monoamine concentration that drives reverse transport. Binding sites on monoamine transporters likely differ for uptake inhibitors and releasers, in agreement with their different mechanisms of action.[43–45] Ultimately, uptake inhibitors enhance activity-dependent monoamine transmission, while releasers primarily promote activity-independent efflux of monoamines.

USING FSCV TO EVALUATE CHANGES IN MONOAMINE DYNAMICS IN RESPONSE TO UPTAKE INHIBITORS AND RELEASERS

Here, we use an uptake inhibitor, cocaine, and a releaser, methamphetamine, to demonstrate the utility of FSCV to measure changes in DA and 5-HT dynamics. Uptake inhibitors and releasers have very different effects on the timecourse of DA overflow curves measured by FSCV. However, both are effective at inhibiting monoamine uptake, which is evidenced by the widening of the clearance portion of the current vs. time curves with increasing concentrations of cocaine and methamphetamine (Figure 4.6).

In addition to preventing monoamine reuptake, uptake inhibitors typically cause an increase in monoamine signal peak height at low concentrations (Figure 4.6, cocaine),[46,47] due to uptake inhibition during stimulation trains and/or diffusion of the monoamine to the electrode from distant sites. The increased peak height of voltammetric signals in response to the diffusion-increasing aspects of uptake inhibition is important because the voltammetric electrode is sampling from a larger anatomical area after the drug than before. The increase in sampling area is generally thought of as a larger-diameter sphere. We generally assume that these increases are not due to actual enhancement of DA release from presynaptic terminals, although a minor contribution cannot be ruled out. At higher concentrations of uptake inhibitors, there is an apparent decrease in release resulting from the increased extracellular monoamine concentrations activating presynaptic autoreceptors (Figure 4.6, cocaine)[48] and obscuring diffusion-related increases in peak height. This effect can be prevented with an autoreceptor antagonist.

The magnitude of change following uptake inhibition is highly correlated with the baseline rate of uptake. For example, in the CPu, where DA uptake is rapid and diffusion is restricted, a large dynamic range of uptake inhibitor effects can be seen; an uptake inhibitor will greatly increase the diffusion distance of DA in the CPu. In contrast, in areas with lower uptake rates, such as the VTA, diffusion is already substantial and uptake inhibition results in a smaller-magnitude effect. For 5-HT the latter case is most common, as there are no brain areas with as great a SERT density as that of DAT in CPu. In the case of cocaine acting at the DAT and SERT (Figure 4.6), the apparent lesser

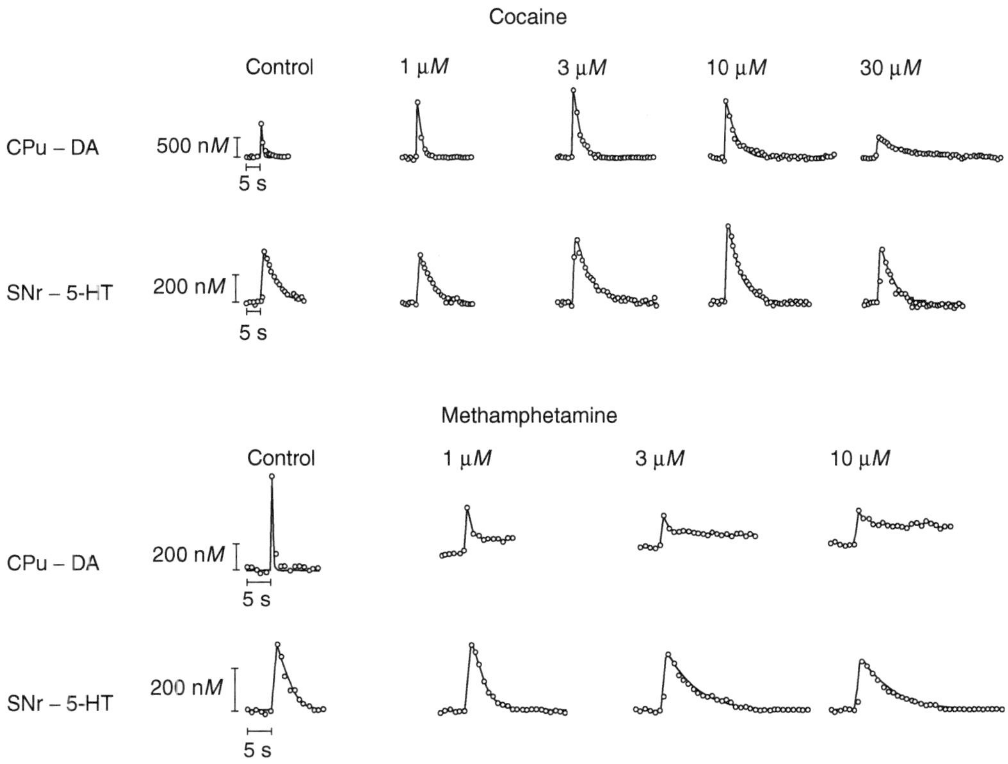

FIGURE 4.6 The effects of an uptake inhibitor, cocaine, and a releaser, methamphetamine, on DA and 5-HT voltammetric signals in wild-type mouse brain slices. Dopamine is measured in caudate-putamen (CPu) (1-pulse, 4 ms stimulations) and 5-HT is measured in SN pars reticulata (SNr) (30-pulse, 30 Hz, 1 ms stimulations). Open circles are raw data (plotted every 10th point for visual clarity). Michaelis–Menten kinetic modeling fits are overlaid (solid line), except in the case of methamphetamine in the CPu, where modeling was not possible because of failure of signals to return to baseline. In this case, dotted lines connect the raw data points. Top panels: cocaine dose-dependently decreased uptake rates, as seen in both brain regions by the widening of the curves. It also altered signal peak heights in a predicted manner (see "Using FSCV to Evaluate Changes in Monoamine Dynamics in Response to Uptake Inhibitors and Releasers"). Bottom panels: methamphetamine dose-dependently widened the curves. The CPu panel shows an example of methamphetamine causing the establishment of a new baseline during the falling phases of the curves after each stimulation. This response occurred frequently, but not always, in the presence of a releaser. The SNr panel shows an example of methamphetamine reducing release and uptake parameters without reestablishment of baselines. Methamphetamine also altered signal peak heights in a predicted manner (see "Using FSCV to Evaluate Changes in Monoamine Dynamics in Response to Uptake Inhibitors and Releasers"). Although baseline changes are depicted in the CPu panel only, there was no difference in the frequency of these types of response between the DA and 5-HT measurements. Representative responses of each type were chosen for illustrative purposes.

effect of cocaine on 5-HT uptake is also related to the greater affinity of cocaine for the DAT than the SERT.[49] While psychostimulants induce a proportionally smaller absolute change in uptake rate (and dialysate monoamine levels) in non-striatal brain regions, there are, nevertheless, still substantial physiological effects of uptake inhibition in these regions.

Current vs. time plots usually return to the baseline established before electrical stimulation after uptake inhibition, albeit slowly. In contrast, after the application of releasers, monoamine levels do not always return to baseline and may appear to "hang-up" or establish a new baseline

(Figure 4.6, methamphetamine effects on DA in the CPu). This is because releasers cause the transporter to operate in two directions, and electrically-stimulated release of monoamine shifts the extracellular concentration of substrate toward a new equilibrium. Amphetamine and its derivatives function as releasers, but these effects can also be caused by endogenous substrates, including phenylethylamine, tyramine, DA, NE, and 5-HT. One must use caution in interpreting the effects of releasers on monoamine clearance curves, as discussed above with regard to exogenous monoamine application (see "Advantages of the Use of Electrical Stimulation"). Michaelis–Menten modeling is useful for amphetamine-altered current vs. time curves that do return to baseline. However, it would be more appropriate to use first-order kinetic decay modeling for those current vs. time plots that do not return to baseline. Also, it should be noted that signals that do not return to baseline (either hanging, drifting, or dropping below baseline) in the absence of transporter substrate are likely being affected by pH, Ca^{2+}, or oxygen changes[50,51] and should not be included in data analysis unless subtraction techniques are available.[51] Also, in contrast to uptake inhibitors, the release-decreasing effects of releasers is only partially attributable to autoreceptor activation.[52–54] Electrically evoked release is dependent on vesicular storage of monoamines; releasers cause displacement of monoamines from vesicles,[38,40] thereby decreasing the releasable pool of monoamines and decreasing evoked release.

CONCLUDING REMARKS

Here, we have described the use of FSCV to measure DA and 5-HT in mouse brain slices. The growing use of genetically defined mouse lines have made mice an important animal model for monoamine research. The utility of FSCV comes primarily from its superb temporal (ms) and spatial (micron) resolution, as well as its reasonable chemical specificity and sensitivity for monoamines. FSCV is a powerful tool that can measure fast changes in extracellular monoamine concentrations in discrete brain regions, both in vitro and in vivo. FSCV in brain slices is particularly useful for characterizing acute, local drug effects; the effects of chronic in vivo drug administration on monoamine systems can be evaluated as well. While brain slice FSCV provides useful information on monoamine dynamics (release, uptake, and diffusion), coupling both in vitro and in vivo data, as well as microdialysis and electrophysiological studies, will provide the most complete description of monoamine functioning in the brain.

ACKNOWLEDGMENTS

We would like to thank Dr Marc G. Caron for the gift of DAT-KO mice. This work was supported by AA11997 and DA016498 to CEJ, and AAO14091, AA013900 and DA018815 to SRJ.

REFERENCES

1. Rice, M. E. et al., Direct monitoring of dopamine and 5-HT release in substantia nigra and ventral tegmental area in vitro, *Exp. Brain Res.*, 100, 395–406, 1994.
2. Jones, S. R. et al., Comparison of dopamine uptake in the basolateral amygdaloid nucleus, caudate-putamen, and nucleus accumbens of the rat, *J. Neurochem.*, 64, 2581, 1995.
3. Bunin, M. A. et al., Release and uptake rates of 5-hydroxytryptamine in the dorsal raphe and substantia nigra reticulata of the rat brain, *J. Neurochem.*, 70, 1077, 1998.
4. Kuhr, W. G. and Wightman, R. M., Real-time measurement of dopamine release in rat brain, *Brain Res.*, 381, 168, 1986.
5. Stamford, J. A., Kruk, Z. L., and Millar, J., Sub-second striatal dopamine release measured by in vivo voltammetry, *Brain Res.*, 381, 351, 1986.
6. Robinson, D. L. et al., Detecting subsecond dopamine release with fast-scan cyclic voltammetry in vivo, *Clin. Chem.*, 49, 1763, 2003.

7. Giros, B. et al., Hyperlocomotion and indifference to cocaine and amphetamine in mice lacking the dopamine transporter, *Nature*, 379, 606, 1996.
8. Cragg, S. J., Hawkey, C. R., and Greenfield, S. A., Comparison of serotonin and dopamine release in substantia nigra and ventral tegmental area: region and species differences, *J. Neurochem.*, 69, 2378, 1997.
9. John, C. E. et al., Neurochemical characterization of the release and uptake of dopamine in the ventral tegmental area and serotonin in substantia nigra of the mouse, *J. Neurochem.*, 2005, in press.
10. Budygin, E. A. et al., Chronic ethanol exposure alters presynaptic dopamine function in the striatum of monkeys: a preliminary study, *Synapse*, 50, 266, 2003.
11. Mateo, Y. et al., Reduced dopamine terminal function and insensitivity to cocaine following cocaine binge self-administration and deprivation, *Neuropsychopharm*, 2005, in press.
12. Kawagoe, K. T., Zimmerman, J. B., and Wightman, R. M., Principles of voltammetry and microelectrode surface states, *J. Neurosci. Methods*, 48, 225, 1993.
13. Kristensen, E. W., Wilson, R. M., and Wightman, R. M., Dispersion in flow injection analysis measured with microvoltammetric electrodes, *Anal. Chem.*, 58, 986, 1986.
14. Xie, T. et al., Effect of temperature on dopamine transporter function and intracellular accumulation of methamphetamine: implications for methamphetamine-induced dopaminergic neurotoxicity, *J. Neurosci.*, 20, 7838, 2000.
15. Wightman, R. M. and Zimmerman, J. B., Control of dopamine extracellular concentration in rat striatum by impulse flow and uptake, *Brain Res. Brain Res. Rev.*, 15, 135, 1990.
16. Jones, S. R. et al., Profound neuronal plasticity in response to inactivation of the dopamine transporter, *Proc. Natl Acad. Sci. U.S.A.*, 95, 4029, 1998.
17. Cragg, S. J. et al., Dopamine-mediated volume transmission in midbrain is regulated by distinct extracellular geometry and uptake, *J. Neurophysiol.*, 85, 1761, 2001.
18. Mateo, Y. et al., Role of serotonin in cocaine effects in mice with reduced dopamine transporter function, *Proc. Natl Acad. Sci. U.S.A.*, 101, 372, 2004.
19. Stamford, J. A. et al., Diffusion and uptake of dopamine in rat caudate and nucleus accumbens compared using fast cyclic voltammetry, *J. Brain Res.*, 448, 381, 1988.
20. Garris, P. A. and Wightman, R. M., Different kinetics govern dopaminergic transmission in the amygdala, prefrontal cortex, and striatum: an in vivo voltammetric study, *J. Neurosci.*, 14, 442, 1994.
21. Cragg, S. J., Rice, M. E., and Greenfield, S. A., Heterogeneity of electrically evoked dopamine release and reuptake in substantia nigra, ventral tegmental area, and striatum, *J. Neurophysiol.*, 77, 863, 1997.
22. Agnati, L. F. et al., Intercellular communication in the brain—wiring versus volume transmission, *Neuroscience*, 69, 711, 1995.
23. Rice, M. E., Cragg, S. J., and Greenfield, S. A., Characteristics of electrically evoked somatodendritic dopamine release in substantia nigra and ventral tegmental area in vitro, *J. Neurophysiol.*, 77, 853, 1997.
24. Zoli, M. et al., Volume transmission in the CNS and its relevance for neuropsychopharmacology, *Trends Pharmacol. Sci.*, 20, 142, 1999.
25. Cragg, S. J. and Rice, M. E., Dancing past the DAT at a DA synapse, *Trends Neurosci.*, 27, 270, 2004.
26. Nirenberg, M. J. et al., The dopamine transporter is localized to dendritic and axonal plasma membranes of nigrostriatal dopaminergic neurons, *J. Neurosci.*, 16, 436, 1996.
27. Nirenberg, M. J. et al., Immunogold localization of the dopamine transporter: an ultrastructural study of the rat ventral tegmental area, *J. Neurosci.*, 17, 5255, 1997.
28. Sesack,, S. R., Aoki, C., and Pickel, V. M., Ultrastructural localization of D2 receptor-like immunoreactivity in midbrain dopamine neurons and their striatal targets, *J. Neurosci.*, 14, 88, 1994.
29. Yung, K. K. et al., Immunocytochemical localization of D1 and D2 dopamine receptors in the basal ganglia of the rat: light and electron microscopy, *Neuroscience*, 65, 709, 1995.
30. Garris, P. A. et al., Efflux of dopamine from the synaptic cleft in the nucleus accumbens of the rat brain, *J. Neurosci.*, 14, 6084, 1994.
31. Heikkila, R. E. et al., Amphetamine: evaluation of D- and L-isomers as releasing agents and uptake inhibitors for 3H-dopamine and 3H-norepinephrine in slices of rat neostriatum and cerebral cortex, *J. Pharmacol. Exp. Ther.*, 194, 47, 1975.
32. Horn, A. S., Dopamine uptake: a review of progress in the last decade, *Prog. Neurobiol.*, 34, 387, 1990.

33. Amara, S. G. and Kuhar, M. J., Neurotransmitter transporters: recent progress, *Annu. Rev. Neurosci.*, 16, 73, 1993.
34. Giros, B. and Caron, M. G., Molecular characterization of the dopamine transporter, *Trends Pharmacol. Sci.*, 14, 43, 1993.
35. Fischer, J. F. and Cho, A. K., Chemical release of dopamine from striatal homogenates: evidence for an exchange diffusion model, *J. Pharmacol. Exp. Ther.*, 208, 203, 1979.
36. Seiden,, L. S., Sabol, K. E., and Ricaurte, G. A., Amphetamine: effects on catecholamine systems and behavior, *Annu. Rev. Pharmacol. Toxicol.*, 33, 639, 1993.
37. Sulzer, D., Maidment, N. T., and Rayport, S., Amphetamine and other weak bases act to promote reverse transport of dopamine in ventral midbrain neurons, *J. Neurochem.*, 60, 527, 1993.
38. Sulzer, D. et al., Amphetamine redistributes dopamine from synaptic vesicles to the cytosol and promotes reverse transport, *J. Neurosci.*, 15, 4102, 1995.
39. Jones, S. R. et al., Profound neuronal plasticity in response to inactivation of the dopamine transporter, *Proc. Natl Acad. Sci. U.S.A.*, 95, 4029, 1998.
40. Mosharov, E. V. et al., Intracellular patch electrochemistry: regulation of cytosolic catecholamines in chromaffin cells, *J. Neurosci.*, 23, 5835, 2003.
41. Mantle, T. J., Tipton, K. F., and Garrett, N. J., Inhibition of monoamine oxidase by amphetamine and related compounds, *Biochem. Pharmacol.*, 25, 2073, 1976.
42. Robinson, J. B., Stereoselectivity and isoenzyme selectivity of monoamine oxidase inhibitors. Enantiomers of amphetamine, *N*-methylamphetamine and deprenyl, *Biochem. Pharmacol.*, 34, 4105, 1985.
43. Sonders, M. S. et al., Multiple ionic conductances of the human dopamine transporter: the actions of dopamine and psychostimulants, *J. Neurosci.*, 17, 960, 1997.
44. Wayment, H., Meiergerd, S. M., and Schenk, J. O., Relationships between the catechol substrate binding site and amphetamine, cocaine, and mazindol binding sites in a kinetic model of the striatal transporter of dopamine in vitro, *J. Neurochem.*, 70, 1941, 1998.
45. Chen, R., Hann, D. D., and Gu, H. H., A triple mutation in the second transmembrane domain of mouse dopamine transporter markedly decreases sensitivity to cocaine and methylphenidate, *J. Neurochem.*, 94, 352, 2005.
46. Schmitz, Y. et al., Presynaptic regulation of dopaminergic neurotransmission, *J. Neurochem.*, 87, 273, 2003.
47. Garris,, P. A. and Rebec, G. V., Modeling fast dopamine neurotransmission in the nucleus accumbens during behavior, *Behav. Brain Res.*, 137, 47, 2002.
48. Wieczorek, W. J. and Kruk, Z. L., A quantitative comparison on the effects of benztropine, cocaine and nomifensine on electrically evoked dopamine overflow and rate of re-uptake in the caudate putamen and nucleus accumbens in the rat brain slice, *Brain Res.*, 657, 42, 1994.
49. Bennett, B. A. et al., Novel 2-substituted cocaine analogs: uptake and ligand binding studies at dopamine, serotonin and norepinephrine transport sites in the rat brain, *J. Pharmacol. Exp. Ther.*, 272, 1176, 1995.
50. Jones, S. R. et al., Interference by pH and Ca^{2+} ions during measurements of catecholamine release in slices of rat amygdala with fast-scan cyclic voltammetry, *J. Neurosci. Methods*, 52, 1, 1994.
51. Venton, B. J., Michael, D. J., and Wightman, R. M., Correlation of local changes in extracellular oxygen and pH that accompany dopaminergic terminal activity in the rat caudate-putamen, *J. Neurochem.*, 84, 373, 2003.
52. Wieczorek, W. J. and Kruk, Z. L., Differential action of (+)-amphetamine on electrically evoked dopamine overflow in rat brain slices containing corpus striatum and nucleus accumbens, *Br. J. Pharmacol.*, 111, 829, 1994.
53. Iravani, M. M. and Kruk, Z. L., Effects of amphetamine on carrier-mediated and electrically stimulated dopamine release in slices of rat caudate putamen and nucleus accumbens, *J. Neurochem.*, 64, 1161, 1995.
54. Schmitz, Y. et al., Amphetamine distorts stimulation-dependent dopamine overflow: effects on D2 autoreceptors, transporters, and synaptic vesicle stores, *J. Neurosci.*, 21, 5916, 2001.

5 High-Speed Chronoamperometry to Study Kinetics and Mechanisms for Serotonin Clearance In Vivo

Lynette C. Daws and Glenn M. Toney

CONTENTS

INTRODUCTION: SEROTONIN AND THE SEROTONIN TRANSPORTER

Serotonin (5-HT) regulates many complex behaviors and physiological functions. These include mood, sleep, feeding and thermoregulation, to name but a few. As a result, dysregulation of 5-HT neurotransmission can have severe consequences. For example, reduced serotonin neurotransmission is thought to underlie such affective disorders as depression and anxiety (Owens and Nemeroff 1994; Nemeroff and Owens 2004) and predispose to addictive disorders such as alcoholism (McBride et al. 1993; Virkkunen and Linnoila 1997). In addition, the hyperserotonergic state resulting from high doses of substituted amphetamines such as 3,4-methylenedioxymethamphetamine (MDMA, "Ecstasy") contributes to the "serotonin syndrome," symptoms of which include agitation, high body temperature, tachycardia, convulsions, coma, and in some cases death (Hegadoren et al. 1999; Green et al. 2003). Given the clear importance of maintaining homeostatic 5-HT neurotransmission, the regulation of extracellular concentration of 5-HT has been the focus of intense study for the past several decades.

The principle mechanism for terminating the neurochemical actions of 5-HT released from nerve terminals is its high affinity uptake from extracellular fluid by the serotonin transporter (5-HTT; SERT). Consequently, since this transporter was first identified in the brain (Shaskan and Snyder 1970; Kuhar et al. 1972), it has been studied extensively (Fuller and Wong 1990; Blakely et al. 1991; Rudnick and Wall 1993; Bengel et al. 1998; Blakely et al. 1998; Kilic et al. 2003). Until relatively recently, active uptake of 5-HT by its transporter has been studied using in vitro techniques such as [^{3}H]5-HT uptake into brain synaptosomes (Shaskan and Snyder 1970; Kokoshka et al. 1998) or into cells transfected with the 5-HTT (Wall et al. 1995; Ramamoorthy and Blakely 1999; Zhu et al. 2005). However, modern voltammetric technologies that stemmed from the pioneering work of Ralph Adams in the 1970s (Kissinger et al. 1973) have allowed 5-HTT function to be measured in vivo. Herein we describe applications of high-speed chronoamperometry, an amperometric technique, to study 5-HTT function in vivo as well as to identify alternative mechanisms for 5-HT clearance from extracellular fluid.

VOLTAMMETRIC TECHNIQUES USED TO STUDY KINETICS OF SEROTONIN CLEARANCE IN BRAIN

As described throughout this book, many forms of voltammetry/amperometry exist (see also Kawagoe et al. 1992; Michael and Wightman 1999; Gerhardt and Burmeister 2000). In simplest terms, they are distinguished by the manner in which a potential is applied to the recording electrode. Two commonly used variations are fast cyclic voltammetry and high-speed chronoamperometry. Applications of fast cyclic voltammetry to measure dopamine and serotonin release and clearance are described elsewhere in this book (see Chapter 4 by Carrie, C. E. and Jones, S. R.; Chapter 8 by Threlfell, S. and Cragg, S. J. Chapter 12 by Garris, P. A. et al.). This chapter details the utility of high-speed chronoamperometry to study mechanisms for 5-HT clearance in the rodent brain.

HIGH-SPEED CHRONOAMPEROMETRY

The principle of high-speed chronoamperometric recording is based on methods originally described by Cottrell (1902), but modern electronics and digital data acquisition have allowed measurements to be made more rapidly. For our studies of 5-HT clearance, we have adopted and modified methods developed by the Gerhardt group for monitoring the extracellular concentration of dopamine in vivo (see Gerhardt et al. 1984; Gerhardt and Hoffman 2001). To measure 5-HT, a carbon fiber electrode (CFE) coated with Nafion (see below) is positioned in suitable buffer solution, tissue section or slice or, as will be detailed in this chapter, in brain in vivo. Figure 5.1 shows the parameters for the square wave potential applied to the CFE. The CFE is held at a zero resting potential vs. a Ag/AgCl reference electrode. Then the potential is stepped to +0.55 V. This voltage step is selected to exceed the peak 5-HT oxidation potential by 0.15 V. The square wave voltage step is held at +0.55 V for 100 ms and this produces a rapid change in the current (typically in a range of 1–3 nA) recorded by the CFE that decays as a function of time. The initial current charges the electrode capacitance and, based on the CFE time constant, is permitted to decay for 20 ms before the oxidation current is sampled and integrated for 80 ms. The square wave voltage step is then returned to 0.0 V and held at this potential for 900 ms. This return to the resting

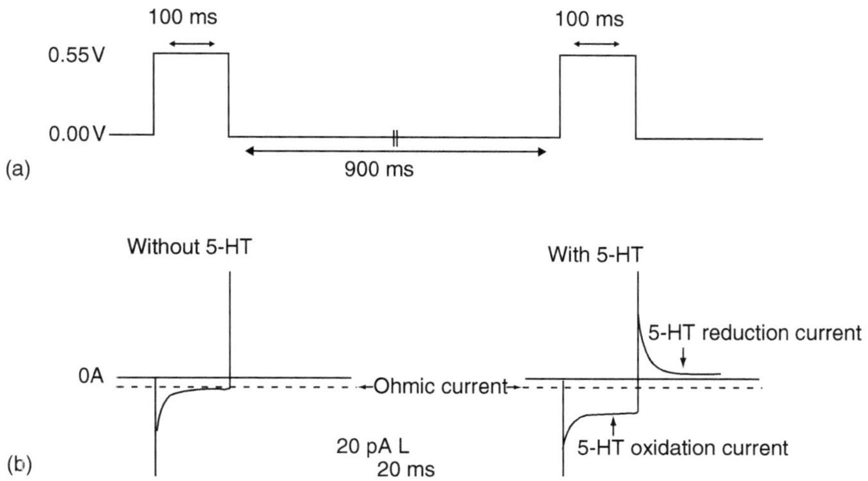

FIGURE 5.1 Square wave voltage pulses applied to the carbon fiber electrode for chronoamperometric recording of 5-HT. B. Ohmic current recorded in the absence of exogenously applied 5-HT (a) is set as "zero" and the change in current produced by oxidation of 5-HT (b) at the electrode is measured as a function of time. The current produced 20–100 ms after the potential is stepped to 0.55 V and then again after it is stepped back to 0.0 V is digitally integrated.

potential gives rise to a reduction current, which for 5-HT and its oxidation products is close to zero. For other neurotransmitters such as dopamine, the applied and resting potentials are typically each 100 ms in duration. However, unlike dopamine, which is readily oxidized and reduced, serotonin is "reluctantly" reduced.

One important implication of this is that the lack of sizable "reverse" current can cause "fouling" of the CFE. Electrogenerated compounds produced from the oxidation of 5-HT can adsorb to the CFE, thereby reducing the active recording area of the CFE and the subsequent ability to detect 5-HT (see also Jackson et al. 1995). We find that leaving a delay of 900 ms between each applied potential greatly increases the lifetime of our CFEs, presumably by decreasing the rate at which adsorptive products are created. CFE potential is therefore stepped at 1 Hz. In theory then, one of the advantages of chronoamperometry is that the high frequency at which recordings are made enhances signal-to-noise because small changes in current due to the detection of 5-HT are not masked by large and potentially varying background current. We presently use the FAST-12 and FAST-16 systems (Quanteon, Lexington, KY) for our chronoamperometric recordings.

CFE Calibration

CFEs are available commercially (e.g., from Quanteon, Lexington, KY) or can be custom made. Details for electrode fabrication can be found in Perez and Andrews (2005) and references therein (Gerhardt 1995; Hoffman and Gerhardt 1998; Gerhardt and Hoffman 2001). Prior to using CFEs for in vivo recordings, they are coated with Nafion and calibrated for 5-HT. Nafion is a perfluronated ion exchange resin that excludes the passage of anions to the CFE (Gerhardt et al. 1984). It has been used with great success to increase the selectivity of CFEs for cations, such as 5-HT, and prevent oxidation of anionic "contaminants" such as the metabolite of 5-HT, 5-hydroxyindoleacetic acid (5-HIAA), uric acid and ascorbic acid, all of which are present in brain in a much higher concentration than 5-HT and have oxidation potentials that overlap that of 5-HT (Crespi et al. 1983; Cespuglio et al. 1986; Crespi et al. 1990; Rivot et al. 1995). The electrode is simply dipped and swirled in

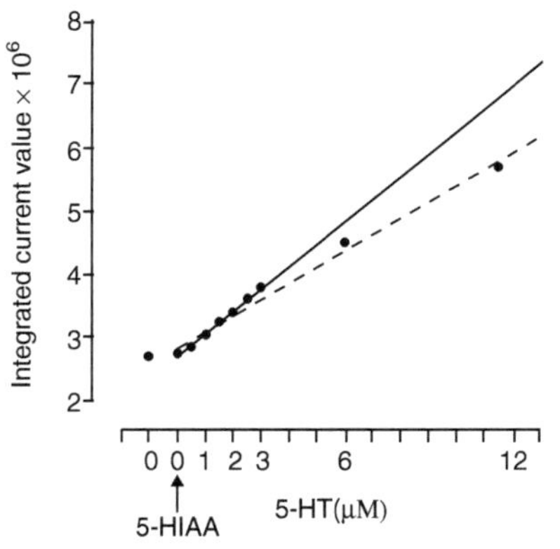

FIGURE 5.2 Calibration of a Nafion-coated carbon fiber electrode in room temperature (22°C) phosphate buffered saline (PBS). To determine the selectivity produced by the Nafion film, the electrode was challenged with 250 μM 5-HIAA before the first addition of 5-HT. The linearity of response to 5-HT over the range of 0.0–3.0 μM is excellent ($r^2 = 0.9956$-solid line) but diminishes as the concentration of 5-HT is increased ($r^2 = 0.9858$-dashed line). The selectivity of this particular sensor for 5-HT against 5-HIAA 3544:1 and the detection limit for 5-HT is 29 nM (signal-noise-ratio = 3).

Nafion (1–2 s) and then dried in an oven at 200°C. This process is repeated 2–5 times and results in a CFE that is highly selective for 5-HT (see also Gerhardt and Hoffman 2001 for methodological considerations). Our criteria is that the CFE must be 500 times more selective for 5-HT than for its metabolite 5-HIAA. To determine this, the CFE is placed in a beaker containing either phosphate buffered saline (PBS) or artificial cerebral spinal fluid (aCSF). Once the current response to stepping the CFE potential stabilizes (within a few minutes), 5-HIAA (250 μM) is added followed by increasing concentrations of 5-HT (typically 0.5 μM increments, ranging from 0.0 to 3.0 μM, see Figure 5.2). In addition to having a 5-HT:5-HIAA selectivity greater than 500:1, electrodes used for our in vivo experiments must have a linear response to 5-HT ($r^2 > 0.90$). We typically find electrodes prepared in this way have a detection limit in the range of 30–60 nM. Figure 5.2 shows the effectiveness of Nafion to prevent detection of 5-HIAA and the excellent linearity of response to increasing concentrations of 5-HT over the range of 0.0–3.0 μM ($r^2 = 0.9956$). Also shown is the response to much higher concentrations of 5-HT (6 and 12 μM). At these higher concentrations, the calibration clearly becomes curvilinear, likely due to adsorption of electrogenerated products to the electrode (Jackson et al. 1995), however, regression analysis continues to yield strong r^2 values (in the range of 0.90–0.98) showing the utility of chronoamperometry to measure relatively high concentrations of 5-HT. We find that if we limit exposure of the electrode to these higher concentrations of 5-HT (1–2 exposures), we can effectively establish values for maximal rates of 5-HT clearance (Vmax) in various brain regions in vivo (Daws et al. 2005). Some of these data are discussed in Section 5.3.2.

High-Speed Chronoamperometry Coupled to Microejection of Serotonin

We have taken the technique of high-speed chronoamperometric recording and coupled it to microinjection of 5-HT and other compounds into discrete brain regions. We developed the use of multibarrelled glass micropipettes (typically 4 or 7 barrels) so as to take full advantage of the excellent spatial resolution afforded by the small size of the active recording area of the CFE (Daws et al. 1997). For our experiments, the CFE electrode (diameter 30 μm) is cut such that 150 μm extends from the pulled glass capillary in which it is housed. By using multibarrel pipettes, we can test the effect of several drugs and/or establish dose-relationships within the same discrete location of brain within each animal. This greatly reduces the number of animals required and importantly, the need to record from multiple sites within a given brain region. The excellent spatial resolution of electrodes used for voltammetric recordings can be a double-edged sword depending on the question being asked. For example, they provide an excellent approach to investigating heterogeneity of response within a given brain structure (e.g., Cass et al. 1993; Cass and Gerhardt 1995; Daws et al. 1998, 2005), but if the research question requires restriction to a confined region (e.g., a distinct set of sterectaxic coordinates), then use of multibarrel pipettes allows considerably greater flexibility in the nature and amount of data that can be obtained from that site without the need to move and/or withdraw the micropipette-electrode assembly (i.e., to exchange or refill the micropipette). Of course, this is a distinct advantage when recording from multiple sites as well. We obtain our 4- and 7-barrel pipettes from Frederick Haer Corp. Inc. (Bowdoinham, ME).

First the CFE is coated with Nafion and calibrated to 5-HT as described above. Then the barrels of the micropipette are filled with the desired solutions, typically 5-HT, vehicle and drugs of interest and a length of PE tubing ($\sim$30 cm) glued with cyanoacrylate adhesive into each of the barrel openings. This tubing serves as the connector to a picospritzer, which is used to deliver solutions into brain by pressure-ejection. The amount of 5-HT or drug delivered can be very accurately quantified by measuring the amount of fluid displaced from the micropipette using a dissection microscope fitted with an eyepiece reticule. This can be simply calculated using the equation $V = \pi r^2 \times l$, where V is the volume displaced, r is the inside radius of the pipette barrel and l is the distance the meniscus falls within the barrel during pressure application. Given that the concentration of each solution contained within the barrel is known, and the volume ejected is readily measured, the number of moles of 5-HT or drug delivered can be easily calculated. A cartoon showing the typical arrangement of our multi-barrel micropipette and electrode assembly is shown in Figure 5.3. The multi-barrel micropipette and CFE are attached just above the taper in the micropipette using sticky wax (Kerr Brand, Emoryville, CA). This is done by clamping the CFE and microelectrode into separate arms of a micromanipulator and carefully aligning the tips in the same plane and separated by a set distance: 200 μm for experiments in mice and 300 μm for

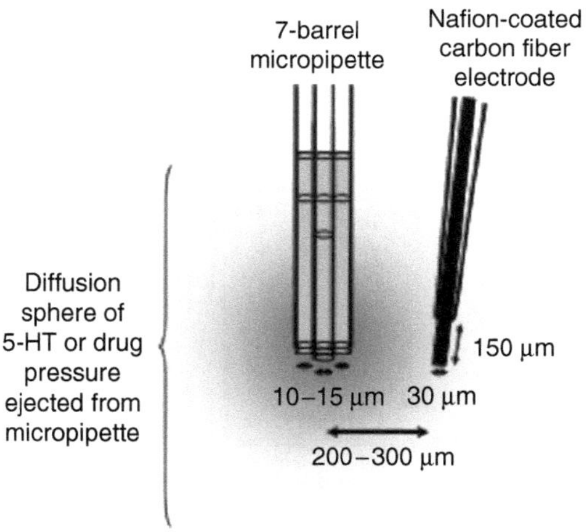

FIGURE 5.3 Cartoon showing only the tips of the carbon fiber electrode and 7-barrel micropipette. Serotonin is pressure-ejected into the brain and diffuses across to the CFE where it is oxidized. This generates a current, which is converted into a micromolar value using a calibration factor determined in vitro. The effects of drugs or vehicle, applied locally, also by pressure-ejection from the multibarrel pipette, or via systemic routes of administration, on the 5-HT signal can then be measured. Note that this is highly idealized cartoon. Many factors, including tortuosity and volume fraction, can influence the diffusion of 5-HT and drug though the brain extracellular matrix. For examples, see Rice and Nicholson (1991) and Nicholson and Syková (1998).

experiments in rats. Tip alignment and separation is confirmed using a dissecting microscope fitted with an eyepiece micrometer. This assembly is implanted, using stereotaxic apparatus, into the desired brain region.

Surgical Procedures

Experiments are routinely performed in anesthetized mice and rats. The procedures are essentially identical with a few minor modifications for the mouse. Rats are anesthetized with an intraperitoneal (ip) injection of chloralose (85 mg/kg) and urethane (850 mg/kg). For mice a mixture of chloralose (35 mg/kg) and urethane (350 mg/kg) is injected at 2 ml/kg body weight to more accurately titrate dose. A stainless steel tube is inserted into the trachea to facilitate breathing and the animal is then placed into the stereotaxic frame. The animal is kept warm by means of a water-circulated heating pad and body temperature monitored using a rectal probe. Body temperature is maintained at $37 \pm 1°C$. The scalp is then incised and reflected and the skull overlying the brain region of interest is removed. Once the dura mater is removed, the electrode–micropipette assembly is lowered into the region of interest. Applications of high-speed chronoamperometry coupled to microinjection of 5-HT are described in the following sections.

KINETICS OF SEROTONIN CLEARANCE

One of the greatest advantages of locally applying 5-HT to brain is that the source and identity of the neurotransmitter analyte is known. Certainly there are many applications where evoking release of endogenous 5-HT is desirable, such as those described by Carrie John and Sara Jones in the previous chapter. However, when the desired measure is that of *transporter function,* specifically in terms of its ability to *remove* neurotransmitter from the extracellular milieu, measuring clearance of exogenously applied 5-HT provides a straightforward and readily interpretable approach.

Serotonin Signal Parameters

Figure 5.4 shows the reproducibility of the signal produced by pressure-ejection of 5-HT into the CA3 region of hippocampus. A minimum of 3 min is allowed to elapse between ejections. We routinely find that 3–5 ejections of serotonin into hippocampus are needed before the signal

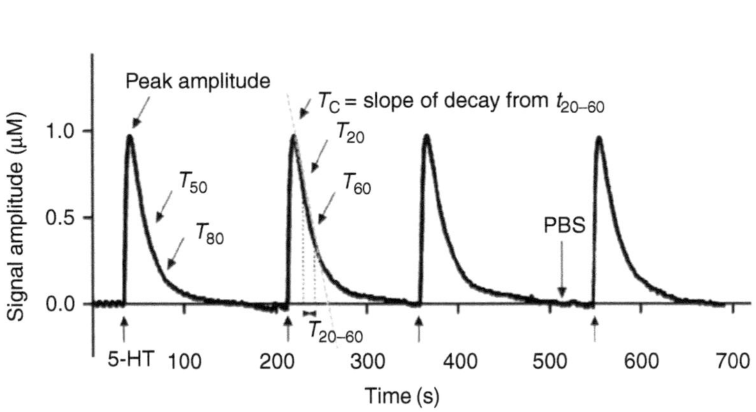

FIGURE 5.4 Signal parameters and reproducibility of electrochemical signals produced by local application of 5-HT in the CA3 region of the anesthetized mouse. Serotonin (2 pmol, indicated by arrows along the lower abscissa) was applied at ~3 min intervals by pressure-ejection. Shown is the oxidation current converted to micromolar concentration using a calibration factor determined in vitro. Once at least 3 reproducible 5-HT signals were obtained, phosphate buffered saline (PBS, a commonly used vehicle for drugs) was applied 60 s prior to the next application of 5-HT. Note the lack of effect of PBS on the 5-HT signal.

becomes reproducible. Before this stable baseline is achieved, the signal amplitude for the same pmol amount of serotonin declines most markedly, but also signal time course decreases modestly. Part of this is no doubt attributable to initial disruption of the extracellular matrix by the bolus ejection of serotonin that could change factors known to influence diffusion (e.g., tortuosity, volume fraction) and hence, clearance. In addition, we believe that trafficking of the 5-HTT to the plasma membrane in response to the rapid increase in extracellular 5-HT also contributes. This phenomenon has been very well characterized by Randy Blakely and his group using in vitro models (Ramamoorthy and Blakely 1999). While experiments are ongoing in our laboratory to address 5-HTT trafficking in vivo, we find that once we have attained stability in the signal produced by repeated pressure-ejection of serotonin into hippocampus, the signal remains remarkably stable for very long periods (h). This provides an excellent preparation for studying the ability of drugs to impact various signal parameters. These are shown in Figure 5.4 and include the peak signal amplitude, rise time (the time for the signal to reach its peak amplitude) the time for the signal to decay by 20, 50, 60, and 80 percent of its peak amplitude (T_{20}, T_{50}, T_{60}, and T_{80}), area under the curve and clearance rate. We and others (e.g., Sabeti et al. 2002) calculate clearance rate in two ways. One method is by calculating the slope of the most linear portion of descending phase of the signal, T_c, defined as the slope between T_{20} and T_{60}. The other is by calculating the velocity, V, for 5-HT clearance by fitting the entire descending portion of the signal to an equation describing one-phase exponential decay. $Y = [\text{5-HT}]_{o,i} \cdot e^{-kt} + [\text{5-HT}]_{o,t}$, where Y is the concentration of 5-HT (μM), $[\text{5-HT}]_{o,i}$ is the initial maximal extracellular concentration of 5-HT (μM), $[\text{5-HT}]_{o,t}$ is the extracellular concentration of 5-HT (μM) at baseline and k is the rate constant (s^{-1}). The half-life of decay is 0.6932/k. Note that because the concentration of 5-HT at baseline (i.e., $[\text{5-HT}]_{o,t}$) equals "zero" (the value given to represent the baseline electrochemical signal *prior* to addition of exogenous 5-HT), this term of the equation reduces to zero and is not used in the analyses. The calculated k value from this equation is then multiplied by the amplitude of the signal to give the velocity of transport (V). These calculations are made for signals of increasing amplitude until an apparent Vmax is reached.

Brain Region Dependency

This approach has been used to demonstrate the relationship between 5-HTT expression and 5-HT clearance (Montañez et al. 2003; Daws et al. 2005). For example, [^{3}H]cyanoimpramine binding to the 5-HTT was carried in the dorsal raphe nucleus (DRN), a cell body region of the 5-HT system, which contains the highest density of 5-HTT in brain (Hensler et al. 1994); the CA3 region of hippocampus, a terminal field region (Hensler et al. 1994), and the corpus callosum, a fiber tract with very few 5-HTTs (Reyes-Haro et al. 2003). Shown in Figure 5.5 are the [^{3}H]cyanoimipramine binding data (Figure 5.5a) and 5-HT clearance profiles (Figure 5.5b). Figure 5.5c shows the remarkable positive correlation between 5-HTT expression and Vmax for 5-HT clearance. The greater the density of 5-HTT, the greater the rate of 5-HT clearance (details can be found in Daws et al. 2005). We have carried out similar experiments in the CA3 region of hippocampus of 5-HTT mutant mice (Montañez et al. 2003). As anticipated, apparent Vmax values for 5-HT clearance in heterozygote mice, which express 50% fewer 5-HTTs than their wild-type counterpart, were approximately half that of wild-type mice. Serotonin clearance rates were substantially reduced in mice lacking the 5-HTT. Indeed, the rate for 5-HT clearance in these mice was comparable to that observed in corpus callosum of rat (see Figure 5.5b); thus, two situations where 5-HTT expression is minimal or absent yield similar profiles (but see also text below and Figure 5.7).

Transporter Promiscuity

The clearance profiles obtained in corpus callosum and 5-HTT KO mice at first glance appear indicative of simple diffusion of 5-HT away from the recording electrode. While this reasonably

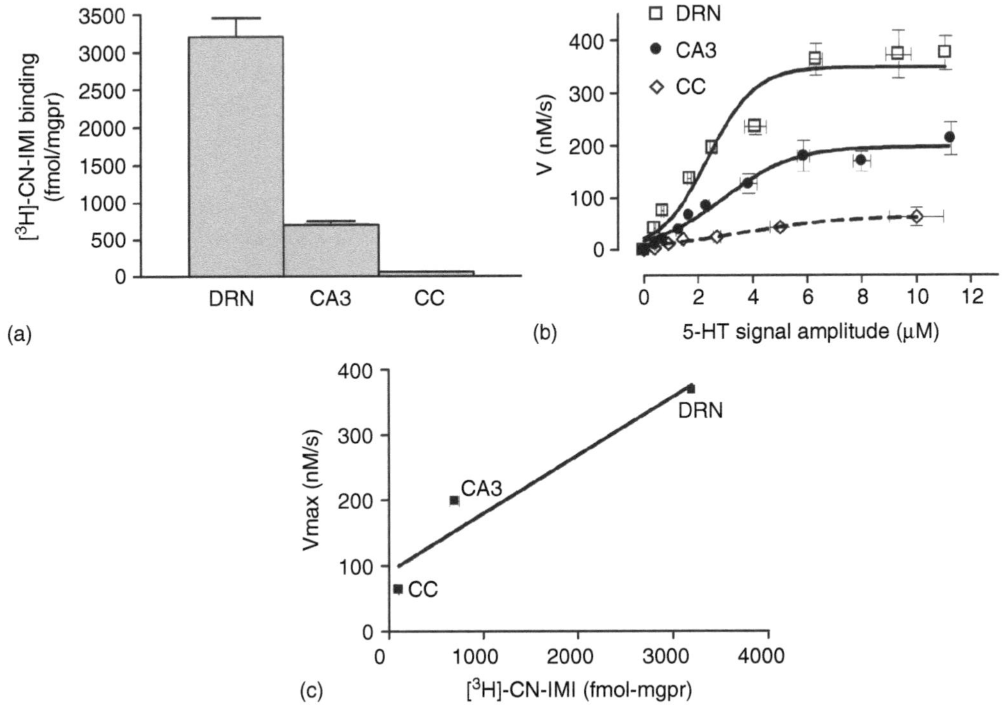

FIGURE 5.5 Relationship between 5-HTT expression and maximal rate (Vmax) of 5-HT clearance in rat brain. (a) Specific binding of [^{3}H]cyanoimipramine (CN-IMI) to 5-HTTs in the dorsal raphe nucleus (DRN), CA3 region of hippocampus and corpus callosum (CC). (b) Velocity of 5-HT clearance in the DRN, CA3 region of hippocampus and corpus callosum (CC) as a function of increasing extracellular concentration of 5-HT. Data are organized according to signals of increasing amplitudes and then averaged into bins according to a pseudo-geometric progression. The horizontal error bar represents the variance around these data bins. Values represent mean $\pm$ SEM from 4 to 8 rats per point. Curves represent fits of a four parameter logistic equation to the data. This equation fits a sigmoid function with variable slope. The minimum value was constrained at zero. Vmax values were determined from this equation and plotted in (c) as a function of [^{3}H]- CN-IMI binding. Regression analysis of data revealed a strong positive correlation between Vmax and 5-HTT density (solid line, $r^2 = 0.9268$). (Adopted from Daws, L. C. et al., *J. Neurosci. Methods*, 143, 49–62, 2005. With permission.)

accounts in large part for the shape of 5-HT clearance profiles in regions of low or no 5-HTT expression, is diffusion alone the only contributor? Our studies in rats treated with 5,7-dihydroxytryptamine (5,7-DHT) to destroy 5-HT neurons (Daws et al. 1998; Daws et al. 2005) and 5-HTT KO mice (Daws et al. 2006) suggest that other processes contribute to 5-HT clearance in the absence of 5-HTTs.

Figure 5.6 shows the clearance profile for 5-HT locally applied to the CA3 region of hippocampus of male Sprague–Dawley rats that had been previously treated with 5,7-DHT (200 μg total; 100 μg in PBS delivered at 10 μL/min bilaterally into the lateral ventricles) or sham operated (for experimental details, see Daws et al. 2005). The success of lesion was confirmed by quantitative autoradiography for [^{3}H]CN–IMI binding to the 5-HTT. Likewise, the success of pretreatment with nomifensine (30 mg/kg, ip) to protect norepinephrine transporters (NET) was also confirmed by [^{3}H]nisoxetine binding to the NET (see Daws et al. 2005). Our expectation was that the Vmax for 5-HT clearance in 5,7-DHT treated rats would be drastically reduced. It was not. As is clear in Figure 5.6, rather than a change in Vmax, there was a clear shift to the right of the concentration effect curve showing that the in vivo affinity value (referred to as K_T) for "active" clearance

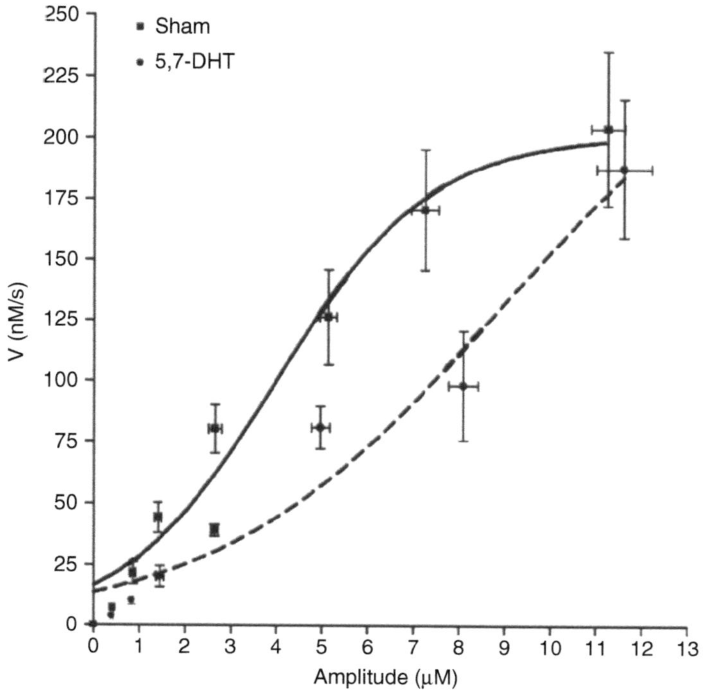

FIGURE 5.6 Maximal velocity of 5-HT clearance in CA3 region of hippocampus is not altered by neurotoxic destruction of 5-HTTs. Velocity of 5-HT clearance in the CA3 region of hippocampus is shown for rats treated with 5,7-DHT (closed circles, dashed line), or vehicle (sham-closed squares, solid line) as a function of increasing extracellular concentration of 5-HT. Values represent mean ± SEM from 4 to 8 rats per point. Curves represent fits of sigmoid function with variable slope (Prism 4.02, GraphPad Software Inc, CA). (Adopted from Daws, L. C. et al., *J. Neurosci. Methods*, 143, 49–62, 2005. With permission.)

mechanisms in the CA3 region of hippocampus had increased (i.e., decreased affinity for 5-HT uptake). These results harkened the early work of Shaskan and Synder (1970) showing that 5-HT uptake was mediated by at least two processes: one high affinity/low capacity the other low affinity but with a high capacity to transporter 5-HT. Because NET is also present in the CA3 region of hippocampus, albeit with expression levels approximately 25% that of 5-HTT, we thought it possible that when 5-HTT was eliminated, the NET may represent another key mechanism for removal of 5-HT from the extracellular milieu. Further investigations in rats where noradrenergic terminals had been destroyed by pre-treatment with 6-hydroxydopamine revealed that the NET is indeed capable of clearing 5-HT from ECF in hippocampus (see Daws et al. 1998, 2005).

One puzzling observation has been why the apparent Vmax for 5-HT clearance in the hippocampus of 5-HTT KO mice is markedly lower than that of wild-type mice. If the NET and/or other transporter is able to maintain Vmax for 5-HT clearance in 5,7-DHT treated rats (i.e., rats with no 5-HTT) at rates similar to those in vehicle-treated rats, why do we not see the same phenomenon in 5-HTT KO mice? One possibility is that in mice, we have simply not yet attained a sufficiently high extracellular concentration of 5-HT to reveal clearance by transporters other than the 5-HTT. Our studies in rat have extended the concentration range to as high as 14 μM (Daws et al. 2005), whereas in mouse, extracellular concentrations of exogenously applied 5-HT have rarely exceeded 8 μM (Montañez et al. 2003). We have begun to study a higher concentration range in 5-HTT mutant mice and we have found, as illustrated in Figure 5.7, that when signal amplitudes produced by exogenous application of 5-HT exceed ~ 15 μM, Vmax for 5-HT, clearance between wild-type

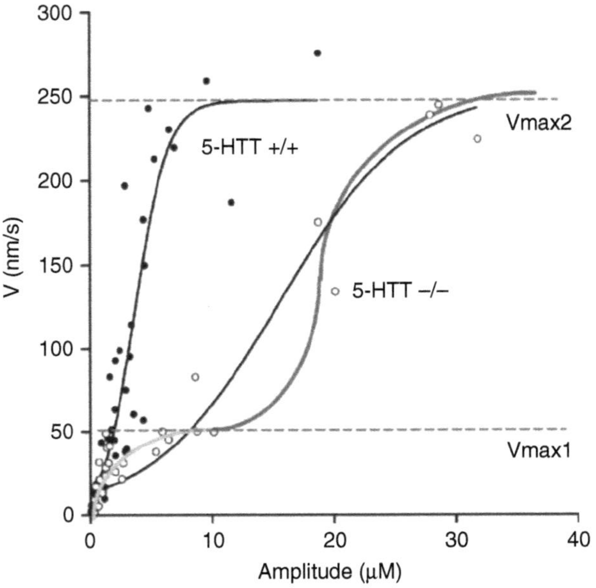

FIGURE 5.7 Evidence for alternative mechanisms for 5-HT clearance in 5-HTT KO mice. Velocity of 5-HT clearance in CA3 region of hippocampus is shown for wild-type (closed circles, $n = 3$) and 5-HTT$-/-$ (open circles, $n = 3$) mice as a function of increasing extracellular concentration of 5-HT. Each symbol represents an individual data point. The black solid lines represent sigmoid fits with variable slope to all data within a genotype (Prism 4.02, GraphPad Software Inc, CA). These fits yielded Vmax values of 231 ± 25 nM/s and 254 ± 2 nM/s for 5-HTT$+/+$ and 5-HTT$-/-$, respectively. K_T values (corrected for volume fraction, $\alpha = 0.2$) were 0.58 ± 0.06 μM and 3.10 ± 0.01 μM for 5-HTT$+/+$ and 5-HTT$-/-$ mice, respectively. The grey dashed lines represent a hypothetical fit revealing two sigmoid functions. The lower "Vmax1" plateau in 5-HTT KO mice may reflect the diffusion component for 5-HT clearance away from the CFE. The higher "Vmax2" plateau may reflect a low affinity, but high capacity transporter for 5-HT such as the NET.

and 5-HTT KO mice becomes indistinguishable (unpublished observations). The black solid lines in Figure 5.7 represent sigmoid fits with variable slope to all data within a genotype. These fits yielded Vmax values of 231 ± 25 nM/s and 254 ± 2 nM/s for 5-HTT$+/+$ and 5-HTT$-/-$, respectively. K_T values (corrected for volume fraction, $\alpha = 0.2$) were 0.58 ± 0.06 μM and 3.10 ± 0.01 μM for 5-HTT$+/+$ and 5-HTT$-/-$ mice, respectively. These fits are consistent with 5-HT clearance in wild-type mice being mediated primarily by the 5-HTT, a high affinity, low capacity transporter for 5-HT, whereas in the 5-HTT KO mice, clearance of 5-HT is mediated by a low affinity but high capacity transporter. The grey dashed lines represent a hypothetical fit revealing two sigmoid functions. The lower plateau "Vmax1" in 5-HTT KO mice (at around 50 nM/s) may reflect the diffusion component for 5-HT clearance away from the CFE. The higher plateau "Vmax2" (at approximately 250 nM/s) may reflect a low affinity, but high capacity transporter for 5-HT such as the NET.

Compelling evidence for the existence of alternative mechanisms for 5-HT clearance in 5-HTT mutant mice come from our studies investigating the effect of ethanol on 5-HT clearance in these mice. We found that ethanol inhibits 5-HT clearance in wild-type mice. Remarkably, however, the ability of ethanol to inhibit 5-HT clearance was significantly potentiated in a 5-HTT genotype-dependent manner, where inhibition was greatest in mice lacking the 5-HTT and intermediate in 5-HTT$\pm$ mice, which express 50% fewer 5-HTTs than wild-type mice (Daws et al., *J. Neuroscience*, 2006). These data clearly demonstrate that (1) while the 5-HTT may be a site of action for ethanol, it is not necessary for ethanol-induced inhibition of 5-HT clearance and (2), that genetic inactivation of the 5-HTT causes a marked upregulation of the principle mechanisms involved. Investigations are ongoing to identify these mechanisms.

Physiological Relevance of Applying Micromolar Concentrations of 5-HT

Of fundamental importance is whether exogenous application of micromolar concentrations of 5-HT or its uptake by catecholaminergic neurons is of physiological relevance. Although studies do not yet provide a complete answer, we believe this approach can yield important insight into the normal physiology and pathophysiology of 5-HT neurotransmission. The current estimate for synaptic concentrations of 5-HT in brain is 6 mM (Bunin and Wightman 1999). This value is consistent with that reported for other neurotransmitters such as glutamate and dopamine (Clements 1996; Cragg and Rice 2004). It has been estimated that 5-HT can diffuse greater than 20 μm away from the synaptic cleft, where the concentration of 5-HT falls to the micromolar and nanomolar range (Bunin and Wightman 1998, 1999). Thus, the micromolar amounts of 5-HT required to reach Vmax for 5-HT clearance in the studies described herein fit well within the current framework of our knowledge. Our data are also in good agreement with those of Bunin and co-workers, in that they support paracrine (or extrasynaptic) transmission as a mode for 5-HT neurotransmission. Ultrastructural studies show that the 5-HTT is located primarily extrasynaptically (on axons and dendrites) (Miner et al. 2000) and also at a distance from potential receptor target sites (Huang and Pickel 2002). The same is true for NET labeling and of particular significance to our studies is evidence that noradrenergic neurons arborize extensively within the hippocampus (Schroeter et al. 2000). Thus, both anatomical and functional data provide strong support that the range of concentrations of 5-HT used in our experiments are physiologically realistic and relevant and further support the involvement of multiple transport mechanisms in regulating 5-HT neurotransmission.

Biological Significance of Transporter Promiscuity

While there is no doubt that the 5-HTT is the "prime-mover" of 5-HT from the extracellular fluid back into the nerve terminal, substantial evidence for uptake of 5-HT by other transporters, including the NET and DAT has been amassed over the past 5 decades (Shaskan and Snyder 1970; Jackson and Wightman 1995; Daws et al. 1998; Pan et al. 2001; Zhou et al. 2002; Callaghan et al. 2005; Daws et al. 2005; Zhou et al. 2005). The data discussed herein suggest that these alternative mechanisms for 5-HT transport play a larger role when extracellular levels of 5-HT are very high and/or when the 5-HTT is compromised or eliminated (genetically or pharmacologically). These findings are particularly intriguing given growing evidence that a low-functioning variant in the promoter region of the human 5-HTT gene is positively associated with risk for emotional disorders and alcoholism. Moreover, they raise important considerations for the use of SSRIs and other therapeutics used in the treatment of disorders such as depression, anxiety and alcoholism. For example, patients who do not respond well to SSRIs may have a greater relative amount of functioning NET than do responders. One could speculate that greater NET function would serve to reduce the overall ability of SSRI treatment to elevate extracellular 5-HT, presumably the first critical step in achieving therapeutic benefits. Taken together with the rapidly expanding body of literature describing interactions between genetically defined expression of the 5-HTT (e.g., 5-HTTLPR polymorphism) (Collier et al. 1996; Smits et al. 2004), environment (Caspi et al. 2003) and response to drug treatment (Eichhammer et al. 2003; Smits et al. 2004), the design of a biological system with built-in functional redundancies makes evolutionary sense in order to maintain homeostasis of serotonergic neurotransmission. Consideration of these alternative mechanisms for 5-HT uptake may lead to the development of improved therapeutics for numerous disorders, in particular those where a hyposerotonergic state is thought to be the key predisposing or exacerbating factor.

RECEPTOR REGULATION OF THE SEROTONIN TRANSPORTER

Another way in which neurotransmitter uptake can be regulated is by receptor control of the neurotransmitter transporter itself. For example, it has been well established that the dopamine2 (D2) autoreceptor can regulate activity of the DAT (Meiergerd et al. 1993; Parsons et al. 1993; Cass and Gerhardt 1994; Wieczorek and Kruk 1995; Dickinson et al. 1999; Mayfield and Zahniser 2001; Gulley and Zahniser 2003). Described in the following section are examples of our application of high-speed chronoamperometry to studying regulation of the 5-HTT by auto- and hetero- receptors.

5-HT$_{1B}$ Receptor

Reports that the D2 autoreceptor could regulate DAT activity turned our attention to the possibility that the 5-HTT could also be under autoreceptor control. The 5-HT$_{1B}$ receptor is located both post-synaptically and also on the pre-synaptic terminal where it functions as an autoreceptor to regulate 5-HT release from the terminal. To test the hypothesis that it also serves as a regulator of 5-HT clearance by influencing the activity of the 5-HTT, we used a pharmacological approach. Cyanopindolol, a 5-HT$_{1B}$ receptor antagonist, was loaded into a 7-barrel glass micropipette. Coupled with a CFE, these multibarrel micropipettes permit the study of dose–response relationships of a locally applied drug on 5-HT clearance at a discrete site within brain. The key advantage is that a different concentration of drug can be loaded into 5 of the 7 barrels, with 5-HT and vehicle loaded into the remaining 2 barrels. In this way, different amounts (typically pmol amounts) of drug can be delivered in equivalent volumes by pressure-ejection. This eliminates any potential confound that might result from pressure-ejection of different volumes of the same barrel concentration of drug to deliver a given pmol amount. It also reduces variability resulting from local heterogeneity of 5-HTT expression.

First, 5-HT was pressure-ejected into the CA3 region of hippocampus until the signal produced became reproducible. This serves as the baseline signal. A given pmol amount of cyanopindolol or vehicle was then pressure-ejected into the same region and 60–90 s. permitted to elapse before again delivering 5-HT. Once all 5-HT was cleared (or 3 min later, whichever was the longer interval), another ejection of 5-HT was made. Pressure-ejection of 5-HT continued according to this schedule until the signal returned to its pre-drug baseline. Once at least two reproducible 5-HT signals were obtained, the next dose of cyanopindolol was locally applied and the protocol repeated.

Cyanopindolol dose dependently inhibited clearance of exogenously applied 5-HT in the CA3 region of hippocampus. Remarkably, the maximal effect (Emax) and the concentration of cyanopindolol producing half Emax (EC_{50}) (42% increase in T_{80} time course and 3.2 pmol respectively) were comparable to that of the SSRI, fluvoxamine (Emax = 38% increase in T_{80} and $EC_{50} = 3.6$ pmol) (see Daws et al. 2000). Similar results were obtained in rats pretreated with parachlorophenylalanine (PCPA) to deplete endogenous stores of 5-HT. The rationale for these experiments was based on the possibility that exogenously applied 5-HT might inhibit release of endogenous 5-HT by stimulating the 5-HT$_{1B}$ autoreceptor. If the autoreceptor is blocked by an antagonist such as cyanopindolol, then this would prevent inhibition of 5-HT release and lead to greater extracellular 5-HT. This in turn could manifest as an increase in signal amplitude and/or time course of the signal produced by exogenous application of 5-HT. The finding that cyanopindolol continued to prolong the time course for clearance of exogenously applied 5-HT in PCPA-treated rats rules out any contribution of endogenously released 5-HT to the signal. In addition, competition binding assays confirmed that inhibition of 5-HT clearance by cyanopindolol was not likely mediated by a direct action at the 5-HTT (*Ki* value > 35,000 nM). Together with data showing that 5-HT moduline and methiothepin, antagonists more selective for the 5-HT$_{1B}$ receptor than cyanopindolol, also inhibited 5-HT clearance (Daws et al. 1999, 2000), these data provide compelling evidence that the 5-HT_{1B} autoreceptor regulates the activity of 5-HTT.

ALPHA$_2$-ADRENOCEPTOR

The 5-HT_{1B} autoreceptor is not the only modulator of 5-HT release. Activation of presynaptic alpha2 (α2)-adrenoceptors have also been demonstrated to regulate 5-HT release (Maura et al. 1992; Gobbi et al. 1993; Numazawa et al. 1995; Gobert et al. 1998; Schneiber et al. 2001). In collaboration with Randy Blakely's group, we explored the possibility that, like the 5-HT_{1B} autoreceptor, the α2-adrenoceptor may also regulate 5-HTT function. Using a similar approach to that described above, we found that in rat parietal cortex, the α2-adrenoceptor agonist, 5-bromo-*N*-[4,5-dihyro-1*H*-imidazol-2-yl]-6-quinoxalinamine (UK14304), inhibited clearance of exogenously applied 5-HT (Ansah et al. 2003). In vitro, yohimbine, an α2 antagonist, augmented [^{3}H]5-HT uptake into mouse forebrain synaptosomes. The augmentation of basal 5-HTT activity by yohimbine may suggest tonic inhibition of 5-HTT by α2-adrenoceptor agonists. In vivo, however, no effect of yohimbine on 5-HT clearance was observed. This apparent lack of effect may be related to dose; as to date, only one dose has been tested in vivo (52 pmol). Alternatively, this discrepancy may be related to the preparations themselves—in vitro versus in vivo. Norepinephrine (NE) was detected by high performance liquid chromatography (HPLC) analysis in the synaptosomal preparation (Ansah et al. 2003); therefore yohimbine could be blocking leaked or released NE in this preparation. In vivo, leakage of NE might be less likely. While studies investigating the mechanism by which this cross-talk between the α2-adrenoceptor and 5-HTT occurs are ongoing, these data provide another example of the utility of in vivo chronoamperometric approaches to identifying the existence of receptor regulation of the 5-HTT outside of the "dish."

ADENOSINE RECEPTOR

Using a rat basophilic leukemia cell line (RBL-2H3), Miller and Hoffman (1994) provided the first evidence of adenosine receptor (AR) regulation of the 5-HTT via a cGMP-dependent mechanism. More recent studies from Randy Blakely's group (e.g, Zhu et al. 2004) have extended these initial observations to reveal the existence of two protein kinase G (PKG)-dependent pathways that support rapid regulation of 5-HTT by the adenosine 3 (A3) receptor. One augments 5-HTT surface trafficking and the other, a p38 mitogen-activated protein kinase (MAPK)-dependent process, increases 5-HTT intrinsic activity (Zhu et al. 2004, 2005). However, to date all of these studies have been carried out in cell lines stably transfected with 5-HTT or synaptosomal preparations. In collaboration with Randy Blakely, we have begun to study regulation of the 5-HTT by these pathways using in vivo chronoamperometric approaches.

First we investigated the dose-dependency for modulation of 5-HT clearance in rat brain by the non-selective AR agonist, 5′-*N*-ethylcarboxamidoadenosine (NECA). Similar approaches to those described above were used. A reproducible baseline 5-HT signal was established and then NECA was locally applied by pressure-ejection 90 s. prior to the next ejection of 5-HT. Consistent with in vitro studies, NECA enhanced clearance of exogenously applied 5-HT in CA3 region of hippocampus, but within a very narrow dose range (Figure 5.8). To investigate a role for PKG and p38 MAPK in mediating this effect of NECA, we then applied either the PKG inhibitor, *N*-[2-(methyamino)ethyl]-5-isoquinoline-sulfonamide (H8), the p38 MAPK inhibitor, 4-(4-fluorophenyl)-2-(4-methylsulfonylphenyl)-5-(4-pyridyl)-1*H*-imidazol (SB203580) or vehicle, 2 min prior to NECA. Pretreatment with either H8 or SB203580 blocked NECA-induced enhancement of 5-HT clearance (Figure 5.9). These studies remain in their very early stages but clearly demonstrate the utility of chronoamperometry coupled to pressure-ejection of exogenous 5-HT to not only identify phenomenological changes in 5-HT clearance, but importantly, to investigate the mechanisms through which these receptor-mediated changes in 5-HT clearance occur in vivo.

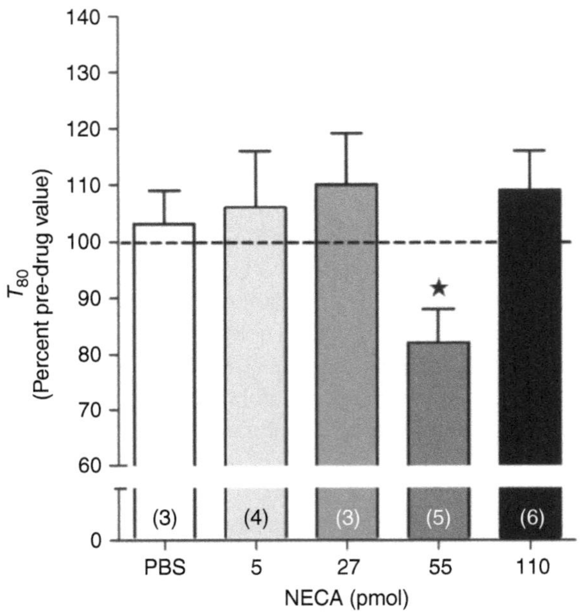

FIGURE 5.8 Dose-dependent effects of the AR agonist, NECA, on 5-HT clearance in rat hippocampus. Data are mean $\pm$ SEM percent change in T_{80} time course parameter 5 min after local application of NECA. Number of rats per "dose" is shown in parentheses.

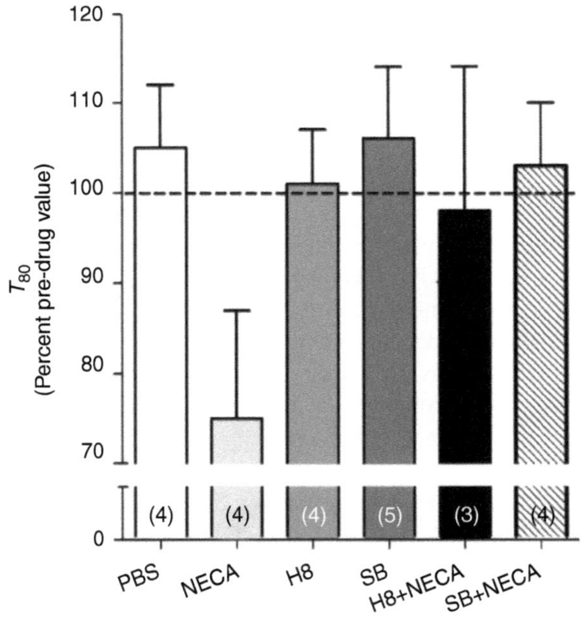

FIGURE 5.9 Inhibitors of PKG and p38 MAPK block AR agonist stimulated clearance of 5-HT. H8 (14 pmol), SB203580 (14 pmol) or equivalent volume of PBS were given either alone, or 2 min. prior to NECA (55 pmol). Data are mean $\pm$ SEM percent change in T_{80} time course parameter 5 min after local application of NECA. Number of rats per "dose" is shown in parentheses.

Significance of Receptor-Mediated Regulation of 5-HTT

Based on a rapidly growing body of in vitro studies, it is now well established that the 5-HTT can be regulated by multiple signaling pathways (Blakely and Bauman 2000; Blakely et al. 2005). The studies described in the previous sections provide initial but convincing evidence that the 5-HTT is also subject to this kind of regulation in the intact brain in vivo. As such, receptor regulation of 5-HTT expression and/or activity presents a new approach to drug development for the treatment of psychiatric disorders linked to dysregulation of 5-HT signaling (e.g., depression and addiction). For example, phosphorylation and sequestration of the 5-HTT is dependent on ligand occupancy (Ramamoorthy and Blakely 1999). Therefore, certain ligands may have superior therapeutic utility by not only preventing 5-HT reuptake but also by allowing kinase-linked signaling pathways to shift the cellular distribution of the 5-HTT. These findings become particularly important given the recent report that individuals with naturally occurring 5-HTT coding variants show altered sensitivity to drugs that target PKG and p38 MAPK (Prasad et al. 2005). In turn, altered 5-HTT regulation by means of PKG and p38 MAPK-linked pathways could influence risk for disorders related to compromised 5-HT neurotransmission.

CONCLUDING REMARKS

Chronoamperometry used to measure clearance of exogenously applied 5-HT offers a unique approach for examining functional changes in 5-HTT activity, in vivo, as they occur in response to genetic aberrations, environmental stimuli and acute or chronic drug treatment. Although in vitro approaches will no doubt continue to provide valuable insights into novel regulatory signaling pathways as well as transport mechanisms for 5-HT uptake, the importance of demonstrating their existence in the whole animal will always remain of paramount importance. In vivo electrochemical approaches, such as those described here, are currently among the most robust methods available to reveal how transporters function in their native environment.

ACKNOWLEDGMENTS

Studies described herein were funded in part by NIH grants MH64489 and DA18992, NARSAD Young Investigator Awards and the Texas Higher Education Coordinating Board Advanced Research Program Award to LCD and NIH grant HL071645 to GMT. The authors gratefully acknowledge the following individuals for their significant contributions toward generating the data contained within: Nicole L. Baganz, Dr. Randy D. Blakely, Alfred S. Calderon, Paul D. Callaghan, Douglas J. Davis, Dr. Alan Frazer, Dr. Greg A. Gerhardt, Dr. Georgianna G. Gould, Rebecca E. Horton, Dr. Sylvia Montañez, Jaclyn L. Munn, W. Anthony Owens, and Susan D. Teicher.

REFERENCES

Ansah, T. A., Ramamoorthy, S., Montañez, S., Daws, L. C., and Blakely, R. D., Calcium-dependent inhibition of synaptosomal serotonin transport by the α2-adrenoceptor agonist 5-bromo-*N*-[4,5-dihydro-1*H*-imidazol-2-yl]-6-quinoxalinamine (UK14304), *J. Pharmacol. Exp. Ther.*, 305, 956–965, 2003.

Bengel, D., Murphy, D. L., Andrews, A. M., Wichems, C. H., Feltner, D., Heils, A., Mossner, R., Westphal, H., and Lesch, K. P., Altered brain serotonin (5-HT) homeostasis and locomotor insensitivity to MDMA ("ecstasy") in 5-HT transporter deficient mice, *Mol. Pharmacol.*, 53, 649–655, 1998.

Blakely, R. D. and Bauman, A. L., Biogenic amine transporters: regulation in flux, *Curr. Opin. Neurobiol.*, 10, 328–336, 2000.

Blakely, R. D., Berson, H. E., Fremeau, R. T. Jr., Caron, M. G., Peek, M. M., Prince, H. K., and Bradley, C. C., Cloning and expression of a functional serotonin transporter from rat brain, *Nature*, 354, 66–70, 1991.

Blakely, R. D., Ramamoorthy, S., Schroeter, S., Qian, Y., Apparsundaram, S., Galli, A., and DeFelice, L. J., Regulated phosphorylation and trafficking of antidepressant-sensitive serotonin transporter proteins, *Biol. Psychiatry*, 44, 169–178, 1998.

Blakely, R. D., DeFelice, L. J., and Galli, A., Biogenic amine neurotransmitter transporters: just when you thought you knew them, *Physiology*, 20, 225–231, 2005.

Bunin, M. A. and Wightman, R. M., Quantitative evaluation of 5-hydroxytryptamine (serotonin) neuronal release and uptake: an investigation of extrasynaptic transmission, *J. Neurosci.*, 18, 4854–4860, 1998.

Bunin, M. A. and Wightman, R. M., Paracrine neurotransmission in the CNS: involvement of 5-HT, *Trends Neurosci.*, 22, 377–382, 1999.

Callaghan, P. D., Irvine, R. J., and Daws, L. C., Differences in the in vivo dynamics of neurotransmitter release and serotonin uptake after acute para-methoxyamphetamine and 3,4-methylenedioxymethamphetamine revealed bychronoamperometry, *Neurochem. Int.*, 47, 350–361, 2005.

Caspi, A., Sugden, K., Moffitt, T. E., Taylor, A., Craig, I. W., Harrington, H., McClay, J., Mill, J., Martin, J., Braithwaite, A., and Poulton, R., Influence of early life stress on depression: moderation by a polymorphism in the 5-HTT gene, *Science*, 301, 386–389, 2003.

Cass, W. A. and Gerhardt, G. A., Direct in vivo evidence that D2 dopamine receptors can modulate dopamine uptake, *Neurosci. Lett.*, 176, 259–263, 1994.

Cass, W. A. and Gerhardt, G. A., In vivo assessment of dopamine uptake in rat medial prefrontal cortex: comparison with dorsal striatum and nucleus accumbens, *J. Neurochem.*, 65, 201–207, 1995.

Cass, W. A., Zahniser, N. R., Flach, K. A., and Gerhardt, G. A., Clearance of exogenous dopamine in rat dorsal striatum and nucleus accumbens: role of metabolism and effects of locally applied uptake inhibitors, *J. Neurochem.*, 61, 2269–2278, 1993.

Cespuglio, R., Sarda, N., Gharib, A., Faradji, H., and Chastrette, N., Differential pulse voltammetry in vivo with working carbon fiber electrodes: 5-hydroxyindole compounds or uric acid detection?, *Exp. Brain Res.*, 64, 589–595, 1986.

Clements, J. D., Transmitter timecourse in the synaptic cleft: its role in central synaptic function, *Trends Neurosci.*, 19, 163–171, 1996.

Collier, D. A., Stober, G., Li, T., Heils, A., Catalano, M., Di Bella, D., Arranz, M. J., et al., A novel functional polymorphism within the promoter of the serotonin transporter gene: possible role in susceptibility to affective disorders, *Mol. Psychiatry*, 1, 453–460, 1996.

Cottrell, F. G., *Z. Physik. Chem.*, 42, 385, 1902.

Cragg, S. J. and Rice, M. E., Dancing past the DAT at a DA synapse, *Trends Neurosci.*, 27, 270–277, 2004.

Crespi, F., Sharp, T., Maidment, N., and Marsden, C. A., Differential pulse voltammetry in vivo evidence that uric acid contributes to the indole oxidation peak, *Neurosci. Lett.*, 43, 203–207, 1983.

Crespi, F., Garratt, J. C., Sleight, A. J., and Marsden, C. A., In vivo evidence that 5-hydroxytryptamine (5-HT) neuronal firing and release are not necessarily correlated with 5-HT metabolism, *Neuroscience*, 35, 139–144, 1990.

Daws, L. C., Toney, G. M., Davis, D. J., Gerhardt, G. A., and Frazer, A., In vivo chronoamperometric measurements of extracellular serotonin clearance in the rat dentate gyrus, *J. Neurosci. Methods*, 78, 139–150, 1997.

Daws, L. C., Toney, G. M., Gerhardt, G. A., and Frazer, A., In vivo chronoamperometric measures of extracellular serotonin clearance in rat dorsal hippocampus: contributions of serotonin and norepinephrine transporters, *J. Pharmacol. Exp. Ther.*, 286, 967–976, 1998.

Daws, L. C., Gerhardt, G. A., and Frazer, A., 5-HT_{1B} antagonists modulate serotonin clearance in the rat dorsal hippocampus, *Neurosci. Lett.*, 266, 165–168, 1999.

Daws, L. C., Gould, G. G., Teicher, S., Gerhardt, G. A., and Frazer, A., 5-HT_{1B} receptor-mediated regulation of serotonin clearance in rat hippocampus in vivo, *J. Neurochem.*, 75, 2113–2122, 2000.

Daws, L. C., Montañez, S., Owens, W. A., Gould, G. G., Frazer, A., Toney, G. M., and Gerhardt, G. A., Transport mechanisms governing clearance in vivo revealed by high-speed chronoamperometry, *J. Neurosci. Methods*, 143, 49–62, 2005.

Daws, L. C., Montañez, S., Munn, J. L., Owens, W. A., Baganz, N. L., Boyce-Rustay, J. M., Millstein, R. A., Wiedholz, L. M., Murphy, D. L., and Holmes, A., Ethanol inhibits clearance of brain serotonin by a serotonin transporter independent mechanism, *J. Neurosci.*, 26, 6431–6438, 2006.

Dickinson, S. D., Sabeti, J., Larson, G. A., Giardina, K., Rubenstien, M., Kelly, M. A., Grandy, D. K., Low, M. J., Gerhardt, G. A., and Zahniser, N. R., Dopamine D2 receptor-deficient mice exhibit decreased dopamine transporter function but no changes in dopamine release in dorsal striatum, *J. Neurochem.*, 72, 148–156, 1999.

Eichhammer, P., Langguth, B., Wiegand, R., Kharraz, A., Frick, U., and Hajak, G., Allelic variation in the serotonin transporter promoter affects neuromodulatory effects of a selective serotonin transporter reuptake inhibitor, *Psychopharmacology*, 166, 294–297, 2003.

Fuller, R. W. and Wong, D. T., Serotonin uptake and serotonin uptake inhibition, *Ann. N. Y. Acad. Sci.*, 600, 68–78, 1990.

Gerhardt, G. A., Rapid chronocoulometric measurements of norepinephrine overflow and clearance in CNS tissues, In *Neuromethods 27*, Boulton, A., Baker, G., and Adams, R. N., Eds., Humana Press, Totowa, NJ, pp. 117–151, 1995.

Gerhardt, G. A. and Burmeister, J., In vivo voltammetry for chemical analysis of the nervous system, In *Encyclopedia of Analytical Chemistry: Instrumentation and Applications*, Meyers, R. A., Ed., Wiley, Chichester, UK, pp. 710–731, 2000.

Gerhardt, G. A. and Hoffman, A. F., Effects of recording media composition on the responses of Nafion-coated carbon fiber microelectrodes measured using high-speed chronoamperometry, *J. Neurosci. Methods*, 109, 13–21, 2001.

Gerhardt, G. A., Oke, A. F., Nagy, G., Moghaddam, B., and Adams, R. N., Nafion-coated electrodes with high selectivity for CNS electrochemistry, *Brain Res.*, 290, 390–394, 1984.

Gobbi, M., Frittoli, E., and Mennini, T., Further studies on $\alpha 2$-adrenoceptor subtypes involved in the modulation of [^{3}H]noradrenaline and [^{3}H]5-hydroxytryptamine release from rat brain cortex synaptosomes, *J. Pharm. Pharmacol.*, 45, 811–814, 1993.

Gobert, A., Rivet, J. M., Audinot, V., Newman-Tancredi, A., Cistarelli, L., and Millan, M. J., Simultaneous quantification of serotonin, dopamine and noradrenaline levels in single frontal cortex dialysates of freely-moving rats reveals a complex pattern of reciprocal auto- and heteroreceptor-mediated control of release, *Neuroscience*, 84, 413–429, 1998.

Green, A. R., Mechan, A. O., Elliott, J. M., O'Shea, E., and Colado, M. I., The pharmacology and clinical pharmacology of 3,4-methylenedioxymethamphetamine (MDMA, "ecstasy"), *Pharmacol. Rev.*, 55, 463–508, 2003.

Gulley, J. M. and Zahniser, N. R., Rapid regulation of dopamine transporter function by substrates, blockers and presynaptic receptor ligands, *Eur. J. Pharmacol.*, 479, 139–152, 2003.

Hegadoren, K. M., Baker, G. B., and Bourin, M., 3,4-Methylenedioxy analogues of amphetamine: defining the risks to humans, *Neurosci. Biobehav. Rev.*, 23, 539–553, 1999.

Hensler, J. G., Ferry, R. C., Labow, D. M., Kovachich, G. B., and Frazer, A., Quantitative autoradiography of the serotonin transporter to assess the distribution of serotonergic projections from the dorsal raphe nucleus, *Synapse*, 17, 1–15, 1994.

Hoffman, A. F. and Gerhardt, G. A., In vivo electrochemical studies of dopamine clearance in the rat substantia nigra: effects of locally applied uptake inhibitors and unilateral 6-hydroxydopamine lesions, *J. Neurochem.*, 70, 179–189, 1998.

Huang, J. and Pickel, V. M., Serotonin transporters (SERTs) within the rat nucleus of the solitary tract: subcellular distribution and relation to 5HT2A receptors, *J. Neurocytol.*, 31, 667–679, 2002.

Jackson, B. P. and Wightman, R. M., Dynamics of 5-hydroxytryptamine released from dopamine neurons in the caudate putamen of the rat, *Brain Res.*, 674, 163–166, 1995.

Jackson, B. P., Dietz, S. M., and Wightman, R. M., Fast-scan cyclic voltammetry of 5-hydroxtryptamine, *Anal. Chem.*, 67, 1115–1120, 1995.

Kawagoe, K. T., Garris, P. A., Wiedemann, D. J., and Wightman, R. M., Regulation of transient dopamine concentration gradients in the microenvironment surrounding nerve terminals in the rat striatum, *Neuroscience*, 51, 55–64, 1992.

Kilic, F., Murphy, D. L., and Rudnick, G., A human serotonin transporter mutation causes constitutive activation of transport activity, *Mol. Pharmacol.*, 64, 440–446, 2003.

Kissinger, P. T., Hart, J. B., and Adams, R. N., Voltammetry in brain tissue: a new neurophysiological measurement, *Brain Res.*, 55, 209–213, 1973.

Kokoshka, J. M., Metzger, R. R., Wilkins, D. G., Gibb, J. W., Hanson, G. R., and Fleckenstein, A. E., Methamphetamine treatment rapidly inhibits serotonin, but not glutamate, transporters in brain, *Brain Res.*, 799, 78–83, 1998.

Kuhar, M. J., Roth, R. H., and Aghajanian, G. K., Synaptosomes from forebrains of rats with midbrain raphe lesions: selective reduction of serotonin uptake, *J. Pharmacol. Exp. Ther.*, 181, 36–45, 1972.

Maura, G., Bonanno, G., and Raiteri, M., Presynaptic α2-adrenoceptors mediating inhibition of noradrenaline and 5-hydroxytryptamine release in rat cerebral cortex: further characterization as different α2-adrenoceptor subtypes, *Naunyn-Schmeideberg's Arch. Pharmacol.*, 345, 410–416, 1992.

Mayfield, R. D. and Zahniser, N. R., Dopamine D2 receptor regulation of the dopamine transporter expressed in Xenopus laevis oocytes is voltage-independent, *Mol. Pharmacol.*, 59, 113–121, 2001.

McBride, W. J., Murphy, J. M., Gatto, G. J., Levy, A. D., Yoshimoto, K., Lumeng, L., and Li, T. K., CNS mechanisms of alcohol administration, *Alcohol*, 2, 463–467, 1993.

Meiergerd, S. M., Patterson, T. A., and Schenk, J. O., D2 receptors may modulate the function of the striatal transporter for DA: kinetic evidence from studies in vitro and in vivo, *J. Neurochem.*, 61, 764–767, 1993.

Michael, D. J. and Wightman, R. M., Electrochemical monitoring of biogenic amine neurotransmission in real time, *J. Pharm. Biomed. Anal.*, 19, 33–46, 1999.

Miller, K. J. and Hoffman, B. J., Adenosine A_3 receptors regulate serotonin transport via nitric oxide and cGMP, *J. Biol. Chem.*, 269, 27351–27356, 1994.

Miner, L. H., Schroeter, S., Blakely, R. D., and Sesack, S. R., Ultrastructural localization of the serotonin transporter in superficial and deep layers of the rat prelimbic prefrontal cortex and its spatial relationship to dopamine terminals, *J. Comp. Neurol.*, 427, 220–234, 2000.

Montañez, S., Owens, W. A., Gould, G. C., Murphy, D. L., and Daws, L. C., Exaggerated effect of fluvoxamine in heterozygote serotonin transporter knockout mice, *J. Neurochem.*, 86, 210–219, 2003.

Nemeroff, C. B. and Owens, M. J., Pharmacologic differences among the SSRIs: focus on monoamine transporters and the HPA axis, *CNS Spectr.*, 9, 23–31, 2004.

Nicholson, C. and Syková, E., Extracellular space structure revealed by diffusion analysis, *Trends Neurosci.*, 21, 207–215, 1998.

Numazawa, R., Yoshioka, M., Matsumoto, M., Togashi, H., Kemmotsu, O., and Saito, H., Pharmacological characterization of α2-adrenoceptor regulated serotonin release in rat hippocampus, *Neurosci. Lett.*, 192, 161–164, 1995.

Owens, M. J. and Nemeroff, C. B., The role of serotonin in the pathophysiology of depression: focus on the serotonin transporter, *Clin. Chem.*, 40, 228–295, 1994.

Pan, Y., Gembom, E., Peng, W., Lesch, K. P., Mossner, R., and Simantov, R., Plasticity in serotonin uptake in primary neuronal cultures of serotonin transporter knockout mice, *Brain Res. Dev. Brain Res.*, 126, 125–129, 2001.

Parsons, L. H., Schad, C. A., and Justice, J. B., Co-administration of the D2 agonist pimozide inhibits up-regulation of dopamine release and uptake induced by repeated cocaine, *J. Neurochem.*, 60, 376–379, 1993.

Perez, X. A. and Andrews, A. M., Chronoamperometry to determine differential reductions in uptake in brain synaptosomes from serotonin transporter knockout mice, *Anal. Chem.*, 77, 818–826, 2005.

Prasad, H. C., Zhu, C. B., McCauley, J. L., Samuvel, D. J., Ramamoorthy, S., Shelton, R. C., Hewlett, W. A., Sutcliffe, J. S., and Blakely, R. D., Human serotonin transporter variants display altered sensitivity to protein kinase G and p38 mitogen-activated protein kinase, *Proc. Natl Acad. Sci.*, 102, 11545–11550, 2005.

Ramamoorthy, S. and Blakely, R. D., Phosphorylation and sequestration of serotonin transporters differentially modulated by psychostimulants, *Science*, 285, 763–766, 1999.

Reyes-Haro, D., García-Alcocer, G., Miledi, R., and García-Colunga, J., Uptake of serotonin by adult rat corpus callosum is partially reduced by common antidepressants, *J. Neurosci. Res.*, 74, 97–102, 2003.

Rice, M. E. and Nicholson, C., Diffusion characteristics and extracellular volume fraction during normoxia and hypoxia in slices of rat neostriatum, *J. Neurophysiol.*, 65, 264–274, 1991.

Rivot, J.-P., Cespuglio, R., Puig, S., Jouvet, M., and Besson, J.-M., In vivo electrochemical monitoring of serotonin in spinal dorsal horn with nafion-coated multi-carbon fiber electrodes, *J. Neurochem.*, 65, 1257–1263, 1995.

Rudnick, G. and Wall, S. C., Non-neurotoxic amphetamine derivatives release serotonin through serotonin transporters, *Mol. Pharmacol.*, 43, 271–276, 1993.

Sabeti, J., Adams, C. E., Burmeister, J., Gerhardt, G. A., and Zahniser, N. R., Kinetic analysis of striatal clearance of exogenous dopamine recorded by chronoamperometry in freely-moving rats, *J. Neurosci. Methods*, 121, 41–52, 2002.

Schneiber, J., Trendelenburg, A. U., Hein, L., and Starke, K., α2-adrenoceptors modulating neuronal serotonin release: a study in α2-adrenoceptor subtype-deficient mice, *Br. J. Pharmacol.*, 132, 925–933, 2001.

Schroeter, S., Apparsundaram, S., Wiley, R. G., Miner, L. H., Sesack, S. R., and Blakely, R. D., Immunolocalization of the cocaine- and antidepressant-sensitive l-norepinephrine transporter, *J. Comp. Neurol.*, 420, 211–232, 2000.

Shaskan, E. G. and Snyder, S. H., Kinetics of serotonin accumulation into slices from rat brain: relationship to catecholamine uptake, *J. Pharmacol. Exp. Ther.*, 175, 404–418, 1970.

Smits, K. M., Smits, L. J. M., Schouten, J. S. A. G., Stelma, F. F., Nelemans, P., and Prins, M. H., Influence of SERTPR and STin2 in the serotonin transporter gene on the effect of selective serotonin reuptake inhibitors in depression: s systematic review, *Mol. Psychiat.*, 9, 433–441, 2004.

Virkkunen, M. and Linnoila, M., Serotonin in early onset alcoholism, *Recent Dev. Alcohol*, 13, 173–189, 1997.

Wall, S. C., Gu, H., and Rudnick, G., Biogenic amine flux mediated by cloned transporters stably expressed in cultured cell lines: amphetamine specificity for inhibition and efflux, *Mol. Pharmacol.*, 47, 544–550, 1995.

Wieczorek, W. and Kruk, Z. L., Influences of neuronal uptake of D2 autoreceptors on regulation of extracellular dopamine in the core, shell and rostral pole of the rat nucleus accumbens, *Brain Res.*, 699, 171–182, 1995.

Zhou, F. C., Lesch, K.-P., and Murphy, D. L., Serotonin uptake into dopamine neurons via dopamine transporters: a compensatory alternative, *Brain Res.*, 942, 109–119, 2002.

Zhou, F.-M., Lang, Y., Salas, R., Zhang, L., De Biasi, M., and Dani, J. A., Corelease of dopamine and serotonin from striatal dopamine terminals, *Neuron*, 46, 65–74, 2005.

Zhu, C. B., Hewlett, W. A., Feoktistov, I., Biaggioni, I., and Blakely, R. D., Adenosine receptor, protein kinase G, and p38 mitogen-activated protein kinase-dependent up-regulation of serotonin transporters involves both transporter trafficking and activation, *Mol. Pharmacol.*, 65, 1462–1474, 2004.

Zhu, C. B., Carneiro, A. M., Dostmann, W. R., Hewlett, W. A., and Blakely, R. D., p38 MAPK activation elevates serotonin transport activity via a trafficking-independent, protein phosphatase 2A-dependent process, *J. Biol. Chem.*, 280, 15649–15658, 2005.

6 Using High-Speed Chronoamperometry with Local Dopamine Application to Assess Dopamine Transporter Function

Joshua M. Gulley, Gaynor A. Larson, and Nancy R. Zahniser

CONTENTS

INTRODUCTION

IMPORTANCE OF DOPAMINE AND THE DOPAMINE TRANSPORTER (DAT)

The catecholamine dopamine (DA) is a neurotransmitter in the central nervous system,[1] where it contributes to normal physiological functions such as motor planning and movement, affect, and reward. Accordingly, DA also plays an important role in diseases such as Parkinson's disease, schizophrenia, and drug addiction. Thus, there are many reasons why neuroscientists and clinicians are interested in understanding DA neurotransmission and how it is regulated.

DA is released from axon varicosities (terminals) of DA neurons and from their dendrites. DA release occurs at low levels spontaneously and at higher levels upon neuronal depolarization, in particular when DA neurons "burst fire." Following diffusion across the synaptic cleft, DA produces its effects by activating two families of postsynaptic DA receptors, D1-like (D1, D5) and D2-like (D2, D3, D4) receptors. DA can also activate presynaptic D2-like receptors localized on the DA neurons themselves (DA autoreceptors), thereby regulating DA neuronal firing rate, DA synthesis, DA release and DA reuptake. All DA receptors are Guanosine triphosphate (GTP) binding protein-coupled receptors that signal through heterotrimeric G proteins to activate or inhibit particular enzymes and ion channels. Because of the receptor signaling cascades involved DA is considered to be a relatively slower acting neurotransmitter, or even a neuromodulator, as compared to the fast neurotransmitters whose receptors are ligand-gated ion channels.

Despite that DA is a relatively slow acting neurotransmitter, similar to all neurotransmitters, it is critical to limit its lifetime in the synapse and its extent of diffusion. In this way, DA neurotransmission can be tightly controlled in both time and space, preventing prolonged receptor stimulation and desensitization. The DA transporter (DAT) is the primary mechanism for clearing extracellular DA. The DAT not only terminates DA neurotransmission by transporting DA back into DA neurons (thus eliminating the possibility for DA to interact with its receptors), but also conserves DA by recycling it into the neuronal cytoplasm, from which it can be taken up into synaptic vesicles for re-release. Additional mechanisms also contribute to eliminating extracellular DA; these include diffusion, intraneuronal metabolism by monoamine oxidase (MAO), and extraneuronal metabolism by catechol-*O*-metyl-transferase (COMT).

DA uptake was first reported more than 30 years ago when it was shown that tritium-labeled DA ([^{3}H]DA) was accumulated in a Na^+- and energy-dependent manner by homogenates of DA-rich brain regions such as striatum.[2,3] Synaptosomes are an in vitro preparation of isolated nerve endings made by a sequence of precise homogenization and centrifugation steps. This preparation is most often used for in vitro uptake studies. When made from striatum, synaptosomes include a relatively high proportion of DA nigrostriatal neuronal terminals; thus, uptake of DA is readily observed in this preparation. DA uptake by DAT, rather than reflecting the activity of a plasma membrane energy-driven pump per se, is dependent upon the electrochemical gradient set up across cellular membranes by the Na^+/K^+-ATPase. The co-transport of 2 Na^+ and 1 Cl^- ions is required to translocate 1 DA^+ molecule (positively charged at physiological pH).[4,5] DAT normally transports DA in an inward direction, i.e., from the extracellular space into DA neuronal dendrites and terminals, from which DA is subsequently either transported into synaptic vesicles by the vesicular monoamine transporter 2 (VMAT2), a H^+-antiporter, or broken down by MAO. However, if the neuronal concentration of free cytoplasmic DA and Na^+ ions becomes locally high in the proximity of DAT, as happens with exposure to amphetamine or tyramine, the DAT can mediate reverse transport of DA out of the DA neuron. This is the basis for amphetamine-induced DA release.[6,7]

Drugs such as cocaine, nomifensine, and mazindol bind to DAT and block [^{3}H]DA uptake with micromolar to nanomolar affinities.[8,9] Although these drugs are all reasonably high affinity inhibitors of DAT, it is important to note that they also have similar (cocaine) or higher (nomifensine and mazindol) affinity for inhibiting the norepinephrine transporter (NET).[10] Cocaine also inhibits the serotonin transporter (SERT or 5-HTT) with an affinity similar to its inhibition of DAT and NET. GBR 12909 is the drug with the highest selectivity between DAT and NET (about fiftyfold)[11] most commonly used. When injected in vivo, all of these DAT inhibitors increase extracellular concentrations of DA, resulting in psychomotor stimulation. Similar to amphetamine and its derivatives, these drugs are self-administered by laboratory animals and have a high abuse and addiction liability in humans.[12] Thus, understanding the function and regulation of DAT is of particular interest to drug abuse researchers.

In the early 1990s, the cloning of DAT revealed that it is a member of the plasma membrane Na^+/Cl^--dependent transporter family, more recently defined as the solute carrier 6 (SLC6) gene family, which includes NET and SERT, as well as transporters for the inhibitory amino acid neurotransmitters GABA (GAT) and glycine (GYT).[13,14] The human gene for DAT is named SLC6A3. Based on primary sequence, DAT and NET are most closely related; thus, some overlap among inhibitors of these two transporters is not surprising, although there are several highly selective (greater than thousandfold) NET inhibitors. Deletion of the DAT gene (DAT knockout; DAT KO) in mice in 1996 provided some of the strongest evidence for the importance of DAT in terminating normal DA neurotransmission.[15,16] DAT KO mice are hyperactive, particularly in a novel environment, despite marked decreases in DA tissue levels, release and receptors—compensatory changes for the elevated extracellular DA concentration. In fact, DA persists 100 times longer in striatum of DAT KO mice than in wild-type mice. Neither cocaine nor amphetamine activates DAT KO mice. However, it came as a surprise that DAT KO mice still self-administer cocaine and exhibit cocaine conditioned place preference. In these mice, SERT inhibition by cocaine is sufficient to mediate cocaine reward.[17]

DAT is found exclusively in DA neurons,[18,19] in which it is functional only when expressed at the plasma membrane. Electron microscopic studies revealed that plasma membrane DAT is localized both near and distant from synaptic junctions.[19] Based on its "perisynaptic/extrasynaptic" localization, one might think that DAT is not all that important. However, the marked behavioral effects of DAT inhibitors and the results with the DAT KO mice argue otherwise. Also, in release assays using rat striatal slices, inclusion of a DAT inhibitor such as nomifensine in the superfusion buffer is required to detect endogenous DA overflow following mild electrical field stimulation.[20]

In the past ten years, it has been appreciated that transporters such as DAT are highly regulated.[21–25] This regulation provides additional evidence for their importance. Regulation may occur rapidly (minutes) or long-term (days to weeks). Mechanisms for rapid regulation include altered DAT kinetics, presumably because of post-translational modifications and conformational changes in the transporter molecule, and DAT trafficking, because of changes in the number of DATs expressed at the plasma membrane. At least in model expression systems, DAT constitutively traffics in a clathrin-mediated manner away from and back to the cell surface, providing an ideal mechanism for rapid regulation.[26] Substrates, inhibitors, ligands acting through presynaptic receptors, and other signaling molecules also rapidly regulate DAT; this often occurs via altered trafficking. One of the main challenges for the future is to develop methods for studying rapid regulation of DAT function in individual neurons, as well as in the intact brain. Disease states and chronic drug administration can also regulate DAT activity over days to weeks. This type of regulation is most often associated with changes in the number of functional DATs; however, whether this is because of alterations in transporter trafficking, turnover (synthesis/degradation rates), or both remains to be elucidated.

Because DA is electroactive, electrochemical methods provide valuable ways of measuring real-time DAT function, most often by determining the clearance of DA by DAT. Before we discuss our electrochemical approach for measuring DA clearance, however, we first mention briefly some of the non-electrochemical approaches used to measure DAT number and function.

NON-ELECTROCHEMICAL METHODS USED TO MEASURE DAT NUMBER AND FUNCTION

Radioligand Binding and Antibodies

The number of DATs in tissue or cell samples is most often measured using radioligand binding studies. DAT binding is determined by calculating specific binding (total binding–nonspecific binding) measured at steady state with a radioligand that has high affinity and reasonable selectivity for DAT, such as the cocaine congener [^{3}H]WIN 35,428 (or [^{3}H]CFT).[27] Binding can be measured in vitro by incubating the radioligand with brain membranes, brain slices, and intact cells or in vivo by injecting the radioligand systemically. Using DAT radioligands adapted to positron emission tomography (PET) or single photon emission tomography (SPET), one can also visualize changes in DAT in the brains of living humans.[28] Radioligands are generally lipophylic and bind to the total population of transporters, viz., transporters associated not only with plasma membranes, but also with intracellular membranes. However, because the binding of [^{3}H]WIN 35,428 is strongly Na^+-dependent and the intracellular concentration of Na^+ is low, in intact cells [^{3}H]WIN 35,428 can selectively detect cell surface binding, which represents the population of functional transporters.[29] When full saturation binding experiments are conducted, the affinity of DAT for the radioligand (K_d) and the maximal number of binding sites (B_{max}) are determined. Comparison of B_{max} values is valuable as an indicator of whether the number of DATs has been altered by a particular manipulation. The advantage of radioligand binding is that it is highly quantitative and, when coupled with image analysis, can yield precise anatomical resolution as well. However, the drawback to this approach is that it measures only the binding interaction of the ligand with the transporter, not transporter function.

DAT antibodies are useful for determining relative amounts of DAT protein and identifying DAT-protein interactions in homogenized tissue or membranes, as well as for localizing DAT protein in fixed tissue and live cell preparations. Although antibody-based methods are less quantitative than radioligand binding, they offer other advantages. For example, assuming equal amounts of protein are loaded in each lane on a gel, detection of bands on a western blot with specific antibodies allows comparison among samples of the relative amounts of DAT protein (not just binding sites). Western blots also reveal potentially different molecular weight forms of DAT, such as dimers and higher weight oligomers.[30]

To assess the number of functional transporters and for trafficking studies, DATs at the cell surface need to be measured. Unfortunately, however, few antibodies are available that interact with the extracellular loop of DAT, thereby detecting exclusively surface DATs, particularly in live cells. Thus, surface DATs have been measured most commonly by chemical cross-linking methods, e.g., with a cell-impermeant biotin reagent in synaptosomes before western blotting.[31]

DATs containing a fluorescent protein-tag have been extremely useful for visualizing DAT trafficking in live cells. New photoswitchable fluorescent-protein derivatives are being developed for use in trafficking and kinetic studies.[32] In general, the drawback has been that these constructs must be overexpressed acutely in cells or neurons. Thus, this approach does not allow native DATs to be studied. In the future, mice expressing fluorescent DATs could be generated. Transgenic mice expressing green fluorescent protein (GFP) under the control of the rat tyrosine hydroxylase gene promoter have already been generated, thus allowing visualization of live DA neurons.[33]

DA Uptake Assays

The oldest, and still an extremely useful, way to assess DAT function is to measure the accumulation of DA by DAT. This approach is limited to in vitro tissue preparations and cells. Both [^{3}H]DA and unlabeled DA have been used to monitor uptake. Rotating disk electrode voltammetry is a highly useful method for studying DA uptake mechanisms in vitro,[34] although [^{3}H]DA is more commonly used to simply measure uptake because of its ease of detection. As with radioligand binding, specific uptake by DAT can only be determined indirectly, by subtracting nonspecific uptake from total uptake. In terms of brain tissue, either synaptosomes or slices can be used. Synaptosomes have lower nonspecific uptake and fewer tissue barriers than slices; however, slices tend to be more viable during longer incubations. Because the rate of uptake is determined, DA accumulation is measured under conditions where it increases linearly with time (minutes). Specific uptake of a single concentration of [^{3}H]DA provides a snapshot comparison whereas a full kinetic curve must be generated to determine if an observed difference between samples is because of a change in affinity (K_m) or maximal velocity (V_{max}) of DAT.

DAT-Associated Currents

As uptake occurs, the inward translocation of 2 Na^+ ions: 1 Cl^- ion: 1 DA^+ molecule with each transport cycle would be expected to generate a small net inward current. Thus, about ten years ago, two-electrode voltage-clamp recording was used to test for such currents in *Xenopus laevis* oocytes expressing the cloned human DAT (hDAT).[35] As predicted, this work revealed small, inward DA-induced currents in oocytes expressing hDATs, but not in control oocytes. The transport-associated currents were concentration-dependent, blocked by pre-exposure to DAT inhibitors, such as cocaine, and magnified at more hyperpolarized potentials. Furthermore, the voltage-dependency of these currents established that DAT is an electrogenic transporter. The results suggested that uptake would be reduced upon depolarization and enhanced upon hyperpolarization. This was confirmed by measuring DA uptake in voltage-clamped hDAT-expressing oocytes. Thus, transport-associated currents provide an alternative real-time (msec) method to assess DAT function and regulation in vitro.

The surprising finding regarding the hDAT currents was that in addition to the transport-associated current, several channel-like leak currents were also associated with DAT expression.[35] One of these conductances was because of a constitutive, nonselective cation leak current; both DAT substrates and inhibitors blocked this leak current, resulting in an outward current. The fact that substrates both induced the transport current and blocked the leak current, whereas inhibitors only blocked the leak current, provided a way to distinguish unequivocally between DAT substrates and blockers. Thus, for example, amphetamine and tyramine were confirmed to be substrates, and cocaine and methylphenidate were confirmed as blockers.

These currents also provide additional approaches for evaluating DAT trafficking and function. For example, capacitance measurements are useful for studying DAT internalization in oocytes.[36] Patch-clamp recording can be used to measure transport-associated currents in mammalian cells, and this has been combined with microamperometric measurements of monoamine flux to provide a more comprehensive picture of NET and DAT function/regulation.[37,38] An interesting, new insight from these studies is that unlike DA, amphetamine has a unique ability to cause DA efflux through a bursting, channel-like mode.[38] On the other hand, whole cell patch clamp and perforated-patch recordings in primary cultures of midbrain DA neurons have revealed that DAT-mediated conductances induced by both DA and amphetamine increase the excitability of DA neurons.[39]

In Vivo Microdialysis

Although samples collected during in vivo microdialysis experiments are often analyzed for monoamines such as DA using high performance liquid chromatography (HPLC) and electrochemical detection, we mention this approach in this section on non-electrochemical methods because it does not use direct electrochemical recording in the brain. Microdialysis experiments are most often conducted in awake, behaving animals—from mice to nonhuman primates. Initially, the animal is anesthetized so that a guide cannula can be implanted under stereotaxic control into the brain region of interest; and this, along with a swivel device for later attachment of the sample tubing to a tether allowing freedom of movement, is then attached securely to the animal's skull.

Following a recovery period, the microdialysis probe is inserted into the guide cannula and slowly perfused with an artificial cerebrospinal fluid (aCSF) buffer. Small molecules such as neurotransmitters, which are present in the extracellular milieu of the brain but not the perfusion buffer, freely diffuse across the dialysis membrane and equilibrate in the perfusion buffer according to their extracellular concentrations. Thus, conventional DA microdialysis experiments measure the extracellular concentration of DA in five to twenty-minute samples; this represents the net result of primarily DA release and uptake processes. To determine absolute levels of basal DA and to quantify changes in DA release and uptake, the "no-net flux" or "zero-net flux" method of quantitative microdialysis is used. Although not without caveats,[40,41] this quantitative microdialysis method has been powerful, e.g., revealing altered basal and cocaine-evoked DAT activity during abstinence from cocaine and how behavioral responsiveness and basal DAT function are related.[42,43] Another in vivo microdialysis method used to assess DAT function selectively relies on the cellular extraction of [^{3}H]MPP$^+$ by DAT and [^{14}C]mannitol as a reference.[44]

Issues with the in vivo microdialysis technique include the amount of tissue damage resulting from relatively large probes ($\sim$200 um diameter) and the need for in vivo calibration of the probe recovery.[45] In general, as noted above, this approach uses relatively long sampling periods, but has the appropriate sensitivity for determining low nanomolar levels of basal extracellular DA. It is also ideal for measuring the relatively slow effects of drugs such as DAT inhibitors on extracellular levels of DA and relating drug-induced changes in DAT function with behavioral changes in freely behaving animals.

> Technological advances involving more rapid sampling and detection methods continue to improve the temporal and spatial resolution of the in vivo microdialysis approach. Analytical methods, such as automated capillary liquid chromatography, also allow simultaneous determination of multiple neurotransmitters in the dialysate. For example, this method has been used to elucidate the effect of the DAT blocker nomifensine on multiple neuroactive amines and amino acids (multi-analyte capability) in a particular brain region.[46]

Electrochemical Methods Used to Measure In Vitro and In Vivo DAT Function

Electrochemical detection of brain catecholamines such as DA, which was initially described more than thirty years ago (for a review, see Adams),[47] has been used extensively by neuroscience researchers interested in elucidating the function of these neurotransmitter systems. Multiple variations of electrochemical detection methods exist, but they all use the facts that electroactive species oxidize in response to an applied voltage and that the resulting movement of electrons can be measured as a change in current. Two commonly used methods are high-speed chronoamperometry (HSC) and fast cyclic voltammetry (FSV). Both HSC and FSV offer high sensitivity with excellent temporal and spatial resolution. Using small diameter (10–30 μm) carbon fibers, detection limits are in the low nanomolar concentration range and measurements can be made within 100 ms. These advantages make electrochemical techniques appealing to those studying the function of neurotransmitters, such as DA, that are characterized by their relatively rapid, synaptic actions.

The primary difference between HSC and FCV is the method of delivering the applied voltage. With HSC, the potential is held constant (e.g., 0.55 V) for a short period of time (typically 100 ms) during which DA at the electrode is oxidized, producing an oxidation (OX) current. This is followed by a return to a resting voltage (typically 0 V) during which the DA is reduced, producing a reduction (RED) current. In most cases, this square-wave pulse is repeated multiple times per second with measurements made at the same time (steady state) during each of the voltage pulses. Because the applied voltage is held at a constant potential and is instantaneously ramped up from the resting value, the measured current is directly proportional to the concentration of the species electrolyzed, and the electrode is sensitive to rapid changes in concentration.

FCV uses a triangular potential, typically with a range of −0.4 to +1.0 V, that is applied rapidly (300 V/sec) to the carbon fiber, with the resulting OX and RED currents measured every 100 ms. An advantage of this voltage-sweep method is that current changes can be resolved over the time and voltage domains, thereby providing an electrochemical "fingerprint" that can be used to aid the identification of the various electroactive species (e.g., DA) that are found in the micro-environment surrounding the electrode. Chemical identification (or verification) is provided, to some degree, with HSC by analysis of the ratio between the RED and OX currents.

USE OF HIGH-SPEED CHRONOAMPEROMETRY (HSC) WITH LOCAL DA APPLICATION TO MEASURE IN VIVO DAT FUNCTION IN DISCRETE BRAIN REGIONS OF ANESTHETIZED RATS AND MICE

In the sections that follow, we describe our typical setup and protocols for measuring DAT function in vivo. We initially chose to measure the clearance of locally applied exogenous DA as an "uncomplicated" indicator of DAT function[48] because this approach eliminates the potential confound of altered DA release. In Summary of Evidence that this Approach Measures DAT Function, we summarize the data showing that this measure does reflect primarily DAT activity. Nonetheless, as the rapid regulation of DAT began to be appreciated, it became apparent that our application of DA could regulate DAT function. This potential confound is discussed in Rapid Regulation of DAT Can Influence Measures of Exogenous DA Clearance. In our experiments, HSC measurements are made using an IVEC-10/FAST-12 system (Quanteon, Nicholasville, Kentucky) that interfaces with an electrode/micropipette assembly. Assemblies are positioned in the brains of rats or mice that are anesthetized and positioned in a stereotaxic apparatus. In Additional Applications of High-speed Chronoamperometry with Local DA Application to Measure DAT Function, we describe modifications that allow us to measure DAT function in brain slices maintained for several hours in a perfusion chamber or in awake, freely behaving rats.

METHODS

HSC

The IVEC-10/FAST-12 system consists of three main components. The first is a low-noise preamplifier, sometimes referred to as a headstage, that interfaces directly with the electrode implanted in the brain. The headstage provides conversion of the measured current into voltage and provides gains of up to 20 picoamps/mV. The second component is a control box that contains secondary amplifiers and a potentiostat for controlling the applied potential. For experiments in which we are interested in measuring the oxidation of DA, we apply square-wave pulses of 0.00–0.55 V (with respect to reference) at a frequency of 5 Hz. The last component is a DOS- or Windows-based computer that contains a National Instruments A/D board. The computer provides a user interface with the control hardware and software and is used for data acquisition, analysis, and storage.

Electrode Construction, Calibration, and Assembly with Micropipettes

Carbon-fiber recording electrodes can be purchased from a number of commercial sources or they can be fabricated easily in the laboratory. We choose the latter approach. For recordings in anesthetized animals, we construct batches of about twenty-five electrodes in three stages, each involving several steps. In the first stage, a 5 cm piece of 4-mm diameter glass tubing (Schott, Elmsford, New York) is pulled to a long, tapered point so that the shank is approximately 2.5 cm long. A carbon fiber (30–33 μm; Textron Systems, Wilmington, Massachusetts) is then inserted through the larger opening of the glass rod and fed as far as possible into the tapered point. The pointed end of the glass rod is then broken by carefully touching it to the surface of a rotating cutting wheel that is powered by a variable speed rotary tool (Dremel, Racine, Wisconsin). When done correctly, the tip will be broken to a diameter a little more than 30 μm and the carbon fiber will protrude through the tip. The fiber is then gently pulled through the small opening so that several millimeters extend outward; a small amount of Epoxylite (The Epoxylite Corp., Irvine, California) is injected into the tip via the larger opening in the glass rod. The electrodes are then cured in a 125°C-oven overnight. Note that if the tip of the glass rod is broken to a diameter much larger than 30 μm, a tight seal will not form between the protruding carbon fiber and the wall of the glass rod. This will result in leakage of the relatively viscous Epoxylite, which will adhere to the surface of the carbon fiber and render it unusable for electrochemical detection.

In the second stage of construction, the cured electrodes are allowed to cool to room temperature and are then loaded with a graphite epoxy paste (Graphpoxy; Dylon Industries, Inc., Cleveland, Ohio). A beveled wooden applicator stick is used to densely pack the paste into the barrel of the glass rod through the larger opening. The tip of the glass rod should be completely filled with paste and contain no pockets or air bubbles. The final step of this stage is to insert an insulated copper wire, with its bottom 5–10 mm scraped bare, into the paste at a sufficient depth so that it forms a continuous contact with the carbon fiber (via the graphite in the paste). The electrodes are then cured in a 125°C-oven overnight.

The final stage of electrode construction is to cut the carbon fiber with a scalpel blade so that 100–200 μm of it are exposed outside the tip of the glass rod. This step is best performed with the assistance of a stereomicroscope. Lastly, and before they are prepared for use in experiments, we typically clean the electrode tips by swirling them for several minutes in a beaker containing an isopropyl alcohol solution. Note that for recording in awake, freely behaving animals, several modifications to the electrode design are necessary, notably substituting fused silica tubing for the glass, and these are described elsewhere[49] (see also In Vivo Chronic Recording in Freely Behaving Rats).

On the day of experimentation, electrodes are initially prepared by coating the exposed carbon fibers with Nafion (5% w/v solution; Sigma–Aldrich, St. Louis, Missouri), which is a perfluorosulfinated polymer that blocks the absorption of potentially interfering anionic species, such as ascorbic acid (AA) and the DA metabolite 3,4-dihydroxyphenylacetic acid (DOPAC).[50] The coating consists of dipping the electrode tips into the Nafion solution, swirling them for several seconds, and then drying them in a 200°C-oven for five minutes. High-temperature drying is an important step that helps ensure electrodes will posses optimal selectivity and response times.[51] Electrodes are then tested for both their selectivity for and response to DA. This is accomplished by placing the electrode in a beaker of 100 mM phosphate buffered saline and adding known concentrations of both AA and DA. We typically test responses to 250 μM AA, followed by 2–12 μM DA in 2 μM-increments. Only those electrodes exhibiting a high selectivity for DA over AA (greater than or equal to 1000:1) and a linear response to DA ($r^2 \geqq 0.997$) are used. Electrodes with a lower selectivity for DA are recoated with Nafion, and the in vitro calibration procedure is repeated, whereas those with an undesirable linear response to DA are not useable (a likely result of a misstep in the construction process). Note that the current response to incremental DA challenges is saved in the computer and later used to convert in vivo measurements of OX current to concentration values.

After the animal has been surgically prepared for the experiment (see below), an electrode/-micropipette assembly is constructed by attaching either a single- or multibarrel pipette to the electrode using sticky wax (Patterson Dental Supply, Aurora, Colorado) that has been liquefied by touching it to a hot metal microspatula. The micropipettes (A–M Systems, Carlsborg, Washington), which are first pulled and broken to a tip diameter of 10–15 μm, are attached so that the tips are parallel and are separated by 150–300 μm (Figure 6.1). The electrode and the micropipette should also be in the same plane as one another (i.e., the tips should be simultaneously

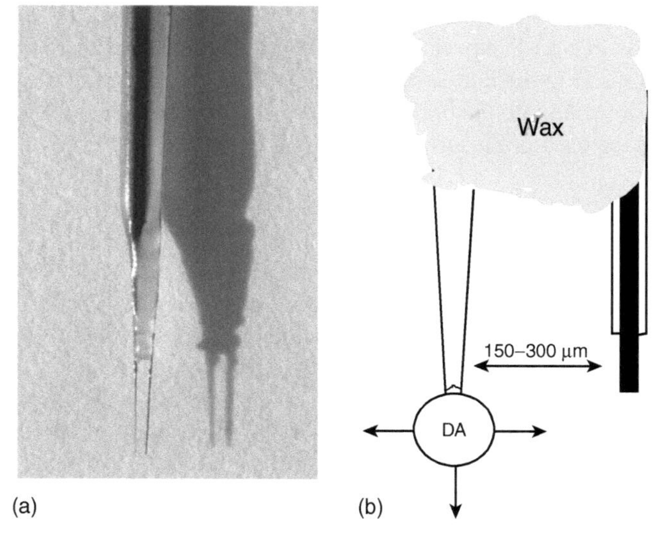

FIGURE 6.1 A picture (a) and schematic diagram (b) of an electrode/micropipette assembly with the micropipette on the left and the carbon fiber electrode on the right. (**a**) The picture shows both the assembly (left) and its shadow (right). The micropipette, which had a tip diameter of 15 μm and was filled with dye to add contrast for the photograph, is positioned 300 μm (center-to-center) away from the electrode. There is a distance of 2650 μm between the edge of the sticky wax and the tips of the assembly. The shadow exaggerates the proportions in the assembly, but it nicely shows that the electrode and micropipette are parallel to one another. (**b**) The schematic diagram (not to scale) represents a close-up view of the assembly tips with a bolus of DA being ejected from the micropipette. The DA diffuses away from the point source (the micropipette); it can then be taken up by nearby DATs, absorbed at the surface of the carbon fiber or continue to diffuse away from the assembly. We find that assemblies that yield the most sensitive and stable recordings are those with center-to-center distances between the micropipette and carbon fiber of 150–300 μm.

in focus under the microscope). Assemblies that do not possess these characteristics typically fail to yield consistent, or even measurable, electrochemical signals once they are implanted in the brain and exogenous DA is applied. A syringe fitted with a flexible microfilament (MicroFil; World Precision Instruments, Sarasota, Florida) is used to fill the single-barrel micropipettes, or one barrel of multibarrel micropipettes, with a DA solution (typically 200 μM in either 154 mM NaCl or 100 mM phosphate buffered saline; pH 7.4, adjusted with NaOH) that also contains 100 μM AA (added to prevent the oxidation of DA in the pipette barrel).

Surgical Procedures

With the exception of survival surgeries in animals fitted with chronic electrochemical recording electrode implants,[52] rats (200–450 g) or mice (15–30 g) are anesthetized with urethane (1.25–1.5 g/kg, i.p.) to a deep surgical level and placed in a stereotaxic frame. We have shown that urethane, unlike chloral hydrate, does not alter exogenous DA clearance, compared to that in an awake, freely behaving rat.[53] Body temperature is maintained at 37°C via a heating pad coupled to a rectal thermometer. A 2-cm incision is made in the scalp under aseptic conditions, which is retracted to expose the surface of the skull. A variable speed rotary tool (Dremel, Racine, Wisconsin) is then used to drill a hole in the skull overlying the brain areas of interest. In our case, these are usually the dorsal striatum (dSTR) and nucleus accumbens (NAc), both of which can be accessed through the single hole in the skull (0.7–1.5 mm anterior and 1.5–2.2 mm lateral to bregma for rats).[54] An additional hole is drilled just anterior to the interaural line for insertion and attachment (via dental acrylic) of a Ag/AgCl reference electrode. The reference electrode is prepared fresh daily by plating a silver wire (A–M Systems, Carlsborg, Washington) with chloride, which can be done with a 12-V power supply and a plating bath of 1 M HCl saturated with NaCl. An opening is then made in the dura with a sharp probe or needle, and the electrode/micropipette assembly is slowly lowered into the brain under stereotaxic control. For the duration of the experiment, the skull and exposed brain surface are bathed in saline.

Experimental Details

After its initial insertion into the brain, the electrode/micropipette assembly is left undisturbed, at a position above the brain area of interest, for about fifteen minutes to allow the electrode and surrounding brain tissue to stabilize. During this time and continuing throughout the experiment, chronoamperometric measurements are made continuously at 5 Hz, with the +0.55 V OX potential applied for 100 ms followed by a 0.00 V resting potential for 100 ms. The resulting OX and RED currents are digitally integrated during the last 70–90 ms (steady state) of each 100-ms pulse. OX current changes are converted to DA concentrations based on the in vitro calibration of the electrode.

When recordings are made in the rat dSTR and NAc, electrode/micropipette assemblies are lowered 4.0–5.0 mm and 6.5–8.0 mm ventral to the skull, respectively.[54] For each recording, a stable background current is established and set to zero before pressure-ejecting DA (5–20 psi for 0.1–5.0 s) at calibrated volumes (10–300 nl). The pressure ejection is controlled by a Picospritzer II (General Valve Corp., Fairfield, New Jersey), and injection volume is verified through a stereomicroscope fitted with an eyepiece reticule to note the movement of fluid in the pipette barrel. Generally, it is desirable to inject as small an amount of DA as possible that will yield signal amplitudes of between 0.75–4 μM. The volume required to achieve this will vary depending upon a number of factors, including the distance between the tips of the electrode and the micropipette, the sensitivity of the electrode, and the number of DATs present in the brain tissue located near the exposed carbon fiber. For most experiments, DA ejections are made at five-minute intervals, with signals used to determine "baseline" established when signal parameters (see Data Analysis) vary by less than or equal to 15% for two consecutive applications. The importance of a five-minute separation between DA ejections is not entirely clear, but we have observed that it leads to DA signals that are stable for long time periods (Figure 6.2). Following baseline measurements, pharmacological manipulations of DAT function are usually performed, with recordings maintained for an additional thirty minutes or more.

After experiments are completed, and while the animal is still deeply anesthetized, a small current is passed for several seconds through the recording electrode to produce a marking lesion at its tip. This can be done by setting both the applied and resting voltages to 2 V, turning off the headstage gain, and then applying the voltage to the carbon fiber. The animal is then euthanized, and the brain is subsequently removed and stored in buffered formalin (4% w/v) for several days. Coronal sections (40 μm) at the level of the brain area of interest are then cut using a microtome, mounted to glass slides, stained with Cresyl Violet, and analyzed under a light microscope to verify the location of the recording site(s).

Data Analysis

A number of electrochemical signal parameters can be analyzed and used to determine in vivo DAT function. We have commonly used two measures that are obtained from DA OX currents: maximal amplitude (A_{max}) and signal decay time (T_{80}; Figure 6.3). A_{max} reflects the maximal extracellular DA concentration detected, whereas T_{80} is the time for the signal to rise to A_{max} and decay by 80%. The T_{80} is chosen because it is in the curvilinear portion of the DA signal curve, takes into account changes in the "tail" of the decay curve where DA concentrations are lower, and is more sensitive to changes induced by DAT uptake inhibitors.

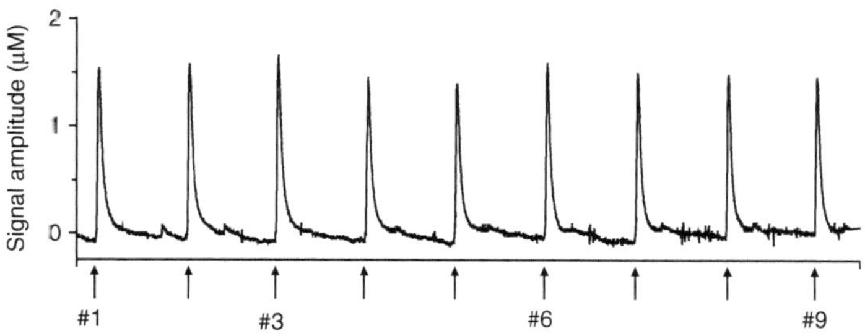

FIGURE 6.2 DA signals measured with HSC in the dSTR of a urethane-anesthetized rat after local pressure ejection of DA (90 nL) every five minutes (solid arrows) are relatively stable after multiple DA ejections. OX currents were converted to DA concentrations based on in vitro calibration of the electrode. Analysis of signal parameters (see Figure 6.3 and Data Analysis for more information) for ejection #1 revealed a maximal amplitude (A_{max}) and decay time (T_{80}) of 1.53 μM and thirty-three seconds, respectively. Forty minutes later, DA ejection #9 resulted in an A_{max} of 1.45 μM and a T_{80} of twenty-nine seconds. Figure modified from data presented elsewhere. (From Gulley, J. M., Doolen, S., and Zahniser, N. R., *J. Neurochem.*, 83, 400, 2002.)

Other options for measuring DA clearance that we have used in previous studies include analyses of the pseudolinear portion of the decay curve ($T_{20–60}$ or $T_{40–60}$)[55,56] and fitting the curve to a first-order exponential decay function.[52] The latter approach allows one to determine the rate constant (k) for DA clearance by approximating from each signal the initial velocity of DA clearance (v) by setting $v = k^*A_{max}$. Following an experiment in which increasing amounts of DA are applied, a plot of v versus the amount DA applied can be generated; and an apparent K_m and V_{max} can be determined from curve fitting of the rectangular hyperbola.[52] Importantly, all of the signal parameters mentioned in this section are affected, to some extent, by DAT inhibitors[52,57] and reflect changes in DA clearance (see Summary of Evidence that this Approach Measures DAT Function). Several other investigators have used similar approaches, while also highlighting other analysis options.[53–62]

Because of individual differences in baseline DAT function measures, we typically normalize data when comparing groups of rats by first obtaining a mean value for parameters during a two- or

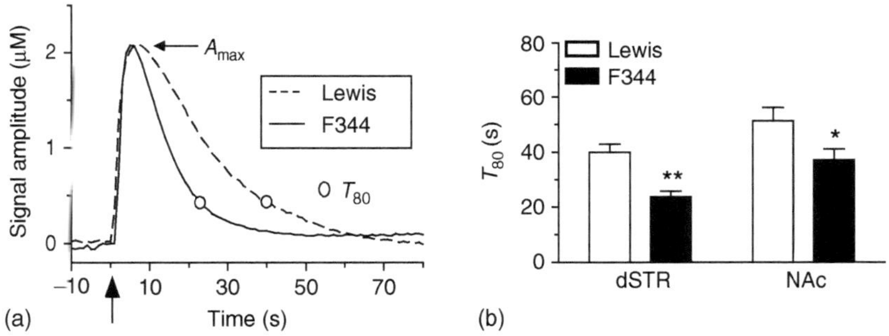

FIGURE 6.3 The baseline capacity of DATs for clearing exogenous DA in dSTR and NAc is reduced in Lewis rats, compared to F344 rats. (**a**) Electrochemical signals measured with HSC in dSTR of a representative Lewis and F344 rat following the local pressure ejection of DA (arrowhead; 20 nL for Lewis, 30 nL for F344). The A_{max} values were similar between strains, but the T_{80} value was more prolonged in the Lewis rat. For the Lewis rat, A_{max} was 2.10 μM and T_{80} was forty seconds; for the F344 rat, A_{max} was 2.07 μM and T_{80} was twenty-seven seconds. (**b**) Mean strain differences in baseline exogenous DA clearance in dSTR and NAc. Brackets indicate SEM ($n = 10–11$ recordings/strain). * $p < 0.05$ and ** $p < 0.01$, compared to Lewis rats; ANOVA and Holm–Sidak *post hoc* test.

three-ejection baseline period, setting this value as 100% and expressing all data (including the baseline data points) as a percentage of baseline. When suitable, or when individual differences are minimal (e.g., studies in inbred rats), non-normalized data are also presented. Appropriate statistical tests are then conducted (e.g., ANOVA with post-hoc Holm-Sidak tests for between-group comparisons at particular data points).

Results

We have used the techniques described above to analyze baseline differences in DAT function, as well, as the effects of drugs that alter DAT activity. For example, in a recent set of experiments, we compared DAT function in the dSTR and NAc of inbred Lewis and Fischer 344 (F344) rat strains, which are known to exhibit differential behavioral responsiveness to psychostimulants such as cocaine.[63] Consistent with a report in the literature that used quantitative autoradiography to reveal Lewis rats had fewer DAT uptake sites in multiple brain regions,[64] we observed that Lewis rats cleared exogenous DA at a slower rate compared to F344 rats (Figure 6.3, panel A). The mean differences in T_{80} values between the two strains were about 70% in dSTR and about 40% in NAc (Figure 6.3, panel B). In these studies, signal amplitudes were maintained at similar levels (1–2 μm) from one experiment to the next by adjusting the amount of DA ejected once every five minutes from the micropipette within each experiment; thus, there were no group differences in A_{max}. We also found no group differences in DA ejection volume. We recently reported a similar finding of baseline strain differences in DAT function in the dSTR of inbred long- and short-sleep mice, which exhibit differential responsiveness to the sedative–hypnotic effects of alcohol and also exhibit diverse behavioral responses to cocaine.[65]

After measures of basal DAT function were obtained, we assessed the effects of systemically-administered *d*-amphetamine sulfate (AMPH) on exogenous DA clearance in the dSTR of Lewis and F344 rats (Figure 6.4). We determined that the baseline clearance responses were stable before a control saline injection was given three minutes before the third DA ejection ($t = 10$ min). The saline was injected intravenously (i.v.) through catheters implanted in either the lateral tail vein or the jugular vein. After two additional DA ejections, rats were injected with 0.5 mg/kg AMPH (i.v.) three minutes before the sixth DA ejection ($t = 25$ min). DA ejections were continued at five-minute intervals for fifteen minutes, followed by a second AMPH injection (0.5 mg/kg; i.v.) three minutes before the tenth DA ejection ($t = 45$ min). DA ejections were then continued at five-minute intervals for fifteen minutes, when the experiment was terminated. Although the i.v. route of drug administration was used here, more convenient systemic routes of drug administration

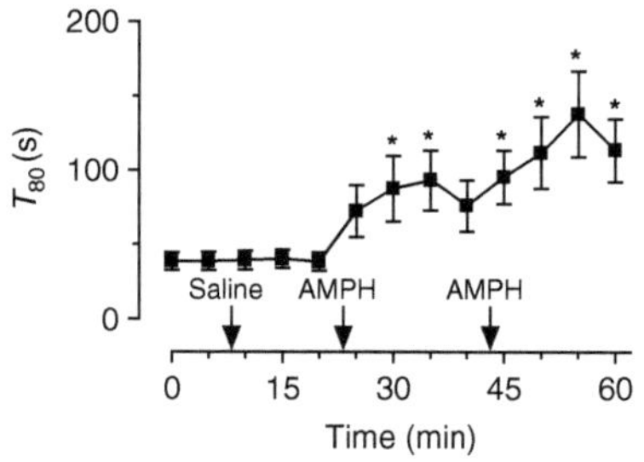

FIGURE 6.4 Prolongation of the DA signal decay time in dSTR of F344 rats following i.v. *d*-amphetamine (AMPH). DA was ejected once every five minutes, and OX currents were measured with HSC as in Figure 6.3. There was no significant change in the T_{80} after saline injection into a catheter positioned in either the lateral tail vein or the jugular vein. In contrast, when rats were given AMPH (0.5 mg/kg), the T_{80} value was significantly increased. Brackets indicate SEM ($n = 5$). $*p < 0.05$ compared to the T_{80} measured from the last DA application before AMPH injection; ANOVA and Holm-Sidak *post hoc* test.

(e.g., subcutaneous or intraperitoneal) can also be used. Furthermore, if assessment of local drug effects is desired, the electrode/micropipette assembly can be constructed with a multibarrel micropipette.[57] DA is loaded into one barrel and the drug solution(s) (pH adjusted to 7.4) is(are) loaded into one or more of the other barrels. Note, however, that local application of drugs must be done with the appropriate controls because particular drugs can alter electrode sensitivity.[66]

Summary of Evidence That This Approach Measures DAT Function

Two fundamental assumptions of the technique described in this chapter are as follows: (1) the changes in the electrochemical OX signal that are observed after local pressure ejection of a DA solution are the result of a change in local DA concentration and (2) the rise and decay of this electrochemical signal represents the clearance of the applied DA solution by DAT. There are several empirical findings that support the validity of these assumptions.

Questions about the first assumption may seem, at least initially, to be easily addressed given that the change in electrochemical signal results within one second after the time of pressure ejection of the DA solution (Figure 6.3). However, because our system simply measures changes in the OX current that result from an applied potential set to a particular voltage (0.55 V in our experiments), any substance that oxidizes at this voltage has the potential to contribute to current changes. Two substances that fit this profile, AA and DOPAC, are found in abundance in brain areas (e.g., dSTR or NAc) where one might assess DAT function. In fact, AA is a component of the DA solution that we pressure eject from our micropipettes. A common solution to this problem is to coat carbon fibers with Nafion, which significantly reduces the OX and/or RED of AA and DOPAC at the electrode surface, so that their influence on the measured signal is greatly attenuated. Through various means, including local infusions of AA or DOPAC via multibarrel micropipettes, we and others have demonstrated that putative DA signals that are measured with Nafion-coated carbon fiber electrodes are, in fact, primarily attributable to DA.[50,56,67–69] Another strategy for identification of the species that compose the electrochemical signal measured with HSC is to monitor the ratio of integrated RED:OX currents.[70] This ratio, which is calculated at the peak of the electrochemical response, is approximately 0.6–0.8 for DA. It can be distinguished from AA, which has a ratio of 0.0 because it lacks a reduction current, and DOPAC (ratio of ~ 1.0).

With respect to the second assumption (i.e., the rise and decay of the electrochemical signal represents the clearance of the applied DA solution), there are several empirical reasons to be confident that it is also reasonable. First, a range of detailed kinetic analyses, performed both in vitro and in vivo, have demonstrated that signal amplitude and decay time primarily reflect transporter activity.[56,60,71–76] Diffusion of applied DA does contribute to certain aspects of the signal parameters,[58,77] but its contribution, in particular in the dSTR where a large number of DATs are present, appears to be relatively minimal.[15,56,78,79] Second, inhibiting DAT activity has consistently been shown to produce changes in these parameters that are indicative of a reduction in DAT function. For example, A_{max}, and even more typically T_{80}, are increased by systemic injection or local infusion of DA uptake inhibitors such as mazindol, nomifensine, and cocaine.[55,57,71,74,80] However, structurally similar compounds that lack significant DAT antagonist properties (e.g., lidocaine) have no effect on A_{max} or T_{80}.[56] When DAT sites are removed from the brain by 6-OHDA or MPTP lesions of DA cell bodies and nerve terminals in the substantia nigra and STR, A_{max}[81] and especially T_{80}[80,81] are increased by up to 130%.

Rapid Regulation of DAT Can Influence Measures of Exogenous DA Clearance

It has been estimated that DATs can be regulated rather rapidly, on the order of several minutes to several hours, upon exposure to both substrates and blockers.[22] In particular, DA has been reported to reduce DAT activity in measures of DAT-associated currents[69] and [^{3}H]DA uptake.[82] D2 receptor agonists have been shown to increase DA uptake by the DAT,[83] whereas D2 receptor antagonists have the opposite effect.[84,85] Indeed, D2 receptor ko mice exhibit reduced exogenous DA clearance in dSTR, suggesting that the locally applied DA normally activates D2 autoreceptors and, thereby, increases DAT surface expression and velocity.[29,86] Potentially, however, local application of DA via micropipettes might alter DAT function either directly, through substrate-induced down-regulation, or indirectly through its action on D2 autoreceptors. In fact, we have reported[69] that frequent (e.g., every two minutes) applications of DA decreases exogenous DA clearance by the DAT, consistent with substrate-induced DAT down-regulation. Although the regulation of DAT is complex, it is noteworthy that DA applications every five minutes do not appear to change DAT function over time (see Figure 6.2).

ADDITIONAL APPLICATIONS OF HSC WITH LOCAL DA APPLICATION TO MEASURE DAT FUNCTION

In Vitro Recording in Acutely Prepared Brain Slices

The analysis of DAT function in vivo provides the important advantage of measurement in the intact organism, but it leads one to give up some aspects of experimental control. Studies in brain slices,[87] which incorporate the same electrode/micropipette assemblies and HSC techniques that we employ in our in vivo studies, provide an in vitro option that regains some of that control. For example, this approach allows for precise anatomical localization of recording electrodes in brain regions of interest. It is relatively easy, using a stereomicroscope, to place electrodes in subregions of the NAc (e.g., the core and shell) that may have different functional populations of DATs. Dose–response analyses of drugs with activity at the DAT are also more practical with in vitro studies because compounds can be bath applied and washed out (at least theoretically) with a relatively rapid time course. Furthermore, multiple brain slices can be tested in a relatively short period of time. For example, the results of an experiment performed in our laboratory demonstrate the effects of different concentrations of cocaine on DAT function in dSTR slices (Figure 6.5).

For in vitro studies, rats are sacrificed by decapitation; their brains are rapidly removed and mounted on a chilled metal block within a small chamber. This chamber, which is filled with cold aCSF, is then attached to a Vibratome (Series 1000; Ted Pella, Inc., Redding, California) that is used

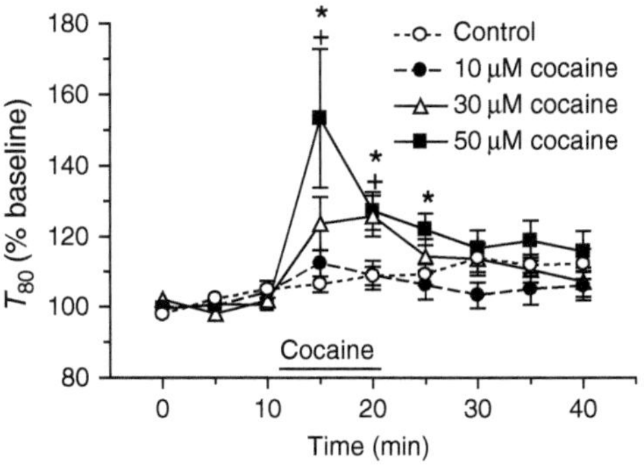

FIGURE 6.5 Consistent with DAT inhibition, in vitro exposure to cocaine increases DA signal decay time in a concentration-dependent manner in slices containing rat dSTR. DA was locally applied at five-minute intervals, and HSC was used to measure OX currents. Cocaine was included for ten minutes (bar) at the times indicated. Each slice was tested with only a single concentration of cocaine. T_{80} values are expressed as percent baseline; baseline values ranged from thirty-two to forty-nine seconds. Mean $\pm$ SEM for $N = 5$–23 slices. $^*p < .05$, 50 μM cocaine vs. control at the same time point; $^+p < .05$, 30 μM cocaine vs. control; ANOVA and Holm–Sidak *post hoc* test.

to cut 400 μm-thick coronal sections at the level of the dSTR and NAc. The brain slices are allowed to recover their energy stores and membrane potentials by incubating them at 22°C in oxygenated aCSF for at least one hour before they are used for HSC. The issue of slice recovery and viability is key to the success of this type of experiment. After recovery, the slices are moved to a small electrophysiological-type slice recording chamber in which they are superfused (2 ml/min) with aCSF and maintained at 32°C–33°C. The electrode/micropipette assembly is then lowered into the desired recording area and, after reproducible baseline signals are obtained, drugs (e.g., DAT inhibitors) are introduced into the chamber. We typically administer drugs at one hundred times the desired final concentration, directly into the aCSF flow via a syringe pump (Model A-99; Razel Scientific Instruments, Stanford, Connecticut) set at 1/100 the normal flow rate (i.e., 20 μl/min). Using this technique, we are able to obtain fairly long lasting (greater than or equal to forty-five minutes) and reproducible HSC signals in response to repeated applications of exogenous DA (Figure 6.5).

In Vivo Chronic Recording in Freely Behaving Rats

Along with our collaborators, we have adapted methods of assessing in vivo DAT function with local DA applications and HSC to studies in awake, freely behaving rats.[49,52] This approach, while technically challenging, offers the opportunity to correlate changes in DAT function with behavior, which is especially useful in studies of chronic drug effects. It also removes potential effects of anesthesia on brain function, although urethane does not appear to influence exogenous DA clearance.[53] Several modifications to the general technique described in this chapter are made, including adjustments to the electrode construction, electrode/micropipette assembly, and analysis of the data. They are described in detail elsewhere[49,52] and are summarized in brief below.

To provide the durability necessary for chronic implantation, recording electrodes are constructed from 30 μm carbon fibers sealed in fused silica tubing. Stainless steel guide cannula are then attached to the electrode assembly; these guide the insertion of fused silica tubing used for rapid infusion of exogenous DA at the recording site, which replaces the glass barrel pipettes used in acute experiments with anesthetized rats. The entire assembly is then lowered via stereotaxic control and cemented to the skull with dental acrylic. After a five- to seven-day recovery period, electrochemical recordings are obtained by inserting the injector and attaching to the electrode assembly a miniature potentiostat/headstage, which interfaces with the IVEC-10 system via a tether cable connected to an electrical commutator. Random movement-generated signal artifacts are eliminated with a low-pass filter (cut-off frequency greater than 0.028 Hz) during fast Fourier transformation analysis.[52]

We have used this technique to study the effects of both acute and repeated administration of cocaine on in vivo DAT function in the dSTR and NAc.[88,89] The effects of cocaine on the clearance of exogenous DA by the DAT depended on the animal's behavioral response to the drug. Following an injection of 10 mg/kg cocaine, a reduction in baseline exogenous DA clearance in dSTR and NAc was observed in outbred male Sprague–Dawley rats with an elevated locomotor response to the drug (i.e., high cocaine responders, or HCRs), but not in those with a reduced locomotor response (i.e., low cocaine responders, or LCRs). After six additional injections with 10 mg/kg cocaine on successive days and a seven-day withdrawal period, LCRs exhibited a sensitized behavioral response to a challenge cocaine injection, whereas HCRs exhibited a response similar to the one they had after the first drug injection. In parallel with the behavioral findings in LCRs was an increased ability of cocaine to inhibit exogenous DA clearance by the DAT in dSTR and NAc. Cocaine-induced inhibition of DAT function was unchanged in HCRs from the last cocaine injection compared to the first. These findings, which were interpreted as evidence for an important role of the DAT in individual differences in cocaine responses, highlight the utility of combining in vivo measures of DAT function with studies in behaving animals.

FUTURE DIRECTIONS IN USING HSC TO MEASURE DAT FUNCTION

The approach of using HSC combined with local ejection of DA has provided a number of insights about DAT function and its regulation. One need for this approach in the future is further miniaturization of the electrochemical electrode/micropipette assemblies. This would facilitate recording in mice (genetic models), as well as in even more circumscribed brain sub-regions. Obviously, any improvements that enhance recording in freely behaving animals will also be important because these will allow simultaneous behavioral and DAT clearance measurements to be compared. This approach should also be useful in helping to understand the physiological and pharmacological relevance of rapid DAT regulation by allowing us to measure transient changes in in vivo DAT function in the brain.

ACKNOWLEDGMENTS

We thank Ms. Melissa Adams for her help in preparing this chapter and acknowledge our grant support from NIDA (DA016485, DA015050, DA004216, DA014204).

REFERENCES

1. Cooper, J. R., Bloom, F. E., and Roth, R. H., *The Biochemical Basis of Neuropharmacology*, 8th ed., Oxford University Press, Oxford, 2003, chap. 9.
2. Snyder, S. H. and Coyle, J. T., Regional differences in H^3-norepinephrine and H^3-dopamine uptake into rat brain homogenates, *J. Pharmacol. Exp. Ther.*, 165, 78, 1969.
3. Harris, J. E. and Baldessarini, R. J., The uptake of [^{3}H]dopamine by homogenates of rat corpus striatum: effects of cations, *Life Sci.*, 13, 303, 1973.
4. Krueger, B. K., Kinetics and block of dopamine uptake in synaptosomes from rat caudate nucleus, *J. Neurochem.*, 55, 260, 1990.
5. McElvain, J. S. and Schenk, J. O., A multisubstrate mechanism of striatal dopamine uptake and its inhibition by cocaine, *Biochem. Pharmacol.*, 43, 2189, 1992.
6. Sulzer, D., Chen, T.-K., Lau, Y. Y., Kristensen, H., Rayport, S., and Ewing, A., Amphetamine redistributes dopamine from synaptic vesicles to the cytosol and promotes reverse transport, *J. Neurosci.*, 15, 4102, 1995.
7. Khosbouei, H., Wang, H., Lechleiter, J. D., Javitch, J. A., and Galli, A., Amphetamine-induced dopamine efflux: a voltage-sensitive and intracellular Na^+-dependent mechanism, *J. Biol. Chem.*, 278, 12070, 2003.
8. Dubocovich, M. L. and Zahniser, N. R., Binding of the dopamine uptake inhibitor [^{3}H]nomifensine to striatal membranes, *Biochem. Pharmacol.*, 34, 1137, 1985.
9. Javitch, J. A., Strittmatter, S. M., and Snyder, S. H., Differential visualization of dopamine and norepinephrine uptake sites in rat brain using [^{3}H]mazindol autoradiography, *J. Neurosci.*, 5, 1513, 1985.
10. Hyttel, J., Citalopram—Pharmacological profile of a specific serotonin uptake inhibitor with antidepressant activity, *Prog. NeuroPsychopharmacol. Biol. Psychiatry*, 6, 277, 1982.
11. Heikkila, R. E. and Manzino, L., Behavioral properties of GBR 12909, GBR 13069 and GBR 13098: specific inhibitors of dopamine uptake, *Eur. J. Pharmacol.*, 103, 241, 1984.
12. Ritz, M. C. and Kuhar, M. J., Cocaine receptors on dopamine transporters are related to self-administration of cocaine, *Science*, 237, 1219, 1987.
13. Amara, S. G. and Kuhar, M. J., Neurotransmitter transporters: recent progress, *Annu. Rev. Neurosci.*, 16, 73, 1993.
14. Chen, N.-H., Reith, M. E. A., and Quick, M. W., Synaptic uptake and beyond: the sodium- and chloride-dependent neurotransmitter transporter family SLC6, *Pflugers Arch. Eur. J. Physiol.*, 447, 519, 2004.

15. Giros, B., Jaber, M., Jones, S. R., Wightman, R. M., and Caron, M. G., Hyperlocomotion and indifference to cocaine and amphetamine in mice lacking the dopamine transporter, *Nature*, 379, 606, 1996.
16. Gainetdinov, R. R. and Caron, M. G., Monoamine transporters: from genes to behavior, *Annu. Rev. Pharmacol. Toxicol.*, 43, 261, 2003.
17. Sora, I., Hall, F. S., Andrews, A. M., Itokawa, M., Li, X.-F., Wei, H.-B., Wichems, C., Lesch, K.-P., Murphy, D. L., and Uhl, G. R., Molecular mechanisms of cocaine reward: combined dopamine and serotonin transporter knockouts eliminate cocaine place preference, *Proc. Natl Acad. Sci. U.S.A.*, 98, 5300, 2001.
18. Ciliax, B. J., Heilman, C., Demchyshyn, L. L., Pristupa, Z. B., Ince, E., Hersch, S. M., Niznik, H. B., and Levey, A. I., The dopamine transporter: immunochemical characterization and localization in brain, *J. Neurosci.*, 15, 1714, 1995.
19. Nirenberg, M. J., Vaughan, R. A., Uhl, G. R., Kuhar, M. J., and Pickel, V. M., The dopamine transporter is localized to dendritic and axonal plasma membranes of nigrostriatal dopaminergic neurons, *J. Neurosci.*, 16, 436, 1996.
20. Gerhardt, G. A., Dwoskin, L. P., and Zahniser, N. R., Outflow and overflow of picogram levels of endogenous dopamine and DOPAC from rat striatal slices: improved methodology for studies of stimulus-evoked release and metabolism, *J. Neurosci. Methods*, 26, 217, 1989.
21. Zahniser, N. R. and Doolen, S., Chronic and acute regulation of Na^+/Cl^-- dependent neurotransmitter transporters: Drugs, substrates, presynaptic receptors and signaling systems, *Pharmacol. Ther.*, 92, 21, 2001.
22. Gulley, J. M. and Zahniser, N. R., Rapid regulation of dopamine transporter function by blockers, substrates and presynaptic receptor ligands, *Eur. J. Pharmacol.*, 479, 139, 2003.
23. Mortensen, O. V. and Amara, S. G., Dynamic regulation of the dopamine transporter, *Eur. J. Pharmacol.*, 479, 159, 2003.
24. Melikian, H. E., Neurotransmitter transporter trafficking: endocytosis, recycling, and regulation, *Pharmacol. Ther.*, 104, 17, 2004.
25. Zahniser, N. R. and Sorkin, A., Rapid regulation of the dopamine transporter: role in stimulant addiction?, *Neuropharmacology*, 47, 80, 2004.
26. Sorkina, T., Hoover, B. R., Zahniser, N. R., and Sorkin, A., Constitutive and protein kinase C-induced internalization of the dopamine transporter is mediated by a clathrin-dependent mechanism, *Traffic*, 6, 157, 2005.
27. Reith, M. E. A. and Coffey, L. L., Cationic and anionic requirements for the binding of 2β- carbomethoxy-3β-(4-fluorophenyl)[^{3}H]tropane] to the dopamine uptake carrier, *J. Neurochem.*, 61, 167, 1993.
28. Volkow, N. D., Wang, G.-J., Fischman, M. W., Foltin, R. W., Fowler, J. S., Abumrad, N. N., Vitkun, S. et al., Relationship between subjective effects of cocaine and dopamine transporter occupancy, *Nature*, 386, 827, 1997.
29. Mayfield, R. D. and Zahniser, N. R., Dopamine D_2 receptor regulation of the dopamine transporter expressed in *Xenopus laevis* oocytes is voltage- independent, *Mol. Pharmacol.*, 59, 113, 2001.
30. Sorkina, T., Doolen, S., Galperin, E., Zahniser, N. R., and Sorkin, A., Oligomerization of dopamine transporters visualized in living cells by fluorescence resonance energy transfer microscopy, *J. Biol. Chem.*, 278, 28274, 2003.
31. Chi, L. and Reith, M. E. A., Substrate-induced trafficking of the dopamine transporter in heterologously expressing cells and in rat striatal synaptosomal preparations, *J. Pharmacol. Exp. Ther.*, 307, 729, 2003.
32. Verkusha, V. V. and Sorkin, A., Conversion of the monomeric red fluorescent protein into a photoswitchable probe, *Chem. Biol.*, 12, 279, 2005.
33. Sawamoto, K., Nakao, N., Kobayashi, K., Matsushita, N., Takahashi, H., Kakishita, K., Yamamoto, A. et al., Visualization, direct isolation, and transplantation of midbrain dopaminergic neurons, *Proc. Natl Acad. Sci. U.S.A.*, 98, 6423, 2001.
34. Schenk, J. O., Wright, C., and Bjorklund, N., Unraveling neuronal dopamine transporter mechanisms with rotating disk electrode voltammetry, *J. Neurosci. Methods*, 143, 41, 2005.

35. Sonders, M. S., Zhu, S.-J., Zahniser, N. R., Kavanaugh, M. P., and Amara, S. G., Multiple ionic conductances of the human dopamine transporter: the actions of dopamine and psychostimulants, *J. Neurosci.*, 17, 960, 1997.
36. Zhu, S. -J., Kavanaugh, M. P., Sonders, M. S., Amara, S. G., and Zahniser, N. R., Activation of protein kinase C inhibits uptake, currents and binding associated with the human dopamine transporter expressed in *Xenopus* oocytes, *J. Pharmacol. Exp. Ther.*, 282, 1358, 1997.
37. Galli, A., Blakely, R. D., and DeFelice, L. J., Patch-clamp and amperometric recordings from norepinephrine transporters: channel activity and voltage-dependent uptake, *Proc. Natl Acad. Sci. U.S.A.*, 95, 13260, 1998.
38. Kahlig, K. M., Binda, F., Khoshbouei, H., Blakely, R. D., McMahon, D. G., Javitch, J. A., and Galli, A., Amphetamine induces dopamine efflux through a dopamine transporter channel, *Proc. Natl Acad. Sci. U.S.A.*, 102, 3495, 2005.
39. Ingram, S. L., Prasad, B. M., and Amara, S. G., Dopamine transporter-mediated conductances increase excitability of midbrain dopamine neurons, *Nat. Neurosci.*, 5, 971, 2002.
40. Tang, A., Bungay, P. M., and Gonzales, R. A., Characterization of probe and tissue factors that influence interpretation of quantitative microdialysis experiments for dopamine, *J. Neurosci. Methods*, 126, 1, 2003.
41. Chen, K. C., Evidence on extracellular dopamine level in rat striatum: implications for the validity of quantitative microdialysis, *J. Neurochem.*, 92, 46, 2005.
42. Chefer, V. I. and Shippenberg, T. S., Changes in basal and cocaine-evoked extracellular dopamine uptake and release in the rat nucleus accumbens during early abstinence from cocaine: quantitative determination under transient conditions, *Neuroscience*, 112, 907, 2002.
43. Chefer, V. I., Zakharova, I., and Shippenberg, T. S., Enhanced responsiveness to novelty and cocaine is associated with decreased basal dopamine uptake and release in the nucleus accumbens: quantitative microdialysis in rats under transient conditions, *J. Neurosci.*, 23, 3076, 2003.
44. Zimmer, I., Kodas, E., Guilloteau, D., Garreau, L., Besnard, J.-C., and Chalon, S., Microdialysis as a tool for in vivo study of dopamine transporter function in rat brains, *J. Neurosci. Methods*, 103, 137, 2000.
45. Borland, L. M., Shi, G., Yang, H., and Michael, A. C., Voltammetric study of extracellular dopamine near microdialysis probes acutely implanted in the striatum of the anesthetized rat, *J. Neurosci. Methods*, 146, 149, 2005.
46. McKenzie, J. A. M., Watson, C. J., Rostand, R. D., German, I., Witowski, S. R., and Kenndy, R. T., Automated capillary liquid chromatography for simultaneous determination of neuroactive amines and amino acids, *J. Chromatogr.*, 962, 105, 2002.
47. Adams, R. N., In vivo electrochemical measurements in the CNS, *Prog. Neurobiol.*, 35, 297, 1990.
48. Zahniser, N. R., Dickinson, S. D., and Gerhardt, G. A., High-speed chronoamperometric electrochemical measurements of dopamine clearance, In *Methods in Enzymology: Neurotransmitter Transporters*, Amara, S. G., Ed., Vol. 296, Academic Press, San Diego, CA, p. 708, 1998.
49. Gerhardt, G. A., Ksir, C., Pivik, C., Dickinson, S. D., Sabeti, J., and Zahniser, N. R., Methodology for coupling local application of dopamine and other chemicals with rapid in vivo electrochemical recordings in freely moving rats, *J. Neurosci. Methods*, 87, 67, 1999.
50. Gerhardt, G. A., Oke, A. F., Nagy, G., Moghaddam, B., and Adams, R. N., Nafion-coated electrodes with high selectivity for CNS electrochemistry, *Brain Res.*, 290, 390, 1984.
51. Gerhardt, G. A. and Hoffman, A. F., Effects of recording media composition on the responses of Nafion-coated carbon fiber microelectrodes measured using high-speed chronoamperometry, *J. Neurosci. Methods*, 109, 13, 2001.
52. Sabeti, J., Adams, C. E., Burmeister, J., Gerhardt, G. A., and Zahniser, N. R., Kinetic analysis of striatal clearance of exogenous dopamine recorded by chronoamperometry in freely moving rats, *J. Neurosci. Methods*, 121, 41, 2002.
53. Sabeti, J., Gerhardt, G. A., and Zahniser, N. R., Chloral hydrate and ethanol, but not urethane, alter the clearance of exogenous dopamine recorded by chronoamperometry in striatum of unrestrained rats, *Neurosci. Lett.*, 343, 9, 2003.
54. Paxinos, G. and Watson, C., *The Rat Brain in Stereotaxic Coordinates*, 4th ed., Academic Press, New York, 1998.

55. Cass, W. A., Gerhardt, G. A., Gillespie, K., Curella, P., Mayfield, R. D., and Zahniser, N. R., Reduced clearance of exogenous dopamine in rat nucleus accumbens, but not in dorsal striatum, following cocaine challenge in rats withdrawn from repeated cocaine administration, *J. Neurochem.*, 61, 273, 1993.
56. Cass, W. A., Zahniser, N. R., Flach, K. A., and Gerhardt, G. A., Clearance of exogenous dopamine in rat dorsal striatum and nucleus accumbens: role of metabolism and effects of locally applied uptake inhibitors, *J. Neurochem.*, 61, 2269, 1993.
57. Zahniser, N. R., Larson, G. A., and Gerhardt, G. A., In vivo dopamine clearance rate in rat striatum: regulation by extracellular dopamine concentration and dopamine transporter inhibitors, *J. Pharmacol. Exp. Ther.*, 289, 266, 1999.
58. Wightman, R. M. and Zimmerman, J. B., Control of dopamine extracellular concentration in rat striatum by impulse flow and uptake, *Brain Res. Rev.*, 15, 135, 1990.
59. Kawagoe, K. T., Garris, P. A., Wiedemann, D. J., and Wightman, R. M., Regulation of transient dopamine concentration gradients in the microenvironment surrounding nerve terminals in the rat striatum, *Neuroscience*, 51, 55, 1992.
60. Suaud-Chagny, M. F., Dugast, C., Chergui, K., Msghina, M., and Gonon, F., Uptake of dopamine released by impulse flow in the rat mesolimbic and striatal systems in vivo, *J. Neurochem.*, 65, 2603, 1995.
61. Schonfuss, D., Reum, T., Olshausen, P., Fischer, T., and Morgenstern, R., Modeling constant potential amperometry for investigations of dopaminergic neurotransmission kinetics in vivo, *J. Neurosci. Methods*, 112, 163, 2001.
62. Daws, L. C., Montañez, S., Owens, W. A., Gould, G. G., Frazer, A., Toney, G. M., and Gerhardt, G. A., Transport mechanisms governing serotonin clearance in vivo revealed by high-speed chronoamperometry, *J. Neurosci. Methods*, 143, 49, 2005.
63. Kosten, T. A. and Ambrosio, E., HPA axis function and drug addictive behaviors: insights from studies with Lewis and Fischer 344 inbred rats, *Psychoneuroendocrinology*, 27, 35, 2002.
64. Flores, G., Wood, G. K., Barbeau, D., Quirion, R., and Srivastava, L. K., Lewis and Fischer rats: a comparison of dopamine transporter and receptors levels, *Brain Res.*, 814, 34, 1998.
65. Hanania, T., Gulley, J. M., Salaz, D. O., Larson, G. A., and Zahniser, N. R., Role of the dopamine transporter in the differential cocaine-induced locomotor activation of inbred long-sleep and short-sleep mice, *Neuropsychopharmacology*, 29, 1814, 2004.
66. Davidson, C., Ellinwood, E. H., Douglas, S. B., and Lee, T. H., Effect of cocaine, nomifensine, GBR 12909 and WIN 35428 on carbon fiber microelectrode sensitivity for voltammetric recording of dopamine, *J. Neurosci. Methods*, 101, 75, 2000.
67. Gratton, A. and Wise, R. A., Drug- and behavior-associated changes in dopamine-related electrochemical signals during intravenous cocaine self-administration in rats, *J. Neurosci.*, 14, 4130, 1994.
68. Garris, P. A., Walker, Q. D., and Wightman, R. M., Dopamine release and uptake rates both decrease in the partially denervated striatum in proportion to the loss of dopamine terminals, *Brain Res.*, 753, 225, 1997.
69. Gulley, J. M., Doolen, S., and Zahniser, N. R., Brief, repeated exposure to substrates down-regulates dopamine transporter function in *Xenopus* oocytes in vitro and rat dorsal striatum in vivo, *J. Neurochem.*, 83, 400, 2002.
70. Gratton, A., Hoffer, B. J., and Gerhardt, G. A., In vivo electrochemical studies of monoamine release in the medial prefrontal cortex of the rat, *Neuroscience*, 29, 57, 1989.
71. Stamford, J. A., Kruk, Z. L., Millar, J., and Wightman, R. M., Striatal dopamine uptake in the rat: In vivo analysis by fast cyclic voltammetry, *Neurosci. Lett.*, 51, 133, 1984.
72. May, L. J., Kuhr, W. G., and Wightman, R. M., Differentiation of dopamine overflow and uptake processes in the extracellular fluid of the rat caudate nucleus with fast-scan *in vivo* voltammetry, *J. Neurochem.*, 51, 1060, 1988.
73. Wightman, R. M., Amatore, C., Engstrom, R. C., Hale, P. D., Kristensen, E. W., Kuhr, W. G., and May, L. J., Real-time characterization of dopamine overflow and uptake in the rat striatum, *Neuroscience*, 25, 513, 1988.
74. Ng, J. P., Menacherry, S. D., Liem, B. J., Anderson, D., Singer, M., and Justice, J. B. Jr., Anomalous effect of mazindol on dopamine uptake as measured by in vivo voltammetry and microdialysis, *Neurosci. Lett.*, 134, 229, 1992.

75. Nicholson, C., Interaction between diffusion and Michaelis–Menten uptake of dopamine after iontophoresis in striatum, *Biophys. J.*, 68, 1699, 1995.
76. Wu, Q., Reith, M. E. A., Wightman, R. M., Kawagoe, K. T., and Garris, P. A., Determination of release and uptake parameters from electrically evoked dopamine dynamics measured by real-time voltammetry, *J. Neurosci. Methods*, 112, 119, 2001.
77. Rice, M. E. and Nicholson, C., Measurement of nanomolar dopamine diffusion using low-noise perfluorinated ionomer coated carbon fiber microelectrodes and high-speed cyclic voltammetry, *Anal. Chem.*, 61, 1805, 1989.
78. Garris, P. A. and Wightman, R. M., Distinct pharmacological regulation of evoked dopamine efflux in the amygdala and striatum of the rat in vivo, *Synapse*, 20, 269, 1995.
79. Schmitz, Y., Lee, C. J., Schmauss, C., Gonon, F., and Sulzer, D., Amphetamine distorts stimulation-dependent dopamine overflow: effects on D2 autoreceptors, transporters, and synaptic vesicle stores, *J. Neurosci.*, 21, 5916, 2001.
80. Hoffman, A. F. and Gerhardt, G. A., In vivo electrochemical studies of dopamine clearance in the rat substantia nigra: effects of locally applied uptake inhibitors and unilateral 6-hydroxydopamine lesions, *J. Neurochem.*, 70, 179, 1998.
81. Gerhardt, G. A., Cass, W. A., Hudson, J., Henson, M., Zhang, Z., Ovadia, A., Hoffer, B. J., and Gash, D. M., In vivo electrochemical studies of dopamine overflow and clearance in the striatum of normal and MPTP-treated rhesus monkeys, *J. Neurochem.*, 66, 579, 1996.
82. Saunders, C., Ferrer, J. V., Shi, L., Chen, J., Merrill, G., Lamb, M. E., Leeb-Lundberg, L. M. F., Carvelli, L., Javitch, J. A., and Galli, A., Amphetamine-induced loss of human dopamine transporter activity: an internalization-dependent and cocaine-sensitive mechanism, *Proc. Natl Acad. Sci. U.S.A.*, 97, 6850, 2000.
83. Meiergerd, S. M., Patterson, T. A., and Schenk, J. O., D2 receptors may modulate the function of the striatal transporter for dopamine: kinetic evidence from studies in vitro and in vivo, *J. Neurochem.*, 61, 764, 1993.
84. Parsons, L. H., Schad, C. A., and Justice, J. B. Jr., Co-administration of the D2 antagonist pimozide inhibits up-regulation of dopamine release and uptake induced by repeated cocaine, *J. Neurochem.*, 60, 376, 1993.
85. Cass, W. A. and Gerhardt, G. A., Direct in vivo evidence that D2 dopamine receptors can modulate dopamine uptake, *Neurosci. Lett.*, 176, 259, 1994.
86. Dickinson, S. D., Sabeti, J., Larson, G. A., Giardina, K., Rubinstein, M., Kelly, M. A., Grandy, D. K., Low, M. J., Gerhardt, G. A., and Zahniser, N. R., Dopamine D_2 receptor-deficient mice exhibit decreased dopamine transporter function but no changes in dopamine release in dorsal striatum, *J. Neurochem.*, 72, 148, 1999.
87. Hoffman, A. F., Lupica, C. R., and Gerhardt, G. A., Dopamine transporter activity in the substantia nigra and striatum assessed by high-speed chronoamperometric recordings in brain slices, *J. Pharmacol. Exp. Ther.*, 287, 487, 1998.
88. Sabeti, J., Gerhardt, G. A., and Zahniser, N. R., Acute cocaine differentially alters accumbens and striatal dopamine clearance in low and high cocaine locomotor responders: behavioral and electrochemical recordings in freely moving rats, *J. Pharmacol. Exp. Ther.*, 302, 1201, 2002.
89. Sabeti, J., Gerhardt, G. A., and Zahniser, N. R., Individual differences in cocaine-induced locomotor sensitization in low and high cocaine locomotor-responding rats are associated with differential inhibition of dopamine clearance in nucleus accumbens, *J. Pharmacol. Exp. Ther.*, 305, 180, 2003.

7 Determining Serotonin and Dopamine Uptake Rates in Synaptosomes Using High-Speed Chronoamperometry

Xiomara A. Perez, Amanda J. Bressler, and Anne Milasincic Andrews

CONTENTS

INTRODUCTION: HISTORY OF INVESTIGATING MONOAMINE NEUROTRANSMITTER UPTAKE

The use of electroanalytical methods to investigate brain chemistry originated in the laboratory of Ralph N. Adams in 1967, with specific reports on the use of chronoamperometry for neurochemical applications beginning to emerge in the literature in the early 1980s.[1–6] Today, the voltammetry techniques most widely used to investigate various aspects of neuronal intercellular signaling, in addition to chronoamperometry, include constant potential amperometry, differential pulse voltammetry, rotating disk electrode voltammetry and fast cyclic voltammetry (for recent reviews on these techniques see[7–13]). All of these methods have most often been applied to the study of the catecholamine neurotransmitter dopamine, due to the importance of dopamine in locomotor and reward-associated behaviors.[14–17] In addition, dense dopamine projections ramify in neuroanatomically discrete brain regions making this electroactive neurotransmitter highly amenable to investigation by voltammetric methods. Serotonin (5-hydroxytryptamine, 5-HT), like dopamine, is oxidized at potentials less than 1 V at the surface of carbon fiber microelectrodes (Figure 7.1a). However, far fewer investigations have focused on serotonergic neurotransmission, partly because sparse serotonin innervation is highly overlapping with other electrochemically active neurotransmitters, complicating analysis by voltammetry.[18–24]

Serotonin neurotransmitter signaling is integral to many types of complex central nervous system function including the regulation of mood and anxiety states, reward-related behavior, and cognitive function.[25–30] The serotonin transporter (SERT) is a presynaptic 68 kD transmembrane protein that regulates the concentration of serotonin in the extraneuronal space by taking serotonin back up into presynaptic neurons after it has been released.[31] SERT is the primary molecular target for many important therapeutics used in the treatment of depression and anxiety disorders (i.e., the tricyclic antidepressants and the serotonin-selective reuptake inhibitors or SRIs),[32–35] as well as recreationally abused substances including cocaine, methamphetamine and 3,4-methylenedioxymethamphetamine (MDMA or Ecstasy).[36–39] SRIs constitute the most widely prescribed class of antidepressant/anti-anxiety drugs and these agents prevent serotonin inward transport from the extracellular space by blocking the transporter.[40–42] Long-term treatment with SRIs (weeks to months) causes sustained reductions in serotonin uptake that likely lead to prolonged increases in extraneuronal serotonin concentrations.[40,43,44] This fundamental homeostatic alteration in serotonin neurotransmission is thought to cause adaptive responses in pre- and postsynaptic neurons, including long-term changes in gene expression, that ultimately underlie the therapeutic efficacy of the SRIs.[29,45,46]

In addition to pharmacologic modulation of the serotonin transporter, genetically controlled alterations in serotonin reuptake occur. A polymorphic element (alternate DNA base sequence) was discovered in the promoter region of the human SERT gene (SLC6A4)[47] that results in 40% reductions in SERT expression in approximately 70% of the normal population.[48] These decreases in SERT have been correlated with increases in anxiety-related personality traits, particularly neuroticism in men and women[48–52] and increased risk for stress-associated depression.[53]

Genetically modified mice have become indispensable tools in biomedical research, increasing our understanding of the contributions of various genes to normal behavior, disease states, and their treatment. Mice have been engineered to express reduced amounts of the serotonin transporter,[54] dopamine transporter,[15,55] or norepinephrine transporter.[56] Like their human counterparts, mice with reduced SERT expression display an increase in anxiety-like behavior.[57–59] Thus, SERT deficient mice, particularly those having 50% reductions in SERT expression, appear to model closely the genetic variability in SERT expression widely occurring in the human population.[60]

Chronoamperometry has been used to investigate changes in serotonin uptake in SERT deficient mice in the hippocampus in vivo.[61] The results of this study showed that serotonin uptake is decreased in mice lacking one copy of the SERT gene (SERT+/− mice) compared to

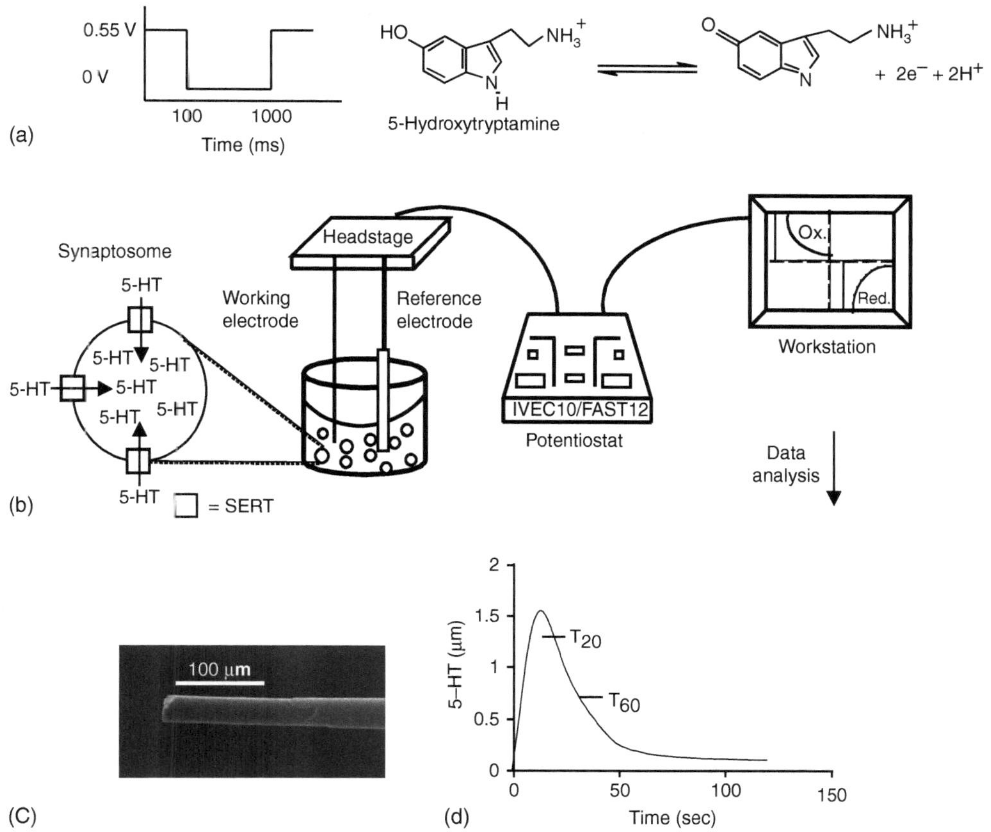

FIGURE 7.1. Chronoamperometry in neuronal synaptosomes. Delayed-mode high-speed chronoamperometry was used to measure uptake rates in synaptosomes. (a) Changes in current due to serotonin oxidation were recorded in response to a 1 Hz square wave step potential. The oxidation potential consisted of a 100-ms pulse at +0.55 V and was followed by a 100-ms reductive pulse at 0 V. The current was maintained at 0 V for an additional 800 ms to prevent fouling of the electrode. (b) Synaptosomes were prepared and brought up in oxygen saturated assay buffer. A carbon fiber microelectrode and Ag/AgCl reference electrode were positioned in the synaptosomal solution, and the square wave step potential was applied. The resulting oxidative and reductive currents, collected before and after addition of analyte, were recorded every second. (c) Transmission electron micrograph of a representative carbon fiber microelectrode. The cylindrical electrodes for these experiments were 30 μm in diameter × 100 μm in length. (d) The faradaic current during the last 80 ms of each pulse was integrated, calibrated against standard solutions of serotonin and plotted on the y-axis versus time on the x-axis. The following parameters were then obtained for data analysis: T_{20} (time for the signal to decay by 20%), T_{60} (time for the signal to decay by 60%), and T_c (the slope of the electrochemical signal between T_{20} and T_{60}). (From Perez, X A. and Andrews, A. M., *Anal. Chem.*, 77, 818–826, 2005. With permission.)

wild-type mice (SERT+/+ mice). However, these findings contradicted the original report on SERT deficient mice where no differences in synaptosomal serotonin uptake rates between SERT+/− and SERT+/+ mice were found using classic radiochemical methods.[38] Based on these discrepancies, we developed chronoamperometry methods specifically aimed at determining serotonin, and later dopamine, uptake rates in synaptosomes. Synaptosomes are neuronal liposomes prepared from brain tissue that typically contain one or more mitochondria, and they retain glycolytic capabilities.[62–64] They also possess active ion and neurotransmitter transport systems across the plasma membrane.

The advantages of using synaptosomes as a simplified model system to investigate serotonin or dopamine uptake kinetics are twofold. First, numerous transport experiments described in the literature have been performed in synaptosomes; thus, data obtained from electroanalytical investigation of synaptosomal uptake can be directly compared to results using radiochemical methods. Second, by employing synaptosomes, the uptake process can be characterized independently of neurotransmitter release and diffusion, processes that also contribute to changes in neurotransmitter concentrations detected at microelectrodes in intact tissue. Under the conditions described here, the chemical identity and concentration of the neurotransmitter under study can be closely controlled while electrochemically monitoring the uptake process. As a result, chronoamperometry studies in synaptosomes lead to straightforward determinations of the kinetics of serotonin or dopamine uptake as they relate to changes in transporter expression or other genetic or pharmacologic manipulations.

The methodology for using chronoamperometry to measure transporter-mediated uptake into synaptosomes is described here. Alterations in uptake in brain tissue from mice lacking one or both (SERT−/−) intact copies of the SERT gene are detailed and compared with previous studies using either radiometric methods in synaptosomes[38] or chronoamperometry in vivo.[61] The factors underlying the discrepancies between the results of radiochemical and chronoamperometric serotonin uptake experiments are also explored. We show that oxygen saturation of the assay buffer greatly influences clearance rates and that a clear incremental relationship exists between serotonin transporter gene expression and serotonin uptake rates.[65,66] We have discovered that mechanical disruption of synaptosomes by stirring or filtration leads to significant loss of transported serotonin resulting in misleadingly low maximal uptake rates (V_{max}) and transporter substrate affinities (K_m), as well as a loss of sensitivity for determining biologically important alterations in uptake rates.[66] Finally, we show how chronoamperometry has been used to determine decreases in dopamine uptake in striatal synaptosomes prepared from a transgenic mouse model expressing a mutant form of human α-synuclein, which is associated with a hyperactive phenotype.[16]

CHRONOAMPEROMETRY METHODS FOR DETERMINING UPTAKE IN SYNAPTOSOMES

MOUSE MODELS OF ALTERED UPTAKE

SERT deficient mice intact were generated at the National Institute of Mental Health (Bethesda, Maryland) on a mixed CD-1×129S6/SVev background.[38,67] Animals used in the present studies were 4- to 6-month-old male mice bred and genotyped at The Pennsylvania State University and were the product of eight to fifteen generations of SERT+/− brother–sister matings.

α-Synuclein transgenic mice expressing mutant human A53T α-synuclein (A53T-tg) or wild-type human (WT-tg) α-synuclein and nontransgenic littermates (non-tg) were generated as previously described at the Johns Hopkins Medical Institutions (Baltimore, Maryland).[68] These mice were derived in a C3H/C57BL6-F1 hybrid, and the established lines were maintained by back-crossing to a C57BL/6 background. Male and female transgenic mice and nontransgenic littermates used for the experiments described here were bred and genotyped at Johns Hopkins and transferred to The Pennsylvania State University. These animals were matched for similar low levels of overexpression of the two forms of human α-synuclein.[16]

All mice were housed in single sex groups of two to five mice per cage on a 12 h light/dark cycle (0600 h lights on) with food and water *ad libitum*. The Pennsylvania State University or Johns Hopkins Medical Institutions Institutional Animal Care and Use Committees approved all animal procedures, which were conducted in strict accordance with National Institutes of Health Animal Care Guidelines.

Carbon Fiber Microelectrode Construction

Cylindrical electrodes were prepared according to the method of Gerhardt and coworkers[69,70] with minor modifications. Microelectrodes were constructed by aspirating 30 μm carbon fibers (Textron Specialty Materials, Lowell, Massachusetts) into 1.2 mm×0.68 mm, 4″ borosilicate glass capillaries. An automatic micropipette puller was used to pull the glass capillaries to create two electrodes. Electrodes were dipped in epoxy (Polysciences Inc., Warrington, Pennsylvania) for five to seven minutes, followed by acetone for five to ten seconds, and heated overnight at 100°C. Carbon fiber tips were cut to a final length of ~100 μm (Figure 7.1c). Nafion (Sigma–Aldrich, St. Louis, Missouri), an anion exchange resin, was coated onto carbon fiber surfaces to increase selectivity and sensitivity for cationic neurotransmitters.[71,72] Electrodes were dipped in Nafion for 10 s and then dried for five minutes at 150°C. This was repeated three to four times with the numbers of Nafion coatings being periodically adjusted to obtain optimal selectivity. Each electrode was filled with 4 M potassium acetate/150 mM potassium chloride to make electrical contact between the carbon fiber and a silver wire to which a gold pin connector (Newark Electronics, Long Island, New York) was soldered.

Preparation of Synaptosomes

Synaptosomes were prepared using a modification of the technique reported by Hyde and Bennett.[73] Brain stem (including tissue containing the dorsal and median raphe nuclei), frontal cortex, and bilateral striata from two mice per genotype were quickly dissected over ice. Tissues were pooled by brain region and genotype, and homogenized by hand in 10 volumes (ml/mg wet tissue weight) Tris-sucrose buffer (0.5 mM Tris–HCl, 0.32 mM sucrose, pH 7.4) with a glass mortar and Teflon pestle. Homogenates were centrifuged at 2000 g for 10 min. The supernatants were carefully removed and centrifuged again at 16,000 g for 10 min. The pellets were then resuspended in 2 mL of assay buffer (124 mM NaCl, 1.80 mM KCl, 1.24 mM KH_2PO_4, 1.40 mM $MgSO_4$, 2.50 mM $CaCl_2$, 26.0 mM $NaHCO_3$, 10.0 mM glucose, saturated with 95% O_2/5% CO_2, pH 7.4 with phosphoric acid). A 50 μL aliquot of each homogenate was reserved for protein analysis according to the method of Lowry et al.[74] Chronoamperometric experiments were conducted using protein concentrations ranging from 0.4 to 0.9 mg/mL protein.

Precalibrated microelectrodes and reference electrodes were placed in synaptosomes in a final volume of 2 mL. A +0.55 V pulsed potential was applied and the current was recorded until a stable baseline was obtained (Figure 7.1b). Serotonin or dopamine (20 μL) was added to each synaptosomal solution and the change in current with respect to time was recorded. Where indicated, synaptosomal preparations from SERT+/+ mice were preincubated with paroxetine (1.0 μM final concentration) or fluoxetine (100 nM final concentration) for 30 min. Before the addition of serotonin for uptake measurements, synaptosomes were centrifuged briefly at 16,000 g and brought up in fresh oxygenated assay buffer.

Chronoamperometry and Electrode Calibration

High-speed chronoamperometry was performed in the delayed-pulse mode.[70] Changes in current were recorded in response to a 1 Hz square wave potential step generated by an IVEC-10/FAST-12 potentiostat (Quanteon Center for Sensor Technology, Lexington, Kentucky). A 100 ms oxidative pulse at +0.55 V was followed by a 100 ms reductive pulse at 0 V (Figure 7.1a). The resulting current was integrated over the last 80 ms of each oxidative and reductive phase. The potential was maintained at 0 V for an additional 800 ms (delayed-mode) to prevent fouling of the electrode.[72,75] Voltage at the carbon fiber working electrode was applied with respect to a Ag/AgCl glass reference electrode (BAS, West Lafayette, Indiana).

Microelectrodes were calibrated over a concentration range of 0.25–1.5 μM for serotonin or 0.25–1.25 μM for dopamine in phosphate buffered saline (pH 7.4) to determine linear responses.

They were challenged with 0.5 μM ascorbate before calibration with serotonin or 0.25 μM 3,4-dihydroxyphenylalanine (DOPAC) prior to dopamine calibration to determine the sensitivity and selectivity of the Nafion coating for cations. Redox ratios between 0.1 and 0.3 for serotonin or 0.5 and 0.7 for dopamine were obtained. Only electrodes with high linear responses ($r^2 > 0.99$) and cation selectivity ratios greater than 700:1 were used in subsequent experiments.

Because electrode sensitivity decreases rapidly by ~60% and then stabilizes as a result of exposure of electrodes to synaptosomes,[65] precalibration was performed to evaluate the sensitivity and linear response of each electrode prior to use in experiments. After uptake experiments were complete, all electrodes were recalibrated and data analysis was performed using postcalibration standard curves. Uptake rates are expressed as pmol 5-HT or DA taken up/mg protein-min. Uptake rates are reported as "not detectable" when T_C (defined below) was less than 0.1 nM/s over the course of 20 min.

HPLC Analysis of Synaptosomal Filtrates

To determine the amount of serotonin lost from synaptosomes during filtration, synaptosomal filtrates were analyzed by high performance liquid chromatography with electrochemical detection (HPLC-ED) using an ESA Coulochem II electrochemical detector ($E_1 = -175$ mV and $E_2 = +220$ mV; Chelmsford, Massachusetts) with minor modifications.[76] After observing complete uptake of 1 μM serotonin into brain stem synaptosomes by chronoamperometry (currents returned to baseline after ten to 20 min depending on SERT genotype), synaptosomal suspensions (2 mL) were filtered (300 mm Hg) through Whatman GF/C filters presoaked in 2% polyethyleneimine using a Millipore sampling manifold (Bedford, Massachusetts) so that filtrates from individual samples could be collected. Samples were then washed with 500 μL of assay buffer. Aliquots of initial filtrates and washes (20 μL each) were injected onto a 100×4.6 mm Spherisorb 3 μm octadecylsulfate reversed-phase chromatography column (Thomson Instruments, Springfield, Virginia) in a mobile phase containing 0.1 M monochloroacetic acid, 8% acetonitrile, 0.5 g/L octanesulfonic acid, 0.3% triethylamine, and 10 μM EDTA at a flow rate of 0.7 ml/min. Serotonin concentrations were quantified by comparing peak areas to a standard curve (0–10 μM 5-HT), and the total amount of serotonin detected in the filtrate from each sample was obtained by adding the amount of serotonin detected in the initial filtrate to that detected in the wash.

Data Analysis and Statistics

For chronoamperometric recordings, the rate of uptake is estimated from T_C, the slope of the electrochemical signal between T_{20} and T_{60} (times needed to transport 20% and 60% of the total neurotransmitter added, respectively; Figure 7.1d).[70] To determine V_{max} and K_m values for 5-HT uptake, data were fit to a one-site hyperbolic function for uptake rates versus initial 5-HT concentrations using Equation 7.1 (GraphPad Prism, San Diego, California). Here, uptake rate is the concentration-dependent T_C, V_{max} is the maximal rate of uptake, K_m is the affinity of the transporter for 5-HT, and [5-HT] is the initial concentration added to synaptosomal suspensions.

$$\text{Uptake rate} = \frac{V_{max} \times [\text{5-HT}]}{K_m + [\text{5-HT}]} \tag{7.1}$$

Data were analyzed statistically by a student's t-test or one-way analysis of variance (ANOVA) with genotype as the independent variable using the Statistical Analysis System (SAS Institute, Carey, North Carolina). A priori differences between groups are indicated by t-test probabilities.

CHARACTERIZATION OF MICROELECTRODE RESPONSES

We assessed carbon fiber microelectrode responses to serotonin over a concentration range of 0.25–1.5 μM. Both oxidative and reductive responses were found to be highly linear before and after the exposure of electrodes to tissue.[65] However, contact of electrodes with solutions containing synaptosomes reduced electrode responsiveness.[65] Upon initial exposure to synaptosomes, baseline currents dropped steadily for about 30 min, after which they soon stabilized. Uptake measurements were made only after baseline currents had stabilized, and data was collected with a particular electrode only as long as the baseline current remained relatively stable. The response of a large group of electrodes ($n = 41$) to tissue exposure was systematically evaluated. The average decrease in the response factor for serotonin was $57 \pm 2\%$ and the mean reduction in the redox ratio was $23 \pm 2\%$. As a result, all electrodes were recalibrated at the end of each experiment, and data analysis was performed using postcalibration values to account for reductions in electrode sensitivity occurring as a result of tissue exposure.

EFFECTS OF OXYGEN ON SYNAPTOSOMAL UPTAKE OF SEROTONIN

The ability of SERT to transport serotonin actively from the extracellular space depends on the co-transport of Na^+, K^+ and Cl^- ions down their respective concentration gradients.[31] The Na^+/K^+-ATPase, which uses the chemical energy stored in ATP to fuel the extrusion of Na^+ from and the accumulation of K^+ into cells, produces these ion gradients. Neurons rely heavily on oxidative respiration to produce large amounts of ATP, and thus, to maintain ion gradients, which are essential for the reuptake of neurotransmitters, as well as for the generation of action potentials. Therefore, normal function of SERT is energetically coupled to the Na^+/K^+-ATPase and depends on mitochondrial respiration and the presence of O_2 to produce high levels of ATP. Synaptosomes, which are sealed presynaptic structures, maintain the ability to take up oxygen and glucose, as well as the capacity to respire and maintain normal membrane potentials.[77–79] Based on this, we evaluated the effects of assay buffer oxygen saturation on uptake rates using chronoamperometry. Serotonin uptake rates were measured in unstirred synaptosomal solutions prepared from SERT+/+ mice.[65] Each sample was divided and half was incubated in nonoxygenated assay buffer, while the other half was incubated in assay buffer saturated for 30 min with 95% O_2/5% CO_2. In the absence of O_2, uptake of 1 μM 5-HT was detected in only three of six samples (mean uptake rate 21 ± 4 pmol/mg protein-min). In the remaining three samples, uptake was not detectable over a twenty-minute period. In the corresponding oxygenated samples, the mean 5-HT uptake rate was 126 ± 8 pmol/mg protein-min ($n = 6$), demonstrating a tenfold increase in uptake rates. These data support the conclusion that oxygen is an essential component in synaptosomal preparations for optimal function of monoamine transporters.

EFFECTS OF STIRRING ON SYNAPTOSOMAL UPTAKE OF SEROTONIN

Rotating disk electrode voltammetry has been used to study neurotransmitter transport kinetics in minced tissue and cultured cells.[80–84] One of the main advantages of this technique is that uptake can be determined under conditions in which the effects of diffusion and mass transport are minimized by the rapid rotation of the electrode.[84] To reduce the effects of mass transport in our system, synaptosomal solutions were stirred and the effects on uptake were determined. However, in contrast to chopped tissue or whole cells, stirring synaptosomal suspensions (2 mL total volume) with a miniature magnetic stirring bar (6 mm length, 2 mm diameter) at the slowest possible stirring plate speed while measuring serotonin clearance by chronoamperometry eliminated measurable uptake.[65] In fact, uptake was not detectable for as long as 90 min after the addition of 5-HT, even in the presence of O_2 in stirred synaptosomes ($n = 3$). We hypothesize that this is because of leakage of transported 5-HT from synaptosomal structures caused by stirring. Unlike intact tissue or cells, synaptosomes have disrupted intracellular cytoskeletal and extracellular matrix elements caused by the tissue homogenization process.

EFFECTS OF UPTAKE INHIBITORS ON SYNAPTOSOMAL UPTAKE OF SEROTONIN

Many drugs that selectively inhibit the serotonin transporter are commonly used in the treatment of mood and anxiety disorders.[85] Paroxetine (trade name Paxil®),[86–88] a compound that belongs to this class of drugs, was used to test the hypothesis that time-dependent reductions in the chronoampero-metric signal in response to the addition of 5-HT to synaptosomes occurs specifically because of serotonin transporter activity. For these experiments, synaptosomal samples were divided. Half of each sample was used to measure serotonin uptake in the absence of paroxetine (saline pre-incubation), while the other half was preincubated with 1 μM paroxetine for 30 min. After preincubation, both sets of samples were briefly centrifuged and brought up in fresh oxygenated assay buffer. Evaluation of serotonin uptake in saline pretreated striatal synaptosomes from SERT+/+ mice yielded a mean uptake rate of 153 ± 12 pmol/mg protein/min ($n = 6$). By contrast, uptake was not detectable in striatal synaptosomes preincubated with paroxetine ($n = 6$).[65]

Uptake was also determined in brain stem synaptosomes preincubated with 100 nM fluoxetine (trade name Prozac®), another selective serotonin uptake inhibitor.[66] No uptake of 5-HT was detected by chronoamperometry in synaptosomes pretreated with fluoxetine (Figure 7.2a).[66] After 20 min of chronoamperometric recording, synaptosomes were filtered to separate them from the surrounding solution and 95% of the serotonin added was detected in the filtrate using

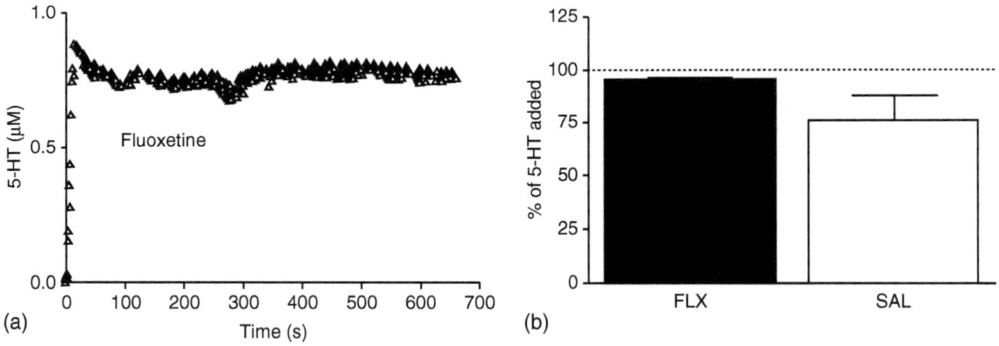

FIGURE 7.2 Effects of fluoxetine and filtration. Synaptosome samples from brain stem ($n = 3$) were divided, and half of each sample was preincubated with either 100 nM fluoxetine in assay buffer or assay buffer alone for 30 min. Samples were then briefly centrifuged at 1600 g and brought up in fresh oxygen saturated assay buffer. (a) Representative current versus time plot for a sample preincubated with fluoxetine showing no decrease in current with respect to time upon addition of 5-HT. This indicates that uptake by SERT was inhibited by fluoxetine. Samples preincubated in assay buffer alone showed robust uptake and currents returned to baseline after 20 min (see Figure 7.3a for a representative current-time trace). (b) In the case of synapto-somes preincubated with fluoxetine, 95% of the serotonin added was detected in the filtrates (error bar is so small that it is almost not visible). This, together with the observation that the chronoamperometric signal did not decline during the monitoring period, indicates that fluoxetine prevents uptake of serotonin by SERT. In the case of samples preincubated with assay buffer alone, 75% of the added serotonin was detected in the filtrates. Here, chronoamperometric monitoring indicated that all of the serotonin had been taken up before filtering. Thus, these data demonstrate that the filtration process disrupts synaptosomal membranes, releasing previously transported 5-HT and resulting in a large underestimation of the amount of serotonin taken up (see also Figure 7.5). (From Perez, X. A., Bianco, L. E., and Andrews, A. M. *J. Neurosci. Methods*, 2006. With permission.)

HPLC (Figure 7.2b). This further indicates that serotonin is not taken up in synaptosomes pretreated with fluoxetine. These results demonstrate that the reduction in the electrochemical signal with respect to time after addition of 5-HT to synaptosomes is attributable to uptake of serotonin by SERT and is not the result of baseline drift, 5-HT oxidation by the microelectrode or uptake by other monoamine neurotransmitters co-expressed in the tissues studied.

INACTIVATION OF SEROTONIN TRANSPORTER EXPRESSION RESULTS IN A GENE DOSE-DEPENDENT REDUCTION IN SEROTONIN UPTAKE

Serotonin-transporter deficient mice lack either one or both copies of the SERT gene constitutively (from conception throughout life). Previous studies have demonstrated that SERT+/− mice have ~50% reductions in serotonin transporter protein levels throughout the brain, including in the striatum, frontal cortex, and brain stem. SERT−/− mice fail to express the serotonin transporter in all brain regions.[54,66] Gene dose-dependent increases in extracellular serotonin levels occur in striatum and frontal cortex in SERT+/− and SERT−/− mice.[89] Furthermore, locomotor responses to the psychomotor stimulant MDMA are attenuated in SERT+/− mice and absent in SERT−/− mice.[38] These observations, and particularly those in SERT+/− mice, indicate that mice lacking one intact copy of the SERT gene are suitable models for investigating alterations in neurochemistry and behavior as they relate to the human SERT promoter polymorphism.

Uptake rates in synaptosomes prepared from the three different genotypes of SERT knockout mice were evaluated using chronoamperometry in synaptosomes prepared from the striatum, frontal cortex, and brain stem.[65,66] Striatum is a brain region that is important for the control of movement, and degeneration of neurons projecting to this brain area is the hallmark of Parkinson's disease.[25] Additionally, ventral striatum (nucleus accumbens) has long been associated with reward-related behavior and drug addiction.[17,90] The frontal cortex is thought to be important in arousal, cognition and emotional states, and the modulation of personality.[91,92] The serotonin neuronal cell bodies are located in the brain stem, and high levels of SERT are also expressed in this brain region.[66] We hypothesized that gene dose-dependent decreases in SERT gene expression in these three brain regions would lead to parallel reductions in serotonin uptake rates and that these changes would be detectable by chronoamperometry.

Representative recordings from brain stem synaptosomes after the addition of serotonin are shown in Figure 7.3. The data indicate that 1 μM (2 nmol) 5-HT is taken up by brain stem synaptosomes from SERT+/+ mice in about 10 min (Figure 7.3a). By contrast, it took about 17 min to take up the same amount of 5-HT into brain stem synaptosomes prepared from SERT+/− mice (Figure 7.3b). No uptake of 5-HT occurred in brain stem synaptosomes from SERT−/− mice (Figure 7.3c).

Table 7.1 shows mean serotonin clearance rates across the three genotypes of SERT deficient mice in different brain regions. In SERT+/− mice, serotonin uptake was significantly reduced by ~60% in the striatum [$t(16) = 9.6$; $p < 0.0001$], frontal cortex [$t(7) = 7.5$; $p < 0.0001$], and brain stem [$t(8) = 5.8$; $p < 0.001$] compared to SERT+/+ mice. Serotonin uptake was not detectable in any brain region in SERT−/− mice. These data support the theory that gene-related reductions in SERT expression result in similar magnitude decreases in the rates of transporter-mediated serotonin clearance.

UPTAKE KINETICS OF SEROTONIN BY CHRONOAMPEROMETRY

Concentrations of 1 μM serotonin were chosen for the uptake experiments described above because this concentration was originally thought to be at least tenfold higher than the K_m (~0.05 μM) of the serotonin transporter for 5-HT based on our previous radiochemical studies of serotonin uptake.[54] However, uptake rates measured by chronoamperometry ended up being much higher

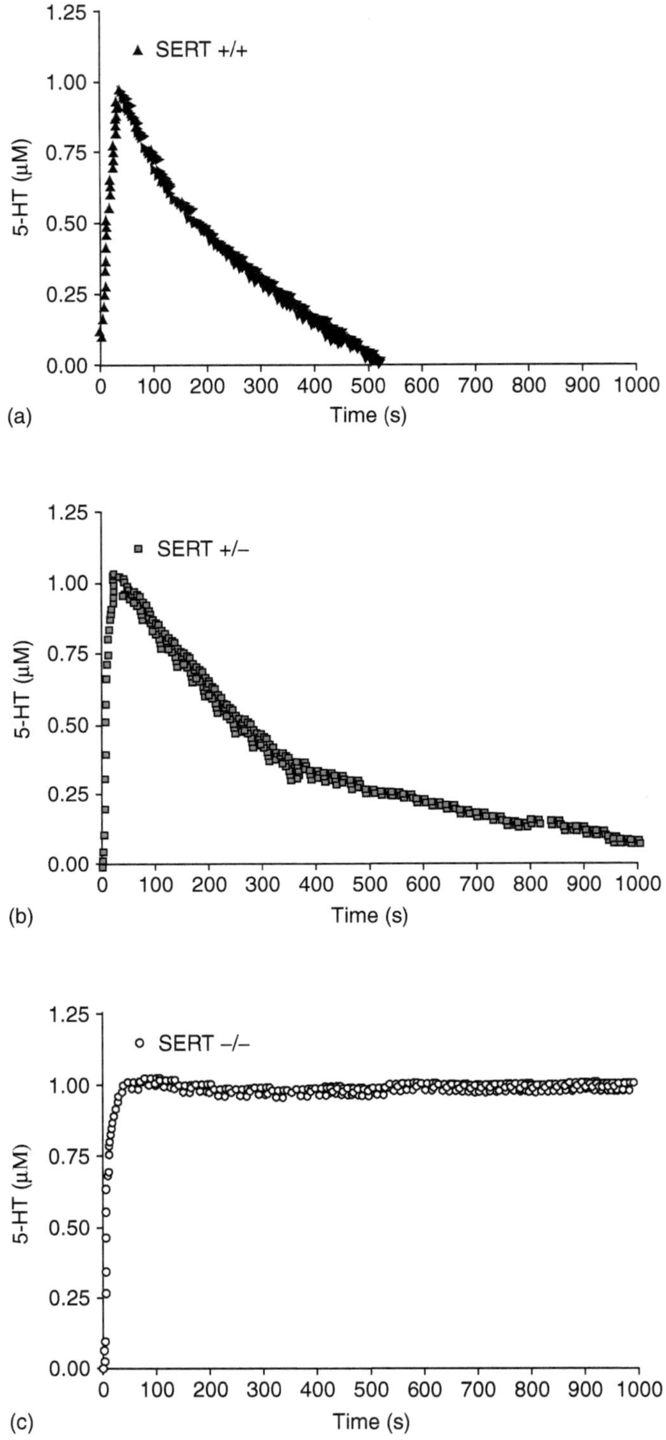

FIGURE 7.3 Serotonin uptake in brain stem synaptosomes from SERT deficient mice. Representative oxidative traces after the addition of 1 μM serotonin to brain stem synaptosomes from (a) SERT+/+ mice, (b) SERT+/− mice and (c) SERT−/− mice. Clearance rates were reduced by 60% in SERT+/− mice, while SERT−/− mice displayed no observable uptake. (This figure is reprinted from Perez, X. A., Bianco, L. E., and Andrews, A. M. *J. Neurosci. Methods*, 2006. With permission.)

TABLE 7.1
Serotonin Uptake Rates Determined by Chronoamperometry in Synaptosomes

	Uptake Rate (pmol/mg protein-min)		
	Brain Stem	Frontal Cortex	Striatum
SERT+/+	161 ± 10	162 ± 10	146 ± 8
SERT+/−	67.7 ± 4^a (60%)	62.0 ± 6^a (62%)	62.7 ± 3^a (57%)
SERT−/−	n.d.	n.d.	n.d.

Uptake rates were determined by chronoamperometry in the presence of 1 μM 5-HT. Average percent decreases with respect to SERT+/+ mice are indicated in parentheses. Rates that were not detectable are indicated as "n.d." Numbers of experiments per genotype are as follows: $n = 13$ per genotype in striatum; $n = 5$ for SERT+/+, $n = 4$ for SERT+/−, $n = 5$ for SERT−/− mice in frontal cortex; $n = 6$ for SERT+/+, $n = 4$ for SERT+/−, and $n = 4$ for SERT−/− mice in brain stem. (From Perez, X.A., Bianco, L.E., and Andrews, A.M., *J. Neurosci. Methods*, 2006. With permission.)

[a] $p < 0.001$ versus SERT+/+ mice.

than those determined by radiometric assay.[54,65,66] For example, V_{max} values reported for radiochemical analysis of serotonin uptake in synaptosomes typically range from 0.2 to 5 pmol/mg protein-min.[54,93–95] Serotonin uptake rates determined by chronoamperometry are ~150 pmol/mg protein-min (Table 7.1). In light of this, the kinetic parameters K_m and V_{max} for serotonin uptake in striatal synaptosomes were reevaluated by high-speed chronoamperometry using concentrations of 5-HT ranging from 0.25 to 1.25 μM.

Figure 7.4 shows the mean kinetic curves for SERT+/+ and SERT+/− mice. Maximal uptake rates determined by nonlinear curve fitting were 198 ± 16 and 118 ± 9.4 pmol/mg protein-min in SERT+/+ and SERT+/− mice, respectively, and these were significantly different from each other [$t(16) = 3.60$; $p < 0.01$]. Affinity constants were 1.1 ± 0.1 and 0.8 ± 0.1 μM for SERT+/+ and SERT+/− mice, respectively, and these were not significantly different [$t(16) = 1.83$; $p = 0.09$]. Kinetic analysis of serotonin uptake in SERT−/− mice was not carried out because it had been demonstrated previously that striatal synaptosomes derived from these mice do not take up serotonin.[65] These data further substantiate that inactivation of one copy of the SERT gene results in a 50% reduction in maximal uptake rates, while affinity of SERT for 5-HT is unchanged. Moreover, they further point to large discrepancies between serotonin uptake rates measured by voltammetric versus radiometric methods.

CHRONOAMPEROMETRY VERSUS RADIOCHEMICAL METHODS: WHY SUCH DISCREPANCIES IN UPTAKE RATES?

One of our major impetuses for developing chronoamperometry methods to measure uptake rates in synaptosomes was to facilitate direct comparisons between kinetic parameters determined by voltammetric versus radiochemical methods using the same tissue preparation. Thus, we attempted to determine why uptake rates determined by radioassay are so much lower than those obtained by chronoamperometry. The first point of consideration is that there are minor differences in the assay conditions between the two types of analyses. The assay buffer that we use for the radiometric assay is similar to that used for chronoamperometry studies with the exception that ascorbic acid and pargyline are included in the former.[54,66] In addition, higher assay temperatures are typically used for radiometric assay (37°C versus 24°C). However, these differences in conditions are expected to maximize rates of uptake in the radiochemical assay by minimizing the oxidation of serotonin, preventing its metabolism by monoamine oxidase, and taking advantage of the temperature dependence of transport rates, respectively. Thus, these minor methodological differences would not

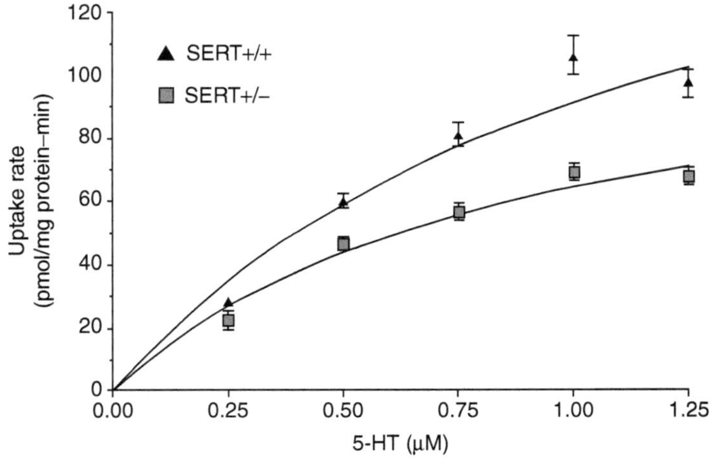

FIGURE 7.4 Kinetic analysis of serotonin uptake in striatal synaptosomes from SERT deficient mice. Data represent mean uptake rates ± SEM from eight to eleven determinations per data point. Data were analyzed by nonlinear curve fitting to a one-site hyperbolic function. Maximal uptake rates were significantly reduced from 198 ± 16 to 118 ± 9.4 pmol/mg protein-min in SERT+/+ versus SERT+/− mice, respectively ($p < 0.01$). No changes were detected in K_m. (This figure is reprinted from Perez, X. A., Bianco, L. E., and Andrews, A. M., *J. Neurosci. Methods*, 2006. With permission.)

account for the greatly lower uptake rates determined by the radiochemical method (compare uptake rates in Table 7.1 with those in Table 7.2).

The second point to consider is that oxygenation of the assay buffer in chronoamperometry experiments increases the rates of serotonin uptake by sevenfold.[65] We assessed whether the same magnitude of effect is observed in radiometric uptake experiments. As discussed above, oxygen is essential for optimal functioning of the monoamine transporters, which depend on the Na^+/K^+ ATPase to maintain the ion gradients necessary for neurotransmitter transport.[96,97] As such, we systematically reevaluated uptake of [^{3}H]5-HT in synaptosomes from SERT+/+ mice using an oxygen saturated assay buffer and standard radiochemical assay conditions.[66] The results of these

TABLE 7.2
Effects of Oxygen Saturation of the Assay Buffer on the Kinetics of [^{3}H]5-HT Uptake

	SERT+/+ Mice			
	V_{max}		K_m	
	No O_2	O_2	No O_2	O_2
Brain Stem	0.54 ± 0.01	0.99 ± 0.07[a]	0.072 ± 0.005	0.056 ± 0.005
Frontal Cortex	0.74 ± 0.04	0.95 ± 0.03[a]	0.043 ± 0.003	0.057 ± 0.006
Striatum	0.91 ± 0.08	1.2 ± 0.07	0.073 ± 0.003	0.067 ± 0.01

Maximal uptake rates (V_{max}) are expressed in pmol/mg protein-min ± SEM. Affinity constants (K_m) are expressed in μM ± SEM. These values were determined by a nonlinear curve fitting to a one-site hyperbolic function. For each experiment, $n = 3$ mice. Each concentration of [^{3}H]5-HT was evaluated in triplicate in three separate experiments. (From Perez, X. A., Bianco, L. E., and Andrews, A. M., *J. Neurosci. Methods*, 2006. With permission.)

[a] $p < 0.01$, absence versus presence of O_2.

studies showed that V_{max} values were only slightly increased in frontal cortex and brain stem synaptosomes, while V_{max} in striatal synaptosomes was not significantly altered (Table 7.2). Oxygenation of the assay buffer had no effect on K_m values, which were still much lower than those measured in chronoamperometry experiments. Thus, despite the fact that the maximal uptake rates determined by the radiometric assay were slightly increased in the presence of oxygen, the magnitude of this increase was considerably smaller than that observed in chronoamperometry experiments. Furthermore, it did not account for the much larger uptake rates calculated by the latter method.

The third point that we considered was the hypothesis that loss of 5-HT during the filtration process used in the radiometric assay accounts for the substantially lower maximal uptake rates and affinity constants measured by this method. As discussed above, stirring of synaptosomal suspensions has detrimental consequences on 5-HT uptake into synaptosomes. Similarly, filtration through a cell harvester or filter manifold requires the use of vacuum pressure to separate synaptosomes and the neurotransmitter they have taken up from the neurotransmitter remaining in the extrasynaptosomal solution. Synaptosomes are further subjected to three to four washes under vacuum pressure with assay buffer before scintillation counting of the filter paper onto which they have been retained.

We have carried out experiments to determine the concentration of serotonin in filtrates (extrasynaptosomal solution) following uptake of 1 μM 5-HT into synaptosomes.[66] Once complete uptake of 5-HT was observed by chronoamperometry (Figure 7.5a), synaptosomal suspensions were filtered through Whatman GF/C filters using a filter manifold (Figure 7.5b). This apparatus permits collection of individual filtrates, which is not feasible using a cell harvester, which is typically used for radiochemical experiments. We hypothesized that if filtration disrupted synaptosomes similar to what we had observed with stirring, then the 5-HT that had been taken up would be subsequently detected in the extrasynaptosomal filtrates (Figure 7.5c). Filtrates were analyzed by HPLC-ED and this confirmed that 75% of the serotonin that had been taken up into synaptosomes was later detected outside of the synaptosomes in the filtrate (see also Figure 7.2b). These observations have lead us to conclude that loss of transported 5-HT during the filtration process used in the radiochemical method leads to: (1) significantly lower uptake rates calculated by this method compared to chronoamperometry; (2) an underestimation of the potency of the effects of oxygen in the radiochemical assay; and (3) a relative lack of sensitivity of the radiometric assay for determining subtle but important alterations in uptake rates predicted by the biology of the system, such as those occurring between SERT+/+ and SERT+/− mice.

DETERMINATION OF DOPAMINE UPTAKE IN α-SYNUCLEIN TRANSGENIC MICE BY CHRONOAMPEROMETRY

Genetic and biochemical abnormalities associated with the abundantly expressed synaptic protein α-synuclein have been linked to the etiology of Parkinson's disease (PD).[98] Two point mutations in α-synuclein have been discovered in rare heritable forms of PD: an alanine to proline substitution at position 30 (A30P), and an alanine to threonine mutation at amino acid 53 (A53T).[99–102] Mutant and normal α-synuclein proteins have been shown to accumulate in Lewy bodies and Lewy neurites, which are characteristic brain pathologies of inherited, as well as the much more commonly occurring forms of sporadic PD. Mice have been genetically engineered to produce normal and mutant forms of human α-synuclein,[68] and expression of A53T mutant human α-synuclein results in a striking adult-onset hyperactive phenotype[16] that is later followed by a fatal motoric syndrome in mice expressing high levels of A53T α-synuclein. Based on this complex phenotype and our observations of reduced striatal dopamine transporter expression in A53T-tg mice, we investigated whether concomitant changes in dopamine uptake occur in these mice.

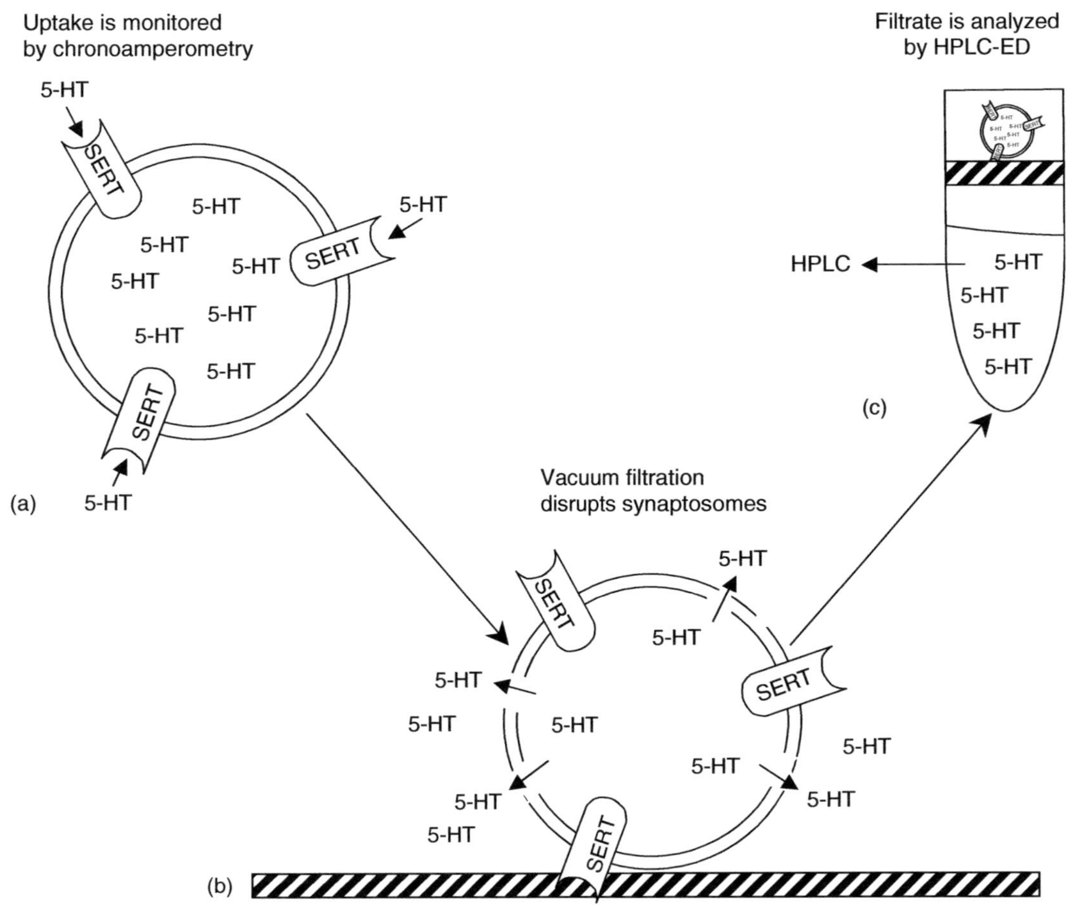

FIGURE 7.5 Measuring serotonin in extrasynaptosomal filtrates. (a) Serotonin (1 μM) was added to each sample and uptake was monitored for 20 min by chronoamperometry. (b) Once uptake was complete, samples were filtered individually using a Millipore filtration manifold. Filtration is hypothesized to disrupt synaptosomal membranes releasing a large portion of the serotonin taken up during the incubation period. (c) Synaptosomal filtrates were analyzed by HPLC-ED.

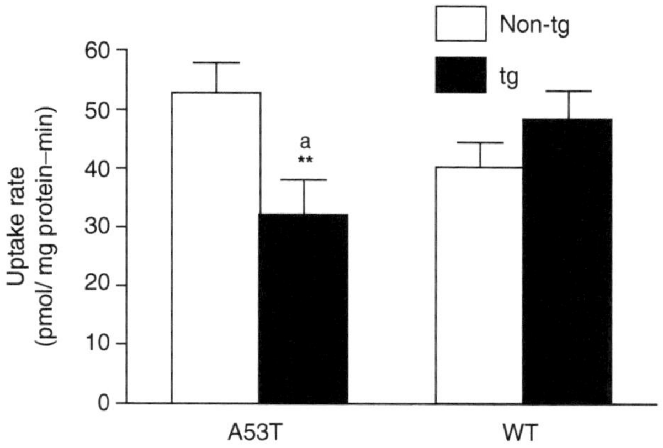

FIGURE 7.6 Dopamine uptake rates measured by chronoamperometry in striatal synaptosomes from α-synuclein transgenic mice. Uptake rates were determined after the addition of 1 μM dopamine to synaptosomes prepared from striatum from A53T-tg, WT-tg, and non-tg mice. A53T-tg mice show a 40% decrease in dopamine uptake compared to the corresponding non-tg littermates ($^{**}p<0.01$). Furthermore, dopamine uptake is reduced in A53T-tg mice compared to WT-tg mice ($^{a}p<0.05$). No significant differences in dopamine uptake were observed between WT-tg and their nontransgenic littermates. (This figure is reprinted from Unger, E. L., Eve, D. J., Perez, X. A., Reichenbach, D. K., Xu, Y., Lee, M. K., and Andrews, A. M., *Neurobiol. Dis.*, 21, 431–443, 2006. With permission.)

Dopamine uptake rates in synaptosomes prepared from striatum from A53T-tg mice, WT-tg mice, and the corresponding non-tg littermate mice were evaluated by high-speed chronoamperometry using the same conditions as those developed for analyzing serotonin uptake. Similar to serotonin, when 1 μM dopamine is added to synaptosomes, the current rises quickly as dopamine is detected at the carbon fiber microelectrode surface. As dopamine is taken up by the dopamine transporter the current decreases and the slope of the current decay curve between the time needed to clear 20% and 60% of the added dopamine is calculated and reported as the dopamine uptake rate in pmol/mg protein-min. One-way ANOVA of uptake data gleaned from α-synuclein transgenic mice demonstrated a significant effect of genotype [$F(3,17)=3.31$, $p<0.05$]. In A53T-tg mice, dopamine clearance rates were 40% lower than those measured in A53T non-tg mice (Figure 7.6). Mean uptake rates were 32 ± 6 and 53 ± 5 pmol/mg protein/min for A53T-tg and A53T non-tg mice, respectively. By contrast, no significant differences in dopamine uptake were detected between WT-tg mice and the corresponding non-tg littermates (49 ± 5 and 42 ± 4 pmol/mg protein/min, respectively). Additionally, uptake rates in A53T-tg mice were significantly decreased compared to WT-tg mice. These data show that the expression of A53T mutant human α-synuclein in mice and the resulting hyperlocomotive phenotype is associated with reduced dopamine uptake in striatum.[16] By contrast, expression of human wild-type α-synuclein does not increase locomotor activity and has no discernable effect on striatal dopamine clearance. From a more global perspective, the results of this study demonstrate that chronoamperometry can be used to measure dopamine, in addition to serotonin, uptake rates and that this technique is sufficiently sensitive to detect modest changes in monoamine neurotransmission in animal models in which the primary genetic alteration may not directly impact transporter expression but in which the behavioral phenotype is indicative of alterations in neurotransmission.

CONCLUSIONS

Traditional radiochemical methods have been used by neurobiologists to measure uptake kinetics for fifty years. These methods require the accumulation of radiolabeled neurotransmitter over the course of minutes. Subsequent determination of uptake rates is based on a single measurement per sample. As such, these types of methods do not offer high temporal resolution of the time course of uptake and, thus, provide significantly less information about transport kinetics. Chronoamperometry, by contrast, yields information on changes in the concentrations of serotonin or dopamine resulting from uptake on a per second basis leading to excellent temporal resolution of uptake rates over the total time period under investigation. The result is improved sensitivity to changes in uptake rates.

The data from the chronoamperometry studies described above support the conclusion that active uptake of serotonin is completely absent in the brains of SERT −/− mice, which concurs with the radiochemical results and SERT protein expression levels published by Bengel et al.[38] In in vivo experiments by Montanez and co-workers, clearance of exogenously applied serotonin was significantly reduced but not abolished in SERT −/− mice.[61] Clearance rates increased linearly with increasing concentrations of 5-HT in SERT −/− mice, in contrast to saturable uptake rates as a function of concentration of 5-HT in SERT +/+ and SERT +/− mice. These data suggest that under normal conditions involving intact tissue, serotonin clearance is a combination of transporter-mediated (saturable) and diffusional (nonsaturable) processes. They also highlight an advantage of using synaptosomes to differentiate between these two processes. Because synaptosomal uptake is measured in finite volumes, changes in the concentration of serotonin can be attributed to active transport, as the studies described above demonstrate, particularly those involving uptake inhibitors or SERT −/− mice. Furthermore, the lack of change in the electrochemical signals in synaptosomes derived from SERT −/− mice after the addition of serotonin is a strong indication that SERT-mediated uptake of serotonin is completely absent in these mice.

In addition to greatly improved temporal resolution, electrochemical studies have revealed that the filtration process used during radiochemical analysis disrupts synaptosomal membranes leading to a loss of transported neurotransmitter. This finding is significant because it explains a number of major discrepancies between uptake rates determined by radiometric versus chronoamperometric methods. Namely, imprecise determination of the amount of neurotransmitter taken up as a function of time accounts for the inability of the radiochemical method to accurately detect reductions in serotonin uptakes rates in with intermediate reductions in SERT expression. Furthermore, it leads to a gross underestimation of the effects of oxygen to increase uptake rates. Finally, these findings explain the considerably lower maximal uptake rates and affinity constants determined radiochemically compared to those measured by chronoamperometry. Specifically, chronoamperometry experiments employing full kinetic analysis indicate that serotonin uptake rates in synaptosomes are approximately 100-fold higher than those commonly reported from radiochemical studies (e.g. ~200 pmol/mg protein-min versus ~2 pmol/mg protein-min, respectively).[54,65,66,93–95] Moreover, the affinity of SERT for 5-HT is calculated to be ~1 μM in the present chronoamperometry experiments in synaptosomes, as well as by in vivo chronoamperometry studies, as opposed to ~0.05 μM reported in radiometric studies.[54,61,66]

In addition to determining serotonin or dopamine uptake as described above, chronoamperometry will also be useful for measuring norepinephrine transport into synaptosomes. We anticipate that future studies aimed at directly comparing kinetic parameters determined by chronoamperometry versus radiochemical methods for dopamine and norepinephrine will yield similar results to those reported for serotonin in that maximal uptake rates and transporter affinities are expected to be greatly underestimated by radiometric analysis.

The methods developed here can also be applied to the investigation of transporter-mediated monoamine efflux or "reverse transport" by preloading synaptosomes with serotonin, dopamine or norepinephrine and monitoring increases in their concentrations with respect to time at carbon fiber microelectrodes upon addition of monoamine releasers, such as the substituted amphetamines.[103] Chronoamperometry methods are also highly amenable to transport studies in cultured cells. The use of smaller Ag/AgCl wire reference electrodes and carbon fiber working electrodes presents further possibilities for studying uptake in reduced volumes and for multiplexing. Microelectrode arrays could be used for the simultaneous determination of uptake in multiple samples in well-plate type assays for high throughput drug screening applications. In all cases, the use of chronoamperometry to investigate monoamine transporter kinetics obviates the need for radioactivity and is expected to lead to more accurate determinations of the kinetic parameters of monoamine transport and to improved sensitivity to changes in V_{max} and/or K_m occurring in response to genetic or pharmacologic manipulations.

ACKNOWLEDGMENTS

This research was supported by a grant from the National Institute of Mental Health (MH64756). Many individuals contributed to the work presented here and the authors would like to express sincere appreciation to Dr. Erica Unger, Dr. Francesca Coppelli, Laura Bianco, Drew Kiraly, Dawn Reichenbach, Dawn Rupp, and Christian Squallanti for their technical assistance; Dr. Dennis Murphy for his generous gift of the SERT deficient mice breeding pairs; Dr. Michael Lee for the α-synuclein transgenic mice; Dr. Jim Schenk for his expert technical advice regarding electrochemical methods; and Drs. Lyn Daws, Andrew Ewing, Greg Gerhardt, and Paul Weiss for their insightful discussions regarding related manuscripts.

REFERENCES

1. Hawley, M. D., Tatawawadi, S. V., Piekarski, S., and Adams, R. N., Electrochemical studies of the oxidation pathways of catecholamines, *J. Am. Chem. Soc.*, 89, 447, 1967.
2. Adams, R. N., Probing brain chemistry with electroanalytical techniques, *Anal. Chem.*, 48, 1126A, 1976.
3. Salamone, J. D., Lindsay, W. S., Neill, D. B., and Justice, J. B., Behavioral observation and intracerebral electrochemical recording following administration of amphetamine in rats, *Pharmacol. Biochem. Behav.*, 17, 445, 1982.
4. Schenk, J. O., Miller, E., Rice, M. E., and Adams, R. N., Chronoamperometry in brain slices: quantitative evaluations of in vivo electrochemistry, *Brain Res.*, 277, 1, 1983.
5. Blakely, R. D. and Duvarney, R. C., A microcomputer controlled system for monitoring multiple voltammetric electrodes in vivo, *Brain Res. Bull.*, 10, 315, 1983.
6. Hefti, F. and Felix, D., Chronoamperometry in vivo: does it interfere with spontaneous neuronal activity in the brain?, *J. Neurosci. Methods*, 7, 151, 1983.
7. Stamford, J. A. and Justice, J. B. Jr., Probing brain chemistry, *Anal. Chem.*, 68, 359A, 1996.
8. Mosharov, E. V. and Sulzer, D., Analysis of exocytotic events recorded by amperometry, *Nat. Methods*, 2, 651, 2005.
9. Schenk, J. O., Wright, C., and Bjorklund, N., Unraveling neuronal dopamine transporter mechanisms with rotating disk electrode voltammetry, *J. Neurosci. Methods*, 143, 41, 2005.
10. Robinson, D. L., Venton, B. J., Heien, M. L., and Wightman, R. M., Detecting subsecond dopamine release with fast-scan cyclic voltammetry in vivo, *Clin. Chem.*, 49, 1763, 2003.
11. Troyer, K. P., Heien, M. L., Venton, B. J., and Wightman, R. M., Neurochemistry and electroanalytical probes, *Curr. Opin. Chem. Biol.*, 6, 696, 2002.
12. Daws, L. C., Montanez, S., Owens, W. A., Gould, G. G., Frazer, A., Toney, G. M., and Gerhardt, G. A., Transport mechanisms governing serotonin clearance in vivo revealed by high-speed chronoamperometry, *J. Neurosci. Methods*, 143, 49, 2005.
13. Michael, D. J. and Wightman, R. M., Electrochemical monitoring of biogenic amine neurotransmission in real time, *J. Pharm. Biomed. Anal.*, 19, 33, 1999.
14. Jones, S. R., Gainetdinov, R. R., and Caron, M. G., Application of microdialysis and voltammetry to assess dopamine functions in genetically altered mice: correlation with locomotor activity, *Psychopharmacology (Berl)*, 147, 30, 1999.

15. Giros, B., Jaber, M., Jones, S. R., Wightman, R. M., and Caron, M. G., Hyperlocomotion and indifference to cocaine and amphetamine in mice lacking the dopamine transporter, *Nature*, 379, 606, 1996.
16. Unger, E. L., Eve, D. J., Perez, X. A., Reichenbach, D. K., Xu, Y., Lee, M. K., and Andrews, A. M., Locomotor hyperactivity and alterations in dopamine neurotransmission are associated with overexpression of A53T mutant human alpha-synuclein in mice, *Neurobiol. Dis.*, 21, 431, 2006.
17. Phillips, P. E., Stuber, G. D., Heien, M. L., Wightman, R. M., and Carelli, R. M., Subsecond dopamine release promotes cocaine seeking, *Nature*, 422, 614, 2003.
18. Daws, L. C., Toney, G. M., Davis, D. J., Gerhardt, G. A., and Frazer, A., In vivo chronoamperometric measurements of the clearance of exogenously applied serotonin in the rat dentate gyrus, *J. Neurosci. Methods*, 78, 139, 1997.
19. Daws, L. C., Toney, G. M., Gerhardt, G. A., and Frazer, A., In vivo chronoamperometric measures of extracellular serotonin clearance in rat dorsal hippocampus: contribution of serotonin and norepinephrine transporters, *J. Pharmacol. Exp. Ther.*, 286, 967, 1998.
20. Daws, L. C., Gerhardt, G. A., and Frazer, A., 5-HT1B antagonists modulate clearance of extracellular serotonin in rat hippocampus, *Neurosci. Lett.*, 266, 165, 1999.
21. Bunin, M. A. and Wightman, R. M., Quantitative evaluation of 5-hydroxytryptamine (serotonin) neuronal release and uptake: an investigation of extrasynaptic transmission, *J. Neurosci.*, 18, 4854, 1998.
22. Bunin, M. A., Prioleau, C., Mailman, R. B., and Wightman, R. M., Release and uptake rates of 5-hydroxytryptamine in the dorsal raphe and substantia nigra reticulata of the rat brain, *J. Neurochem.*, 70, 1077, 1998.
23. Jackson, B. P. and Wightman, R. M., Dynamics of 5-hydroxytryptamine released from dopamine neurons in the caudate putamen of the rat, *Brain Res.*, 674, 163, 1995.
24. Jackson, B. P., Dietz, S. M., and Wightman, R. M., Fast-scan cyclic voltammetry of 5-hydroxytryptamine, *Anal. Chem.*, 67, 1115, 1995.
25. Hornykiewicz, O., Biochemical aspects of Parkinson's disease, *Neurology*, 51, S2, 1998.
26. Sharma, S. K. and Carew, T. J., The roles of MAPK cascades in synaptic plasticity and memory in Aplysia: facilitatory effects and inhibitory constraints, *Learn Mem.*, 11, 373, 2004.
27. Lesch, K. P., Zeng, Y., Reif, A., and Gutknecht, L., Anxiety-related traits in mice with modified genes of the serotonergic pathway, *Eur. J. Pharmacol.*, 480, 185, 2003.
28. Murphy, D. L., Li, Q., Engel, S., Wichems, C., Andrews, A., Lesch, K. P., and Uhl, G., Genetic perspectives on the serotonin transporter, *Brain Res. Bull.*, 56, 487, 2001.
29. Murphy, D. L., Andrews, A. M., Wichems, C. H., Li, Q., Tohda, M., and Greenberg, B., Brain serotonin neurotransmission: an overview and update with an emphasis on serotonin subsystem heterogeneity, multiple receptors, interactions with other neurotransmitter systems, and consequent implications for understanding the actions of serotonergic drugs, *J. Clin. Psychiatry*, 59, 4, 1998.
30. Uhl, G. R., Hall, F. S., and Sora, I., Cocaine, reward, movement and monoamine transporters, *Mol. Psychiatry*, 7, 21, 2002.
31. Blakely, R. D., De Felice, L. J., and Hartzell, H. C., Molecular physiology of norepinephrine and serotonin transporters, *J. Exp. Biol.*, 196, 263, 1994.
32. Sheehan, D. V., Current concepts in the treatment of panic disorder, *J. Clin. Psychiatry*, 60 (suppl. 18), 16, 1999.
33. Liebowitz, M. R., Update on the diagnosis and treatment of social anxiety disorder, *J. Clin. Psychiatry*, 60 (suppl. 18), 22, 1999.
34. Goodman, W. K., Obsessive-compulsive disorder: diagnosis and treatment, *J. Clin. Psychiatry*, 60 (suppl. 18), 27, 1999.
35. Keller, M. B., The long-term treatment of depression, *J. Clin. Psychiatry*, 60 (suppl. 17), 41, 1999.
36. Gough, B., Imam, S. Z., Blough, B., Slikker, W. Jr., and Ali, S. F., Comparative effects of substituted amphetamines (PMA, MDMA, and METH) on monoamines in rat caudate: a microdialysis study, *Ann. N. Y. Acad. Sci.*, 965, 410, 2002.

37. Callaway, C. W., Wing, L. L., and Geyer, M. A., Serotonin release contributes to the locomotor stimulant effects of 3,4-methylenedioxymethamphetamine in rats, *J. Pharmacol. Exp. Ther.*, 254, 456, 1990.
38. Bengel, D., Murphy, D. L., Andrews, A. M., Wichems, C. H., Feltner, D., Heils, A., Mossner, R., Westphal, H., and Lesch, K. P., Altered brain serotonin homeostasis and locomotor insensitivity to 3,4-methylenedioxymethamphetamine ("Ecstasy") in serotonin transporter-deficient mice, *Mol. Pharmacol.*, 53, 649, 1998.
39. Sora, I., Hall, F. S., Andrews, A. M., Itokawa, M., Li, X. F., Wei, H. B., Wichems, C., Lesch, K. P., Murphy, D. L., and Uhl, G. R., Molecular mechanisms of cocaine reward: combined dopamine and serotonin transporter knockouts eliminate cocaine place preference, *Proc. NatlAcad. Sci. U.S.A.*, 98, 5300, 2001.
40. Frazer, A., Antidepressants, *J. Clin. Psychiatry*, 58 (suppl. 6), 9, 1997.
41. Frazer, A., Serotonergic and noradrenergic reuptake inhibitors: prediction of clinical effects from in vitro potencies, *J. Clin. Psychiatry*, 62 (suppl. 12), 16, 2001.
42. Benmansour, S., Owens, W. A., Cecchi, M., Morilak, D. A., and Frazer, A., Serotonin clearance in vivo is altered to a greater extent by antidepressant-induced downregulation of the serotonin transporter than by acute blockade of this transporter, *J. Neurosci.*, 22, 6766, 2002.
43. Fuller, R. W., Uptake inhibitors increase extracellular serotonin concentration measured by brain microdialysis, *Life Sci.*, 55, 163, 1994.
44. Nestler, E. J., Antidepressant treatments in the 21st century, *Biol. Psychiatry*, 44, 526, 1998.
45. Nibuya, M., Nestler, E. J., and Duman, R. S., Chronic antidepressant administration increases the expression of cAMP response element binding protein (CREB) in rat hippocampus, *J. Neurosci.*, 16, 2365, 1996.
46. Blier, P. and de Montigny, C., Current advances and trends in the treatment of depression, *Trends Pharmacol. Sci.*, 15, 220, 1994.
47. Heils, A., Teufel, A., Petri, S., Stober, G., Riederer, P., Bengel, D., and Lesch, K. P., Allelic variation of human serotonin transporter gene expression, *J. Neurochem.*, 66, 2621, 1996.
48. Lesch, K. P., Bengel, D., Heils, A., Sabol, S. Z., Greenberg, B. D., Petri, S., Benjamin, J., Muller, C. R., Hamer, D. H., and Murphy, D. L., Association of anxiety-related traits with a polymorphism in the serotonin transporter gene regulatory region, *Science*, 274, 1527, 1996.
49. Greenberg, B. D., Li, Q., Lucas, F. R., Hu, S., Sirota, L. A., Benjamin, J., Lesch, K. P., Hamer, D., and Murphy, D. L., Association between the serotonin transporter promoter polymorphism and personality traits in a primarily female population sample, *Am. J. Med. Genet.*, 96, 202, 2000.
50. Schinka, J. A., Busch, R. M., and Robichaux-Keene, N., A meta-analysis of the association between the serotonin transporter gene polymorphism (5-HTTLPR) and trait anxiety, *Mol. Psychiatry*, 9, 197, 2004.
51. Osher, Y., Hamer, D., and Benjamin, J., Association and linkage of anxiety-related traits with a functional polymorphism of the serotonin transporter gene regulatory region in Israeli sibling pairs, *Mol. Psychiatry*, 5, 216, 2000.
52. Sen, S., Burmeister, M., and Ghosh, D., Meta-analysis of the association between a serotonin transporter promoter polymorphism (5-HTTLPR) and anxiety-related personality traits, *Am. J. Med. Genet.*, 127B, 85, 2004.
53. Caspi, A., Sugden, K., Moffitt, T. E., Taylor, A., Craig, I. W., Harrington, H., McClay, J., Mill, J., Martin, J., Braithwaite, A., and Poulton, R., Influence of life stress on depression: moderation by a polymorphism in the 5-HTT gene, *Science*, 301, 386, 2003.
54. Bengel, D., Murphy, D. L., Andrews, A. M., Wichems, C. H., Feltner, D., Heils, A., Mossner, R., Westphal, H., and Lesch, K. P., Altered brain serotonin homeostasis and locomotor insensitivity to 3,4-methylenedioxymethamphetamine ("Ecstasy") in serotonin transporter-deficient mice, *Mol. Pharmacol.*, 53, 649, 1998.
55. Jones, S. R., Gainetdinov, R. R., Jaber, M., Giros, B., Wightman, R. M., and Caron, M. G., Profound neuronal plasticity in response to inactivation of the dopamine transporter, *Proc. NatlAcad. Sci. U.S.A.*, 95, 4029, 1998.
56. Xu, F., Gainetdinov, R. R., Wetsel, W. C., Jones, S. R., Bohn, L. M., Miller, G. W., Wang, Y. M., and Caron, M. G., Mice lacking the norepinephrine transporter are supersensitive to psychostimulants, *Nat. Neurosci.*, 3, 465, 2000.

57. Holmes, A., Lit, Q., Murphy, D. L., Gold, E., and Crawley, J. N., Abnormal anxiety-related behavior in serotonin transporter null mutant mice: the influence of genetic background, *Genes Brain Behav.*, 2, 365, 2003.
58. Holmes, A. and Hariri, A. R., The serotonin transporter gene-linked polymorphism and negative emotionality: placing single gene effects in the context of genetic background and environment, *Genes Brain Behav.*, 2, 332, 2003.
59. Holmes, A., Murphy, D. L., and Crawley, J. N., Abnormal behavioral phenotypes of serotonin transporter knockout mice: parallels with human anxiety and depression, *Biol. Psychiatry*, 54, 953, 2003.
60. Murphy, D. L., Uhl, G. R., Holmes, A., Ren-Patterson, R., Hall, F. S., Sora, I., Detera-Wadleigh, S., and Lesch, K. P., Experimental gene interaction studies with SERT mutant mice as models for human polygenic and epistatic traits and disorders, *Genes Brain Behav.*, 2, 350, 2003.
61. Montanez, S., Owens, W. A., Gould, G. G., Murphy, D. L., and Daws, L. C., Exaggerated effect of fluvoxamine in heterozygote serotonin transporter knockout mice, *J. Neurochem.*, 86, 210, 2003.
62. Ghijsen, W. E., Leenders, A. G., and Lopes da Silva, F. H., Regulation of vesicle traffic and neurotransmitter release in isolated nerve terminals, *Neurochem. Res.*, 28, 1443, 2003.
63. Raiteri, L. and Raiteri, M., Synaptosomes still viable after 25 years of superfusion, *Neurochem. Res.*, 25, 1265, 2000.
64. Joyce, O. J., Farmer, M. K., Tipton, K. F., and Porter, R. K., Oxidative phosphorylation by in situ synaptosomal mitochondria from whole brain of young and old rats, *J. Neurochem.*, 86, 1032, 2003.
65. Perez, X. A. and Andrews, A. M., Chronoamperometry to determine differential reductions in uptake in brain synaptosomes from serotonin transporter knockout mice, *Anal. Chem.*, 77, 818, 2005.
66. Perez, X. A., Bianco, L. E., and Andrews, A. M., Filtration disrupts synaptosomes during radiochemical analysis of serotonin uptake: Comparison with chronoamperometry in SERT knockout mice, *J. Neurosci. Methods*, 154 (1–2), 245–255, 2006.
67. Heils, A., Wichems, C., Mossner, R., Petri, S., Glatz, K., Bengel, D., Murphy, D. L., and Lesch, K. P., Functional characterization of the murine serotonin transporter gene promoter in serotonergic raphe neurons, *J. Neurochem.*, 70, 932, 1998.
68. Lee, M. K., Stirling, W., Xu, Y., Xu, X., Qui, D., Mandir, A. S., Dawson, T. M., Copeland, N. G., Jenkins, N. A., and Price, D. L., Human alpha-synuclein-harboring familial Parkinson's disease-linked Ala-53 → Thr mutation causes neurodegenerative disease with alpha-synuclein aggregation in transgenic mice, *Proc. NatlAcad. Sci. U.S.A.*, 99, 8968, 2002.
69. Hoffman, A. F. and Gerhardt, G. A., In vivo electrochemical studies of dopamine clearance in the rat substantia nigra: effects of locally applied uptake inhibitors and unilateral 6-hydroxydopamine lesions, *J. Neurochem.*, 70, 179, 1998.
70. Gerhardt, G. A., Rapid chronocoulometric measurements of norepinephrine overflow and clearance in CNS tissues, In *Neuromethods*, Boulton, A., Baker, G., and Adams, R. N., Eds., Humana Press, Inc., Totowa, NJ, p. 117, 1995.
71. Gerhardt, G. A., Oke, A. F., Nagy, G., Moghaddam, B., and Adams, R. N., Nafion-coated electrodes with high selectivity for CNS electrochemistry, *Brain Res.*, 290, 390, 1984.
72. Kawagoe, K. T., Zimmerman, J. B., and Wightman, R. M., Principles of voltammetry and microelectrode surface states, *J. Neurosci. Methods*, 48, 225, 1993.
73. Hyde, C. E. and Bennett, B. A., Similar properties of fetal and adult amine transporters in the rat brain, *Brain Res.*, 646, 118, 1994.
74. Lowry, O. H., Rosebrough, N. J., Farr, A. L., and Randall, R. J., Protein measurement with the Folin phenol reagent, *J. Biol. Chem.*, 193, 265, 1951.
75. Cass, W. A. and Gerhardt, G. A., In vivo assessment of dopamine uptake in rat medial prefrontal cortex: comparison with dorsal striatum and nucleus accumbens, *J. Neurochem.*, 65, 201, 1995.
76. Numis, A. L., Unger, E. L., Sheridan, D. L., Chisnell, A. C., and Andrews, A. M., The role of membrane and vesicular monoamine transporters in the neurotoxic and hypothermic effects of 1-methyl-4-(2′-aminophenyl)-1,2,3,6-tetrahydropyridine (2′-NH(2)-MPTP), *Mol. Pharmacol.*, 66, 718, 2004.
77. Whittaker, V. P., Thirty years of synaptosome research, *J. Neurocytol.*, 22, 735, 1993.

78. Scott, I. D. and Nicholls, D. G., Energy transduction in intact synaptosomes Influence of plasma-membrane depolarization on the respiration and membrane potential of internal mitochondria determined in situ, *Biochem. J.*, 186, 21, 1980.
79. Erecinska, M., Nelson, D., and Silver, I. A., Metabolic and energetic properties of isolated nerve ending particles (synaptosomes), *Biochim. Biophys. Acta*, 1277, 13, 1996.
80. Earles, C. and Schenk, J. O., Multisubtrate mechanism for the inward transport of dopamine by the human dopamine transporter expressed in HEK cells and its inhibition by cocaine, *Synapse*, 33, 230, 1999.
81. Earles, C. and Schenk, J. O., Rotating disk electrode voltammetric measurements of dopamine transporter activity: an analytical evaluation, *Anal. Biochem.*, 264, 191, 1998.
82. Chen, N. and Justice, J. B. Jr., Cocaine acts as an apparent competitive inhibitor at the outward-facing conformation of the human norepinephrine transporter: kinetic analysis of inward and outward transport, *J. Neurosci.*, 18, 10257, 1998.
83. Povlock, S. L. and Schenk, J. O., A multisubstrate kinetic mechanism of dopamine transport in the nucleus accumbens and its inhibition by cocaine, *J. Neurochem.*, 69, 1093, 1997.
84. Earles, C., Wayment, H., Green, M., and Schenk, J. O., Resolution of biogenic amine transporter kinetics by rotating disk electrode voltammetry: methodology and mechanistic interpretations, *Methods Enzymol.*, 296, 660, 1998.
85. Richelson, E., Interactions of antidepressants with neurotransmitter transporters and receptors and their clinical relevance, *J. Clin. Psychiatry,*, 64 (suppl. 13), 5, 2003.
86. Habert, E., Graham, D., Tahraoui, L., Claustre, Y., and Langer, S. Z., Characterization of [3H]paroxetine binding to rat cortical membranes, *Eur. J. Pharmacol.*, 118, 107, 1985.
87. Hrdina, P. D., Foy, B., Hepner, A., and Summers, R. J., Antidepressant binding sites in brain: autoradiographic comparison of [3H]paroxetine and [3H]imipramine localization and relationship to serotonin transporter, *J. Pharmacol. Exp. Ther.*, 252, 410, 1990.
88. Bolden-Watson, C. and Richelson, E., Blockade by newly-developed antidepressants of biogenic amine uptake into rat brain synaptosomes, *Life Sci.*, 52, 1023, 1993.
89. Mathews, T. A., Fedele, D. E., Coppelli, F. M., Avila, A. M., Murphy, D. L., and Andrews, A. M., Gene dose-dependent alterations in extraneuronal serotonin but not dopamine in mice with reduced serotonin transporter expression, *J. Neurosci. Methods*, 140, 169, 2004.
90. Wise, R. A., Neurobiology of addiction, *Curr. Opin. Neurobiol.*, 6, 243, 1996.
91. Damasio, H., Grabowski, T., Frank, R., Galaburda, A. M., and Damasio, A. R., The return of Phineas Gage: clues about the brain from the skull of a famous patient, *Science*, 264, 1102, 1994.
92. Coccaro, E. F. and Murphy, D. L., In *Serotonin in Major Psychiatric Disorders*, Spiegel, D., Ed. *Progress in Psychiatry*, American Psychiatric Press, Inc, Washington, DC, 1990.
93. O'Reilly, C. A. and Reith, M. E., Uptake of [3H]serotonin into plasma membrane vesicles from mouse cerebral cortex, *J. Biol. Chem.*, 263, 6115, 1988.
94. Reyes-Haro, D., Garcia-Alcocer, G., Miledi, R., and Garcia-Colunga, J., Uptake of serotonin by adult rat corpus callosum is partially reduced by common antidepressants, *J. Neurosci. Res.*, 74, 97, 2003.
95. Fernandez, F., Sarre, S., Launay, J. M., Aguerre, S., Guyonnet-Duperat, V., Moisan, M. P., Ebinger, G., Michotte, Y., Mormede, P., and Chaouloff, F., Rat strain differences in peripheral and central serotonin transporter protein expression and function, *Eur J. Neurosci.*, 17, 494, 2003.
96. Blakely, R. D. and Bauman, A. L., Biogenic amine transporters: regulation in flux, *Curr. Opin. Neurobiol.*, 10, 328, 2000.
97. Zahniser, N. R. and Doolen, S., Chronic and acute regulation of Na^+/Cl^--dependent neurotransmitter transporters: drugs, substrates, presynaptic receptors, and signaling systems, *Pharmacol. Ther.*, 92, 21, 2001.
98. Kruger, R., Muller, T., and Riess, O., Involvement of alpha-synuclein in Parkinson's disease and other neurodegenerative disorders, *J. Neural. Transm.*, 107, 31, 2000.
99. Mezey, E., Dehejia, A., Harta, G., Papp, M. I., Polymeropoulos, M. H., and Brownstein, M. J., Alpha synuclein in neurodegenerative disorders: murderer or accomplice?, *Nat. Med.*, 4, 755, 1998.
100. Polymeropoulos, M. H., Lavedan, C., Leroy, E., Ide, S. E., Dehejia, A., Dutra, A., Pike, B. et al., Mutation in the alpha-synuclein gene identified in families with Parkinson's disease, *Science*, 276, 2045, 1997.

101. Spillantini, M. G., Crowther, R. A., Jakes, R., Hasegawa, M., and Goedert, M., Alpha-Synuclein in filamentous inclusions of Lewy bodies from Parkinson's disease and dementia with lewy bodies, *Proc. NatlAcad. Sci. U.S.A.*, 95, 6469, 1998.
102. Kruger, R., Kuhn, W., Muller, T., Woitalla, D., Graeber, M., Kosel, S., Przuntek, H., Epplen, J. T., Schols, L., and Riess, O., Ala30Pro mutation in the gene encoding alpha-synuclein in Parkinson's disease, *Nat. Genet.*, 18, 106, 1998.
103. Rothman, R. B. and Baumann, M. H., Targeted screening for biogenic amine transporters: potential applications for natural products, *Life Sci.*, 78, 512, 2005.

8 Using Fast-Scan Cyclic Voltammetry to Investigate Somatodendritic Dopamine Release

Sarah Threlfell and Stephanie J. Cragg

CONTENTS

WHAT IS SOMATODENDRITIC DOPAMINE RELEASE?

MIDBRAIN DOPAMINE NEURONS RELEASE DOPAMINE FROM THEIR SOMATODENDRITES

Midbrain dopamine (DA) neurons of the substantia nigra (SN, "A9") and adjacent ventral tegmental area (VTA, "A10") are critical to a range of CNS functions, including motor facilitation by the basal ganglia and the regulation of motivation by natural rewards as well as by drugs of addiction. A characteristic shared by DA cells in the SN and VTA is that they release DA locally from somatodendritic regions[1–4] as well as from their axonal projections. There is evidence for release from soma[5] as well as from dendrites.[2,6] Somatodendritic release of neurotransmitter is not restricted to DA neurons; rather, neurons found throughout the brain can signal via the somatodendritic release of neurotransmitters, including GABA and glutamate as well as neuropeptides.[7,8] Somatodendritic neurotransmission operates both at a synaptic level and by more paracrine/autocrine-like modes to offer neuronal cross-talk as well as self- or auto-feedback control.[7,8] This chapter will focus specifically on the somatodendritic release of DA within the midbrain and how voltammetric methods, particularly fast-scan cyclic voltammetry (FCV), have been used to explore its characteristics.

THE FUNCTIONS AND SITES OF ACTION OF SOMATODENDRITIC DOPAMINE

What are the functions of somatodendritic DA? Somatodendritically released DA acts on somatodendritic DA D_2-like autoreceptors to regulate subsequent somatodendritic DA release in the SN pars compacta (SNc),[9] DA neuron firing activity,[10–13] as well as downstream axonal release in striatum.[14,15] Moreover, somatodendritically released DA in both SN and VTA can act at DA D_1-like receptors presynaptic on GABAergic and glutamatergic terminals to modulate release of those transmitters.[16–18] For example, activation of D_1 receptors on striatonigral terminals increases GABA inhibitory transmission to SN pars reticulata (SNr) output neurons, which decreases the inhibitory SNr output to the thalamus;[17] this would reinforce motor activation by the striatonigral pathway. In fact, the critical role of the nigrostriatal pathway in movement, as demonstrated by the motor deficits of Parkinson's disease that accompany loss of nigrostriatal DA,[19] may result from a loss of DA in striatal and midbrain regions because both axon terminal and somatodendritic release have been shown to be required for basal ganglia-mediated movement.[20–23]

Where are DA receptors ultrastructurally located? There are few dendro-dendritic synapses in the SN/VTA.[24,25] DA receptors are, however, densely located at extrasynaptic sites throughout the length of the somatodendritic plasmamembrane and on afferent input membranes in SNc and SNr.[12,26] Moreover, the DA uptake transporter (DAT) is expressed at extrasynaptic sites remote from vesicles in the SN.[27] Together, the emerging picture of somatodendritic DA signaling is one where, unlike classical synaptic transmission (e.g., intrasynaptic release and receptor activation by glutamate), transmission by DA is mediated primarily by extrasynaptic receptors.[28–30] In fact, DA in both midbrain and striatum must act at its receptors via volume transmission.[28,31] As a consequence, a full understanding of DA transmission by somatodendritic DA release requires an understanding of local diffusion characteristics. This particular issue is discussed later in this chapter.

METHODOLOGICAL APPROACHES TO THE STUDY OF SOMATODENDRITIC DOPAMINE RELEASE

ELECTROCHEMICAL AND OTHER APPROACHES TO DOPAMINE DETECTION

A wide range of techniques have been used to investigate somatodendritic DA release. Initial studies used detection of [^{3}H]-DA in vitro using midbrain slices[2,32] and in vivo using push–pull perfusion.[4,33] Development of more sensitive off-line detection methods, especially HPLC with electrochemical detection, permitted monitoring of *endogenous* DA release from midbrain slices in vitro.[34] Another advance was seen with the introduction of in vivo microdialysis, which permits evaluation of extracellular levels of either exogenous or endogenous DA when coupled with an appropriate off-line analytical method.[14,35–39] Microdialysis measurements have been helpful in elucidating factors that influence somatodendritic DA release. Like other in vivo methods, microdialysis affords the opportunity to study somatodendritic DA release after systemic drug administration or during behavior in freely moving animals.[40] Another feature of microdialysis is that the off-line dialysate analysis, usually after an HPLC separation step, permits selective detection of DA as well as the concurrent monitoring of DA metabolites or other transmitters. These strengths come with caveats, however. Firstly, the temporal and spatial resolutions of microdialysis and corresponding probes are limited. For example, control of release by discrete action potentials or bursts of patterned activity cannot be detected or explored on a subsecond basis. Spatially, neurotransmitter release cannot readily be attributed to discrete functional subterritories within any given nucleus, e.g., dorsal or ventral tiers of DA cells. Secondly, interpretation of in vivo studies to address the mechanism or factors regulating somatodendritic DA release are complicated by the unavoidable influence of the overall circuitry governing DA cell activity in the SN or VTA.

The use of voltammetric recording in vitro at carbon-fiber microelectrodes (CFMs) with fast-scan cyclic voltammetry (FCV) was introduced for the study of somatodendritic DA release over a decade ago.[6] FCV at CFMs is a high-speed, high-spatial-resolution detection method that is ideal for monitoring release of DA or other biogenic amines by discrete stimuli in discrete brain nuclei, including the SNc and VTA. Many insights into somatodendritic DA release in midbrain over the last decade have been obtained using this method. Major advantages of this method over those already described are that DA release can be monitored within and between discrete nuclei, e.g., SNc vs. VTA, with sub-second resolution, i.e., in "real-time," and relatively independently from whole-brain circuitry.

Although considered a high-speed (or "real-time") detection technique, the spatial resolution of FCV is finite. The types of scans used with FCV, typically bi-, tri- or quadri-phasic waveforms that are designed to include oxidizing as well as reducing scans, limit the number of FCV scans per second. In comparison, some other high-speed electrochemical detection methods suitable at microelectrodes, e.g., constant potential amperometry, offer the possibility to scan continuously at a fixed potential and thus offer continuous, temporally more precise data. And yet, temporal precision with constant potential amperometry can be at a cost of identification of electroactive species. Scan potentials fixed at the DA oxidation potential ($\sim$500–600 mV vs. Ag/AgCl) will detect DA but cannot readily discern DA from the possible multiple detectable events that accompany somatodendritic DA release in midbrain in situ, including changes in pH and local extracellular Ca^{2+} concentration,[6,41,42] as well as the release of other monoamines, including 5-HT. Appreciation of the complex nature of these signals is imperative when undertaking electrochemical recordings in midbrain, where these other current sources are significant.[6,43–45] The use of a scanning waveform that applies a reducing as well as an oxidizing phase, as used in FCV but not in constant potential amperometry, can circumvent some of these problems because these different components can be resolved, and, thus, the contribution that is selectively DA can be dissociated.

Co-Detection of DA and 5-HT in Midbrain

FCV Waveform Choice

FCV offers an ideal approach for the study of somatodendritic DA release. Nonetheless, its own limitations must also be appreciated. In particular, voltammetric studies with FCV of somatodendritic DA release in the SN of some rodents, including rats[45–48] and mice[49] have been hindered by the concomitant or even predominant detection of 5-HT. The SN receives a major 5-HT innervation

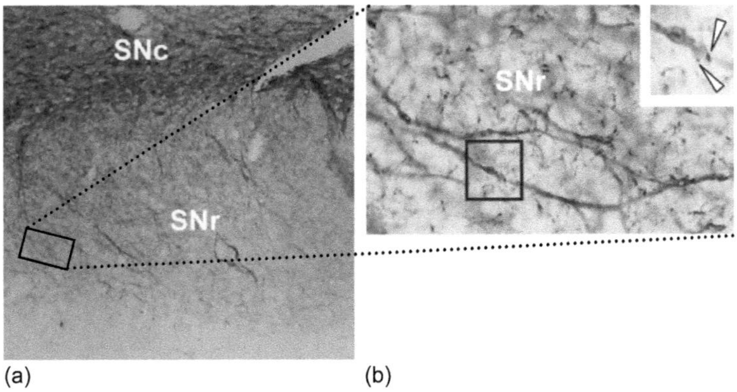

FIGURE 8.1 **(See color insert following page 272.)** Dopamine and 5-HT immunoreactivity in substantia nigra. (a) Low-power image of tyrosine hydroxylase-immunoreactivity (TH-IR) (grey/brown profiles) of soma and dendrites in SN pars compacta (SNc) and dendrites in SN pars reticulata (SNr) combined with serotonin transporter-IR (particulate black profiles) prominent on fine processes in SNr of rat midbrain. (b) An enlarged image of the boxed area (in a) reveals dense expression of the 5-HT marker on fine axonal processes throughout SNr. Serotonin transporter-IR (black profiles) and TH-IR (brown profiles) in the SNr; the boxed area (in b) is shown at higher power as an inset (arrowheads indicate varicosities with serotonin transporter-IR in the vicinity of dendritic processes containing TH-IR). Scales: (a) black box length, 85 μm; (b) black box width, 15 μm. (Figure reproduced in part from Threlfell, S., Cragg, S. J., Kallo, I. et al. *J. Neurosci.*, 24, 8704, 2004. With permission.)

density (see Figure 8.1) from raphe nuclei projections, which provide direct, asymmetric synaptic inputs to both dopaminergic and non-dopaminergic dendrites in SNc and SNr in primates and rodents (SNr > SNc > VTA).[50–54] Despite the theoretical ability to resolve DA and 5-HT on the basis of their reduction currents during a reducing scan (Figure 8.2), this task is not always simple analytically. The problem is compounded by the typical 3- to 10-fold greater sensitivity of CFMs to 5-HT than DA.[45]

Historically, different scan waveforms have been used for the preferential detection of DA or 5-HT. A biphasic triangular waveform has largely been used for detection of DA release and a triphasic "N" waveform for 5-HT. Triangular waveforms beginning and ending at negative potentials that include the reduction potentials for DA are typically less problematic for DA than 5-HT, due to the relatively lower propensity of the catecholamine DA to adsorb to the recording electrode and foul the surface. Furthermore, the triangular waveform has fewer changes in polarity of dV/dt and, therefore, fewer artifacts that can frequently occur with these switches in polarity, particularly upon electrical stimulation of tissue.[6,43] These artifacts can cause problems when quantifying neurotransmitter oxidation/reduction peaks, especially if they overlap with or distort a peak of interest. The indoleamines, such as 5-HT, have a much greater propensity than DA to adsorb to the electrode and foul the surface, potentially leading to artificially enhanced oxidation peaks.[55] For example, on calibration of 5-HT, using the triangular waveform and a resting potential of −0.4 V, the concentration-time profile is not rectangular as expected from a species that does not adsorb but displays an ever-rising profile characteristic of species that do adsorb.[55] To minimize this adsorption, a higher rest potential (between scans) for the triangular waveform can be employed, typically 0 or +0.2 V. To continue to be able to observe the reduction peak, which allows discrimination of the substance as 5-HT vs. DA, the reducing scan must be ramped below this potential before returning to rest; hence, the triphasic "N" waveform, e.g., scanning from (i) +0.2 to +1.0 V, (ii) +1.0 to −0.1 V, and (iii) −0.1 back to +0.2 V. Net adsorption is thought to be reduced by this procedure while the reduction peak essential for identification of the signal is retained.[55]

Alternatively, our procedure of choice when recording in midbrain, where DA and 5-HT can both occur, is to use a biphasic triangular waveform (from −0.7 V) that uses a minimal number of switches in polarity of d*V*/d*t* and which is switched out of circuit (or to 0 V) between scans to minimize adsorption. The scan range we use contains potentials for oxidation and reduction of both DA and 5-HT and thus leads to a relative ease of signal identification. Using this waveform, 5-HT exhibits dual reduction peaks and DA a single reduction peak at a potential between those for 5-HT (Figure 8.2), while the fewer changes in polarity of d*V*/d*t* (compared to the "N") generate less interference from capacitance-associated electrical artifacts. We have taken this approach in our studies for the detection of DA and 5-HT in SN.[43–45,56]

Attempts to improve the distinction between DA and 5-HT currents have recently been made in related FCV studies in striatum[57] based on the temporal profiles of current decay due to a more rapid reuptake of DA by DATs and more rapid clearance of DA than 5-HT from the CFM. A similar approach has not been explored in midbrain where current decay is already difficult due to both slower DA uptake[58] and the accompanying current shifts that usually accompany any release-evoking stimulus.[43] One of the most convincing ways to differentiate between 5-HT and DA in midbrain remains to be the use of the extended triangular waveform (−0.7 to +1.3 V) switched out of circuit between scans, with companion studies using pharmacological tools for confirmation.

Species Variation in 5-HT Interference

Contamination of somatodendritic DA voltammograms with interference from 5-HT varies with the species of rodent. The guinea pig SN receives a less dense 5-HT innervation than the rat SN so that pure somatodendritic DA release can be monitored in guinea pig SNc but not in rat SNc.[45] Interestingly, there are species differences in 5-HT receptor binding profiles as well, e.g., $5HT_4$ receptors,[59] with the pattern in guinea pig better resembling that in human SN. Thus, the guinea pig

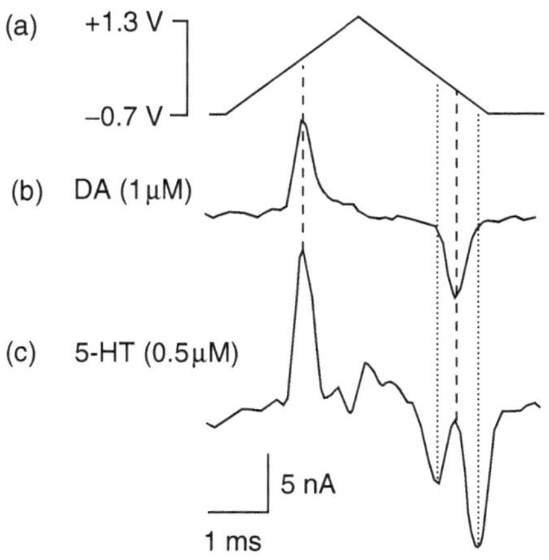

FIGURE 8.2 Voltammograms for dopamine (DA) and 5-HT. (a) Fast-scan cyclic voltammetry voltage waveform, 800 V/s scan rate, applied against an Ag/AgCl reference electrode. Scan switched out of circuit between scans. (b) DA voltammogram indicates oxidation and reduction peak potentials at +530 and −180 mV, respectively, vs. Ag/AgCl (dashed lines) for this electrode. Calibration factor was 6 nA/μM for DA in aCSF at 32°C. (c) 5-HT voltammogram indicates an oxidation peak potential similar to that for DA (+530 mV), but subsequent reduction generates two peaks (~0 and −560 mV vs. Ag/AgCl, represented by dotted lines) giving 5-HT a distinctly different voltammogram from that of DA. Calibration factor was 14 nA/μM for 5-HT. (Figure adapted from Rice, M. E., Cragg, S. J., and Greenfield, S. A., *J. Neurophysiol.*, 77, 853, 1997. With permission.)

is the rodent species of choice for the characterization of somatodendritic DA release in SN and VTA using FCV[45] while the rat is preferable over the guinea pig for exploring nigral 5-HT.[56] The problematic co-detection of 5-HT is not a concern for microdialysis because of the separate off-line separation step usually used for DA detection. Nonetheless, this problem is worse yet for amperometric detection methods because, without a reducing scan, no attempt at distinction between DA and 5-HT can be made at all.

Characteristic Voltammetric and Pharmacological Differences

Several groups have studied 5-HT release in SN using FCV.[6,45,47,48,56] Release of 5-HT in SN may exhibit a different characteristic frequency response from that of DA. Reports to date suggest that DA release in SNc is frequency-dependent up to 10 Hz, beyond which no further increase in $[DA]_o$ is observed,[43] whereas 5-HT release in SNr is frequency dependent up to 50 Hz.[47,48,56] Nonetheless, train lengths were not constant across these studies; thus, frequency responses have not yet been systematically compared within a given study using a single experimental paradigm.

Is it possible to differentiate between 5-HT and DA pharmacologically in SN? There are indeed a few key features of the pharmacology of these two species in SN that differ. Firstly, evoked extracellular concentrations of 5-HT are extremely sensitive to blockade of the 5-HT uptake transporter (SERT) but not to blockade of the DAT,[47,48,56] whereas evoked extracellular concentrations of DA are modified by the DAT but not the SERT.[44] Also apparent is the control by DA D_2 receptors of somatodendritic release of DA but not 5-HT release in SNr.[47]

Reproducibility of 5-HT and DA voltammetric signals over time also aids in species identification in the SN. Once DA has been detected in a single site, it can be difficult to evoke concentrations of the same magnitude again from the same site, even after a considerable pause after initial stimuli. In contrast to the release of DA, 5-HT release in SN is extremely robust,[56] and single sites can reproducibly release for up to eight hours (Threlfell and Cragg, unpublished observation). Thus, strong re-releasability from a recording site is likely to further indicate signal identity as 5-HT.

Further points worth noting are empirical observations of the profiles of the oxidation peaks and the concentrations of species expected in the SN. In our hands, using a triangular waveform (-0.7 to $+1.3$ V), 5-HT oxidation peaks are notably narrower (span a smaller range of potentials) than those for DA that tend to have a much broader profile. Detected concentrations of each electroactive species in the SN vary significantly from one another by up to one order of magnitude ($[DA]_0 >$ $[5\text{-HT}]_0$), e.g., evoked [DA] in SNc can reach $\sim$500 nM while evoked [5-HT] (in rat SN) are in the range of 50–120 nM. However, the enhanced sensitivity of the CFMs to 5-HT (typically three to five times greater than that for DA) can lead to similar overall detected currents. Where DA and 5-HT occur together at these levels, e.g., in rat SNc,[45] reduction currents comprise multiple irresolvable peaks, which precludes accurate signal identification and quantification.

In Vitro Methods for FCV in Midbrain

Slice Preparations and Species

To date, successful voltammetric approaches to explore somatodendritic DA release have used coronal brain slice preparations (300–400 μm) from a range of rodent species, including guinea pig, rat, and mouse. Acute slices are prepared by conventional methods used for electrophysiology preparations, e.g., over ice using commercially available vibratomes. Coronal slices offer a choice of recording sites in VTA, SNc, and SN pars reticulata (SNr), with readily identifiable anatomical landmarks to delineate each region, e.g., the myelinated accessory optic tract neatly divides the VTA from the SNc. The coordinates of midbrain slices containing the SN/VTA in guinea pig correspond to A8.0–A8.7 mm anterior (A) to the interaural line, according to the atlas of Smits et al.[60] In adult rat, the equivalent slice range corresponds to the coordinates A3.2–A4.2 mm, according to the atlas of Paxinos and Watson,[61] i.e., between rostral interpeduncular nucleus and caudal mammillary bodies. Somatodendritic DA release can be detected from early postnatal ages onwards (e.g., postnatal day two) in guinea pig and rat[45] at ages prior to the development of innervation by 5-HT projection axons. If dissecting multiple brain regions for voltammetry from one brain, e.g., midbrain and forebrain, optimal success with midbrain recordings in our hands has been achieved when the midbrain block is sectioned first, i.e., given priority over any other. On a further technical note, the large network of blood vessels present on the ventral aspect of the adult rat brain must be removed as quickly as possible. These elastic vessels may otherwise be poorly cut by even the best commercially available sectioning instruments and thus cause tissue snagging, stretching, and subsequent damage that limits slice viability.

Solutions

To date, published methods for preparing animals for midbrain slices for FCV have used inhalation or intraperitoneal anesthesia (halothane or pentobarbital, respectively), and, after brain removal, coronal slices have typically been cut on vibratomes in ice-cold, HEPES-buffered artificial CSF (ACSF) containing (in mM): 120 NaCl, 5 KCl, 1.25 KH_2PO_4, 20 $NaHCO_3$, 6.7 HEPES acid, 3.3 HEPES salt, 2 $CaCl_2$, 2 $MgSO_4$, and 10 glucose (saturated with 95% O_2/5% CO_2).[43,44,62] After cutting, slices are allowed to recover in HEPES-buffered ACSF for at least one hour at room temperature before being transferred to a submersion recording chamber. Once in the recording chamber, slices are equilibrated for an additional thirty minutes with a physiological bicarbonate-buffered aCSF, which usually contains (in mM): 124 NaCl, 3.7 KCl, 26 $NaHCO_3$, 1.5–2.4 $CaCl_2$, 1.3 MgSO4, 1.3 KH_2PO_4, and 10 glucose (saturated with 95% O_2/5% CO_2). Chamber temperatures used to date are typically 32°C.

During recording, viewing through a low-power binocular microscope is usually sufficient to locate VTA, SNc, and SNr within a coronal slice; higher magnification optics and infra red difference interference contrast (IR-DIC) microscopy can offer more precise electrode placements to tiers of DA neurons within the SNc, if required. Carbon-fiber electrodes are only 7–8 μm in diameter and thus sample from highly discrete regions; as a consequence, the accurate placement of electrodes to within specific nuclei is important. Nonetheless, the typical electrode tip lengths (e.g., 30–100 μm tip length) required to generate currents of appropriate signal to noise ratios for voltammetry entail that electrodes sample from a local *population* rather than single release sites. In turn, the concentration of DA release detected will thus depend on the local packing density of somatodendritic release sites and correlates well with the density of immunoreactivity to tyrosine hydroxylase (TH-ir) (Figure 8.3). Other factors that govern the detected concentration of DA release include stimulation protocol, Ca^{2+} availability, DA uptake, extracellular volume fraction, and autoreceptor and heteroreceptor regulation.

Scan Waveforms

A variety of scan waveforms, scan rates, and sample frequencies have successfully been used to detect and experiment with somatodendritic DA release in VTA and SNc, ranging from so-called "N"-shape waveforms (scans from 0.0 to +1.1 V to −0.5 to 0.0 V vs. Ag/AgCl) at 660 V/s using sample frequencies of 2–4 Hz[6] to "triangular" waveforms (scans from −0.7 to +1.3 V to −0.7 V vs. Ag/AgCl, switched out of circuit between scans) at 800 V/s at 8–10 Hz[9,43,62] where the scan duration is ~5 ms. Arguably, the biphasic triangular scan is more simple to interpret than a triphasic N scan due to the fewer number of switches in scan rate polarity and consequent artifacts that are associated with each switch; these can be particularly problematic during stimulated release in SN and VTA.[6,43,44] The scan potentials for the oxidation and reduction of DA are well within such scan potential ranges (occurring at ~500 to −600 mV and −200 to −300 mV), and these generous scan potential ranges have been used in midbrain to increase the separation of oxidation/reduction

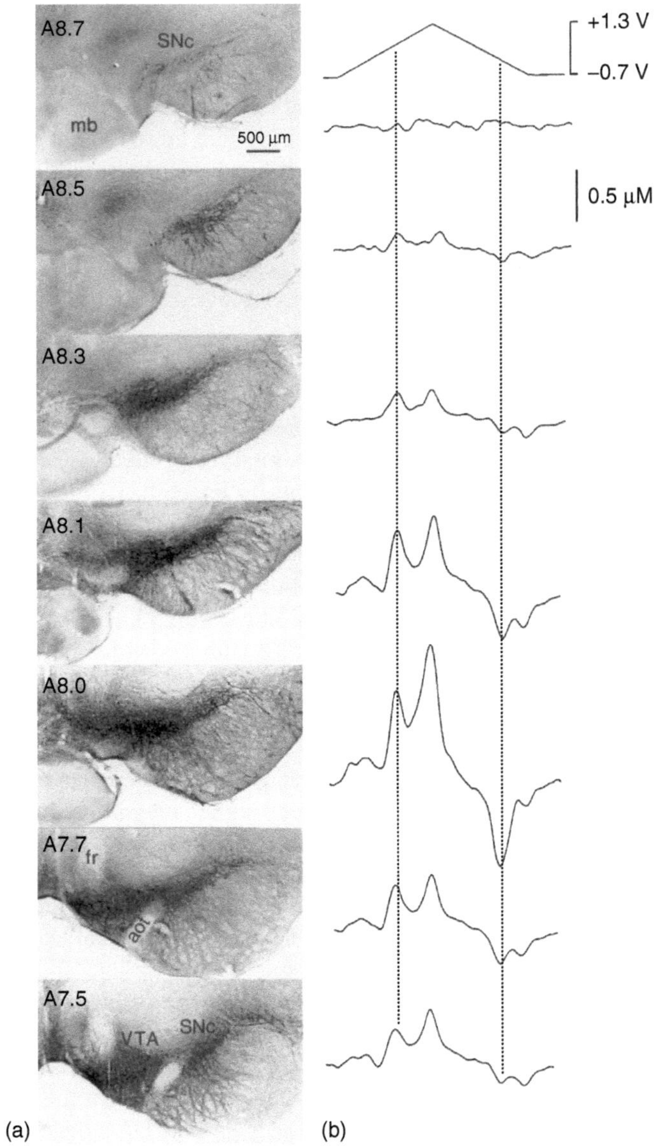

FIGURE 8.3 Rostral-caudal range of SN and corresponding voltammograms obtained in SNc after electrical stimulation. (a) Coronal sections extending from rostral (A8.7 mm) to mid-caudal SN (A7.5) stained for tyrosine hydroxylase immunoreactivity (TH-ir). (Coordinates are taken from the atlas of Smits, R. P. J., Steinbusch, H. W. M., and Mulder, A. H., *J. Chem. Neuroanat.*, 3, 101, 1990). Notation: *mb*, mammillary bodies; *fr*, fasciculus retroflexus; *aot*, accessory optic tract; *VTA*, ventral tegmental area. (b) Representative voltammograms obtained during electrical stimulation (10 Hz for five seconds) from medial SNc in each slice range corresponding to co-ordinates in (a). Evoked extracellular DA concentrations ranged from 50 to 700 nM. Dotted lines indicate dopamine (DA) oxidation potentials obtained during postcalibration of the electrode. Other prominent currents are capacitance-related currents occurring at the switch in scan rates from $+dV/dt$ to $-dV/dt$ and from $-dV/dt$ to $dV/dt = 0$ at the temporary holding potential (-0.7 V). (Figure adapted from Cragg, S. J., Rice, M. E., and Greenfield, S. A., *J. Neurophysiol.*, 77:863, 1997. With permission.)

currents from any artifacts occurring at the switches between $+\delta V/\delta t$ and $-\delta V/\delta t$. Both DA and 5-HT can be resolved with such scans (Figure 8.2). Given the strong innervation density of 5-HT and DA in midbrain (Figure 8.1), it is imperative to use a scan that has the capability to resolve both amines.

Carbon-Fiber Microelectrodes: Preparation, Calibration

Carbon-fiber microelectrodes (CFMs) for use with FCV in the study of somatodendritic DA release have typically been prepared from 7- to 8-μm diameter untreated carbon fibers in a 2 mm glass capillary tube pulled to a taper. The electrodes we prepared ourselves, as well as those purchased from MPB electrodes (London), were prepared to final tip lengths of 30–60 μm (length of carbon-fiber exposed beyond the glass insulation), using either cutting or spark-etching followed by beveling to a point of tip diameter of 2–4 μm.[63] Full instructions for electrode preparation using the method of Millar and colleagues as used by us, are detailed at http://www.qmw.ac.uk/~physiol/makingCFE.html.

Electrode calibrations for sensitivity to DA (or 5-HT) were performed in the recording chamber at 32°C using known concentrations of DA (or where appropriate 5-HT), made up in aCSF solutions to amine concentrations similar to those seen in situ (0.5–2 μM). Electrodes should respond linearly to concentration over this range (data not illustrated). Stock amines (1000–2500x) were kept in solutions of 0.1 M $HClO_4$ and made up in aCSF buffer immediately before use. Electrode sensitivity during recordings was assessed by calibration immediately after recording. Typically, electrodes had sensitivities to DA in the order of 2–10 nA/μM, and detection limits (~2x noise) were 30–50 nM.

Choice of Stimulation Parameters

Because electrochemical current responses are proportional to the concentration of electroactive species, optimal signal strength of DA will occur following synchronous release from a population of local release sites. With FCV to date, data acquisition has been most successful using electrical stimulation at local bipolar electrodes.[9,43,44,62,64–66] Spontaneous somatodendritic release has not been reported with FCV. However, this could be due to limitations of sensitivity rather than a paucity of spontaneous DA release events because DA neurons are spontaneously active within midbrain slice preparations and, furthermore, amperometry measurements at single cells cleared of adjacent tissue by micropipette suction do reveal spontaneous exocytotic-like events from DA-like neurons, albeit at very low frequencies (~0.3 to 3 events per minute).[5] Nonetheless, if there is spontaneous somatodendritic DA release in slices, it is not sufficient to activate release-regulating autoreceptors.[9]

Prior to the introduction of electrical stimulation to study discrete and repeated observations of DA release with FCV, the chemical secretagogue veratrine (mixed alkaloids)[6,43] or high potassium[6] were used to evoke DA release. Voltammetric signals following veratrine superfusion were attributable to DA but were frequently accompanied by the progressive distortion of the DA voltammogram by 5-HT as well as changes in background currents in VTA and SNc.[6] These artifacts were particularly problematic with high potassium.[6] These additional shifts can correspond to changes in Ca^{2+}, pH, O_2 tension, and "switching artifacts."[6] Repeat measures in a single slice cannot readily be achieved following veratrine superfusion, thus limiting data collection with this approach. We have previously experimented with iontophoresis of high K^+ concentrations or glutamate (Cragg and Rice, unpublished data) but with notably less success than electrical stimulation.

Chemical stimulation of the subthalamic nucleus (STN) has been used to remotely activate the SN in a study using amperometry.[13] In that study, changes in electroactive species were not detected following electrical stimulation of the STN, but, rather, local pressure application

(36 s, 20 psi) of carbachol (10 mM) was apparently the most effective means of stimulating currents in SN. However, the speculated identity of the signals detected in SN as DA in that study remains unresolved and has not been explored with FCV.

By using electrical stimulation at local bipolar electrodes, we have found FCV recordings within SN or VTA to have a higher data yield, to be spatially discrete, repeatable at several loci within a slice, and less confounded by accompanying stimulation artifacts than with local veratrine,[43,44] as well being more readily controlled in intensity, duration, and location than was possible with the depolarization induced by veratrine. However, electrically eliciting DA release from the SN can prove difficult, as noted by others,[46] and is not reliably repeatable. Typically, we have constructed bipolar stimulating electrodes from Teflon-coated platinum wire (Clark Electromedical Instruments) of either (1) bare diameter 125 μm, coated diameter 175 μm, and tip separation of 100 μm or (2) bare diameter 50 μm, coated diameter 75 μm, tip separation 50 μm. However, stimuli generated with the larger electrodes readily evoked release in a manner that was independent of voltage-gated Na^+ channels, i.e., TTX-insensitive,[43,44] whereas stimulation with the smaller electrodes (presumably higher resistance and lower applied currents) permitted DA release via recruitment of voltage-gated Na^+ channels, i.e., TTX-sensitive release.[9] In all cases, we superficially position the tips of the bipolar electrode for surface stimulation and insert the CFM 50–100 μm into the slice between the stimulator poles. Constant voltage (10–20 V, 0.1 ms pulses)[9] and constant current (0.4–0.8 mA, 0.1–1.0 ms pulses)[62,64] stimuli have been used to evoke TTX-sensitive somatodendritic DA release. Commercially available, e.g., concentric bipolar electrodes, would likely be equally successful.

Several factors will govern the choice of electrical stimulation parameters for studies of neurotransmitter release. Experimenters should decide whether they want to optimize signal to noise ratio (relative signal amplitude), local synaptic input by other neurotransmitters, independence from other neurotransmitter inputs, or repeatable release at a single loci. For example, long stimulation trains will have the disadvantages (or advantages) of poor reproducibility in a short timeframe and engaging the neuromodulation of somatodendritic DA release by other neurotransmitters released during the stimulus that will subsequently control DA release at successive pulses. In contrast, release by single stimulus pulses may be lower in concentration but more reproducible (see below) and will be governed only by factors that regulate intrinsic somatodendritic release probability as well as any spontaneously active sources of neuromodulatory inputs.

In SNc and VTA, the most routinely reported stimulation paradigms are trains of pulses at 10 Hz (e.g., 0.1 ms pulses in three second trains)[9,62,64,65] (although up to ten seconds have been used)[43,44], or single pulses (when longer pulse durations are required, 1 ms).[62,66] In our hands, 10 Hz trains for three seconds most reliably evokes the greatest single increase in $[DA]_0$.[43] Typically, however, electrical stimulation evokes the greatest $[DA]_0$ during the *first* of any stimulation train at a given SNc or VTA recording site.[43,44,62] Although detectable release can again be elicited after several minutes, subsequent stimulations are rarely as effective. This failure at successive stimuli has been assessed most rigorously following 10 Hz stimulation trains (of 3–10 s duration) when maximal detected $[DA]_0$ decline significantly at subsequent trains, even after 20–30 min recovery between trains.[43,44] However, we have found that stimulation trains of shorter duration and higher frequency (e.g., ten pulses of 0.1 ms duration at 100 Hz)[9,67] can be relatively successful at evoking reproducible $[DA]_0$ at a given recording site when using 5–10 min sampling intervals.[67] Nonetheless, concentrations of DA detected with this paradigm are low, in the order of ~100–200 nM DA, which are at least five-fold less than concentrations detected at single 10 Hz trains only.

Because optimal $[DA]_0$ are released by longer pulse trains, after which release is not fully reproducible at any given recording site, a suitable experimental design that can be used to assess the control of somatodendritic release before and after manipulation of a parameter of interest (ion channel, neurotransmitter receptor, ligand, etc.), is one involving paired observations in two hemispheres. For example, single measurements can be taken at several recording sites in VTA and SNc, then the parameter of interest can be suitably modulated (e.g. by drug or ligand application), and then measurements can be taken in the opposite hemisphere by single measurements at matching, i.e. paired recording sites in VTA and SNc. Caution should be taken if currents are reproducible at a given recording site because 5-HT release is readily reproducible at recording sites in rat SNc and SNr; e.g.,[56] the identity of the reproducible component of these voltammograms should be appropriately analyzed.

Another useful consideration when using electrical stimulation in conjunction with any electrochemical scan is to prevent the stimulus from occurring during the limited duration of the scan in order to avoid distortion of the electrochemical currents with a stimulus artifact. This can be achieved by either electronically blanking stimuli during the voltammetric scans or by the introduction of a simple electronic timing circuit to offset the depolarizing stimulus at a fixed interval out of phase with FCV scans.

Concerns in Signal Identification

Among the advantages of FCV as a technique for real-time detection of somatodendritic DA release is its ability to discriminate between DA and the other sources of apparent currents that constant potential scans or non-cyclic scans cannot. Nonetheless, these other sources of currents can either obscure redox currents due to DA or can contribute to the current measured at the DA oxidation potential. Thus, due caution should be taken in attributing solely to DA the electrochemical currents at the DA oxidation potential. Where necessary, baseline adjustments may need to be made to correctly determine currents due to DA. Such interferents can also greatly hinder the apparent decay of currents at the DA oxidation potential after release, thereby often precluding determination of uptake rates.[44] This issue is particularly important in the midbrain where these other current sources, e.g., due to changes in pH, extracellular Ca^{2+}, O_2 tension[6,41,42,68] and 5-HT seem to be particularly significant.[6,43–45]

As with the use of FCV anywhere in the brain, the identity of DA-like electrochemical signals must be validated by supplementary data in addition to electrochemical information provided by the scan, e.g., through anatomic verification and pharmacological manipulations. Clearly, a comparison of the electrochemical signals with calibration voltammograms should indicate that the released substance is a catechol, like DA or DOPAC, rather than the indole 5-HT. The contribution of electroactive DOPAC must be considered; nonetheless the low sensitivity of our electrodes for DOPAC (1:70, DOPAC:DA) suggest that any contribution of DOPAC to the catechol signal will likely not be appreciable. We have confirmed this experimentally: the monoamine oxidase inhibitor pargyline does not modify the amplitude of the release signal in SNc.[6] Another possible interferent, ascorbic acid (ascorbate), is also unlikely to contribute; previous studies indicate that 95% of tissue ascorbate content is lost from slices.[69] Even when present, the ascorbate voltammogram is readily distinguished from that of a catechol.[70] Importantly, DA-like signals in SN and VTA should be anatomically specific, i.e., restricted to regions that are positive for TH-immunoreactivity[43,44] with, furthermore, a general correlation between the amplitude of the responses and density of TH-ir at these locations[44] (Figure 8.3). In addition, these signals should be indistinguishable (except in amplitude) from those obtained in striatum, a region for which DA release has been extensively characterized with FCV.

Additional pharmacological manipulations can identify whether Ca^{2+}-free media reversibly eliminate DA-like signals in an expected manner and which uptake transporter (DAT, NET, or SERT) plays the most prominent role in regulating amine-like signals. Primarily, somatodendritic DA is governed by the DAT rather than the NET or SERT,[44,58] but monoaminergic neurotransmitter uptake systems can be highly promiscuous with respect to neuroamine substrates.[44,58,71–73] Thus, care should be taken in identifying amine-like signals in SN and VTA based solely on the effects of uptake transport inhibitors. Although the voltammogram of DA is similar to that of the catecholamine norepinephrine (NE), anatomical observations of the restricted localization of immunoreactivity to the NE synthesizing enzyme, DA-β-hydroxylase, in conjunction with the limited effects of NE uptake inhibitors on the catechol signals, together suggest that NE does not contribute to the release signal in VTA or SNc[44].

WHAT IS THE MECHANISM OF RELEASE?

ANATOMICAL EVIDENCE

Whether or not somatodendritic DA release is exocytotic remains a subject of continuing speculation and investigation.[13,43,44,62,66] Anatomical evidence for exocytosis is certainly limited in SN; synaptic sites where vesicle fusion might occur are rare.[25] Vesicles are also rare in DA cells of the SNc. Whereas vesicle density is high in DA terminals in striatum,[27,74] there are few vesicles in DA somata or dendrites,[24,25,27] implying a limited source for exocytotic release.[27]

Storage of somatodendritic DA has been proposed to be in saccules of smooth endoplasmic reticulum[75,76] as well as in vesicles.[24,25] Consistent with dual storage sites, the vesicular monoamine transporter, VMAT2, is expressed in saccules (so-called tubulovesicles) and, less commonly in vesicles.[77] Whether both storage sites contribute to the releasable pool of DA is not known. Both would be susceptible to DA depletion by reserpine, an irreversible inhibitor of VMAT2, which limits the information provided by reserpine sensitivity about vesicular release or otherwise.[37]

MECHANISTIC EVIDENCE

The non-axonal source for somatodendritic DA release and paucity of dendritic synapses has led to speculation over whether the mechanism of somatodendritic release might also differ from that seen from axons. The limited numbers of vesicles and abundance of DATs in DA dendrites led several groups to speculate that somatodendritic release might be mediated by vesicle-independent reverse DAT transport.[24,27,77,78] Indeed, this is a primary mode for release following certain pharmacological manipulations, including veratridine-induced depolarization[34] or amphetamine-induced DA displacement from intracellular stores.[79–81] Somatodendritic DA release elicited by veratridine requires voltage-dependent Na^+ channel opening and subsequent Na^+ loading to activate reverse DA transport; release can subsequently be blocked by the DAT inhibitor GBR-12909.[34] Under normal conditions, it has been suggested that cytoplasmic DA concentrations are not sufficient to reverse the DAT unless increased by pharmacological agents, e.g., by amphetamine.[80] However, Falkenburger et al.[13] reported electrophysiological evidence that carbachol stimulation of the subthalamic nucleus could also induce dendritic DA release in the SN by a mechanism that could be prevented by low concentrations of GBR-12909. This finding suggests that under some physiological conditions, somatodendritic release might be mediated by reversal of the DAT. It is not yet clear how the data of Falkenburger et al. can be reconciled with other in vitro and in vivo studies of somatodendritic DA release, including those with FCV that typically find an increase, not a decrease, in $[DA]_0$ in the presence of an uptake blocker.[14,44,58,62,82–85]

Apart from the study by Falkenburger et al.[13] there is little evidence to contradict the original proposal by Geffen et al.[2] that DA release in the SN is vesicular and mediated by exocytosis, as it is in axon terminals in striatum. Consistent with the ionic and pharmacologic characteristics of exocytosis, dendritic DA release is depolarization-, Na^+ channel-, and Ca^{2+}-dependent,[2,6,9,33,43,62] is sensitive to DA depletion by reserpine,[6,37] and is not prevented by DAT blockers.[14,44,62,82,84] Additionally, other manipulations that alter $[DA]_0$ in striatum generally also modify $[DA]_0$ in SN[9,14,37,44,62,86] although there is evidence for less similarity between DA release in VTA vs. nucleus accumbens.[9,87] As detected by FCV, the dependence of somatodendritic DA release on voltage-gated Na^+ channels (TTX-sensitivity)[9,62] is in keeping with the presence of regenerative Na^+ currents in DA neuron dendrites,[88] as in axons. And yet, how somatodendritic DA release depends on action potential vs. other regulation of membrane excitability, e.g., local afferent input, remains unknown. Thus, it remains unclear how somatodendritic DA release reflects the characteristic repertoire of tonic and phasic firing patterns of DA neuron activity.[89–92]

Evidence for Ca^{2+} dependence is often taken as confirmatory of vesicular release because Ca^{2+} entry is typically required for exocytosis.[93] Voltammetric studies of somatodendritic DA release in SN have been particularly informative. Somatodendritic DA release in SN requires Ca^{2+}; incubation in Ca^{2+}-free media with EGTA inhibits evoked DA release by >90%.[6,43] However, in contrast to striatal release, DA release in SNc persists in low-Ca^{2+} media;[6,43,62,86] inclusion of EGTA is required to eliminate release,[62] suggesting a limited regulation of somatodendritic DA release by Ca^{2+} entering the neuron.[66] Furthermore, blockers of voltage-gated Ca^{2+}-channels (N-, P/Q-, T-, R-, L-types) have undetectable to modest effects on $[DA]_0$ in the SN contrasting with their various, strong roles in the regulation of DA release in striatum.[66] Together, these findings suggest a difference in requirement for Ca^{2+} entry, intracellular Ca^{2+} availability, Ca^{2+} sensitivity or some other related aspect in the mechanisms of somatodendritic vs. axonal release. For example, ready somatodendritic release in low $[Ca^{2+}]_0$ might reflect differential sensitivity of Ca^{2+}-dependent release mechanisms, including fusion proteins, between these compartments. For example, the varying forms of synaptotagmin, a key Ca^{2+}-sensing protein responsible for triggering *synaptic* transmitter release, can differ in Ca^{2+} affinity by 10- to 20-fold.[94,95] Differential expression of fusion proteins among cell compartments is not without precedent; Bergquist et al.[39] recently suggested that different synaptobrevin isoforms underlie axonal DA release in striatum vs. somatodendritic release in the SN.

A possible confounding factor in voltammetric studies of the Ca^{2+}-dependence of somatodendritic DA release was that the pulse-train stimulation used to evoke DA release simultaneously elicits release of other local transmitters, including GABA and glutamate, that strongly modulate somatodendritic DA release in the SNc.[65] Furthermore, comparison with striatal DA release is complicated by the strong dependence of striatal release on endogenous ACh.[96] This raises the concern that the true Ca^{2+}-dependence of somatodendritic and striatal DA release might be masked by the Ca^{2+}-dependence of GABA, glutamate, and ACh release in each region that could subsequently differently govern evoked $[DA]_0$. This concern has been addressed to some extent in SNc by studies that reexamined the Ca^{2+}-dependence of DA release using single-pulse stimulation, which elicits DA release independently of ionotropic glutamate or GABA receptor modulation.[66,97] Importantly, DA release varied with $[Ca^{2+}]_0$ in a manner similar to that observed with pulse trains[97] indicating that the Ca^{2+}-dependence reported previously was intrinsic rather than extrinsic. Full comparison of the Ca^{2+}-dependence, as well as voltage-gated Ca^{2+} channel control of somatodendritic compared to striatal DA release during conditions similarly independent of afferent neurotransmitter input (e.g., during ACh receptor block), has not yet been assessed and requires further delineation.

REGULATION OF SOMATODENDRITIC DOPAMINE SIGNALS BY EXTRACELLULAR GEOMETRY AND UPTAKE

VOLUME TRANSMISSION AND UPTAKE

Since DA transmission is mediated primarily by extrasynaptic receptors[28–30] and thus must act via volume transmission,[28,31] a full understanding of DA transmission by somatodendritic DA release requires an understanding of local extracellular diffusion characteristics. Key features that define extracellular concentrations of DA and its extracellular diffusion are the extracellular volume fraction, extracellular tortuosity, and plasmamembrane re-uptake.[30,98] Together, these factors define the sphere of influence of a released neurotransmitter.

We have defined these variables using FCV as well as using ion-selective microelectrode recordings. Strikingly, the extracellular volume fraction (α) available for DA diffusion in the SN and VTA is 0.30,[58] compared to values of ~0.20 that are typical for forebrain structures, including striatum.[99] This means that the extracellular concentration of DA ($[DA]_0$) after release of a given number of molecules will be >30% lower in the SN/VTA than in striatum in the absence of other regulatory mechanisms. This has obvious implications for concentration-dependent receptor activation in SN/VTA vs. striatum as well as for experimental observations of $[DA]_0$ in these regions. The tortuosity factor, λ, which governs the apparent diffusion coefficient of a diffusing substance, is similar between midbrain and striatum.[58]

Electrophysiological studies indicate a physiological role for uptake in the modulation of somatodendritic DA signaling.[100,101] In keeping with this and with the high density of DAT-expressing cells in these areas,[27,102–104] in vitro experiments with FCV (and high-speed chronoamperometry) indicate that there is uptake of DA via the DAT after evoked release or application of exogenous DA in SNc or VTA.[44,58,84] Nonetheless, the DAT is not the only transporter that can transport somatodendritic DA. Uptake of DA by the NE transporter (NET), albeit modest, is more prominent in VTA than SNc[44,58] and appears to be mediated by a few sparsely packed, en passant rostral norepinephrinergic processes.[44] Uptake of DA in SNr is much less avid than in SNc or VTA enabling DA to diffuse over larger distances without encountering uptake sites.[58]

An important point to be made about uptake is in the context of DA as a volume transmitter. Dopamine transporters (DATs) are often considered to participate in this process by "gating" spillover from release sites. Indeed, although there is competition between DATs and diffusion in sculpting extracellular DA transients after release,[30] it should be noted that a role for the DAT or any uptake process is not at odds with the notion that a neurotransmitter acts by volume transmission, in contrast to the recent publication of this notion.[85] DATs are expressed only by dopaminergic neurons, at perisynaptic and non-synaptic sites on DA dendrites and axons.[27,105] Neither striatal nor somatodendritic DATs are empowered anatomically or kinetically to prevent initial spillover from a release site.[30] This is in contrast to the situation seen for glutamate at glutamate synapses that are enriched in synaptic and extrasynaptic glutamate uptake transporters (GluTs) on multiple presynaptic and postsynaptic cells, including astrocytes, and not only occur at densities at least two orders of magnitude greater than for DATs surrounding DA release sites but, furthermore, operate with transport numbers one order of magnitude greater than for DA at DATs.[30] Thus, GluTs are better positioned than DATs, anatomically, numerically, and kinetically, to limit spillover and active lifetime of transmitter.

Rather, DATs particularly govern the sphere of influence, lifetime, and, thus, net extracellular concentration of DA beyond a release site.[30] This can readily be demonstrated both experimentally and theoretically even for axonally released DA where DAT density is at its highest. There is abundant experimental evidence from voltammetric studies at extrasynaptic CFMs that DA signals readily appear in the extracellular space following evoked action potentials[106] and during motivational and reward-associated behaviors.[107–109] Furthermore, the extrasynaptic sphere of influence of DA can be modeled using a modified version of the equation for three-dimensional diffusion in brain tissue;[30,58,110] this model has been used to predict the sphere of influence of DA at its receptors in the striatum where DA uptake kinetics are at least one order of magnitude greater than in midbrain.[30,58] Such modeling readily predicts that in striatum, despite avid DA uptake, e.g., V_{max} of 4–5 μM/s defined from studies with FCV,[111] the release of a single vesicle of 2000–14000 DA molecules would reach an estimated 20–100 synapses, or a volume approaching 2000 μm^3, at concentrations greater than the EC_{50} value for D_2 receptors.[30] These spheres of influence and synapse numbers would be up to twice these values with lower rates of DA uptake.[30] Thus, volume transmission readily occurs for DA whether released from somatodendrites or striatal synapses.

Regional Kinetic Distinctions

Within the midbrain, there is a regional variation in the regulation of $[DA]_0$ by uptake. The role of the DAT on $[DA]_0$ after somatodendritic release may be more marked in SNc than VTA,[44] but see [58] reflecting differential DAT expression in these regions. Ventral tier DA neurons (found in the SNc only) have greater mRNA and protein levels of DA transporter (and D_2 DA receptor) than dorsal tier DA cells (dorsal SNc, VTA).[102,103,112–114] In particular, in FCV studies of evoked somatodendritic DA release, a role for DA uptake was less apparent in VTA than SNc.[44] Subsequently, we used FCV in conjunction with exogenously applied DA to eliminate the contribution of local depolarization as a possible hindrance to DA uptake by the electrogenic DA transporter; in turn, uptake of somatodendritic DA was also identified in VTA.[58] It is relevant to note that ventral tier SNc cells are more susceptible to degeneration in Parkinson's disease than dorsal tier cells in either VTA or SNc.[115–117] This pattern of susceptibility is paralleled in the vulnerability to the DA uptake substrate and toxin, 1-methyl-4-phenyl-1,2,3,6-tetrahydropyridine (MPTP),[118] which can be prevented by DA uptake inhibition. Together, such findings have implicated DA uptake activity or other regionally specific DA handling mechanisms as possible risk factors in Parkinsonian degeneration.[119–121] DA uptake within the midbrain, therefore, is crucial not only for normal DA neuron physiology but may also contribute to the differential vulnerability of DA cells to pathophysiology.

Although uptake via the DAT is an important mechanism of regulating $[DA]_0$ in midbrain,[44,58] it plays an apparently lesser role on the regulation of $[DA]_0$ in the midbrain than in striatum,[44] as shown empirically by a greater increase in $[DA]_0$ during local electrical stimulation in striatum than in SNc when the DAT is blocked. Approximations to linear uptake constants for each region have been estimated at $\sim 0.1\ s^{-1}$ in SNc[58] and $\sim 20\ s^{-1}$ in striatum.[30] This difference is in keeping with the lower density of DAT protein in somatodendritic than axon terminal regions.[102,103,122] It can be speculated that a consequence of this apparently less avid regulation of dendritic compared to axon terminal $[DA]_0$ will be a greater sphere of influence of dendritic than axonal DA.[30,58]

NEUROMODULATION OF SOMATODENDRITIC RELEASE

AUTORECEPTOR REGULATION

Inhibition of DA release from striatal axon terminals via presynaptic D_2-like autoreceptors plays a powerful role in the regulation of striatal $[DA]_0$.[9,123–125] D_2 receptors are also expressed and located directly on DA dendrites in VTA and SN.[12,26,113,126] However, particular care with experimental protocol is needed when using FCV to evaluate whether evoked neurotransmitter release is controlled by neurotransmitter receptors. Without a large background or basal level of $[DA]_0$, the control of DA release by a tonic DA activation of D_2 receptors is unlikely to be detected. The detection of autoreceptor activation by discretely released endogenous neurotransmitter requires that concentrations of release as well as durations of pulse trains are appropriate. D_2-receptor tone may need to be generated by stimulated release (e.g., in slices where spontaneous release is minimal). Furthermore, a time delay for activation may occur: short stimulus trains (<100 ms) are unable to reveal DA D_2-receptor control of DA release at either somatodendrites or axon terminals in striatum.[9,106,125,127] Rather, longer stimulus trains (>100 ms) are required to permit sufficient time for DA released early in the train to activate D_2 autoreceptors such that the effects of D_2 antagonists can be observed later in the train.[9] With due consideration to this issue, D_2-receptor control of somatodendritic DA release has been documented in SNc indicating that D_2 receptors can operate as an autoinhibitory mechanism to regulate somatodendritic release.[9] The modulation of $[DA]_0$ by D_2 receptors is less marked in VTA than in SNc, consistent with the higher expression and protein levels of the D_2 receptor in ventral tier (SNc only) DA neurons.[113] As for differential DAT function discussed above, regional variation in the management of somatodendritic DA transmission in VTA and SNc by autoreceptor regulation might also contribute to differential susceptibility of these cell groups to degeneration in Parkinson's disease.

Notably, the degree of control of somatodendritic $[DA]_0$ by D_2 autoreceptors is less avid than for striatal DA. Less effective autoinhibition of release, in conjunction with less avid DA reuptake via the DAT[44,128] and other Ca^{2+}-dependent processes, could result in altogether different regulation of $[DA]_0$ in somatodendritic and axon terminal regions.

Heteroreceptor and Other Neuromodulatory Regulation

DA neurons receive significant non-DA synaptic input and concomitantly express somatodendritic receptors for corresponding major classes of neurotransmitters, including GABA, glutamate, 5-HT, and acetylcholine (ACh), as well as co-transmitters like dynorphin and ATP. The role of each of these inputs in the direct heteroreceptor regulation of DA release within the SN and VTA is increasingly being revealed.

Strong inhibitory control of somatodendritic DA release by GABA has been identified in experiments with FCV, especially in the SNc where both $GABA_A$ and $GABA_B$ receptors control release.[65] This control is particularly evident during somatodendritic release elicited by pulse trains (10 Hz for three seconds), rather than single pulses.[97] This difference, therefore, presumably reflects insufficient spontaneous GABA release in coronal midbrain slices to modify DA release as measured at an isolated stimulus but that GABA released by an electrical stimulus can subsequently modulate DA release by ongoing depolarization, i.e., in a stimulus train. This issue readily highlights an important methodological strength of FCV. FCV can be used to dissociate intrinsic vs. extrinsic (input) control of neurotransmitter release; its sophisticated time resolution permits resolution of release by instantaneous (single pulse) as well as longer events (bursts of action potentials). In turn, a related point must be made that the effects of the local stimulation on the release of other non-electroactive neurotransmitters must always be considered in evaluating the regulation of the neurotransmitter of study (here, somatodendritic DA). If the focus of experimental investigation is the regulation of somatodendritic DA release by mechanisms that are intrinsic to DA neurons and not those that concomitantly regulate the release of other local neurotransmitters then either appropriate short stimulus trains and/or appropriate synaptic blockers should be incorporated into the experimental design.

The role of GABA in the VTA is less clear. Although measurements in vivo using microdialysis show that $GABA_B$ receptors tonically inhibit DA release in VTA,[129] in vivo data can be complicated by concomitant activation of long-loop circuits. Local measurements in slices with FCV suggest that $GABA_A$ and $GABA_B$ receptor control of release in VTA is limited and may depend on the strength and timing of other synaptic input.[65]

Glutamate receptor regulation of somatodendritic DA release has been assessed with several methods and can take the form of direct excitation by ionotropic receptors,[65,130–133] particularly in the VTA,[65] or heterosynaptic inhibition via AMPA-receptor activation on GABA afferents in the SNc.[65] As discussed previously for GABA, evaluation of glutamate receptor control with FCV requires careful consideration of experimental protocol to ensure local glutamate synapses are activated by the stimulation protocol.[97] Possible differences between SNc and VTA are consistent with the dominance of glutamate input to VTA DA neurons (70% of axodendritic synapses) and the dominance of GABA input to SNc DA cells (70% of axodendritic synapses).[134] Whether local glutamate input enhances or inhibits somatodendritic DA release (via GABA afferents) therefore depends on the balance between excitatory and inhibitory inputs at a given moment[65] and, as such, can vary according to brain region, the pattern of local input circuitry, and experimental conditions.

In keeping with the expression of nicotinic ACh receptors by DA neurons (primarily $\alpha 4\beta 2$-subunit-containing),[135,136] nicotine/ACh facilitatory control of somatodendritic DA release has been identified in VTA using microdialysis following systemic nicotine administration[137] and more directly in a dendrosomal preparation from SN/VTA using a $[^3H]$DA release assay.[138] Although the effect of nicotinic ACh receptor activation on DA neuron activity is typically excitatory, inhibitory effects can also occur by subsequent activation of Ca^{2+}-activated K^+ channels,[139] and a complex multisynaptic activation can occur via nAChR desensitization on synaptic inputs.[140,141] However, measurements with FCV have not yet been exploited to dissect the roles of direct and indirect actions of ACh on any control of endogenous somatodendritic DA release.

Somatodendritic DA release may also be regulated by other classes of molecules, including reactive oxygen species; FCV has been used to reveal the real-time control of somatodendritic DA by endogenously generated hydrogen peroxide (H_2O_2). H_2O_2 can suppress DA release in the SNc, although not in the VTA,[64] on a time course of a few seconds. This feature is not unique to dendrites; striatal synaptic DA release is similarly governed by the generation of endogenous peroxide.[62,142] Whether other inputs or neuromodulatory molecules regulate somatodendritic DA release, e.g., as suggested by microdialysis studies for ATP,[143] has yet to be explored directly with the superior spatial and temporal resolution offered by FCV at CFMs.

Thus, in addition to the role that major synaptic inputs play in the regulation of DA neuron activity, these neuromodulatory inputs can also powerfully and variously gate the release of somatodendritic DA. As a consequence, these inputs will govern the action of somatodendritic DA release at D_1 and D_2 receptors in VTA and SN on the final signal integration and the ultimate activity of all dopaminoceptive output neurons of the VTA, SNc, and SNr.

SUMMARY AND CONCLUSIONS

FCV at CFMs offers a powerful approach for the study of endogenous somatodendritic DA release in real-time. With attention to appropriate scan and stimulation waveforms, pharmacological characterizations, and choice of species, DA can readily be detected with sub-second resolution and also discerned from possible contaminants and artifacts (including 5-HT, NE, DOPAC, pH, Ca^{2+}, O_2) that other electrochemical measurements with greater temporal (and possible spatial) resolution cannot, e.g., constant potential amperometry. Despite the greater possible resolutions of amperometry, FCV is actually better suited to electrochemical detection of somatodendritic DA release in the slice environment due to the likelihood of interference with DA signals by artifacts and contaminants that amperometry would not readily discern from DA.

Evidence from FCV has not contradicted the original proposal by Geffen et al.[2] that DA release in the SN is mediated by vesicular exocytosis as it is in axon terminals in striatum. In fact, somatodendritic DA release, like synaptic release in striatum, is regulated by depolarization, Ca^{2+}, D_2-like autoreceptors, and DA uptake via the DAT (and occasionally also the NET) as well as local neurotransmitters and other neuromodulators. Nonetheless, the extent or manner to which these features regulate somatodendritic release can differ from the control of synaptic DA release in striatum; however, whether these differences are merely quantitative or more qualitative remains to be clarified. Locally within the midbrain, the good spatial resolution of FCV at a CFM has also permitted comparison of the regulation of somatodendritic DA release in SNc with the adjacent cell body region, the VTA. Several discrete differences have already been noted between these two key DA cell groups (e.g., control by afferent input, degree of regulation by D_2 receptors, uptake transporters), reinforcing the principle that these two regions and projections differ functionally and should always be considered separately. Moreover, the many functional differences in DA signals between these regions may yet have ramifications for the differential vulnerability of DA cells to pathophysiology, e.g., in Parkinson's disease, where regionally specific DA handling mechanisms may pose differential risks to neurodegeneration.

Several key questions regarding the mechanisms that regulate somatodendritic DA remain to be resolved. Fundamental mechanisms underlying release continue to be questioned. Not only is it unclear what might constitute, ultrastructurally, the presynaptic active zone or other site for release or which "presynaptic" proteins are localized to dendritic plasmamembranes but also the possible conditions under which alternative mechanisms proposed for release to Ca^{2+}-dependent exocytosis, i.e., DAT reversal, could operate. It is also still unknown how somatodendritic DA is released in response to the tonic vs. phasic firing patterns and frequencies of activity of DA neurons and/or whether dendritic release of DA is under discrete control by the local excitability of remote dendrites or whether sub-second processes akin to short-term synaptic dynamics seen at DA synapses[144] operate in dendrites.

Somatodendritic DA release in the SN and VTA serves a number of critical functions in the regulation of animal behavior, from motivation to motion. Released DA in midbrain acts at predominantly extrasynaptic DA receptors to mediate autoreceptor control of DA cell activity and heteroreceptor regulation of synaptic input that is also reciprocal. In the thirty years since somatodendritic DA release was proposed, much has been learned about its role and how it is regulated, as summarized in this chapter. Excitingly, there are many remaining questions to address.

ACKNOWLEDGMENTS

We acknowledge support from Novartis Pharma AG (ST) and the Paton Fellowship, University of Oxford (SJC).

REFERENCES

1. Bjorklund, A. and Lindvall, O., Dopamine release in dendrites of substantia nigra neurons: Suggestions for a role in dendritic terminals, *Brain Res.*, 83, 531, 1975.
2. Geffen, L. B., Jessell, T. M., Cuello, A. C. et al., Release of DA from dendrites in rat SN, *Nature*, 260, 258, 1976.
3. Groves, P. M., Wilson, C. J., Young, S. J. et al., Self-inhibition by dopaminergic neurons, *Science*, 190, 522, 1975.
4. Nieoullon, A., Cheramy, A., and Glowinski, J., Release of DA in vivo from cat SN, *Nature*, 266, 375, 1975.
5. Jaffe, E. H., Marty, A., Schulte, A. et al., Extrasynaptic vesicular transmitter release from the somata of substantia nigra neurons in rat midbrain slices, *J. Neurosci.*, 18, 3548, 1998.
6. Rice, M. E., Richards, C. D., Nedergaard, S. et al., Direct monitoring of dopamine and 5-HT release in substantia nigra and ventral tegmental area in vitro, *Exp. Brain Res.*, 100, 395, 1994.
7. Ludwig, M. and Pittman, Q. J., Talking back: Dendritic neurotransmitter release, *Trends Neurosci.*, 26, 255, 2003.
8. Ludwig, M., *Dendritic Neurotransmitter Release*, Springer Science–Business Media, Inc., New York, 2005.
9. Cragg, S. J. and Greenfield, S. A., Differential autoreceptor control of somatodendritic and axon terminal dopamine release in substantia nigra, ventral tegmental area, and striatum, *J. Neurosci.*, 17, 5738, 1997.
10. Lacey, M. G., Mercuri, N. B., and North, R. A., On the potassium conductance increase activated by GABAB and dopamine D2 receptors in rat substantia nigra neurones, *J. Physiol.*, 401, 437, 1992.
11. Chiodo, L. A., Dopamine autoreceptor signal transduction in the DA cell body: A "current view", *Neurochem. Int.*, 20, 81S, 1992.
12. Yung, K. K. L., Bolam, J. P., Smith, A. D. et al., Immunocytochemical localization of D1 and D2 dopamine receptors in the basal ganglia of the rat: Light and electron microscopy, *Neuroscience*, 65, 709, 1995.
13. Falkenburger, B. H., Barstow, K. L., and Mintz, I. M., Dendrodendritic inhibition through reversal of dopamine transport, *Science*, 293, 2465, 2001.
14. Santiago, M. and Westerink, B. H., Characterization and pharmacological responsiveness of dopamine release recorded by microdialysis in the substantia nigra of conscious rats, *J. Neurochem.*, 57, 738, 1991.
15. Kalivas, P. W. and Duffy, P., A comparison of axonal and somatodendritic DA release using in vivo dialysis, *J. Neurochem.*, 56, 961, 1991.
16. Miyazaki, T. and Lacey, M. G., Presynaptic inhibition by dopamine of a discrete component of GABA release in rat substantia nigra pars reticulata, *J. Physiol.*, 513 (Pt 3), 805, 1998.

17. Radnikow, G. and Misgeld, U., Dopamine D1 receptors facilitate GABAA synaptic currents in the rat substantia nigra pars reticulata, *J. Neurosci.*, 18, 2009, 1998.
18. Koga, E. and Momiyama, T., Presynaptic dopamine D2-like receptors inhibit excitatory transmission onto rat ventral tegmental dopaminergic neurones, *J. Physiol.*, 523 (Pt 1), 163, 2000.
19. Wichmann, T. and DeLong, M. R., Functional and pathophysiological models of the basal ganglia, *Curr. Opin. Neurobiol.*, 6, 751, 1996.
20. Robertson, G. S. and Robertson, H. A., Evidence that L-dopa-induced rotational behavior is dependent on both striatal and nigral mechanisms, *J. Neurosci.*, 9, 3326, 1989.
21. Timmerman, W. and Abercrombie, E. D., Amphetamine-induced release of dendritic dopamine in substantia nigra pars reticulata: D1-mediated behavioral and electrophysiological effects, *Synapse*, 23, 280, 1996.
22. Crocker, A. D., The regulation of motor control: an evaluation of the role of dopamine receptors in the substantia nigra, *Rev. Neurosci.*, 8, 55, 1997.
23. Bergquist, F., Shahabi, H. N., and Nissbrandt, H., Somatodendritic dopamine release in rat substantia nigra influences motor performance on the accelerating rod, *Brain Res.*, 973, 81, 2003.
24. Groves, P. M. and Linder, J. C., Dendro-dendritic synapses in substantia nigra: Descriptions based on analysis of serial sections, *Exp. Brain Res.*, 49, 209, 1983.
25. Wilson, C. J., Groves, P. M., and Fifkova, E., Monoaminergic synapses, including dendro-dendritic synapses in the rat substantia nigra, *Exp. Brain Res.*, 30, 161, 1977.
26. Sesack, S. R., Aoki, C., and Pickel, V. M., Ultrastructural localization of D2-receptor-like immunoreactivity in midbrain dopamine neurons and their striatal targets, *J. Neurosci.*, 14, 88, 1994.
27. Nirenberg, M. J., Vaughan, R. A., Uhl, G. R. et al., The dopamine transporter is localized to dendritic and axonal plasma membranes of nigrostriatal dopaminergic neurons, *J. Neurosci.*, 16, 436, 1996.
28. Rice, M. E., Distinct regional differences in dopamine-mediated volume transmission, In *Volume Transmission Revisited*, Agnati, L. F., Fuxe, K., Nicholson, C., and Sykova, E., Eds., Elsevier Science, Amsterdam, The Netherland, p. 275, 2000.
29. Pickel, V. M., Extrasynaptic distribution of monoamine transporters and receptors, *Prog. Brain Res.*, 125, 267, 2000.
30. Cragg, S. J. and Rice, M. E., Dancing past the DAT at a DA synapse, *Trends Neurosci.*, 27, 270, 2004.
31. Fuxe, K. and Agnati, L. F., *Volume Transmission in the Brain*, Raven Press, New York, 1991.
32. Tagerud, S. E. O. and Cuello, A. C., DA release from the rat SN in vitro. Effect of Raphe lesions and veratridine stimulation, *Neuroscience*, 4, 2021, 1979.
33. Cheramy, A., Leviel, V., and Glowinski, J., Dendritic release of dopamine in the substantia nigra, *Nature*, 289, 537, 1981.
34. Elverfors, A., Jonason, J., Jonason, G. et al., Effects of drugs interfering with sodium channels and calcium channels on the release of endogenous dopamine from superfused substantia nigra slices, *Synapse*, 26, 359, 1997.
35. Elverfors, A. and Nissbrandt, H., Reserpine-insensitive dopamine release in the substantia nigra? *Brain Res.*, 557, 5, 1991.
36. Robertson, G. S., Damsa, G., and Fibiger, H. C., Characterisation of DA release in the SN by in vivo microdialysis in freely moving rats, *J. Neurosci.*, 11, 2209, 1991.
37. Heeringa, M. J. and Abercrombie, E. D., Biochemistry of somatodendritic DA release in the substantia nigra: An in vivo comparison with striatal dopamine release, *J. Neurochem.*, 65, 192, 1995.
38. Gerhardt, G. A., Cass, W. A., and Huettl, P., GDNF improves dopamine function in the substantia nigra but not the putamen of unilateral MPTP-lesioned rhesus monkeys, *Brain Res.*, 817, 163, 1999.
39. Bergquist, F., Niazi, H. S., and Nissbrandt, H., Evidence for different exocytosis pathways in dendritic and terminal dopamine release in vivo, *Brain Res.*, 950, 245, 2002.
40. Bergquist, F. and Nissbrandt, H., Dopamine release in substantia nigra: Release mechanisms and physiological function in motor control, In *Dendritic Neurotransmitter Release*, Ludwig, M., Ed., Springer Science + Business Media, Inc., New York, p. 85, 2005.
41. Jones, S. R., Mickelson, G. E., Collins, L. B. et al., Interference by pH and Ca2+ ions during measurements of catecholamine release in slices of rat amygdala with fast-scan cyclic voltammetry, *J. Neurosci. Methods*, 52, 1, 1994.

42. Rice, M. E. and Nicholson, C., Diffusion and ion shifts in the brain extracellular microenvironment and their relevance for voltammetric measurements: The brain is not a beaker, In *Neuromethods, Voltammetric Methods in Brain Systems*, Boulton, A., Baker, G., and Adams, R. N., Eds., Vol. 27, Humana Press, Totowa, NJ, p. 27, 1995.
43. Rice, M. E., Cragg, S. J., and Greenfield, S. A., Characteristics of electrically evoked somatodendritic dopamine release in substantia nigra and ventral tegmental area in vitro, *J. Neurophysiol.*, 77, 853, 1997.
44. Cragg, S. J., Rice, M. E., and Greenfield, S. A., Heterogeneity of electrically evoked dopamine release and reuptake in substantia nigra, ventral tegmental area, and striatum, *J. Neurophysiol.*, 77, 863, 1997.
45. Cragg, S. J., Hawkey, C. R., and Greenfield, S. A., Comparison of serotonin and dopamine release in substantia nigra and ventral tegmental area: region and species differences, *J. Neurochem.*, 69, 2378, 1997.
46. Stamford, J. A., Palij, P., Davidson, C. et al., Simultaneous "real-time" electrochemical and electrophysiological recording in brain slices with a single carbon-fibre microelectrode, *J. Neurosci. Methods*, 50, 279, 1993.
47. Iravani, M. M. and Kruk, Z. L., Real-time measurement of stimulated 5-hydroxytryptamine release in rat substantia nigra pars reticulata brain slices, *Synapse*, 25, 93, 1997.
48. Bunin, M. A. and Wightman, R. M., Quantitative evaluation of 5-hydroxytryptamine (serotonin) neuronal release and uptake: An investigation of extrasynaptic transmission, *J. Neurosci.*, 18, 4854, 1998.
49. John, C. E., Budygin, E. A., Mateo, Y. et al., Voltammetric characterization of monoamines in mouse midbrain, In *Monitoring Molecules in Neuroscience 2003*, Kehr, J., Fuxe, K., Ungerstedt, U., Svensson, T. H. et al., Eds., Karolinska University Press, Stockholm, p. 490, 2003.
50. Steinbusch, H. W. M., Distribution of serotonin-immunoreactivity in the central nervous system of the rat- cell bodies and terminals, *Neuroscience*, 6, 557, 1981.
51. Nedergaard, S., Bolam, J. P., and Greenfield, S. A., Facilitation of a dendritic calcium conductance by 5-HT in the substantia nigra, *Nature*, 333, 174, 1988.
52. Lavoie, B. and Parent, A., Immunohistochemical study of the serotonergic innervation of the basal ganglia in the squirrel monkey, *J. Comp. Neurol.*, 299, 1, 1990.
53. Corvaja, N., Doucet, G., and Bolam, J. P., Ultrastructure and synaptic targets of the raphe-nigral projection in the rat, *Neuroscience*, 55, 417, 1993.
54. Moukhles, H., Bosler, O., Bolam, J. P. et al., Quantitative and morphometric data indicate precise cellular interactions between serotonin terminals and postsynaptic targets in rat substantia nigra, *Neuroscience*, 76, 1159, 1997.
55. Jackson, B. P., Dietz, S. M., and Wightman, R. M., Fast-scan cyclic voltammetry of 5-HT, *Anal. Chem.*, 67, 1115, 1995.
56. Threlfell, S., Cragg, S. J., Kallo, I. et al., Histamine H3 receptors inhibit serotonin release in substantia nigra pars reticulata, *J. Neurosci.*, 24, 8704, 2004.
57. Zhou, F. M., Liang, Y., Salas, R. et al., Corelease of dopamine and serotonin from striatal dopamine terminals, *Neuron*, 46, 65, 2005.
58. Cragg, S. J., Nicholson, C., Kume-Kick, J. et al., Dopamine-mediated volume transmission in midbrain is regulated by distinct extracellular geometry and uptake, *J. Neurophysiol.*, 85, 1761, 2001.
59. Waeber, C., Sebben, M., Nieoullon, A. et al., Regional distribution and ontogeny of 5-HT4 binding sites in rodent brain, *Neuropharmacology*, 33, 527, 1994.
60. Smits, R. P. J., Steinbusch, H. W. M., and Mulder, A. H., Distribution of DA-immunoreactive cell bodies in the guinea-pig brain, *J. Chem. Neuroanat.*, 3, 101, 1990.
61. Paxinos, G. and Watson, C., *The Rat Brain in Stereotaxic Coordinates*, Academic Press, London, 1982.
62. Chen, B. T. and Rice, M. E., Novel Ca^{2+} dependence and time course of somatodendritic dopamine release: Substantia nigra versus striatum, *J. Neurosci.*, 21, 7841, 2001.
63. Millar, J., Connor, J. J., Trout, S. et al., CSCV: A new high-speed electrochemical method for monitoring neuronal dopamine release, *J. Neurosci. Methods*, 43, 109, 1992.

64. Chen, B. T., Avshalumov, M. V., and Rice, M. E., Modulation of somatodendritic dopamine release by endogenous H(2)O(2): Susceptibility in substantia nigra but resistance in VTA, *J. Neurophysiol.*, 87, 1155, 2002.
65. Chen, B. T. and Rice, M. E., Synaptic regulation of somatodendritic dopamine release by glutamate and GABA differs between substantia nigra and ventral tegmental area, *J. Neurochem.*, 81, 158, 2002.
66. Chen, B. T., Moran, K. A., Avshalumov, M. V., et al., Limited regulation of somatodendritic dopamine release by voltage-sensitive Ca^{2+} channels contrasted with strong regulation of axonal dopamine release, *J. Neurochem.*, 96, 645, 2006.
67. Cragg, S. J. and Greenfield, S. A., Somatodendritic dopamine release is not modulated by a dopamine D_3 receptor agonist (PD 128,907) in either substantia nigra or ventral tegmental area, In *Soc. Neurosci. Abstr.*, Vol. 25., Washington, DC, p. 1214, 1999.
68. Venton, B. J., Michael, D. J., and Wightman, R. M., Correlation of local changes in extracellular oxygen and pH that accompany dopaminergic terminal activity in the rat caudate-putamen, *J. Neurochem.*, 84, 373, 2003.
69. Rice, M. E. and Cammack, J., Anoxia-resistant turtle brain maintains ascorbic acid content in vitro, *Neurosci. Lett.*, 132, 141, 1991.
70. Stamford, J. A., Kruk, Z. L., and Millar, J., Regional differences in extracellular ascorbic acid levels in the rat brain determined by high speed cyclic voltammetry, *Brain Res.*, 299, 289, 1984.
71. Kelly, E., Jenner, P., and Marsden, C. D., Evidence that [3H]-dopamine is taken up and released from non-dopaminergic nerve terminals in the rat substantia nigra in vitro, *J. Neurochem.*, 45, 137, 1985.
72. Carboni, E., Tanda, G. L., Frau, R. et al., Blockade of the noradrenaline carrier increases extracellular dopmaine concentrations in the prefrontal cortex: Evidence that dopamine is taken up in vivo by noradrenergic terminals, *J. Neurochem.*, 55, 1060, 1990.
73. Pozzi, L., Invernizzi, R., Cervo, L. et al., Evidence that extracellular concentrations of dopamine are regulated by noradrenergic neurons in the frontal cortex of rats, *J. Neurochem.*, 63, 195, 1994.
74. Nirenberg, M. J., Chan, J., Liu, Y. et al., Vesicular monoamine transporter-2: Immunogold localization in striatal axons and terminals, *Synapse*, 26, 194, 1997.
75. Mercer, L., del Fiacco, M., and Cuello, A. C., The smooth endoplasmic reticulum as a possible storage site for dendritic dopamine in substantia nigra neurones, *Experientia*, 35, 103, 1979.
76. Wassef, M., Berod, A., and Sotelo, C., Dopaminergic dendrites in the pars reticulata of the rat substantia nigra and their striatal input. Combined immunocytochemical localization of tyrosine hydroxylase and anterograde degeneration, *Neuroscience*, 6, 2125, 1981.
77. Nirenberg, M. J., Chan, J., Liu, Y. et al., Ultrastructural localization of the vesicular monoamine transporter-2 in midbrain dopaminergic neurons: Potential sites for somatodendritic storage and release of dopamine, *J. Neurosci.*, 16, 4135, 1996.
78. Leviel, V., The reverse transport of DA, what physiological significance? *Neurochem. Int.*, 38, 83, 2001.
79. Sulzer, D., Chen, T. K., Lau, Y. Y. et al., Amphetamine redistributes dopamine from synaptic vesicles to the cytosol and promotes reverse transport, *J. Neurosci.*, 15, 4102, 1995.
80. Jones, S. R., Gainetdinov, R. R., Wightman, R. M. et al., Mechanisms of amphetamine action revealed in mice lacking the dopamine transporter, *J. Neurosci.*, 18, 1979, 1998.
81. Torres, G. E., Gainetdinov, R. R., and Caron, M. G., Plasma membrane monoamine transporters: Structure, regulation and function, *Nat. Rev. Neurosci.*, 4, 13, 2003.
82. Elverfors, A. and Nissbrandt, H., Effects of d-amphetamine on dopaminergic neurotransmission: a comparison between the substantia nigra and the striatum, *Neuropharmacology*, 31, 661, 1992.
83. Engberg, G., Elverfors, A., Jonason, J. et al., Inhibition of dopamine re-uptake: Significance for nigral dopamine neuron activity, *Synapse*, 25, 215, 1997.
84. Hoffman, A. F., Lupica, C. R., and Gerhardt, G. A., Dopamine transporter activity in the substantia nigra and striatum assessed by high-speed chronoamperometric recordings in brain slices, *J. Pharm. Exp. Ther.*, 287, 487, 1998.
85. Beckstead, M. J., Grandy, D. K., Wickman, K. et al., Vesicular dopamine release elicits an inhibitory postsynaptic current in midbrain dopamine neurons, *Neuron*, 42, 939, 2004.

86. Hoffman, A. F. and Gerhardt, G. A., Differences in pharmacological properties of dopamine release between the substantia nigra and striatum: An in vivo electrochemical study, *J. Pharmacol. Exp. Ther.*, 289, 455, 1999.
87. Iravani, M. M., Muscat, R., and Kruk, Z. L., Comparison of somatodendritic and axon terminal dopamine release in the ventral tegmental area and the nucleus accumbens, *Neuroscience*, 4, 1025, 1996.
88. Stuart, G., Spruston, N., Sakmann, B. et al., Action potential initiation and backpropagation in neurons of the mammalian CNS, *Trends Neurosci.*, 20, 125, 1998.
89. Grace, A. A. and Bunney, B. S., The control of firing pattern in nigral dopamine neurons: Burst firing *J. Neurosci.*, 4, 2877, 1984.
90. Grace, A. A. and Bunney, B. S., The control of firing pattern in nigral dopamine neurons: Single spike firing, *J. Neurosci.*, 4, 2866, 1984.
91. Schultz, W., Getting formal with dopamine and reward, *Neuron*, 36, 241, 2002.
92. Hyland, B. I., Reynolds, J. N. J., Hay, J. et al., Firing modes of midbrain dopamine cells in the freely moving rat, *Neuroscience*, 114, 475, 2002.
93. Douglas, W. W. and Rubin, R. P., The mechanism of catecholamine release from the adrenal medulla and the role of calcium in stimulation-secretion coupling, *J. Physiol.*, 167, 288, 1963.
94. Sudhof, T. C. and Rizo, J., Synaptotagmins: C2-domain proteins that regulate membrane traffic, *Neuron*, 17, 379, 1996.
95. Sudhof, T. C., Synaptotagmins: Why so many? *J. Biol. Chem.*, 277, 7629, 2002.
96. Rice, M. E. and Cragg, S. J., Nicotine amplifies reward-related dopamine signals in striatum, *Nat. Neurosci.*, 7, 583, 2004.
97. Moran, K. A., Chen, B. T., Tao, L., et al., Calcium dependence of dopamine release in the VTA reveals both somatodendritic and synaptic characteristics. Program number 461.16, In *2003 Abstract Viewer and Itinerary Planner*, Society for Neuroscience, Washington, DC, Online, 2003. http://sfn.scholarone.com/itin2003/.
98. Nicholson, C. and Sykova, E., Extracellular space structure revealed by diffusion analysis, *Trends Neurosci.*, 21, 207, 1998.
99. Rice, M. E. and Nicholson, C., Diffusion characteristics and extracellular volume fraction during normoxia and hypoxia in slices of rat neostriatum, *J. Neurophysiol.*, 65, 264, 1991.
100. Lacey, M. G., Mercuri, N. B., and North, R. A., Actions of cocaine on rat dopaminergic neurones in vitro, *Br. J. Pharmacol.*, 99, 731, 1990.
101. Mercuri, N. B., Calabresi, P., Bernardi, G., Dopamine uptake inhibition potentiates the effects of L-DOPA on rat substantia nigra zona compacta neurons, *Neurosci. Lett.*, 126, 79, 1991.
102. Ciliax, B. J., Heilman, C., Demchyshyn, L. L. et al., The dopamine transporter: Immunocytochemical characterization and localization in brain, *J. Neurosci.*, 15, 1714, 1995.
103. Freed, C., Revay, R., Vaughan, R. A. et al., Dopamine transporter immunoreactivity in rat brain, *J. Comp. Neurol.*, 359, 340, 1995.
104. Nirenberg, M. J., Chan, J., Pohorille, A. et al., The dopamine transporter: Comparative ultrastructure of dopaminergic axons in limbic and motor compartments of the nucleus accumbens, *J. Neurosci.*, 17, 6899, 1997.
105. Hersch, S. M., Yi, H., Heilman, C. J. et al., Subcellular localization and molecular topology of the dopamine transporter in the striatum and substantia nigra, *J. Comp. Neurol.*, 388, 211, 1997.
106. Schmitz, Y., Benoit-Marand, M., Gonon, F. et al., Presynaptic regulation of dopaminergic neurotransmission, *J. Neurochem.*, 87, 273, 2003.
107. Rebec, G. V., Christensen, J. R., Guerra, C. et al., Regional and temporal differences in real-time dopamine efflux in the nucleus accumbens during free-choice novelty, *Brain Res.*, 776, 61, 1997.
108. Garris, P. A., Kilpatrick, M., Bunin, M. A. et al., Dissociation of dopamine release in the nucleus accumbens from intracranial self-stimulation, *Nature*, 398, 67, 1999.
109. Phillips, P. E., Stuber, G. D., Heien, M. L., et al., Subsecond dopamine release promotes cocaine seeking, *Nature*, 422, 614, 2003.
110. Nicholson, C., Diffusion from an injected volume of a substance in brain tissue with arbitrary volume fraction and tortuosity, *Brain Res.*, 333, 325, 1985.
111. Cragg, S. J., Hille, C. J., and Greenfield, S. A., Dopamine release and uptake dynamics within nonhuman primate striatum in vitro, *J. Neurosci.*, 20, 8209, 2000.

112. Blanchard, V., Raisman-Vozari, R., Vyas, S. et al., Differential expression of tyrosine hydroxylase and membrane dopamine transporter genes in subpopulations of dopaminergic neurons of the rat mesencephalon, *Mol. Brain Res.*, 22, 29, 1994.
113. Hurd, Y. L., Pristupa, Z. B., Herman, M. M. et al., The DA transporter and DA D2 receptor mRNAs are differentially expressed in limbic- and motor-related subpopulations of human mesencephalic neurons, *Neuroscience*, 63, 357, 1994.
114. Sanghera, M. K., Manaye, K. F., Liang, C. L. et al., Low dopamine transporter mRNA levels in midbrain regions containing calbindin, *Neuroreport*, 5, 1641, 1994.
115. Yamada, T., McGeer, P. L., Baimbridge, K. G. et al., Relative sparing in Parkinson's disease of substantia nigra dopamine neurons containing calbindin-D28k, *Brain Res.*, 526, 303, 1990.
116. Fearnley, J. M. and Lees, A. J., Ageing and Parkinson's disease: Substantia nigra regional selectivity, *Brain*, 114, 2283, 1991.
117. Gibb, W. R G. and Lees, A. J., Anatomy, pigmentation, ventral and dorsal subpopulations of the substantia nigra, and differential cell death in Parkinson's disease, *J. Neurol. Neurosurg. Psych.*, 54, 388, 1991.
118. German, D. C., Dubach, M., Askari, S. et al., MPTP-induced Parkisonian syndrome in Macaca fascularis: Which midbrain dopaminergic neurons are lost? *Neuroscience*, 24, 161, 1988.
119. Javitch, J. A., D'Amato, R. J., Strittmatter, S. M. et al., Parkisonism-inducing neurotoxin, MPTP: Uptake of the metabolite *N*-methly-4-phenly pyridine by dopamine neurons explains selective toxicity, *Proc. Natl Acad. Sci. U.S.A.*, 82, 2173, 1985.
120. Pifl, C., Giros, B., and Caron, M. G., Dopamine transporter expression confers cytotoxicity to low does of the parkinsonism-inducing neurotoxin MPTP, *J. Neurosci.*, 13, 4246, 1993.
121. Sundstrom, E., Goldstein, M., and Jonsson, G., Uptake inhibition protects nigro-striatal dopamine neurons from the neurotoxicity of 1-methyl-4-phenylpyridine (MPP+) in mice, *Eur. J. Pharmacol.*, 131, 289, 1986.
122. Donnan, G. A., Kaczmarczyk, S. J., Paxinos, G. et al., Distribution of catecholamine uptake sites in human brain as determined by quantitative [3H] mazindol autoradiography, *J. Comp. Neurol.*, 304, 419, 1991.
123. Limberger, N., Trout, S. J., Kruk, Z. L. et al., "Real-time" measurement of endogenous DA release during short trains of pulses in slices of rat neostriatum and nucleus accumbens: Role of autoinhibition, *Naunyn Schmiedebergs Arch. Pharmacol.*, 344, 623, 1991.
124. Trout, S. J. and Kruk, Z. L., Differences in evoked dopamine efflux in rat caudate putamen, nucleus accumbens and tuberculum olfactorium in the absence of uptake inhibition: Influence of autoreceptors, *Br. J. Pharmacol.*, 106, 452, 1992.
125. Benoit-Marand, M., Borrelli, E., and Gonon, F., Inhibition of dopamine release via presynaptic D2 receptors: Time course and functional characteristics in vivo, *J. Neurosci.*, 21, 9134, 2001.
126. Bouthenet, M. -L., Martres, M. P., Sales, N. et al., A detailed mapping of dopamine D2 receptors in rat central nervous system by autoradiography with [125I]-iodosulpiride, *Neuroscience*, 20, 117, 1987.
127. Phillips, P. E., Hancock, P. J., and Stamford, J. A., Time window of autoreceptor-mediated inhibition of limbic and striatal dopamine release, *Synapse*, 44, 15, 2002.
128. Nissbrandt, H., Engberg, G., and Pileblad, E., The effects of GBR 12909, a dopamine re-uptake inhibitor, on monoaminergic neurotransmission in rat striatum, limbic forebrain, cortical hemspheres and substantia nigra, *Naunyn Schmiedebergs Arch. Pharmacol.*, 344, 16, 1991.
129. Giorgetti, M., Hotsenpiller, G., Froestl, W. et al., In vivo modulation of ventral tegmental area dopamine and glutamate efflux by local GABA(B) receptors is altered after repeated amphetamine treatment, *Neuroscience*, 109, 585, 2002.
130. Araneda, R. and Bustos, G., Modulation of dendritic release of dopamine by *N*-methyl-D-aspartate receptors in rat substantia nigra, *J. Neurochem.*, 52, 962, 1989.
131. Westerink, B. H., Santiago, M., and Vries, J. B., The release of dopamine from nerve terminals and dendrites of nigrostriatal neurons induced by excitatory amino acids in the conscious rat, *Naunyn Schmiedebergs Arch. Pharmacol.*, 345, 523, 1992.
132. Gauchy, C., Desban, M., Glowinski, J. et al., NMDA regulation of dopamine release from proximal and distal dendrites in the cat substantia nigra, *Brain Res.*, 635, 249, 1994.

133. Rosales, M. G., Flores, G., Hernandez, S. et al., Activation of subthalamic neurons produces NMDA receptor-mediated dendritic dopamine release in substantia nigra pars reticulata: A microdialysis study in the rat, *Brain Res.*, 645, 335, 1994.
134. Bolam, J. P. and Smith, Y., Characterisation of the synaptic inputs to DA-ergic neurones in the rat SN, In *The Basal Ganglia III*, Bernardini, G., Ed., Plenum Press, New York, 1990.
135. Picciotto, M. R., Zoli, M., Rimondini, R. et al., Acetylcholine receptors containing the beta2 subunit are involved in the reinforcing properties of nicotine, *Nature*, 391, 173, 1998.
136. Champtiaux, N., Gotti, C., Cordero-Erausquin, M. et al., Subunit composition of functional nicotinic receptors in dopaminergic neurons investigated with knock-out mice, *J. Neurosci.*, 23, 7820, 2003.
137. Rahman, S., Zhang, J., and Corrigall, W. A., Effects of acute and chronic nicotine on somatodendritic dopamine release of the rat ventral tegmental area: In vivo microdialysis study, *Neurosci. Lett.*, 348, 61, 2003.
138. Reuben, M., Boye, S., and Clarke, P. B., Nicotinic receptors modulating somatodendritic and terminal dopamine release differ pharmacologically, *Eur. J. Pharmacol.*, 393, 39, 2000.
139. Fiorillo, C. D. and Williams, J. T., Cholinergic inhibition of ventral midbrain dopamine neurons, *J. Neurosci.*, 20, 7855, 2000.
140. Pidoplichko, V. I., DeBiasi, M., Williams, J. T. et al., Nicotine activates and desensitizes midbrain dopamine neurons, *Nature*, 390, 401, 1997.
141. Mansvelder, H. D., Keath, J. R., and McGehee, D. S., Synaptic mechanisms underlie nicotine-induced excitability of brain reward areas, *Neuron*, 33, 905, 2002.
142. Avshalumov, M. V., Chen, B. T., Marshall, S. P. et al., Glutamate-dependent inhibition of dopamine release in striatum is mediated by a new diffusible messenger, H_2O_2, *J. Neurosci.*, 23, 2744, 2003.
143. Krugel, U., Kittner, H., Franke, H. et al., Stimulation of P2 receptors in the ventral tegmental area enhances dopaminergic mechanisms in vivo, *Neuropharmacology*, 40, 1084, 2001.
144. Cragg, S. J., Variable dopamine release probability and short-term plasticity between functional domains of the primate striatum, *J. Neurosci.*, 23, 4378, 2003.

9 From Interferant Anion to Neuromodulator: Ascorbate Oxidizes Its Way to Respectability

George V. Rebec

CONTENTS

INTRODUCTION

Ascorbate, the deprotonated form of ascorbic acid, or vitamin C, is an ideal substance for detection by in vivo voltammetry. As an antioxidant, ascorbate readily gives up electrons, a fundamental requirement for voltammetric detection. In addition, the principal target of voltammetric studies is the extracellular fluid of the brain where ascorbate exists in plentiful amounts. In fact, the extracellular concentration of brain ascorbate is much higher than that of most other molecules of interest, making it relatively easy to detect and to monitor over time. A related point is that the level of extracellular brain ascorbate is not static but fluctuates with behavior, and there is mounting evidence that these fluctuations are related to the release of glutamate, an amino acid transmitter responsible for most neuronal excitations. In short, an understanding of ascorbate fluctuations is likely to reveal important information about how the brain operates.

From the very beginning, however, voltammetry was aimed at studying the catecholamine and indoleamine transmitters. Not only were they, like ascorbate, easily oxidizable but also they appeared to play critical roles in such fundamental functions as attention, cognition, mood, motivation, and motor control. Indeed, these transmitters, collectively known as the monoamines, have been the center of neurochemical attention since the 1950s. And so, when Ralph Adams and colleagues first used voltammetric techniques to study brain chemistry, the monoamines were a key target.[1,2] Ascorbate, far from being a molecule of interest, gained recognition, instead, as an interferent anion. Its high extracellular concentration, at more than 1000 times that of the monoamines, complicated monoamine detection. Strategies soon were developed to improve monoamine selectivity and, when applied carefully and appropriately, as described in other chapters in this volume, they helped launch a new appreciation of the complex role that the monoamines play in shaping brain function.

But ascorbate, too, has had a renaissance in neuroscience because of voltammetry. Improvements in electrochemical detection have made it possible to study ascorbate independently of other oxidizable molecules and have suggested functions well beyond those of a simple antioxidant. Other experimental approaches have begun to confirm an expanded role for brain ascorbate. Slowly but steadily, ascorbate has been gaining respect as a versatile molecule linked to the extracellular modulation of glutamate and, coming full-circle, perhaps the monoamines as well. This chapter focuses on how voltammetry helped to shape the ascorbate renaissance.

BASIC CHEMISTRY

ISOLATION AND TISSUE AVAILABILITY

Ascorbate is a hexuronic sugar acid, first isolated from adrenal glands and cabbages by Szent-Györgyi in 1928. A few years later, King identified the same molecule in orange juice. It was named ascorbate when it became clear that in humans this molecule could prevent scurvy, a condition whose symptoms include weak and swollen limbs, bleeding gums, hemorrhages under the skin, and a failure of wounds to heal. *Scorbutic* is the term used to describe scurvy, and thus an ascorbutic molecule (i.e., ascorbate) prevents or counteracts this condition. Although the symptoms of scurvy have been known for at least a thousand years, the underlying cause was first studied systematically in eighteenth century British sailors who had no access to fresh fruits and vegetables during long sea voyages.[3] The introduction of lemon juice into the diet of the British Navy virtually eliminated scurvy in this population.

Humans are vulnerable to this disease because, unlike most animals, they are unable to synthesize ascorbate from glucose. The problem is a non-functional gene for gulonolactone oxidase, the enzyme that catalyzes the final step in this process. Without ascorbate, collagen, the most abundant fibrous protein in the body, becomes malformed and gives rise to fragile connective tissue and blood vessels. The subsequent identification of scurvy in the guinea pig, another animal that fails to synthesize ascorbate, allowed various foods to be tested for ascorbutic potency. The most effective foods, citrus fruits and red and green vegetables, are rich sources of ascorbate.

Although humans and guinea pigs belong to a select group of animals, including other primates, fruit bats, and passeriform birds, that must forage for ascorbate, its concentration in various tissues is strikingly similar to that reported for animals that synthesize it. Dietary ascorbate is absorbed from the gut, but animals equipped with the appropriate synthetic machinery produce ascorbate either in the liver (mammals) or kidneys (reptiles). Regardless of the source, ascorbate enters the bloodstream where it can be taken up and stored by all relevant tissues. Apart from the adrenals, the brain contains the highest tissue level of ascorbate. Blood-borne ascorbate enters brain tissue from the ventricular system.[4] A selective transport mechanism in the choroid plexus ensures a high ventricular concentration even when blood ascorbate levels are low, making it virtually impossible to create an ascorbate deficiency in an otherwise healthy animal. Even in the scorbutic guinea pig, brain ascorbate declines very slowly relative to ascorbate in other organs. On the other hand, when blood ascorbate levels are unusually high, ascorbate can enter the brain by simple diffusion.[5] Neurons accumulate ascorbate via a sodium-dependent transporter[6] subsequently identified as SVCT2.[7] Because glial cells lack this transporter,[8] they may not be a major ascorbate storage site,[9] but a glial role in ascorbate release is still possible (see below).

The distribution of ascorbate in adult mammalian brain is remarkably heterogeneous. In forebrain structures such as the striatum, a region of the basal ganglia that integrates information from the entire cortical mantle, the tissue level of ascorbate is as much as 50% higher than in some midbrain and hindbrain areas.[10] Most ascorbate is found intracellularly, but the level of ascorbate in extracellular fluid is maintained at the expense of intracellular stores.[11] In striatum, interestingly, extracellular ascorbate is highest in dorsal and ventral regions, suggesting heterogeneity in release.[12]

Oxidation

Because it has two dissociable protons with pKa values of 4.2 and 11.8, ascorbate is a monovalent anion at physiological pH. The most striking chemical property of ascorbate is ease of oxidation, which arises from its enediol structure conjugated to a carbonyl. This capacity to donate electrons allows ascorbate to function as a co-factor for several enzymes that contain either copper or iron at their active sites. Two of these iron-containing enzymes, proline hydoxylase and lysine hydroxylase, are essential for the synthesis of collagen.[13] A well-known copper-containing enzyme, dopamine-β-hydroxylase, converts dopamine into norepinephrine.[14]

As an electron donor, ascorbate also protects against the damaging effects of free radicals generated by normal biological activity.[15–17] Most of these free radicals have one unpaired electron and thus can be easily neutralized by the one-electron oxidation of ascorbate. This reaction forms semi-dehydroascorbate or the ascorbyl radical. Loss of a second electron forms dehydroascorbate, the final oxidation product. Both forms of oxidized ascorbate, semi-dehydroascorbate and dehydroascorbate, can be returned to their original, reduced form by glutathione (GSH), a low-molecular-mass, thiol-containing tri-peptide, and by dehydroascorbate reductase, a GSH-dependent enzyme found in brain. Because ascorbate is water soluble, it works in the aqueous phase of tissue to scavenge hydroxyl- and peroxyl-radicals, peroxynitrite, singlet oxygen, and superoxide. Ascorbate also can stop the oxidation of cellular membranes by recycling or preventing the oxidation of vitamin E (α-tocopherol), which is lipid soluble. Ascorbate, therefore, appears to be a critical part of the cellular antioxidant network.

It is also the case, however, that ascorbate combined with copper or iron will promote free-radical production and thus act as a pro-oxidant. Combining a high level of ascorbate with either of these metals in the stomach, for example, generates hydroxyl radicals. Thus, ascorbate in some vitamin pills may have a pro-oxidant effect if they dissolve in the presence of copper or iron ions.[18] A key issue, therefore, is the availability of these ions in vivo. Fortunately, in most cellular systems, these ions are largely sequestered and unable to catalyze free radical reactions. In fact, the presence of ascorbate may actually facilitate iron sequestration. Moreover, extracellular fluid, which contains abundant amounts of ascorbate, is almost completely devoid of copper and iron ions. Thus, although an increase in the normal extracellular level of either ascorbate or one of these reactive metals could elevate the risk of free radical damage, this outcome is unlikely under normal conditions. It could occur, however, in certain pathological states such as iron-overload or tissue injury, both of which increase the availability of transition metal ions and thus increase the likelihood of an interaction with ascorbate.[19]

ASCORBATE ELECTROCHEMISTRY

In essence, electrochemistry is a form of chemical measurement that occurs at a surface-solution interface. Voltammetry is a special case in which the variable of interest is the current recorded at the surface of a working electrode. The current is generated by an applied potential, which causes oxidation or reduction of a chemical species in solution. The amount of recorded current is directly proportional to the concentration of the chemical species. The identification of ascorbate is based on current-voltage curves and their position along the voltage axis. This section focuses on the oxidative voltammetry of ascorbate in brain extracellular fluid and what this and other experimental approaches have revealed about the mechanisms underlying ascorbate release.

METHODOLOGY

Classical electroanalytical methodology uses various types of carbon-based electrodes to measure oxidation current in response to an applied potential. Separating the ascorbate oxidation signal from that generated by the monoamines became a significant challenge when voltammetry was first used to study brain chemistry.[2] Signals attributed to dopamine, for example, likely included a major contribution from ascorbate. Only when ascorbate was detected independently of other oxidizable substances and shown to fluctuate with behavior did it become clear that ascorbate deserved further study in its own right.[20] The first step in this direction came with development of carbon-paste electrodes that could distinguish ascorbate from other easily oxidized compounds in the brain.[21] Used in conjunction with linear sweep voltammetry, these electrodes revealed a distinct oxidation peak at $\sim +50$ mV when a potential was applied from -100 to $+500$ mV versus a Ag/AgCl reference. Ascorbate appeared to be the main component of this peak since the infusion of additional ascorbate in the immediate vicinity of the carbon-paste electrode selectively increased the signal. Confirmation came from evidence that local infusions of ascorbate oxidase, a dimeric copper-requiring enzyme that removes ascorbate by oxidizing it to dehydroascorbate, had the opposite effect. In fact, because dehydroascorbate is electrochemically inert, ascorbate oxidase significantly attenuated the signal.

Carbon-paste electrodes are made by packing carbon paste into a Teflon® (E.I. Du Pont de Nemours & Company, Inc., Wilmington, DE) tube or pulled-glass capillary. A similar but slightly more robust electrode is made from carbon epoxy. In either case, diameters typically range from 50 to 200 μm. Although recipes for carbon-paste and carbon-epoxy mixtures are available,[22,23] the relatively large size of these electrodes presents certain electrochemical and physiological complications that can be avoided with the use of electrodes whose active recording area is the exposed tip of a small-diameter carbon fiber.[24]

Formed from the high-temperature pyrolysis of polyacrylonitrile (PAN) or, in some cases, pitch, carbon fibers come in large bundles for use in a wide range of commercial applications. PAN-based fibers, for example, are used in composites to make extremely strong yet lightweight materials, such as fishing rods and airplane wings. Pitch-based fibers are not used for critical structural applications because of their low mechanical properties. Fiber diameters of 5–30 μm are preferred for voltammetry because such a small size minimizes perturbation of the brain environment and also permits a high degree of spatial resolution. Thornel P-55 carbon fibers, which are PAN-based and supplied by BP Amoco Chemicals, have individual diameters of ~ 10 μm and are perhaps the most commonly used. For electrochemistry, a single fiber is inserted in a pulled-glass capillary and positioned to extend beyond the glass tip by 150–200 μm. Sealing the fiber in the glass can be accomplished by injecting an epoxy resin into the non-pulled end of the capillary; micro-filaments in the glass draw the resin into the pulled tip where it hardens and forms an effective seal. The resin is made of Shell EPON Resin 828 (distributed by Miller-Stephenson) mixed with a small amount of *ortho*-phenylenediamine (Mallinckrodt, St. Louis, MO) to ensure reasonably rapid hardening; if the resin sets too rapidly, which occurs when meta-phenylenediamine is used, the seal will not reach the tip, thus exposing a large segment of the fiber within the glass capillary to extracellular fluid. After the fiber is sealed in the glass tip, conductive metal is inserted into the non-pulled end of the glass to establish electrical contact between the fiber and the recording equipment. Although mercury has been used for this purpose, safety concerns have largely replaced this metal with a bismuth alloy (Small Parts, Inc.), which melts at >70°C. Thus, heating the glass capillary liquefies the alloy and allows a connecting wire to be inserted. As the capillary cools, the alloy hardens, and the electrode is ready for use. A typical carbon-fiber electrode is shown schematically in Figure 9.1.

To become selective for ascorbate, the exposed fiber requires electrochemical pretreatment in phosphate-buffered saline, a procedure pioneered by Gonon et al.[25] Although several variations of the pretreatment protocol are now in use, the basic procedure is to apply a 70-Hz triangular wave from zero to up to +2.5 V. The result is a shift in the oxidation potential of ascorbate to a lower value distinct from that for catechols such as dopamine and its metabolite, 3,4-dihydroxyphenylacetic acid (DOPAC). The higher the triangular-wave voltage, the greater the ascorbate shift away from the catechols, but a voltage > +2.5 V appears to move ascorbate oxidation toward other interfant signals. A related point is that the longer the duration of the triangular wave application during pretreatment, the longer the electrode appears to maintain its selectivity for ascorbate and DOPAC in vivo. In light of these considerations, a recently developed protocol includes two applications of the 70-Hz triangular wave (versus a saturated calomel reference and stainless steel auxiliary): first, for 140 s from zero to +2.1 V and then for 5 s from zero to 2.5 V. Each application, moreover, is followed by 20 s of a constant potential at 1.5 V. In striatum, where most assessments of extracellular ascorbate have taken place, DOPAC is the dominant catechol, accounting for >95% of the catechol peak.[26] Thus, for electrode testing prior to in vivo use,

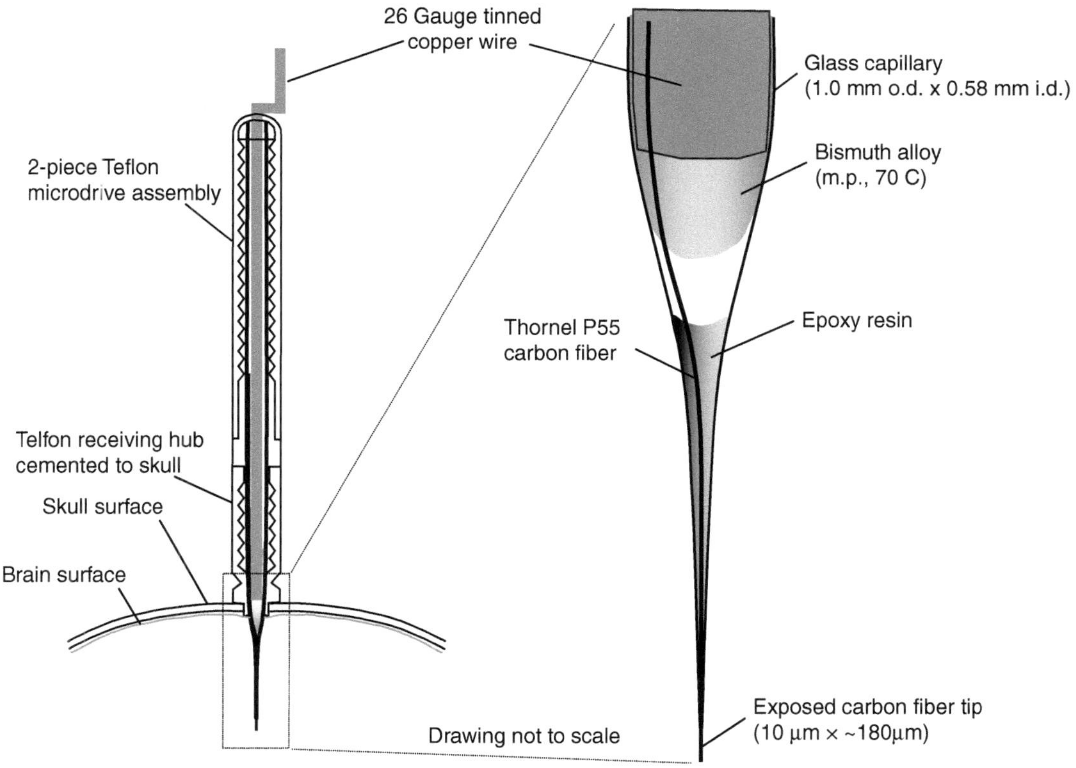

FIGURE 9.1 Schematic representation (not to scale) of the working electrode carbon fiber installed in the head-mounted microdrive assembly (left) and an expanded view of the electrode tip (right). The microdrive assembly mates with a teflon hub secured to the skull during stereotaxic surgery. The carbon fiber is sealed in a pulled-glass capillary. Bismuth alloy (heated to >50°C and subsequently cooled) ensures electrical contact between the fiber and the copper wire connection to the voltammetry system. See text for details.

ascorbate and DOPAC are added to the buffer at concentrations that reflect their relative physiological proportions. Buffer content also may affect the extent to which the electrochemical pretreatment separates the ascorbate and DOPAC signals. Phosphate-buffered saline, for example, may require slightly altered time and voltage applications.

The effects of electrochemical pretreatment are evident in Figure 9.2, which depicts voltammograms obtained in citrate-phosphate buffer containing 100 μM ascorbate and 20 μM DOPAC. Note that the untreated electrode is unable to distinguish ascorbate from DOPAC such that both are included in a single oxidation peak; the small size of the peak, moreover, indicates minimal sensitivity. When the same electrode is tested again after application of the triangular wave—even for a few seconds—sensitivity improves and the previously homogeneous oxidation peak moves to a lower potential as separate peaks for ascorbate and DOPAC begin to form. Very distinct peaks emerge when the fiber is exposed to the full pretreatment protocol; a difference of at least 150 mV is evident between each peak. Although it is unclear why electrochemical pretreatment has this effect, a change in the surface properties of the carbon is a likely explanation.[27]

Variations on the electrochemical pretreatment protocol have been used to improve the selectivity of carbon-paste and carbon-epoxy electrodes.[27,28] For each electrode type, the goal of electrochemical pretreatment is to shift the oxidation potential of ascorbate away from that for catechols. Another effect of electrochemical pretreatment is the emergence of a third in vivo peak at applied potentials >200 mV that represents the combined oxidation of serotonin, 5-hydroxyindoleacetic acid (5-HIAA, the serotonin metabolite), and uric acid.

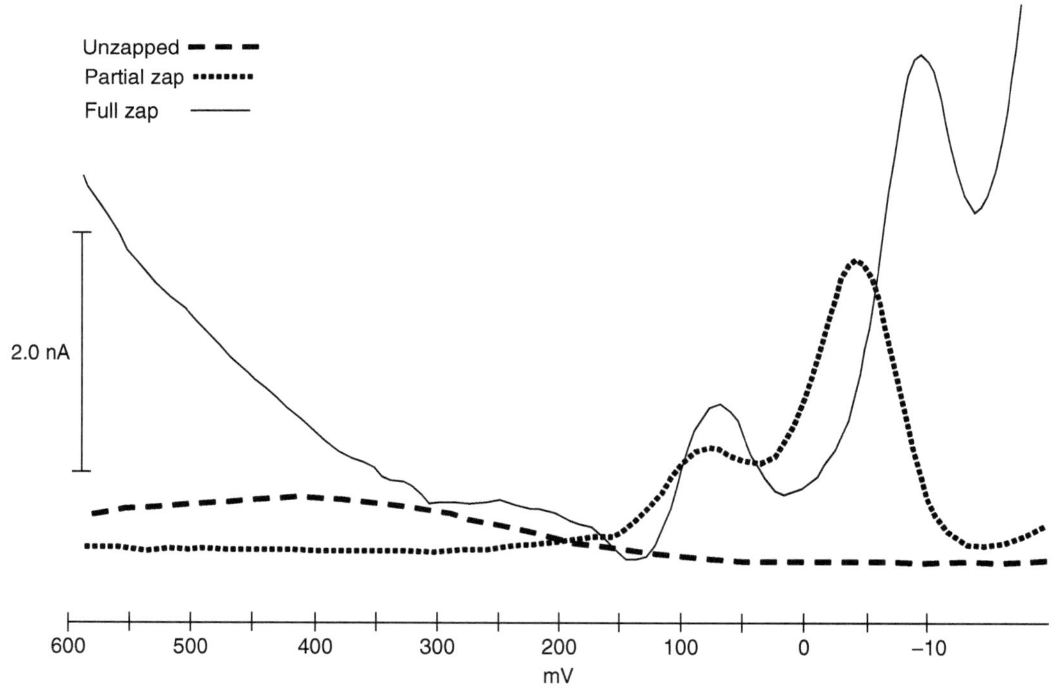

FIGURE 9.2 Individual voltammograms obtained in citrate-phosphate buffer containing 100 μM ascorbate and 20 μM DOPAC. Voltammograms were obtained before electrochemical pretreatment (Unzapped), after a brief pretreatment exposure (Partial zap: 3 s at 2.1 V), and after the full pretreatment procedure (Full zap: 140 s at 2.1 V and 5 s at 2.5 V). Note the broad, relatively small undifferentiated oxidation peak of the unzapped electrode compared to the much sharper, higher magnitude, and well-differentiated peaks of the fully zapped electrode. See text for details.

Although the irreversible nature of ascorbate oxidation allows it to be distinguished from catechols at untreated carbon-fiber electrodes used for fast-scan cyclic voltammetry,[29] electrochemical monitoring of ascorbate in brain extracellular fluid is typically performed with electrochemically pretreated carbon fibers scanned at a relatively slow rate (<100 mV/s). The applied potential typically sweeps from -200 to $+600$ mV for 30 s at a rate of 20–30 mV/s. Slow scanning provides a stable oxidation signal for extracellular ascorbate that can be monitored continuously on a minute-by-minute basis for more than two hours without significant loss of sensitivity. A common procedure for slow-scan voltammetry is to apply the potential in a stepwise or staircase fashion while calculating a local differentiation of the oxidation wave by recording the difference of the oxidation currents sampled just before and after each step. The entire scan lasts for 60 s as the potential is applied for 30 s in each direction. For in vivo recording, the reference electrode is typically a Ag/AgCl wire in contact with brain extracellular fluid.

EXTRACELLULAR CHANGES

Rodents have been the animals of choice for voltammetric monitoring of brain extracellular ascorbate, largely because they have been so valuable in revealing the extracellular dynamics (i.e., the release, reuptake, and diffusion properties) of the monoamine transmitters. In fact, because the monoamines appear to play such a key role in behavior, it was a natural extension of this line of research to make parallel assessments of extracellular ascorbate. Most of these investigations, moreover, have focused on striatum. Although dopamine and glutamate are known to play key roles in striatal function, it is difficult to ignore evidence that the extracellular level of striatal ascorbate changes with behavioral state. In resting rats, for example, the level typically hovers at $\sim 300\ \mu M$ but may double during periods of motor activation such as the nocturnal phase of the circadian cycle[30] or following administration of amphetamine or other dopamine agonists.[31,32] Such data, however, do not establish a causal link between ascorbate release and behavioral activation. Although additional support comes from evidence that scorbutic guinea pigs suffer from behavioral depression, this effect is confounded by the poor health of such animals.[33] To make a direct assessment of the relationship between endogenous striatal ascorbate release and behavior, we infused ascorbate oxidase or vehicle bilaterally into the striatum of rats tested in an open-field arena equipped with novel objects and populated with other rats.[34] Motor activation patterns were assessed along with the frequency of approach of novel objects and social interactions. The extent of ascorbate depletion was monitored simultaneously with slow-scan voltammetry, which also was used to measure DOPAC.

Intra-striatal infusion of ascorbate oxidase caused a rapid and pronounced decline in extracellular ascorbate. Within 20 min after infusion onset, a $>50\%$ decline was evident in all tested animals, and this deficit persisted throughout the recording session, which lasted for 60 min. Vehicle infusions had no effect on striatal ascorbate. The amplitude of the DOPAC signal was largely unchanged after infusion of either ascorbate oxidase or vehicle. Because ascorbate oxidase uses ascorbate as a cofactor in the enzymatic conversion of molecular oxygen to water, there is the possibility of oxygen depletion and subsequent cell loss due to hypoxic injury. In some cases, therefore, histological assessment of striatal tissue was carried out one or two days after ascorbate oxidase or vehicle infusion. No evidence of cell damage other than that typically observed along the infusion tract was found.

Loss of striatal ascorbate led to a near-total inhibition of all recorded behavior, including open-field locomotion, approach of novel objects, and social interaction with other rats. Episodes of simple movement, when they occurred, were limited to brief (<5 s) bouts of head bobbing or turning. No locomotion or rearing was observed in any of these animals as soon as 25 min after ascorbate oxidase infusion. Vehicle controls, in contrast, maintained baseline levels of responsiveness throughout the recording period. Because ascorbate oxidase may disrupt striatal functioning as a result of its relatively large size as a glycoprotein (140,000 kDa), an inactive form of the enzyme was infused as a second control. The inactive enzyme mimicked the effect of vehicle, ruling out an effect of molecular size. To determine if a loss of striatal ascorbate simply changed the dynamics of extracellular ascorbate throughout the brain and thus altered behavior through an extra-striatal mechanism, some rats received bilateral infusions of ascorbate oxidase into dorsal hippocampus, another forebrain area with a high level of extracellular ascorbate. These animals, however, maintained normal behavioral activity, further implicating the loss of striatal ascorbate in the behavioral deficit.

Systemic administration (sc) of 1.0 mg/kg D-amphetamine immediately after the intra-striatal infusion of ascorbate oxidase in separate animals reversed the decline in ascorbate and restored the extracellular level to pre-infusion values. Concomitantly, behavioral activation in these animals returned to baseline and surpassed it by the end of the recording session. Although removal of ascorbate from striatal extracellular fluid may increase susceptibility to oxidative damage, this effect cannot explain the behavioral deficits induced by ascorbate oxidase given their rapid onset and their equally rapid reversal by amphetamine. In fact, our behavioral results with amphetamine rule out the possibility that ascorbate oxidase inhibits behavior non-selectively by binding to dopamine receptors or the dopamine molecule. A related point is that ascorbate oxidase failed to alter striatal DOPAC, arguing against a loss of dopamine. It appears, therefore, that normal behavioral output depends on a critical level of ascorbate in striatal extracellular fluid.[34] Insight into the neural mechanisms controlling the release of ascorbate has emerged from work on glutamate transmission.

Glutamate-Related Mechanisms of Striatal Ascorbate Release

The glutamate projection to the striatum arises primarily from cerebral cortex.[35–37] All cortical areas contribute to this projection, which originates bilaterally from both supragranular and infragranular layers. Corticostriatal neurons comprise the major source of extracellular ascorbate. Cortical ablations, for example, lower the level of extracellular ascorbate in striatum by $>70\%$,[38,39] whereas intra-striatal infusion of kainic acid, which destroys intrinsic cells, has no such effect.[40] Conversely, increases in striatal glutamate transmission enhance the release of ascorbate.[41,42] Taken together, these lines of evidence suggest that the level of extracellular ascorbate is closely linked to activity in glutamate afferents arising from cerebral cortex. Evidence also suggests that ascorbate is released by a heteroexchange mechanism involving the uptake of glutamate.[20,21,43] According to this model, glutamate uptake triggers the outward transport of ascorbate. Thus, stimuli that activate corticostriatal neurons and release glutamate also activate glutamate transport, which, in turn, increases ascorbate release. The model also suggests that a high level of extracellular ascorbate would oppose further ascorbate release and thus prevent glutamate uptake. Indeed, a 500 μM increase in striatal ascorbate not only elevated basal extracellular glutamate but also significantly prolonged the disappearance of glutamate release evoked by cortical stimulation.[44] Because glia, like neurons, express glutamate transporters,[45–47] the heteroexchange model also may apply to one or more types of glial cells, but this is uncertain.[41] Regardless of the mechanism, however, activation of the corticostriatal pathway is mainly responsible for ascorbate release in striatum.

Although drugs that enhance dopamine transmission also promote striatal ascorbate release,[32,48,49] this effect is due to activation of corticostriatal afferents rather than a direct change in striatal dopamine signaling. This conclusion comes mainly from evidence that neurotoxic loss of striatal dopamine terminals has little or no effect on striatal ascorbate release, whereas prevention of cortical activation by disruption of thalamo-cortical afferents blocks this process.[50–53] It also is interesting that other dopamine-rich forebrain areas, such as the nucleus accumbens and medial prefrontal cortex, show relatively little behavior- or drug-related change in ascorbate release, even though the basal level of ascorbate in these areas matches that in striatum.[54,55] Ascorbate release from corticostriatal afferents, therefore, may play a unique behavioral role.

Identifying the functional significance of this release is likely to shed new light on basal ganglia operations in health and disease.

Ascorbate Modulation of Glutamate-Evoked Neuronal Signaling

In general terms, the striatum processes cortical information and routes it via both mono- and multisynaptic connections to basal ganglia output structures to influence motor control. Thus, the striatum is strategically positioned to shape motor output, and, not surprisingly, the striatum plays key roles in motor memory as well as habit learning and retrieval.[56–58] As the main information-processing unit of the striatum, the medium spiny projection neuron responds to cortical activation with a glutamate-mediated excitation. In fact, cortical input is necessary to maintain striatal neurons in a depolarized or "up" state required for impulse activity.[59] Dopamine-containing afferents, arising from the ventral midbrain, modulate cortical input by adjusting the strength of the glutamate response. When applied iontophoretically at low concentrations in resting rats, for example, dopamine enhances glutamate-induced excitations by increasing the magnitude of the glutamate signal relative to background activity.[60] From striatum, information is routed via direct and indirect projections to the entopeduncular nucleus (internal globus pallidus in primates) and substantia nigra reticulata (SNr), which together form the basal ganglia output to thalamo-cortical and downstream motor pathways.

Preliminary experiments with ascorbate iontophoresis also showed a potentiation of the excitatory effects of glutamate on striatal neurons.[61,62] To determine if this effect, like that of dopamine, could be explained by a change in the glutamate signal-to-noise ratio, a detailed assessment of the ascorbate-glutamate interaction was performed.[63] First, the effects of ascorbate iontophoresis were tested on spontaneously active units as well as on silent or sporadically active units driven by glutamate. To control for antioxidant effects, D-*iso*-ascorbate was substituted for the naturally occurring isomer in some cases. The data revealed that brief (20 s) applications of either ascorbate or D-*iso*-ascorbate at low ejection currents (5–40 nA) typically had no effect on either spontaneously active or glutamate-driven units. When ascorbate was applied at these currents for two to four minutes, however, glutamate-induced excitations were altered dramatically. For example, prolonged applications of 5–40 nA ascorbate, which often failed to alter basal activity, significantly increased the excitatory response to brief pulses of glutamate. Further analysis revealed that ascorbate enhanced both the onset and absolute magnitude of the glutamate response. In contrast, prolonged application of D-*iso*-ascorbate at the same ejection currents never potentiated the action of glutamate. In fact, D-*iso*-ascorbate often attenuated the glutamate response. Interestingly, ascorbate had an attenuating effect only at high ejection currents (>80 nA), which have been shown by concomitant voltammetric measurements to increase extracellular concentrations above 1.0 mM.[61] Thus, at extremely high concentrations, ascorbate appears to antagonize glutamate transmission through a non-specific antioxidant effect.[64] This interpretation is supported by recent evidence that systemic injection of high ascorbate doses (500 mg/kg and 1000 mg/kg) in freely behaving rats not only elevates striatal ascorbate beyond 1.0 mM but also suppresses the responsiveness of striatal neurons to sensorimotor stimulation.[65] In fact, treatment with 1000 mg/kg ascorbate reversed many stimulation-evoked excitations to inhibitions.

It also is interesting that the modulatory role of ascorbate differs from that reported for dopamine. For example, although iontophoretic dopamine can enhance the glutamate response, this effect is manifest as an increase in the glutamate-evoked signal relative to background activity rather than an increase in the absolute magnitude of the signal.[66] Ascorbate, therefore, appears to play a unique role in modulating the strength of glutamate-induced excitations. Because behavior-related increases in striatal ascorbate parallel the ascorbate increases achieved with low-to-moderate iontophoretic doses, behaviorally relevant fluctuations in extracellular ascorbate are likely to exert a critical influence on the flow of glutamate-mediated information through striatal circuits.

ASCORBATE AND HUNTINGTON'S DISEASE

In an interesting extension of evidence that striatal ascorbate is essential for normal behavioral output,[34] mice that model Huntington's disease (HD), a genetic disorder characterized by striatal degeneration and severe behavioral deficits, suffer from a loss of ascorbate in striatal extracellular fluid.[67] Even more intriguing is evidence that some of the neurological motor signs that these mice display improve upon reversal of the ascorbate deficit.[68] Elevating striatal ascorbate also reverses abnormal electrophysiological activity in HD striatum.[69] This section reviews emerging evidence implicating abnormal striatal ascorbate in HD genetic mouse models.

HD AND TRANSGENIC MICE

HD is an autosomal dominant condition caused by an expanded tri-nucleotide (CAG) repeat, which results in the increased length of a poly-glutamine (polyQ) tract in the huntingtin protein.[70] Although normal huntingtin function can tolerate up to thirty-five or thirty-six repeats, just one more is enough to cross the pathogenic threshold. Like other polyQ disorders, juvenile-onset HD is associated with longer polyQ tracts than adult-onset cases. Typical onset occurs in middle-age. Huntingtin is widely expressed throughout the nervous system, and, accordingly, all brain regions show some atrophy in HD. The most pronounced neuronal pathology, however, strikes the medium spiny neurons of the striatum.

The function of the huntingtin protein is far from understood, but it may have anti-apoptotic properties. In HD, mutated huntingtin appears to acquire a neurotoxic function, which may be related to the formation of intra-nuclear inclusions, a disruption of gene transcription, or abnormal interactions with other proteins.[71,72] At the synaptic level, the polyQ expansion is likely to result in aberrant transmission that causes toxic insults to vulnerable neuronal populations. These and other insights into HD pathology have come from genetic mouse models. The most extensively characterized are the R6 lines, which contain an exon-1 transgene.[73] The behavioral phenotype and neuropathology have been studied most thoroughly in the R6/2 line, which carries ~150 CAG repeats. These animals develop motor abnormalities (e.g., tics, stereotyped and choreiform movements, and tremor) as early as five or six weeks of age[74] and show intra-nuclear inclusions of huntingtin conjugated to ubiquitin.[75] Although the R6/2 striatum does not show marked cell loss, there is evidence of late-onset neuronal degeneration, striatal atrophy, and decreased expression of striatal signaling genes.[71,76] We have been using this line to assess possible HD-related abnormalities in the regulation of extracellular striatal ascorbate.[67]

ASCORBATE DYSREGULATION IN HD STRIATUM

Recording voltammetric signals from behaving mice requires some modification of the head-mounted assembly that has been used successfully in studies of behaving rats.[77–79] A key component is a lightweight (1.5 g) bilateral micro-drive system fashioned from two nylon rods as described elsewhere.[80] The bilateral design allows for a working electrode to be lowered into either the left or right target structure (e.g., striatum), while a reference electrode is positioned on the brain surface of the contralateral hemisphere. As shown in Figure 9.3, the recording system fits conveniently on the head to permit noise-free voltammetry. Because the working electrode is lowered acutely on the recording day, we avoid the loss of sensitivity and other complications associated with chronically implanted voltammetric electrodes.[81]

At the beginning of each voltammetry session, all mice are anesthetized to allow electrode placement in striatum. While under anesthesia (30–60 min), both wild-type and transgenic mice show no difference in signal amplitude for either ascorbate or DOPAC. As recording continues over the next 60–90 min, however, clear ascorbate differences emerge as both groups of mice become behaviorally active. Thus, whereas wild-type animals show the expected behavior-related increase in ascorbate release, HD mice respond with an ascorbate decrease that in some cases represents a

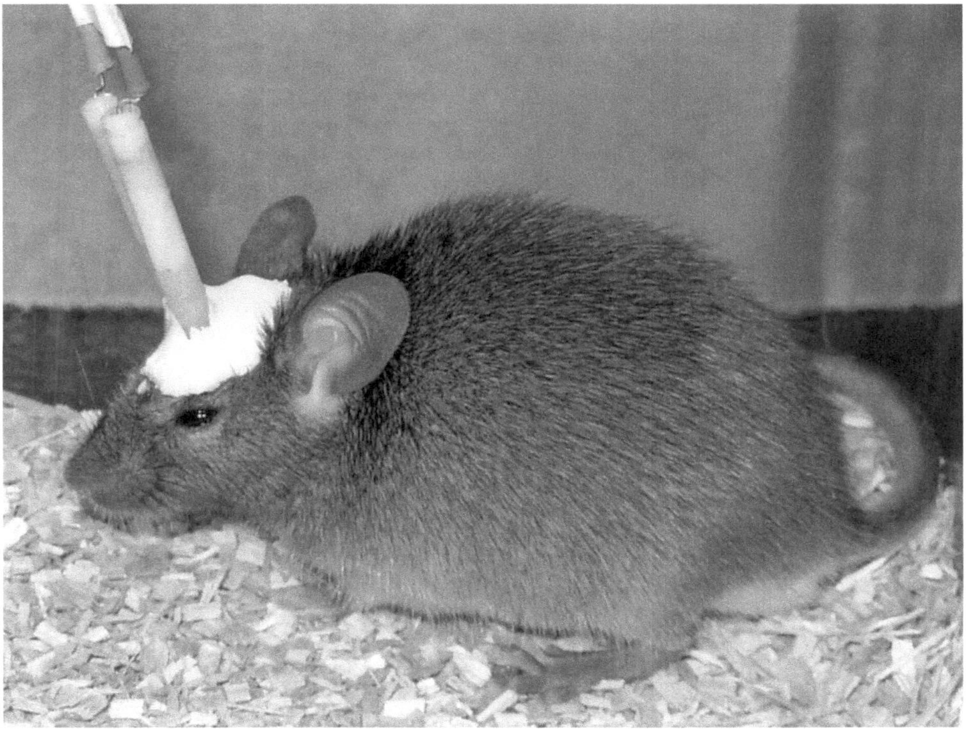

FIGURE 9.3 Digital photograph of a freely behaving mouse prepared for voltammetric recording. The bilateral microdrive system allows for lowering the working electrode into the target structure on the recording day, while the reference electrode is positioned at brain surface. The entire head-mounted system is lightweight (<2 g) and provides noise-free recording (see Figure 9.4).

decline of >50% from anesthesia baseline.[67] Figure 9.4 illustrates this difference recorded from representative wild-type and HD mice during and after recovery from anesthesia. This behavior-related divergence in the ascorbate signal is evident at six weeks of age—the earliest recording age to date—and is present two or three weeks later, emphasizing the persistence of the effect.[67]

Behavioral assessments indicate that relative to wild-type mice, HD transgenics show a more restricted range of motor responses.[68] HD mice, for example, move in fixed, stereotypic patterns and spend less time interacting with the environment. These animals also show hind-paw flicks, which appear as an aborted grooming maneuver but are considered an HD neurological sign.[74] Repeated daily injections of ascorbate (300 mg/kg, ip), however, restore the loss of behavior-related striatal ascorbate release and improve motor behavior.[68] Ascorbate-treated HD mice, for example, significantly increase turning probability in a plus maze, a measure of motor flexibility, and decrease hind-paw flicks compared to vehicle-treated transgenics. These groups did not differ, however, on total locomotion arguing against an ascorbate-induced suppression of overall motor activity. It also is interesting that treatment of wild-type mice with ascorbate had no significant effect on either striatal ascorbate release or behavior indicating that under normal conditions this dose of ascorbate does not easily penetrate brain tissue.

At the neuronal level, HD striatum appears hyperactive. An assessment of the distribution of striatal firing rates in R6/2 and wild-type mice revealed that transgenics had significantly more fast-firing units.[69] Although both groups modulated unit activity with behavioral state such that anesthesia slowed and subsequent behavioral recovery elevated unit firing, R6/2 neurons

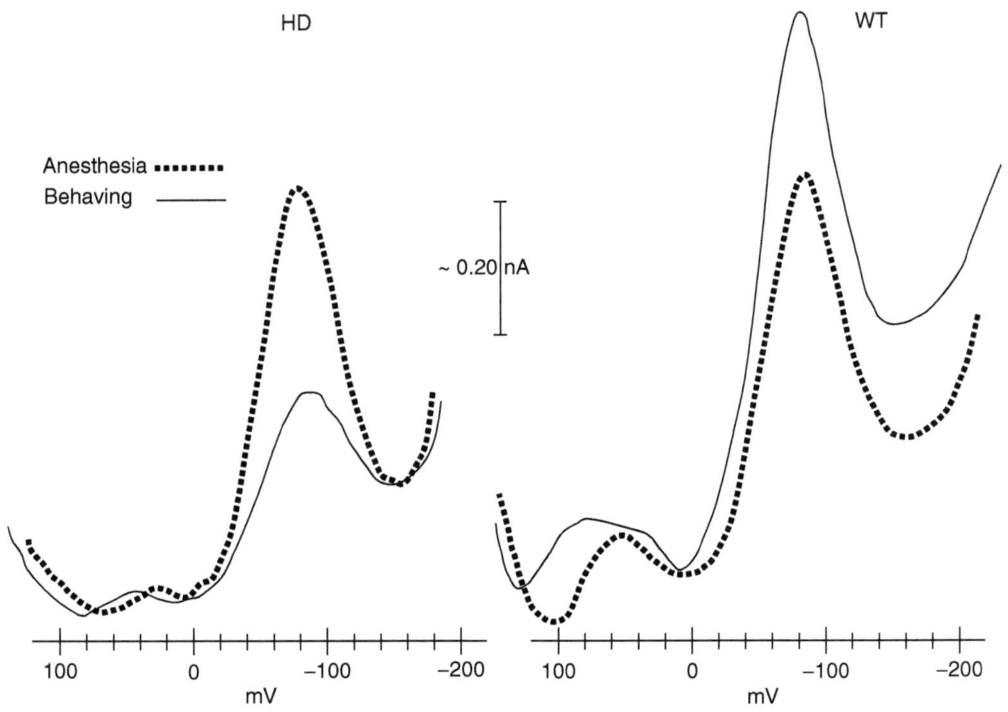

FIGURE 9.4 Individual voltammograms obtained from the striatum of a transgenic (HD) and a wild-type (WT) mouse during anesthesia and subsequent waking (behaving). Note the typical increase in the ascorbate signal associated with behavioral arousal in the WT mouse but a pronounced behavior-related ascorbate decline in the HD animal. These results are representative of a low level of extracellular ascorbate in HD mouse striatum relative to that in wild-type controls.

consistently discharged above the wild-type rate. Again, however, multiple injections of ascorbate (300 mg/kg, ip), which restore extracellular ascorbate levels in R6/2 striatum, attenuated this difference.

Although these lines of evidence suggest that a deficit in striatal ascorbate release plays a key role in HD, the mechanism underlying this effect is unclear. In view of evidence that ascorbate release is linked to glutamate uptake (see above), one could argue that low extracellular ascorbate simply indicates a glutamate uptake problem. In fact, there is evidence for decreased mRNA levels of the major astroglial glutamate transporter (GLT1 or EAAT2) accompanied by decreased glutamate uptake in the striatum of transgenic mice.[82] Thus, a defect in astrocytic glutamate uptake may contribute to striatal neuropathology. This hypothesis is consistent with rapidly firing striatal neurons in the R6/2 line because over-activation of glutamate receptors can result in excitotoxicity. As the most abundant glutamate transporter in the brain, GLT1 is critically important for maintaining low resting levels of extracellular glutamate (<1 μM) by concentrating the transmitter across the cell membrane. Because excitotoxicity includes the production of cell-damaging free radicals, e.g.,[83–85] an accompanying low level of extracellular ascorbate may simply exacerbate the problem by lowering anti-oxidant capacity.

Although a failure of glutamate uptake and the accompanying loss of extracellular ascorbate may comprise critical components of the neuropathological cascade in HD, it is important to recognize that a restoration of striatal ascorbate attenuates both behavioral and firing-rate abnormalities in R6/2 mice. Conceivably, therefore, impaired ascorbate release could play a primary role in

HD pathology. One of these roles may involve an anti-oxidant effect. In fact, GLT1, like other glutamate transporters, is vulnerable to the action of biological oxidants.[86] Thus, without the oxidant-fighting capacity of extracellular ascorbate, a progressive positive feedback loop may develop in which glutamate uptake declines, leading to less ascorbate release and further impairing glutamate uptake. The addition of ascorbate may lower the oxidant load on GLT1 and resume the removal of glutamate. The result would be a decline in striatal firing rate and, thus, improved striatal processing of behavior-related neuronal signals. Alternatively, restoring extracellular ascorbate could alter striatal function directly by interfering with the binding of glutamate to the *N*-methyl-D-aspartate (NMDA) receptor. At high extracellular concentrations ($\sim$1 mM), for example, ascorbate occupies an NMDA redox site that antagonizes the action of glutamate.[64] Because striatal NMDA receptors appear to be abnormally sensitive to glutamate activation in HD mice,[87] ascorbate treatment may stabilize striatal physiology and, thus, behavior by preventing NMDA receptor over-activation. It is clear that pre- and postsynaptic modulation of glutamate transmission by ascorbate deserves further assessment in the ongoing effort to understand and treat HD.

SUMMARY AND CONCLUSION

The ability of ascorbate to give up electrons not only accounts for most of its physiological functions but also makes it accessible to voltammetric detection. Although ascorbate has never approached the level of interest generated by the monoamines, voltammetry has been largely responsible for evidence implicating ascorbate release in the striatum as a modulator of neuronal information processing and behavior. This modulatory role seems closely tied to glutamate through a direct influence on both its uptake and postsynaptic action. In fact, a deficit in striatal ascorbate release in mouse genetic models of HD suggests several ways in which glutamate may contribute to this and, perhaps, other neurodegenerative diseases. Pursuit of these suggestions may identify key targets in the search for treatment strategies. As research on ascorbate continues, it is likely that voltammetry will remain a critical part of this effort.

ACKNOWLEDGMENTS

Portions of the research described in this chapter were supported by U.S. Public Health Service grants from the National Institute on Drug Abuse, (DA02451), and the National Institute of Neurological Disease and Stroke (NS35663). Support from the Hereditary Disease Foundation is also acknowledged. Mr. Scott J. Barton and Mr. Paul E. Langley provided expert technical support; Ms. Faye Caylor assisted with editing and manuscript organization.

REFERENCES

1. Adams, R. N., Probing brain chemistry with electroanalytical techniques, *Anal. Chem.*, 48, 1128A, 1976.
2. Adams, R. N., In vivo electrochemical measurements in the CNS, *Prog. Neurobiol.*, 35, 297, 1990.
3. Carpenter, K. J., *The History of Scurvy and Vitamin C*, Cambridge, New York, 1986.
4. Spector, R., Micronutrient homeostasis in mammalian brain and cerebrospinal fluid, *J. Neurochem.*, 53, 1667, 1989.
5. Tolbert, L. C., Thomas, T. N., Middaugh, L. D., and Zemp, J. W., Effect of ascorbic acid on neurochemical behavioral and physiological systems mediated by catecholamines, *Life Sci.*, 25, 2189, 1979.
6. Rice, M. E. and Nicholson, C., Interstitial ascorbate in turtle brain is modulated by release and extracellular volume change, *J. Neurochem.*, 49, 1096, 1987.

7. Tsukaguchi, H., Tokui, T., Mackenzie, B., Berger, U. V., Chen, X. Z., Wang, Y., Brubaker, R. F., and Hediger, M. A., A family of mammalian Na+-dependent L-ascorbic acid transporters, *Nature*, 399, 70, 1999.
8. Berger, U. V. and Hediger, M. A., The vitamin C transporter SVCT2 is expressed by astrocytes in culture but not in situ, *Neuroreport*, 11, 1395, 2000.
9. Rice, M. E. and Russo-Menna, I., Differential compartmentalization of brain ascorbate and glutathione between neurons and glia, *Neuroscience*, 82, 1213, 1998.
10. Oke, A. F., May, L., and Adams, R. N., Ascorbic acid distribution patterns in human brain, In *Beyond Deficiency: New Views on the Function and Health Effects on Vitamins (Annals of the New York Academy of Sciences)*, New York Academy of Sciences, New York, p. 1, 1992.
11. Schenk, J. O., Miller, E., Gaddis, R., and Adams, R. N., Homeostatic control of ascorbate concentration in CNS extracellular fluid, *Brain Res.*, 253, 353, 1982.
12. Basse-Tomusk, A. and Rebec, G. V., Regional distribution of ascorbate and 3,4-dihydroxyphenylacetic acid (DOPAC) in rat neostriatum, *Brain Res.*, 538, 29, 1991.
13. Englard, S. and Seifter, S., The biochemical functions of ascorbic acid, *Ann. Rev. Nutr.*, 6, 365, 1986.
14. Diliberto, E. J. and Allen, P. L., Mechanism of dopamine-beta-hydroxylation. Semidehydroascorbate as the enzyme oxidation product of ascorbate, *J. Biol. Chem.*, 256, 3385, 1981.
15. Halliwell, B., Oxygen radicals as key mediators in neurological disease: fact or fiction, *Ann. Neurol.*, 32, S10, 1992.
16. Niki, E., Action of ascorbic acid as a scavenger of active and stable oxygen radicals, *Am. J. Clin. Nutr.*, 54, S1119, 1991.
17. Rice, M. E., Ascorbate regulation and its neuroprotective role in the brain, *Trends Neurosci.*, 23, 209, 2000.
18. Halliwell, B. and Gutteridge, J. M. C., *Free Radicals in Biology and Medicine*, 3rd ed., Oxford University Press, New York, 1999.
19. Connor, J. R., Evidence for iron mismanagement in the brain in neurological disorders, In *Metals and Oxidative Damage in Neurological Disorders*, Connor, J. R., Ed., Plenum Press, New York, p. 23, 1997.
20. Rebec, G. V. and Pierce, R. C., A vitamin as neuromodulator: ascorbate release into the extracellular fluid of the brain regulates dopaminergic and glutamatergic transmission, *Prog. Neurobiol.*, 43, 537, 1994.
21. Fillenz, M., O'Neill, R. D., and Grunewald, R. A., Changes in extracellular brain ascorbate concentration as an index of excitatory aminoacid release, In *Monitoring Neurotransmitter Release During Behaviour*, Joseph, M. H., Fillenz, M., MacDonald, A., and Marsden, C. A., Eds., Ellis Norwood, Chichester, p. 144, 1986.
22. Kissinger, P. T., Hart, J. B., and Adams, R. N., Voltammetry in brain tissue: a new neurophysiological measurement, *Brain Res.*, 55, 209, 1973.
23. Conti, J. C., Strope, E., Adams, R. N., and Marsden, C. A., Voltammetry in brain tissue: chronic recording of stimulated dopamine and 5-hydroxytryptamine release, *Life Sci.*, 23, 2705, 1978.
24. Justice, J. B. Jr., Introduction to in vivo voltammetry, In *Voltammetry in the Neurosciences, Principles, Methods, and Applications*, Justice, J. B. Jr., Ed., Humana Press, Clifton, New Jersey, USA, p. 3, 1987.
25. Gonon, F. G., Frombarlet, C. M., Buda, M. J., and Pujol, J. F., Electrochemical treatment of pyrolytic carbon fiber electrodes, *Anal. Chem.*, 53, 1386, 1981a.
26. Gonon, F., Buda, M., Cespuglio, R., Jouvet, M., and Pujol, J. F., Voltammetry in the striatum of chronic freely moving rats: detection of catechols and ascorbic acid, *Brain Res.*, 223, 69, 1981b.
27. Falat, L. and Cheng, H. Y., Voltammetric differentiation of ascorbic acid and dopamine at an electrochemically treated graphite/epoxy electrode, *Anal. Chem.*, 54 (12), 2108, 1982.
28. Rice, M. E., Galus, Z., and Adams, R. N., Graphite paste electrodes. Effects of paste composition and surface states on electron-transfer rates, *J. Electroanal. Chem.*, 143, 89, 1983.
29. Stamford, J. A., Kruk, Z. L., and Miller, J., Regional differences in extracellular ascorbic levels in the rat brain determined by high speed cyclic voltammetry, *Brain Res.*, 299, 289, 1984.
30. O'Neill, R. D., Grunewald, R. A., Fillenz, M., and Albery, W. J., Linear sweep voltammetry with carbon paste electrodes in the rat striatum, *Neuroscience*, 7, 1945, 1982.
31. Mueller, K., Repeated administration of high doses of amphetamine increases release of ascorbic acid in caudate but not nucleus accumbens, *Brain Res.*, 494, 30, 1989.

32. Pierce, R. C. and Rebec, G. V., Stimulation of both D1 and D2 receptors increases behavioral activation and ascorbate release in the neostriatum of freely moving rats, *Eur. J. Pharmacol.*, 191, 295, 1990.
33. Kaufmann, P., Wiens, W., Dirks, M., and Krehbiel, D., Changes in social behavior and brain catecholamines during the development of ascorbate deficiency in guinea pigs, *Behav. Process.*, 13, 13, 1986.
34. Rebec, G. V. and Wang, Z., Behavioral activation in rats requires endogenous ascorbate release in striatum, *J. Neurosci.*, 21, 668, 2001.
35. Bolam, J. P., Hanley, J. J., Booth, P. A., and Bevan, M. D., Synaptic organisation of the basal ganglia, *J. Anat.*, 196, 527, 2000.
36. Groves, P. M., A theory of the functional organization of the neostriatum and the neostriatal control of voluntary movement, *Brain Res. Rev.*, 5, 109, 1983.
37. Parent, A., Cote, P. Y., and Lavoie, B., Chemical anatomy of primate basal ganglia, *Prog. Neurobiol.*, 46, 131, 1995.
38. Basse-Tomusk, A. and Rebec, G. V., Corticostriatal and thalamic regulation of amphetamine-induced ascorbate release in the neostriatum, *Pharmacol. Biochem. Behav.*, 35, 55, 1990.
39. O'Neill, R. D., Grunewald, R. A., Fillenz, M., and Albery, W. J., The effect of unilateral cortical lesions on the circadian changes in rat striatal ascorbate and homovanillic acid levels measured *in* vivo using voltammetry, *Neurosci. Lett.*, 42, 105, 1983.
40. Pierce, R. C., Miller, D. M., Reising, D., and Rebec, G. V., Unilateral neostriatal kainate but not 6-OHDA lesions block dopamine agonist-induced ascorbate release in the neostriatum of freely-moving rats, *Brain Res.*, 597, 138, 1992.
41. Cammack, J., Ghasemzadeh, B., and Adams, R. N., The pharmacological profile of glutamate-evoked ascorbic acid efflux measured by in vivo electrochemistry, *Brain Res.*, 565, 17, 1991.
42. Pierce, R. C. and Rebec, G. V., Intraneostriatal administration of glutamate antagonists increases behavioral activation and decreases neostriatal ascorbate via non-dopaminergic mechanisms, *J. Neurosci.*, 13, 4272, 1993.
43. O'Neill, R., The measurement of brain ascorbate in vivo and its link with excitatory amino acid neurotransmission, In *Neuromethods Vol. 27: Voltammetric Methods in Brain Systems*, Boulton, A., Baker, G., and Adams, R. N., Eds., Humana Press Inc, Clifton, New Jersey, USA, p. 221, 1995.
44. Rebec, G. V., Witowski, S. R., Sandstrom, M. I., Rostand, R. D., and Kennedy, R. T., Extracellular ascorbate modulates cortically evoked glutamate dynamics in rat striatum, *Neurosci. Lett.*, 378, 166, 2005.
45. Balazs, R., Bridges, R. J., and Cotman, C. W., *Excitatory Amino Acid Transmission in Health and Disease*, Oxford University Press, New York, 2006.
46. Kanai, Y., Trotti, D., Nussberger, S., and Hediger, M. A., The high-affinity glutamate transporter family: structure function and physiological relevance, In *Neurotransmitter Transporters—Structure Function and Regulation*, Reith, M. E. A., Ed., Humana Press, Totowa, New Jersey, USA, p. 171, 1997, (Contemporary Neuroscience).
47. Torp, R., Danbolt, N. C., Babaie, E., Bjoras, M., Seeberg, E., Storm-Mathisen, J., and Otterson, O. P., Differential expression of two glial glutamate transporters in the rat brain: an in situ hybridization study, *Eur. J. Neurosci.*, 6, 936, 1994.
48. Mueller, K. and Haskett, C., Effects of haloperidol on amphetamine-induced increases in ascorbic acid as determined by voltammetry in vivo, *Pharmacol. Biochem. Behav.*, 27, 231, 1987.
49. Zetterstrom, T., Wheeler, D. B., Boutelle, M. G., and Fillenz, M., Striatal ascorbate and its relationship to dopamine receptor stimulation and motor activity, *Eur. J. Neurosci.*, 3, 940, 1991.
50. Christensen, J. C., Wang, Z., and Rebec, G. V., GABA infusion in substantia nigra pars reticulata inhibits ascorbate release in ipsilateral striatum, *Neurosci. Lett.*, 280, 191, 2000.
51. Kamata, K., Wilson, R. L., Alloway, K. D., and Rebec, G. V., Multiple amphetamine injections reduce the release of ascorbic acid in the neostriatum of the rat, *Brain Res.*, 362, 331, 1986.
52. Pierce, R. C., Clemens, A. J., Grabner, C. P., and Rebec, G. V., Amphetamine promotes neostriatal ascorbate release via a nigro-thalamo-corticoneostriatal loop, *J. Neurochem.*, 63, 1499, 1994.
53. Wilson, R. L. and Wightman, R. M., Systemic and nigral application of amphetamine both cause an increase in extracellular concentration of ascorbate in the caudate nucleus of the rat, *Brain Res.*, 339, 219, 1985.

54. Phebus, L. A., Roush, M. E., and Clemens, J. A., Effect of direct and indirect dopamine agonists on brain extracellular ascorbate levels in the striatum and nucleus accumbens of awake rats, *Life Sci.*, 47, 1317, 1990.
55. Yount, S. E., Kraft, M. E., Pierce, R. C., Langley, P. E., and Rebec, G. V., Acute and long-term amphetamine treatments alter extracellular ascorbate in neostriatum but not nucleus accumbens of freely moving rats, *Life Sci.*, 49, 1237, 1991.
56. Calabresi, P., Demurtas, M., and Bernardi, G., The neostriatum beyond the motor function: experimental and clinical evidence, *Neuroscience*, 78, 39, 1997.
57. Graybiel, A. M., The basal ganglia and chunking of action repertoires, *Neurobiol. Learn. Mem.*, 70, 119, 1998.
58. Middleton, F. A. and Strick, P. L., Basal ganglia output and cognition: evidence from anatomical behavioral and clinical studies, *Brain Cogn.*, 42, 183, 2000.
59. Bennett, B. D. and Wilson, C. J., Synaptology and physiology of neostriatal neurones, In *Brain Dynamics and the Striatal Complex*, Miller, R. and Wickens, J., Eds., Harwood Academic Publishers, Australia, 2000.
60. Kiyatkin, E. A. and Rebec, G. V., Dopaminergic modulation of glutamate-induced excitations of neurons in the neostriatum and nucleus accumbens of awake unrestrained rats, *J. Neurophysiol.*, 75 142, 1996.
61. Gardiner, T. W., Armstrong-James, M., Caan, A. W., Wightman, R. M., and Rebec, G. V., Modulation of neostriatal activity by iontophoresis of ascorbic acid, *Brain Res.*, 344, 181, 1985.
62. Pierce, R. C. and Rebec, G. V., Iontophoresis in the neostriatum of awake unrestrained rats: differential effects of dopamine glutamate and ascorbate on motor- and nonmotor-related neurons, *Neuroscience*, 67, 313, 1995.
63. Kiyatkin, E. A. and Rebec, G. V., Ascorbate modulates glutamate-induced excitations of striatal neurons, *Brain Res.*, 812, 14, 1998.
64. Majewska, M. D., Bell, J. A., and London, E. D., Regulation of the NMDA receptor by redox phenomena—inhibitory role of ascorbate, *Brain Res.*, 537, 328, 1990.
65. Cortright, J. J. and Rebec, G. V., Ascorbate modulation of sensorimotor processing in striatum of freely moving rats, *Brain Res.*, 1092, 108, 2006.
66. Kiyatkin, E. A. and Rebec, G. V., Modulation of striatal neuronal activity by glutamate and GABA: iontophoresis in awake, unrestrained rats, *Brain Res.*, 822, 88, 1999.
67. Rebec, G. V., Barton, S. J., and Ennis, M. D., Dysregulation of ascorbate in the striatum of behaving mice expressing the Huntington's disease gene, *J. Neurosci.*, 22, RC202:1-6, 2002.
68. Rebec, G. V., Barton, S. J., Marseilles, A. M., and Collins, K., Ascorbate treatment attenuates the Huntington behavioral phenotype in mice, *Neuroreport*, 14 (9), 1263, 2003.
69. Rebec, G. V., Conroy, S. K., and Barton., S. J., Hyperactive striatal neurons in symptomatic Huntington R6/2 mice: variations with behavioral state and repeated ascorbate treatment, *Neuroscience*, 137, 327, 2006.
70. Huntington's Disease Collaborative Group, A novel gene containing a trinucleotide repeat that is expanded and unstable on Huntington's disease chromosomes, *Cell*, 72, 971, 1993.
71. Luthi-Carter, R., Strand, A., Peters, N. L., Solano, S. M., Hollingsworth, Z. R., Menon, A. S., Frey, A. S. et al., Decreased expression of striatal signaling genes in a mouse model of Huntington's disease, *Hum. Mol. Genet.*, 9, 1259, 2000.
72. Panov, A. V., Gutekunst, C. A., Leavitt, B. R., Hayden, M. R., Burke, J. R., Strittmatter, W. J., and Greenamyre, J. T., Early mitochondrial calcium defects in Huntington's disease are a direct effect of polyglutamines, *Nat. Neurosci.*, 5, 731, 2002.
73. Mangiarini, L., Sathasivam, K., Seller, M., Cozens, B., Harper, A., Hetherington, C., Lawton, M. et al., Exon 1 of the HD gene with an expanded CAG repeat is sufficient to cause a progressive neurological phenotype in transgenic mice, *Cell*, 87, 493, 1996.
74. Carter, R. J., Lione, L. A., Humby, T., Mangiarini, L., Mahal, A., Bates, G. P., Dunnett, S. B., and Morton, A. J., Characterization of progressive motor deficits in mice transgenic for the human Huntington's disease mutation, *J. Neurosci.*, 19, 3248, 1999.
75. Davies, S. W., Turmaine, M., Cozens, B. A., DiFiglia, M., Sharp, A. H., Ross, C. A., Scherzinger, E., Wanker, E. E., Mangiarini, L., and Bates, G. P., Formation of neuronal intranuclear inclusions underlies the neurological dysfunction in mice transgenic for the HD mutation, *Cell*, 90, 537, 1997.

76. Klapstein, G. J., Fisher, R. S., Zanjani, H., Cepeda, C., Jokel, E. S., Chesselet, M. F., and Levine, M. S., Electrophysiological and morphological changes in striatal spiny neurons in R6/2 Huntington's disease transgenic mice, *J. Neurophysiol.*, 86, 2667, 2001.
77. Rebec, G. V., Langley, P. E., Pierce, R. C., Wang, Z., and Heidenreich, B. A., A simple micromanipulator for multiple uses in freely moving rats: electrophysiology, voltammetry, and simultaneous intracerebral infusions, *J. Neurosci. Methods*, 47, 53, 1993.
78. Rebec, G. V., Christensen, J. R. C., Guerra, C., and Bardo, M. T., Regional and temporal differences in dopamine efflux in the nucleus accumbens during free-choice novelty, *Brain Res.*, 776, 61, 1997.
79. Garris, P. A., Christensen, J. R. C., Rebec, G. V., and Wightman, R. M., Real-time measurement of electrically evoked extracellular dopamine in the striatum of freely moving rats, *J. Neurochem.*, 68, 152, 1997.
80. Rebec, G. V. and Wang, Z., Microtechnology for monitoring real-time electrical and chemical events at the neuronal level in freely behaving animals, *Proc. IEEE-EMBS Asia-Pacific Conf. Biomed. Eng.*, 2, 511, 2000.
81. Yavich, L., A new technique for measuring the temporal characteristics of the carbon fibre microelectrodes in in vivo voltammetry at millisecond time intervals, *J. Neurosci. Methods*, 84, 29, 1998.
82. Lievens, J.-C., Woodman, B., Mahal, A., Spasic-Boscovic, O., Samuel, D., Kerkerian-Le Goff, L., and Bates, G. P., Impaired glutamate uptake in the R6 Huntington's disease transgenic mice, *Neurobiol. Dis.*, 8, 807, 2001.
83. Andersen, J. K., Oxidative stress in neurodegeneration: cause or consequence, *Nat. Rev. Neurosci. Neurodegeneration*, 5, S18, 2004, http://www.nature.com/focus/neurodegen/.
84. Barnham, K. J., Masters, C. L., and Bush, A. I., Neurodegenerative diseases and oxidative stress, *Nat. Rev. Drug Discov.*, 3, 205, 2004.
85. Coyle, J. T. and Puttfarcken, P., Oxidative stress, glutamate, and neurodegenerative disorders, *Science*, 262, 689, 1993.
86. Trotti, D., Danbolt, N. C., and Volterra, A., Glutamate transporters are oxidant-vulnerable: a molecular link between oxidative and excitotoxic neurodegeneration, *Trends Pharmacol. Sci.*, 19, 328, 1998.
87. Levine, M. S., Klapstein, G. J., Koppel, A., Gruen, E., Cepeda, C., Vargas, M. E., Jokel, E. S. et al., Enhanced sensitivity to *N*-methyl-D-aspartate receptor activation in transgenic and knockin mouse models of Huntington's disease, *J. Neurosci. Res.*, 58, 515, 1999.

10 Biophysical Properties of Brain Extracellular Space Explored with Ion-Selective Microelectrodes, Integrative Optical Imaging and Related Techniques

Sabina Hrabětová and Charles Nicholson

CONTENTS

INTRODUCTION: EXTRACELLULAR SPACE

The extracellular space (ECS) may be imagined as a system of interconnected channels demarcated by cellular membranes (Figure 10.1) and filled with an ionic solution, primarily NaCl, along with macromolecules of the extracellular matrix, negatively-charged proteoglycans and glycosaminoglycans (Margolis and Margolis 1993; Ruoslahti 1996; Novak and Kaye 2000). The ECS mediates intercellular communication (Nicholson 1979) and the transport of nutrients and metabolites (Syková et al. 2000), and it forms a reservoir of ions that establishes the resting potentials of cells and mediate fluxes across the membranes. It also may serve to deliver therapeutic substances to cells (Ulbrich et al. 1997; Saltzman 2001). All these processes are primarily mediated by diffusion. The ECS diffusive properties are thus critical for neurotransmission over short and

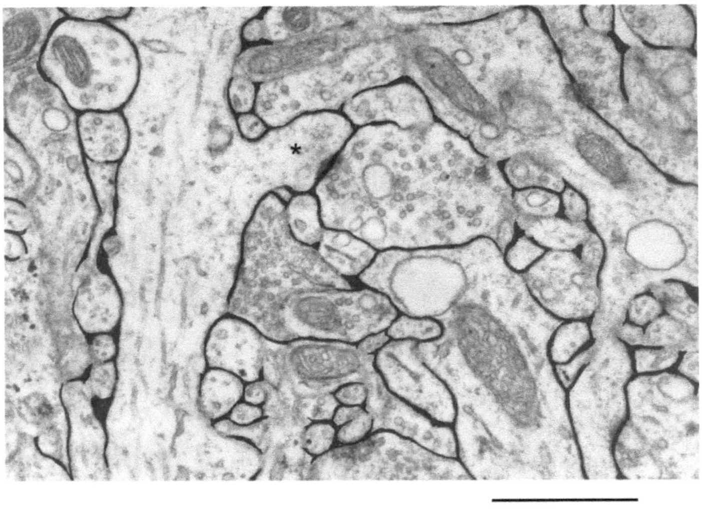

FIGURE 10.1 Electron micrograph of a small region of the cerebral cortex of a rat with a prominent synapse. The black areas between cells indicate the ECS, which may have been reduced in size as a consequence of the histological processing. The asterisk indicates an extension of a dendritic profile into a dendritic spine. The spine is joined by a synaptic cleft to a pre-synaptic terminal containing several small round vesicles filled with transmitter molecules. The presynaptic membrane and vesicles, cleft and immediate post-synaptic membrane constitute the synapse. Some of the round profiles abutting the dendrite are probably unmyelinated axons cut in cross-section. The scale bar represents a distance of about 1 μm. (Reproduced from Nicholson, C., *Rep. Prog. Phys.*, 64, 815–884, 2001.)

long distances (the latter being called volume transmission, Fuxe and Agnati 1991; Agnati et al. 2000), neurotrophic effects, general electrical activity, basic cellular homeostasis, and drug delivery. This chapter will introduce the brain ECS, describe techniques and preparations most commonly used to measure diffusion properties together with relevant theory and software for data analysis, review selected findings, demonstrate the utility of mathematical modeling and computer simulations, and outline some future directions for research.

Brief History

The middle years of the 20th century witnessed great controversy about the ECS. Experiments that measured the distribution of sodium and chloride in the brain, both regarded as having a predominantly extracellular distribution, indicated that the ECS might occupy as much as 40% of brain volume while emerging electron microscopy of the CNS indicated that the spaces between cells were so small that the ECS might only occupy 5% of the brain volume or even be absent altogether (Horstmann and Meves 1959; Gonzalez-Anguilar 1969; Johnston and Roots 1972).

After it was recognized that sodium and chloride had an appreciable intracellular distribution, investigators turned to radiolabeled sucrose, inulin and other membrane impermeant markers to measure the ECS volume fraction, resulting in values of 15%–20% (Fenstermacher and Kaye 1988). At the same time it became evident that the ischemia that occurred during fixation and also the subsequent chemical processing associated with conventional electron microscopy cause a drastic reduction of ECS volume (Van Harreveld 1972). Later studies (ibid.) employed freeze substitution, which to some extent remedies the shortcomings of conventional fixation, and also arrived at an ECS volume fraction of 15%–20%. These estimates are close to the values typically obtained with the modern diffusion methods that will be described in detail in this chapter. The existence of a substantial ECS then became an accepted concept and, the structure of the ECS was likened to the water phase of a foam (Kuffler and Potter 1964).

Even while some questioned the existence of the ECS, others debated its content. It was recognized early it its history that the ECS could not merely be filled with just a facsimile of cerebrospinal fluid (CSF) but also likely contained a "ground substance" composed of "mucopolysacharrides" and related molecules (Schmitt and Samson 1969; Johnston and Roots 1972). Today the reality of what is now called the extracellular matrix (ECM) is accepted but the composition, amount, and distribution are still debated.

Volume Fraction and Width

The macroscopic properties of the ECS are summarized by two aggregate parameters: the volume fraction and the tortuosity (discussed in the next section). The volume fraction α represents the proportion of tissue volume (V) occupied by the ECS; $\alpha = V_{ECS}/V_{tissue}$. In healthy isotropic brain, $\alpha \approx 0.2$ (Nicholson and Syková 1998; Nicholson 2001), which, as noted in the previous section, means that typically the ECS occupies 20% of brain tissue. From a practical point of view, α determines the dilution of substances released to the ECS.

In addition to α, the width and shape of a two-dimensional (2D) cross section of individual ECS channels can be estimated from electron micrographs. The gaps between the cells were estimated to be 10–20 nm wide (Brightman 1965; Johnston and Roots 1972; Van Harreveld 1972) when brain tissue was processed using freeze substitution or other techniques that were thought to preserve the dimensions. However, it is becoming evident that electron micrographs underestimate the width of ECS channels. Recent diffusion studies employing large ECS probes, such as the polysaccharide Ficoll (hydrodynamic diameter 41 nm, Hrabětová and Nicholson 2002) in rat neocortex in vitro or quantum dots (hydrodynamic diameter 34 nm, Thorne and Nicholson 2006) in rat neocortex in vivo suggest that at least some interstitial spaces are sufficiently wide and interconnected to permit these large entities to diffuse in the ECS. Regarding the shape of ECS channels, electron micrographs uncovered many aspects such as widely distended regions, "lakes" (Van Harreveld et al. 1965) and intercellular gaps of uneven width (Brightman 1965; Bondareff and Pysh 1968; Cragg 1979), although it cannot be excluded that some might be artifacts caused by ice crystal formation during the freeze substitution phase of the fixation procedure. Electron microscopy also revealed regions of the ECS surrounded by glial processes (Špaček 1985; Kosaka and Hama 1986; Grosche et al. 1999). The functional significance of these morphological variations of the ECS channels remains unknown.

Hindrance and Tortuosity

Molecular transport in brain ECS is hindered by the structure of the tissue compared to a free medium. Diffusion analysis of small extracellular marker molecules quantifies this hindrance, expressed as tortuosity $\lambda = (D/D^*)^{1/2}$, where D is the free diffusion coefficient and D^* is the effective diffusion coefficient in tissue (Nicholson and Phillips 1981; Nicholson 2001). In isotropic healthy brain, λ is ≈ 1.6 (Nicholson and Syková 1998; Nicholson 2001). From a practical point of view, λ determines the ease of spread of molecules in the ECS.

Unlike α, the nature of λ is incompletely understood; it has proven difficult to ascertain which structural elements of the brain determine λ. Cells and their processes create obstacles that diffusing molecules have to navigate around; thus geometry of the ECS, as determined by the shape of cells undoubtedly contributes to the tortuosity of brain tissue. The open question is whether ECS geometry alone can account for the value of this parameter or whether other factors, such as viscosity of the extracellular fluid arising from the presence of the ECM, come into play. We shall return to this issue using computer simulation techniques in the Section: "Monte Carlo Simulation of Diffusion in 3D Media."

Bulk Flow

It has been postulated from time to time that there is a bulk flow of interstitial fluid within the ECS, but conclusive evidence is still lacking. The most probable origin of bulk flow would be an osmotically-balanced fluid that is secreted across the blood–brain barrier (BBB), although some flow might be generated by cells when water is produced through metabolism (Abbott 2004). The fluid then flows into the CSF or enters the lymphatics (ibid.). The actual route for the flow was indicated by Cserr and colleagues (Ichimura et al. 1991) who concluded that bulk flow was restricted to the Virchow–Robin perivascular space surrounding the capillaries, rather than the whole ECS. The flow velocity has been estimated at only about 10 μm min^{-1} towards the ventricle in white matter and may not occur in grey matter at all under normal conditions (Rosenberg et al. 1980). These conclusions make it unlikely that bulk flow affects the diffusion measurements to be described in this chapter.

Volume Transmission and Drug Delivery

Electrically active neurons cause voltage changes in the immediate ECS, and synchronized populations can lead to quite widespread potentials; indeed this is the basis of the familiar EEG diagnostic recording. The invention of ion-selective microelectrodes (ISMs) established that the electrical fluctuations in the ECS were often accompanied by ionic variation, especially in extracellular potassium, which is typically at a low concentration of about 3 mM. Thus the idea arose that the electrical or ionic changes might transmit signals between brain cells along pathways, which were independent of the conventional synaptic connections, so the ECS reasonably might be regarded as a communication channel (Nicholson 1979). Little evidence has been found to support a role for these signal mediators in normal brain function, perhaps because they are non-specific, but the idea has been generalized into the concept of volume transmission (Fuxe and Agnati 1991) involving more complex molecules and receptors. The signaling domain can range from the local spillover of transmitter at a synapse, enabling the molecules to reach extrasynaptic receptors on the same or adjacent cell, to long distance communication using the cerebral ventricles as conduits for hormones (Agnati et al. 2000).

Volume transmission involves endogenous substances, either released by cells or entering from the ventricles. Obviously the ECS is also capable of distributing exogenous substances—drugs—so long as they can be introduced into the brain without undue damage (Saltzman 2001). Like volume transmission, drug delivery will rely on diffusion to transport substances but this is not effective over long distances. For this reason convection enhanced delivery (CED) is gaining popularity (Morrison et al. 1994). This uses pressure injection from a cannula to force a drug through the brain; after the pressure injection ends, the drug distribution is again dominated by diffusion (ibid.).

TOOLS AND METHODS TO STUDY EXTRACELLULAR SPACE IN REAL TIME

As mentioned above, two independent approaches have been adopted to study the ECS in the past: morphometric and diffusion methods. The morphometric method employs electron micrographs that directly visualize the interstitial channels in a small 2D region of fixed tissue, typically containing a few cells. The diffusion method uses the distribution of extracellular probe molecules or markers to quantify the aggregate parameters of the ECS from a larger 3D region, typically containing hundreds of cells. Individual interstitial channels are below the resolution of current diffusion methods. Diffusion methods offer two important advantages over the morphometric approach. First, diffusion methods can yield both α and λ, while the morphometric method quantifies only α (and, as noted, can be subject to systematic error caused by tissue fixation techniques). Second, and more importantly, contemporary real-time diffusion methods can be exploited to study dynamic changes of the ECS during normal and pathological conditions in living tissue, while morphometric methods are limited to fixed tissue.

Diffusion Measurements Reveal Extracellular Space Properties

The diffusion methods were pioneered by a group around Fenstermacher and Patlak (Rall et al. 1962, Levin et al. 1970) using radiotracers, while the contemporary, real-time techniques, which do not employ radiotracers, were developed by Nicholson and coworkers (Nicholson and Philips 1981; Nicholson 1993; Rice and Nicholson 1995). In principle, in all diffusion methods, a probe molecule is delivered to the ECS, and its subsequent distribution is measured and analyzed using the appropriate solution of the diffusion equation. The marker can be any molecule that stays predominantly in the ECS and can be quantitatively detected. In the most reliable early studies, radiolabeled sucrose or inulin were delivered into the ventricle and their distributions were obtained post mortem from sections of the caudate nucleus, the region adjacent to ventricles. The parameters measured in several species were in the range 15%–20% for α and 1.52–1.64 for λ (Fenstermacher and Kaye 1988). Although the diffusion of a radiotracer occurred in living tissue, the quantification was performed in post mortem tissue yielding a single time point. This drawback was overcome by Nicholson who introduced new methods based on release of a probe ion or molecule from a point source (Figure 10.2) and detection with ion-selective microelectrodes, voltammetry or fluorescence microscopy. In these techniques, the diffusion of probe molecules in the ECS of living tissue is detected in a real time and this facilitates experiments involving tissue manipulation. This chapter will largely focus on these contemporary real-time diffusion techniques.

The Diffusion Equation and Its Meaning

Typically, diffusion measurements are made over distances of the order of 100 μm and this ensures that the local anatomical properties are averaged, enabling a modification of the classical diffusion

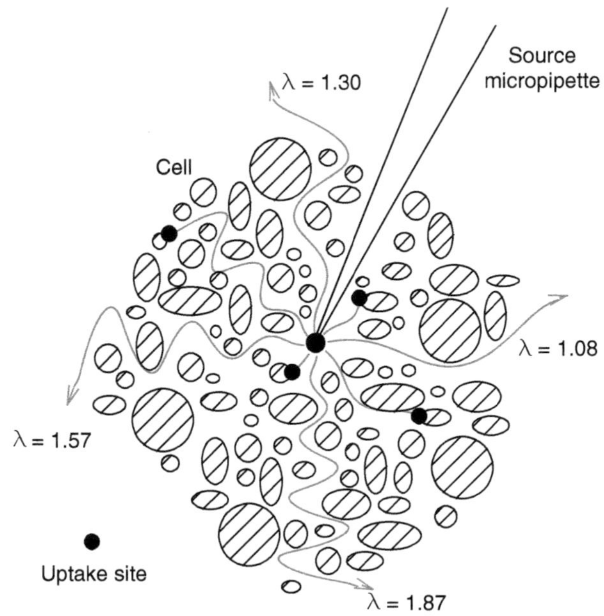

FIGURE 10.2 Point source paradigm for diffusion studies. Glass micropipette (*source micropipette*) is filled with extracellular probe molecules that are released to the ECS by passing current (iontophoretic delivery of small charged molecules) or by pressure injection (any molecule). The diameter of micropipette tip is small, ≈3–6 μm; the micropipette therefore is a good approximation of point source (*large black circle*). Released molecules diffuse through the ECS (*lines with arrowheads*) around cellular elements and possibly other diffusion barriers. Some molecules may be removed from the ECS at uptake sites (*black dots*). The diffusion of molecules is recorded by appropriate sensors (see text for details). (Reproduced from Nicholson and Syková, E., *Trends Neurosci*., 21, 207–215, 1998.)

equation of Adolf Fick to be used (Nicholson and Phillips 1981; Nicholson 2001):

$$\frac{\partial C}{\partial t} = \frac{D}{\lambda^2}\nabla^2 C + \frac{Q}{\alpha} - \mathbf{v}\cdot\nabla C - \frac{f(C)}{\alpha}. \tag{10.1}$$

The concentration, C, is that which exists in the narrow interstitial space; it is a function both of time t and position, represented by x, y, z in a rectangular coordinate system, or the single coordinate r in spherically symmetric coordinate system centered on the point source, as is frequently assumed here. The symbols ∇ and ∇^2 symbolize the first and second spatial derivatives in the appropriate coordinate system. The averaged tissue structure comes into Equation 10.1 through the two nondimensional parameters mentioned above, the volume fraction (α) and the tortuosity (λ).

The various terms of Equation 10.1 may be interpreted as follows. The term on the left of the equality sign represents the way the interstitial concentration changes with time at a given location. The first term on the right of the equality sign is the contribution of diffusion itself. The free diffusion coefficient is D and the effective diffusion coefficient in brain, $D^* = D/\lambda^2$, indicating the role of tortuosity in reducing the free diffusion coefficient. The next term is the source-term, Q, that may be used to describe iontophoresis from a micropipette (Nicholson and Phillips 1981; Nicholson 1992), or release of molecules by pressure ejection from a micropipette (Nicholson 1985, 1992). The third term represents the possible contribution of bulk flow. Flow is characterized by a velocity vector $\mathbf{v}$ that forms a scalar product with the concentration gradient ∇C. The final term, $f(C)$, represents irreversible loss or clearance of material from the ECS. This may occur through uptake into cells, loss across the BBB, or simply by the destruction of the molecule. Often clearance is proportional to local concentration and we can write $f(C) = -k'\alpha C$. On the other hand, if $f(C)$ represents uptake that is governed by, say, Michaelis–Menten kinetics, Equation 10.1 becomes non-linear and the analysis is more complex (Nicholson 1995). Equation 10.1 is quite general and must be solved for specific experimental situations. Some of the elementary solutions will be detailed in later sections.

Choice of Molecular Probes for Extracellular Space

The aim of the diffusion study determines the type of molecular probe employed. In the first type of study, we can quantify two macroscopic parameters, λ and α, that encapsulate the properties of the ECS. For this purpose, the ideal probe should fulfill the following criteria: water solubility; small size; no interaction with the extracellular environment; no membrane permeability; and, for in vivo studies, no transport across the BBB. This probe would visit all channels of the ECS and report true, probe-independent, ECS parameters. In the second type of study, a probe significantly violating one or more criteria outlined above is used. The use of such non-ideal probes allows us to ask questions about: (1) the width of the ECS channels; (2) interactions (non-specific and specific binding); and (3) loss from the ECS (transport into cells or across the BBB). Such probes might not visit all channels of the ECS or might be detained in, or lost from, the ECS compartment. The experimental outcome combines the basic properties of the ECS with one or more probe-specific elements. In the following paragraphs, some examples will be given for the two categories of molecular probes; a more comprehensive overview has been given by Nicholson (2001).

The truth is that an ideal probe molecule that would belong to the first group does not exist and molecules used in this context do not fully comply with criteria for an ideal probe. Molecular probes most commonly used in real-time diffusion techniques (Table 10.1) are small but of finite size; may carry a charge, thus introducing the possibility of interactions with negatively charged molecules of the ECM; and are lost in small amounts from the ECS. Based on studies reported to date, however, they appear to be a good approximation of an ideal probe. For example, α-naphthalenesulfonate, the largest ion listed in Table 10.1, has a hydrodynamic diameter (calculation based on a free diffusion coefficient) of only ≈ 1.0 nm, about 10 times less than the lower estimate of the width of an ECS channel from electron micrographs (10–20 nm, Brightman 1965; Johnston and Roots 1972; Van Harreveld 1972). Because both monovalent cations and anions measure similar λ (Nicholson and Phillips 1981), no charge-based interactions are manifested. Finally, non-specific clearance (also called a non-specific uptake), which quantifies the loss of a probe from the ECS (Nicholson 1992,

TABLE 10.1
Selected Measurements of Tortuosity and Volume Fraction of Brain ECS Measured with Molecules Approximating Ideal Probe

Probe	M_r	Charge	λ	α	Reference
TMA	74	1+	1.49	0.18	Nicholson and Phillips (1981)[a]
TEA	130	1+	1.48	0.21	Nicholson and Phillips (1981)[a]
α-NS	207	1−	1.54	0.18	Nicholson and Phillips (1981)[a]
AsF_6	189	1−	1.68	0.26	Nicholson and Phillips (1981)[a]
Sucrose	342	Neutral	1.60	0.21	Fenstermacher and Kaye (1988)[b]

[a] Rat cerebellar molecular layer in vivo.
[b] Rabbit caudate nucleus in vivo.

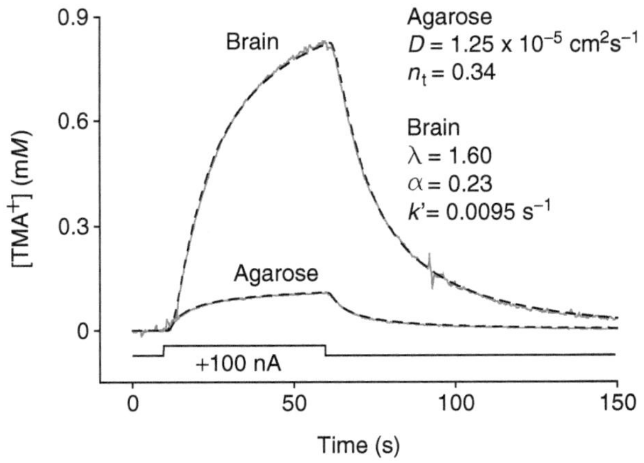

FIGURE 10.3 Diffusion of tetramethylammonium (TMA) detected with an ion-selective microelectrode (ISM) in rat neocortex in vitro. Tetramethylammonium (*MW* 74) was delivered by an iontophoretic pulse (+100 nA, 50 s) and the concentration was measured with a TMA^+-ISM positioned 120 μm away. The measurements were done at 34°C. Recorded curves (*solid grey lines*) are superimposed with the theoretical curves (*short dashed lines*). In agarose, the TMA^+ diffusion curve is used to estimate the free diffusion coefficient (D cm^2 s^{-1}) and the transport number, n_t, using the known spacing between two microelectrodes r. In brain, the effective diffusion coefficient (D^* cm^2 s^{-1}), volume fraction (α), and non-specific clearance (k' s^{-1}) was extracted using r and n_t. The tortuosity (λ) was then calculated as $(D/D^*)^{1/2}$.

2001), is small in both in vitro and in vivo preparations ($k' \approx 10^{-3}$ s^{-1}), or is even absent when brain slices are held in an interface rather than a submersion chamber (Nicholson et al. 2000). In summary, small ions can be used as extracellular probes in the diffusion analysis of ECS properties; in isotropic healthy brain tissue, $\alpha \approx 0.2$ and $\lambda \approx 1.6$ (Figure 10.3, Nicholson 2001; Nicholson and Syková 1998). These estimates of α and λ agree well with radiotracer studies employing sucrose, a small (*MW* 342), neutral, membrane impermeable molecule (Table 10.1; Fenstermacher and Kaye 1988).

The studies employing "imperfect" probes from the second category have utility even for investigators not primarily interested in the structure of the ECS. Using inert globular molecules of increasing molecular weight, such as dextrans, violates the criterion of small size. We observe that as the hydrodynamic diameter increases, λ increases as the hindrance to diffusion rises (Figure 10.4). Note that the largest molecule shown in Figure 10.4 has a diameter approaching the average electron microscopic estimate of the width of an ECS channel. When even larger molecules are used, such as Ficoll or quantum dots (Hrabětová and Nicholson 2002; Thorne and Nicholson 2006) with hydrodynamic diameter significantly larger than the upper estimate of width of the ECS (20 nm), one is surprised that they diffuse in brain at all. This finding indicates that the width of at least some channels of the ECS is larger than previously thought, and that these wide channels are sufficiently interconnected to permit macroscopic diffusion processes. In summary, using probes of large sizes reveals new information about the biophysical properties of the ECS and has practical implications for effective design of macromolecular drug carriers and viral vectors underpinning gene therapy.

Examples of an interaction between the probe molecule, specific receptors and the extracellular environment include measurement of the diffusion of neurotransmitters (Rice et al. 1985; Cragg et al. 2001; Cragg and Rice 2004—see Section: "Diffusion Measurements Using Dual Probe Microdialysis Probes"); second messengers such as calcium (Morris and Krnjević 1981; Nicholson and Rice 1987; Hrabětová and Nicholson 2004); and trophic factors (Krewson et al. 1995; Stroh et al. 2003; Thorne et al. 2004).

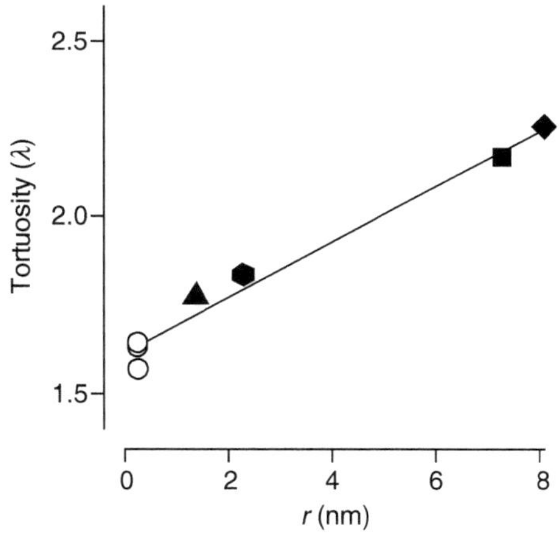

FIGURE 10.4 Tortuosity as a function of hydrodynamic radius. Diffusion of TMA^+ (*MW* 74, *white circles*) in a rat neocortex in vivo and in vitro was quantified using real-time iontophoresis (RTI). The diffusion of fluorophore-labeled dextran molecules (*MW* 3,000, *black triangle*; *MW* 10,000, *black hexagon*; *MW* 40,000, *black square*; *MW* 70,000, *black diamond*) in a rat neocortex in vitro was quantified using Integrative Optical Imaging (IOI). The hydrodynamic radius was calculated from Einstein's formula, $R_H = kT/(6\pi\eta D)$, where k is Boltzmann's constant, T is absolute temperature (K), η is viscosity (mN s m^{-2}), and D is the free diffusion coefficient. The TMA^+ and dextran molecules fitted quite well to a linear regression, with an intercept for $R_H = 0$ of $\lambda = 1.63$. (Reproduced from Tao, L. and Nicholson, C., *Neuroscience*, 75, 839–847, 1996.)

Finally, the older studies with radiotracers showed that non-specific uptake of a molecular probe into cells can be measured when mannitol or creatinin are used, and loss across the BBB can be determined when ethylene glycol or urea are employed (reviewed in Nicholson 2001). In midbrain, specific uptake of neurotransmitter dopamine (DA) can be quantified when the diffusion is measured in the absence and presence of dopamine uptake blockers (Figure 10.5; Cragg et al. 2001).

By combining studies, the biophysical properties of the ECS can be explored using diffusion analysis. True, probe non-specific, diffusion parameters can be measured with small exogenous ions. In contrast, probe molecules of large size, strong charge, containing binding motifs or those that can penetrate cellular membranes or move across the BBB, explore molecular interactions with extracellular environment. Next, we will discuss the methods of delivery and the detection; a schematic of the recording setup used for diffusion measurements is shown in Figure 10.6.

Delivery of Molecular Probes: Iontophoresis and Pressure Ejection

Two methods of delivery are employed in point source real-time diffusion techniques: iontophoresis and pressure ejection, both from a glass micropipette. In iontophoretic delivery, charged molecules are released by passing a current (on the order of tens to hundreds of nA) through the micropipette. The advantage is that ions (not the solution) can be released in a precise amount by controlling the iontophoretic current and this amount can be quantified with an appropriate detector. The pressure ejection technique has broader applicability because solutions of both charged and neutral molecules can be delivered; the disadvantage, however, is the injected volume is hard to control and it varies with the medium into which the ejection is made.

The release by iontophoresis or pressure injection can be triggered manually, but is typically controlled by the iontophoretic current generator or pressure ejection system from a MATLAB-based data acquisition program (Wanda, see Section: "Specialized Software for Point-Source

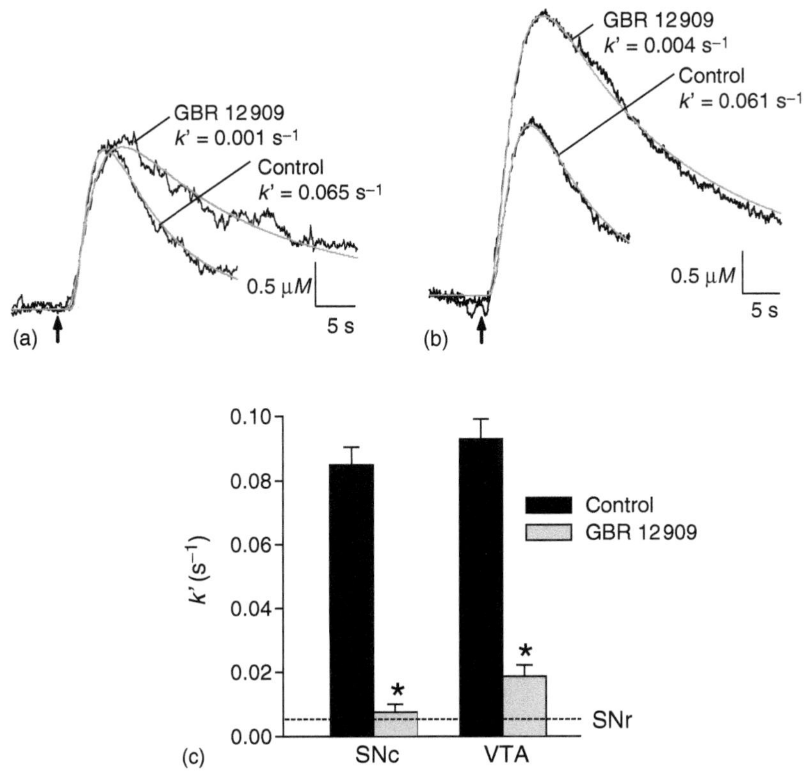

FIGURE 10.5 Diffusion analysis determines specific uptake of neurotransmitter dopamine (DA) in midbrain. Dopamine was released by pressure injection and detected with a carbon fiber microelectrode (CFM) using fast-scan cyclic voltammetry (FCV). To quantify the specific uptake by DA transporters, DA diffusion was measured under control conditions (*control*) and in the presence of the DA uptake blocker, *GBR 12909*. Representative diffusion curves of DA in substantia nigra pars compacta (SNc; a) and ventral tegmental area (VTA; b). (c) Summary. Inhibition of DA transporters significantly (*asterisks*) reduces DA uptake in SNc and VTA. For comparison, the low level of clearance in substantia nigra reticulata (SNr), a region with sparse DA processes, is shown (dashed line). (Reproduced from Cragg, S. J. et al., *J. Neurophysiol.*, 85, 1761–1771, 2001. With permission.)

Diffusion Analysis") running on a PC equipped with an A/D and D/A converter (model PCI-MIO-16E-4; National Instruments Corp., Austin, TX), as the timing of delivery is critical for data analysis.

Iontophoresis

For iontophoresis, a micropipette is pulled from a borosilicate glass capillary (single barrel capillaries from A-M Systems, Inc., Carlsborg, WA, catalog # 617000 or double barrel capillaries from Harvard Apparatus Ltd, Holliston, MA, catalog # 30-0117) and the tip is broken (by bumping against the edge of a glass slide under a microscope with 100× total magnification) to have an outer diameter (o.d.) of 3–6 μm. We prefer tips of this size to a larger diameter because they create less damage to the tissue upon insertion, and because with a bigger tip, the iontophoretic current carries large amounts of ions creating uncontrollable flow. The micropipette is back-filled with a concentrated solution of the ion (100 m*M*—1 *M* made in distilled water), a chloridized silver wire is inserted into the barrel and immersed in the filling solution. The barrel is then sealed with a hard, fast-setting wax (Sticky Wax, KerrLab Corporation, Orange, CA, catalog # 00625) to

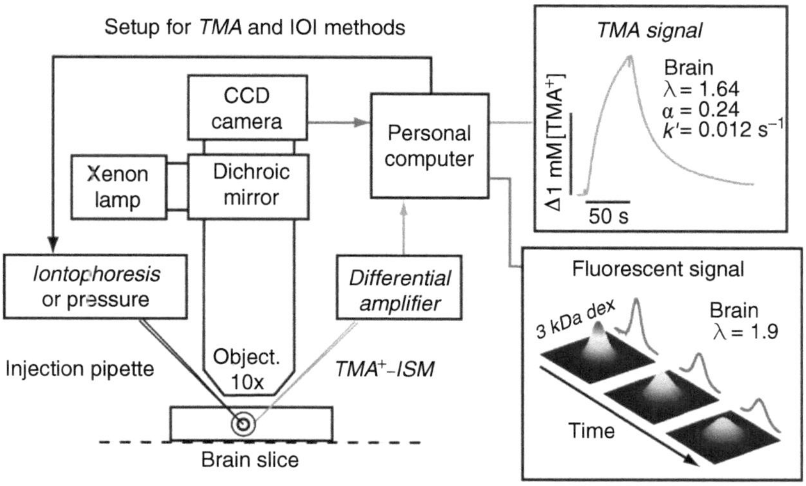

FIGURE 10.6 Recording setup for diffusion measurements using real-time iontophoretic (RTI) and integrative optical imaging (IOI) methods. The specimen (dilute agarose gel, brain slice, or living animal) is placed on the stage of a compound microscope. For RTI, an extracellular probe, most commonly TMA^+, is released from a glass micropipette (*Injection pipette*) by iontophoresis and detected with an ion-selective microelectrode (*TMA^+-ISM*) positioned about 100 μm away. The resulting voltage versus time curve is amplified, digitized and converted to a diffusion curve (concentration versus time) on a computer enabling an appropriate diffusion equation to be fitted to the data (Nicholson, 2001). In diluted agarose gel, the free diffusion coefficient, D ($cm^2\ s^{-1}$), and the transport number, n_t, are measured. In brain, effective diffusion coefficient, D^* ($cm^2\ s^{-1}$), volume fraction, α, and non-specific clearance (uptake), k' (s^{-1}), are measured. The tortuosity, λ, is calculated as $(D/D^*)^{1/2}$. For IOI, a fluorescent molecule, here 3000 *MW* dextran labeled with Texas Red fluorophore, is pressure injected and a time series of images captured using a charge-coupled device (CCD) camera. The appropriate solution of the diffusion equation is fitted to the intensity profiles extracted along selected image axes (Nicholson, 2001). The diffusion coefficients, D and D^*, are extracted from measurements in dilute agarose gel and brain, respectively. The tortuosity, λ, is calculated as $(D/D^*)^{1/2}$. As shown in this figure, RTI and IOI measurements can be performed simultaneously.

prevent evaporation and mechanical artifacts from wire movement. The iontophoretic pipette is attached via the silver wire to an accurate source of constant current (e.g., an iontophoretic current generator ION-100 made by Dagan Corporation, Minneapolis, MN). Positively and negatively charged ions are ejected by passing positive and negative current, respectively. When a sensor selective for a particular ion (see below) is used, transport number (n_t) of the iontophoretic electrode can be quantified in a free solution (where $\alpha = 1$). For this purpose, dilute agarose gel (0.3% in 150 m*M* NaCl; NuSieve GTG, FMC BioProducts, Rockland, ME) rather than water or isoosmotic pH-balanced solution is used to avoid the influence of thermal convection on the measurement. In theory, n_t should be about 0.5 if the dissociated cation and anion have similar mobility, because half of the current passing through the micropipette should be carried by ions of the same polarity leaving the micropipette through the tip, while the other half is carried by ions of opposite charge entering the tip. In practice, however, various factors lower n_t, such as geometry of the tip and interaction with the glass.

Following are several practical tips based on many years of experience. First, even when a current generator indicates that the current is passing, sometimes only ions of the opposite charge are entering the tip of the micropipette but no ions are moving out. In general, without detecting released molecules with a selective sensor, no assumption can be made about whether there is release or how much (unfortunately this is equally true when microiontophoresis is used in other contexts to release transmitter substances or their analogues). Second, it is a common practice in other work (e.g., iontophoresis of transmitters) to use a retaining current of the opposite polarity at the end of the delivery pulse to hold the charged substance in the barrel. While this strategy fulfills its purpose, this can allow the solution around the micropipette to enter and dilute the tip. Subsequently, the transport number of the micropipette changes, resulting in a different amount of ions being released by the next pulse. For diffusion measurements, the dilution is prevented by constantly passing a small forward bias current of the same polarity as the main ejection current (Nicholson and Phillps 1981). Third, iontophoresis can be used to deliver large charged molecules (proteins, dextrans); however, it is less reliable. Typically, a large current has to be applied and these molecules are most probably carried out by a water flow (electroosmosis) rather than true iontophoresis.

The solution to Equation 10.1 for iontophoresis is (Nicholson and Philips 1981; Nicholson 2001):

$$C(t) = \frac{Q}{8\pi D^{*}\alpha r}\left[\operatorname{erfc}\left(\frac{r}{2\sqrt{D^{*}t}}+\sqrt{k't}\right)\exp\left(r\sqrt{\frac{k'}{D^{*}}}\right)+\operatorname{erfc}\left(\frac{r}{2\sqrt{D^{*}t}}-\sqrt{k't}\right)\exp\left(-r\sqrt{\frac{k'}{D^{*}}}\right)\right]. \tag{10.2}$$

Here, erfc is the complementary error function. Equation 10.2 represents the behavior of the concentration C at distance r from the source, after the iontophoretic pulse is switched on and maintained, i.e., the current is a step function. To account for what happens when the iontophoretic current is switched off, e.g., to make the step function into a pulse of duration t_p, a delayed version of Equation 10.2 is subtracted from the original function:

$$C(t > t_p) = C(t) - C(t - t_p). \tag{10.3}$$

It is important to note that Equation 10.2 can be extended to take account of anisotropy in the tissue; the details may be found in (Rice et al. 1993; Nicholson 2001).

For iontophoresis, the magnitude of the source Q is given by

$$Q = In_t/zF \tag{10.4}$$

where I is the magnitude of the iontophoresis current, n_t the transport number of the iontophoresis electrode, z the valency of the ion and F is Faraday's electrochemical equivalent.

Pressure Ejection

For pressure ejection, a micropipette is pulled from a borosilicate glass capillary (single barrel capillaries, A–M Systems, Inc., catalog # 617000) and the tip is again broken to have an o.d. of 3–6 μm. Because sub-nanoliter volumes are typically injected (Cragg et al. 2001), only the very tip of the micropipette needs to be filled with solution. Extreme care must be exercised to remove all air bubbles from the solution by gently tapping the micropipette because, when compressed during the pressure pulse, air bubbles prevent the ejection of the solution from the micropipette. A Teflon® tube of suitable diameter (Small Parts Inc, Miami Lakes, FL, STT-30–50 PTFE Tube) is inserted in the open end of the micropipette that is then sealed with melted wax (as used for iontophoresis electrodes; Teflon® tubing is used so that the hot wax does not melt it) and connected to a pressure ejection system (e.g., Picospritzer III, Parker Hannifin Corp., Pneutronics Division, General Valve Operation., Fairfield, NJ). Compressed nitrogen is used in our laboratory but any non-toxic compressed gas would do.

The appropriate solution to Equation 10.1 for pressure ejection is for the concentration C at distance r from the source (Nicholson 1993, 2001):

$$C(t) = \frac{Q}{\alpha} \frac{1}{(4D^* t\pi)^{3/2}} \exp\left(-\frac{r^2}{4D^* t} - k' t\right) \quad (10.5)$$

This assumes that the pressure pulse is very brief in duration and confined to a point but still releases a finite amount of substance, i.e., it is an impulse or mathematical delta-function in both time and space. While this latter assumption is obviously violated, it is generally an adequate approximation because measurements are made at some distance and time from the source. When the point-source assumption breaks down, a more realistic solution for a finite volume is available (Nicholson 1985, 1993, 2001).

For pressure ejection described by Equation 10.5, the source is defined as

$$Q = UC_{\mathrm{f}} \quad (10.6)$$

where U is the volume ejected and C_{f} is the concentration of the ejected solution. Unfortunately, U is generally not known accurately, and because α is also unknown it is evident from Equation 10.5 that only the ratio U/α can be estimated. Indeed, C_{f} may not be well defined either because of dilution of the solution in the tip of the ejecting electrode (a problem that is overcome with a bias current in iontophoresis).

Detection of Molecular Probes: Ion-Selective Microelectrodes, Integrative Optical Imaging and Carbon-Based Microelectrodes

Three detection systems have been used: ion-selective microelectrodes (ISMs), integrative optical imaging (IOI) employing fluorescence microscopy and a (charge-couple device CCD) camera and carbon-based sensors combined with chronoamperometry or voltammetry.

The ISM is limited to detecting charged molecule for which a selective ion exchanger exists in a formulation that works in a micropipette (for overview see, Ammann 1986; Nicholson 1993). These probe molecules may be delivered by pressure ejection or iontophoresis; when an ISM (or occasionally a carbon-based sensor) is used in a tandem with iontophoresis, the technique is called the real-time iontophoretic (RTI) method. The IOI method quantifies the diffusion of neutral or charged fluorophore-labeled substances, usually in the form of macromolecules; these probe molecules are almost always delivered by pressure ejection. Carbon-sensor methods quantify the diffusion of electroactive substances. In following three subsections, ISMs, IOI and carbon-based sensors are described briefly; detailed descriptions of the techniques and theory can be found elsewhere (Nicholson and Phillips 1981; Nicholson and Rice 1988; Nicholson 1992, 1993, 2001; Nicholson and Tao 1993; Tao and Nicholson 1995).

Ion-Selective Microelectrodes

To fabricate an ISM, a micropipette is pulled from double-barreled capillaries (Harvard Apparatus Ltd, catalog # 30-0117). The capillaries are cleaned internally by flushing with acetone and dried by forcing compressed gas, pulled and the tip broken to have an o.d. of 3–6 μm. The reference barrel is filled with 150 mM NaCl and a chloridized silver wired sealed in with wax (see "Iontophoresis"). The barrel to be used for ion sensing is filled with a 100–150 mM solution of the ion to be sensed and a Teflon® tube (see "Pressure Ejection") is temporarily sealed with wax into the back of the barrel. The inner surface of the tip of the ion-detecting barrel is then coated with a solution of chlorotrimethylsilane (Sigma-Aldrich, Inc., St Louis, MO, catalog # C72854; 4% v/v in xylene) by applying pressure and suction (a disposable syringe may be used) through the Teflon® tube to make the surface hydrophobic and organophilic and prepare it to retain the ion-exchanger. After silanization, the ion exchanger, a liquid-membrane containing carrier molecules selective for the ion (Ammann 1986; Nicholson 1993) is drawn in and finally the Teflon® tube is removed and a chloridized silver wire inserted in this barrel. The length of the exchanger column in the tip determines the resistance of the barrel and this resistance can be made small by inserting another capillary; this will shorten the response time of the ISM (Ujec et al. 1979). For example, the response time of an ion exchanger column ≈200 μm long might be on the order of hundreds of milliseconds (Nicholson 1993). However, when the length of the exchanger column is reduced to 10–20 μm, the response time is shortened below 10 ms (Ujec et al. 1979). The ion-selective barrel detects a potential corresponding to the concentration of the ion of interest together with the local potential in the tissue at the tip of the electrode while the reference barrel detects the local potential only. Consequently, a potential that exclusively reflects the ion concentration is obtained by continuously subtracting the potential measured by the reference barrel from the signal measured on the ion-selective barrel using a dual channel microelectrode preamplifier (model IX2-700; Dagan Corp., Minneapolis, MN). Each ISM is calibrated in a set of standard solutions in which the concentration of the ion of interest is increased by doubling its amount against background of 150 m*M* NaCl (the Fixed Interference Method, Nicholson 1993). Calibration parameters are used to obtain the slope and the interference of the electrode by fitting the data to the Nikolsky Equation (Ammann 1986; Nicholson 1993). The slope and interference are used for potential-to-concentration conversion. A detailed description of ISM fabrication can be found in (Nicholson and Rice 1988; Nicholson 1993).

For diffusion measurements in agarose gel or brain, the iontophoretic microelectrode and the ISM are glued together to make a fixed array with inter-tip spacing 100–200 μm (Nicholson 1993). Alternatively, when robotic micromanipulators (such as MP 285; Sutter Instrument Co., Novato, CA) are available, the two microelectrodes can be held and moved independently (Hrabětová and Nicholson 2000, 2004). In the second scenario, the recording chamber has to be in a setup equipped with optics that allows the spacing between the tips of the electrodes to be set under visual control. In our laboratory we use a compound microscope (model BX61WI, Olympus America, Melville, NY) equipped with a water-immersion objective (Olympus, UMPlanFl, 10×, NA 0.3) and infrared (IR) differential interference contrast (DIC) optics. Independent microelectrodes offer great advantages when records are taken along *x*-, *y*-, and *z*-axes to test for anisotropy (i.e., when the diffusion is different along three orthogonal axes) of the region because the microelectrodes can be repositioned easily.

During diffusion measurements, a continuous bias current of 20 nA, with polarity appropriate to the ion, is applied from a constant-current, high-impedance source (see above). This iontophoretic current is stepped up to 60–100 nA for 50–100 s to obtain diffusion curves in dilute agarose gel and in brain. As mentioned above, the ion signal is extracted by continuous subtraction of the potential on the reference barrel from the potential on the ion-detecting barrel using a dual-channel microelectrode amplifier (model IX2-700; Dagan Corp., Minneapolis, MN). The headstage buffer amplifier connected to the ISM must have adequate impedance and a sufficiently low leakage current to accommodate the impedance of the ISM, which equals or exceeds 1000 MOhms. The ion and DC signals are amplified, low-pass filtered (6 Hz, 8 pole Bessel filter) using a CyberAmp 320 (Axon Instruments, Inc.), and continuously monitored on a chart recorder. The ion diffusion curves are digitized using an A/D converter installed in a personal computer and saved using a MATLAB-based (MathWorks, Natick, MA) program Wanda and analyzed with the MATLAB-based program Walter, both described in "Specialized Software for Point-Source Diffusion Analysis" in this chapter. Typical examples of records obtained in dilute agarose and brain are shown (Figure 10.3).

Integrative Optical Imaging

In the Integrative Optical Imaging (IOI) method (Nicholson and Tao 1993), a molecule of interest with an attached fluorescent marker is pressure ejected from a glass micropipette into agar or brain and its distribution is detected optically over a sequence of intervals using an epifluorescent microscope equipped with a CCD camera. The distributions are fitted to the diffusion equation to obtain an estimate of the effective diffusion coefficient (D^*). Various fluorophore-labeled molecules have been used: dextrans (*MW* 3 000–70 000; Nicholson and Tao 1993; Tao and Nicholson 1996; Tao 1999; Hrabětová et al. 2003; Thorne et al. 2004; Hrabětová 2005; Thorne and Nicholson 2006), proteins (albumins and EGF; Tao and Nicholson 1996; Prokopová-Kubinová et al. 2001; Thorne et al. 2004), and synthetic polymers (PHPMA (Prokopová-Kubinová et al. 2001); polyethylene glycol (PEG) (Thorne et al. 2005)). These macromolecules, labeled with Texas Red®, tetramethylrhodamine, fluorescein, or Oregon Green® 514 fluorophores, were obtained from Molecular Probes, Sigma, Nektar Therapeutics (PEG) or custom made (PHPMA) at the Institute of Macromolecular Chemistry, Academy of Sciences of the Czech Republic. The working concentration was 0.1–1 m*M* in 150 m*M* NaCl (dextrans and PHPMAs) or in PBS (proteins and PEG) with the exceptions of EGF, which was used at the concentration of 0.6×10^{-3} m*M*, and PEG, which was used at concentrations of 5–50 m*M*. In our experience, the small dextrans and PHPMAs are the easiest to work with. Tao and Nicholson (1996) reported that albumins tend to block the pipette, and Thorne et al. (2004) noted that solutions of proteins have to be carefully checked for aggregation.

The fluorophore-labeled probes are injected into dilute agarose gel or brain by a brief pulse (typically 10–200 ms, 10–20 psi) of compressed nitrogen applied to the micropipette containing the probe solution. The injected volume is less than 1.0 nL (Nicholson and Tao 1993) and typically 25–50 pL (Cragg et al. 2001). For fluorescence imaging, the excitation light from a 75 W xenon or 100 W halogen source is directed to the specimen using a dichroic mirror with a filter set corresponding to the fluorophore used. A shutter (Uniblitz®, Vincent Associates, Rochester, NY) is located in the light path to prevent any photobleaching of the fluorophore that can occur when a set of images is acquired over a long period and the fluorophore is prone to photobleaching, e.g., fluorescein. The light emitted by the diffusing molecules is imaged by a CCD camera (CH350 or Cool-Snap HQ Monochrome, Photometrics, Tuscon, AZ are used in our laboratory). A set of 11 or more images is obtained at preset intervals for each injection. The first image, taken before the injection, serves as a background that is subtracted from subsequent images. The image acquisition is controlled by the program Vida, written in the V++ language (Digital Optics Ltd, Auckland, New Zealand) and the overall control of the experiment and analysis of the data is implemented in the MATLAB-based program Ida (see "Software for Integrative Optical Imaging-Vida and Ida").

Given that the pressure ejection paradigm underlies the IOI method, Equation 10.5 with $k' = 0$ forms the basis for interpreting the results. Ideally, the 3D distribution of molecules described by this equation forms a spherical cloud that is projected onto a plane in the camera attached to the fluorescent microscope with the intensity distribution modified by the defocussed point spread function of the microscope objective (Nicholson and Tao 1993; Tao and Nicholson 1995). Although the details are complicated, the intensity distribution (I_i) can be represented by the simple expression for the *i*th image at time t_i after the initial injection of the fluorescent molecules:

$$I_i(r,\gamma_i) = E_i\exp(-(r/\gamma_i)^2) \tag{10.8}$$

where r is the distance from the point injection site and

$$\gamma_i^2 = 4D^*(t_i + t_0). \tag{10.9}$$

The offset t_0 represent a virtual time origin prior to the actual time of injection to approximate the fact that the initial injection will have a finite volume (Prokopová-Kubinová et al. 2001).

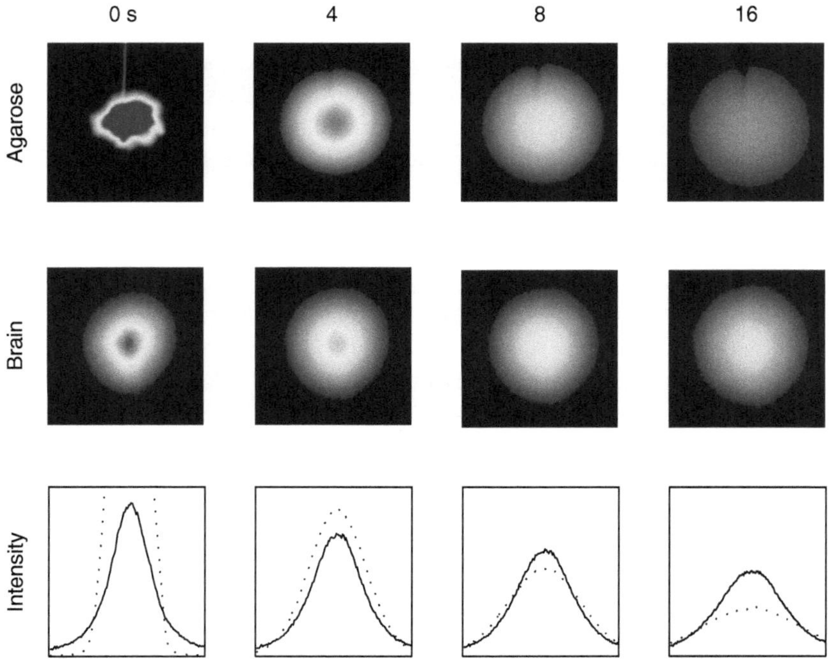

FIGURE 10.7 Diffusion of fluorophore-labeled dextran measured with IOI in dilute agarose gel and in rat neocortex in vitro. At time zero (0 *s*), a small volume of dextran (*MW* 3000) labeled with the fluorophore Texas Red® was pressure ejected from a glass micropipette into either 0.3% agarose gel (*Agarose*) or a rat cortical slice (*Brain*). The cloud of fluorescent molecules was viewed through an epifluorescent microscope and imaged with a cooled CCD camera. Images are shown at four, eight, and 16 s after the injection. The dextran diffuses away more rapidly in the agarose (free medium) than in the cortex (hindered medium). The lowest panel shows concentration (*Intensity*) profiles along a horizontal line through the center of the cloud of diffusing molecules. The profiles confirm the more rapid dispersion in the free medium (*dotted line*) compared with the hindered medium (*full line*), and the characteristic bell-shaped curve of the concentration distribution. Appropriate analysis of the curves enables D, D^*, and λ to be obtained. Scale size: side of image, 282 μm. (Reproduced from Nicholson, C., *Rep. Prog. Phys.*, 64, 815–884 2001.)

The analysis of a set of IOI images is shown in Figure 10.7. The images are 2D projections of the diffusing molecules and the program fits an appropriate solution of the diffusion equation to intensity curves obtained along horizontal, vertical, and two diagonal lines running through the center of each image (Hrabětová et al. 2003). Diffusion coefficients (D for agarose gel and D^* for brain tissue) were obtained by (a) fitting Equation 10.8 to each image to obtain a sequence of γ_i for each t_i and then (b) performing a linear regression based on Equation 10.9 to obtain D or D^* and t_0 (Nicholson and Tao 1993; Prokopová-Kubinová et al. 2001). Unlike the RTI method, the IOI method does not determine α and k'. Potentially, a finite k' would affect the accuracy of λ but it is anticipated that most macromolecules cannot easily leave the ECS so $k' \approx 0$. Eventually some endocytocis may be expected.

Carbon-Based Microelectrodes

Because there have been few diffusion measurements using electroactive compounds, this section will discuss both methods and results. Carbon epoxy electrodes or carbon fiber microelectrodes (CFMs) can sense electroactive substances and function much like an ISM. Such electrodes can therefore be used as sensors in the RTI or pressure ejection method, however, the theory and technology are distinctly different from those associated with ISMs; they are described in detail in Chapter 11 by Rice and colleagues.

The earliest diffusion measurements using electroactive probes combined with a point-source paradigm were described by Dayton et al. (1983) who employed CFMs, 7–10 μm diameter, and pressure ejected ascorbic acid (AA), dihydroxyphenylacetic acid (DOPAC) and α-methyldopamine (α-MeDA) from a 20 gauge needle (0.7 mm diameter) some 0.5–1.0 mm distant from the CFM in the rat cortex. The resulting signals were analyzed by chronoamperometry and Equation 10.5 with $k' = 0$ was utilized to interpret the data. Generally, the investigators relied on t_{max}, the time of the peak of the diffusion curve, to estimate D^* rather than attempting detailed curve fitting. The relationship for this is (Nicholson 2001):

$$D^*_{k'=0} = r^2/6t_{max}. \tag{10.7}$$

For the three substances studied, the values of D^* are consistent with tortuosities of ~1.8. Rice et al. (1985) used a more extensive range of electroactive molecules in a similar approach but with graphite epoxy sensing electrodes some 100–250 μm in diameter, a pressure ejection micropipette of 5 μm diameter, and injection and sensing locations spaced 0.4–0.8 mm apart. Measurements were made in the rat caudate nucleus (striatum) and, when diffusion of DA or α-MeDA was measured, cellular uptake was blocked pharmacologically. Chronoamperometry was again used, and the interpretation of the data based on Equation 10.5 and Equation 10.7. Rice et al. (1985) found that, for anionic electroactive compounds, including AA and DOPAC, they obtained results similar to those of Dayton et al. (1983) but for cationic compounds, including DA and α-MeDA, D^* was about three times less than for the anions. The authors attributed these results to differential interaction with the negatively charged ECM but the results have remained controversial and other factors, such as incomplete blockage of cellular uptake, lack of detailed curve fitting and large sensing electrode may have distorted the results, although the systematic differences between anions and cations remain striking.

These point-source paradigms were refined by Rice and Nicholson (1989) who used CFMs of 10–12 μm diameter and an iontophoretic micropipette some 150–200 μm away, combined with improved CFM fabrication techniques, signal processing and fast-scan cyclic voltammetry (FCV) to make accurate diffusion measurements of DA at concentrations as low as 35 nM in agarose gels. The analysis was based on fitting the experimental curves to Equation 10.2 through Equation 10.4 with $k' = 0$. The techniques were also reviewed by Rice and Nicholson (1995).

These developments led to a more sophisticated measurement of DA diffusion in brain tissue with the point-source paradigm using CFMs and a micropipette pressure source by Cragg et al. (2001). Here, the full version of Equation 10.5 was used to determine both λ and k'. A value for α was obtained from iontophoretic measurements with TMA^+, along with λ for that ion. Moreover, 3000 MW fluorescent dextran was included with the DA to enable the ejected volume to be estimated (using the TMA^+ data) and controlled; the ejected volume was maintained in the range 25–50 pL. These combined measurements led to a detailed analysis of the diffusion properties of the guinea pig midbrain. The TMA^+ measurements revealed that the substantia nigra compacta (SNc), substantia nigra reticulata (SNr) and ventral tegmental tract (VTA) had tortuosities in the range 1.58–1.69, quite comparable to the value in the guinea pig cortex (1.59) but the volume fraction in all three midbrain regions was a surprising 0.3, almost 50% greater than that seen in the cortex (0.22). The values of λ measured with DA (range 1.68–1.80) were slightly higher than those measured with TMA^+ and the reason is not clear, however, they were nowhere near as high as the value obtained in rat caudate in the earlier work of Rice et al. (1985).

The studies of Cragg et al. (2001) also revealed distinct DA clearance mechanisms (Figure 10.5). In contrast to the striatum (caudate) where pressure-injected DA is rapidly removed (Van Horne et al. 1992) and Michaelis–Menten kinetics dominate (Nicholson 1995), in the midbrain simple clearance kinetics proportional to extracellular DA concentration were adequate to describe the data but the value of k' was 0.085 s^{-1} in SNc and 0.093 s^{-1} in VTA, in marked contrast to the value of 0.006 s^{-1} in SNr. Indeed, the DA clearance value in SNr was comparable to the non-specific loss measured for TMA^+ ($k' = 0.005–0.007\ s^{-1}$).

One other study (Xin and Wightman 1997) employed modified CFMs coupled to an enzyme to attempt to measure the diffusion of choline and acetylcholine in rat brain slices, using a point source paradigm with pressure ejection. Although the study indicated some potential for the approach, the slow response of the sensors and a separation between sensor and source comparable to the thickness of the brain slice (see Section: "Effects of Boundary Conditions on Diffusion Measurements in Brain Tissue") compromised the study. Finally we note that the diffusion of the electroactive neurotransmitter serotonin (5-hydroxytryptamine) can be measured in brain tissue with the RTI method, using the same ion-exchanger as is employed to measure TMA^+ (Rice and Nicholson 1986).

Diffusion Measurements Using Dual Microdialysis Probes

Because of the size of a microdialysis probe (typically 4 mm long, 0.24 mm o.d., Höistad et al. 2002) diffusion measurements with two such probes violates the point source/point sensor paradigm described above; however, the underlying concept is similar and this approach is briefly described here. Note that in order to make a correct quantitative assessment of microdialysis results in brain tissue in general, it is essential to take into account diffusion but these issues will not be addressed here; some discussion is given by Nicholson (2001). In this section we will only be concerned with the use of microdialysis to make actual diffusion measurements.

In principle it is possible to measure in vivo diffusion coefficients using two microdialysis probes (Höistad et al. 2000; Kehr et al. 2000) but there is only one study where an effective diffusion coefficient has been measured (Chen et al. 2002; Höistad et al. 2002). The potential advantages are that any substance that can be detected by microdialysis could be measured. The disadvantages are that the microdialysis probes occupy quite a large volume and, because they remove an appreciable amount of the substance being measured, they perturb the diffusion field. These two issues greatly complicate the theoretical analysis of the diffusion problem, however, they are not insurmountable given the availability of personal computers (Chen et al. 2002). A further disadvantage is that the time-scale of diffusion measurements with dual probes is long so they are unsuitable for measuring transient changes in diffusion conditions. Finally, the insertion of the probes leads to the formation of glia scar tissue around them over hours and days and this may alter the local environment in ways that have yet to be characterized fully.

In the study by Höistad et al. (2002), a radiolabeled tracer, ^{3}H-mannitol, was continuously infused at different concentrations via a probe acutely implanted into the striatum of anesthetized male rats. Control measurements were made in a dilute agar gel. Samples were collected by a second probe 1 mm away from the first, and the recovered ^{3}H-mannitol was measured by liquid scintillation counting. In the striatum, the delivery of ^{3}H-mannitol was counteracted by its removal from the ECS by passive uptake into cells and clearance into the microcirculation, causing the diffusion profile to approach quasi steady-state levels within two hours. Diffusion data from brain and agar were analyzed using a mathematical model (Chen et al. 2002). The effective diffusion coefficient for ^{3}H-mannitol was $D^* = 2.9 \times 10^{-6}\ cm^2\ s^{-1}$, the effective volume fraction $\alpha^* = 0.30$ and the clearance rate constant $\kappa = 2.3 \times 10^{-5}\ s^{-1}$. From these data, the tortuosity value was $\lambda =$ 1.81, and penetration distance, $\Gamma = 4.2$ mm, where $\Gamma = (D^*/\kappa)^{0.5}$ (Bungay et al. 1990). Mannitol has been widely used in various brain studies, including radiotracer diffusion measurements (Patlak and Fenstermacher 1975), but there is evidence that this substance enters cells to some degree and sucrose might be a better choice of probe substance for future diffusion measurements using dual probes (see, Höistad et al. 2002, for further discussion).

CHOICE OF BRAIN TISSUE PREPARATION

Two classes of brain preparation have been used in past diffusion studies: tissue slices and slabs (in vitro preparations) and living animals (in vivo preparation). The major advantages and disadvantages for diffusion studies are outlined.

TISSUE SLICES AND SLABS

Brain slices, either 400 μm thick representing normal tissue or 1000 μm thick representing a model of ischemia, prepared from adult animals have been used extensively in diffusion studies. Detailed protocols are given by Rice and Nicholson (1991); Pérez-Pinzón et al. (1995). Slices are prepared according to a standard procedure and then kept at room temperature submerged in artificial cerebrospinal fluid (ACSF). Normal, 400 μm thick, slices are left to recover for at least an hour prior to recording. For the diffusion measurement, a single slice is transferred to the recording chamber where the temperature usually is maintained at 32°C–34°C. This temperature is below the typical physiological temperature of in vivo preparation (37°C) to prolong the survival of brain slices prepared from adult animals. Some studies, where the diffusion measurements were combined with patch clamp recordings, have been done at room temperature (Vargová et al. 2001).

Two types of tissue slab preparations have been used: isolated turtle cerebellum (Rice et al. 1993; Križaj et al. 1996) and isolated spinal cord from juvenile rats (P4-P21; Syková et al. 1999). Diffusion measurements were done at room temperature, which in the case of the turtle is the physiological temperature. Both these slab preparations exhibit a high resilience to ischemia so that a volume of healthy tissue containing intact 3D neuronal circuits can be obtained.

In general, in vitro preparations are employed when substantial manipulation of ACSF composition is required or when prompt and uniform drug delivery is needed. In the tissue chamber, one or both tissue surfaces (see below) are exposed to flowing ACSF to which substances can be added for rapid delivery to the tissue. Another advantage is region accessibility. In the intact brain, deep brain structures constitute a challenge for the RTI method. In typical use, the spacing between the tips of an array, comprising an ISM and a source micropipette, is stabilized by putting a drop of dental cement about 500 μm from the tips (Nicholson 1993). Such a strategy has to be abandoned when placing the array deep in the brain; consequently the spacing of the tips is likely to change as the microelectrodes are lowered into the tissue. Using a slice, however, almost any chosen brain region can be selected and it is possible to use two independent micromanipulators to further guarantee accuracy of placement for the micropipettes. Finally, brain slices are easier to work with than anesthetized animals. Once a successful dissection is done, slices are kept in the holding chamber and are used throughout the experimental day without need for further care.

It is important to keep in mind, however, that even the best prepared and maintained slices have limitations. Brain slices undergo an ischemic insult during preparation and, in addition, the tissue is traumatized by slicing. Consequently, the protoplasmic extensions of many cells are severed and the cut surface of the slice contains dying cells. It is pertinent to mention here that some of the recently introduced vibrating-blade microtomes are designed to have minimal movement perpendicular to the cutting plane, which reduces surface damage. Oxygen tension in the isolated preparation is usually un-physiological and the ACSF certainly lacks some chemical ingredients that are present in the real CSF (Artru 1999). Working with the slice preparation is always a race against time as the tissue slowly deteriorates. Finally, physiological questions that involve blood flow cannot be addressed in a slice.

EFFECTS OF BOUNDARY CONDITIONS ON DIFFUSION MEASUREMENTS IN BRAIN TISSUE

Diffusion measurements made close to the interface between the tissue and the environment will be influenced by the conditions at the boundary. Some of these issues were addressed previously in the context of bath-application of pharmacological agents to slices (Nicholson and Hounsgaard 1983;

Lipinski and Bingmann 1987). In this chapter, mathematical modeling has been used to investigate the effect of boundary conditions on diffusion measurements. The outcome is most relevant to brain slices but it is applicable also to brain slabs and in vivo preparations when the measurements are close to the tissue surface.

Diffusion measurements in slice preparations are performed in the center of the slice (i.e., for 400 μm thick slices at a depth of 200 μm). In the recording chamber, the brain slice is usually supported on a mesh with ACSF flowing underneath. The upper surface of the slice is either also exposed to flowing ACSF or it is exposed to a humidified and pre-warmed atmosphere (95% O_2/5% CO_2; Figure 10.8). The first type of chamber is called a *submersion* chamber, because the slice is totally submerged in flowing ACSF; the second type of chamber is called an *interface* chamber, because the moist upper surface of the slice forms an interface with the humidified gas. The two different scenarios at the upper surface result in distinct boundary conditions. In one case, the flowing solution interface functions as a sink that takes molecules away from the slice, while in the other case, the tissue–gas interface functions as an impermeable boundary where diffusing molecules within the slice get reflected back. These two scenarios were simulated to investigate their effect on diffusion measurements.

This analysis will be restricted to iontophoresis. Starting with Equation 10.2 and assuming the loss of probe ion is negligible ($k' = 0$), Equation 10.2 reduces to:

$$C(t) = \frac{Q}{4\pi D^{*} \alpha r} \operatorname{erfc}\left(\frac{r}{2\sqrt{D^{*} t}}\right). \tag{10.10}$$

Again, the falling phase of the concentration curve after the iontophoresis ceases after a pulse of duration t_p, will be described by Equation 10.3.

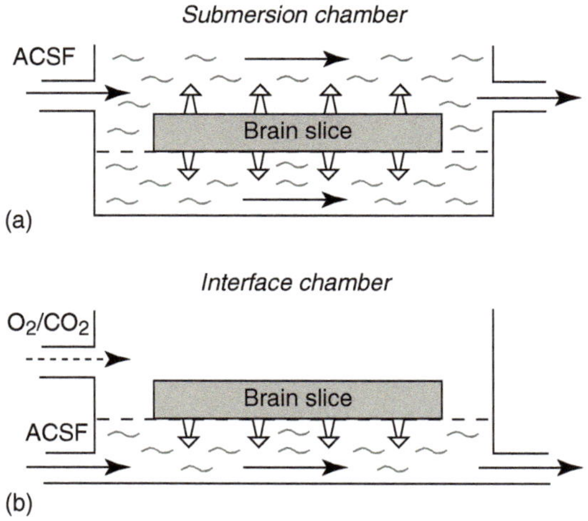

FIGURE 10.8 Submersion and interface tissue chambers. (a) In a submersion tissue chamber, the brain slice rests on a supporting mesh and the ACSF flows both underneath and over the slice. The slice is held down with a weight in the form of a "harp" or simply a silver ring placed on the slice to prevent movement (not shown). Molecules released into the tissue gradually diffuse to the slice surfaces and are washed away (*thick arrows*). (b) In an interface chamber, the brain slice also rests on a supporting mesh but the ACSF flows only underneath the tissue. The upper surface of the slice is exposed to pre-warmed and humidified gas mixture (95% O/5% CO). Molecules released into the tissue gradually diffuse to the slice surfaces and are washed away at the lower surface of the slice (*thick arrows*) but get reflected back into the tissue when reaching upper surface that functions as a non-permeable boundary.

The derivation of Equation 10.10 assumes that the source and measuring electrode are both far from any boundaries, i.e., they are embedded in an infinite medium with effective diffusion coefficient D^*. If the source and measuring electrodes are moved close to, say, the upper boundary then the influence of this boundary may be accounted for in two limiting cases: (a) all diffusing crossing the boundary are swept away by the superficial flow of bathing medium; this is the absorbing boundary; or (b) all molecules reaching the boundary are reflected back into the tissue so the boundary is non-permeable. Both these conditions can be modeled by placing a virtual point source above the location of the boundary as a mirror image of the actual source and then removing the boundary so there is an infinite medium with two sources. When the virtual source is $-Q$ in magnitude, equal and opposite to the real source, then on the boundary the concentration will be zero, equivalent to the loss of all particles, i.e., case (a) the absorbing boundary. When the virtual source is equal to $+Q$, then the flux across the boundary will be zero, so there will be no loss of molecules equivalent to case (b), an impermeable boundary.

In the simulations we have taken Equation 10.10 with images, calculated the concentrations and then analyzed the results by fitting Equation 10.2, which embodies a loss term but returns to an infinite medium representation, to determine whether the effects of a bounded medium without loss are equivalent to an infinite medium with a "virtual" loss.

The equation with an image source has a contribution from the actual source at distance r_{direct} and the virtual, image, source at r_{image} (See figure with Table 10.2).

$$C(t) = \frac{Q}{4\pi D^* \alpha r}\left[\operatorname{erfc}\left(\frac{r_{\text{direct}}}{2\sqrt{D^* t}}\right) \pm \operatorname{erfc}\left(\frac{r_{\text{image}}}{2\sqrt{D^* t}}\right)\right] \tag{10.11}$$

where

$$r_{\text{direct}} = \sqrt{(a-b)^2 + h^2} \quad \text{and} \quad r_{\text{image}} = \sqrt{(a+b)^2 + h^2} \tag{10.12}$$

and a is the vertical distance of the source from the brain/external medium interface, equal to the distance from the interface to the image source or sink on the opposite side of the interface, b is the vertical distance of the sensor from the brain/external medium interface and h is the horizontal distance between source and sensor in the plane of the surface (see figure with Table 10.2).

When the source and sensor electrodes are lowered to the same depth, a, (i.e., $a = b$) with fixed spacing h, Equation 10.12 reduces to

$$r_{\text{direct}} = h \quad \text{and} \quad r_{\text{image}} = \sqrt{4a^2 + h^2}. \tag{10.13}$$

Table 10.2 depicts the results of calculations for a variety of configurations of source and sensor with respect to the interface. In all cases it is assumed that the other, lower, interface of the slice is so distant that the medium can be regarded as being semi-infinite. The procedure was (a) use Equation 10.10 through Equation 10.12 and Equation 10.3 to compute a concentration verses time curve; and (b) use the program Walter (described in "Software for RTI and Pressure Ejection-VOLTORO, Wanda and Walter") to try to fit Equation 10.2 and Equation 10.3 to this curve. This means that we tried to fit the solution to the diffusion for a source in an infinite medium with loss of ions to the curves derived for a source in a semi-infinite medium with two types of boundary condition (absorbing and non-permeable) without loss. We required a "good" fit but not a perfect one.

The outcome of modeling is summarized in Table 10.2 and examples of simulated diffusion curves shown in Figure 10.9. Before describing the results in detail, we would like to stress that the two types of boundary conditions modeled here mimic extreme conditions and at the same time over-simplify the typical experimental situation. In an experimental setting, a slice in the

TABLE 10.2
Simulations of the Effect of Absorbing and Non-Permeable Boundary on λ, α, k'.
Determined by Curve-Fitting for Input Parameters: $\lambda_{input} = 1.6$, $\alpha_{input} = 0.20$, $k'_{input} = 0\ s^{-1}$. Molecule is TMA^+.

Simulation number	Boundary conditions	*a* (μm)	*b* (μm)	*h* (μm)	*r* (μm)	λ	α	k' (s^{-1})
1	Absorbing	200	200	50	50	1.65	0.204	5.68×10^{-3}
2	Absorbing	200	200	100	100	1.66	0.196	6.72×10^{-3}
3	Absorbing	200	200	150	150	1.65	0.186	7.17×10^{-3}
4	Absorbing	200	150	0	50	1.65	0.203	8.02×10^{-3}
5	Absorbing	200	100	0	100	1.63	0.187	1.29×10^{-2}
6	Absorbing	200	50	0	150	*1.49*	*0.194*	*1.72×10^{-2}*
7	Absorbing	1	1	50	50	*1.12*	N/A	*17.5×10^{-2}*
8	Absorbing	1	1	100	100	*1.16*	N/A	*5.4×10^{-2}*
9	Absorbing	1	1	150	150	*1.19*	N/A	*2.9×10^{-3}*
10	Non-permeable	200	200	50	50	1.68	0.215	0
11	Non-permeable	200	200	100	100	*1.73*	*0.218*	0
12	Non-permeable	200	200	150	150	*1.77*	*0.213*	0
13	Non-permeable	1	1	50	50	1.60	0.10	0
14	Non-permeable	1	1	100	100	1.60	0.10	0
15	Non-permeable	1	1	150	150	1.60	0.10	0

S, sensor (ion selective microelectrode); *Q*, source (iontophoretic microelectrode); $\pm Q_{image}$, image source (+non-permeable boundary) or sink (−absorbing boundary); *a*, distance of the source from the boundary; *b*, distance of sensor from the boundary; *h*, horizontal distance between source and sensor; r_{direct}, distance between source and sensor; r_{image}, distance between image source (or sink) and sensor; λ, tortuosity; α, volume fraction; k' (s^{-1}), non-specific clearance; *N/A*, not applicable.

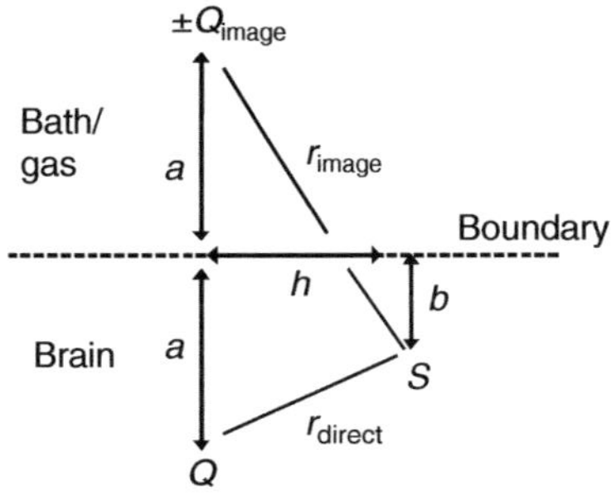

submersion chamber has a stagnant or unstirred layer of solution (Lipinski and Bingmann 1987) on the surface rather then laminar flow of the ACSF and this leads to incomplete absorption, while a slice in the interface chamber has a thin layer of the ACSF resulting in incomplete reflection of diffusing molecules. In addition, another complexity not taken into account in the present modeling is that the boundary conditions at the lower surface may affect the measurements in the center of the slice under some conditions. Therefore the models provide trends rather than a precise description.

In the simulation mimicking an absorbing boundary, the sink effect of the boundary conditions had minimal effect on the extracellular diffusion measured 200 μm underneath the boundary in a plane parallel with the surface of the slice (Table 10.2, Simulation number 1–3; Figure 10.9a). The tortuosity derived from curve fitting was slightly higher ($<+4\%$) for spacing $r=50$–150 μm while α was slightly lower (-7%) only for $r=150$ μm. The sink effect of absorbing boundary was to be found in the apparent value of k' (recall that the model had $k'=0$).

We also tested a special case, where the source microelectrode was kept in place but the detector was moved between the source and the boundary to measure the diffusion in a plane perpendicular to the surface of the tissue. This configuration is used in experiments testing for

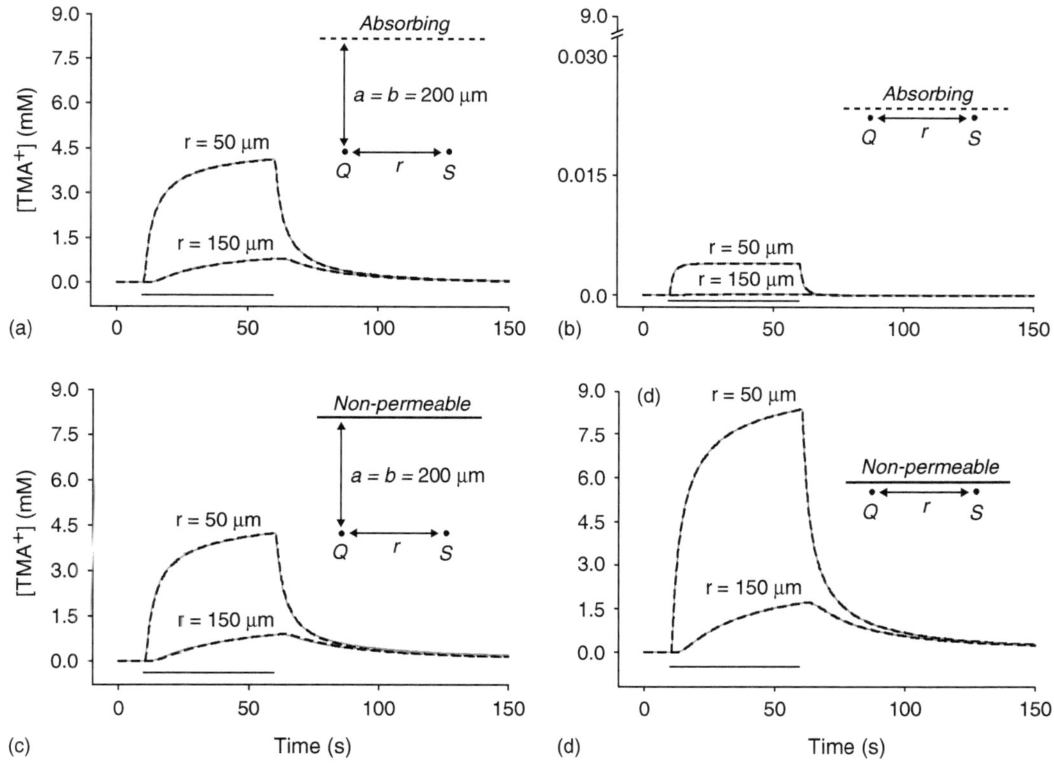

FIGURE 10.9 Effect of boundary conditions at tissue-environment interface on the diffusion measurements. The diffusion curves (*solid grey line*) were generated (see text for details) with the following input parameters: $n_t=0.3$, $r=50$, 100 or 150 μm, $\lambda_{input}=1.6$, $\alpha_{input}=0.2$, $k'_{input}=0\ s^{-1}$. Simulated curves were then fitted (*black dashed lines*) with MATLAB fitting program Walter used to analyze experimental data and λ, α, and k' (s^{-1}) were extracted. The results are summarized in Table 10.2; selected curves are shown in panels (a–d). (a) The source (*Q*) and the sensor (*S*) are 200 μm below the absorbing boundary. (b) The source and the sensor are 1 μm below the absorbing boundary. (c) The source and the sensor are 200 μm below the non-permeable boundary. (d) The source and the sensor are 1 μm below the non-permeable boundary. Notice the difference in shape and amplitude for curves obtained in near proximity to the tissue–environment interface (b, d).

tissue anisotropy, where the diffusion is measured along three perpendicular axes (Rice et al. 1993). Our simulations show (Table 10.2, Simulations 4–6) that for $r = 50$ and 100 μm, the boundary conditions influence the diffusion measurements only slightly. The results are similar to those obtained in the parallel plane. However, when the spacing was increased to 150 μm (i.e., the detector is only 50 μm away from the tissue surface, Simulation 6), λ was significantly reduced and an increase in k' was observed, attributable to an accentuated sink effect close to the boundary.

Next, we tested a case when both microelectrodes were placed just beneath the slice surface (Table 10.2, Simulation 7–9; Figure 10.9b). The sink effect of an absorbing boundary clearly dominated. The diffusion curves were greatly distorted both in shape and amplitude. A large drop in λ and an increase in k' were observed while the values of α were meaningless ($\alpha > 1.0$).

In the simulation mimicking a non-permeable boundary, a reasonable result for the plane parallel with the tissue surface (depth of 200 μm) was obtained only when r was small (Table 10.2, Simulation 10–12; Figure 10.9c). At $r \geq 100$ μm, the tortuosity was increased by more than 8%.

Finally, when both source and measuring electrodes were placed close to the non-permeable boundary, a correct result for λ and k' was obtained (Table 10.2, Simulation 13–15; Figure 10.9d). To get a correct value of α, the volume fraction extracted from such records has to be doubled; this is because of the reflective properties of the tissue–gas interface causing the sensor to detect twice as many molecules as would have been detected in an infinite medium.

It is obvious that great caution has to be taken when designing and carrying out diffusion experiments in brain slices to avoid possible effects from conditions at tissue–environment interfaces. When using a submersion tissue chamber, the iontophoretic microelectrode is typically positioned at a depth of 200 μm, i.e., the source of ions is equidistant from the upper and lower surface of the slice, for all measurements. The ISM is positioned typically 100 μm away along x-, y-, and z-axes. The recordings are thus obtained from the central portion of the slice where the influence of boundary conditions is minimal.

In Vivo Preparations

Many diffusion studies have been done in in vivo preparations. In fact, the very first major study using the TMA^+ RTI method investigated diffusion properties of the ECS in rat cerebellum in vivo (Nicholson and Phillips 1981). Subsequently, large numbers of in vivo studies have been done in Professor Syková's laboratory in Prague where the research topics include the changes in the ECS during development, learning and memory, aging, as well as in animal models of various pathological conditions such as ischemia and terminal anoxia, stab wound, and transgenic models of Alzheimer's disease (for a review of these topics see Syková 2004). Other important in vivo diffusion studies using point source diffusion methods include a study of Ca^{2+} diffusion using a pressure source (Nicholson and Rice 1987), a study of volume regulation (Cserr et al. 1991), a study of changes in diffusion parameters during development (Lehmenkühler et al. 1993), and studies of epileptic foci (Lehmenkühler et al. 1988; Schwindt et al. 1997). In vivo diffusion studies with CFMs have been reviewed in Section: "Carbon-Based Microelectrodes".

For diffusion measurements, animals are anesthetized, placed in a stereotaxic apparatus and the cortex or other brain region is exposed, then the dura mater is carefully removed. Body temperature is maintained at 37°C with a heating pad and the exposed cortex is superfused with ACSF warmed to 37°C and bubbled with 5% CO_2/95% O_2 to maintain pH at 7.4. The precise details of the preparation will vary with the animal and laboratory and original papers should be consulted. Diffusion measurements are done using a glued array of an iontophoretic or pressure ejection electrode and ISM. The array is calibrated in agar or agarose gel first and then it is lowered into the brain to measure the diffusion properties of the ECS. More detail is provided by Nicholson and Phillips (1981); Cserr et al. (1991); Lehmenkühler et al. (1993), and Voříšek and Syková (1997), among others. Recently the IOI technique has been adapted to the in vivo preparation (Thorne and Nicholson 2006).

The living animal does not have the same limitations as brain slices and the brain tissue remains in a framework of the whole organism. Its necessity is undisputed when questions involving BBB, bulk flow or diffusion over long distances are addressed. In addition, while in vitro models of various pathological conditions exist, they are usually an approximation of in vivo conditions. However, difficulties with drug application, inaccessibility of deep brain tissues, and finally a need for non-trivial surgery combined with the necessity to maintain careful control of the preparation throughout the experiment can make use of in vivo preparation truly challenging.

SPECIALIZED SOFTWARE FOR POINT-SOURCE DIFFUSION ANALYSIS

Software for Iontophoresis and Pressure Ejection—VOLTORO, Walter and Wanda

The earliest curve fitting was accomplished on hand-held electronic calculators (Nicholson and Phillips 1981) and did not allow for a non-zero clearance parameter (k'). When personal computers became available it was possible to adopt a more sophisticated approach using the non-linear simplex approach to fit Equation 10.2 and Equation 10.3 or Equation 10.5 to data that were acquired with a digital oscilloscope. A succession of personal computers were programmed by Dr C. Nicholson using the language Pascal, and this phase reached its final sophistication in the program VOLTORO that accommodated iontophoresis, pressure ejection, ISMs and CFMs combined with FCV. The first version of VOLTORO ran on a DEC PDP 11 computer and subsequently on an Apple II. In its final incarnation the program ran under the DOS operating system on the IBM PC. It is still possible to run the program under Windows XP but support for the programming language has long ceased and VOLTORO can no longer be extended to operate contemporary data acquisition boards or many features of modern PC hardware.

To take full advantage of the Windows operating system, the analysis part of VOLTORO was re-written in the MATALAB (MathWorks, Natick, MA) programming environment by Dr Nicholson and became the program Walter. Walter was considered a means of reading in substantial numbers of data files generated by VOLTORO and re-analyzing them, still using the simplex curve fitting, but removing any operator bias by randomizing starting parameters. The program also allowed better visualization of the data and storage of all data and fitting parameters in an Excel spreadsheet.

Initially, Walter lacked any means of acquiring data because MATLAB did not support this function. This was remedied by Dr C. Chen who used the LabView language (National Instruments) to write a data acquisition program called Wanda which generated files that could be read by Walter. Some years later the Data Acquisition Toolbox became available for MATLAB and Wanda was re-written in MATLAB by Dr L. Tao. Today both Wanda and Walter run under MATLAB and support a variety of data acquisition boards from National Instruments (and some other manufacturers). The programs are available from Dr C. Nicholson.

Software for Integrative Optical Imaging—Vida and Ida

The IOI technique was introduced more than 10 years after the RTI method, when PCs were in general use, and the first version of the accompanying program was written in the C language by Drs L. Tao and C. Nicholson. This program both operated the CCD camera and fitted the resulting image data to Equation 10.8 and Equation 10.9.

After the experience with MATLAB and the Walter program it was also decided to move the IOI program to this environment. Because MATLAB lacked, and still lacks, any toolbox to control a Photometrics CCD camera, the programming language V++ (formerly V-Pascal) from Digital Optics, Auckland, New Zealand, was adopted for this part of the package. The main part of the program, including data analysis, was written in MATLAB by Dr C. Nicholson and called Ida;

and the camera operation program, called Vida, was written in V++, also by Dr Nicholson. Vida runs under control of Ida using the DDE (Dynamic Data Exchange) facilities of the two programming environments. This arrangement has the disadvantage that two program environments are required (V++ and MATLAB) but it has the advantage that the camera can be run independently under the extensive turnkey V++ package for other image analysis tasks. Ida and Vida are available from Dr C. Nicholson.

DIFFUSION PROPERTIES OF BRAIN EXTRACELLULAR SPACE

Here, selected examples of results of the diffusion studies in brain tissue in vivo and in vitro are given. More comprehensive reviews may be found elsewhere (Syková 1997; Nicholson and Syková 1998; Nicholson 2001; Syková 2004).

BRAIN UNDER PHYSIOLOGICAL CONDITIONS

The first major RTI study quantified the diffusion of small cations and anions in rat cerebellum in vivo (Nicholson and Phillips 1981). These investigators found that all ions tested gave similar results and the values were pooled to $\alpha = 0.21$ and $\lambda = 1.55$. It may be noted that this early study looked for anisotropy in the molecular layer of the cerebellum but failed to find it, primarily because the data analysis techniques were not sensitive enough; later studies did reveal anisotropy both in the cerebellum and elsewhere (see below). Extensive subsequent studies in in vivo and in vitro preparations of rat, mice, guinea pig, and turtle brain revealed that most commonly, λ is about 1.6 and α is about 0.2 in isotropic tissue under physiological conditions (Nicholson and Syková 1998; Nicholson 2001). However, there are exceptions. For example, α is sometimes larger, about 0.3 in in vitro substantia nigra in guinea pigs (Cragg et al. 2001) and about 0.4 in in vivo neocortex of rat pups (age 2–3 days postnatal; Lehmenkühler et al. 1993); interestingly, λ remains close to 1.6 in these preparations (1.60 in SNr VTA, 1.69 in SNc, 1.59–1.68 in layers III–VI of the somatosensory neocortex). In contrast, a low tortuosity, $\lambda = 1.46$, is observed in in vitro preparations of the CA1 region of hippocampus (Hrabětová 2005, but see Mazel et al. 1998), the region vulnerable to ischemia and epilepsy (Traynelis and Dingeldine 1988; Kawasaki et al. 1990; Schmidt-Kastner and Freund 1991). We may speculate that the tissue hindrance corresponding to $\lambda \approx 1.6$ is optimal for intercellular communication and maintenance of extracellular environment, and that smaller values (as seen in the hippocampus) or higher values (as seen in many pathological conditions) have negative consequences for brain function. The maintenance of stable tissue hindrance for diffusion might be achieved by various, as yet unknown, regulatory mechanisms.

Tissue anisotropy, which signifies that diffusion is different in different axes, can be studied with RTI methods. This was first demonstrated by Rice et al. (1993) in the molecular layer of cerebellum of the turtle (where $\lambda_x = 1.44$, $\lambda_y = 1.95$, $\lambda_z = 1.58$), followed by studies in corpus callosum (Voříšek and Syková 1997), hippocampus (Mazel et al. 1998 but see Hrabětová 2005), and auditory cortex (Voříšek et al. 2002). Diffusion anisotropy most often correlates with a directional arrangement of fibers (Beaulieu 2002).

The studies employing IOI explored how macromolecules are hindered in the ECS. Nicholson and Tao (1993) found that the tortuosity increases as the diameter of globular dextran increases (Figure 10.4). In contrast, long-chain flexible synthetic polymers (PHPMA) of molecular weight ranging from 7800 to 1 057 000 are hindered similarly to the small TMA cation (Prokopová-Kubinová et al. 2001), although their actual effective diffusion coefficient is small. The mechanism of low hindrance for long-chain polymers is unknown. Finally, studies employing proteins show that albumins with *MW* ranging from 14 500 to 66 000 are hindered more ($\lambda = 2.24$–2.50; Tao and Nicholson 1996) than the smaller epidermal growth factor, *MW* 6600 ($\lambda = 1.8$; Thorne et al. 2004).

Brain during Reversible Osmotic Challenge

The ECS is not a static structure but one in which properties change during brain activity and under various challenges. One might conjecture that there is a relationship between λ and α; that λ will increase as α gets smaller and vice versa. However, while this is true under some circumstances it is not a universal fact and generally there is no simple relation between λ and α in brain (Figure 10.10). This was demonstrated using the TMA–RTI method by Kume-Kick et al. (2002) in neocortical slices where α can be reversibly reduced by hypotonic stress (150 mosmol kg^{-1}) to as low as 0.12. As might be expected, λ increases in value to reach 1.86. But enlarging α from a control value of 0.24–0.42 has no effect on λ, which remains, even in the most severe hypertonic stress (500 mosmol kg^{-1}), close to the control value of 1.68. Tao (1999) measured qualitatively the same behavior of λ with fluorophore-labeled *MW* 3000 dextran (see, Kume-Kick et al. 2002, for discussion).

Similar trends in osmotic behavior were also seen in a prior study on the isolated turtle cerebellum using the TMA–RTI method (Križaj et al. 1996). Both this and the Kume-Kick et al. (2002) study combined TMA$^+$ diffusion measurements with determinations of wet-weight and dry-weight of tissue to arrive at a description of both the extra- and intracellular compartments. This approach had been pioneered in an earlier in vivo study of hypernatremia and brain volume regulation (Cserr et al. 1991).

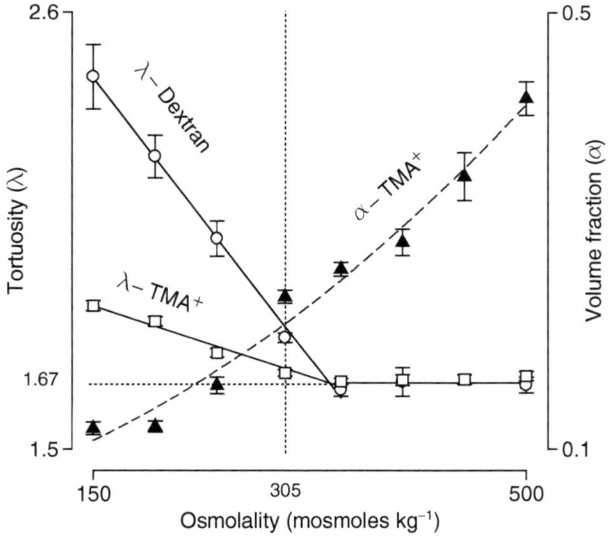

FIGURE 10.10 Independence of extracellular tortuosity and volume fraction during osmotic stress in rat neocortex in vitro. Diffusion of TMA$^+$ was measured using RTI; diffusion of fluorophore-labeled 3000 *MW* dextran was measured using IOI. Osmotic stress was induced by changing the NaCl content to alter osmolality of the bathing media for rat neocortical slices. Tortuosity measured with both molecules increases linearly as osmolality decreases below (i.e., becomes hypotonic) control value (305 mosmol kg^{-1}), but the slope measured with the larger molecule (dextran) is greater than that measured with the smaller (TMA$^+$). In hypertonic ACSF, the value of λ measured with both molecules quickly reaches a constant value, $\lambda = 1.67$ even though the osmolality of the medium continues to increase. In contrast to the behavior of λ, α varies smoothly and monotonically with osmolality, decreasing in hypotonic and increasing in hypertonic medium. (Reproduced from Kume-Kick, J., et al., *J. Physiol.*, 542, 515–527, 2002.)

Brain in Pathological States

The dissociation of α and λ can become even more pronounced under pathological situations (Syková et al. 2000) suggesting that a combination of factors is in play. During terminal anoxia in rat cortex, α is reduced to as low as 0.05, while X-irradiation increases α to 0.50. Yet, the tortuosity of tissue under both these pathological conditions is elevated (to 2.0 in terminal anoxia and to 1.8 in X-irradiated tissue).

Diffusion is significantly hindered and the volume of the ECS is reduced in many neuropathological states associated with cellular edema (Nicholson and Syková 1998; Syková et al. 2000). Several studies, employing in vitro and in vivo models of ischemia, showed that the tortuosity increases and the volume fraction decreases (Lundbæk and Hansen 1992; Syková et al. 1994; Pérez-Pinzón et al. 1995; Hrabětová and Nicholson 2000). In thick-slices (1000 μm), which model aspects of ischemia in vitro (Newman et al. 1988, 1989, 1991), λ measured with TMA^+ increases to 1.9 while α is reduced to 0.14 (Hrabětová and Nicholson 2000; Hrabětová et al. 2002). It has been proposed that the large hindrance observed during ischemia arises from formation of dead-space microdomains in the ECS, dead-end pores, formed by swelling cells, where diffusing molecules are delayed (Hrabětová and Nicholson 2000; Hrabětová et al. 2003). In in vivo models of anoxia and ischemia, λ measured with TMA^+ increases to about 2.0 while α is reduced to 0.05 (Syková et al. 1994; Voříšek and Syková 1997). Such dramatic changes in diffusion parameters occur even in rat pups (P4-6) but it takes longer for the changes to develop (Voříšek and Syková 1997). Changes in α seen in in vivo ischemic models are more pronounced than in in vitro studies. This difference may reflect the fact that, in vivo, the spatial expansion of edematous ischemic tissue is restricted by the skull, but no such constraint is imposed on a slice (Hrabětová et al. 2002). Among the many consequences of the impaired transport in ischemia is the disruption of nutrient and metabolic trafficking; such disruption augments brain dysfunction and delays recovery during vascular reperfusion.

A large alteration in diffusion parameters is also observed during astrogliosis induced in rat spinal cord (Syková et al. 1999; Vargová et al. 2001), in anaplastic astrocytoma in humans (Vargová et al. 2003), and in experimental autoimmune encephalomyelitis that is used as a model for multiple sclerosis (Šimonová et al. 1996). Under all these, as well as many other conditions, the ECS environment changes, altering the diffusion of substances and the homeostasis of the extracellular environment.

MONTE CARLO SIMULATION OF DIFFUSION IN 3D MEDIA

Over the years, diffusion measurements have generated lots of interesting but often counterintuitive data. Some striking examples include the findings that α and λ are independent parameters (Chen and Nicholson 2000; Kume-Kick et al. 2002, see above); and the discovery that, while the hindrance for globular molecules increases with the molecular weight (Nicholson and Tao 1993; Tao and Nicholson 1996), PHPMA linear polymers (*MW* 7800–1 057 000) have the same tortuosity as the small TMA cation (*MW* 74; Prokopová-Kubinová et al. 2001). As a final example it was established that the hindrance of ischemic tissue is actually reduced when large molecules are introduced into the ECS (Hrabětová and Nicholson 2000; Hrabětová et al. 2003). The interpretation of these results is challenging because the ECS parameters obtained with diffusion analysis are the result of macroscopic averaging and consequently they do not provide information about the features of individual ECS channels. We know very little about the local structure of these channels, whether they are homogeneous or heterogeneous and how they are organized in 3D. Microscopic heterogeneity would imply the existence of regions with distinct physiological characteristics. Although λ was originally thought of as a reflection of how molecules navigate around round cells, other biophysical features of the ECS may contribute to hindrance in brain tissue (Chen and Nicholson 2000; Hrabětová et al. 2003; Hrabě et al. 2004; Tao et al. 2005).

To test ideas and to facilitate data interpretation, modeling has come to our aid. Here we will focus on the parameters α and λ although the modeling approach has also been applied in other contexts to describe, e.g., the interaction between diffusion and Michaelis–Menten mediated uptake of dopamine from the ECS (Nicholson 1995; Cragg et al. 2001; Cragg and Rice 2004).

MCell and DReAMM

The Monte Carlo Simulation of Cellular Neurophysiology (MCell) software package was developed to numerically simulate the events at the neuromuscular junction. These include: acetylcholine release, diffusion, and binding to receptors as well as channel kinetics (Stiles et al. 1996; Stiles et al. 1999; Stiles and Bartol 2001). This software (available at www.mcell.psc.edu) was designed for biologically oriented users and therefore it uses a simple programming language or model description language (MDL) to specify simulation components such as 3D surfaces and reaction mechanisms.

Tao and Nicholson (2004) adopted the MCell software package to simulate the diffusion of probe molecules in virtual models of brain tissue. Instead of using brain ECS reconstructed from electron micrographs, Tao and Nicholson (2004), and later others (Hrabě et al. 2004; Tao et al. 2005), built simple virtual models where the features of ECS structure can be fully controlled. During Monte Carlo simulation of diffusion, probe molecules are released into the artificial 3D media from a point source representing conditions of the biological experiment (pressure pulse release of molecules) and they perform random walks. During one time step, each diffusing molecule moves in a randomly chosen direction and travels a randomly chosen step. When a marker encounters the cell boundary, it is reflected at the appropriate angle, just like a billiard ball colliding with a smooth wall. The positions of individual molecules are recorded over time. The effective diffusion coefficient, D^*, is calculated from the distribution of molecules as a function of time and distance, and λ derived (Tao and Nicholson 2004). The results are visualized using the Data Explorer-based program DReAMM (www.mcell.pcs.edu).

All simulations so far are based on simple 3D models that test the effect of certain geometrical features on the diffusion of a point particle. Interaction with the extracellular matrix, transport across the cellular wall or effects caused by finite size of the marker are neglected.

Three-Dimensional Media Composed of Convex Cells

In the first study, the ECS was depicted as interconnected spaces of uniform size that surround convex elements (Figure 10.11; Tao and Nicholson 2004). The simplest 3D media was built from cubes (Figure 10.11a). Two additional geometries were generated (Tao and Nicholson 2004): one built from truncated octahedra and the other built from three elements of different size, rhombicuboctahedra, cubes and tetrahedra (Figure 10.11b). All three ensembles of geometrical objects can pack 3D space completely without gaps (in contrast to spheres) so α can be varied from 0 to 1. Ensembles of cubes have through-channels that might be supposed to decrease tortuosity but the two additional geometries did not have such channels and the last geometry also did not have just one type of "cell". Surprisingly, all three geometries gave identical results, a simple relationship between the two ECS parameters, λ_g defined as the tortuosity arising only from the geometrical constraints, and α, the total volume fraction

$$\lambda_g(\alpha) = \sqrt{(3-\alpha)/2} \qquad (10.14)$$

giving an upper limit of 1.225 for λ_g (Figure 10.11). For $\alpha = 0.2$ (a typical value in brain), $\lambda_g = 1.183$ (Figure 10.11c). Interestingly, Maxwell (1954) derived the same formula from a study of electric conductivity of a medium containing a dilute suspension of spheres. Tao and Nicholson (2004) proposed that this result is generally applicable to 3D media composed of

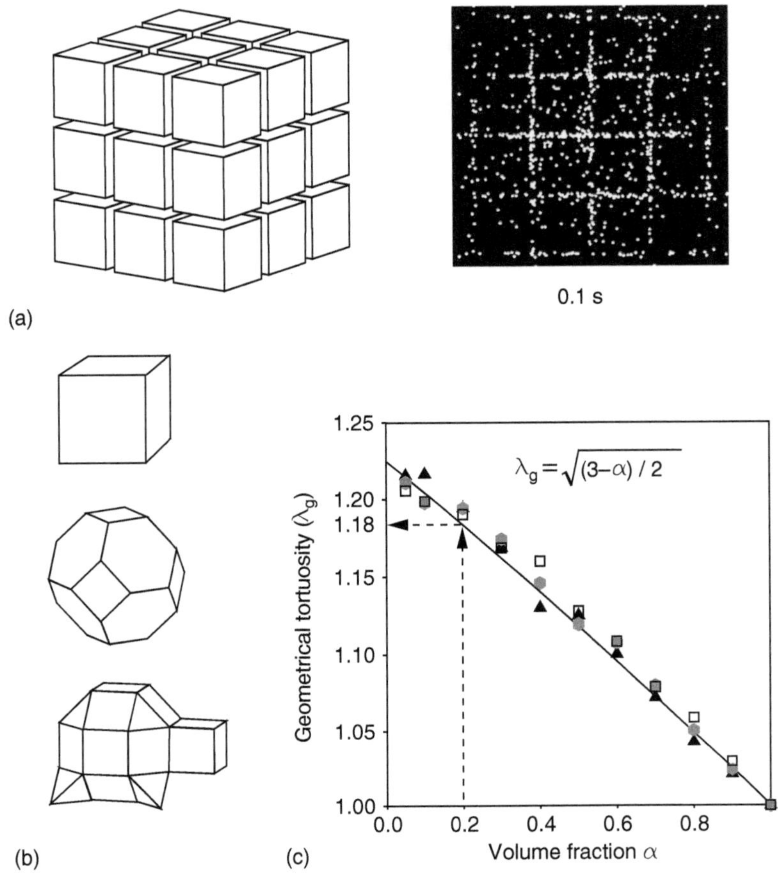

FIGURE 10.11 Computer simulation of diffusion in virtual 3D media composed of convex cells. (a) ECS model constructed with uniformly spaced cubic cells (*left*). Projection of the 3D distribution of the diffusing particles onto the x–y plane recorded at time 0.1 s (*right*). (b) Cellular convex elements used to build three ECS models. Cube (*top*), truncated octahedra (*middle*), combination of one rhombicuboctahedron, one cube and two tetrahedra (*bottom*). (c) Summary of simulations in three ECS models (cubes *squares*, truncated octahedra *hexagon*, three differently-shaped convex cells *triangles*). In all three virtual 3D media, the volume fraction was varied from 0.05 to 0.90, D^* was calculated from the particle counts and the tortuosity was calculated. (Reproduced from Tao, L. and Nicholson, C., *J. Theor. Biol.*, 229, 59–68, 2004.)

convex elements separated by uniform spaces. Indeed, simulations in a 3D model composed of random convex polyhedra confirmed this prediction (Figure 10.12; Hrabě et al. 2004). In brain, however, α and λ are independent parameters, and $\lambda > \lambda_g$. The 3D models built from convex cells thus did not capture all the critical structural features of brain ECS.

Dwell-Time Diffusion Theory

The surprising result of Tao and Nicholson (2004) together with the experimental finding that the introduction of background macromolecules to ischemic tissue decreases extracellular tortuosity (Hrabětová and Nicholson 2000; Hrabětová et al. 2003) led to the development of a simple semi-quantitative model of tortuosity based on diffusion dwell-times (Hrabětová et al. 2003; Hrabě et al. 2004). It is assumed there that the ECS is composed both of well-connected spaces where molecules move with a hindrance given by simple geometric considerations and it also contains dead-spaces

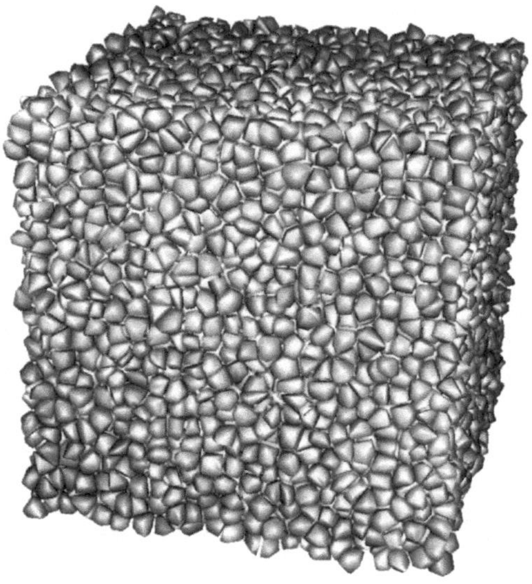

FIGURE 10.12 Virtual 3D media composed of random convex polyhedra. The gaps between the polyhedra are uniform. (Reproduced from Hrabě, J., et al. *Biophys. J.*, 87, 1606–1617, 2004. With permission.)

where diffusing molecules are delayed (Figure 10.13a). This model offers an explanation as to why the limiting tortuosity of 3D media composed of convex elements is 1.225 and shows how the addition of dead-spaces increases λ. Furthermore, the model predicts a relationship between λ and α in isotropic media containing dead spaces:

$$\lambda_g = \lambda_0 \sqrt{\alpha/\alpha_0}, \tag{10.15}$$

where λ_0 and α_0 are the values associated with the well-connected space under normal conditions. The volume fraction α here is the sum of volume fraction occupied by dead-spaces, α_d, and well-connected spaces, α_0. Dead-space volume fraction is thus a proportion of the ECS, forming the space inside dead-end pores, leaving the remainder of the ECS to constitute the well-connected space (Figure 10.13a). The dwell-time diffusion model supports the interpretation of experimental data. When the background macromolecule, 70,000 *MW* dextran, was added to thick slice models of ischemia, α decreased from 0.14 to 0.10 and λ decreased from 1.9 to 1.54; the model predicts a decrease to 1.61 (Hrabětová et al. 2003). The model also allows us to make predictions about the distribution of volume fraction between well-connected spaces and dead-spaces based on λ measured in brain (Hrabětová et al. 2003; Hrabětová and Nicholson 2004). Note that this model assumes that molecules exchange rapidly between dead-space and well-connected space.

Three-Dimensional Media Containing Concave Cells and Lakes

To confirm the dwell time diffusion model by Monte Carlo simulations, concave elements were incorporated into the virtual 3D media (Figure 10.13a; Hrabě et al. 2004). The ECS of such media is composed of well-connected spaces (α_0) and dead-spaces within concavities (α_d) that delay diffusing molecules by transiently trapping them. Pathways between the elements were kept narrow to comply with assumption of the dwell-time diffusion model that ECS width is significantly smaller than the side of an element. Dead-spaces were mimicked by introducing slits into the faces of cubes (Figure 10.13a). Monte Carlo numerical simulations showed that dead-spaces can increase the tortuosity to, and even well beyond, values measured in brain; for a given α, the increase in λ was dependent on the volume occupied by dead-spaces (Figure 10.13b). Monte Carlo numerical simulations thus fully support the dwell-time diffusion model.

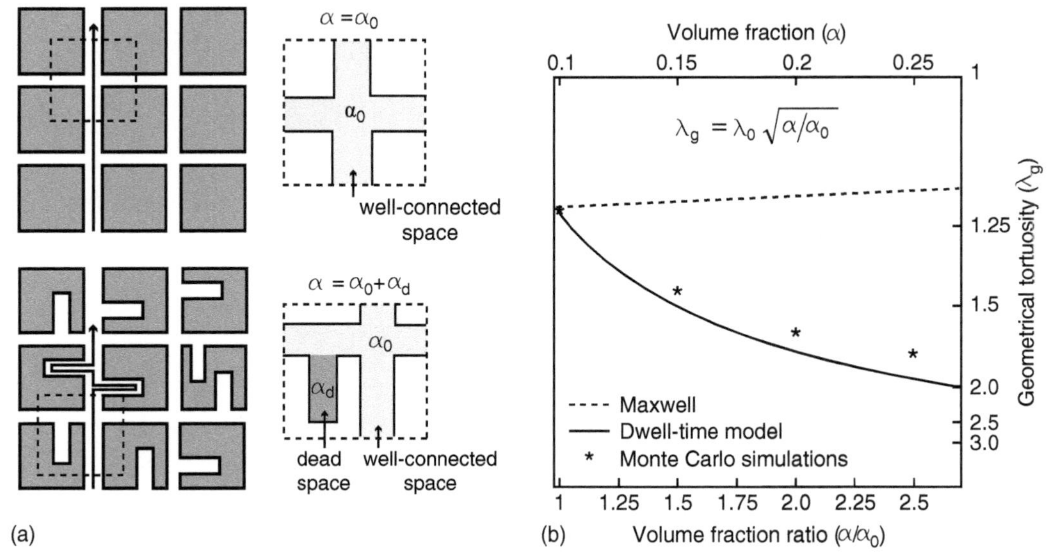

FIGURE 10.13 Computer simulation of diffusion in virtual 3D media containing concave slit-like dead spaces. (a) Two-dimensional sections through 3D model of the ECS. In the media composed of uniformly spaced convex elements, the ECS is composed of well-connected spaces (α_0; *top*). In contrast, when concave elements are introduced, the ECS is composed of both well-connected spaces (α_0) and dead-spaces residing in concavities (α_d). (b) Addition of dead-spaces increases the tortuosity. A dwell-time diffusion model predicts a relationship between λ and $\alpha/\alpha_0 = (\alpha_d + \alpha_0)/\alpha_0$ (*solid line*) that was confirmed by Monte Carlo simulations of diffusion process (*asterisks*). In virtual 3D media containing slit-like dead-spaces, α was varied from 0.10 to 0.25 with α_d equal 0.05, 0.1, and 0.15. Values of D^* were obtained from the particle counts and tortuosities calculated. Maxwell relationship derived from simulations in media containing convex cell is shown (*dotted line*) for a comparison. (Reproduced from Hrabě, J. et al., *Biophys. J.*, 87, 1606–1617, 2004. With permission.)

A related study (Tao et al. 2005) employed a set of 3D models composed of cubes where dead-space microdomains were mimicked by open rectangular cavities inserted either into the face of a cube or at the vertex (Figure 10.14). At the vertex, the dead-spaces do not have the form of narrow slits but rather form a residual extracellular space volume or lake. Such models have physiological relevance; it has been proposed that such lakes are formed during hyperosmotic stress and function as transient traps for diffusing molecules, keeping values of λ constant while α increases (Chen and Nicholson 2000; Kume-Kick et al. 2002). In addition, dilated regions of the ECS have been observed in electron micrographs of brain tissue (Brightman 1965; Van Harreveld et al. 1965; Cragg 1979). Simulations (Tao et al. 2005) revealed that in such ECS models, λ can equal or exceed the typical experimental value of about 1.6. The equation derived from dwell-time diffusion theory no longer fits the data because the assumption of narrow spaces is violated here (Tao et al. 2005). To accommodate wide dead-spaces, λ_0 was replaced with Maxwell's expression and a reasonably good fit was obtained (Figure 10.13a). In addition, Tao et al. (2005) introduced an empirical factor β that correlated with the ease with which a molecule could leave a cavity and its vicinity (Figure 10.13b; Tao et al. 2005).

FUTURE DIRECTIONS

Research done over past 25 years gives a solid picture of the macroscopic biophysical properties of the ECS (Nicholson and Syková 1998; Nicholson 2001). We have learned that the ECS

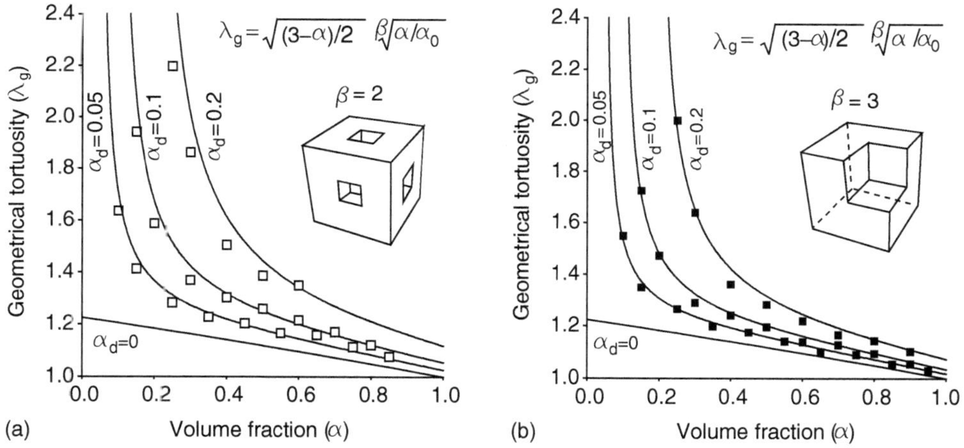

FIGURE 10.14 Computer simulation of diffusion in virtual 3D media containing concave lake-like dead spaces. (a) A model of the ECS was built from cubes containing cavities in the center of each face; lake-like dead-space is surrounded by two cubes. (b) A model of the ECS was built from cubes where a cavity was placed at a vertex; lake-like dead space is surrounded by four cubes. In both virtual 3D media, the volume fraction was varied from 0.05 to 0.90 with α_d equal 0.05, 0.10, and 0.20. The D^* was obtained from the particle counts and the tortuosity was calculated. Summary of the simulations are shown (*squares*) together with equations extracted from fitting the data (*curved lines*). Modified Maxwell relationship derived from simulations in media containing convex cells is shown ($\alpha_d = 0$) for a comparison. (Reproduced from Tao et al. *J. Theor. Biol.*, 234, 525–536, 2005.)

occupies about 20% of tissue volume and that inert molecules are hindered so that the diffusion coefficient, D, is reduced by about 2.6 (i.e., λ^2). We know now that these values change-sometimes independently-during brain activity and in various pathological states. In addition, when the size, shape, and charge of the probe molecule are altered, additional information reflecting the interaction of the probe with the extracellular environment, including the matrix, may be obtained. However, an unequivocal interpretation of all the results has proved elusive. It is clear that better understanding of ECS microstructure, in terms of geometry and molecular composition, would facilitate the interpretation of data. Diffusion studies in combination with modeling offer a unique exploratory tool. Future results will be highly relevant for understanding the transport of substances, including neurotransmitters, neuromodulators and exogenous drugs. The work to date does establish, however, that the ECS is an essential component of brain function.

Supported by NIH NINDS grants NS47557 (S.H.) and NS28642 (C.N.).

REFERENCES

Abbott, N. J., Evidence for bulk flow of brain interstitial fluid: Significance for physiology and pathology, *Neurochem. Int.*, 45, 545–552, 2004.

Agnati, L. F., Fuxe, K., Nicholson, C., and Syková, E., *Volume Transmission Revisited (Prog. Brain Res. 125)*, Elsevier, Amsterdam, 2000.

Ammann, D., *Ion-Selective Microelectrodes*, Springer, Berlin, 1986.

Artru, A. A., Spinal cerebrospinal fluid chemistry and physiology, In *Spinal Drug Delivery*, Yaksh, T. L., Ed., Elsevier, Amsterdam, pp. 177–237, 1999.

Beaulieu, C., The basis of anisotropic water diffusion in the nervous system—a technical review, *NMR Biomed.*, 15, 435–455, 2002.

Bondareff, W. and Pysh, J. J., Distribution of the extracellular space during postnatal maturation of rat cerebral cortex, *Anat. Rec.*, 160, 773–780, 1968.

Brightman, M. W., The distribution within the brain of ferritin injected into cerebrospinal fluid compartments. II. Parenchymal distribution, *Am. J. Anat.*, 117, 193–219, 1965.

Bungay, P. M., Morrison, P. F., and Dedrick, R. L., Steady-state theory for quantitative microdialysis of solutes and water in vivo and in vitro, *Life Sci.*, 46, 105–119, 1990.

Chen, K. C. and Nicholson, C., Changes in brain cell shape create residual extracellular space volume and explain tortuosity behavior during osmotic challenge, *Proc. Natl. Acad. Sci. U.S.A*, 97, 8306–8311, 2000.

Chen, K. C., Höistad, M., Kehr, J., Fuxe, K., and Nicholson, C., Quantitative dual-probe microdialysis: mathematical model and analysis, *J. Neurochem.*, 81, 94–107, 2002.

Cragg, B., Brain extracellular space fixed for electron microscopy, *Neurosci. Lett.*, 15, 301–306, 1979.

Cragg, S. J. and Rice, M. E., DAncing past the DAT at a DA synapse, *Trends Neurosci.*, 27, 270–277, 2004.

Cragg, S. J., Nicholson, C., Kume-Kick, J., Tao, L., and Rice, M. E., Dopamine-mediated volume transmission in midbrain is regulated by distinct extracellular geometry and uptake, *J. Neurophysiol.*, 85, 1761–1771, 2001.

Cserr, H. F., DePasquale, M., Nicholson, C., Patlak, C. S., Pettigrew, K. D., and Rice, M. E., Extracellular volume decreases while cell volume is maintained by ion uptake in rat brain during acute hypernatremia, *J. Physiol.*, 442, 277–295, 1991.

Dayton, M. A., Ewing, A. G., and Wightman, R. M., Diffusion processes measured at microvoltammetric electrodes in brain tissue, *J. Electroanal. Chem.*, 146, 189–200, 1983.

Fenstermacher, J. D. and Kaye, T., Drug "diffusion" within the brain, *Ann. N. Y. Acad. Sci.*, 531, 29–39, 1988.

Fuxe, K. and Agnati, L. F., *Volume Transmission in the Brain*, Raven Press, New York, 1991.

Gonzalez-Anguilar, F., Extracellular space in the rat brain after fixation with 12 *M* formaldehyde, *J. Ultrastruct. Res.*, 29, 76–85, 1969.

Grosche, J., Matyash, V., Möller, T., Verkhratsky, A., Reichenbach, A., and Kettenmann, H., Microdomains for neuron–glia interaction: parallel fiber signaling to Bergmann glial cells, *Nat. Neurosci.*, 2, 139–143, 1999.

Höistad, M., Kehr, J., Andbjer, B., Jansson, A., and Fuxe, K., Intracerebral infusion of ^{3}H-dopamine and ^{3}H-mannitol in the striatum of halothane-anaesthetized male rats. A dual-probe microdialysis study of long-distance diffusion, *Eur. J. Neurosci.*, 12, 2505–2514, 2000.

Höistad, M., Chen, K. C., Nicholson, C., Fuxe, K., and Kehr, J., Quantitative dual-probe microdialysis: Evaluation of [^{3}H]mannitol diffusion in agar and rat striatum, *J. Neurochem.*, 81, 80–93, 2002.

Horstmann, E. and Meves, H., Die Feinstrucktur des moleculären Rindengraues und ihre physiologisches Bedeutung, *Z. Zellforsch.*, 49, 569–604, 1959.

Hrabě, J., Hrabětová, S., and Segeth, K., A model of effective diffusion and tortuosity in the extracellular space of the brain, *Biophys. J.*, 87, 1606–1617, 2004.

Hrabětová, S., Extracellular diffusion is fast and isotropic in the stratum radiatum of hippocampal CA1 region in rat brain slices, *Hippocampus*, 15, 441–450, 2005.

Hrabětová, S. and Nicholson, C., Dextran decreases extracellular tortuosity in thick-slice ischemia model, *J. Cerebral Blood Flow Metab.*, 20, 1306–1310, 2000.

Hrabětová, S. and Nicholson, C., Do electron micrographs underestimate gaps between brain cells? *Soc. Neurosci. Abs. Wash*, 28, 505.17, 2002.

Hrabětová, S. and Nicholson, C., Contribution of dead-space microdomains to tortuosity of brain extracellular space, *Neurochem. Int.*, 45, 467–477, 2004.

Hrabětová, S., Chen, K. C., Masri, D., and Nicholson, C., Water compartmentalization and spread of ischemic injury in thick-slice ischemia model, *J. Cerebral Blood Flow Metab.*, 22, 80–88, 2002.

Hrabětová, S., Hrabě, J., and Nicholson, C., Dead-space microdomains hinder extracellular diffusion in rat neocortex during ischemia, *J. Neurosci.*, 23, 8351–8359, 2003.

Ichimura, T., Fraser, P. A., and Cserr, H. F., Distribution of extracellular tracers in perivascular spaces of the rat brain, *Brain Res.*, 545, 103–113, 1991.

Johnston, P. V. and Roots, B. I., *Nerve Membranes. A Study of the Biological and Chemical Aspects of Neuron–Glia Relationships*, Pergamon, Oxford pp. 73–92 1972.

Kawasaki, K., Traynelis, S. F., and Dingledine, R., Different responses of CA1 and CA3 regions to hypoxia in rat hippocampal slice, *J. Neurophysiol.*, 63, 385–394, 1990.

Kehr, J., Höistad, M., and Fuxe, K., Diffusion of radiolabeled dopamine, its metabolites and mannitol in the rat striatum studied by dual-probe microdialysis, In *Volume Transmission Revisited (Prog. Brain Res. 125)*, Agnati, L. F., Fuxe, K., Nicholson, C., and Syková, E., Eds., Elsevier, Amsterdam, pp. 179–190, 2000.

Kosaka, T. and Hama, K., Three-dimensional structure of astrocytes in the rat dentate gyrus, *J. Comp. Neurol.*, 249, 242–260, 1986.

Krewson, C. E., Klarman, M. L., and Saltzman, W. M., Distribution of nerve growth factor following direct delivery to brain interstitium, *Brain Res.*, 680, 196–206, 1995.

Križaj, D., Rice, M. E., Wardle, R. A., and Nicholson, C., Water compartmentalization and extracellular tortuosity after osmotic changes in cerebellum of *Trachemys scripta*, *J. Physiol.*, 492, 887–896, 1996.

Kuffler, S. W. and Potter, D. D., Glia in the leech central nervous system: Physiological properties and neuron–glia relationship, *J. Neurophysiol.*, 27, 290–320, 1964.

Kume-Kick, J., Mazel, T., Voříšek, I., Hrabětová, S., Tao, L., and Nicholson, C., Independence of extracellular tortuosity and volume fraction during osmotic challenge in rat neocortex, *J. Physiol.*, 542, 515–527, 2002.

Lehmenkühler, A., Kersting, U., and Nicholson, C., Diffusion of penicillin in agar and cerebral cortex of the rat, *Brain Res.*, 444, 181–183, 1988.

Lehmenkühler, A., Syková, E., Svoboda, J., Zilles, K., and Nicholson, C., Extracellular space parameters in the rat neocortex and subcortical white matter during postnatal development determined by diffusion analysis, *Neuroscience*, 55, 339–351, 1993.

Levin, V. A., Fenstermacher, J. D., and Patlak, C. S., Sucrose and inulin space measurements of cerebral cortex in four mammalian species, *Am. J. Physiol.*, 219, 1528–1533, 1970.

Lipinski, H.-G. and Bingmann, D., Diffusion in slice preparations bathed in unstirred solutions, *Brain Res.*, 437, 26–34, 1987.

Lundbæk, J. A. and Hansen, A. J., Brain interstitial volume fraction and tortuosity in anoxia. Evaluation of the ion-selective micro-electrode method, *Acta. Physiol. Scand.*, 146, 473–484, 1992.

Margolis, R. K. and Margolis, R. U., Nervous tissue proteoglycans, *Experientia*, 49, 429–446, 1993.

Maxwell, J. C., In *A Treatise on Electricity and Magnetism*, Vol. 1, Dover; New York, 1954. (1891, Article 314).

Mazel, T., Šimonová, Z., and Syková, E., Diffusion heterogeneity and anisotropy in rat hippocampus, *Neuroreport*, 9, 1299–1304, 1998.

Morris, E. and Krnjević, K., Slow diffusion of Ca^{2+} in the rat's hippocamus, *Can. J. Physiol. Pharm.*, 59, 1022–1025, 1981.

Morrison, P. F., Laske, D. W., Bobo, H., Oldfield, E. H., and Dedrick, R. L., High-flow microinfusion: tissue penetration and pharmacodynamics, *Am. J. Physiol.*, 266, R292–R305, 1994.

Newman, G. C., Hospod, F. E., and Wu, P., Thick brain slices model the ischemic penumbra, *J. Cerebral Blood Flow Metab.*, 8, 586–597, 1988.

Newman, G. C., Hospod, F. E., and Wu, P., Glucose utilization of ischemic hippocampal slices, *J. Neurosci. Meth.*, 28, 23–34, 1989.

Newman, G. C., Hospod, F. E., and Schissel, S. L., Ischemic brain slice glucose utilization: effects of slice thickness, acidosis, and K^+, *J. Cerebral Blood Flow Metab.*, 11, 398–406, 1991.

Nicholson, C., Brain cell microenvironment as a communication channel, In *The Neurosciences Fourth Study Program*, Schmitt, F. O. and Worden, F. G., Eds., MIT Press, Cambridge, MA, pp. 457–476, 1979.

Nicholson, C., Diffusion from an injected volume of a substance in brain tissue with arbitrary volume fraction and tortuosity, *Brain Res.*, 333, 325–329, 1985.

Nicholson, C., Quantitative analysis of extracellular space using the method of TMA^+ iontophoresis and the issue of TMA^+ uptake, *Can. J. Physiol. Pharm.*, 70 (Suppl.), S314–S322, 1992.

Nicholson, C., Ion-selective microelectrodes and diffusion measurements as tools to explore the brain cell microenvironment, *J. Neurosci. Methods.*, 48, 199–213, 1993.

Nicholson, C., Interaction between diffusion and Michaelis–Menten uptake of dopamine after iontophoresis in striatum, *Biophys. J.*, 68 (5), 1699–1715, 1995.

Nicholson, C., Diffusion and related transport properties in brain tissue, *Rep. Prog. Phys.*, 64, 815–884, 2001.

Nicholson, C. and Hounsgaard, J., Diffusion in the slice microenvironment and implications for physiological studies, *Fed. Proc.*, 42, 2865–2868, 1983.

Nicholson, C. and Phillips, J. M., Ion diffusion modified by tortuosity and volume fraction in the extracellular microenvironment of the rat cerebellum, *J. Physiol.*, 321, 225–257, 1981.

Nicholson, C. and Rice, M. E., Calcium diffusion in the brain cell microenvironment, *Can. J. Physiol. Pharm.*, 65, 1086–1091, 1987.

Nicholson, C. and Rice, M. E., Use of ion-selective microelectrodes and voltammetric microsensors to study brain cell microenvironment, In *Neuromethods: The Neuronal Microenvironment*, Boulton, A. A., Baker, G. B., and Walz, W., Eds., Humana, Clifton, NJ, pp. 247–361, 1988.

Nicholson, C. and Syková, E., Extracellular space structure revealed by diffusion analysis, *Trends Neurosci.*, 21, 207–215, 1998.

Nicholson, C. and Tao, L., Hindered diffusion of high molecular weight compounds in brain extracellular microenvironment measured with integrative optical imaging, *Biophys. J.*, 65, 2277–2290, 1993.

Nicholson, C., Chen, K. C., Hrabětová, S., and Tao, L., Diffusion of molecules in brain extracellular space: Theory and experiment, In *Volume Transmission Revisited (Prog. Brain Res. 125)*, Agnati, L. F., Fuxe, K., Nicholson, C., and Syková, E., Eds., Elsevier, Amsterdam, pp. 129–154, 2000.

Novak, U. and Kaye, A. H., Extracellular matrix and the brain: components and function, *J. Clin. Neurosci.*, 7, 280–290, 2000.

Patlak, C. S. and Fenstermacher, J. D., Measurements of dog blood–brain transfer constants by ventriculocisternal perfusion, *Am. J. Physiol.*, 229, 877–884, 1975.

Pérez-Pinzón, M. A., Tao, L., and Nicholson, C., Extracellular potassium, volume fraction, and tortuosity in rat hippocampal CA1, CA3, and cortical slices during ischemia, *J. Neurophysiol.*, 74, 565–573, 1995.

Prokopová-Kubinová, S., Vargová, L., Tao, L., Ulbrich, K., Šubr, V., Syková, E., and Nicholson, C., Poly[*N*-(2-hydroxypropyl)methacrylamide] polymers diffuse in brain extracellular space with same tortuosity as small molecules, *Biophys. J.*, 80, 542–548, 2001.

Rall, D. P., Oppelt, W. W., and Patlak, C. S., Extracellular space of brain as determined by diffusion of inulin from the ventricular system, *Life Sci.*, 2, 43–48, 1962.

Rice, M. E. and Nicholson, C., Serotonin migration in the neuronal microenvironment, *Ann. N. Y. Acad. Sci.*, 481, 381–382, 1986.

Rice, M. E. and Nicholson, C., Measurement of nanomolar dopamine diffusion using low-noise perfluorinated ionomer coated carbon fiber microelectrodes and high-speed cyclic voltammetry, *Anal. Chem.*, 61, 1805–1810, 1989.

Rice, M. E. and Nicholson, C., Diffusion characteristics and extracellular volume fraction during normoxia and hypoxia in slices of rat neostriatum, *J. Neurophysiol.*, 65, 264–272, 1991.

Rice, M. E. and Nicholson, C., Diffusion and ion shifts in the brain extracellular microenvironment and their relevance for voltammetric measurements, In *Voltammetric Methods in Brain Systems. Neuromethods 27*, Boulton, A. A., Baker, G. B., and Adams, R. N., Eds., Humana, Totowa, NJ, pp. 27–79, 1995.

Rice, M. E., Gerhardt, G. A., Hierl, P. M., Nagy, G., and Adams, R. N., Diffusion coefficients of neurotransmitters and their metabolites in brain extracellular fluid space, *Neuroscience*, 15, 891–902, 1985.

Rice, M. E., Okada, Y. C., and Nicholson, C., Anisotropic and heterogeneous diffusion in the turtle cerebellum: Implications for volume transmission, *J. Neurophysiol.*, 70, 2035–2044, 1993.

Rosenberg, G. A., Kyner, W. T., and Estrada, E., Bulk flow of brain interstitial fluid under normal and hyperosmolar conditions, *Am. J. Physiol.*, 238, F42–F49, 1980.

Ruoslahti, E., Brain extracellular matrix, *Glycobiology*, 6, 489–492, 1996.

Saltzman, W. M., *Drug Delivery: Engineering Principles for Drug Therapy*, Oxford University Press, New York, 2001.

Schmidt-Kastner, R. and Freund, T. F., Selective vulnerability of the hippocampus in brain ischemia, *Neuroscience*, 40, 599–636, 1991.

Schmitt, F. O. and Samson, F. E., The brain cell microenvironment, *Neurosci. Res. Prog. Bull.*, 7, 277–417, 1969.

Schwindt, W., Nicholson, C., and Lehmenkühler, A., Critical volume of rat cortex and extracellular threshold concentration for a pentylenetetrazol-induced epileptic focus, *Brain Res.*, 753, 86–97, 1997.

Šimonová, Z., Svoboda, J., Orkand, P., Bernard, C. C. A., Lassmann, H., and Syková, E., Changes of extracellular space volume and tortuosity in the spinal cord of Lewis rats with experimental autoimmune encephalomyelitis, *Physiol. Res.*, 45, 11–22, 1996.

Špaček, J., Three-dimensional analysis of dendritic spines. III. Glial sheath, *Anat. Embryol. (Berl.)*, 171, 245–252, 1985.

Stiles, J. R. and Bartol, T. M., Monte Carlo methods for simulating realistic synaptic microphysiology using MCell, In *Computational Neuroscience: Realistic Modeling for Experimentalists*, De Schutter, E., Ed., CRC Press, London, pp. 87–127, 2001.

Stiles, J. R., Van Helden, D., Bartol, T. M., Salpeter, E. E., and Salpeter, M. M., Miniature endplate current rise times <100 μs from improved dual recordings can be modeled with passive acetylcholine diffusion from a synaptic vesicle, *Proc. Natl. Acad. Sci. U.S.A*, 93, 5747–5752, 1996.

Stiles, J. R., Kovyazina, I. V., Salpeter, E. E., and Salpeter, M. M., The temperature sensitivity of miniature endplate currents is mostly governed by channel gating: evidence from optimized recordings and Monte Carlo simulations, *Biophys. J.*, 77, 1177–1187, 1999.

Stroh, M., Zipfel, W. R., Williams, R. M., Webb, W. W., and Saltzman, W. M., Diffusion of nerve growth factor in rat striatum as determined by multiphoton microscopy, *Biophys. J.*, 85, 581–588, 2003.

Syková, E., The extracellular space in the CNS: Its regulation, volume and geometry in normal and pathological neuronal function, *Neuroscientist*, 3, 28–41, 1997.

Syková, E., Diffusion properties of the brain in health and disease, *Neurochem. Int.*, 45, 453–466, 2004.

Syková, E., Svoboda, J., Polák, J., and Chvátal, A., Extracellular volume fraction and diffusion characteristics during progressive ischemia and terminal anoxia in the spinal cord of the rat, *J. Cerebral Blood Flow Metab.*, 14, 301–311, 1994.

Syková, E., Vargová, L., Prokopová, and Šimonová, Z., Glial swelling and astrogliosis produce diffusion barriers in the rat spinal cord, *Glia*, 25, 56–70, 1999.

Syková, M. E., Mazel, T., Vargová, L., Voříšek, I., and Prokopová-Kubinová, S., Extracellular space diffusion and pathological states, In *Volume Transmission Revisited (Prog. Brain Res. 125)*, Agnati, L. F., Fuxe, K., Nicholson, C., and Syková, E., Eds., Elsevier, Amsterdam, pp. 155–178, 2000.

Tao, L., Effects of osmotic stress on dextran diffusion in rat neocortex studied with integrative optical imaging, *J. Neurophysiol.*, 81, 2501–2507, 1999.

Tao, L. and Nicholson, C., The three-dimensional point spread functions of a microscope objective in image and object space, *J. Microsc.-Oxford*, 178, 267–271, 1995.

Tao, L. and Nicholson, C., Diffusion of albumins in rat cortical slices and relevance to volume transmission, *Neuroscience*, 75, 839–847, 1996.

Tao, L. and Nicholson, C., Maximum geometrical hindrance to diffusion in brain extracellular space surrounding uniformly spaced convex cells, *J. Theor. Biol.*, 229, 59–68, 2004.

Tao, A., Tao, L., and Nicholson, C., Cell cavities increase tortuosity in brain extracellular space, *J. Theor. Biol.*, 234, 525–536, 2005.

Thorne, R. G. and Nicholson, C., In vivo diffusion analysis with quantum dots and dextrans predicts the width of brain extracellular space, *Proc. Natl. Acad. Sci. U.S.A.*, 103, 5567–5577, 2006.

Thorne, R. G., Hrabětová, S., and Nicholson, C., Diffusion of epidermal growth factor in rat brain extracellular space measured by integrative optical imaging, *J. Neurophysiol.*, 92, 3471–3481, 2004.

Thorne, R. G., Hrabětová, S., and Nicholson, C., Diffusion measurements for drug design, *Nat. Mater.*, 4, 713, 2005.

Traynelis, S. F. and Dingledine, R., Potassium-induced spontaneous electrographic seizures in the rat hippocampal slice, *J. Neurophysiol.*, 59, 259–276, 1988.

Ujec, E., Keller, O., Machek, J., and Pavlík, V., Low impedance coaxial K^+ selective microelectrodes, *Pflügers Arch.*, 382, 189–192, 1979.

Ulbrich, K., Pechar, M., Strohalm, J., Šubr, V., and Říhová, B., Synthesis of biodegradable polymers for controlled drug release, *Ann. N. Y. Acad. Sci.*, 831, 47–56, 1997.

Van Harreveld, A., The extracellular space in the vertebrate central nervous system, In *The Structure and Function of Nervous Tissue*, Bourne, G. H., Ed., Academic Press, New York, pp. 447–511, 1972.

Van Harreveld, A., Crowell, J., and Malhotra, S. K., A study of extracellular space in central nervous tissue by freeze-substitution, *J. Cell Biol.*, 25, 117–137, 1965.

Van Horne, C., Hoffer, B. J., Strömberg, I., and Gerhardt, G. A., Clearance and diffusion of locally applied dopamine in normal and 6-hydroxydopamine-lesioned rat striatum, *J. Pharmacol. Exp. Ther.*, 263, 1285–1292, 1992.

Vargová, L., Chvátal, A., Anděrová, M., Kubinová, S., Žiak, D., and Syková, E., Effect of osmotic stress on potassium accumulation around glial cells and extracellular space volume in rat spinal cord slices, *J. Neurosci. Res.*, 65, 129–138, 2001.

Vargová, L., Homola, A., Zámečník, J., Tichý, M., Beneš, V., and Syková, E., Diffusion parameters of the extracellular space in human gliomas, *Glia*, 42, 77–88, 2003.

Voříšek, I. and Syková, E., Evolution of anisotropic diffusion in the developing rat corpus callosum, *J. Neurophysiol.*, 78, 912–919, 1997.

Voříšek, I., Hájek, M., Tintěra, J., Nicolay, K., and Syková, E., Water ADC, extracellular space volume, and tortuosity in the rat cortex after traumatic injury, *Mag. Res. Med.*, 48, 994–1003, 2002.

Xin, Q. and Wightman, R. M., Transport of choline in rat brain slices, *Brain Res.*, 776, 126–132, 1997.

11 Hydrogen Peroxide as a Diffusible Messenger: Evidence from Voltammetric Studies of Dopamine Release in Brain Slices

Margaret E. Rice, Marat V. Avshalumov, and Jyoti C. Patel

CONTENTS

INTRODUCTION

Reactive oxygen species (ROS) are often considered to be toxic byproducts of cell metabolism. Increased ROS production and oxidative stress contribute to cell death after acute brain injury, as well as in slowly progressing neurodegenerative disorders, including Parkinson's disease [1,2]. However, this view of ROS is rapidly evolving in light of increasing evidence that ROS also act as cellular messengers that can modulate processes from short-term ion-channel activation to gene transcription. Hydrogen peroxide (H_2O_2) is a particularly intriguing candidate as a signaling molecule because it is neutral and membrane-permeable [3], and therefore can diffuse freely from a site of generation, like the established diffusible messengers, nitric oxide and carbon monoxide. Moreover, H_2O_2 has no unpaired electrons and is therefore not a free radical, in contrast to superoxide ($\cdot O_2^-$) and the hydroxyl radical ($^{\cdot}OH$), and so does not readily cause oxidative damage [4].

We have recently discovered that endogenously generated H_2O_2 regulates the release of dopamine (DA). DA is an important neurotransmitter and neuromodulator that participates in a wide range of brain functions, including learning and cognition [5], control of movement [6,7], and mediation of desire and reward [8]. Consequently, dysfunction of dopaminergic transmission has been implicated in several significant brain disorders, including the psychoses of schizophrenia, the movement deficits of Parkinson's disease, and addiction to substances like alcohol and cocaine. Studies of DA release regulation using voltammetric methods with carbon-fiber microelectrodes have made significant contributions to the present understanding of DA regulation and the role of DA in normal and pathological conditions. A number of key mechanistic insights have been obtained using these methods in brain slices in vitro, which readily permits voltammetric detection of DA release in discrete brain regions, without complicating factors inherent to in vivo studies, including animal behavior, anesthesia, and indirect effects from distant structures via long pathways.

In this chapter, we will discuss various methodological considerations for the use of in vitro voltammetric recording in brain slices, which will indicate the types of questions that these methods are best suited to address. In addition, we will review evidence that H_2O_2 is a diffusible messenger that mediates glutamate-dependent inhibition of DA release in dorsolateral striatum, based on data obtained using fast-scan cyclic voltammetry (FCV) with carbon-fiber microelectrodes in brain slices [9,10]. These studies also indicate the mechanism of release suppression, which is the activation of ATP-sensitive potassium (K_{ATP}) channels. Companion electrophysiological studies show that H_2O_2 also regulates the spontaneous activity of DA neurons in the substantia nigra pars compacta (SNc) via K_{ATP} channels [11], consistent with the apparently abundant expression of these channels throughout the nigrostriatal pathway [12–14]. Together, these studies illustrate how voltammetric methods in brain slices, both alone and in combination with companion techniques, can be used to address important questions in neuroscience.

OVERVIEW

Physiology of DA Release: Why It Can Be Detected in the Extracellular Space

Dopaminergic cell body regions within the central nervous system (CNS) are subdivided into cell groups A8–A17 based on the classification system of Dahlström and Fuxe [15]. The principal DA neuronal pathways arise from midbrain nuclei designated as the A9 and A10 cell groups, which are the SNc and ventral tegmental area (VTA). These project ipsilaterally via the medial forebrain bundle (MFB) to various forebrain structures. Dopaminergic neurons in the SNc send axon projections via the nigrostriatal pathway to the dorsolateral striatum (caudate putamen, CPu), which is critically involved in the control of movement [6,7]. DA neurons in the VTA project via the mesolimbic pathway to the ventromedial striatum (nucleus accumbens, NAc), which is involved in motivation and reward [8]. Additionally, DA neurons of the VTA project via the mesocortical pathway to the amygdala and prefrontal cortex, which have been implicated in the control of emotion and the pathophysiology of schizophrenia, respectively.

Synthesis of DA occurs in the cytosol of DA cell bodies and axons from the amino acid tyrosine; the first step mediated by tyrosine hydroxylase (TH) is rate limiting. Cytosolic DA is taken up into intracellular storage vesicles via the vesicular monoamine transporter 2 (VMAT2). Release of DA from axon terminals, like that of the primary excitatory transmitter glutamate and the primary inhibitory transmitter GABA (γ-aminobutyric acid), occurs via two main mechanisms: impulse-dependent (vesicular) and carrier-mediated release. With impulse-dependent DA release, Na^+-dependent action potentials trigger Ca^{2+} entry via the opening of voltage-gated Ca^{2+}-channels [16]. In DA axons, this catalyzes a process in which VMAT2-expressing vesicles that contain DA fuse with the presynaptic membrane to release their contents into a synapse or the extracellular environment via exocytosis (Figure 11.1). The amplitude and time course of an increase in extracellular DA concentration ($[DA]_o$), after stimulation of a population of DA cells or axons, are determined by the competing processes of release and uptake. Uptake of DA is mediated by the dopamine transporter (DAT), which is found only on the plasma membrane of DA cells and axons [17–20] (Figure 11.1). However, conditions that cause the cytoplasmic concentration of DA to exceed that in the extracellular environment can lead to carrier-mediated DA release via reversal of the DAT, which is a Ca^{2+}-independent process. Under normal physiological conditions, impulse-dependent exocytosis is the primary mechanism of axonal DA release.

In contrast to fast, synaptically acting neurotransmitters like glutamate, DA acts on a slower time scale as a neuromodulator that sets a tone for local excitability. Consequently, the actions and

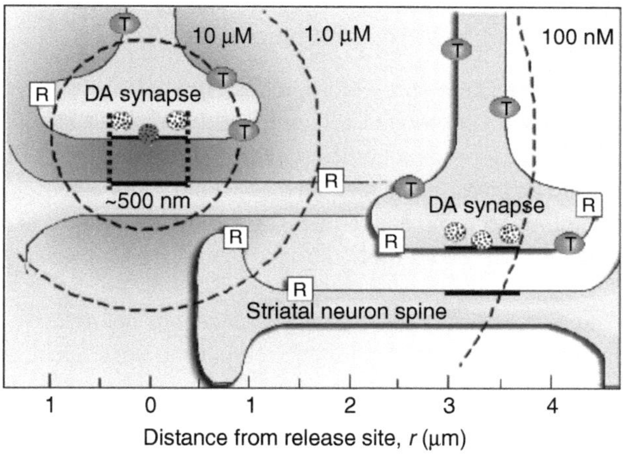

FIGURE 11.1 Relationship between distance and DATs on $[DA]_o$ transients following quantal release in striatum. Schematic representations of a releasing DA synapse and a neighboring synapse on dendritic spines of striatal medium spiny neurons showing extrasynaptic DA receptors (R) and DATs (T) and the maximum distances (dashed arcs) at which peak $[DA]_o$ reaches 10 μM, 1 μM, and 100 nM after quantal release with normal uptake via the DAT. The concentration and distances were derived from simulations shown in Cragg, S.J. and Rice, M. E., *Trends Neurosci.*, 27, 270–277, 2004, in which the quantal content of a synaptic vesicle, Q, is 9800 DA molecules and the average distance between DA synapses is ~ 3.5 μm based on a mean of 0.05 synapses μm^{-3}. Within 1 μm of a release site, DA concentration falls entirely by diffusion to <10% of the peak within milliseconds. Thus, on reaching an adjacent synapse, released DA would be diluted by $> 10^6$, from 250 mM in a vesicle of 25 nm radius to ~100 nm. This peak $[DA]_o$ is sufficient to activate all DA receptors with affinities in the nanomolar range and thereby readily mediate crosstalk. DA molecules will only be removed by DAT uptake on reaching an adjacent DA axon. (Modified from Cragg, S. J. and Rice, M. E., *Trends Neurosci*, 27, 270–277, 2004.)

regulation of DA differ in range and time course from those of "conventional" transmitters [20]. Released DA activates two broad classes of DA receptors, classified as D1-like and D2-like [21,22]. Both classes of DA receptors are G-protein coupled, with response times exceeding 100 ms [21]. By contrast, glutamate primarily mediates fast synaptic transmission via ionotropic receptors that generate excitatory postsynaptic potentials (EPSPs) lasting only 1–2 ms [23,24]. Moreover, the locations of DA and glutamate receptors differ. Both classes of DA receptors are located predominantly outside a synapse (extrasynaptic) [17,18], whereas glutamate receptors are primarily found within a synapse (subsynaptic) [23,24].

The extrasynaptic location of DA receptors implies that DA must escape a synapse and diffuse to its receptors to mediate local effects [20,25,26]. Consistent with this functional requirement, the DAT is also found extrasynaptically on DA axons [27]. Immediately after release, a DA molecule diffusing in three-dimensional extracellular space will be removed only when it encounters a DA axon (Figure 11.1). By contrast, glutamate synapses are literally surrounded by glutamate transporters, which are located not only on pre- and postsynaptic sites within a synapse, but also on glial membranes that ensheath glutamate synapses [28,29]. This three-dimensional barrier strongly limits glutamate "spillover" from one synapse to another.

Overall, DA neurotransmission is designed to occur primarily in the extracellular space. This makes DA an ideal transmitter substance to monitor with sensors that greatly exceed the <10 nm diameter of a synaptic cleft, e.g., 8–30 μm carbon-fiber microelectrodes. Although there continues to be interest in further electrode miniaturization to enable DA detection within a synapse, this is not really "where the action is." Rather, changes in $[DA]_o$ monitored with voltammetric methods provide a direct indication of DA transmission, since the extent of activation of extrasynaptic DA receptors will be governed by the amplitude and duration of a given increase in $[DA]_o$.

A Brief History of In Vivo and In Vitro Voltammetry

The use of voltammetric methods to study biological substances in vivo was first described in the mid-1950s, when Leland Clark introduced a membrane covered Pt electrode to monitor oxygen levels in blood and other tissues using cathodic reduction of oxygen [30]. Other electrode materials, including glassy carbon [31], were introduced later to examine the anodic oxidation of exogenous marker substances, including ascorbic acid and hydrogen. Subsequently, Ralph Adams introduced carbon paste as an electrode material with better characteristics for oxidation reactions than either Pt or glassy carbon [32].

The first report of anodic voltammetry to study *endogenous* neuroactive substances *in vivo* was by Adams and his colleagues, who used miniaturized carbon paste electrodes to examine neurotransmission in the rat CPu [33]. Although the initial goal of those studies was to monitor DA, the primary substance detected in the absence of stimulation was ascorbate. Interference from ascorbate held back progress in voltammetric studies of DA release in vivo for several years while researchers explored a variety of methods to overcome this problem. Several other applications of voltammetry for in vivo monitoring were introduced during this time, including detection of activity-dependent increases in DA metabolites in cerebrospinal fluid [34]. Eventually, two successful approaches were developed to eliminate interference from ascorbate. First, Gonon et al. [35,36] introduced a method for anodic pretreatment of carbon fibers that modifies an electrode surface to permit detection of DA and ascorbate oxidation at distinct potentials. At unmodified carbon surfaces, there is normally an overpotential for irreversible ascorbate oxidation that raises the required potential to roughly that required for DA oxidation (typically 0.3–0.5 V versus Ag/AgCl with slow scan methods or amperometry), despite the lower thermodynamic oxidation potential for ascorbate versus DA. Anodic pretreatment decreases the ascorbate overpotential by ~ 200 mV so that ascorbate and DA can be monitored simultaneously, but independently. The second approach for eliminating the ascorbate interference was to coat electrodes with Nafion®, a perfluorosulfonated ionomer that excludes anionic ascorbate, as well as the DA metabolite DOPAC

(3,4-dihydroxyphenylacetic acid), while extracting cationic DA [37,38]. Coating with Nafion® not only improves the selectivity of voltammetric electrodes for DA, but also protects the electrode surface during tissue recording, which helps maintain sensitivity for DA detection.

Three additional developments in the late 1970s and early 1980s revolutionized the use of voltammetric methods to examine DA release and have led to widespread acceptance of these methods in the neuroscience community today. The first of these was the introduction of carbon-fiber microelectrodes for in vivo voltammetry by two groups working independently [35,36,39,40]. The carbon fibers used were typically 7–10 μm in diameter, which produced much smaller microelectrodes than the 150–300-μm carbon paste-based electrodes first introduced for in vivo recording (e.g., [33,37]). Carbon-fiber microelectrodes approached physiological dimensions and caused minimal damage when implanted in living tissue; indeed recent studies indicate that there is little cellular damage at distances >7 μm from the center of a carbon-fiber microelectrode track [41]. The second development was the discovery that a pure DA release signal in the striatum could be elicited by electrical stimulation of the dopaminergic MFB pathway [42–44]. This observation was critical because it eliminated the need for electrode pretreatment or Nafion® coating, although both methods are still used for some applications.

The third key development was the introduction of in vitro voltammetric monitoring of neuroactive substances, initially ascorbate, in viable brain slices [45,46]. This methodology was soon extended to the detection of evoked DA release in acute striatal slices [47–49], and eventually to measurements of spontaneous DA release in organotypic slice culture [50]. A creative advance that followed from these first in vitro studies was to use voltammetric methods, specifically amperometry, to monitor quantal release of catecholamines from isolated chromaffin cells [51,52]. Amperometry was subsequently applied to the study of DA release from cultured CNS neurons, including carotid glomus cells [53] and midbrain DA neurons [54], and later to quantal release of DA from substantia nigra neurons in acute midbrain slices [55].

Together, these developments demonstrated the value and feasibility of using voltammetric methods to study DA release in vitro, as well as in vivo. Reflecting this, the titles of most voltammetric studies published in the last decade confidently state the result, rather than the methods used, indicating that these electrochemical techniques have become an accepted part of state-of-the-art methods in neuroscience.

Advantages of Brain Slices to Study DA Release

Brain slices provide a valuable experimental preparation in which intact, but isolated, brain regions containing ensembles of neurons and glia can be readily identified and studied under controlled conditions. In brain slice preparations, DA release elicited by local stimulation, whether the stimulus is chemical (e.g., high K^+) or electrical, is a "population response" such that monitored $[DA]_o$ reflects the sum of DA released from the local population of all DA axons, or cell bodies and dendrites, within the stimulated region. Stimulation procedures are discussed in "Stimulating and Recording DA Release in Brain Slices" in this chapter.

Key factors that can be controlled in brain slices include recording position, composition of the extracellular fluid, and the concentration of applied drugs or other agents. The ability to identify a region of interest is of obvious importance. For voltammetric recording in vivo, electrodes are inserted through the skull via burr holes made over the region to be studied, then lowered according to coordinates from a stereotaxic atlas of the brain for the species examined, often a rat (e.g., [56]). Using these procedures, electrode placement in vivo is essentially blind and requires post-experiment confirmation of electrode position. Although electrode placement is usually not a problem for large brain structures like the striatum, accurate placement in smaller, deeper structures can be difficult. For example, voltammetric recording in midbrain slices has been successfully used to study somatodendritic DA release in the SNc, which is a small DA cell body region in the midbrain [65,95,100]; however, comparable in vivo studies in the SNc have not been reported. A related advantage of brain slice preparations is that measurements of DA release can be combined with other methods for which ready visualization is beneficial, like electrophysiological recording [57–59], or essential, like fluorescence imaging [60,61]. Indeed, with appropriate instrumentation, a carbon-fiber microelectrode used for voltammetry can also be used for simultaneous electrophysiological recording in slices [57].

The ability to control the extracellular microenvironment in a brain slice is another advantage of this in vitro preparation. The composition of the artificial cerebrospinal fluid (aCSF) used to maintain brain slice viability (discussed in "aCSF Composition for Healthy Slices" in this chapter) typically mimics the ionic composition of CSF in vivo. However, aCSF composition can be varied in vitro, for example to examine the Ca^{2+} dependence of DA release [62], whereas normal ion homeostasis in the intact brain would preclude addressing such questions in vivo. Additionally, known concentrations of pharmacological agents can readily be added to aCSF. Such direct application circumvents two major problems with in vivo drug administration. First, with systemic administration of a given drug, actual drug concentration in a brain region of interest is difficult to pinpoint because of peripheral metabolism, variable blood–brain barrier permeability, and/or other pharmacokinetic complications. Second, the site(s) of action is often unclear with systemic drug administration. For example, if a decrease in evoked DA release in striatum were seen following systemic administration of a DA autoreceptor agonist in vivo, it would not be obvious, without further experimentation, whether this reflected a direct inhibitory effect within the striatum, an indirect effect due to autoreceptor-dependent suppression of DA neuron excitability in the SNc, or some combination of both. On the other hand, the question of whether DA autoreceptor activation has a direct, local effect on DA release in the striatum can be readily addressed using striatal slices [63–69].

It is important to consider the limitations of brain slice preparations, as well. Most obviously, the use of isolated brain tissue in vitro precludes evaluation of DA release induced by animal behavior and how this might be affected by systemic drug administration. Additionally, some questions are best addressed by considering the influence of distant structures connected by long pathways that cannot easily be preserved in slice preparations. Even then, however, slices can provide important mechanistic information to complement in vivo studies. For example, Phillips et al. [70] examined DA release and uptake in brain slices from rats that had received prenatal exposure to cocaine in order to assess long-term consequences on DA regulation when these animals reached adulthood.

Overall, the strengths of brain slice preparations outweigh the weaknesses for mechanistic studies of DA release in discrete brain regions without interference from systemic interactions that can complicate data interpretation in vivo. Although cultured or freshly dissociated neurons can also be used for such studies at the single cell level, slices provide a reduced preparation in which the geometric structure of brain tissue is preserved, including intact local circuitry and maintained neuron–glial interactions. This structural preservation facilitates the study of the regulation of DA release by other endogenous neurotransmitters and modulators, including those discussed later in "Regulation of Striatal DA Release in Hydroden Peroxide". Lastly, direct, in situ voltammetric detection of $[DA]_o$ in slice preparations offers clear advantages over other methods that require efflux into the medium outside a slice (for review see [71]).

VOLTAMMETRIC METHODS

COMPARISON OF VOLTAMMETRIC TECHNIQUES

A range of voltammetric methods have been used to monitor DA release in vitro, as well as in vivo; the three most commonly used are amperometry, chronoamperometry, and FCV. Of these, FCV has been most widely used in brain slice studies; however, discussion of these methods will begin with the conceptually simplest, which is constant potential amperometry.

Constant Potential Amperometry

For amperometric detection of DA, a potential sufficient for DA oxidation is applied constantly to the working electrode. The potential required depends on the electrode used and must be determined experimentally; a range of +0.3 to +0.8 V versus Ag/AgCl can be found in the literature [55,59,69,72]. Given that the electrode potential is always sufficient to oxidize all molecules of DA (or other electroactive substances oxidizing at the same or lower potential) in contact with the electrode surface, the integrated current monitored during a release event is directly proportional to the number of molecules oxidized. An advantage of this method is that continual DA oxidation curtails release signals by enhancing DA "clearance" as DA molecules are electrochemically consumed, which can improve the resolution of two signals separated by <1 s. Therefore, amperometry is well-suited for detection and quantification of the release of DA from single vesicles, i.e., quantal release [54,55]. Additionally, because of the constant applied potential, amperometric measurements have little contribution from residual non-Faradaic current (capacitative or charging current), and are not subject to interference from pH or divalent ions like Ca^{2+} (J. Patel and M. Harrell, unpublished observations), which can be a problem with FCV, as discussed below.

On the other hand, there are several aspects of amperometric monitoring that limit its usefulness for the study of population-based DA release. First, amperometry alone provides little information about signal identity. Other electroactive substance with oxidation potentials similar to that of DA include ascorbate, the DA metabolite DOPAC, and the related monoamine transmitters, norepinephrine (NE) and serotonin (5-hydroxytryptamine; 5-HT). Consequently, amperometric measurements require the use of an independent method, like FCV [69,72], to confirm that the monitored substance is DA. Without this, the identity of the monitored substance(s) remains uncertain. Second, this technique cannot readily be used to estimate $[DA]_o$ after a release event because electrode calibration factors obtained using amperometry are flow-rate dependent [72]; again, however, FCV has been used as an adjunct method for quantitative assessment of a release response. Third, because constant-potential amperometry irreversibly consumes DA, this complicates assessment of DA uptake from release records obtained using this method [73,74]. Thus, although amperometry is conceptually simple, data interpretation can be complex.

Chronoamperometry

The earliest voltammetric method used for DA detection in brain slices was chronoamperometry [47], in which the potential is stepped briefly from a non-oxidizing holding potential to a voltage sufficient for DA oxidation, then returned to the holding potential. When first introduced, the duration of the voltage step for chronoamperometry was 1 s delivered at regular intervals of 20–30 s. However, more recent studies have used shorter pulses of 100 ms applied at 200 ms intervals, leading to the term "high-speed" chronoamperometry [75–78]. Like the closely related technique of amperometry, chronoamperometry permits DA detection with minimal interference from pH and Ca^{2+} [77], which offers particular advantages for studies of the ionic dependence of DA release, e.g., the Ca^{2+} dependence [76]. A typical voltage step in chronoamperometry is from 0 to +0.55 V versus Ag/AgCl and back, which allows for reduction of oxidized DA after the pulse. The signal-to-noise ratio is usually optimized by integrating the current detected during the last 70 ms of the oxidation step, and again during the reduction step. The reduction step not only minimizes sampling-induced DA depletion, but also provides assistance in the identification of a monitored substance, since the ratio of reduction/oxidation current (or S2/S1) for a given signal tends to be substance-specific. For example, the S2/S1 ratio for DA is typically >0.5 in calibration, albeit >1.0 in slices. By contrast, the ratio for 5-HT is <0.1 [75,76], which permits these substances to be distinguished when only one or the other is released. The variability of the S2/S1 ratio for DA in calibration versus in slices precludes 100% certainty of signal identification, such that the term "DA-like" is sometimes used in reports in which chronoamperometry was used (e.g., [75,76]).

Fast-Scan Cyclic Voltammetry

FCV is more widely used for monitoring DA release than amperometry or chronoamperometry for three main reasons: (1) reliable signal identification; (2) reduction of oxidized DA; and (3) a short sampling time that is commensurate with physiological events. The simplest voltage waveform used for FCV is a triangle wave (Figure 11.2a), with an initial positive-going ramp that starts from a typically negative initial potential, rises to potential that is more positive than the DA oxidation potential, then reverses as a negative-going ramp that returns to the initial potential, which is typically more negative than the potential required for reduction of oxidized DA. Other voltage waveforms have also been used for FCV measurements; indeed, the waveform first introduced began with an initial negative (cathodic) ramp, followed by a conventional triangle wave that returned to the most negative potential, with a final anodic ramp back to the starting point, so that the overall quadraphasic waveform was like a "W" [40]. More dramatically altered

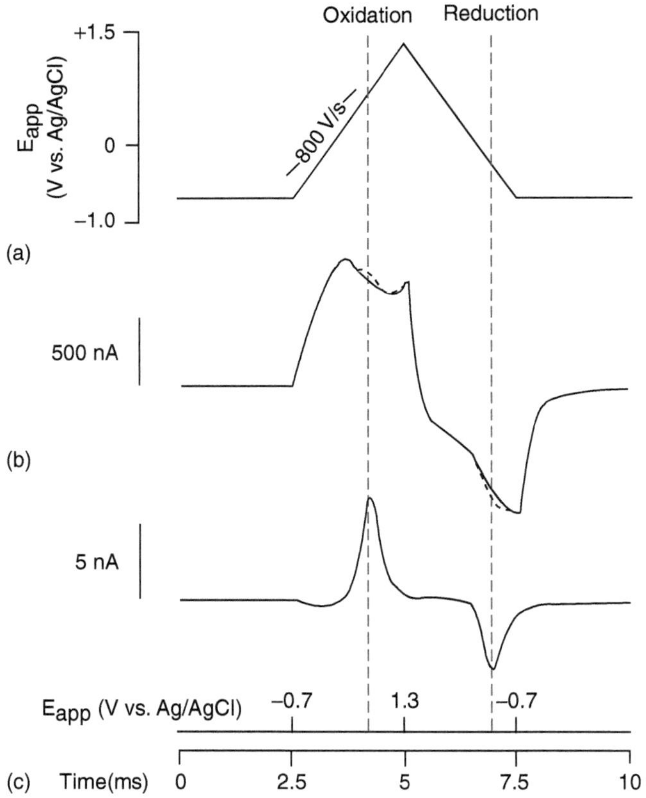

FIGURE 11.2 Fast-scan cyclic voltammetry (FCV) voltage waveform and dopamine (DA) voltammograms. (a) Typical triangular voltage waveform applied to a carbon-fiber electrode with FCV. (b) Background current generated at the carbon-fiber electrode with FCV in the absence (solid-line) and presence (broken line) of 5 μ*M* DA. The electrolyte was recording aCSF (as described in "aCSF Composition for Healthy Slices"); scan rate was 800 V/s. (c) Voltammogram for 5 μ*M* DA obtained by subtracting the background current in aCSF alone from that obtained in the presence of DA in panel (b). Electrode sensitivity to DA in this example was 7 nA/μ*M*; oxidation and reduction peak potentials for DA were +580 mV and −190 mV versus Ag/AgCl, respectively.

waveforms have also been used to facilitate the detection of a substance or multiple substances. For example, to examine DA release and oxygen consumption simultaneously, Kennedy et al. [79] used a scan that rose from a holding potential of 0.0 V versus Ag/AgCl to +1.0 V, reversed to −1.3 V, then returned to the holding potential, which encompassed potentials required for both DA oxidation and oxygen reduction.

Regardless of the waveform used, the DA oxidation peak in FCV scans typically occurs around +0.6 V versus Ag/AgCl, with a single reduction peak around −0.2 V (Figure 11.2). Although the voltammogram of 5-HT obtained with FCV has a similar oxidation peak potential to that of DA, the 5-HT voltammogram has two reduction peaks, neither of which coincide with the potential of the single DA reduction peak [49,80,81]. Thus, the characteristic DA voltammogram can be used as a "fingerprint" for signal identification (Figure 11.2c). The reduction of oxidized DA during the cathodic part of the FCV scan allows sampling of $[DA]_o$ following a release event, with minimal DA consumption, such that FCV is less destructive than amperometry. This characteristic helps ensure that evaluation of DA uptake and release parameters from release records obtained using FCV reflect physiological characteristics and not consequences of oxidative loss of DA from the detection method used [73,74].

Like high-speed chronoamperometric voltage steps, FCV scans are typically repeated at subsecond intervals, ranging from 250 ms (4 Hz) to 100 ms (10 Hz). However, in the case of FCV, the term "fast" (or sometimes "high-speed") refers to the scan rate, typically 300–900 V/s [40,49,65,82], rather than the sampling interval. With these scan rates, a forward and backward sweep over a 2 V potential range, e.g., −0.7 to +1.3 V, takes ~5 ms (Figure 11.2), which only slightly exceeds the duration of action potentials in the SNc [11].

Overall, the positive characteristics of FCV have made it the electrochemical method of choice for in vitro studies of endogenous DA release. There are some caveats for this method, however. In contrast to the low non-Faradaic current component of amperometry or chronoamperometry, there is a high non-Faradaic background current in all FCV scans [40,82] (Figure 11.2b). An unfortunate physicochemical fact is that the background current generated with any linear scanning method is directly proportion to scan rate, whereas the Faradaic oxidation current for a given concentration of analyte is proportional to the square root of scan rate [83]. This creates an analytical challenge, since the background current at a carbon-fiber microelectrode is usually ~1 μA with FCV (Figure 11.2b), whereas a physiologically relevant increase in $[DA]_o$ of 1 μM at such an electrode would generate a DA oxidation current in the low nA range (Figure 11.2c). One strategy to address this problem is to improve electrode sensitivity for DA by adjusting FCV scan parameters. For example, increasing the anodic limit from +1.0 to +1.4 V versus Ag/AgCl can enhance electrode sensitivity several fold, thereby lowering the detection limit for DA [84].

To monitor the time course of a release response with FCV, current detected at the DA oxidation peak potential can be followed using a sample-and-hold circuit; and the change above background can then be amplified as needed. Signal identification, however, requires the storage of a background scan before a DA increase (Figure 11.2b) that can be digitally subtracted from subsequent scans to provide background-subtracted voltammograms of the monitored substance (Figure 11.2c). Background subtraction is often done off-line; in early studies, this was accomplished using a digital oscilloscope with subtraction capabilities that permitted identification of only the peak DA signal [43,82]. Subsequently, Julian Millar developed an instrument, the Millar Voltammeter, which stores a background scan that is subtracted from subsequent scans, so the Faradaic voltammogram corresponding to changes in $[DA]_o$ can be monitored directly in real time. The Millar Voltammeter was, until recently, commercially produced by PD Systems International, U.K., but now is available by special request to Dr Millar (Queen Mary, University of London, U.K.). More recently, improved software for use with 12- to 16-bit D/A boards (e.g., National Instruments, Austin, Texas) has been developed that permits off-line subtraction of all scans [85].

Electrodes

As noted earlier in this chapter, the first electrodes developed to monitor DA release in brain tissue were made using carbon paste, including carbon paste mixed with epoxy to improve electrode stability. These were superseded by carbon-fiber microelectrodes, which are now used in almost all in vitro or in vivo voltammetric studies. A number of methods for the fabrication of carbon-fiber microelectrodes suitable for brain slice recording have been described in the literature [39,55,77,82,86–91]. In all of these, the fundamental design consideration has been to achieve the highest signal-to-noise ratio and lowest detection limit possible. The basic procedure often involves inserting a carbon fiber in a glass capillary tube, then pulling it on a conventional micro-electrode puller to provide a seal and insulate the fiber with glass, although other coatings have been used to provide insulation. Every attempt is then made to ensure a tight seal, because small gaps or cracks that draw solution by capillary action can lead to electrode noise.

The design of carbon-fiber microelectrodes to be used for brain slice recording should include an active surface that can be contained completely within a slice, which has a typical thickness of $\leq$400 μm. Such electrode surfaces can be prepared by beveling or cutting, or by chemical or electrical etching of the carbon-fiber tip. The resultant working surface is either a disk that is flush with surrounding insulation (shown for a beveled electrode in Figure 11.3a), the surface of a cylinder that extends $\leq$50 μm beyond the insulation (Figure 11.3b), or that of a slightly-extended fiber with a conical shape (Figure 11.3c). Longer electrodes, like the 500 μm cylinders introduced by Gonon et al. [36], work well in vivo, but are not practical for use in slices. Selection of electrode design depends on the application and the carbon fibers available. Beveled and conical electrodes, which have tip diameters of typically 2–4 μm (e.g., [91]), minimize tissue damage at the recording site. Cylindrical and conical electrodes have larger surface areas than disk electrodes, and therefore produce more current for a given DA concentration. This can improve absolute detection limits, with the caveat that background currents are also proportional to surface area, so that signal-to-noise ratio may not necessarily improve. An additional point is that some carbon fibers are "sized," which means that they have a coating that may prevent electron transfer on the surface of the cylinder; in some cases, this can be removed by soaking fibers in an appropriate solvent, like acetone or dimethyl sulfoxide [91]. Also, isopropanol has also been used to clean electrode surfaces, whether sized or not [89].

One final issue in making carbon-fiber microelectrodes is creating electrical contact between a 5–10 μm carbon fiber and the instrumentation required for voltammetric recording (Figure 11.3d). Poor contact with the fiber can be a major source of electrical noise. Materials used include Wood's, metal, silver epoxy, silver paint, carbon paste, and conductive solutions like potassium acetate mixed with KCl, each of which can link the fiber to a larger diameter wire that is then used for connection to an appropriate headstage.

BRAIN SLICE METHODOLOGY

The use of brain slices as model CNS preparations were first explored in the 1950s by McIlwain (e.g., [92]). Over time, optimal conditions and incubation media were established to prepare and maintain viable slices; many of the issues evolved, as well as a range of uses for brain slice preparations are discussed in the comprehensive book *Brain Slices* edited by Dingledine [93]. An important parameter for brain slices is slice thickness; this should not exceed 400 μm, which was determined experimentally to be the maximum thickness at which the core of the slice can receive adequate oxygen and glucose. On the other hand, the minimal thickness of acute slices is ~200 μm, because the cut surfaces of a slice contains a layer of damaged cells that can extend 50 μm into the tissue, such that a 200 μm slice provides at least 100 μm of viable tissue within the slice where DA release can be studied.

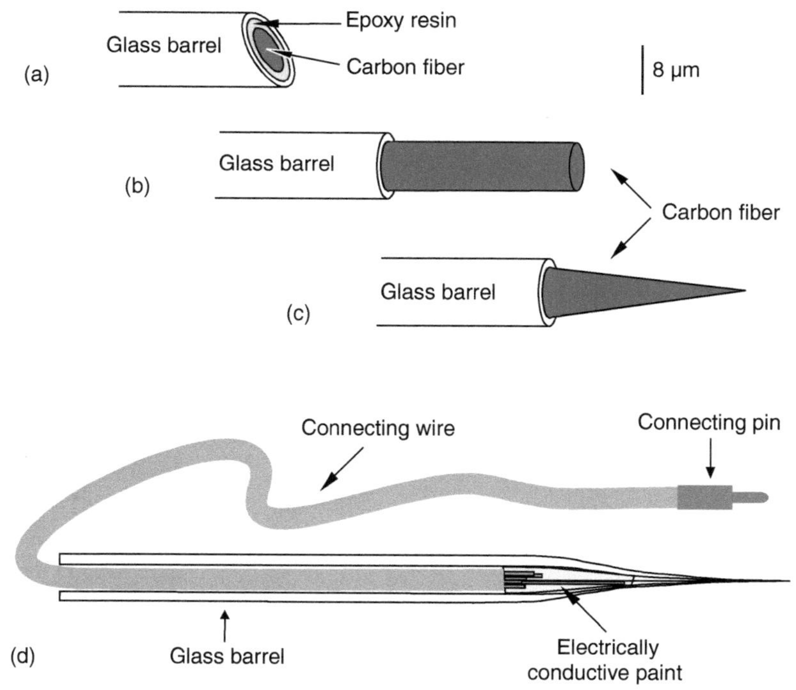

FIGURE 11.3 Carbon-fiber microelectrodes used for voltammetric detection of dopamine (DA) release. Carbon-fiber microelectrodes are typically prepared by inserting a carbon fiber into a glass capillary, then pulling in a conventional electrode puller to melt the glass, which seals and insulates the fiber. Electrode tips usually have one of three shapes: (a) disk, in which the carbon fiber is cut and or beveled flush with the surrounding insulation; epoxy resin is often used to provide mechanical stability for this procedure; (b) cylindrical, in which the carbon fiber protruding beyond the glass insulation is trimmed to a desired length (typically less than 50 μm); or (c) conical, in which the protruding carbon fiber is etched/or beveled to obtain a final tip diameter of 2–4 μm. (d) Final carbon-fiber electrode product in which electrical contact is made between a fine wire and the carbon fiber via electrically conductive material (e.g., [91]). The wire is stripped of insulation for 1–2 cm and inserted into the stem of the electrode barrel. The other end of the wire is soldered to a suitable pin for connection to the headstage of a voltammeter.

Brain slices are typically prepared from deeply anesthetized animals; after decapitation, the brain is carefully removed into ice-cold aCSF (composition is discussed in "aCSF Composition for Healthy Slices") for 1–2 min to minimize on-going metabolism and possible damage from anoxia. In early studies, McIlwain and others prepared slices using a "chopper," which is composed of a weighted arm with attached blade to cut slices quickly; McIlwain tissue choppers are commercially available (Brinkmann Instruments, Westbury, New York). Later designs for tissue slicing instrumentation incorporated a vibrating blade to cut slices submerged in ice-cold aCSF; these instruments include the now classic Vibratome (Ted Pella, Inc., Redding, California) and the more recently introduced Leica vibrating microtome (Leica Microsystems, Bannockburn, Illinois). The idea behind a vibrating microtome is that a plane of solution and foam generated at the edge of the vibrating blade precedes the blade and separates the slice from the tissue block, which minimizes compression injury to the tissue. Once slices are prepared, they are usually allowed to recover for at least an hour in appropriate media at room temperature before experimentation. The slices can be held in individual vials or in a larger group holding chamber (e.g., a large Petri dish); with continuous oxygenation, slices can remain viable for eight to ten hours at room temperature. The number of slices prepared from a given animal depends on the region of interest, with more slices possible from the striatum than from smaller structures like the SNc. If DA release is to be examined in more than one region from a given animal, optimal viability will be obtained if more than one slicer is available for tissue preparation; otherwise, the quality of the second region may be degraded from lack of oxygenation during slicing of the first tissue block.

For voltammetric recording, it is preferable to use a submersion chamber in which slices (and electrode tips) are continuously immersed in aCSF, rather than an interface chamber in which the slice surface is exposed to a humidified atmosphere of 95% O_2/5% CO_2. Removal of electrodes from the lipid environment of brain tissue into air can alter electrode sensitivity and response time. The use of a superfusion chamber also helps ensure the accuracy of electrode calibration data used to calculate evoked $[DA]_o$, because electrodes can be readily calibrated immediately after recording by raising the electrode from the tissue into the superfusing aCSF, then adding a known concentration of DA. A related technical point is that electrodes calibrated in the recording chamber should be positioned near the superfusion inlet to avoid dilution of the calibration solution with DA-free aCSF in the surrounding chamber; such dilution can give an erroneously low calibration factor thereby causing overestimation of evoked $[DA]_o$ in a tissue slice.

aCSF Composition for Healthy Slices

The composition of modified aCSF solutions used for brain slice studies is critical for slice viability. These solutions differ from in vivo CSF in several ways. First, aCSF is hyperglycemic: the glucose concentration of CSF is 1–3 m*M*, whereas aCSF typically contains 10 m*M* glucose. Second, aCSF is hyperoxic: the medium surrounding a brain slice is continually saturated with 95% O_2/5% CO_2 to provide normal O_2 levels in the interior of the slice. Third, the superfusing aCSF is hypothermic: the temperature used in most studies of DA release in brain slices is 31–32°C, such that the tissue is somewhat below normal core body temperature of 37°C. This mild hypothermia helps preserve slice viability during recording periods that can last for several hours. In initial studies of electrically-evoked DA release in brain slices at 37°C, Bull and colleagues [49] found that evoked $[DA]_o$ in striatal slices was below detection limits; however DA release was readily detected when the temperature was lowered to 32°C. This reflected both enhanced temperature-dependent DA uptake via the DAT at 37°C, as well as poorer slice viability: even in the presence of a DAT inhibitor, stable DA release is difficult to maintain at 37°C [49]. A further decrease in temperature is required for brain slice preparation; as noted in the introduction to this section, slices prepared on a vibrating microtome are submerged in ice-cold aCSF to preserve slice viability during cutting.

A variety of media changes have also been introduced for slice preparation and for use in a recovery or holding chamber. Many of these are indeed beneficial, with the caveat that some changes like decreasing Ca^{2+} levels or adding glutamate receptor antagonists can have lasting effects on cell excitability. One particularly helpful approach is to use a HEPES-bicarbonate-buffered aCSF for slice preparation, which improves slice quality [94] without obvious effects on tissue physiology. This solution contains (in m*M*): NaCl (120); $NaHCO_3$ (20); glucose (10); HEPES acid (6.7); KCl (5); HEPES sodium salt (3.3); $CaCl_2$ (2); $MgSO_4$ (2), equilibrated with 95% O_2/5% CO_2 [81,95,96]. Slices are then maintained in this solution at room temperature until they are used (at least 1 h, but up to 10 h). HEPES was originally included to enhance buffering capacity when the medium was not bubbled with 95% O_2/5% CO_2 during slicing; it was later found that inclusion of HEPES improves slice quality by preventing excessive water gain (edema) during the holding period [94].

In addition to being hyperglycemic, hyperoxic, and hypothermic, the aCSF used for brain slice studies also often has slightly elevated concentrations of K^+ and/or Ca^{2+} compared to normal levels in CSF (3 m*M* for K^+ and 1.2 m*M* for Ca^{2+}) to enhance slice excitability and/or elevate Ca^{2+}-dependent transmitter release. An example of aCSF composition is (in m*M*): NaCl (124); KCl (3.7); $NaHCO_3$ (26); $CaCl_2$ (2.4); $MgSO_4$ (1.3); KH_2PO_4 (1.3); and glucose (10), equilibrated with 95% O_2/5% CO_2 [62,81,95]. An advantage of in vitro studies is that the aCSF composition can be readily altered to suit the experiment; for example, Ca^{2+} concentration can be varied to examine the Ca^{2+} dependence of DA release [62,76], oxygen and glucose can be lowered to study the effects of ischemia on DA release [96], or Mg^{2+} can be omitted to enhance glutamatergic NMDA (*N*-methyl-D-aspartate) receptor activation [97].

Species Selection

Selection of the species to be studied, like the other methodological options already discussed, depends on the questions asked. Many fundamental properties of DA release regulation have been examined in rats, which are often a natural choice simply because of the large body of literature available on many aspects of DA physiology, anatomy, biochemistry, and pharmacology in this species. On the other hand, the increasing availability of transgenic mice developed to address aspects of DA regulation, like DAT or D_2 receptor knockouts, make the mouse the species of choice for some applications. The use of other species can offer specific advantages, as well. For example, the study of somatodendritic DA release in the SNc of small rodents is hindered by significant 5-HT innervation [98], such that 5-HT rather than DA can be the primary substance detected in rat and mouse midbrain slices [57,99–102]. The pattern of 5-HT innervation in guinea pig SNc differs, however, such that pure DA release can be detected [62,95,100]. Thus, guinea pigs are the species of choice for studies of somatodendritic DA release in the SNc, although rats or mice also offer advantages for studying 5-HT in this region.

The use of brain slices also permits species comparisons of DA release regulation in animals for which in vivo recording is technically challenging, including non-human primates and birds. For example, Cragg and colleagues [103,104] examined axonal DA release probability and uptake in motor versus limbic regions of striatal slices from the marmoset; these studies revealed a different and more extensive pattern of DA regulation and plasticity in primate brain than seen in rodent striatum. Gale and Perkel [105] recently examined DA release in slices containing the songbird striatal analog, Area X, to assess the extent to which avian Area X is like mammalian striatum. Intriguingly, regulation of evoked $[DA]_o$ by uptake and by DA autoreceptors in Area X is surprisingly similar to that seen in rodent striatum.

Plane of Slicing

Brain slices prepared from either midbrain or forebrain are most commonly cut in the coronal plane (Figure 11.4a). In the midbrain, the lateral projecting dendrites of DA neurons in the SNc lie in this plane (Figure 11.4b, left), such that the use of coronal slices avoids having to sever DA cell dendrites, which can compromise cell viability. In axon terminal regions like dorsal striatum (CPu) or NAc, DA axons fan out after leaving the MFB, such that most axons will be perpendicular to the coronal plane (Figure 11.4b, right). When local electrical or chemical stimulation is used, axon orientation should not matter significantly, because the stimulus has a field effect that elicits DA release from all sites for which the field strength is sufficient to cause action potential generation and Ca^{2+} entry. Such stimulation also causes the release of other local transmitters, which can be an advantage or disadvantage depending on the goals of a given experiment (see "Regulation of Striatal DA Release in H_2O_2" in this chapter). It is sometimes desirable to elicit DA release by pathway stimulation. Because of the non-planar path of the MFB, only a few millimeters can be contained in a given slice, even when cut in a parasagittal plane. Nonetheless, we have found that preservation of 1–2 mm of DA axons in guinea pig striatum cut in the parasagittal plane is sufficient to permit examination of pure axonal DA release [61,106]. With the smaller mouse brain, horizontal slices can be prepared in which the MFB connections between midbrain and ventral striatum are sufficiently intact to generate spontaneous axonal DA release in ventral striatum arising from action potentials generated in midbrain DA neurons [107].

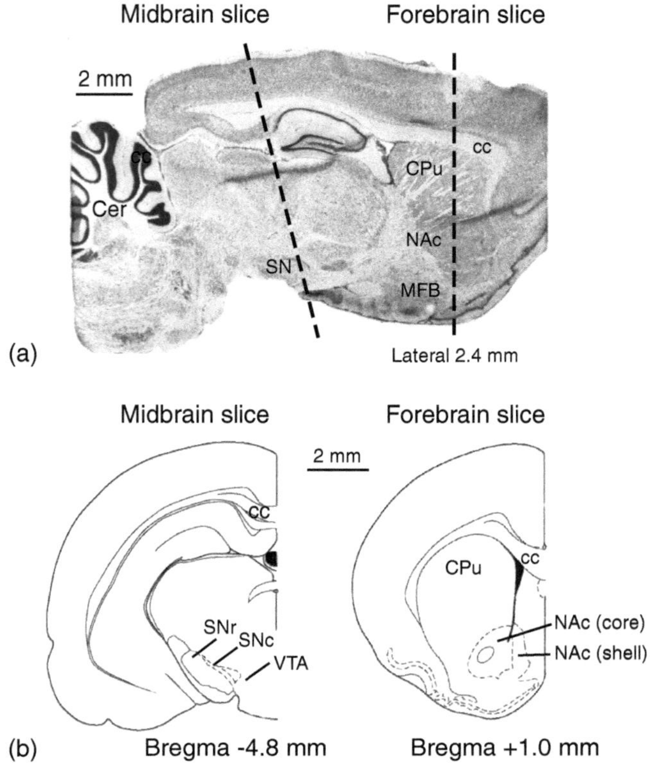

FIGURE 11.4 (**See color insert following page 272.**) Rat brain sections showing major midbrain and forebrain dopamine (DA) regions. (a) Sagittal section of rat brain, stained with cresyl violet, showing main dopaminergic midbrain (somatodendritic) and forebrain (axon terminal) regions for DA release studies. Dashed lines show the relative positions at which coronal midbrain or forebrain slices are typically made. (Modified from Paxinos, G. and Watson, C., *The Rat Brain, in Stereotaxic Coordinates*, 4th ed., Academic Press, Inc, New York, 1998). Lateral coordinate indicates distance from the central suture, a landmark on rat skull. (b) Crossection of coronal midbrain (left panel) and forebrain (right panel) slices at the levels indicated in (a). Note that at some levels, it is possible it is possible to record somatodendritic DA release from the SNc (A9 region), as well as the VTA (A10 region) in the same midbrain slice. Similarly, at appropriate forebrain levels, it is possible to record axonal DA release from the dorsal striatum (CPu) as well as the ventral striatum (NAc, core and shell). Bregma coordinates indicate distance anterior (+) and posterior (−) to this landmark on the skull. (Modified from Paxinos, G. and Watson, C., *The Rat Brain, in Stereotaxic Coordinates*, 4th ed., Academic Press, Inc, New York, 1998). *Abbreviations*: Cer, cerebellum; cc, corpus callosum; CPu, caudate putamen; MFB, medial forebrain bundle; NAc, nucleus accumbens; SN, substantia nigra; SNc, SN pars compacta; SNr, SN pars reticulata; VTA, ventral tegmental area.

STIMULATING AND RECORDING DA RELEASE IN BRAIN SLICES

Release can be elicited by various means, including chemical (e.g., high K^+), electrical, or pharmacological (e.g., amphetamine) stimulation. In this section, we will focus on stimulation paradigms that are used to elicit exocytotic release, especially high K^+ and local electrical stimulation. It should be noted that local electrical stimulation can also be used in combination with other releasing agents to address specific questions. For example, the effect of amphetamine on DA release via DAT reversal and its consequence on impulse-dependent DA release has been used to examine mechanisms of vesicular storage [72,108–111].

Regardless of the stimulation method used, recording electrodes are typically inserted ~50–100 μm into the slice [49,62,95]. When local electrical stimulation is used to evoke DA release, the carbon-fiber electrode is typically positioned within 100 μm of the stimulating electrode. A technical note about recording in striatal slices is that a number of myelinated fiber bundles pass through the striatum, giving it a striated appearance (hence its name). Recording (and stimulating) electrodes should be placed in gray matter, away from these non-dopaminergic fibers [49].

The optimal stimulation paradigm for a given experiment, once again, depends on the question addressed. When the question is about factors that regulate DA release (or DA uptake), a physiological, or depolarizing, stimulus is often preferable to a pharmacological stimulus. Neuronal membranes are normally polarized to a negative intracellular potential (often -60 to -70 mV with respect to a grounded reference electrode) (discussed in [112]). Transmitter release occurs when a Na^+-dependent action potential propagates to a synapse (or other release sites) to cause Ca^{2+} entry through voltage-sensitive Ca^{2+} channels, vesicle fusion, and diffusion of transmitter into the synaptic cleft and/or extracellular space (see "Physiology of DA Release" in this chapter). This process can be initiated artificially by a variety of methods that cause membrane depolarization. The simplest method is to superfuse a slice with aCSF made with a high K^+ concentration (e.g., 30 m*M*) for a brief period of time [76] or to apply veratridine, which is a pharmacological agent that opens Na^+ channels directly [81,113]. The disadvantage of these methods is that they can produce non-physiological DA release signals that last several minutes and are accompanied by large ion shifts that can be problematic for measurements with FCV [112]. Moreover, application of a depolarizing agent to the entire slice causes the uncontrolled release of many other transmitters, as well as DA. An additional concern about veratridine is that it can lead to Na^+ overload and consequent DA release via reversal of the DAT, rather than by Ca^{2+}-dependent exocytosis [113].

Better protocols to elicit DA release involve more discrete stimulation. Local electrical stimulation using insulated bipolar electrodes is used most widely [49,62,66,67,69,79,85,95,103,111, 114–117] although local application of high K^+ via pressure ejection has also been reported [76]. Both single electrical pulses and trains of multiple-pulses have been used to elicit DA release. Each offers advantages for specific experiments. For example, single pulses are used to indicate whether or not a regulatory factor affects DA axons directly, whereas indirect effects involving local microcircuitry can be revealed using pulse-train stimulation (see "Regulation of Striatal DA Release in H_2O_2" in this chapter). It is also sometimes informative to vary the frequency and number of applied electrical pulses to mimic tonic versus phasic firing patterns of DA neurons [66,106,115–117].

Stimulating electrodes are typically bipolar and can be made from tungsten, platinum, or platinum–iridium, and can be made locally or purchased commercially. The stimulating electrode is usually placed on the slice surface, rather than implanted, with both poles of the electrode in the same plane. In general, narrow gauge wire (e.g., 75 μm diameter bare) is recommended for construction of twisted bipolar electrodes; tip separation is usually ~100 μm. Commercially available electrodes should also be selected on the basis of size, based on the balance between small size and adequate field of depolarization. The parameters required to elicit DA release depend on the specific electrode used, as well as the brain region examined. Usually, stimulation pulse duration is 100–1000 μs, which is shorter than a typical action potential, with similar parameters used for both single- and multiple-pulse stimulation. Although large gauge electrodes used with high currents can elicit release that is not action potential dependent (i.e., that cannot be prevented by the Na^+ channel blocker tetrodotoxin (TTX) [95]), this is not ideal. Rather, stimulation parameters should be optimized to produce consistent release that is TTX sensitive [62,103].

Signal versus Interference

The earliest voltammetric studies designed to monitor DA release in vivo were hindered by overwhelming interference from ascorbate and from the anionic DA metabolite, DOPAC, as discussed earlier. In vivo, extracellular ascorbate concentration is ~400 μ*M* and DOPAC is 20–30 μ*M* [118,119]. These levels are initially preserved in acute slices made with a tissue chopper and recorded in an interface chamber (see "Brain Slice Methodology" in this chapter) [46]. After only thirty minutes of in vitro incubation, ascorbate is lost from its intracellular stores in the absence of added ascorbate in the incubation medium, resulting in intra-and extracellular depletion [46]. Striatal DOPAC is lost even more quickly, since it is predominantly extracellular and readily washed out of slices [46,81]. Thus, neither of these electroactive species interfere with DA detection in superfused brain slices [81].

Slice recording is not devoid of interference, however, as several electroactive and non-electroactive substances can interfere with DA detection. The most commonly reported electroactive interferents are 5-HT and NE. The degree of interference depends on the relative innervation of each transmitter versus DA in a given region. For example, in dorsal striatum, DA is the only electroactive substance detected because of the high density of DA axons, as well as the low level of 5-HT axons arising from 5-HT neurons in midbrain. By contrast, in the SNc and the SN pars reticulata, DA content is ~10-fold lower than in striatum [81] with specific 5-HT innervation of DA dendrites, such that 5-HT can be the predominant substance identified, especially in rat and mouse midbrain [57,99–102], as noted in "Species Selection." The extent of 5-HT interference is often amplified by the typically 2-4-fold higher sensitivity of carbon-fiber microelectrodes for 5-HT than for DA. Consequently, even if DA were 4-fold higher than 5-HT, both substances would contribute equally to the current monitored. When FCV is used, 5-HT interference is immediately indicated by the characteristic 5-HT voltammogram, which can be readily distinguished from that of DA (see "Species Selection" in this chapter) [57,80,81,100]. With amperometry or chronoamperometry, however, pharmacological methods (e.g., uptake blockade) must be used to confirm signal identity. In contrast to the cyclic voltammogram for 5-HT, the voltammogram for NE is largely indistinguishable from that for DA, especially in tissue, since the redox characteristics of DA and NE are defined primarily by their identical catechol ring structures; electrode sensitivity for NE is typically similar to or less than that for DA. Consequently, in regions with possible NE interference, additional pharmacological and/or anatomical evidence must be obtained to confirm that DA is the substance monitored, regardless of the electrochemical method used. For example, both anatomical and pharmacological data were used to confirm that DA was the primary substance detected in the VTA (A10 DA cell body region) and in the subthalamic nucleus, both of which have particularly low DA innervation, as well as NE fibers passing through [120,121].

The most important non-electroactive interferents are Ca^{2+} and pH (H^+). These factors are most problematic for FCV, which has a large non-Faradaic background that is sensitive to the ionic microenvironment (as discussed in "Fast-Scan Cyclic Voltammetry" in this chapter). Possible interference can arise from ion-dependent background shifts [82,114] and from the fact that changes in ionic composition, especially in the divalent cations Ca^{2+} and Mg^{2+}, can alter electrode sensitivity to DA [122]. Understanding these interferents is important because the ionic microenvironment of brain cells is dynamic, with decreases in extracellular Ca^{2+} concentration and activity-dependent alkaline and acid shifts in the extracellular space accompanying all depolarizing stimuli (see [112] for review).

The sensitivity of electrode background in FCV to changes in ionic composition was first shown for pH, with a decrease in pH (increase in acidity) reflected in an anodic surface wave with a peak around 0.3 V versus Ag/AgCl [82,112,114]. The potential of this background wave can readily be distinguished from the DA oxidation potential (~ +0.6 V versus Ag/AgCl in FCV at most electrodes); however, this can interfere with DA quantitation in brain regions with low DA innervation like the amygdala [111]. Stimulated Ca^{2+} entry (required for exocytotic release of DA) and the associated transient decrease in extracellular Ca^{2+} concentration is reflected in a negative deflection in current output at roughly the same potential as the pH shift [114]. Heien, Wightman and colleagues recently reported the use of principal component regression coupled with appropriate calibration data to resolve contributions from substances that give overlapping voltammetric waveforms, including DA and pH [85].

A less obvious source of ionic interference arises from interactions of divalent cations with an electrode surface, which can alter electrode sensitivity to DA [122]. The calibration factor for

DA using FCV can be two-fold lower in media containing Ca^{2+} and Mg^{2+} than in divalent-ion free solutions, presumably because divalent ions associate strongly with anionic functionalities on electrode surfaces, thereby decreasing DA adsorption and, thus, DA oxidation current. This is relevant for two practical reasons. First, electrode calibration in many previously reported studies was done using HEPES- or phosphate-buffered saline that lack divalent ions; this means that when these data are used to calculate evoked $[DA]_o$ for tissue bathed in normal aCSF (see "aCSF Composition for Healthy Slices" in this chapter), $[DA]_o$ can be underestimated by at least 50%. Second, studies of the Ca^{2+}- or Mg^{2+}- dependence of DA release will require that electrode sensitivity to DA be determined in all media tested to ensure quantitative accuracy (e.g., [62]).

One final set of potentially interfering substances is the agents used to manipulate DA release. We have found that a number of such substances are electroactive, including the DA-receptor blocker haloperidol, the DAT inhibitor nomifensine, and the catalase inhibitor 3-aminotriazole. An advantage of in vitro methods is that testing whether a drug agent is electroactive, and calibrating electrodes in the presence and absence of the agent, can be done readily in the slice chamber at the same temperature used for DA release studies.

REGULATION OF STRIATAL DA RELEASE BY H_2O_2

Research in our laboratory over the past decade has focused on two converging areas of interest. The first is the neuroprotective role of low molecular weight antioxidants, including ascorbate; the second is DA release regulation by endogenous neurotransmitters and neuromodulators. These topics converged when we began to examine DA release regulation under conditions of oxidative stress. In our initial studies, we examined the effect of exogenously applied H_2O_2 on electrically evoked $[DA]_o$ in guinea pig striatal slices, monitored using carbon-fiber microelectrodes and FCV. Transient exposure to exogenous H_2O_2 (15 min, 1.5 m*M*) causes a reversible 30%–40% decrease in pulse-train evoked $[DA]_o$ in dorsolateral striatum and is not accompanied by a change in tissue DA content or evidence of oxidative damage [123]. In the course of these studies, we also discovered that endogenous H_2O_2 plays an important role in the physiological regulation of axonal DA release in striatum. This finding shifted our focus from viewing H_2O_2 as a potential mediator of oxidative damage, to viewing it as a neuromodulator. In this section we summarize studies that have been done to determine the source of modulatory H_2O_2, the neuronal circuitry involved, and the mechanism of DA release suppression. This body of work not only provides a good example of the power of voltammetric detection of DA release, but also demonstrates an advantage of in vitro brain slice preparation studies, which is that they facilitate the use of companion techniques, including whole-cell physiological recording and fluorescence imaging.

ENDOGENOUS H_2O_2 INHIBITS AXONAL DA RELEASE

We first examined a possible regulatory role for endogenously generated H_2O_2 by manipulating slice peroxidase activity using exogenously applied catalase and/or by inhibiting endogenous glutathione (GSH) peroxidase (as noted in "Signal Versus Interference," the catalase inhibitor, 3-aminotriazol is electroactive and could not be tested in these experiments). For example, amplification of endogenous H_2O_2 levels by GSH-peroxidase inhibition with mercaptosuccinate (MCS) causes a similar decrease in pulse-train evoked $[DA]_o$ (30 pulses, 10 Hz) in dorsolateral striatum to that seen with exogenously applied H_2O_2 [9,124]. Again, no change in DA content is seen and DA release suppression is fully reversible upon MCS washout or when the slice is superfused with catalase in the continued presence of MCS [9]. Regulation of DA release by endogenous H_2O_2 also occurs in the shell of the NAc and in the SNc [124]. Importantly, MCS has no effect on $[DA]_o$ evoked by a single stimulus pulse [9], implying that even under conditions of impaired GSH peroxidase activity, basal levels of H_2O_2 in striatum are insufficient to inhibit DA release. Instead, modulatory H_2O_2 must be generated dynamically during the first pulse of a stimulus-train and inhibits DA released by subsequent pulses. Additional studies show that H_2O_2 generation is derived from concomitant glutamate release and AMPA receptor activation (see "Regulation of Axonal DA Release by Glutamate Acting at AMPA Receptors Requires H_2O_2" in this chapter). The time-scale of H_2O_2-dependent DA release suppression is rapid, occurring within a few hundred milliseconds after initiation of a 10 Hz pulse train [9].

Potential Sources of H_2O_2 Generation

The major source of ROS in neurons, as in all cells, is mitochondrial respiration [125], although other superoxide ($\cdot O_2^-$) generating processes, notably NADPH oxidase [126], also contribute. During mitochondrial respiration, ($\cdot O_2^-$) is formed from the one electron reduction of O_2; H_2O_2 is generated from ($\cdot O_2^-$) by superoxide dismutase (SOD), as well as by spontaneous dismutation [4,127] (Figure 11.5). The amount of H_2O_2 produced by mitochondria is 1%–2% of O_2 metabolized [127], such that the concentration of H_2O_2 within the restricted volume of a dendrite, for example, could transiently reach m*M* levels [123]. Absolute levels of H_2O_2 at a given point in time, however, will depend on the balance between H_2O_2 generation and cellular peroxidase activity [4,11] (Figure 11.5).

Is mitochondrial respiration the primary source of modulatory H_2O_2 in the striatum? Our recent studies indicate that the answer is yes: the effect of MCS on DA release in striatum is lost when mitochondrial complex I is inhibited by rotenone in the presence of the complex II substrate succinate to maintain tissue ATP content [128]. These data also argue against significant involvement of other $\cdot O_2^-/H_2O_2$ generating pathways, including NADPH oxidase. Additional experiments confirm a lack of involvement of monoamine oxidases (MAOs), which produce one molecule of H_2O_2 for each molecule of DA metabolized [4]: a cocktail of MAO inhibitors does not alter evoked $[DA]_o$ versus control or prevent the effect of MCS [128].

Regulation of Axonal DA Release by Glutamate Acting at AMPA Receptors Requires H_2O_2

Given the close apposition of mitochondria to presynaptic sites in DA axons [129], we intially assumed that modultory H_2O_2 generation in the striatum might occur directly in DA axons, such that activity-dependent H_2O_2 would provide a feedback signal to decrease DA release and thereby "augment the effect of D_2-autoreceptor-mediated inhibition" [123]. However, our subsequent studies disprove DA axons as primary sites of generation. The most direct evidence comes from comparisons of the effect of GSH peroxidase inhibition on DA release evoked by local versus distal stimulation in parasagittal slices of striatum, in which DA-axon tracts are preserved for 1–2 mm.

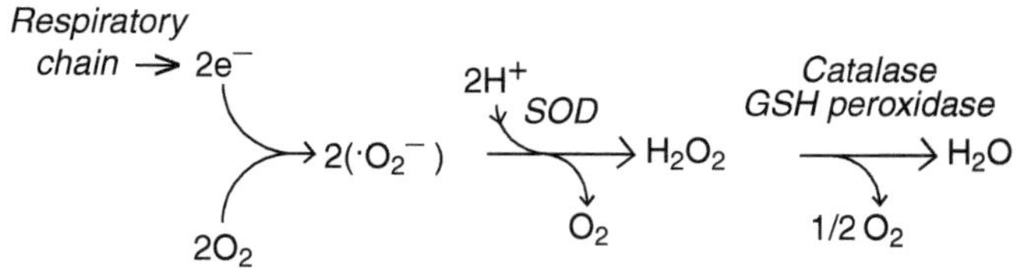

FIGURE 11.5 Production of H_2O_2 by the mitochondrial respiratory chain. The primary source of H_2O_2 generation in all cells is mitochondrial respiration during which $\cdot O_2^-$ is formed from the single-electron reduction of O_2. H_2O_2 is then generated from $\cdot O_2^-$ by superoxide dismutase (SOD), as well as by spontaneous dismutation. Absolute levels of H_2O_2 depend on the balance between H_2O_2 generation and the cellular activity of the H_2O_2 scavenging enzymes, catalase and glutathione (GSH) peroxidase.

In this preparation, as in our usual coronal slices, local stimulation evokes the release of glutamate, GABA, and other transmitters, as well as that of DA, whereas distal pathway stimulation selectively activates DA axons to elicit relatively pure DA release. Under these conditions, inhibiting GSH peroxidase with MCS suppresses $[DA]_o$ evoked by local stimulation, as seen previously, but not by distal stimulation [61]. Also arguing against H_2O_2 generation in DA axons is the finding that H_2O_2-dependent inhibition of DA release requires glutamatergic AMPA receptors [9], which are apparently absent from DA axons [130,131].

Thus, H_2O_2-mediated inhibition of DA release in striatum is not an auto-regulatory process, but rather represents potent external regulation by glutamatergic input to the striatum. Indeed, when striatal AMPA receptors are blocked by a selective antagonist, GYKI-52466, pulse-train evoked $[DA]_o$ increases by 100% [9]; this effect is completely prevented by catalase, confirming that H_2O_2 is required (Figure 11.6a and Figure 11.6b). Thus, glutamate released under physiological conditions inhibits action potential-dependent DA release in striatum via H_2O_2. Moreover, H_2O_2 amplification by MCS has no effect on DA release in the presence of GYKI-52466, indicating that glutamate acting at AMPA receptors is required for generation of modulatory H_2O_2 in striatum [9]. Consistent with the non-DA axon origin of AMPA-receptor dependent H_2O_2 generation, blockade of AMPA receptors with GYKI-52466 enhanced $[DA]_o$ evoked locally, but had no effect on that evoked distally [61]. The question of how, or even if, glutamate regulates striatal DA release has been a long-standing source of controversy. Our discovery that glutamate inhibits DA release via H_2O_2 generation resolves this conundrum.

Our studies show that regulation of DA release in the dorsolateral striatum by GABAergic input is also unconventional: blockade of $GABA_A$ receptors by picrotoxin causes a ~50% decrease in evoked $[DA]_o$ (Figure 11.6c), whereas $GABA_B$-receptor blockade with saclofen has no effect [9], indicating that GABA, acting at $GABA_A$ receptors, normally enhances DA release. The influence of GABA on DA release, like that of glutamate, must be indirect, since DA axons in striatum do not

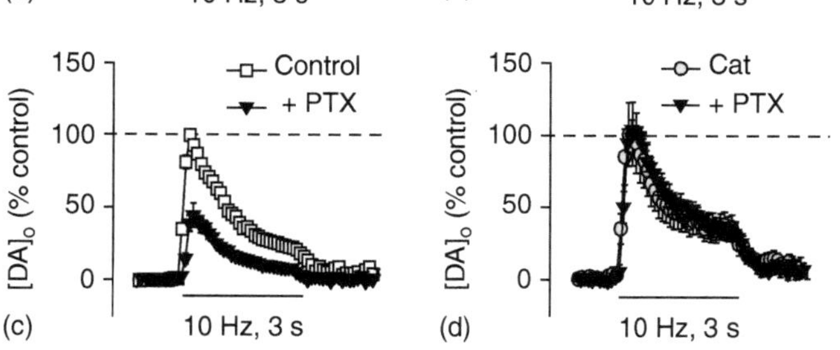

FIGURE 11.6 Regulation of striatal DA release by glutamate and GABA requires H_2O_2. (a) AMPA-receptor blockade by GYKI-52466 (GYKI; 50 μ*M*) causes a ~100% increase in evoked $[DA]_o$ in striatum ($p<0.001$, $n=6$). (b) The effect of AMPA-receptor blockade is prevented by catalase (Cat). (c) $GABA_A$ receptor blockade by picrotoxin (PTX; 100 μ*M*) causes a ~50% decrease in evoked $[DA]_o$ ($p<0.001$, $n=6$). (d) Catalase abolishes the effect of picrotoxin. Responses in the presence of heat-inactivated catalase were the same as control. Data are means ± SEM, shown as percentage of same-site control (Modified from Avshalumov, M. V., Chen, B. T., Marshall, S. P., Peña, D. M., and Rice, M. E., *J. Neurosci.*, 23, 2744–2750, 2003. With permission.)

express $GABA_A$ receptors [132]. The mediator again proved to be H_2O_2, since the effect of picrotoxin is prevented by catalase (Figure 11.6d). Moreover, picrotoxin has no effect on DA release when AMPA receptors are blocked by GYKI-52466 [9], indicating that glutamate and GABA act on the same pool of AMPA-receptor dependent H_2O_2.

Activity-Dependent H_2O_2 Generation in Striatal Medium Spiny Neurons

Increasing evidence from our laboratory implicates striatal medium spiny neurons as the primary cellular source of modulatory H_2O_2 in the striatum. Not only are these the most abundant striatal neurons (90%–95%; [133]), but the pattern of sensitivity of DA release to glutamate and GABA antagonists [9] mirrors the electrophysiological responsiveness of medium spiny neurons to these agents [134,135]. Moreover, glutamate synapses are closely apposed to DA synapses on medium spiny neuron dendrites [130,131,136], such that they are ideally positioned to modulate DA release via postsynaptically generated H_2O_2. Medium spiny neurons also express $GABA_A$ receptors at dendritic sites near spines [132], which would facilitate GABAergic opposition of glutamatergic excitation and H_2O_2 generation.

Consistent with this indirect evidence, we have recently confirmed that medium spiny neurons are a key cellular source of modulatory H_2O_2 in striatum. In these studies, we used a method for H_2O_2 detection that we developed for simultaneous whole-cell recording and fluorescence imaging using the H_2O_2-sensitive dye H_2DCF (dihydro-dichlorofluorescein) in brain slices; this dye becomes fluorescent DCF when oxidized by ROS [11]. Under control conditions, basal DCF fluorescence is seen in all medium spiny neurons, reflecting a basal ROS tone (Figure 11.7). Local stimulation (with the same 30 pulse, 10 Hz pulse trains used to elicit DA release) causes a ~30% increase in DCF fluorescence in medium spiny neurons, which is further enhanced by inhibition of GSH peroxidase with MCS [61]. Importantly, AMPA-receptor blockade by GYKI-52466 prevented stimulus-induced action potentials, as well as activity-dependent H_2O_2 generation in MSNs [61].

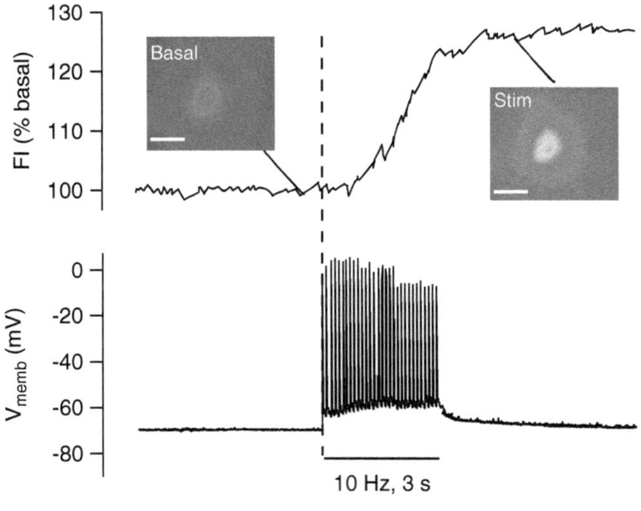

FIGURE 11.7 Activity-dependent H_2O_2 generation in a striatal medium spiny neuron. H_2O_2-dependent DCF fluorescence intensity (FI) under control conditions and after local stimulation (10 Hz, 30 pulses) with simultaneously recorded membrane voltage (V_{memb}); FI plateau reflects irreversible DCF activation by H_2O_2. Images are basal and stimulated (Stim) FI obtained at the times indicated (scale bar is 20 μm).

H_2O_2 Acts via K_{ATP} Channels to Inhibit DA Release

The mechanism by which H_2O_2 regulates the nigrostriatal pathway is H_2O_2-dependent opening of K_{ATP} channels [9,10]. These channels are multimeric proteins composed of four inwardly rectifying pore-forming units, typically Kir 6.2 in neurons [137], and four sulfonylurea-binding subunits (SUR1 or SUR2) [138]. Previous physiological studies demonstrated that *exogenous* H_2O_2 can cause membrane hyperpolarization by activating a K^+ conductance in a variety of cell types, including pancreatic β-cells [139] and CA1 hippocampal neurons [140]. Our studies of striatal DA release provided the first evidence that *endogenous* H_2O_2 can activate K_{ATP} channels. Blocking K_{ATP} channels with tolbutamide or glibenclamide leads to a significant increase in evoked $[DA]_o$ in dorsolateral striatum, indicating that during normal stimulation, K_{ATP}-channel activation causes the inhibition of DA release; these sulfonylureas also prevent the usual pattern of H_2O_2-dependent modulation by MCS, GYKI-52466, and picrotoxin [9,10], demonstrating that K_{ATP} channels are required for modulation of DA release by H_2O_2, glutamate, and GABA. Conversely, striatal DA release is decreased by ~40% by either diazoxide, a SUR1-selective K_{ATP}-channel opener, or cromakalim, a SUR2-selective opener. However, diazoxide prevents the effects of MCS, GYKI-52466, and picrotoxin, whereas cromakalim does not, demonstrating that glutamate-dependent H_2O_2 preferentially opens SUR1-based K_{ATP} channels (Avshalumov and Rice 2003) [10].

As mentioned in the introduction to this chapter, K_{ATP} channels are expressed abundantly throughout the nigrostriatal pathway [12–14]. We recently reported that DA neurons in the SNc are also responsive to endogenously generated H_2O_2: 50% of these neurons respond to moderate H_2O_2 elevation during GSH peroxidase inhibition with membrane hyperpolarization and loss of spontaneous firing, whereas all cells respond in this way to greater H_2O_2 elevation when catalase is inhibited [11]. These effects are reversible and are blocked by glibenclamide, confirming a role for K_{ATP} channels. However, the more sensitive "responders" are those that express SUR1-containing K_{ATP} channels, which is the channel subtype involved in mediating AMPA-receptor/H_2O_2-dependent DA release suppression in the striatum [10]. Whether regulation of K_{ATP} channels by H_2O_2 is direct or indirect in DA neurons remains to be elucidated, although previous studies using inside-out membrane patches from cardiac cells have shown a direct, effect of H_2O_2 on K_{ATP} channel opening by decreasing channel sensitivity to ATP [141].

SUMMARY AND FUTURE DIRECTIONS FOR STUDIES OF H_2O_2 AS A NEUROMODULATOR

In this chapter, we have presented rationale, methods, and results for the study of DA release in brain slices. This background should give readers a general understanding of the strengths and weaknesses of brain slices as experimental preparations, as well as an overview of the voltammetric methods and carbon-fiber microelectrodes currently used to monitor DA release. There is much left to discover about the regulation and function of DA in the CNS. Future studies are likely to exploit the ease of combining other methods with voltammetry in brain slices. Fluorescence imaging and whole-cell recording of single neuron activity are especially promising companion techniques, as illustrated in "Regulation of Striatal DA Release by H_2O_2" in this chapter.

Our exploration of H_2O_2 as an endogenous neuromodulator indicates that activity-dependent H_2O_2 generation is predominantly mitochondrial. In dorsolateral striatum, H_2O_2 is generated downstream from AMPA-receptor activation in medium spiny neurons, rather than DA axons, and must therefore diffuse to DA synapses where it inhibits DA release via opening of SUR1-based K_{ATP} channels. Our working model of H_2O_2-dependent regulation of striatal DA release involves a triad of DA, glutamate, and GABA synapses, which are separated by a few micrometers on the dendrites of medium spiny neurons, but bound together functionally by diffusible H_2O_2 [9,142].

A noted in the beginning of this chapter, H_2O_2 and other ROS can be neurotoxic. Thus, neuromodulation by H_2O_2 can be a double-edged sword: oxidative stress resulting from an imbalance between H_2O_2 generation and regulation could lead to cellular damage. Indeed, oxidative stress has been implicated in nigrostriatal degeneration in Parkinson's disease, which selectively targets the nigrostriatal pathway [1,2,4]. Thus, loss of normal H_2O_2 regulation could contribute to nigrostriatal pathophysiology and to the motor deficits of Parkinson's disease. Further exploration of H_2O_2 as a diffusible messenger in the migrostriatal pathway should provide new insights into both normal and pathophysiological roles of this novel neuromodulators.

ACKNOWLEDGMENTS

The authors are grateful for support from NIH grants NS-36362 and NS-45325 and the National Parkinson Foundation.

REFERENCES

1. Olanow, C. W. and Tatton, W. G., Etiology and pathogenesis of Parkinson's disease, *Annu. Rev. Neurosci.*, 22, 123–144, 1999.
2. Zhang, Y., Dawson, V. L., and Dawson, T. M., Oxidative stress and genetics in the pathogenesis of Parkinson's disease, *Neurobiol. Dis.*, 7, 240–250, 2000.
3. Ramasarma, T., Generation of H_2O_2 in biomembranes, *Biochem. Biophys. Acta*, 694, 69–93, 1982.
4. Cohen, G., Enzymatic/nonenzymatic sources of oxyradicals and regulation of antioxidant defenses, *Ann. N.Y. Acad. Sci.*, 738, 8–14, 1994.
5. Castner, S. A., Goldman-Rakic, P. S., and Williams, G. V., Animal models of working memory: insights for targeting cognitive dysfunction in schizophrenia, *Psychopharmacology (Berl.)*, 174, 111–125, 2004.
6. Carlsson, A., Treatment of Parkinson's with L-DOPA. The early discovery phase, and a comment on current problems, *J. Neural. Transm.*, 109, 777–787, 2002.
7. Evans, A. H. and Lees, A. J., Dopamine dysregulation syndrome in Parkinson's disease, *Curr. Opin. Neurol.*, 17, 393–398, 2004.
8. Cannon, C. M. and Bseikri, M. R., Is dopamine required for natural reward?, *Physiol. Behav.*, 81, 741–748, 2004.
9. Avshalumov, M. V. and Rice, M. E., Activation of ATP-sensitive K^+ (K_{ATP}) channels by H_2O_2 underlies glutamate-dependent inhibition of striatal dopamine release, *Proc. Natl Acad. Sci. U.S.A.*, 100, 11729–11734, 2003.
10. Avshalumov, M. V., Chen, B. T., Marshall, S. P., Peña, D. M., and Rice, M. E., Glutamate-dependent inhibition of dopamine release in striatum is mediated by a new diffusible messenger, H_2O_2, *J. Neurosci.*, 23, 2744–2750, 2003.
11. Avshalumov, M. V., Chen, B. T., Koós, T., Tepper, J. M., and Rice, M. E., Endogenous H_2O_2 regulates the excitability of midbrain dopamine neurons via K_{ATP} channels, *J. Neurosci.*, 25, 4222–4231, 2005.
12. Mourre, C., Smith, M. L., Siesjo, B. K., and Lazdunski, M., Brain ischemia alters the density of binding sites for glibenclamide, a specific blocker of ATP-sensitive K^+ channels, *Brain Res.*, 526, 147–152, 1990.
13. Xia, Y. and Haddad, G. G., Major differences in CNS sulfonylurea receptor distribution between the rat (newborn, adult) and turtle, *J. Comp. Neurol.*, 314, 278–289, 1991.
14. Liss, B., Bruns, R., and Roeper, J., Alternative sulfonylurea receptor expression defines sensitivity to K-ATP channels in dopaminergic midbrain neurons, *EMBO J.*, 18, 833–846, 1999.
15. Dahlström, A. and Fuxe, K., Evidence of the existence of monoamine-containing neurons in the central nervous system. I: demonstration of monoamines in the cell bodies of brain stem neurons, *Acta Physiol. Scand.*, 62, 1–55, 1964.

16. Meir, A., Ginsburg, S., Butkevich, A., Kachalsky, S. G., Kaiserman, I., Ahdut, R., Demirogoren, S., and Rahamimoff, R., Ion channels in presynaptic terminals and control of transmitter release, *Physiol. Rev.*, 793, 1019–1088, 1999.
17. Sesack, S. R., Aoki, C., and Pickel, V. M., Ultrastructural localization of D2-receptor-like immunoreactivity in midbrain dopamine neurons and their striatal targets, *J. Neurosci.*, 14, 88–106, 1994.
18. Yung, K. K. L., Bolam, J. P., Smith, A. D., Hersch, S. M., Ciliax, B. J., and Levey, A. I., Immunocytochemical localization of D1 and D2 dopamine receptors in the basal ganglia of the rat: light and electron microscopy, *Neuroscience*, 65, 709–730, 1995.
19. Torres, G. E., Gainetdinov, R. R., and Caron, M. G., Plasma membrane monoamine transporters: structure, regulation and function, *Nat. Rev. Neurosci.*, 4, 13–25, 2003.
20. Cragg, S. J. and Rice, M. E., Dancing past the DAT at a DA synapse, *Trends Neurosci.*, 27, 270–277, 2004.
21. Neve, K. A. and Neve, R. L., Molecular biology of dopamine receptors, In *The Dopamine Receptors*, Neve, K. A and Neve, R. L., Eds., Humana Press, Totowa, NJ, pp. 27–76, 1997.
22. Missale, C., Nash, S. R., Robinson, S. W., Jaber, M., and Caron, M. G., Dopamine receptors: from structure to function, *Physiol. Rev.*, 78, 189–225, 1998.
23. Rusakov, D. A., Kullmann, D. M., and Stewart, M. G., Hippocampal synapses: do they talk to their neighbours? *Trends Neurosci.*, 22, 382–388, 1999.
24. Barbour, B., An evaluation of synapse independence, *J. Neurosci.*, 21, 7969–7984, 2001.
25. Garris, P. A., Ciolkowski, E. L., Pastore, P., and Wightman, R. M., Efflux of dopamine from the synaptic cleft in the nucleus accumbens of the rat brain, *J. Neurosci.*, 14, 6084–6093, 1994.
26. Rice, M. E., Distinct regional differences in dopamine-mediated volume transmission, *Prog. Brain Res.*, 125, 277–290, 2000.
27. Nirenberg, M. J., Vaughan, R. A., Uhl, G. R., Kuhar, M. J., and Pickel, V. M., The dopamine transporter is localized to dendritic and axonal plasma membranes of nigrostriatal dopaminergic neurons, *J. Neurosci.*, 16, 436–447, 1996.
28. Bergles, D. E., Diamond, J. S., and Jahr, C. E., Clearance of glutamate inside the synapse and beyond, *Curr. Opin. Neurobiol.*, 9, 293–298, 1999.
29. Danbolt, N. C., Glutamate uptake, *Prog. Neurobiol.*, 65, 1–105, 2001.
30. Clark, L. C., Monitor and control of blood and tissue oxygen tensions, *Trans. Am. Soc. Artif. Intern. Organs*, 2, 41–45, 1956.
31. Clark, L. C. and Lyons, C., Studies of a glassy carbon electrode for brain polarography with observations of the effects of carbonic anhydrase inhibition, *Ala. J. Med. Sci.*, 2, 353–359, 1965.
32. Adams, R. N., Carbon paste electrodes, *Anal. Chem.*, 30, 1576, 1958.
33. Kissinger, P. T., Hart, J. B., and Adams, R. N., Voltammetry in brain tissue—a new neurophysiological measurement, *Brain Res.*, 55, 209–213, 1973.
34. Wightman, R. M., Strope, E., Plotsky, P. M., and Adams, R. N., Monitoring of transmitter metabolites by voltammetry in cerebrospinal fluid following neural pathway stimulation, *Nature*, 262, 145–146, 1976.
35. Gonon, F., Cespuglio, R., Ponchon, J. L., Buda, M., Jouvet, M., Adams, R. N., and Pujol, J.-F., *CR. Acad. Sci. Hebd. Seances*, 286, 1203–1206, 1978.
36. Gonon, F. G., Buda, M., Cespuglio, R., Jouvet, M., and Pujol, J.-F., In vivo electrochemical detection of catechols in the neostriatum of anesthetized rats: DA or DOPAC? *Nature*, 286, 902–904, 1980.
37. Gerhardt, G. A., Oke, A. F., Nagy, G., Moghaddam, B., and Adams, R. N., Nafion-coated electrodes with high selectivity for CNS electrochemistry, *Brain Res.*, 290, 390–394, 1984.
38. Nagy, G., Gerhardt, G. A., Oke, A. F., Rice, M. E., Adams, R. N., Moore, R. B., III, Szentirmay, M. N., and Martin, C. R., Ion exchange and transport of neurotransmitters in Nafion films on conventional and microelectrode surfaces, *J. Electroanal. Chem.*, 188, 85–94, 1985.
39. Armstrong-James, M. and Millar, J., Carbon fibre microelectrodes, *J. Neurosci. Methods*, 1, 279–287, 1979.
40. Armstrong-James, M., Millar, J., and Kruk, Z. L., Quantification of noradrenaline iontophoresis, *Nature*, 288, 181–183, 1980.
41. Peters, J. L., Miner, L. H., Michael, A. C., and Sesack, S. R., Ultrastructure at carbon fiber microelectrode implantation sites after acute voltammetric measurements in the striatum of anesthetized rats, *J. Neurosci. Methods*, 137, 9–23, 2004.

42. Ewing, A. G., Bigelow, J. C., and Wightman, R. M., Direct in vivo monitoring of dopamine released from two striatal compartments in the rat, *Science*, 221, 169–171, 1983.
43. Millar, J., Stamford, J. A., Kruk, Z. L., and Wightman, R. M., Electrochemical, pharmacological and electrophysiological evidence of rapid dopamine release and removal in the rat caudate nucleus following electrical stimulation of the median forebrain bundle, *Eur. J. Pharmacol.*, 109, 341–348, 1985.
44. Stamford, J. A., Kruk, Z. L., and Millar, J., Ascorbic acid does not modulate stimulated dopamine release: in vivo voltammetric data in the rat, *Neurosci. Lett.*, 60, 357–362, 1985.
45. Schenk, J. O., Miller, E., Gaddis, R., and Adams, R. N., Homeostatic control of ascorbate concentration in CNS extracellular fluid, *Brain Res.*, 253, 353–356, 1982.
46. Schenk, J. O., Miller, E., Rice, M. E., and Adams, R. N., Chronoamperometry on brain slices: quantitative evaluations of in vivo electrochemistry, *Brain Res.*, 277, 1–8, 1983.
47. Rice, M. E., Oke, A. F., Bradberry, C. W., and Adams, R. N., Simultaneous voltammetric and chemical monitoring of dopamine release in situ, *Brain Res.*, 340, 151–155, 1985.
48. Kelly, R. S. and Wightman, R. M., Detection of dopamine overflow and diffusion with voltammetry in slices of rat brain, *Brain Res.*, 423, 79–87, 1987.
49. Bull, D. R., Palij, P., Sheehan, M. J., Millar, J., Stamford, J. A., Kruk, Z. L., and Humphrey, P. P., Application of fast cyclic voltammetry to measurement of electrically evoked dopamine overflow from brain slices in vitro, *J. Neurosci. Methods*, 32, 37–44, 1990.
50. Cragg, S. J., Holmes, C., Hawkey, C. R., and Greenfield, S. A., Dopamine is released spontaneously from developing midbrain neurons in organotypic culture, *Neuroscience*, 84, 325–330, 1998.
51. Leszczyszyn, D. J., Jankowski, J. A., Viveros, O. H., Diliberto, E. J. Jr., Near, J. A., and Wightman, R. M., Nicotinic receptor-mediated catecholamine secretion from individual chromaffin cells. Chemical evidence for exocytosis, *J. Biol. Chem.*, 265, 14736–14737, 1990.
52. Wightman, R. M., Jankowski, J. A., Kennedy, R. T., Kawagoe, K. T., Schroeder, T. J., Leszczyszyn, D. J., Near, J. A., Diliberto, E. J. Jr., and Viveros, O. H., Temporally resolved catecholamine spikes correspond to single vesicle release from individual chromaffin cells, *Proc. Natl Acad. Sci. U. S. A.*, 88, 10754–10758, 1991.
53. Urena, J., Fernandez-Chacon, R., Benot, A. R., Alvarez de Toledo, G. A., and Lopez-Barneo, J., Hypoxia induces voltage-dependent Ca^{2+} entry and quantal dopamine secretion in carotid body glomus cells, *Proc. Natl. Acad. Sci. U. S. A*, 91, 10208–10211, 1994.
54. Pothos, E., Desmond, M., and Sulzer, D., L-3,4-dihydroxyphenylalanine increases the quantal size of exocytotic dopamine release in vitro, *J. Neurochem.*, 66, 629–636, 1996.
55. Jaffe, E. H., Marty, A., Schulte, A., and Chow, R. H., Extrasynaptic vesicular transmitter release from the somata of substantia nigra neurons in rat midbrain slices, *J Neurosci.*, 18, 3548–3553, 1998.
56. Paxinos, G. and Watson, C., *In The Rat Brain, in Stereotaxic Coordinates*, 4th ed., Academic Press, Inc, New York, 1998.
57. Stamford, J. A., Palij, P., Davidson, C., and Millar, J., Simultaneous "real-time" electrochemical and electrophysiological recording in brain slices with a single carbon–fibre microelectrode, *J. Neurosci. Methods*, 50, 279–290, 1993.
58. Hoffman, A. F., Zahniser, N. R., Lupica, C. R., and Gerhardt, G. A., Voltage-dependency of the dopamine transporter in the rat substantia nigra, *Neurosci. Lett.*, 260, 105–108, 1999.
59. Falkenburger, B. H., Barstow, K. L., and Mintz, I. M., Dendrodendritic inhibition through reversal of dopamine transport, *Science*, 293, 2465–2470, 2001.
60. Mundorf, M. L., Joseph, J. D., Austin, C. M., Caron, M. G., and Wightman, R. M., Catecholamine release and uptake in the mouse prefrontal cortex, *J. Neurochem.*, 79, 130–142, 2001.
61. Patel, J., Avshalumov, M. V., and Rice, M. E., Glutamate-dependent inhibition of striatal dopamine release is mediated by H_2O_2 generated in medium spiny neurons but not in dopamine terminals, *Abstract Viewer/Itinerary Planner*, Society for Neuroscience, Washington, DC, p.46.6, 2004.
62. Chen, B. T. and Rice, M. E., Novel Ca^{2+} dependence and time course of somatodendritic dopamine release: substantia nigra vs. striatum, *J. Neurosci.*, 21, 7841–7847, 2001.
63. Palij, P., Bull, D. R., Sheehan, M. J., Millar, J., Stamford, J. A., Kruk, Z. L., and Humphrey, P. P. A., Presynaptic regulation of dopamine release in corpus striatum monitored in vitro in real time by fast cyclic voltammetry, *Brain Res.*, 509, 172–174, 1990.

64. Bull, D. R. and Sheehan, M. J., Presynaptic regulation of electrically evoked dopamine overflow in nucleus accumbens: a pharmacological study using fast cyclic voltammetry in vitro, *Naunyn-Schmiedebergs Arch. Pharmacol.*, 343, 260–265, 1991.
65. Kennedy, R. T., Jones, S. R., and Wightman, R. M., Dynamic observation of dopamine autoreceptor effects in rat striatal slices, *J. Neurochem.*, 59, 449–455, 1992.
66. Trout, S. J. and Kruk, Z. L., Differences in evoked dopamine efflux in rat caudate putamen, nucleus accumbens and tuberculum olfactorium in the absence of uptake inhibition: influence of autoreceptors, *Br. J. Pharmacol.*, 106, 452–458, 1992.
67. Patel, J., Trout, S. J., Palij, P., Whelpton, R., and Kruk, Z. L., Biphasic inhibition of stimulated endogenous dopamine release by 7-OH-DPAT in slices of rat nucleus accumbens, *Br. J. Pharmacol.*, 115, 421–426, 1995.
68. Cragg, S. J. and Greenfield, S. A., Differential autoreceptor control of somatodendritic and axon terminal dopamine release in the substantia nigra, ventral tegmental area, and striatum, *J. Neurosci.*, 17, 5738–5746, 1997.
69. Phillips, P. E. M., Hancock, P. J., and Stamford, J. A., Time window of autoreceptor-mediated inhibition of limbic and striatal dopamine release, *Synapse*, 44, 15–22, 2002.
70. Phillips, P. E. M., Johns, J. M., Lubin, D. A., Budygin, E. A., Gainetdinov, R. R., Lieberman, J. A., and Wightman, R. M., Presynaptic dopaminergic function is largely unaltered in mesolimbic and mesostriatal terminals of adult rats that were prenatally exposed to cocaine, *Brain Res.*, 961, 63–72, 2003.
71. Gardener, E. L., Chen, J., and Paredes, W., Overview of chemical sampling techniques, *J. Neurosci. Methods*, 48, 173–197, 1993.
72. Schmitz, Y., Lee, C. J., Schmauss, C., Gonon, F., and Sulzer, D., Amphetamine distorts stimulation-dependent dopamine overflow: effects on D2 autoreceptors, transporters, and synaptic vesicle stores, *J. Neurosci.*, 21, 5916–5924, 2001.
73. Schönfuß, D., Reum, T., Olshausen, P., Fischer, T., and Morgenstern, R., Modelling constant potential amperometry for investigations of dopaminergic neurotransmission kinetics in vivo, *J. Neurosci. Methods*, 112, 163–172, 2001.
74. Venton, B. J., Zhang, H., Garris, P. A., Phillips, P. E. M., Sulzer, D., and Wightman, R. M., Real-time decoding of dopamine concentration changes in the caudate-putamen during tonic and phasic firing, *J. Neurochem.*, 87, 1284–1295, 2003.
75. Hoffman, A. F., Lupica, C. R., and Gerhardt, G. A., Dopamine transporter activity in the substantia nigra and striatum assessed by high-speed chronoamperometric recordings in brain slices, *J. Pharmacol. Exp. Ther.*, 287, 487–496, 1998.
76. Hoffman, A. F. and Gerhardt, G. A., Differences in pharmacological properties of dopamine release between the substantia nigra and striatum: an in vivo electrochemical study, *J. Pharmacol. Exp. Ther.*, 289, 455–463, 1999.
77. Gerhardt, G. A. and Hoffman, A. F., Effects of recording media composition on the responses of Nafion-coated carbon fiber microelectrodes measured using high-speed chronoamperometry, *J. Neurosci. Methods*, 109, 13–21, 2001.
78. Partridge, J. G., Apparsundaram, S., Gerhardt, G. A., Ronesi, J., and Lovinger, D. M., Nicotinic acetylcholine receptors interact with dopamine in induction of striatal long-term depression, *J. Neurosci.*, 22, 2541–2549, 2002.
79. Kennedy, R. T., Jones, S. R., and Wightman, R. M., Simultaneous measurement of oxygen and dopamine: coupling of oxygen consumption and neurotransmission, *Neuroscience*, 47, 603–612, 1992.
80. O'Connor, J. J. and Kruk, Z. L., Fast cyclic voltammetry can be used to measure stimulated endogenous 5-hydroxytyrptamine release in untreated rat brain slices, *J. Neurosci. Methods*, 38, 25–33, 1991.
81. Rice, M. E., Richards, C. D., Nedergaard, S., Hounsgaard, J., Nicholson, C., and Greenfield, S. A., Direct monitoring of dopmine and 5-HT release from substantia nigra and ventral tegmental area in vitro, *Exp. Brain Res.*, 100, 395–406, 1994.
82. Rice, M. E. and Nicholson, C., Measurement of nanomolar dopamine diffusion using low-noise perfluorinated ionomer coated carbon fiber microelectrodes and high-speed cyclic voltammetry, *Anal. Chem.*, 61, 1805–1810, 1989.
83. Adams, R. N., *Electrochemistry at Solid Electrodes*, Marcel Dekker, New York, p. 124, 1969.

84. Hafizi, S., Kruk, Z. L., and Stamford, J. A., Fast cyclic voltammetry: improved sensitivity to dopamine with extended oxidation scan limits, *J. Neurosci. Methods*, 33, 41–49, 1990.
85. Heien, M. L., Johnson, M. A., and Wightman, R. M., Resolving neurotransmitters detected by fast-scan cyclic voltammetry, *Anal. Chem.*, 76, 5697–5704, 2004.
86. Armstrong-James, M., Fox, K., and Millar, J., A method for etching the tips of carbon fibre microelectrodes, *J. Neurosci. Methods*, 431–432, 1980.
87. Ewing, A. G., Wightman, R. M., and Dayton, M. A., In vivo voltammetry with electrodes that discriminate between dopamine and ascorbate, *Brain Res.*, 249, 361–370, 1982.
88. Kawagoe, K. T., Garris, P. A., Wiedemann, D. J., and Wightman, R. M., Regulation of transient dopamine concentration gradients in the microenvironment surrounding nerve terminals in the rat striatum, *Neuroscience*, 51, 55–64, 1992.
89. Cahill, P. S., Walker, Q. D., Finnegan, J. M., Mickelson, G. E., Travis, E. R., and Wightman, R. M., Microelectrodes for the measurement of catecholamines in biological systems, *Anal. Chem.*, 68, 3180–3186, 1996.
90. Koh, D. S. and Hille, B., Rapid fabrication of plastic-insulated carbon-fiber electrodes for microamperometry, *J. Neurosci. Methods*, 88, 83–91, 1999.
91. Millar, J. and Pelling, C. W., Improved methods for construction of carbon fibre electrodes for extracellular spike recording, *J. Neurosci. Methods*, 110, 1–8, 2001.
92. McIlwain, H., Thomas, J., and Bell, J. L., The composition of isolated cerebral tissues: ascorbic acid and cozymase, *Biochem. J.*, 64, 332–335, 1956.
93. Dingledine, R., *Brain Slices*, Plenum Publishing Corporation, New York, 1984.
94. MacGregor, D. G., Chesler, M., and Rice, M. E., HEPES prevents edema in rat brain slices, *Neurosci. Lett.*, 303, 141–144, 2001.
95. Rice, M. E., Cragg, S. J., and Greenfield, S. A., Characteristics of electrically evoked somatodendritic dopamine release in substantia nigra and ventral tegmental area in vitro, *J. Neurophysiol.*, 77, 853–862, 1997.
96. Toner, C. C. and Stamford, J. A., 'Real time' measurement of dopamine release in an in vitro model of neostriatal ischaemia, *J. Neurosci. Methods*, 67, 133–140, 1996.
97. Iravani, M. M. and Kruk, Z. L., Real-time effects of *N*-methyl-D-aspartic acid on dopamine release in slices of rat caudate putamen: a study using fast cyclic voltammetry, *J. Neurochem.*, 66, 1076–1085, 1996.
98. Moukhles, H., Bosler, O., Bolam, J. P., Vallee, A., Umbriaco, D., Geffard, M., and Doucet, G., Quantitative and morphometric data indicate precise cellular interactions between serotonin terminals and postsynaptic targets in rat substantia nigra, *Neuroscience*, 76, 1159–1171, 1997.
99. Iravani, M. M. and Kruk, Z. L., Real-time measurement of stimulated 5-hydroxytryptamine release in rat substantia nigra pars reticulata brain slices, *Synapse*, 25, 93–102, 1997.
100. Cragg, S. J., Hawkey, C. R., and Greenfield, S. A., Comparison of serotonin and dopamine release in substantia nigra and ventral tegmental area: region and species differences, *J. Neurochem.*, 69, 2378–2386, 1997.
101. Bunin, M. A. and Wightman, R. M., Quantitative evaluation of 5-hydroxytryptamine (serotonin) neuronal release and uptake: an investigation of extrasynaptic transmission, *J. Neurosci.*, 18, 4854–4860, 1998.
102. John, C. E., Budygin, E. A., Mateo, Y., and Jones, S. R., Voltammetric characterization of monoamines in mouse midbrain, In *Monitoring Molecules in Neuroscience*, Kehr, J., Fuxe, K., Ungerstedt, U., and Svensson, T.H, Eds., Karolinska University Press, Stockholm, pp. 490–491, 2003.
103. Cragg, S. J., Hille, C. J., and Greenfield, S. A., Dopamine release and uptake dynamics within nonhuman primate striatum in vitro, *J. Neurosci.*, 20, 8209–8217, 2000.
104. Cragg, S. J., Variable dopamine release probability and short-term plasticity between functional domains of the primate striatum, *J. Neurosci.*, 23, 4378–4385, 2003.
105. Gale, S. D. and Perkel, D. J., Properties of dopamine release and uptake in the songbird basal ganglia, *J. Neurophysiol.*, 93, 1871–1879, 2005.
106. Rice, M. E. and Cragg, S. J., Nicotine amplifies reward-related dopamine signals in striatum, *Nat. Neurosci.*, 7, 583–584, 2004.
107. Zhou, F. M., Liang, Y., and Dani, J. A., Endogenous nicotinic cholinergic activity regulates dopamine release in the striatum, *Nat. Neurosci.*, 4, 1224–1229, 2001.

108. Wieczorek, W. J. and Kruk, Z. L., Differential action of (+)-amphetamine on electrically evoked dopamine overflow in rat brain slices containing corpus striatum and nucleus accumbens, *Br. J. Pharmacol.*, 111, 829–836, 1994.
109. Iravani, M. M. and Kruk, Z. L., Effects of amphetamine on carrier-mediated and electrically stimulated dopamine release in slices of rat caudate putamen and nucleus accumbens, *J. Neurochem.*, 64, 1161–1168, 1995.
110. Jones, S. R., Joseph, J. D., Barak, L. S., Caron, M. G., and Wightman, R. M., Dopamine neuronal transport kinetics and effects of amphetamine, *J. Neurochem.*, 73, 2406–2414, 1999.
111. Patel, J., Mooslehner, K. A., Chan, P. M., Emson, P. C., and Stamford, J. A., Presynaptic control of striatal dopamine neurotransmission in adult vesicular monoamine transporter 2 (VMAT2) mutant mice, *J. Neurochem.*, 85, 898–910, 2003.
112. Rice, M. E. and Nicholson, C., Diffusion and ion shifts in the brain extracellular microenvironment and their relevance for voltammetric measurements: the brain is not a beaker, In *Neuromethods* Vol. 27, *Voltammetric Methods in Brain Systems*, Boulton, A. A., Baker, G. B., and Adams, R. N., Eds., Humana Press, Totowa, NJ, pp. 27–79, 1995.
113. Elverfors, A., Jonason, J., Jonason, G., and Nissbrandt, H., Effects of drugs interfering with sodium channels and calcium channels on the release of endogenous dopamine from superfused substantia nigra slices, *Synapse*, 26, 359–369, 1997.
114. Jones, S. R., Mickelson, G. E., Collins, L. B., Kawagoe, K. T., and Wightman, R. M., Interference by pH and Ca^{2+} ions during measurements of catecholamine release in slices of rat amygdala with fast-scan cyclic voltammetry, *J. Neurosci. Methods*, 52, 1–10, 1994.
115. Patel, J., Trout, S. J., and Kruk, Z. L., Regional differences in evoked dopamine efflux in brain slices of rat anterior and posterior caudate putamen, *Naunyn-Schmiedeberg's Arch. Pharmacol.*, 346, 267–276, 1992.
116. Phillips, P. E. and Stamford, J. A., Differential recruitment of N-, P- and Q-type voltage-operated calcium channels in striatal dopamine release evoked by 'regular' and 'burst' firing, *Brain Res.*, 884, 139–146, 2000.
117. Zhang, H. and Sulzer, D., Frequency-dependent modulation of dopamine release by nicotine, *Nat. Neurosci.*, 7, 581–582, 2004.
118. Stamford, J. A., Kruk, Z. L., and Millar, J., Regional differences in extracellular ascorbic acid levels in the rat brain determined by high speed cyclic voltammetry, *Brain Res.*, 299, 289–295, 1984.
119. Basse-Tomusk, A. and Rebec, G. V., Regional distribution of ascorbate and 3,4-dihydroxyphenylacetic acid (DOPAC) in rat striatum, *Brain Res.*, 538, 29–35, 1991.
120. Cragg, S. J., Rice, M. E., and Greenfield, S. A., Heterogeneity of electrically-evoked dopamine release and uptake in substantia nigra, ventral tegmental area, and striatum, *J. Neurophysiol.*, 77, 863–873, 1997.
121. Cragg, S. J., Baufreton, J., Xue, Y., Bolam, J. P., and Bevan, M. D., Synaptic release of dopamine in the subthalamic nucleus, *Eur. J. Neurosci.*, 20, 1788–1802, 2004.
122. Kume-Kick, J. and Rice, M. E., Dependence of dopamine calibration factors on media Ca^{2+} and Mg^{2+} at carbon-fiber microelectrodes used with fast-scan cyclic voltammetry, *J. Neurosci. Methods*, 84, 55–62, 1998.
123. Chen, B. T., Avshalumov, M. V., and Rice, M. E., H_2O_2 is a novel, endogenous modulator of synaptic dopamine release, *J. Neurophysiol.*, 85, 2468–2476, 2001.
124. Chen, B. T., Avshalumov, M. V., and Rice, M. E., Modulation of somatodendritic dopamine release by endogenous H_2O_2: susceptibility in substantia nigra but resistance in VTA, *J. Neurophysiol.*, 87, 1155–1158, 2002.
125. Finkel, T. and Holbrook, N. J., Oxidants, oxidative stress and the biology of ageing, *Nature*, 408, 239–247, 2000.
126. Serrano, F., Kolluri, N. S., Wientjes, F. B., Card, J. P., and Klann, E., NADPH oxidase immunoreactivity in the mouse brain, *Brain Res.*, 988, 193–198, 2003.
127. Boveris, A. and Chance, B., The mitochondrial generation of hydrogen peroxide, general properties and effect of hyperbaric oxygen, *Biochem. J.*, 134, 707–716, 1973.
128. Bao, L. and Rice, M. E, Mitochondrial H_2O_2 mediates glutamate-dependent inhibition of striatal dopamine release, *2004 Abstract Viewer/Itinerary Planner*, Soc. Neuroscience, Washington, DC, p. 46.7, 2004.

129. Nirenberg, M. J., Chan, J., Liu, Y., Edwards, R. H., and Pickel, V. M., Vesicular monoamine transporter-2: immunogold localization in striatal axons and terminals, *Synapse*, 26, 194–198, 1997.
130. Bernard, V. and Bolam, J. P., Subcellular and subsynaptic distribution of the NR1 subunit of the NMDA receptor in the neostriatum and globus pallidus of the rat: colocalization at synapses with the GluR2/3 subunit of the AMPA receptor, *Eur. J. Neurosci.*, 10, 3721–3738, 1998.
131. Chen, Q., Veenman, L., Knopp, K., Yan, Z., Medina, L., Song, W. J., Surmeier, D. J., and Reiner, A., Evidence for the preferential localization of glutamate receptor-1 subunits of AMPA receptors to the dendritic spines of medium spiny neurons in rat striatum, *Neuroscience*, 83, 749–761, 1998.
132. Fujiyama, F., Fritschy, J. M., Stephenson, F. A., and Bolam, J. P., Synaptic localization of $GABA_A$ receptor subunits in the striatum of the rat, *J. Comp. Neurol.*, 416, 158–172, 2000.
133. Kemp, J. M. and Powell, T. P., The structure of the caudate nucleus of the cat: light and electron microscopy, *Philos. Trans. R. Soc. Lond. B Biol. Sci.*, 262, 383–401, 1971.
134. Jiang, Z. G. and North, R. A., Membrane properties and synaptic responses of rat striatal neurones in vitro, *J. Physiol. (Lond.)*, 443, 533–553, 1991.
135. Kita, H., Glutamatergic and GABAergic postsynaptic responses of striatal spiny neurons to intrastriatal and cortical stimulation recorded in slice preparations, *Neuroscience*, 70, 925–940, 1996.
136. Smith, A. D. and Bolam, J. P., The neural network of the basal ganglia as revealed by the study of synaptic connections of identified neurons, *Trends Neurosci.*, 13, 259–265, 1990.
137. Ashcroft, F. M. and Gribble, F. M., Correlating structure and function in ATP-sensitive K^+ channels, *Trends Neurosci.*, 21, 288–294, 1998.
138. Aguilar-Bryan, L., Clement, J. P. T., Gonzalez, G., Kunjilwar, K., Babenko, A., and Bryan, J., Toward understanding the assembly and structure of K_{ATP} channels, *Physiol. Rev.*, 78, 227–245, 1998.
139. Krippeit-Drews, P., Kramer, C., Welker, S., Lang, F., Ammon, H. P. T., and Drews, G., Interference of H_2O_2 with stimulus-secretion coupling in mouse pancreatic β-cells, *J. Physiol. (Lond.)*, 15, 471–481, 1999.
140. Seutin, V., Scuvee-Moreau, J., Masotte, L., and Dresse, A., Hydrogen peroxide hyperpolarizes rat CA1 pyramidal neurons by inducing an increase in potassium conductance, *Brain Res.*, 683, 275–278, 1995.
141. Ichinari, K., Kakei, M., Matsuoka, T., Nakashima, H., and Tanaka, H., Direct activation of the ATP-sensitive potassium channel by oxygen free radicals in guinea-pig ventricular cells: its potentiation by MgADP, *J. Mol. Cell. Cardiol.*, 28, 1867–1877, 1996.
142. Rice, M. E. and Avshalumov, M. V., Somatodendritic H_2O_2 from medium spiny neurons inhibits axonal dopamine release, In *Dendritic Neurotransmitter Release*, Ludwig, M., Ed., Springer, New York, pp. 301–313, 2005.

12 In Vivo Voltammetry with Telemetry

Paul A. Garris, Phillip G. Greco, Stefan G. Sandberg, Greg Howes, Sirinun Pongmaytegul, Byron A. Heidenreich, Joseph M. Casto, Robert Ensman, John Poehlman, Andy Alexander, and George V. Rebec

CONTENTS

INTRODUCTION

OVERVIEW

A new instrument for the monitoring and control of neural activity is described in this chapter. The novel characteristic, wireless communication, confers several advantages. Most revolve around freeing the animal from the cable tether linking animal and recording equipment in traditional, hardwired set-ups. The new instrument, called real-time animal telemetry or RAT, was initially developed to support fast-scan cyclic voltammetry (FSCV) at a carbon-fiber microelectrode (CFM). Hence, a large portion of the background section of this chapter is dedicated to this voltammetric microsensor technique for chemical monitoring. We are also working towards the goal of incorporating other microelectrode approaches into RAT, such as, electrophysiology for measuring bioelectric signals and potentiometry for measuring ions. In theory, RAT could accommodate any sensing technology that transduces the input signal into a voltage signal. While RAT was also developed to support electrical stimulation for focally controlling neural activity, it should be possible to incorporate other methods for this purpose as well. Before the actual description of the wireless instrument, a brief primer on telemetry is presented, along with an abbreviated review of existing wireless instruments for neural monitoring and control.

MONITORING NEURAL ACTIVITY

As described in Figure 12.1, neural activity in the brain is fundamentally composed of chemical and bioelectrical signals. Consequently, one approach for characterizing brain function is to sample these signals. Several techniques have been developed over the past decades to achieve this important goal. Among the most widely used is microdialysis [1–3], which removes chemicals from the brain for subsequent analysis by methods, such as, electrochemical detection, fluorescence, capillary electrophoresis, and mass spectroscopy. Two general microelectrode techniques for monitoring neural activity, voltammetry and electrophysiology, are also well established. Shown diagrammatically in Figure 12.1, voltammetry monitors the chemistry of the brain, whereas electrophysiology measures its bioelectrical activity. Like microdialysis, voltammetry and electrophysiology have been applied to animals that are awake, rendering these techniques well suited for investigating the neural substrates of behavior. Also shown diagrammatically in Figure 12.1 is potentiometry. This electrochemical technique is used with ion selective microelectrodes [4].

As described in detail in the "Principles of FSCV at a CFM" and the "Applications of FSCV at a CFM" sections of this chapter, we are working towards the goal of incorporating voltammetry, electrophysiology and potentiometry into a wireless instrument. The rationale is that combined, the use of these microelectrode techniques affords a broad-based, integrative study of neuronal activity in the fast time and small space domains in which neurons function during important types of behavior. The initial thrust of instrument development was focused on FSCV at a CFM. Wightman establishes four criteria for sampling brain chemistry faithfully [5]: specificity, sensitivity, temporal resolution, and spatial resolution. It is our conviction that this voltammetric microsensor technique meets these stringent criteria for chemical monitoring in the brain quite well under suitable experimental conditions (see "Voltammetry" and "Principles of FSCV at a CFM" sections of this chapter).

Developing an instrument for neural monitoring begs the question of how the study of brain function could be enhanced. The primary answer is rooted in the nature of the connection between the animal and recording equipment in existing set-ups: the cable tether. Clearly, hardwire set-ups have afforded elegant studies investigating neural substrates of behavior, with the cable tether providing a robust connection. However, the cable tether is also restrictive, as the animal is never completely free to behave, and recordings in natural environmental settings, such as in a

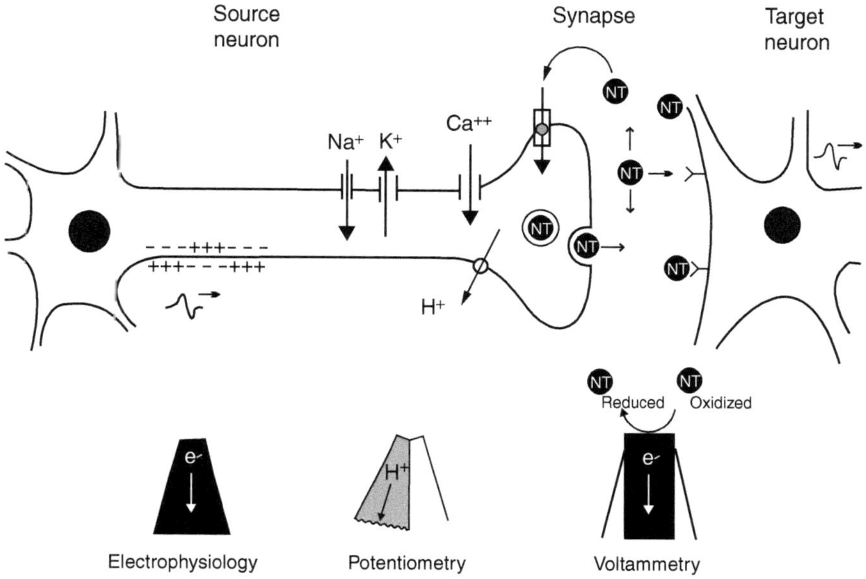

FIGURE 12.1 Monitoring neuronal activity. The function of the brain is to control behavior, and the functional cells of the brain are called neurons. (From Kandel, E. R., Schwartz, J. H., and Jessell, T. M., *Principles of Neural Science*, McGraw-Hill, New York, 2000; Zigmond, M. J., Bloom, F. E., Landis, S. C., Roberts, J. L., and Squire, L. R., *Fundamental Neuroscience*, Academic Press, San Diego, CA, 1999.) These cells are highly specialized for rapid signaling over large distances. During neurotransmission, an electrical signal called an action potential is conducted along the length of the axon to the synapse, where it elicits the release of a chemical neurotransmitter from the source neuron. Released neurotransmitter diffuses in brain extracellular space to a target neuron. Although shown immediately across the synaptic cleft, the target neuron may also be several microns away from the source neuron, when the neurotransmitter signals via non-synaptic communication. (From Vizi, E. S., *Pharmacological Reviews*, 52 (1), 63–89, 2000; Zoli, M., Torri, C., Ferrari, R., Jansson, A., Zini, I., Fuxe, K., and Agnati, L. F., *Brain Research Reviews*, 26 (2–3), 136–147, 1998.) Regardless of the spatial arrangement, subsequent binding of the neurotransmitter to a specific protein receptor elicits a change in the physiological status of the target neuron. Receptor binding often manifests as an alteration in the generation of action potentials. Mediating many of the processes of neurotransmission is the flux of ions between intracellular and extracellular compartments. For example, Na^+ and K^+ underlie the ion fluxes associated with the action potential, and Ca^{2+} is intimately involved in exocytosis, the mechanism releasing neurotransmitter into the extracellular compartment. Intracellular pH, which changes as a result of neural activity, is tightly controlled via several mechanisms transporting H^+ across the neuronal membrane, including the amiloride-sensitive Na^+/H^+ exchanger. Playing an integral role in neuronal activity is the brain cell microenvironment, the physical and chemical milieu immediately surrounding neurons. (From Nicholson, C. and Rice, M. E., *Volume Transmission in the Brain: Novel Mechanisms for Neural Transmission*, Raven Press, New York, 1991.) In addition to an information pathway, the dynamic brain cell microenvironment maintains ionic homeostasis and provides a conduit for metabolic substrates supporting the proper functioning of neurons.

social housing condition or in an enriched environment that demands species-specific responses (such as crawling in or around obstacles), are impossible. Yet, it is in these settings that we are most likely to gain the most insight into behaviorally relevant neuronal operations. Another limitation related to investigating social interaction is monitoring in a single animal only. Indeed, monitoring neural activity in more than one animal, simultaneously, would result in a tangled mess of tethers. In short, the answer to the question posed above is a wireless connection, which would offer an elegant solution to the limitations imposed by the cable tether.

Voltammetry

Voltammetry directly monitors chemicals in brain's extracellular fluid. Although many different methodologies have been developed, the basis of each is that the chemical is either oxidized or reduced at the sensor surface by an applied potential. The measured faradaic current is then proportional to the concentration of chemical in the region of the sensor [6–8]. Several characteristics make voltammetry attractive for chemical monitoring. These include the presence of important, easily oxidized neurotransmitters in the brain (e.g., dopamine, norepinephrine, epinephrine, and histamine), fast kinetics and measurement of redox reactions, and ready fabrication of voltammetric microsensors. Voltammetric monitoring of dopamine has clearly received the most attention. This neurotransmitter is found in high concentrations in the brain and subserves important functions related to motor control, cognition, motivation, and endocrine control [9–11].

Early attempts to monitor brain dopamine voltammetrically, pioneered by Adams and co-workers in the 1970s [12,13], were hindered by the high concentrations of electroactive interferents in brain extracellular fluid [8]. Even today, selectivity represents the most contentious issue of the voltammetric technique, especially for measurements in the behaving animal [14–16]. Second generation techniques utilized modified electrodes and more complex waveforms to improve selectivity [7,8,17], enabling reliable measurements of neurotransmitter metabolites in ambulant animals with minute temporal resolution [18–20]. Continued advances have resulted in the development of three in vivo voltammetric techniques for the real-time monitoring of neurotransmitters: high-speed chronoamperometry [21], FSCV [22–24] and amperometry [25,26]. These techniques sample dopamine with sub-second temporal resolution at a CFM.

No attempt is made here to contrast the various voltammetric techniques developed and applied to investigate brain function. Such comparisons are available in the literature [6,8,17,27] and, if interested, the reader is encouraged to pursue this direction. Rather, this chapter focuses on FSCV at a CFM, which is incorporated into the wireless instrument. The rationale for incorporation is straightforward: this technique affords millisecond temporal and micron spatial resolution for chemical monitoring, along with provision for identifying the analyte detected in the form of a background-subtracted cyclic voltammogram. Moreover, FSCV at a CFM has been used recently in significant neurobiological experiments; thus, appears to be well positioned for contributing prominently to neurobiology research for many years to come. In the following two sections, the principles and applications of this voltammetric microsensor technique are described in more detail.

Principles of FSCV at a CFM

Figure 12.2 describes the technique of FSCV at a CFM. Typically, a triangle waveform is applied intermittently to the CFM at 100 ms intervals. In the example shown in Panel A, voltage is linearly ramped from a resting potential of -400 mV to a peak of 1000 mV and back. At a rate of 300 V/s, the duration of the voltage scan is approximately 9.3 ms. This scan rate supports monitoring at a temporal resolution (100 ms) that is suitable for characterizing brain dopamine dynamics, as the scan interval should be about ten times the scan length. The approximately 90 ms in between scans also allows ample time for dopamine to adsorb to the CFM, thus increasing sensitivity. During the time in between scans, the CFM rests at -400 mV. A single triangle waveform is shown in Panel B (top). Immediately below (middle) are two recordings obtained at the same CFM during single voltage scans. The black line denotes current collected in the striatum of an anesthetized rat, in the absence of electrical stimulation. Notice the large background or charging current that changes direction at the switching potential of the scan. The red line denotes current collected at the same CFM, but during electrical stimulation of dopaminergic fibers to elicit dopamine release. Notice that two additional, but smaller, features appear on top of the background current around 4 and 8 ms, corresponding to approximately 750 and -100 mV, respectively. Because the background current is stable, it can be removed by simple subtraction to reveal these smaller features more

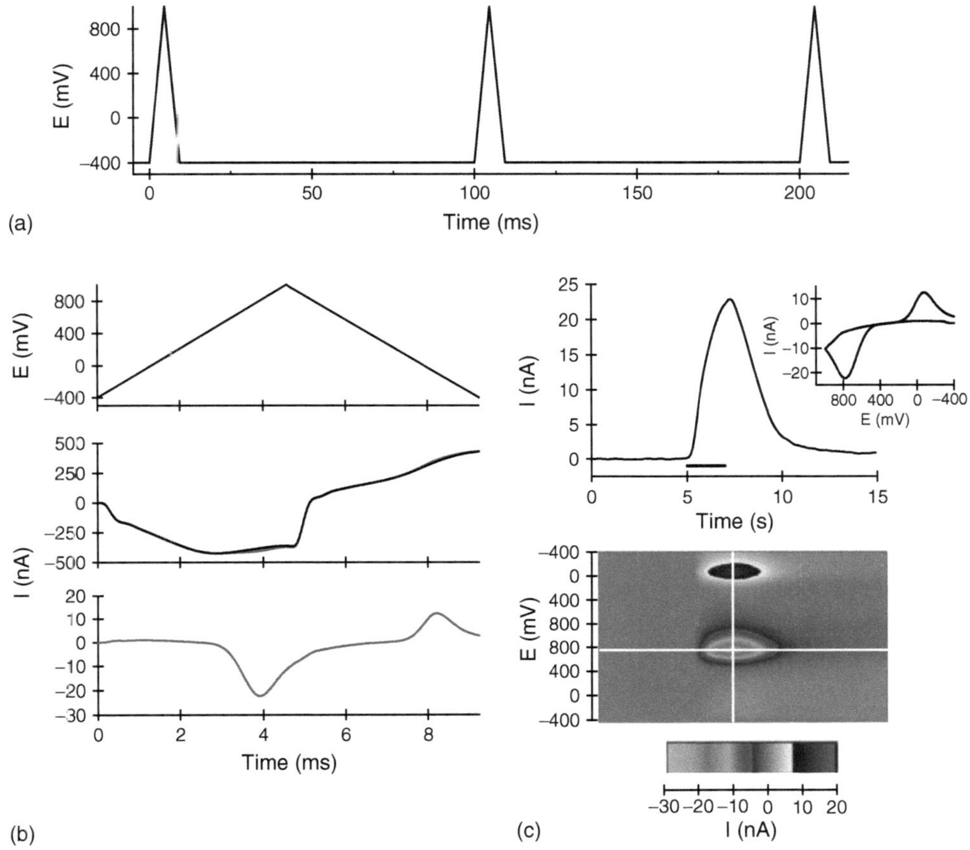

FIGURE 12.2 **(See color insert following page 272.)** Fast-scan cyclic voltammetry at a carbon-fiber microelectrode. (a) application of the triangle wave form at 100 ms intervals; (b) current monitored at a CFM during a single voltage scan; (c) voltammetric recording of electrically evoked dopamine release in the striatum of a urethane-anesthetized rat.

clearly (bottom). These features are due to faradaic current, when dopamine oxidizes to a quinone (the early negative current) and when the electroformed quinone is reduced back to dopamine (the later positive current).

The faradaic current collected during a scan is typically shown as a voltammogram, which is a plot of current as a function of applied potential. Because charging current is removed, the voltammogram shown in Panel C, INSET, is called a *background-subtracted* voltammogram. This voltammogram was constructed from the recording shown in Panel B (bottom). Notice the same negative and positive peaks around 750 and −100 mV, respectively. These peaks are indicative of dopamine; thus, the background-subtracted voltammogram is used as a chemical signature for identifying the analyte detected. Additionally, the change in dopamine-per-time is determined by plotting the current obtained for each scan at the oxidation potential for dopamine. For example, Panel C (top) shows the current monitored at approximately 750 mV in the rat striatum before, during, and after electrical stimulation. The stimulus train is demarcated by the solid line underneath the voltammetric recording, between 5 and 7 s.

While informative, the limitation of the voltammogram is that it represents a change in analyte(s) between only two time points. In the present case, pre-stimulus voltammograms were averaged and subtracted from the average of voltammograms collected around the peak of the electrically evoked signal. However, the entire electrochemical record can be viewed by successively plotting all of the background-subtracted voltammograms with time [28]. Such a plot is shown in Panel C (bottom), with time as the x-axis, applied potential as the y-axis, and current (in color) as the z-axis. Notice the same two, prominent features displayed in the color plot at around 750 and −100 mV, coincident with electrical stimulation. Indeed, a plot of the current recorded at approximately 750 mV versus time, and corresponding to the horizontal white line, yields the current versus time trace shown in the top of Panel C. Moreover, a plot of the current recorded at approximately 7 s versus applied potential, and corresponding to the vertical white line, yields the voltammogram shown in the INSET at the top of Panel C. The color plot clearly demonstrates that dopamine is the only analyte detected at the CFM under these conditions.

Applications of FSCV at a CFM

Carbon-fiber microelectrodes were originally used for electrophysiological measurements (see Figure 12.4) [29], and FSCV was originally developed for calibrating the iontophoretic application of dopamine with a CFM [30,31]. Iontophoresis uses current to apply charged chemicals directly into the brain via a micropipette. However, the first use of FSCV at a CFM for monitoring endogenous dopamine was in the anesthetized rat during electrical stimulation [32]. Although a relatively simple signal (e.g., see Figure 12.2, top), the evoked measurement is replete with information about the fundamental mechanisms regulating extracellular dopamine levels in the brain, such as, release, uptake, and diffusion [22,33,34]. Hence, the anesthetized experiment with electrical stimulation, and its analogous experiment in vitro [35–37], have proven to be workhorse approaches for investigating dopamine neurochemistry and neuropharmacology.

A twist on the evoked experiment with FSCV at a CFM is, instead of using electrical stimulation, focally applying pharmacological agents to alter brain dopamine levels [38,39]. Evaluation of the factors regulating dopaminergic tone, the ambient concentration of dopamine in extracellular fluid, is possible with this later approach, which is a prospect not considered amenable to real-time voltammetric techniques. More recently, coupling FSCV with a CFM to principal component analysis [40] has been used to quantify the increase in basal dopamine levels following cocaine administration [41].

Towards the end of the 1990s, FSCV at a CFM was applied to unanesthetized animals, paving the way for a new era of more behaviorally oriented experiments. The first measurements in freely moving animals were dopamine evoked by electrical stimulation [42]. This development enabled pharmacological experiments much like those performed in the anesthetized animal, but without the concern for anesthesia artifacts, and accompanied by the simultaneous assessment of drug-altered behavior [43–45]. It was also possible to investigate the role of dopamine in the classic paradigm of intracranial self-stimulation on a sub-second temporal level [46–50].

Perhaps the most impressive application of FSCV at a CFM in unanesthetized animals is the measurement of behaviorally evoked changes in brain dopamine levels [51]. The evolution of this application has been driven by steadily improving detection limits, achieved through advances in instrumentation, sensors and waveforms [52–56]. As a result, it is now possible to monitor the sub-second dopamine changes presumably elicited by burst firing of dopaminergic neurons during phasic signaling [57,58] and associated with sociosexual interaction [59,60], and food [61] and drug [62,63] seeking.

Figure 12.3 shows a recording collected by FSCV at a CFM in an unanesthetized animal. The trace described by the dotted line in the left panel was electrically evoked by a pulse train applied at 5 s to the medial forebrain bundle. Animals lever-press to obtain this same pulse train (60 Hz, 0.4 s, 125 μA) during intracranial self-stimulation [46,47]. The black line in the INSET at the top is the background-subtracted cyclic voltammogram obtained at the peak of the evoked signal. The large, downward (i.e., oxidative) peak around 700 mV and smaller, upward (i.e., reductive) peak around −300 mV clearly indicated dopamine. Underneath the evoked trace is the color plot showing all of the voltammograms collected during this time. Coincident with the electrical stimulation and the

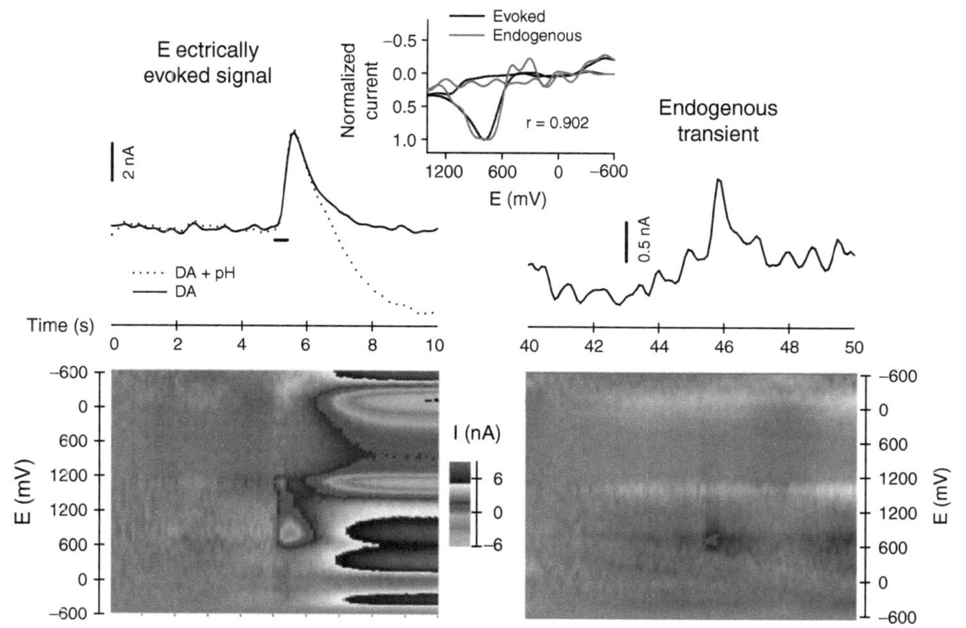

FIGURE 12.3 **(See color insert following page 272.)** Voltammetric measurement of dopamine in a freely moving rat. For this recording, the CFM was placed in the dorsomedial caudate–putamen, and the stimulating electrode was placed in the medial forebrain bundle. Voltammograms shown in the INSET were normalized to peak oxidative current.

increase in the evoked trace is a greenish dot at approximately 5 s and 700 mV, corresponding to the oxidative peak in the dopamine voltammogram. Also notice several features emerging in the color plot just after the dopamine peak. These features reflect changes in brain pH associated with the electrical stimulation [64,65] and cause the evoked trace (dotted line) to plummet between 7 and 10 s. However, the pH interference is readily removed by differential subtraction, resulting in a pure signal describing the dopamine changes (solid line).

A 10 s portion of the same recording, but beginning 35 s after the electrical stimulation, is shown in the right panel. The same CFM location at which an evoked dopamine signal was recorded also yielded a non-electrically evoked, transient signal. Although smaller in amplitude, the endogenous signal exhibited comparable dynamics to the electrically evoked dopamine recording, suggestive of similar origins. The INSET at the top overlays the background-subtracted cyclic voltammogram obtained from electrical stimulation (black line) and the endogenous transient (red line). Voltammograms are similar ($r = 0.902$), and indicate that the endogenous transient is a dopamine signal. Indeed, the color plot below shows a clearly demarcated purplish dot at about 45 s and 700 mV, corresponding to the transient signal and the oxidative peak of the voltammogram. As identified above, this transient dopamine concentration spike is thought to result from a burst of action potentials during phasic dopaminergic signaling [16,66].

Carbon-fiber microelectrodes also serve as excellent sensors for electrophysiological measurements [29]. Figure 12.4 compares extracellular single units (i.e., action potentials) recorded in the striatum of an anesthetized rat by a saline-filled glass capillary (A) and a CFM (B). At least in our hands, CFMs perform as well as or exceed glass capillaries for this purpose [67]. The ability to measure both single unit activity and dopamine was exploited by Millar and co-workers in developing quasi-simultaneous electrophysiology and voltammetry at a CFM [68,69]. In this elegant technique, the CFM is switched in the time between voltage scans from a current-to-voltage transducer circuit for measuring voltammetry to a voltage-follower circuit for measuring

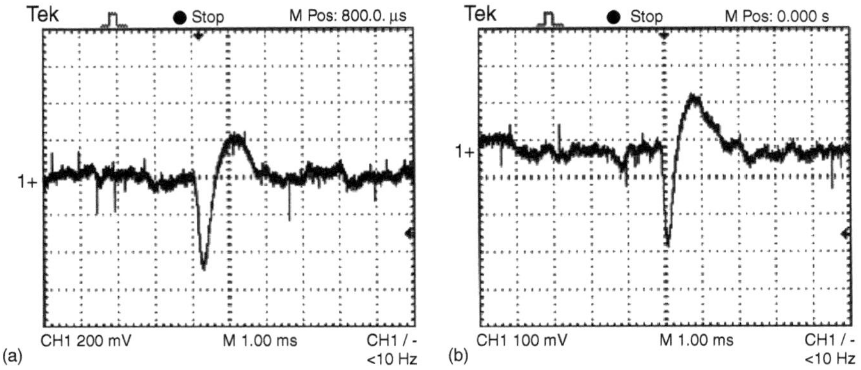

FIGURE 12.4 Comparison of electrodes for measuring electrophysiology. (a) Saline-filled glass micropipette; (b) CFM. Both recordings were collected in the dorsomedial caudate–putamen of a urethane-anesthetized rat.

electrophysiology. The beauty of this technique is that dopamine and the target cell response to dopamine are measured at approximately the same time and in approximately the same vicinity. More recently, quasi-simultaneous electrophysiology and voltammetry at a CFM has been applied to the freely moving rat [70]. Although not attempted yet, it should be possible to incorporate this dual sensing technique into a wireless instrument.

PRINCIPLES OF TELEMETRY

RESOURCES

Several telemetry resources are available for both the technically and less technically inclined. For the more general reader, two excellent texts are *The Essential Guide to RF and Wireless* by Carl Weisman (2002, Prentice Hall) [71] and *The Essential Guide to Wireless Communications Applications* by Andy Dornan (2002, Prentice Hall) [72]. Focusing primarily on consumer oriented wireless devices, both provide readable reviews of basic telemetry principles and technologies. Much of the information provided in the next two sections was obtained from these two texts. Although some information is understandably dated now, it nevertheless is interesting to compare the various predictions as to the direction it was believed wireless technology would take, with the direction it has actually taken, particularly regarding Wireless Fidelity (Wi-Fi) and Bluetooth, as both have enjoyed rapid developments in the last three years. More up to date information about wireless devices is found in *PC Magazine Wireless Solutions* by Neil Randall and Barrie Sosinksy (2005, Wiley) [73].

THEORY AND BASIC INSTRUMENTATION

Radio waves, a form of electromagnetic radiation, allow telemetry devices to transmit data wirelessly [72]. Typically conceptualized as waves, electromagnetic radiation is characterized by frequency (which is the number of cycles per second and inversely related to wavelength) and amplitude (which describes the power of transmission). The term "radio" was originally given to electromagnetic radiation used in communication, although modern communication technologies also use waves with much greater frequencies, such as, microwaves and infrared. Radio waves range in frequency between a lower limit of around 5 kHz to an upper limit of around 300 GHz, where they overlap with the lower end of microwaves. The higher the frequency, the shorter is the range. This is due to the greater attenuation (i.e., loss of energy due to collisions) of higher

TABLE 12.1
Example Radio Frequencies (in Hz)

AM radio	530,000
FM radio	88,000,000
TV	746,000,000
Wireless LAN	2,400,000,000
Satellite	4,200,000,000
Radar	9,000,000,000

Source: From Weisman, C. J., *The Essential Guide to RF and Wireless*, Upper Saddle River, Prentice Hall, NJ, 2002.

frequency waves. The amount of data transmitable, hence the speed of data transmission, is also related to frequency. More specifically, the larger the bandwidth, the frequency range or channel allocated to transmission, the greater the amount of information that can be sent. Example frequencies of radio applications are shown in Table 12.1.

Telemetry devices share common principles (Figure 12.5) [71]. The original data signal is converted by a transmitter from an electrical signal into an electromagnetic wave, which is sent through the air. If everything goes smoothly, the receiver then converts the airborne wave back into the original data signal. The transmitter is comprised of several components including an amplifier, a filter and an antenna. Two additional and critical components are the oscillator and mixer. During wireless transmission, the data signal is not sent directly by the transmitter. Rather, the data signal is "piggy-backed" onto a carrier signal, e.g., a sine wave of a particular frequency and generated by the oscillator. The data signal is added to the carrier signal by the mixer in a process called modulation. The receiver operates in the opposite fashion, removing the original data signal from the carrier signal in a process called demodulation.

Telemetry is distinguished on the basis of several characteristics, including the frequency of the carrier wave, the type of modulation, and whether the original data signal is analog or digital [71]. Example frequencies for various radio devices were described above in Table 12.1. Two general types of modulation are amplitude modulation (AM) and frequency modulation (FM). During AM,

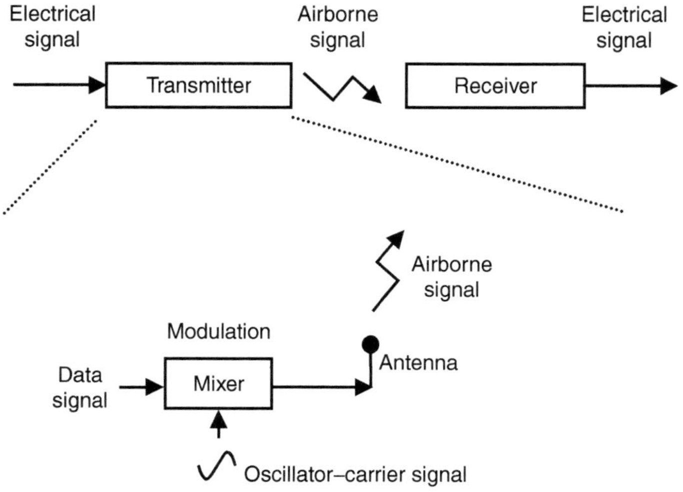

FIGURE 12.5 Principles of telemetry. (Modified from Weisman, C. J., *The Essential Guide to RF and Wireless*, Prentice Hall, Upper Saddle River, NJ, 2002.)

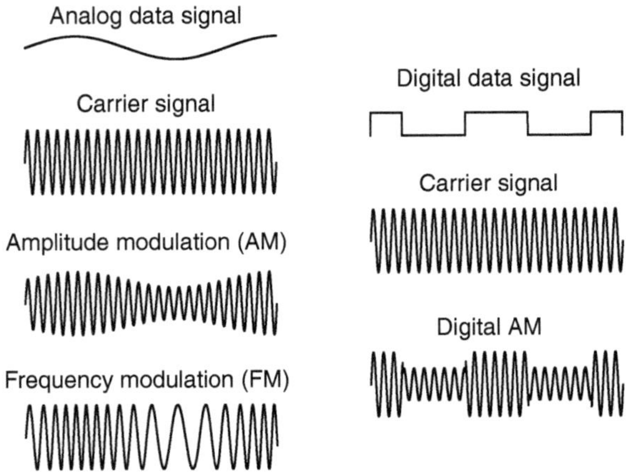

FIGURE 12.6 Modulation. (Modified from Weisman, C. J., *The Essential Guide to RF and Wireless*, Prentice Hall, Upper Saddle River, NJ, 2002.)

the amplitude of the carrier signal is modified by the data signal, whereas frequency is modified during FM. The examples found in the left panel of Figure 12.6 show AM and FM transmitting an analog, or continuous, data signal. More recently, digital, or binary, signals have been sent wirelessly. The zeroes and ones of a digital signal are readily observed on a carrier wave as "highs" and "lows" after digital AM in the stylized example shown in right panel of Figure 12.6. Phase modulation, which adds the original signal onto the carrier wave by adjusting its phase, is commonly used in digital applications. There are various forms of amplitude, frequency and phase modulation and, as described in more detail in the next section, additional modulation is used when different signals share the same carrier frequency.

Although analog signals were the first to be sent wirelessly, the subsequent rise in digital telemetry and the great fanfare surrounding its introduction, suggest important advantages to the wireless transmission of digital signals. Indeed, digital telemetry confers several key advantages [72]: noise reduction, reliability, transmission efficiency, security, and timing. Perhaps the easiest advantage to conceptualize is noise reduction. Whereas a receiver cannot distinguish between noise and signal in an analog wave, the highs and lows of a digital signal are distinct states, more easily separated from noise (see Figure 12.6). Reliability is enhanced, because digital signals are encoded with extra data permitting error correction. Although error correction reduces capacity, the increased reliability compensates by affording more efficient use of the same bandwidth. Increased spectral efficiency is also achieved by the compression of digital data. Security has always plagued the wireless transmission of analog signals, and while possible, encryption is unreliable. On the other hand, digital signals are readily encrypted to various degrees and without the loss of capacity during transmission. Finally, because digital signals are easily stored, they can be divided into packets and sent at any time to enhance efficiency. The internet is based on timing in the form of packet-switching.

Are there any advantages to analog telemetry? Yes, in fact, there are several. For example, analog devices are cheaper, simpler, and smaller. Consequently, because they are sufficient for many applications, analog telemetry can be the right choice.

Digital Telemetry at 2.4 GHz

Two types of digital telemetry have gained prominence recently, Wi-Fi and Bluetooth [73]. Both operate in the internationally allocated band of 2.4 GHz. While this unlicensed band is only one of several designated by the International Telecommunications Union for industrial, scientific, and medical (ISM) purposes, it is the only ISM band available in every country. Wireless fidelity has emerged as the standard in the United States for wireless local area networks (LANs). Although sometimes confused as a competitor, Bluetooth was developed to support personal area networks (PANs), characterized by mobile access points (compare with the fixed access point in LAN), and piconets, a collection of several Bluetooth devices. Interestingly, another 2.4 GHz technology, called HomeRF, was developed specifically for the home networking and entertainment market [71]. This technology combines features of Wi-Fi and cordless phones, and was originally designed to be more affordable and simpler to implement than Wi-Fi. Since then, Wi-Fi has become relatively inexpensive, and many home networks are now based on this technology. Table 12.2 compares salient features of Wi-Fi and Bluetooth.

Wireless fidelity is based on the Institute of Electrical and Electronic Engineers (IEEE) 802.11 standard [73]. To date, the two most successful protocols are 802.11b and 802.11g. Back compatible, 802.11g is newer and faster (11 versus 54 Mbps). As with any protocol operating in the unlicensed 2.4 GHz band, Wi-Fi must use a second type of modulation, in addition to piggybacking the data signal onto a carrier wave, to combat interference from other signals existing at the same time and in the same place [71]. One such general technology is spread spectrum, which essentially hides the signal in the noise, so it becomes undetected. This is accomplished by "spreading" the carrier wave across a wider frequency spectrum. For example, the direct sequence spread spectrum (DSSS) used by 802.11b multiplies the digital signal by a pseudo random noise (PN) code, effectively converting the signal into apparent noise. The signal is readily uncovered again if the PN code is known. The newer 802.11g uses a modulation called orthogonal frequency division multiplexing (OFDM), which breaks the bandwidth into separate and distinct sub-bands. More complex than DSSS, OFDM is especially valuable for minimizing multipath (i.e., indirect signals caused by reflections).

Like the first 802.11 protocol, Bluetooth uses frequency hopping spectrum spread (FHSS) to navigate the crowded 2.4 GHz airwaves [71]. In this spread spectrum technology, the frequency of the carrier wave is jumped between 79 1 MHz bands, again according to a PN code. In contrast to the FHSS of the original 802.11, the dwell time (the time at which the wave rests at one frequency) is very short, as Bluetooth hops frequencies at a rate up to 3200/s (compare with 2.5 hops/s). The fast hopping rate makes Bluetooth well suited for crowded 2.4 GHz environments and relatively resistant to multipath. On the downside, it limits the data transmission rate (see next paragraph) and can interfere with other 2.4 GHz signals. For this reason, Bluetooth has been called a "rude radio" [72].

Although transmitting more slowly (less than 1 versus greater than or equal to 11 Mbps) and across shorter distances (9 versus 90 m maximally) than Wi-Fi, Bluetooth enjoys several advantages, including reliability (see preceding paragraph), small size, low power requirements and low price [71]. As described below in the "Real-Time Animal Telemetry" section of this chapter,

TABLE 12.2
Comparison of Wi-Fi and Bluetooth

Characteristic	802.11b	802.11g	Bluetooth
Frequency	2.4 GHz	2.4 GHz	2.4 GHz
Application	LAN	LAN	PAN, piconet
Technology	DSSS	OFDM	FHSS
Range	90 m (300 ft.)	90 m (300 ft.)	9 m (30 ft.)
Data rate	11 Mbps	54 Mbps	1 Mbps

Source: From Weisman, C. J., *The Essential Guide to RF and Wireless*, Prentice Hall, Upper Saddle River, NJ, 2002; Dornan, A., *The Essential Guide to Wireless Communications Applications*, Prentice Hall, Upper Saddle River, NJ, 2002.

similar advantages make Bluetooth well suited for developing a wireless instrument supporting FSCV at a CFM. As one might expect, these advantages are also consistent with the original goal of Ericsson (Sweden) in developing Bluetooth: a protocol supporting PANs and piconets. In a PAN, one wireless device becomes the access point for several other wireless devices. A Bluetooth piconet, in turn, enables eight devices to be linked together, with one device serving as the master to the other seven slaves. Bluetooth piconets are scalable, as a slave can be a master to a different set of slaves. To make the concept of PANs and piconets work most successfully, it is speculated that Ericsson must provide Bluetooth devices at the low cost of $5 or less [72]. Dating back to 1994, Bluetooth was initially received as the "next best thing." Although applications were slow in coming to the marketplace, which led many industry analysts to wonder whether the technology was a bust, it appears now that Bluetooth products are everywhere, from the computer mouse and key board, to the camera, the PDA, the cell phone and even the car [73]. The next generation of Bluetooth technology is predicted to send signals at 20 Mbps to distances of up to 90 m, transmission characteristics similar to 801.11b [71].

WIRELESS NEURAL MONITORING AND CONTROL

Scope

No attempt is made here to exhaustively review the literature on wireless neural monitoring and control; the use of telemetry for electrophysiological recording and electrical stimulation in experimental animals is not new; analog systems supporting these applications have been available for decades [74–77] and similar systems, but based on digital technology, have been introduced in the past few years [78–81]. Rather, this section will highlight the most recent developments related to our wireless instrument. On a related front, it should be mentioned here that the use of wireless devices in medicine is rapidly expanding [82–85] and that this phenomenon will, no doubt, have a great impact on the future development of wireless neural monitoring and control. The availability of a turnkey, commercial system will ultimately drive more widespread applications. Such wireless instruments, though limited to the monitoring of physiological parameters, such as, heart rate, blood pressure, ECG, EEG, temperature and locomotor activity, are currently available from Data Sciences International (St Paul, MN) and Minimitter (Sunriver, OR).

Electrophysiology

Analog FM technology has a suitable data transmission rate for capturing single unit activity in the brain. For this reason, several FM-based systems have been developed over the years for the wireless monitoring of electrophysiology in freely behaving animals, including the rat [86], rabbit [86], toad [87], monkey [88], song bird [89], and owl [90]. As mentioned above in the "Theory and Basic Instrumentation" section of this chapter, analog devices are readily miniaturized, hence their suitability for smaller laboratory animals. Although stereo-FM has been also used to record single unit activity from two implanted electrodes simultaneously [90], there is a general trend in electrophysiology to move towards multiwire array electrodes in order to record tens of units at once [91–94]. Three wireless devices have been described recently to support these state-of-the-art, multiwire array applications. Nicolelis and co-workers have developed a device based on 802.11b to record single unit activity in the monkey from 16 electrodes sampled at 30 kS/s with 12-bit resolution [95]. This system weighs 260 g and measures 14×8.3×4 cm, about the size of a young adult rat. However, a smaller wireless array system has been specifically developed for the rat. This analog device, available commercially from Triangle BioSystems (Durham, NC, U.S.A.), comes in a 7- and 15-channel package and is compatible with Plexon (Dallas, TX, U.S.A.) hardware and software [96]. Plexon makes hardwired systems for multiwire electrophysiology. A smaller digital system is also in the developmental stages [97]. This device operates at a frequency of

433 MHz using frequency-shift keying modulation and samples 88 channels simultaneously, with 10-bit resolution at 15 kS/s. It is worth noting that this device also uses a 2.64 MHz inductive link to power the unit.

Voltammetry

Perhaps related to the fact that there are a smaller number of users of voltammetry compared with the number of users of electrophysiology, there are fewer descriptions of wireless voltammetry. A slow scan electrochemical technique called *differential normal pulse voltammetry*, which samples electroactive neurotransmitter metabolites with minute temporal resolution, has been coupled with infrared telemetry [98–101]. This example highlights another advantage of wireless for microelectrode applications: reducing electrical cross talk when combining voltammetry with other techniques such as electroencephalography, electrooculography, and electromyography. This optoelectronic system was later modified to support direct current amperometry (i.e., pulsed amperometry or chronoamperometry) for sub-second measurements of the neurotransmitter serotonin at a CFM [102]. A company called Pinnacle Technology (Lawrence, KS, U.S.A.) now offers a digital wireless instrument supporting chronoamperometry [103,104]. This same company has also developed a wireless instrument for an enzyme-linked microsensor for monitoring the neurotransmitter glutamate [105]. Because glutamate is not electroactive, an enzyme, glutamate oxidase, is used to produce hydrogen peroxide from glutamate, which can then be detected amperometrically (i.e., at a constant potential). The sampling rates for the infrared device supporting direct current amperometry (5 S/s) and for both Pinnacle Technology devices (0.5 S/s) are far too slow to support FSCV, which requires sampling rates in excess of 20 kS/s [106].

Electrical Stimulation

As with electrophysiology, analog systems for wireless electrical stimulation have been available for decades [75]. In fact, electrophysiology, in some form, and electrical stimulation are often combined into the same device [76,77]. Some wireless devices for electrical stimulation are powered inductively [87,107]. Most appear to stimulate with constant voltage. However, by correcting for the different impedances of stimulating electrodes, constant current is preferable. One of the more interesting applications of wireless electrical stimulation appeared in the journal *Nature* in 2002. In this so called robot rat example, stimulating electrodes implanted bilaterally in the somatosensory cortex were wirelessly controlled and used to steer a rat in real time across a variety of terrains [108]. Another stimulating electrode was implanted in the medial forebrain bundle for activating the brain reward system, similar to the intracranial self-stimulation experiment described above in the "Principles of FSCV at a CFM" section of this chapter. This rewarding stimulation was used to train animals to turn in the appropriate direction depending upon whether the right or left somatosensory cortex was activated. The wireless device for electrical stimulation is described in more detail in a later paper [109]. Constant voltage stimulation was controlled by an onboard micro-processor and remotely triggered by a UHF (approximately 420 MHz) transmitter. The device also supported intracranial self-stimulation, with a delay between lever press and pulse train application of only 45 ms.

REAL-TIME ANIMAL TELEMETRY

Instrument Overview

We are developing wireless instrumentation, called Real-time Animal Telemetry or RAT, for both monitoring and controlling neuronal activity in laboratory animals [106]. As described later in "Future Directions," the ultimate goal is a wireless instrument supporting a wide variety of microsensors. The initial strategy was to implement FSCV at a CFM. The rationale is that, because FSCV is technically more challenging than electrophysiology or potentiometry, for example, expanding RAT functionality to include these other sensing technologies at a later time would not be as difficult once FSCV is established. The general design for the RAT instrument supporting FSCV at a CFM is shown in Figure 12.7 [106]. RAT consists of a remote unit, which is affixed to the animal, and a home-base unit, for interfacing with the experimenter. The remote unit contains three components: (1) voltammetry analog front-end, i.e., a miniature potentiostat for controlling electrochemistry; (2) micro-processor, for controlling the potentiostat, data acquisition and telemetry; and (3) transmitter–receiver, for two-way communication with the home-base unit. A desktop PC computer and transmitter–receiver comprise the home-base unit. Front-end software for this unit has provision for initiating data collection, storing data to a file, and data viewing and analysis.

Wireless communication between home-base and remote units is performed by Bluetooth digital telemetry. Bluetooth was selected for several reasons, in particular fidelity, small size and low power requirements. The small size and low power requirements of Bluetooth are essential for developing a wireless device supporting measurements in small laboratory animals such as rodents. Efforts by the IEEE to develop an industry standard based on Bluetooth (802.15) and the success of Bluetooth applications in the marketplace will drive continued development of this wireless technology. Therefore, one additional advantage is that upgrading the wireless capabilities of RAT will be relatively straightforward by incorporating the latest Bluetooth offering. The selection of a digital telemetry to support FSCV at a CFM is also vital, because the voltammogram requires high-resolution acquisition and is very susceptible to transmission artifacts. Indeed, the rapid conversion of an analog signal to digital decreases the chance of degradation in signal because of noise and loss of signal integrity, as further signal conditioning stages are used (e.g., amplifiers, external antenna noise adding to signals, modulation/demodulation, filtering, etc.). In contrast to an analog system, signal integrity can be maintained beyond 10–12 bits with digital telemetry, a necessity for the color plot of FSCV (Figure 12.2 and Figure 12.3) [28].

The transmitter–receivers of the home-base and remote units are identical. This Bluetooth module, a developmental version (ROK 101008/21, Ericsson, Stockholm, Sweden), is accessed via a high-speed serial or USB interface. The former was chosen, because of our expertise and the

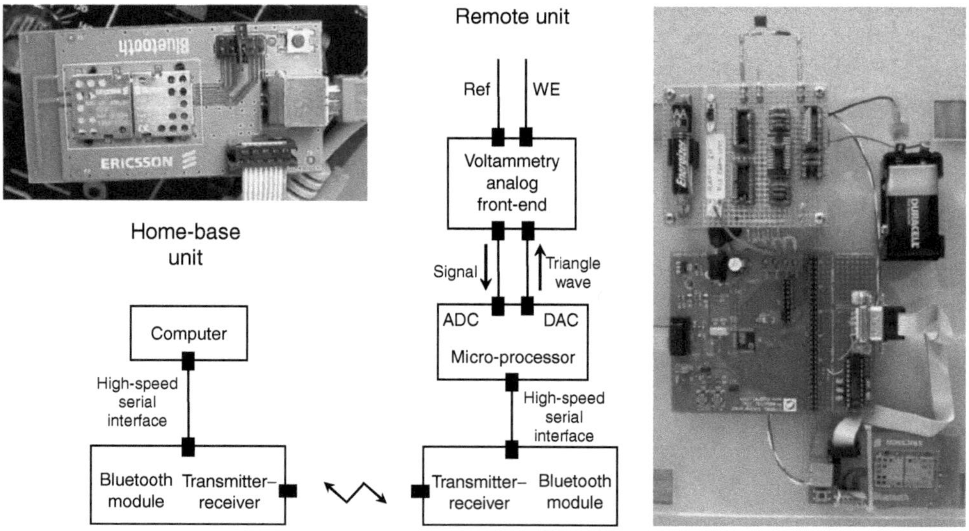

FIGURE 12.7 Prototype RAT. This large prototype was used to establish proof of principle for the wireless transmission of FSCV at a CFM using RAT. Abbreviations: Ref (reference electrode); WE (working electrode or CFM).

serial-communication support built into LabVIEW (National Instruments, Austin, TX, U.S.A.), the platform for the home-base, front-end software. While the USB interface is faster, the data transmission rate (460 Kbps) of the serial interface (232PCI1A, B&B Electronics, Ottawa, IL, U.S.A.) is sufficiently matched to Bluetooth (approximately 700 Kbps) such that additional speed is not necessary. Of course, a USB, PCI or other interface would be required to exploit fully proposed improvements that would increase Bluetooth's transmission rate to 20 Mbps. Whereas we initially developed our own driver, LabVIEW now supports Bluetooth. The left photograph in Figure 12.7 shows the home-base Bluetooth module with a serial connection (lower right). A USB connection (center right) is also shown, but it is only used to provide 5 V for powering the unit.

The micro-processor plays a key role in RAT, by generating the triangle waveform for FSCV via a digital-to-analog converter (DAC), digitizing the microsensor signal via an analog-to-digital converter (ADC), and directing wireless transmission. To perform these functions, we selected the C8051F007DK micro-processor from Silicon Laboratories (Austin, TX, U.S.A.), which features a 12-bit ADC with configurable gain and multiple inputs, two 12-bit DACs, and a high-speed serial interface. Software stored in FLASH memory generates the voltage scan, controls data acquisition, and communicates with the Bluetooth module. An approximately 15 MHz crystal supports digitization at a rate of 100 kS/s. Digitized data, stored temporarily in internal RAM, is wirelessly sent to the home-base unit during the time in between voltage scans. Overall, RAT characteristics for data acquisition are very similar to hardwired systems, which typically digitize at rates between 20 and 100 kS/s at a resolution of either 12 or 16 bits [106].

The analog circuit of the voltammetry front-end contains three amplifiers and is configured such that the reference electrode is held at circuit common potential, while the working electrode (i.e., the CFM) is scanned. The operational amplifier serves as a current-to-voltage converter with gain. One differential amplifier subtracts the voltage scan from the measured signal, and the second differential amplifier is used to generate potentials for the subtraction and reference electrode. The potentiostat circuit is operated in *two-electrode mode* (i.e., a working and a reference electrode), as there is no provision for an auxiliary electrode. However, this mode is made possible by virtue of the small currents recorded at the CFM, thereby minimizing IR voltage errors.

Establishment of Proof of Principle

A prototype RAT instrument was built to establish the wireless transmission of FSCV at a CFM [106]. Although too large to be affixed to a laboratory rat, the size of the prototype expedited construction, modification and testing. The potentiostat electronics are shown at the top of the right picture in Figure 12.7. A dummy cell, a resistor and capacitor circuit mimicking the electrical properties of the CFM, is connected to the working and reference electrode inputs. The printed circuit board in the middle of the picture contains the micro-processor, the component near the center. A serial cable connects the micro-processor board with the Bluetooth transmitter–receiver. The prototype is powered by standard 9.0 V radio and 1.5 V AAA batteries.

Several dry and wet tests were performed on the prototypte RAT instrument [106]. Dummy cell recordings were sent wirelessly for distances up to 16 m with line of sight, in real time, without data loss, and with high fidelity and stability. Wireless measurements of dopamine in the brain of an anesthetized rat during electrical stimulation and collected at a CFM with flow injection analysis, compared favorably to a conventional hardwired voltammetry system. Collectively, these data establish proof of principle for the wireless transmission of FSCV at a CFM. Of considerable interest was that measurements collected by the prototype RAT were largely immune to ambient electrical noise, while measurements with a conventional hardwired system need to be performed inside a grounded Faraday cage to minimize external interference, typically from 60 Hz noise generated from power lines and AC-powered equipment. Because only the analog signal is affected by the ambient noise, we reasoned that the reduced interference was due to digitization of the analog signal near the measurement site. This finding indicates that FSCV at a CFM can be performed outside of the Faraday cage, another advantage of digital telemetry for neuronal monitoring.

Preliminary Findings in Freely Moving Animals

To support measurements in freely moving animals, a miniature RAT instrument was built based on the successful design establishing proof of principle (Greco et al., in preparation). The miniature RAT, shown in Figure 12.8, consists of four components: (1) custom printed circuit board housing the micro-processor and voltammetry circuit, (2) transmitter–receiver, (3) battery, and (4) analog circuit for electrical stimulation. Each module is roughly the same size (approximately 4.0×2.0 cm). When sandwiched together, the height is 1.5 cm with a weight of 17 g. The micro-processor and Bluetooth module are identical to those used in the large prototype RAT. A marked reduction in overall size was achieved by the use of surface mount electronic components, careful design of the printed circuit board, and stripping away unnecessary components on the large prototype.

In contrast to the main board, the stimulus generator was constructed on breadboard with conventional electronic devices. It also uses a separate, isolated power supply. The stimulus generator creates biphasic stimulus pulses, which are applied in between voltage scans to minimize stimulus artifact. The operational amplifier in the circuit is configured for constant-current output. Digital outputs from the micro-processor generate positive and negative phases of the pulses, which are optically isolated. The stimulus generator circuit is passive and can be detached without altering the function of the other components. After detachment, the remaining components of the miniature RAT weigh 10 g.

Tests similar to those described above establishing proof of principle were also performed on the miniature RAT instrument. Overall, the miniature RAT compared favorably with the large prototype RAT and a hardwired voltammetry system (Greco et al., in preparation). Due to a shorter antenna, the transmission distance of approximately 8 m was less than the large prototype, but near the Bluetooth specification of 9 m. As anticipated, the miniature RAT was as insensitive to ambient electrical noise as the large prototype. Figure 12.9 compares the functions of FSCV and electrical stimulation for the miniature RAT and a conventional hardwired system. Panel A shows evoked

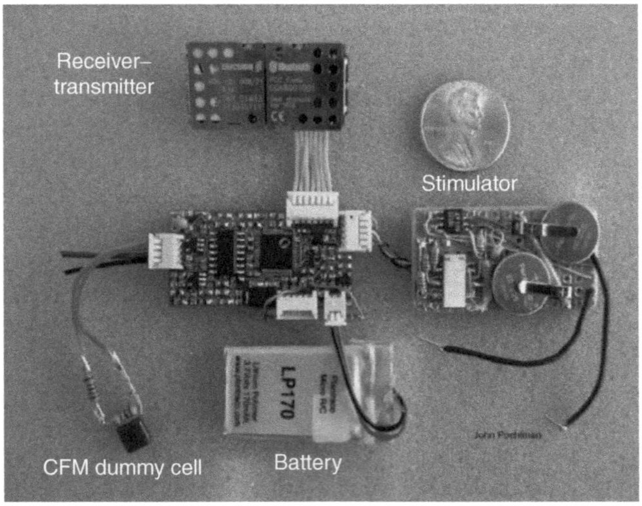

FIGURE 12.8 Miniature RAT. This small RAT device was used for monitoring electrically evoked dopamine in a freely moving rat.

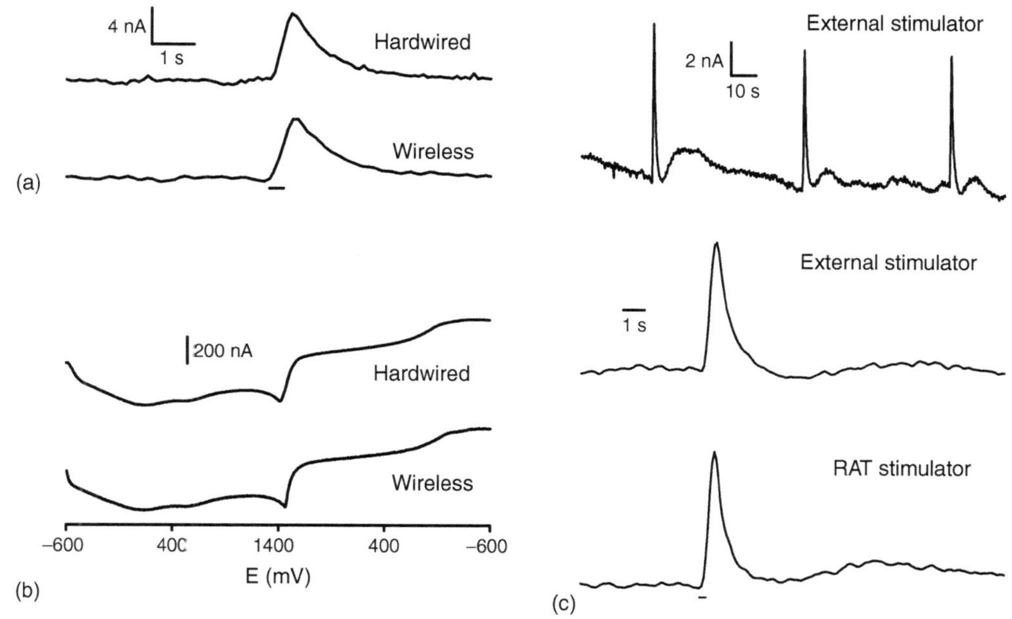

FIGURE 12.9 Comparison of hardwired and wireless systems for FSCV at a CFM and electrical stimulation: (a) dopamine levels were recorded at the same CFM by either the hardwired or wireless system (electrical stimulation was computer controlled and provided by the hardwired system as indicated); (b) background cyclic voltammograms were recorded at the same CFM by either the hardwired or wireless system as indicated; (c) dopamine levels were measured at the same CFM by the miniature RAT, but evoked either by the wireless device or an external stimulator (S88 Grass Stimulator, Grass-Telefactor, West Warwick, RI, U.S.A.) as indicated. All wireless measurements and electrical stimulation were performed by the miniature RAT shown in Figure 12 8. All recordings were collected in a urethane-anesthetized rat. The CFM was implanted in the dorsomedial caudate–putamen, and the stimulating electrode was implanted in the medial forebrain bundle. Electrical stimulation consisted of a 0.4 s, 60 Hz pulse train delivered at 125 μA.

traces recorded in the striatum of an anesthetized rat at the same CFM and collected by either the hardwired system (top) or the miniature RAT (bottom). Electrical stimulation, which elicited the signal increase, was generated by the computer of the hardwired system. Background currents for these recordings are found in Panel B. Overall, responses recorded by the hardwired system and miniature RAT are very similar. Panel C shows recordings collected in another anesthetized rat and at the same CFM by the miniature RAT. In this case, electrical stimulation was provided either by a stand alone unit (i.e., external stimulator) or the stimulus generator module of the miniature RAT. Again, similar responses were recorded by the two stimulation units.

The miniature RAT was also tested in a freely moving rat. The four components of this device were bundled together, wrapped in parafilm® (American National Can, Chicago, IL, U.S.A.), and attached using Velcro® tape to a home-made vest worn by the animal. This vest fit over the front paws, akin to a backpack. Animals were previously habituated to the vest and a mock RAT made with Play-Doh® (Hasbro, Pawtucket, RI, U.S.A.), of comparable size and weight. Figure 12.10 shows an electrically evoked trace recorded in the striatum of an ambulatory rat using the miniature RAT. A 60 Hz, 24 pulse train, applied to the medial forebrain bundle at the arrow, elicits a robust increase in the voltammetric signal. The background subtracted voltammogram (INSET) and color plot (below) identify the signal increase as originating from dopamine. The quality of the signal recorded with the miniature RAT is comparable to a hardwired voltammetry system (e.g., see Figure 12.4). Taken together, these results support the utility of RAT for the wireless monitoring of FSCV at a CFM and control of electrical stimulation in freely moving animals.

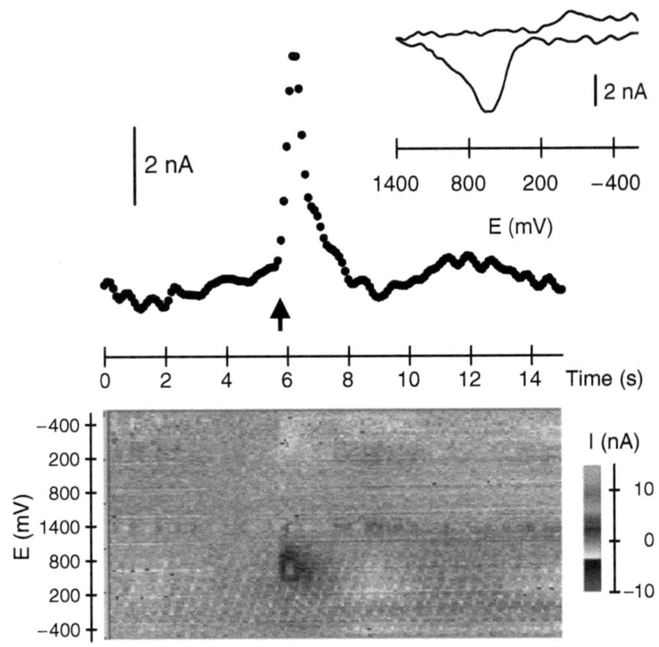

FIGURE 12.10 **(See color insert following page 272.)** Wireless measurement of electrically evoked dopamine levels in a freely moving rat. The recording was collected by the miniature RAT shown in Figure 12.8. The CFM was implanted in the dorsomedial caudate–putamen, and the stimulating electrode was implanted in the medial forebrain bundle.

FUTURE DIRECTIONS

The Smaller the Better

While the miniature RAT is manageable for an adult laboratory rat, a reduction in size will be beneficial by making the instrument less intrusive to the animal. Reduced size is additionally necessary for applying RAT to smaller rodents, such as the hamster and mouse. Hamsters are well suited models for studies on circadian rhythms [110], aggression [111], and sexual behavior [112]. Development of a wireless instrument for use in mice would be significant, because of the capability for generating transgenic animals. Indeed, transgenic mice have revolutionized biomedical research, making the mouse the model system of choice in several fields of neuroscience [113]. Currently, all electronic components on the main board and the Bluetooth module are now available in a profile approximately half the size. In addition, the newer Bluetooth devices have lower power requirements. Because the transmitter–receiver is the most power hungry of the electronic components on RAT, the size of the battery can be reduced as well. Another space saving modification is moving the stimulus generator circuit to the main board and using surface mount components for its fabrication. Thus, further miniaturizing RAT is imminently possible, even without a major change in engineering strategy.

Multifunctional Wireless Instrument

The ultimate goal of RAT is a multifunctional wireless instrument accommodating several technologies for neuronal monitoring and control. A module instrument is one strategy for realizing this goal. In such a design, the main module would consist of the micro-processor and Bluetooth telemetry. Plugged into the main module, and depending upon the application, is second a module supporting a microsensor(s). This design is based on the notion that, regardless of the sensing technology, the measured signal is eventually converted into voltage and digitized. Front-end software for the home-base unit would also be developed for conditioning, viewing and analysis of each data type. In addition to the stimulus generator of the miniature RAT, it may also be possible to develop an iontophoresis module for focal delivery of drugs. Iontophoresis is routinely coupled with electrophysiology [114,115] and to a lesser extent, voltammetry (see below) [69,116].

As a first step to a multifunctional wireless instrument, we have already built, on the main board of the miniature RAT, a voltage-follower circuit with gain and filtering for extracellular recording of unit activity (Figure 12.8). The two black wires just above the dummy cell are the leads, and electrophysiological signals could be digitized by a free ADC channel on the micro-processor. In this manner, the miniature RAT could be used to measure voltammetry and electrophysiology simultaneously at separate microelectrodes. Additional switching circuitry between the voltammetry and electrophysiology circuits would support quasi-simultaneous monitoring of chemical and bioelectrical signals at the same CFM [68,69].

The impedance of the voltage-follower circuit is also sufficiently high to accommodate ion selective microelectrodes. Software changes for incorporating the latter sensing technology are straightforward to implement. Because changes in extracellular ion concentrations are relatively slow, a measurement would only need to be collected once every voltammetric scan (i.e., 10 Hz or every 100 ms). The data packet format used to send the voltammogram wirelessly easily accommodates this additional datum, thus affording simultaneous voltammetric and potentiometric measurements at separate microelectrodes.

Portable Microsensor Measurement System

In addition to measurements in freely moving animals, RAT will uniquely support other types of applications by functioning as a portable microsensor measurement system (PMMS). Ideally, the so called PMMS is a mobile instrument that is readily moved from place to place for operation. In this capacity, RAT confers at least four distinct advantages. First, the small size of the instrument permits convenient incorporation into existing, instrument-crowded set-ups. Second, RAT comes with its own data acquisition system. Thus, beyond the home-base Bluetooth module and serial card, no additional hardware is required. Our experience also indicates that the serial card does not interfere with the operation of standard data acquisition and control boards interfacing to the PCI-bus of the same computer, suggesting that RAT is compatible with most set-ups. If not, RAT can be simply run by a separate computer. The latter alternative, moreover, highlights another advantage of RAT: wireless communication allows the home-base unit to be situated several feet from the remote unit, further supporting convenient incorporation into existing set-ups. In the age of massive tangles of cables, even for home computer operations but especially in the laboratory, wireless connections are most welcome. Third, the relative insensitivity to ambient electrical noise permits applications of RAT in experiments that do not utilize a Faraday cage or are difficult to perform in a Faraday cage. And fourth, some applications, perhaps due to harsh reaction conditions, may require physically separating the sensor from other components of the set-up, and wireless communication may be the best way to achieve this separation.

Although there are several potential applications for RAT as a PMMS, two are discussed here. Figure 12.11a shows an electrophysiological record of extracellular unit activity in the striatum of the anesthetized rat (Sandberg et al. 2005). These multiple units were activated by glutamate and then inhibited by dopamine, both administered by iontophoresis. Because the units were recorded by a CFM (e.g., see Figure 12.4), it was possible to measure the iontophoresed dopamine profile subsequent to the electrophysiological record using RAT (Figure 12.11b). Such a determination is critical for calibrating iontophoresis, as the same applied current may yield different dopamine concentrations depending upon pipette fabrication. In this example, RAT was incorporated into a very crowded electrophysiology set-up containing various instruments, including an amplifier,

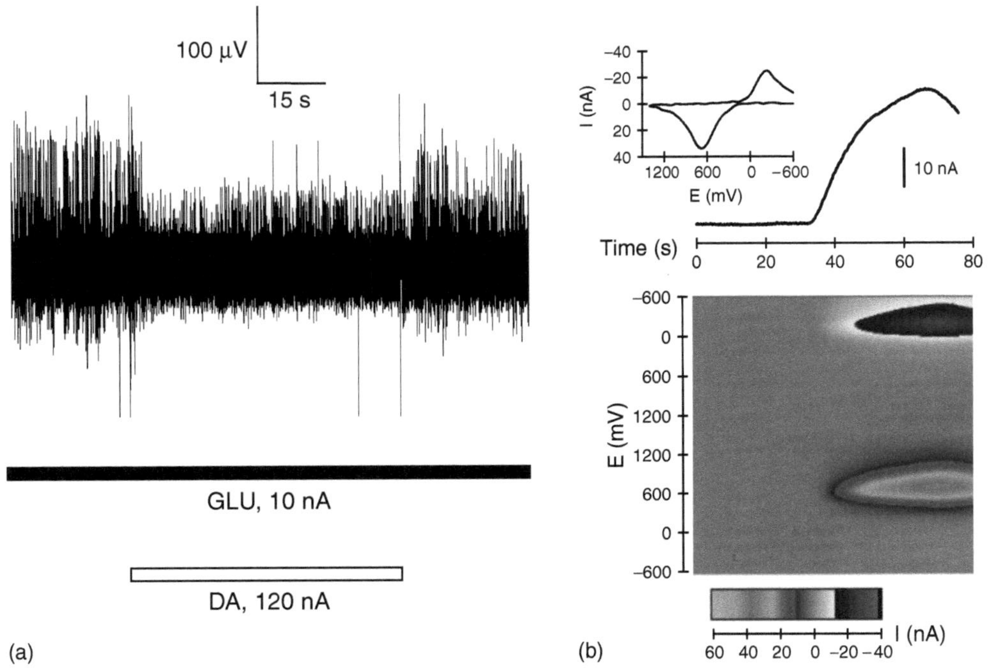

FIGURE 12.11 (**See color insert following page 272.**) Electrophysiological and voltammetric measurements at the same CFM. The CFM was implanted in the dorsomedial caudate–putamen of the urethane-anesthetized rat. Glutamate and dopamine were administered by iontophoresis. Electrophysiology was recorded by a standard hardwired set-up (negative voltage up), and dopamine was monitored by FSCV using RAT.

a filter, an iontophoresis unit, a stimulus generator, and an oscilloscope. RAT signals were also collected by a separate computer, for the simple reason that installing the home-base serial card in the computer interfaced to the electrophysiology set-up was inconvenient. Of course, an alternative and more elegant approach is quasi-simultaneous electrophysiology and voltammetry as discussed above. However, this approach requires special hardware and software and may not be absolutely necessary for all experiments.

A second application of RAT as a PMMS is shown in Figure 12.12. This figure shows a RAT measurement of dopamine in the striatum of an anesthetized rat during cerebral ischemia [117]. Focal ischemia was caused by the intraluminal thread model of middle cerebral artery occlusion [118]. In this model, a fine nylon thread is inserted into the internal carotid artery and advanced until the middle cerebral artery, which feeds the striatum and adjacent cortex, is unilaterally occluded. In agreement with previous work [119,120], ischemia elicited a massive increase in extracellular dopamine. The advantage of FSCV in these ischemia experiments is the better temporal resolution compared to the microdialysis and voltammetry approaches used earlier.

As a PMMS, the ischemia example highlights another advantage of RAT: measurements outside of a Faraday cage. Indeed, the ischemia experiment requires a microsurgical scope and accessory instruments, e.g., for blood pressure and heart rate monitoring and temperature control, that are difficult to fit inside a Faraday cage. Moreover, the rat lies in a supine position during experimental ischemia, but pronely for voltammetric measurements in a stereotaxic apparatus. To overcome this latter problem, the animal was first surgically prepared for dopamine monitoring under freely moving conditions (see Figure 12.3 and Figure 12.10). For the subsequent ischemia experiment, the rat, with an implanted CFM and stimulating electrode, was then placed supinely on a raised platform with a cutout portion through which the hardware mounted on the head could

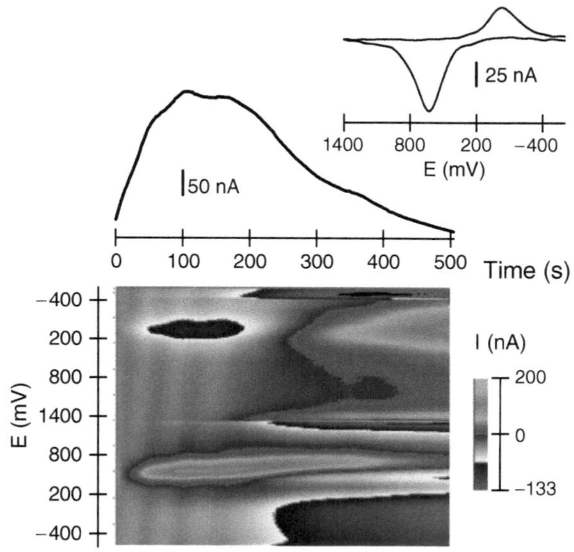

FIGURE 12.12 **(See color insert following page 272.)** Voltammetric measurement of dopamine during focal cerebral ischemia. The CFM was implanted in the dorsomedial caudate–putamen of the isoflurane-anesthetized rat. The middle cerebral artery was occluded for 20 min, beginning approximately 120 s prior to the onset of the voltammetric recording.

safely pass. The implanted electrodes were then connected to RAT, which transmitted data to the home-base unit located several feet away from the crowded ischemia set-up.

CONCLUSIONS

We describe a new instrument called RAT, developed for wireless neural monitoring and control. Presently, RAT supports FSCV at a CFM and electrical stimulation, and is suitably sized for adult rats. Extending the functionality of the existing RAT to electrophysiology and potentiometry should be straightforward. Further miniaturization of RAT for use in smaller rodents such as hamsters and mice should be possible with available smaller components and without major re-engineering. In addition to experiments with freely behaving animals, RAT also has utility as a PMMS.

ACKNOWLEDGMENTS

This research supported by NSF (DBI-0138011) and DOD (USAMRMC 03281055).

REFERENCES

1. Justice, J. B. J., Quantitative microdialysis of neurotransmitter, *Journal of Neuroscience Methods*, 48 (3), 263–276, 1993.
2. Plock, N. and Kloft, C., Microdialysis—theoretical background and recent implementation in applied life-sciences, *European Journal of Pharmaceutical Sciences*, 25 (1), 1–24, 2005.
3. Parrot, S., Bert, L., Mouly-Badina, L., Sauvinet, V., Colussi-Mas, J., Lambas-Senas, L., Robert, F. et al., Microdialysis monitoring of catecholamines and excitatory amino acids in the rat and mouse brain: recent developments based on capillary electrophoresis with laser-induced fluorescence detection—a mini-review, *Cell and Molecular Neurobiology*, 23 (4–5), 793–804, 2003.

4. Sykova E., Ion-selective microelectrodes, In *Monitoring Neuronal Activity: the Pratical Approach*, Stamford J., Ed., Oxford University Press, Oxford, pp. 261–282, 1992.
5. Wightman, R. M., Brown, D. S., Kuhr, W. G., and Wilson, R. L., Molecular specificity of *in vivo* eletrochemical measurements, In *Voltammetry in the Neurosciences*, Justice, J.B., Ed., Human Press Clifton, NJ, pp. 103–138, 1987.
6. Justice, J.B., Introduction to *in vivo* voltammetry, In *Voltammetry in the Neurosciences*, Justice, J.B., Ed., Human Press Clifton, NJ, pp. 3–101, 1987.
7. Adams, R. N., *In vivo* electrochemical measurements in the CNS, *Progress in Neurobiology*, 35 (4), 297–311, 1990.
8. Marsden, C. A., Joseph, M. H., Kruk, Z. L., Maidment, N. T., O'Neill, R. D., Schenk, J. O., and Stamford, J. A., *In vivo* voltammetry—present electrodes and methods, *Neuroscience*, 25 (2), 389–400, 1988.
9. Bjorklund, A., and Lindvall, O., Dopamine-containing systems in the CNS, Bjorklund, A. and Hokfelt, T., Eds., Elsevier, New York, pp. 55–122, 1984.
10. Le Moal, M. and Simon, H., Mesocorticolimbic dopaminergic network: functional and regulatory roles, *Physiological Reviews*, 71 (1), 155–234, 1991.
11. Ben-Jonathan, N., Dopamine: a prolactin-inhibiting hormone, *Endocrine Reviews*, 6 (4), 564–589, 1985.
12. Kissinger, P. T., Hart, J. B., and Adams, R. N., Voltammetry in brain tissue—a new neurophysiological measurement, *Brain Research*, 55 (1), 209–213, 1973.
13. Adams, R. N., Probing brain chemistry with electroanalytical techniques, *Analytical Chemistry*, 48 (14), 1126A–1138A, 1976.
14. Blaha, C. D. and Phillips, A. G., A critical assessment of electrochemical procedures applied to the measurement of dopamine and its metabolites during durg-induced and species-typical behaviours *Behavioural Pharmacology*, 7, 675–708, 1996.
15. Joseph, M. H., Can voltammetric electrodes measure brain extracellular dopamine in the behaving animal?, *Behavioural Pharmacology*, 7 (7), 709–713, 1996.
16. Wightman, R. M. and Robinson, D. L., Transient changes in mesolimbic dopamine and their association with 'reward', *Journal of Neurochemistry*, 82 (4), 721–735, 2002.
17. Stamford, J. A., *In vivo* voltammetry: some methodological considerations, *Journal of Neuroscience Methods*, 17 (1), 1–29, 1986.
18. Gonzalez-Mora, J. L., Guadalupe, T., Fumero, B., and Mas, M., Mathematical resolution of mixed *in vivo* voltammetry signals, *Journal of Neuroscience Methods*, 39 (3), 231–244, 1991.
19. Pierce, R. C. and Rebec, G. V., Dopamine-, NMDA- and sigma-receptor antagonists exert differential effects on basal and amphetamine-induced changes in neostriatal ascorbate and DOPAC in awake, behaving rats, *Brain Research*, 579 (1), 59–66, 1992.
20. Brose, N., O'Neill, R. D., Boutelle, M. G., Anderson, S. M., and Fillenz, M., Effects of an anxiogenic benzodiazepine receptor ligand on motor activity and dopamine release in nucleus accumbens and striatum in the rat, *Journal of Neuroscience*, 7 (9), 2917–2926, 1987.
21. Gerhardt, G. A., Rapid chronocoulometric measurements of norepinephrine overflow and clearance in CNS tissues, In *Neuromethods: Voltammetric Methods in Brain Systems,* Boulton, A., Baker, G., and Adams, R. N., Eds., Humana Press Inc., Totowa, NJ, pp. 117–151, 1995, Chap. 27.
22. Garris, P. A. and Wightman, R. M., Regional differences in dopamine release, uptake, and diffusion measured by fast-scan cyclic voltammetry, In *Neuromethods: Voltammetric Methods in Brain Systems,* Boulton, A., Baker, G., and Adams, R. N., Eds., Humana Press Inc., Totowa, NJ, pp. 179–220, 1995, Chap. 27.
23. Rice, M. E. and Nicholson, C., Diffusion and ion shifts in the brain extracellular microenvironment and their relevance for voltammetric measurements, In *Neuromethods: Voltammetric Methods in Brain Systems,* Boulton, A., Baker, G., and Adams, R. N., Eds., Humana Press Inc., Totowa, NJ, pp. 27–81, 1995, Chap. 27.
24. Stamford, J. A., Fast cyclic voltammetry in brain slices, In *Neuromethods: Voltammetric Methods in Brain Systems,* Boulton, A., Baker, G., and Adams, R. N., Eds., Humana Press Inc., Totowa, NJ, pp. 179–220, 1995, Chap. 27.
25. Gonon, F., Monitoring dopamine and noradrenaline release in central and peripheral nervous systems with treated and untreated carbon-fiber electrodes, In: *Neuromethods: Voltammetric Methods in Brain Systems,* Boulton, A., Baker, G., and Adams, R. N., Eds., Humana Press Inc., Totowa, NJ, pp. 153–177, 1995, Chap. 27.

26. Benoit-Marand, M., Borrelli, E., and Gonon, F., Inhibition of dopamine release via presynaptic D2 receptors: time course and functional characteristics *in vivo*, *Journal of Neuroscience*, 21 (23), 9134–9141, 2001.
27. Kawagoe, K. T., Zimmerman, J. B., and Wightman, R. M., Principles of voltammetry and microelectrode surface states, *Journal of Neuroscience Methods*, 48 (3), 225–240, 1993.
28. Michael, D., Travis, E. R., and Wightman, R. M., Color images for fast-scan CV measurements in biological systems, *Analytical Chemistry*, 70 (17), 586A–592A, 1998.
29. Armstrong-James, M. and Millar, J., Carbon fibre microelectrodes, *Journal of Neuroscience Methods*, 1 (3), 279–287, 1979.
30. Armstrong-James, M., Fox, K., Kruk, Z. L., and Millar, J., Quantitative ionophoresis of catecholamines using multibarrel carbon fibre microelectrodes, *Journal of Neuroscience Methods*, 4 (4), 385–406, 1981.
31. Millar, J., Armstrong-James, M., and Kruk, Z. L., Polarographic assay of iontophoretically applied dopamine and low-noise unit recording using a multibarrel carbon fibre microelectrode, *Brain Research*, 205 (2), 419–424, 1981.
32. Millar, J., Stamford, J. A., Kruk, Z. L., and Wightman, R. M., Electrochemical, pharmacological and electrophysiological evidence of rapid dopamine release and removal in the rat caudate nucleus following electrical stimulation of the median forebrain bundle, *European Journal of Pharmacology*, 109 (3), 341–348, 1985.
33. Wightman, R. M., Amatore, C., Engstrom, R. C., Hale, P. D., Kristensen, E. W., Kuhr, W. G., and May, L., Real-time characterization of dopamine overflow and uptake in the rat striatum, *Journal of Neuroscience*, 25 (2), 513–523, 1988.
34. Peters, J. L. and Michael, A. C., Changes in the kinetics of dopamine release and uptake have differential effects on the spatial distribution of extracellular dopamine concentration in rat striatum, *Journal of Neurochemistry*, 74 (4), 1563–1573, 2000.
35. Schmitz, Y., Lee, C. J., Schmauss, C., Gonon, F., and Sulzer, D., Amphetamine distorts stimulation-dependent dopamine overflow: effects on D2 autoreceptors, transporters, and synaptic vesicle stores, *Journal of Neuroscience*, 21 (16), 5916–5924, 2001.
36. Jones, S. R., Gainetdinov, R. R., Jaber, M., Giros, B., Wightman, R. M., and Caron, M. G., Profound neuronal plasticity in response to inactivation of the dopamine transporter, *Proceedings of the National Academy of Sciences of the United States of America*, 95 (7), 4029–4034, 1998.
37. Phillips, P. E., Hancock, P. J., and Stamford, J. A., Time window of autoreceptor-mediated inhibition of limbic and striatal dopamine release, *Synapse*, 44 (1), 15–22, 2002.
38. Borland, L. M. and Michael, A. C., Voltammetric study of the control of striatal dopamine release by glutamate, *Journal of Neurochemistry*, 91 (1), 220–229, 2004.
39. Kulagina, N. V., Zigmond, M. J., and Michael, A. C., Glutamate regulates the spontaneous and evoked release of dopamine in the rat striatum, *Neuroscience*, 102 (1), 121–128, 2001.
40. Heien, M. L., Johnson, M. A., and Wightman, R. M., Resolving neurotransmitters detected by fast-scan cyclic voltammetry, *Analytical Chemistry*, 76 (19), 5697–5704, 2004.
41. Heien, M. L., Khan, A. S., Ariansen, J. L., Cheer, J. F., Phillips, P. E., Wassum, K. M., and Wightman, R. M., Real-time measurement of dopamine fluctuations after cocaine in the brain of behaving rats, *Proceedings of the National Academy of Sciences of the United States of America*, 102 (29), 10023–10028, 2005.
42. Garris, P. A., Christensen, J. R., Rebec, G. V., and Wightman, R. M., Real-time measurement of electrically evoked extracellular dopamine in the striatum of freely moving rats, *Journal of Neurochemistry*, 68 (1), 152–161, 1997.
43. Budygin, E. A., Kilpatrick, M. R., Gainetdinov, R. R., and Wightman, R. M., Correlation between behavior and extracellular dopamine levels in rat striatum: comparison of microdialysis and fast-scan cyclic voltammetry, *Neuroscience Letters*, 281 (1), 9–12, 2000.
44. Budygin, E. A., Phillips, P. E., Robinson, D. L., Kennedy, A. P., Gainetdinov, R. R., and Wightman, R. M., Effect of acute ethanol on striatal dopamine neurotransmission in ambulatory rats, *Journal of Pharmacology and Experimental Therapeutics*, 297 (1), 27–34, 2001.
45. Garris, P. A., Budygin, E. A., Phillips, P. E., Venton, B. J., Robinson, D. L., Bergstrom, B. P., Rebec, G. V., and Wightman, R. M., A role for presynaptic mechanisms in the actions of nomifensine and haloperidol, *Neuroscience*, 118 (3), 819–829, 2003.

46. Garris, P. A., Kilpatrick, M., Bunin, M. A., Michael, D., Walker, Q. D., and Wightman, R. M., Dissociation of dopamine release in the nucleus accumbens from intracranial self-stimulation, *Nature*, 398 (6722), 67–69, 1999.
47. Kilpatrick, M. R., Rooney, M. B., Michael, D. J., and Wightman, R. M., Extracellular dopamine dynamics in rat caudate-putamen during experimenter-delivered and intracranial self-stimulation, *Neuroscience*, 96 (4), 697–706, 2000.
48. Kruk, Z. L., Cheeta, S., Milla, J., Muscat, R., Williams, J. E., and Willner, P., Real time measurement of stimulated dopamine release in the conscious rat using fast cyclic voltammetry: dopamine release is not observed during intracranial self stimulation, *Journal of Neuroscience Methods*, 79 (1), 9–1?, 1998.
49. Yavich, L. and Tiihonen, J., *In vivo* voltammetry with removable carbon fibre electrodes in freely-moving mice: dopamine release during intracranial self-stimulation, *Journal of Neuroscience Methods*, 104 (1), 55–63, 2000.
50. Yavich, L. and Tiihonen, J., Patterns of dopamine overflow in mouse nucleus accumbens during intracranial self-stimulation, *Neuroscience Letters*, 293 (1), 41–44, 2000.
51. Rebec, G. V., Christensen, J. R., Guerra, C., and Bardo, M. T., Regional and temporal differences in real-time dopamine efflux in the nucleus accumbens during free-choice novelty, *Brain Research*, 776 (1–2), 61–67, 1997.
52. Michael, D. J., Joseph, J. D., Kilpatrick, M. R., Travis, E. R., and Wightman, R. M., Improving data acquisition for fast-scan cyclic voltammetry, *Analytical Chemistry*, 71 (18), 3941–3947, 1999.
53. Cahill, P. S., Walker, Q. D., Finnegan, J. M., Mickelson, G. E., Travis, E. R., and Wightman, R. M., Microelectrodes for the measurement of catecholamines in biological systems, *Analytical Chemistry*, 68 (18), 3180–3186, 1996.
54. Rice, M. E. and Nicholson, C., Measurement of nanomolar dopamine diffusion using low-noise perfluorinated ionomer coated carbon fiber microelectrodes and high-speed cyclic voltammetry, *Analytical Chemistry*, 61 (17), 1805–1810, 1989.
55. Wiedemann, D. J., Kawagoe, K. T., Kennedy, R. T., Ciolkowski, E. L., and Wightman, R. M., Strategies for low detection limit measurements with cyclic voltammetry, *Analytical Chemistry*, 63 (24), 2965–2970, 1991.
56. Heien, M. L., Phillips, P. E., Stuber, G. D., Seipel, A. T., and Wightman, R. M., Overoxidation of carbon-fiber microelectrodes enhances dopamine adsorption and increases sensitivity, *Analyst*, 128 (12), 1413–1419, 2003.
57. Overton, P. G. and Clark, D., Burst firing in midbrain dopaminergic neurons, *Brain Research. Brain Research Reviews*, 25 (3), 312–334, 1997.
58. Schultz, W., Predictive reward signal of dopamine neurons, *Journal of Neurophysiology*, 80 (1), 1–27, 1998.
59. Robinson, D. L., Phillips, P. E., Budygin, E. A., Trafton, B. J., Garris, P. A., and Wightman, R. M., Sub-second changes in accumbal dopamine during sexual behavior in male rats, *Neuroreport*, 12 (11), 2549–2552, 2001.
60. Robinson, D. L., Heien, M. L., and Wightman, R. M., Frequency of dopamine concentration transients increases in dorsal and ventral striatum of male rats during introduction of conspecifics, *Journal of Neuroscience*, 22 (23), 10477–10486, 2002.
61. Roitman, M. F., Stuber, G. D., Phillips, P. E., Wightman, R. M., and Carelli, R. M., Dopamine operates as a subsecond modulator of food seeking, *Neuroscience*, 24 (6), 1265–1271, 2004.
62. Stuber, G. D., Roitman, M. F., Phillips, P. E., Carelli, R. M., and Wightman, R. M., Rapid dopamine signaling in the nucleus accumbens during contingent and noncontingent cocaine administration, *Journal Neuropsychopharmacology*, 30 (5), 853–863, 2005.
63. Phillips, P. E., Stuber, G. D., Heien, M. L., Wightman, R. M., and Carelli, R. M., Subsecond dopamine release promotes cocaine seeking, *Nature*, 422 (6932), 614–618, 2003.
64. Venton, B. J., Michael, D. J., and Wightman, R. M., Correlation of local changes in extracellular oxygen and pH that accompany dopaminergic terminal activity in the rat caudate-putamen, *Journal of Neurochemistry*, 84 (2), 373–381, 2003.
65. Jones, S. R., Mickelson, G. E., Collins, L. B., Kawagoe, K. T., and Wightman, R. M., Interference by pH and Ca^{2+} ions during measurements of catecholamine release in slices of rat amygdala with fast-scan cyclic voltammetry, *Journal of Neuroscience Methods*, 52 (1), 1–10, 1994.

66. Carelli, R. M. and Wightman, R. M., Functional microcircuitry in the accumbens underlying drug addiction: insights from real-time signaling during behavior, *Current Opinion in Neurobiology*, 14 (6), 763–768, 2004.
67. Sandberg, G., Heidenreich, B. A., Garris, P. A., A methodology for *in vivo* detection of the effects of dopamine transients on glutamate-evoked striatal activity using electrophysiology, fast-scan cyclic voltammetric, and microiontophoretic techniques, Program No. 568.3. *2005 Abstract Viewer/Itinerary Planner*. Washington, DC: Society for Neuroscience, 2005. Online.
68. Williams, G. V. and Millar, J., Concentration-dependent actions of stimulated dopamine release on neuronal activity in rat striatum, *Neuroscience*, 39 (1), 1–16, 1990.
69. Williams, G. V. and Millar, J., Differential actions of endogenous and iontophoretic dopamine in rat striatum, *European Journal of Neuroscience*, 2 (7), 658–661, 1990.
70. Cheer, J. F., Heien, M. L., Garris, P. A., Carelli, R. M., and Wightman, R. M., Simultaneous electrochemical and single-unit recordings in the nucleus accumbens reveal GABA-mediated responses: implications for intracranial self-stimulation, *Proc. Natl. Acad. Sci. U.S.A.*, 102 (52), 19150–19155, 2005.
71. Weisman, C. J., *The Essential Guide to RF and Wireless*, Prentice Hall, Upper Saddle River, NJ, 2002.
72. Dornan, A., *The Essential Guide to Wireless Communications Applications*, Prentice Hall, Upper Saddle River, NJ, 2002.
73. Randall, N. and Sosinksy, B., *PC Magazine Wireless Solutions*, Wiley Publishing, Indianapolis, IN, 2005.
74. Patterson, M. M., Instrumentation and techniques in physiological psychology, *American Psychologist*, 30 (3), 249–256, 1975.
75. Robinson, B. W., Brain telestimulation in primates, *American Psychologist*, 24 (3), 248–250, 1969.
76. Sperry, C. J., Bach, L. M., Happel, L. T., and Cabaniss, J. S., Implantable stimulator and transmitter for telemetry of evoked potentials during defensive behavior, *Biomedical Sciences Instrumentation*, 4, 119–124, 1968.
77. Upson, R. B., King, F. A., and Roberts, L., An improved remote brain stimulator and EEG transmitter for small animals, *Biomedical Sciences Instrumentation*, 4, 164–170, 1968.
78. Rollins, D. L., Killingsworth, C. R., Walcott, G. P., Justice, R. K., Ideker, R. E., and Smith, W. M., A telemetry system for the study of spontaneous cardiac arrhythmias, *IEEE Transactions on Biomedical Engineering*, 47 (7), 887–892, 2000.
79. Harada, E., Yonezawa, Y., Caldwell, W. M., and Hahn, A. W., A microcontroller-based telemetry system for sympathetic nerve activity and ECG measurement, *Biomedical Sciences Instrumentation*, 35, 63–67, 1999.
80. Drewes, A. M., Andreasen, A., Assentoft, J. E., and Nagel, O., Seven-channel digital telemetry system for monitoring and direct computer capturing of biological data, *Journal of Biomedical Engineering*, 15 (5), 435–440, 1993.
81. Sundstrom, L. E., Sundstrom, K. E., and Mellanby, J. H., A new protocol for the transmission of physiological signals by digital telemetry, *Journal of Neuroscience Methods*, 77 (1), 55–60, 1997.
82. Jasemian, Y. and Arendt-Nielsen, L., Evaluation of a realtime, remote monitoring telemedicine system using the Bluetooth protocol and a mobile phone network, *Journal of Telemedicine and Telecare*, 11 (5), 256–260, 2005.
83. Jasemian, Y. and Nielsen, L. A., Design and implementation of a telemedicine system using Bluetooth protocol and GSM/GPRS network, for real time remote patient monitoring, *Technology and Health Care*, 13 (3), 199–219, 2005.
84. Pandolfino, J. E., Schreiner, M. A., Lee, T. J., Zhang, Q., Boniquit, C., and Kahrilas, P. J., Comparison of the Bravo wireless and Digitrapper catheter-based pH monitoring systems for measuring esophageal acid exposure, *American Journal of Gastroenterology*, 100 (7), 1466–1476, 2005.
85. Arshak, A., Arshak, K., Waldron, D., Morris, D., Korostynska, O., Jafer, E., and Lyons, G., Review of the potential of a wireless MEMS and TFT microsystems for the measurement of pressure in the GI tract, *Medical Engineering and Physics*, 27 (5), 347–356, 2005.
86. Eichenbaum, H., Pettijohn, D., Deluca, A. M., and Chorover, S. L., Compact miniature microelectrode-telemetry system, *Physiology and Behavior*, 18 (6), 1175–1178, 1977.

87. Pinkwart, C. and Borchers, H. W., Miniature three-function transmitting system for single neuron recording, wireless brain stimulation and marking, *Journal of Neuroscience Methods*, 20 (4), 341–352, 1987.
88. O'Byrne, K. T., Thalabard, J. C., Grosser, P. M., Wilson, R. C., Williams, C. L., Chen, M. D., Ladendorf, D., Hotchkiss, J., and Knobil, E., Radiotelemetric monitoring of hypothalamic gonadotropin-releasing hormone pulse generator activity throughout the menstrual cycle of the rhesus monkey, *Endocrinology*, 129 (3), 1207–1214, 1991.
89. Nieder, A. and Klump, G. M., Time course of simultaneous masking in the starling's auditory forebrain, *Experimental Brain Research*, 124 (3), 311–320, 1999.
90. Nieder, A., Miniature stereo radio transmitter for simultaneous recording of multiple single-neuron signals from behaving owls, *Journal of Neuroscience Methods*, 101 (2), 157–164, 2000.
91. Carelli, R. M., Ijames, S., Konstantopoulos, J., and Deadwyler, S. A., Examination of factors mediating the transition to behaviorally correlated nucleus accumbens cell firing during cocaine self-administration sessions in rats, *Behavioural Brain Research*, 104 (1–2), 127–139, 1999.
92. Chang, J. Y., Janak, P. H., and Woodward, D. J., Neuronal and behavioral correlations in the medial prefrontal cortex and nucleus accumbens during cocaine self-administration by rats, *Neuroscience*, 99 (3), 433–443, 2000.
93. Nicolelis, M. A., Dimitrov, D., Carmena, J. M., Crist, R., Lehew, G., Kralik, J. D., and Wise, S. P., Chronic, multisite, multielectrode recordings in macaque monkeys, *Proceedings of the National Academy of Sciences of the United States of America*, 100 (19), 11041–11046, 2003.
94. Gulley, J. M., Kosobud, A. E., and Rebec, G. V., Behavior-related modulation of substantia nigra pars reticulata neurons in rats performing a conditioned reinforcement task, *Neuroscience*, 111 (2) 337–349, 2002.
95. Obeid, I., Nicolelis, M. A. L., and Wolf, P. D., A multichannel neural telmetry system, Program No. 405.17. *2002 Abstract Viewer/Itinerary Planner*. Washington, DC: Society for Neuroscience, 2002. Online.
96. Morizio, J., Vinson, G., Jack, P., Steven, G., Steven, M., Lucas, S., Gary, L., and Miguel, N., Fifteen-channel wireless headstage system for singl-unit rat recordings, Program No. 68.4. *2005 Abstract Viewer/Itinerary Planner*. Washington, DC: Society for Neuroscience, 2005. Online.
97. Harrison, R. R., Watkins, P. T., Kier, R. J., Lovejoy, R. O., Black, D. J., Solzbacher, F., and Normann, R. A., A low-power microchip for a wireless 100-electrode neural recording system, Program No. 707.7. *2005 Abstract Viewer/Itinerary Planner*. Washington, DC: Society for Neuroscience, 2005. Online.
98. Imeri, L., De Simoni, M. G., Giglio, R., Clavenna, A., and Mancia, M., Changes in the serotonergic system during the sleep-wake cycle: simultaneous polygraphic and voltammetric recordings in hypothalamus using a telemetry system, *Neuroscience*, 58 (2), 353–358, 1994.
99. Imeri, L., Gemma, C., De Simoni, M. G., Opp, M. R., and Mancia, M., Hypothalamic serotonergic activity correlates better with brain temperature than with sleep-wake cycle and muscle tone in rats, *Neuroscience*, 89 (4), 1241–1246, 1999.
100. De Simoni, M. G., De Luigi, A., Imeri, L., and Algeri, S., Miniaturized optoelectronic system for telemetry of *in vivo* voltammetric signals, *Journal of Neuroscience Methods*, 33 (2–3), 233–240, 1990.
101. Annovazzi-Lodi, V. and Donati, S., An optoelectronic interconnection for bidirectional transmission of biological signals, *IEEE Transactions Biomedical Engineering*, 35, 595–606, 1988.
102. Crespi, F., Dalessandro, D., Annovazzi-Lodi, V., Heidbreder, C., and Norgia, M., *In vivo* voltammetry: from wire to wireless measurements, *Journal of Neuroscience Methods*, 140 (1–2), 153–161, 2004.
103. Johnson, D. A., Aillon, D. V., Johnson, D. A., Palamarchouk, V., and Smagin, G., *In vitro* study of a new carbon fiber microelectroe for real-time measurement of dopamine with digital wireless transmisison, Program No. 453.20. *2005 Abstract Viewer/Itinerary Planner*. Washington, DC: Society for Neuroscience, 2005. Online.
104. Palamarchouk, V. S, Song, D., Aillon, D., Johnson, D. A., Johnson, D. Goeders, N. E., and Smagin, G. N., *In vivo* real time measurement of dopamine with digital wireless data transmission: application to studies in behaving animals, Program No. 453.21. *2005 Abstract Viewer/Itinerary Planner*. Washington, DC: Society for Neuroscience, 2005. Online.

105. Johnson, D. A., Burch, M., Gabbert, S., Aillon, D. V., Wilson, R., Gifford, R., and Xiaohong, C., *In vitro* and *in vivo* real-time wireless monitoring system for glutamate, Program No. 326.7. *2003 Abstract Viewer/Itinerary Planner*. Washington, DC: Society for Neuroscience, 2003. Online.
106. Garris, P. A., Ensman, R., Poehlman, J., Alexander, A., Langley, P. E., Sandberg, S. G., Greco, P. G., Wightman, R. M., and Rebec, G. V., Wireless transmission of fast-scan cyclic voltammetry at a carbon-fiber microelectrode: proof of principle, *Journal of Neuroscience Methods*, 140 (1–2), 103–115, 2004.
107. Deutsch, S., Seven-electrode EER stimulator-monitor for ambulatory monkeys, *Biotelemetry and Patient Monitoring*, 7 (3–4), 231–241, 1980.
108. Talwar, S. K., Xu, S., Hawley, E. S., Weiss, S. A., Moxon, K. A., and Chapin, J. K., Rat navigation guided by remote control, *Nature*, 417 (6884), 37–38, 2002.
109. Xu, S., Talwar, S. K., Hawley, E. S., Li, L., and Chapin, J. K., A multi-channel telemetry system for brain microstimulation in freely roaming animals, *Journal of Neuroscience Methods*, 133 (1–2), 57–63, 2004.
110. Duffield, G. E., Hastings, M. H., and Ebling, F. J., Investigation into the regulation of the circadian system by dopamine and melatonin in the adult Siberian hamster (Phodopus sungorus), *Journal of Neuroendocrinology*, 10 (11), 871–884, 1998.
111. Sterner, M. R., Meisel, R. L., and Diekman, M. A., Forebrain sites of estradiol-17 beta action on sexual behavior and aggression in female Syrian hamsters, *Behavioral Neuroscience*, 106 (1), 162–171, 1992.
112. Shelley, D. N. and Meisel, R. L., The effects of mating stimulation on c-Fos immunoreactivity in the female hamster medial amygdala are region and context dependent, *Hormones and Behavior*, 47 (2), 212–222, 2005.
113. Hofker, M. H., Introduction: the use of transgenic mice in biomedical research, *Methods in Molecular Biology*, 209, 1–8, 2003.
114. Heidenreich, B. A., Trytek, E. S., Schroeder, D. M., Sengelaub, D. R., and Rebec, G. V., A methodology for determining the patch-matrix compartmental location of extracellular single-unit recordings in the striatum of freely moving rats, *Journal of Neuroscience Methods*, 52 (2), 169–174, 1994.
115. Kiyatkin, E. A. and Rebec, G. V., Heterogeneity of ventral tegmental area neurons: single-unit recording and iontophoresis in awake, unrestrained rats, *Neuroscience*, 85 (4), 1285–1309, 1998.
116. Kiyatkin, E. A., Kiyatkin, D. E., and Rebec, G. V., Phasic inhibition of dopamine uptake in nucleus accumbens induced by intravenous cocaine in freely behaving rats, *Neuroscience*, 98 (4), 729–741, 2000.
117. Casto, J. M., Greco, P. G., Pongmayteegul, S., Howes, G. A., Kattner, K. A., and Garris, P. A., Rapid increases in extracellular dopamine following ischemia/hypoxia as measured by fast-scan cyclic voltammetry, Program No. 100.22. *2005 Abstract Viewer/Itinerary Planner*. Washington, DC: Society for Neuroscience, 2005. Online.
118. Koizumi, J., Yoshida, Y., and Nakazawa, T., Experimental studies of ischemic brain edema: a new experimental model in which recirculation can be reintroduced, *Japanese Journal of Stroke*, 8, 1–8, 1986.
119. Globus, M. Y., Busto, R., Dietrich, W. D., Martinez, E., Valdes, I., and Ginsberg, M. D., Effect of ischemia on the *in vivo* release of striatal dopamine, glutamate, and gamma-aminobutyric acid studied by intracerebral microdialysis, *Journal of Neurochemistry*, 51 (5), 1455–1464, 1988.
120. Ahn, S. S., Blaha, C. D., Alkire, M. T., Wood, E., Gray-Allan, P., Marrocco, R. T., and Moore, W. S., Biphasic striatal dopamine release during transient ischemia and reperfusion in gerbils, *Stroke*, 22 (5), 674–679, 1991.
121. Kandel, E. R., Schwartz, J. H., and Jessell, T. M., *Principles of Neural Science*, McGraw-Hill, New York, 2000.
122. Zigmond, M. J., Bloom, F. E., Landis, S. C., Roberts, J. L., and Squire, L. R., *Fundamental Neuroscience*, Academic Press, San Diego, CA, 1999.
123. Vizi, E. S., Role of high-affinity receptors and membrane transporters in nonsynaptic communication and drug action in the central nervous system, *Pharmacological Reviews*, 52 (1), 63–89, 2000.
124. Zoli, M., Torri, C., Ferrari, R., Jansson, A., Zini, I., Fuxe, K., and Agnati, L. F., The emergence of the volume transmission concept, *Brain Research Reviews*, 26 (2–3), 136–147, 1998.
125. Nicholson, C. and Rice, M. E., Diffusion of ions and transmitters in the brain cell microenvironment, In *Volume Transmission in the Brain: Novel Mechanisms for Neural Transmission,* Fuxe, A. and Agnati, L. F., Eds., Raven Press, New York, 1991.

13 Oxidative Stress at the Single Cell Level

Christian Amatore and Stéphane Arbault

CONTENTS

INTRODUCTION

OXIDATIVE STRESS PROCESSES

Aerobic cells derive most of their energy from the controlled oxidation of CH bonds via oxygen-atom transfer catalyzed by metalloenzymes. In particular, these metalloenzymes are central actors of the respiratory chain in mitochondria. However, the same enzymes are generally good reducing agents, prone to open a side route (6–8% yield), leading to oxygen reduction into superoxide ion ($O_2^{\bullet -}$, see Figure 13.1), the precursor of a whole series of hazardous species for living systems coined the "reactive oxygen species" (ROS).[1–4]

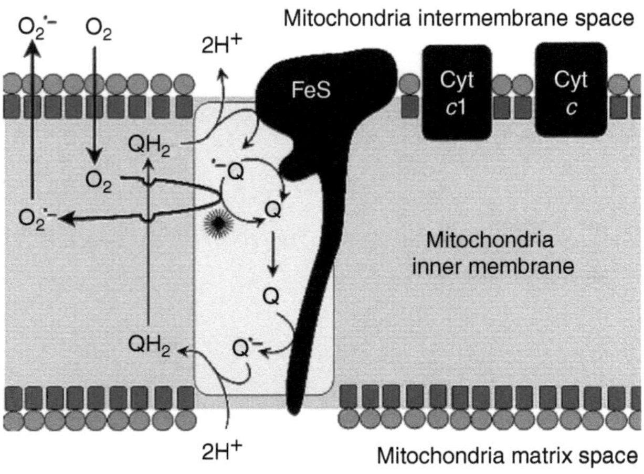

FIGURE 13.1 Scheme describing the formation of superoxide anion as a side reaction of oxygen reduction in the respiratory chain of mitochondria. (Adapted from Raha, S. and Robinson, B. H., *Trends in Biochemical Sciences,* 25, 10, 2000.)

At physiological pH, the superoxide ion is readily scavenged through its disproportionation ($k = 2\times10^5\ M^{-1}\ s^{-1}$, $T = 25°C$, pH = 7.4) into hydrogen peroxide and oxygen:

$$O_2^{\bullet-} + H^+ \rightleftharpoons HOO^{\bullet} \quad (K_{pH7} \ll 1) \tag{13.1}$$

$$HOO^{\bullet} + O_2^{\bullet-} \rightarrow HOO^- + O_2 \quad \text{(fast)} \tag{13.2}$$

$$HOO^- + H^+ \rightarrow H_2O_2 \quad \text{(fast)} \tag{13.3}$$

which may be even more rapid in living cells when catalyzed ($k = 2\times10^9\ M^{-1}\ s^{-1}$, $T = 25°C$, pH = 7.4) by superoxide dismutases (SOD):

$$2O_2^{\bullet-}(+2H^+) \rightarrow H_2O_2 + O_2 \tag{13.4}$$

Two main types of SOD are present in mammalian aerobic cells: a Mn-based enzymatic center form (Mn SOD) located in mitochondria and a Cu/Zn-based form (Cu/Zn SOD) located in cytoplasm in specific granules, and possibly in membranes.[5–11] These enzymes are prone to protect cells from superoxide oxidation of R–H bonds, though this reaction is generally considered as slow, and also from the reaction of superoxide with other radicals such as nitric oxide that may generate more hazardous compounds (Figure 13.2 and explanations below).

H_2O_2 formed by $O_2^{\bullet-}$ disproportionation is an even more potent cytotoxic chemical than superoxide, since its lifetime is sufficient to allow its diffusion to almost any cellular compartment, where it may act as a potential source of hydroxyl radicals $OH^{\bullet}$ (Figure 13.2) through the Fenton reaction in the presence of iron-II free ions or metalloenzymes.[1,2] Because hydroxyl radicals are among the best hydrogen atom acceptors, they are prone to induce a large variety of biological modifications, among which are the peroxidation of cell membrane bilipids through the Haber–Weiss reaction, DNA bases, and backbone oxidation and hydroxylation.[1,2,12]

In living cells, H_2O_2 is scavenged primarily by catalase and secondly by peroxidases (glutathione peroxidase), which catalyze respectively its fast disproportionation into oxygen and water (or into oxidized organic substrates and water), so that its concentration is usually maintained at nanomolar steady-state levels and the lethal routes mentioned above prevented.[3,13–16] In spite of its considerable

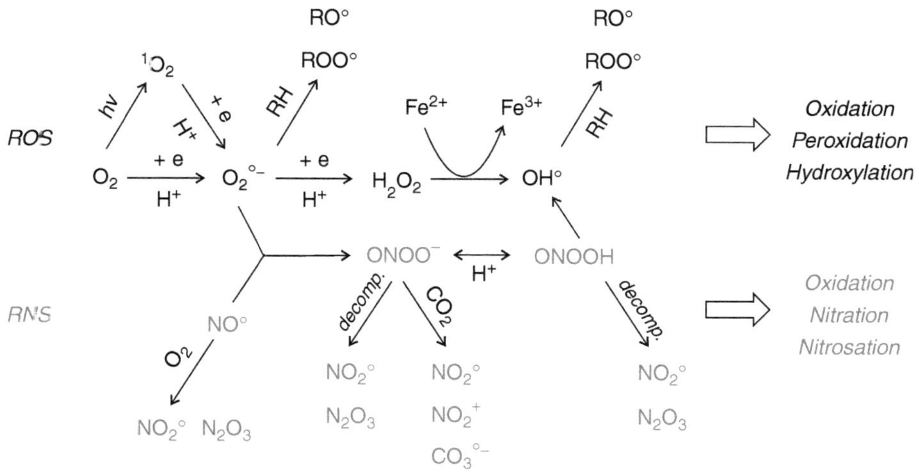

FIGURE 13.2 Main reactive oxygen species (ROS) and reactive nitrogen species (RNS) derived from the biological conversion of oxygen into superoxide ion ($O_2^{\bullet-}$) and nitric oxide ($NO^{\bullet}$).

efficiency at the nanomolar and sub-nanomolar physiological concentrations of hydrogen peroxide, catalase is inhibited by its substrate at higher concentrations. More active scavengers designed to perform in such higher concentration ranges may be produced by a cell through protein expression, but this requires at least half an hour after signaling of the ROS increase.[1,17,18] Thus, aerobic cells might remain unprotected against H_2O_2 and other ensuing ROS whenever its concentration rises rapidly at levels significantly beyond normal physiological ones.

Besides, the production of superoxide and other ROS may not only result from side-routes of the normal oxidative or energetic metabolism of cells. Specific processes are used purposefully by many cell types for communication, for the control of redox sensitive mechanisms (gene expression) leading eventually to cell death by apoptosis, or for defense against xenobiotics (bacteria, virus, etc.) in the case of phagocytes (macrophages and neutrophils). Indeed, a cell may change its normal oxygen metabolism so as to produce deliberately important quantities of superoxide through the involvement of specific enzymes, such as NADPH oxidases, which catalyze the following reaction[19–22]:

$$2O_2 + NADPH \rightarrow 2O_2^{\bullet-} + NADP^+ + H^+ \tag{13.5}$$

NADPH oxidase (from the NOX family) activity is of particular importance in the phagocytosis process of foreign bacteria and viruses by macrophages or neutrophils, since the bactericidal activity is provided essentially by ROS (and RNS). However, excess of ROS production by active or passive mechanisms will amplify cell damages with time. These damages might be prevented by enzymes (SOD, catalase, and peroxidases mentioned above), soluble antioxidant molecules (vitamins, glutathione, uric acid, etc.), or repaired by specific enzymes (DNA repair systems, lysosomes fusion for membrane repair, etc.). This critical situation termed *oxidative stress* basically describes an acute or chronic imbalance between the production of ROS and antioxidant capacities of living cells and organisms. Oxidative stress processes are suspected to be involved directly or indirectly (apoptosis) in many human pathologies (aging, cancers, Parkinson and Alzheimer diseases, auto-immune pathologies, arthritis, etc.) and have stressed in the last two decades, the interest of many research groups in several disciplines of Biology and Medicine.

From recent studies, the imprecation of ROS metabolism with that of nitric oxide ($NO^{\bullet}$) in oxidative stress processes appears ubiquitous in many physiological or pathological situations.

Nitric oxide (or nitrogen monoxide) displays very different effects (positive or negative) in organisms as a function of its concentration and of the local conditions in vivo. $NO^{\bullet}$ mostly originates from the activation of enzymes, the NO synthases (three different mammalian types), which catalyze the reaction[23–26]:

$$\text{L-arginine} + 2\text{NADPH} + 2O_2 + 2H^+ \rightarrow NO^{\bullet} + \text{citrulline} + 2\text{NADP}^+ + 2H_2O \quad (13.6)$$

Nitric oxide is a weak oxidant by itself and should not cause oxidative stress. In fact, it is at most, a weak reductant associated with the strong oxidizing species NO^+.[27] Nevertheless, if $O_2^{\bullet-}$ and $NO^{\bullet}$ are produced in the same cell environment, or in closely related ones, they may couple at a rate close to the diffusion limit (approximately $3\text{–}19\times10^9\ M^{-1}\ s^{-1}$)[27–29] to form the potent peroxynitrite anion:

$$O_2^{\bullet-} + NO \rightarrow ONOO^- \quad (13.7)$$

Peroxynitrite was generally ignored in the past. Contrastingly, it is now suspected to be the source of many oxidative, nitrating, or nitrosating processes involving several cell components and has focused a broad research in the last decade. At physiological pH, peroxynitrite may decompose through the intermediate formation of its conjugated acid ONOOH ($pK_a \approx 6.8$)[30–34]:

$$ONOO^- + H_3O^+ \rightleftharpoons ONOOH + H_2O \quad (13.8)$$

ONOOH undergoes two facile decomposition routes. One goes through the formation of the very reactive radicals $NO_2^{\bullet}$ and $OH^{\bullet}$ and is considered as biologically feasible, especially in non-polar environments (e.g., membranes). Besides, peroxynitrite anion ONOO may react in biological conditions with carbon dioxide (overall constant $k = 3\text{–}6\times10^4\ M^{-1}\ s^{-1}$, Equation 13.9 through Equation 13.12)[32,35–37]:

$$ONOO^- + CO_2 \rightleftharpoons [ONOOCO_2] \quad (13.9)$$

$$[ONOOCO_2] \rightarrow CO_2 + NO_3^- \quad (70\%) \quad (13.10)$$

$$[ONOOCO_2^-] \rightarrow NO_2^{\bullet} + CO_3^{\bullet-} \quad (30\%) \quad (13.11)$$

$$NO_2^{\bullet} + CO_3^{\bullet-} \rightarrow CO_2 + NO_3^- \quad (13.12)$$

This mechanism competes with the direct decomposition of $ONOO^-$, since CO_2 concentration in vivo is quite high (12–30 mM in blood). In this process, it appears that a significant fraction of the short time living species $ONOOCO_2^-$ ends into the potent nitrating and caboxylating radicals $NO_2^{\bullet}$ and $CO_3^{\bullet-}$.

Similarly, when diffusing over significant distances from its source, $NO^{\bullet}$ is expected to react significantly ($k = 2\times10^6\ M^{-1}s^{-1}$)[38,39] with oxygen (Equation 13.13) since this species is present at rather important concentrations (mean of 0.24 mM) in extracellular fluid or in blood:

$$4NO^{\bullet} + O_2 + 2H_2O \rightarrow 4NO_2^- + 4H^+ \quad (13.13)$$

The side reactions of nitric oxide oxidation may form species such as $NO_2^{\bullet}$ and N_2O_3, which are also very efficient nitrating and nitrosating compounds. Finally, though the in vitro decomposition of nitric oxide and peroxynitrite leads ultimately to very stable nitrite and nitrate ions (which, in vivo, may accumulate in cells and their environment), these processes generate very reactive intermediates, such as $NO_2^{\bullet}$, N_2O_3 or NO^+ (suspected to be at the origin of nitrated, nitrosated derivatives of amine acids, thiols, and other important biological components).

The existence of such a series of complex and intricately interconnected pathways, which necessarily follow any simultaneous production of superoxide and nitrogen monoxide by living cells, explains why the very nature of the ROS and/or RNS compounds has not yet been fully characterized in the majority of oxidative stress processes. The origin and mechanism of these processes has been mostly hypothesized based upon the observation of cell metabolites (requiring anywhere from hours to years to appear) or upon effects on cellular activation mechanisms such as gene expression (hours to days). Despite the apparent ubiquity and importance in aerobic cells of the ROS and RNS, few analytical methods have been developed that allows for direct monitoring of their production by living cells, i.e., for investigating the processes at the very origin of oxidative stress and its severe consequences. Electrochemistry on microelectrodes became a method of choice in this regard during the last 15 years owing to the electroactivity of several important ROS and RNS. While, initial studies were conducted on groups of cells or cell monolayers, they have finally evolved to allow direct measurements at the single cell level.

Electroactive Species Implicated in Oxidative Stress

The ability of the electrochemical methods to detect in biological conditions, some of the ROS and RNS mentioned above (or presented in Figure 13.1) obviously depends on two main factors: the half-wave potential of the redox couple (implicating a given ROS or RNS), and the time-life of this compound when produced by a living cell, since this latter commands its permanence and diffusion into cells or tissues. Moreover, with the ultimate goal here being the analysis on a single aerobic living cell, measurements have to be conducted in aqueous medium (buffer) in the presence of oxygen (solutions in equilibrium with normal atmosphere). This condition mostly precludes the use of the reduction domain for electrochemical analysis, given that the reduction of oxygen takes place at low reductive potentials on most electrode surface, e.g. -0.1 V vs. (saturated sodium calomel electrode SCE) on platinum microelectrodes, and oxygen concentration in the close environment of the cell may vary during oxidative stress processes due to its consumption. Thus, direct reductive analysis of any other electroactive species would be generally convoluted with oxygen detection.

Hydroxyl radical is among the most reactive radicals found in biochemistry because it can abstract a hydrogen atom to any CH bond. This species is, thus, definitively not able to diffuse from its generation point in the cell toward the electrode surface without reacting with other molecules, such as the cell membrane constituents. The eventuality of its detection would require positioning the electrode surface at a nanometric distance from its source, which is outside of our present experimental limits in electrochemistry at living cells. Nitrogen dioxide, $NO_2^{\bullet}$, time-life is reported to be significantly longer under biological conditions than that of hydroxyl radical (a microsecond domain versus a nanosecond domain, respectively). This would be compatible with a direct electrochemical detection, though it still represents an analytical challenge. However, the standard potential of the redox couple for the oxidation of $NO_2^{\bullet}$ into NO_2^+ has been measured as $E'(NO_2^+/NO_2^{\bullet}) = +1.56$ V vs. NHE,[27,28,40,41] a high potential close to the water oxidation wall. These conditions strongly limit the probability for a direct detection of $NO_2^{\bullet}$ through its oxidation, and to the best of our knowledge, such results have never been reported into biological conditions.

Conversely, the primary radical species which lead to the complex family of ROS and RNS, i.e., superoxide and nitric oxide, are certainly an easier target for their electrochemical detection by oxidation. The thermodynamic data providing the following standard potential of the redox couple for the oxidation of $NO^{\bullet}$ ($E'(NO^+/NO^{\bullet}) = +1.21$ V vs. NHE) and a standard potential of the redox couple for the oxidation of $O_2^{\bullet-}$ ($E'(O_2/O_2^{\bullet-}) = -0.33$ V vs. NHE) do not preclude a priori detection.[27,28,40] From a kinetic point of view, the life span of each of these primary species strongly depends on the local conditions in cells for their production. As mentioned above, superoxide disappearance by disproportionation depends on the species concentration (order two kinetic) and is much accelerated in presence of SOD. Similarly, the nitric oxide auto-oxidation process is dependent on the species concentration and controls the availability of $NO^\bullet$ in conjunction with many possible reactions with biological targets in cells. Yet, nitric oxide at nano- to micromolar local concentration may diffuse over micrometric distances, thus its detection in the extracellular space is feasible. Moreover, the high lipophilic character of $NO^\bullet$ and the slight lipophilic character of $O_2^{\bullet -}$ allow their diffusion through cell membranes and their detection by a microelectrode placed outside the cell. Besides, we may consider the possibility that these species can be produced by enzymatic systems located in cell membranes, such as the NADPH oxidases (NOX family) for superoxide, and constitutive NO synthases for $NO^\bullet$. Under such conditions, the local geometry disfavors possible reactions between $NO^\bullet$ or $O_2^{\bullet -}$ and the cell components, thus providing a better probability for their observation.

Several other neutral or anionic ROS and RNS are also good candidates for their detection and the characterization of oxidative stress at living cells. Concerning ROS, hydrogen peroxide is probably the most stable compound among this family of species; its time-life in vivo depends essentially on its possible reaction with free metallic ions such as iron(II) or catalase present in the medium (see also detailed explanations about diffusion of H_2O_2 in vivo in Chapter 11). The redox potential of its couple [$E'(H_2O_2/O_2) = -0.146$ V vs. NHE] makes it a good candidate for direct detection on cells. Nitrite is a very stable compound in vivo, and the redox potential of its couple [$E'(NO_2^\bullet/NO_2^-) = +0.99$ V vs. NHE] is a priori compatible for direct oxidation. Nitrite is essentially the end product of nitric oxide metabolism, and is classically used as an indirect marker of oxidative or nitrosative stress with the problem that its basal levels may be quite high. Nevertheless, a large area of studies defining our present understanding of $NO^\bullet$ physiology has been based on the detection of nitrite and nitrate by Griess reaction. Nitrate ions are difficult to oxidize [$E'(NO_3^\bullet/NO_3^-) = +2.50$ V vs. NHE] in absence of an enzymatic transducing system and are also stable end products of RNS.[27,28,40] These two compounds did not focus much interest for their direct analysis on single cells. Conversely, the intermediate and reactive species, peroxynitrite anion, is an ideal target for electrochemical measurements. Other methods reported in the literature for peroxynitrite detection are routed on fluorimetric and chemiluminescence analysis, though they are indirect and not effectively selective.[42–45] Quantization and kinetic profiles of species release by cells is also impossible by such methods. In contrast, electrochemistry at microelectrodes may offer the considerable advantage of a direct detection of peroxynitrite by its oxidation, since its formal redox potential is quite low ($E'(ONOO^\bullet/ONOO^-) = +0.20$ V vs. NHE), and its decomposition kinetic is in the sub-second range.[27,40,46] Consequently, considering an electrochemical sensor positioned close to the cell source of peroxynitrite, i.e., under some conditions equivalent to the ones used for $NO^\bullet$ or $O_2^{\bullet -}$ detection (the actual species that lead to $ONOO^-$), it may be possible to detect and monitor that species before it reacts with its biological targets or before being decomposed into its daughter species.

Direct Electrochemical Detection of Oxidative Release on Single Cells

The average amount of oxidants that may be released by a single cell during an induced oxidative stress could be estimated from studies on cell populations (that number in the millions), by submiting the whole colony to a chemical or immunological stimulation. Then following with spectrophotometric analysis or fluorimetric assays.[45,47,48] For instance, neutrophils in culture may produce about 0.1 μmol of $O_2^{\bullet -}$/h/10^6 cells, and macrophages in culture may produce about 1 μ*M* of RNS/h/10^6 cells. These values lead to the estimation of a rate of release in the range of 10^{-16}–10^{-13} mol/s/cell. Such infinitely small quantities represent a priori, a considerable challenge compared with the actual analytical standards. The difficulty is even greater whenever the whole event is completed within a very short time, and the timing is "chosen" by the cell itself. Both factors emphasize the huge difficulty of the analytical detection in terms of signal-to-noise ratio, and easily explain why, despite their extreme biological importance and huge medical consequences they have. the study of oxidative stress primary events has seldom been investigated by biologists at the single cell level.

The quality of any analytical information requires a sufficient signal-to-noise ratio. Taking for granted that the analytical device has been selected to adequately perform, both thermodynamically and kinetically, a sufficient signal-to-noise ratio imposes a sufficient concentration of the chemical to be detected given that a minimal noise level cannot be avoided. It is important to stress here that a large *concentration* does not mean a large *quantity,* since a concentration is a quantity of chemicals divided by the sampling volume. Thus, an infinitely small number of molecules may be easily converted into a large concentration by restricting the volume in which these molecules are released. This is the simple arithmetic solution retained by nature in its biological synapses, so as to detect kinetically (viz., concentration), with a high signal-to-noise ratio, the release of zepto- to attomoles of chemical messengers on a millisecond time-scale.[49,50]

This principle was first adapted to electroanalytical purposes by Wightman's group and our group.[50–53] Indeed, whenever the released molecules are electroactive and identifiable electrochemically, placing an electrode at micrometric or sub-micrometric distances from a living cell ensures an adequate signal-to-noise ratio by restricting the extracellular volume in which the molecules produced by the cell are released. For example, one femtomole delivered in a volume of a thousand μm^3 creates a millimolar concentration rise, i.e., produces a concentration variation which can be monitored kinetically via electrochemical techniques with an excellent signal-to-noise ratio, provided that the electrode does not pick up too much noise through its surface.

Ultimately, electrodes pick up electrical noise through their capacitance, viz., through their overall conducting surface area, while the analytical information (viz., the Faradaic current) may arise only from the surface area exposed to the cell release. Thus, using an electrode with an active surface matching that of the examined cell decreases the noise, while not affecting the quality and intensity of the analytical information. This increases the signal-to-noise ratio and ensures, simultaneously, a quantitative collection efficiency, since the artificial synaptic cleft covers the whole cell emitting surface. It is understood that positioning an ultramicroelectrode at micrometric distances from an isolated living cell (Figure 13.3), cells being of micrometric dimensions, ensures the most adequate signal-to-noise ratio and guarantees that the release of the collection of electroactive chemicals at the surface of a single living cell is total.[54–56]

Based on these analytical principles, studies were initiated by several groups at the end of eighties and beginning of the nineties in order to detect either superoxide and its first derivatives hydrogen peroxide, or nitric oxide released by a single cell or a small group of cells. More recent studies have led to the fine analysis of cross-reaction products, such as peroxynitrite, between the two primary compounds ($NO^\bullet$ and $O_2^{\bullet-}$). The following examples have been chosen in order to illustrate different types of applications and different technologies for the microelectrode preparation and their sensing properties.

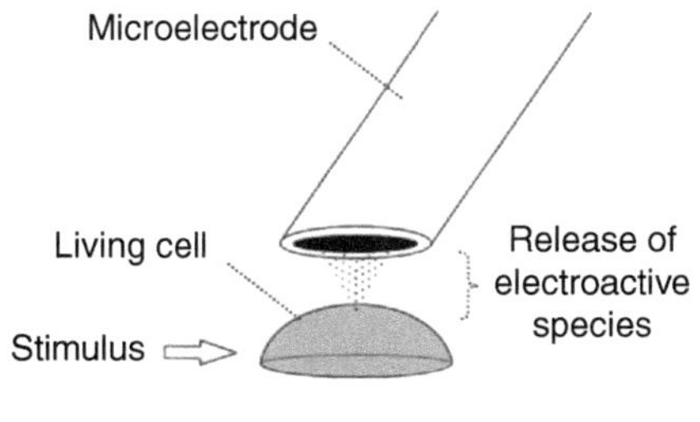

FIGURE 13.3 Principle of analysis by a microelectrode of a single cell's secretion of electroactive species.

MICROELECTRODE DETECTION OF SUPEROXIDE AND HYDROGEN PEROXIDE RELEASED BY LIVING CELLS

DETECTION OF SUPEROXIDE BY CARBON AND GOLD MICROELECTRODES ON PHAGOCYTES AND VASCULAR CELLS

Pioneer work in the field of detection of species implicated in oxidative stress processes was reported by the group of Hill (Oxford University, U.K.) in the mid eighties. This group worked previously on SOD activity and investigated the possibility to detect superoxide anion release by human blood neutrophils.[57–59] Two series of results were reported. In the first one, they proposed to detect superoxide anion by its oxidation wave at low potential (+50 mV vs. Ag/AgCl) on the surface of a millimetric pyrolytic graphite electrode. Then, the electrode was "opsonized", i.e., its surface was modified by a deposit of an immunological factor, the immunoglobulin G (IgG). This was obtained through a simple procedure, dipping the electrode (2 min) into a solution of IgG (30 mg/mL^{-1} in Hanks buffer, 37°C). Measurements were made into a buffer solution containing neutrophils previously purified and used in suspension. The neutrophils that met the electrode surface were activated by the deposit of IgG and rapidly initiated a respiratory burst response. Such typical response of neutrophils and macrophages during immunological processes has been known for decades to be a key factor of phagocytosis. The increase of oxygen consumption (the respiratory burst) by activated phagocytes correlates with the activation of their NADPH oxidases (NOX-2 type) that catalyze the formation of superoxide and hydrogen peroxide. Hill et al. have shown that $O_2^{\bullet -}$ released by neutrophils in their environment was detected on the graphite electrode surface. The selectivity for $O_2^{\bullet -}$ of their analysis was supported by the observation of a strong decrease of response when SOD or when *N*-ethyl-maleimide (NEM, a non selective inhibitor of NADPH oxidases) was injected into the medium.

In a second series of experiments, a gold microelectrode (10 μm in diameter) was used instead of the classic pyrolytic graphite millimetric electrode.[58] The microelectrode surface was opsonized by IgG and placed into a suspension of fresh neutrophils, as described previously. The contact between cells and the microelectrode surface induced very low variations of current, in the range of 1–1.5 pA, with a spike-shaped kinetic profile (see Figure 13.4). Control experiments in the presence of SOD or NEM (see above) showed that superoxide was responsible for this oxidation current, or at least was related to it. Measurements of the oxygen consumption rate by the neutrophils and its evaluation for a single cell led the authors to conclude that the oxidative bursts detected on their opsonized microelectrode originated in most of cases from the activation of a single neutrophil being in contact with the electrode surface.

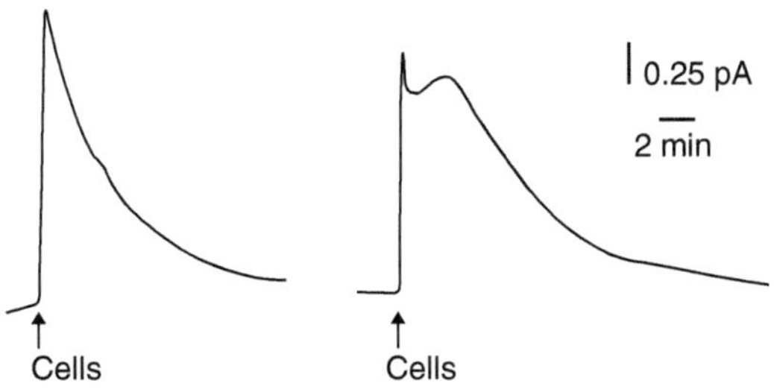

FIGURE 13.4 Detection on an opsonized (IgG modified) gold microelectrode of superoxide release by a single neutrophil. At time *t*(Neuts.), neutrophils were added to the buffer solution in which the microelectrode was placed with the intention that one of them would encounter the microelectrode surface and be activated. Two examples of cell response are presented. (Adapted from Hill, H. A. O., Tew, D. G., and Walton, N. J., *FEBS Letters*, 191, 1985.)

These seminal results seem to be the first demonstration of the perfect suitability of electrochemical methods for the detection of ROS species implicated in oxidative stress at the level of one living cell.

In 1996, Tanaka et al. reported studies[60,61] that confirmed and refined the work of Hill et al. Superoxide production by phagocytes in culture (HL-60 cell line) was detected on a 10 μm-diameter carbon fiber microelectrode. The microelectrode surface was opsonized by dipping it into a solution of human IgG (80 mg mL^{-1} in PBS, 3 dips of 15 min at 37°C), a procedure akin to that of Hill's group, or modified by adsorption of an activator, the phorbol ester PMA (4-phorbol-12-β myristate-13-acetate at 80 n*M* in PBS, 3 dips of 15 min at 37°C). Cells were immobilized on a collagen modified coverslip and consequently, easy to localize for analysis without relying on a hypothetical interaction between cells in suspension and the electrode surface. When the microelectrode surface was put in contact with the surface of a phagocyte, the cell shape was changed and a phagocytosis process was initiated. As reported in Figure 13.5, amperometric currents detected at +0.1 V vs. Ag/AgCl, a potential sufficient to oxidize superoxide into oxygen on the carbon surface, provided slow developing signals in both condition of cell activation. Their maximum amplitude was in the range of a few picoampers. The authors assumed that the collection efficiency of superoxide secretion from the cell attached to the electrode surface was total, so that they could estimate the maximum rate and total amount of release: 1.2 fmol min^{-1} and 15 fmol total for IgG activation; and 0.4 fmol min^{-1} and 30 fmol total for PMA. The very low noise level of their background current, i.e., about 0.1 pA, led them to draw the conclusion that the oscillations on the cell signal analyzed for one to three hours originated from variations of NADPH during cell response. Indeed, this may be used by cells to modulate their NADPH oxidase activity or other enzymes activity also dependent on NADPH metabolism.

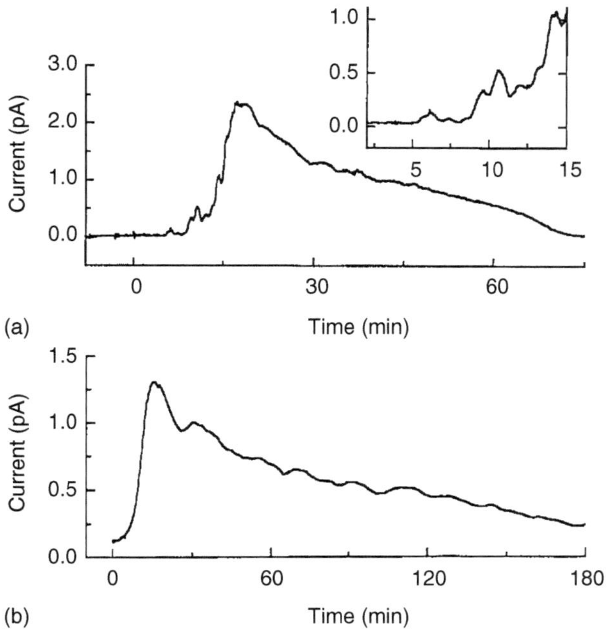

FIGURE 13.5 Detection on (a) an opsonized (IgG modified) or (b) PMA-modified carbon microelectrode of superoxide release by single phagocytes (of the HL-60 cell line). The microelectrode surface was positioned in contact with one isolated cell at $t = 0$ min. (Adapted from Isogai, Y., Tsuyama, T., Osada, H., Iizuka, T., and Tanaka, K., *FEBS Letters*, 380, 3, 1996.)

Based on the work of Tanaka et al., several groups have used carbon fiber microelectrodes to monitor superoxide release by living cells into biological situations different from the one exposed above.[62–65] In particular, Privat et al.[66] have reported measurements of $O_2^{\bullet-}$ produced by vascular cells (endothelial and smooth muscle cells). Indeed, oxidative stress following overproduction of superoxide (and possibly nitric oxide) at the vascular level may lead to local dysfunctions. This has been recognized as a key factor in atherosclerosis, ischemia/reperfusion injury, etc. The authors have carefully calibrated their carbon fiber microelectrodes (7 μm in diameter) against superoxide solutions prepared from dissolved KO_2 over a wide range of pHs (7.4–14). The electrode surfaces were electrochemically pretreated into PBS by cycling the potential, and measurements were conducted by differential pulse amperometry between 0 and +0.1 V vs. SCE (40 ms step length and sampling for 6 ms). This procedure offered an excellent sensitivity for $O_2^{\bullet-}$ (0.8 nM/pA^{-1}) while the low potential applied provided selectivity. In these experiments, the sensing property of the electrodes was separated from the cell stimulating step without including adsorption of chemical or immunological agents on the electrode surface. Vascular cells in culture were stimulated by a biological agent, the cytokine IL-1β, after positioning (via a micromanipulator) the microelectrode surface at a close distance to the cells, supposedly a few tens of microns. The configuration ensured an efficient collection of superoxide release, without a priori any significant loss of response due to the formation of hydrogen peroxide by disproportionation or due to reaction with $NO^{\bullet}$, which was not supposed to be produced in these conditions. These hypotheses were supported by the low local concentrations, in the range of tens of nanomolar of superoxide detected (see Figure 13.6). This means that the second order kinetics of reactions Equation 13.4 and Equation13.8 are slow enough and henceforth do not affect $O_2^{\bullet-}$ concentration before these molecules reach the electrode surface. Yet, it may be noted that this may be a circular argument, given that if superoxide were to be produced at larger amounts, most of it would be consumed during the time-of-flight required to reach the electrode surface. Then, only a minimal fraction would survive and be detected, i.e., that for which spontaneous disproportionation becomes negligible.

Detection of Superoxide by Cytochrome *C* Modified Gold Electrodes on Neuronal Cells

Let us present, now, another seminal study from the group of C. J. McNeil (University of Newcastle upon Tyne, U.K.). Starting in 1993 these authors have proposed a new concept for electrode preparation in order to improve their selectivity toward superoxide. These studies were based on millimetric electrodes and did not apply to single cells but instead to confluent cell layers.

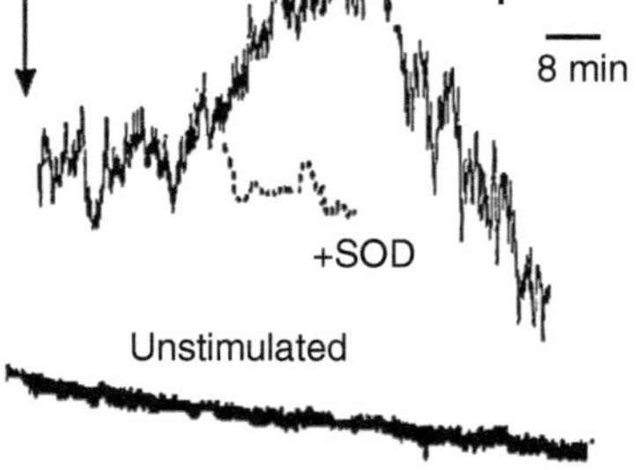

FIGURE 13.6 Detection with a carbon fiber microelectrode of superoxide release from unstimulated and IL-1 β-stimulated endothelial vascular cells. SOD was injected in another experiment to verify the signal origin. (Adapted from Privat, C., Stepien, O., David-Dufilho, M., Brunet, A., Bedioui, F., Marche, P., and Devynck, J., *Free Radical Biology and Medicine*, 27, 1999.)

We believe, nevertheless, that the present principle of electrode modification may also be used for the preparation of microelectrodes and that single cell analysis would be possible with such sensors. Hence, the authors have developed millimetric gold electrodes modified by a deposit of ferricytochrome *c*,[67–70] a protein classically used for indirect spectrophotometric measurements of superoxide produced by living cells in culture since the reaction between the two compounds is very fast ($k = 2\times10^6\ M^{-1}\ s^{-1}$):[47]

$$\text{ferricytochrome } c(Fe^{3+}) + O_2^{\bullet-} \rightarrow \text{ferrocytochrome } c(Fe^{2+}) + O_2 \quad (13.14)$$

Ferricytochrome *c* was covalently immobilized by a thiol derivative (DTSSP) linker interacting with the protein and the gold surface (SAM formation through Au–S bonds). This was decided to provide a better selectivity for $O_2^{\bullet-}$, particularly against $NO^{\bullet}$, than that offered by simple carbon or gold electrodes. A complete in vitro study with solutions of superoxide generated enzymatically by a system of xanthine/xanthine oxidase, demonstrated the interest of the concept based on the electro-oxidation of reduced cytochrom *c*, thus providing a catalytic effect on the response. Consequently, these modified microelectrodes offered a high sensitivity for $O_2^{\bullet-}$, e.g., 2 p*M*/pA^{-1}.[68,69] Each electrode was used in conjunction with a second electrode, a commercial NO sensor, for simultaneous measurements of $O_2^{\bullet-}$ and $NO^{\bullet}$ produced by neuronal cells in culture (different cell lines were analyzed).

The data presented in Figure 13.7 were obtained from simultaneous measurements with the two types of sensor over a layer of adherent glioblastoma cells from the A172 cell line.[68] Their oxidative response was stimulated by addition of PMA (a phorbol ester activating NADPH oxidases, as explained before) or (lipopolysaccharide- LPS α prepared from fragments of bacteria walls, known to activate NO synthases) to the buffer solution. Based on their in vitro tests, the authors assumed no possible interference between the two sensors, i.e., superoxide was not detected by the NO sensor, and the $O_2^{\bullet-}$ sensor was insensitive to $NO^{\bullet}$. However, simultaneous measurements gave them the ability to observe the cross reaction between $O_2^{\bullet-}$ and $NO^{\bullet}$ in their experimental conditions. Following addition of the stimulus (LPS in Figure 13.7), a rapid increase of $NO^{\bullet}$ was detected, and even possibly amplified, under conditions of NO synthases pre-activation (inducible NOS induction by the cytokine interferon-γ), while superoxide was not detected.

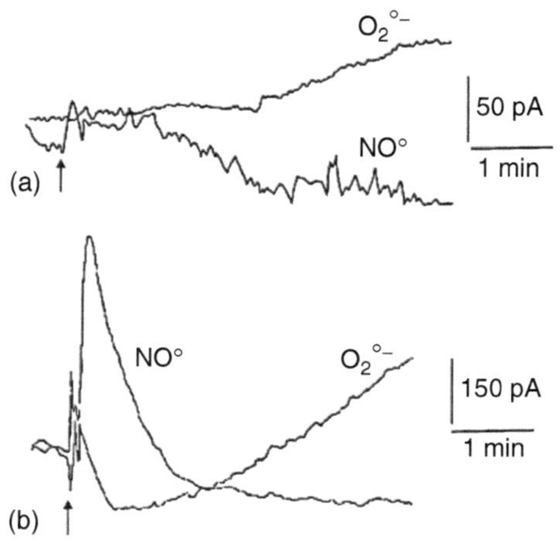

FIGURE 13.7 Detection of $O_2^{\bullet-}$ on a ferricytochrome c modified gold electrode and $NO^{\bullet}$ on a commercial selective electrode at a sub-millimetric distance from a layer of 10^6 glioblastoma cells (of the A172 line). The LPS stimulation was applied to control cells (a) or cells pre-activated by IFN-γ for 24 h (b). (Adapted from Manning, P., Calum, J., Cooper, J. C., and Hillhouse, E., *Free Radical Biology and Medicine*, 24, 1998.)

Then, when the production of $NO^{\bullet}$ decreased, superoxide started to be observed showing that in the initial phase of the cell response $O_2^{\bullet -}$ was titrated by excess of $NO^{\bullet}$ to form $ONOO^-$. Consequently, these studies reported one of the first examples of direct and selective measurement of primary species of oxidative stress at living cells and the observation of reaction between these species at origin of RNS formation. Nevertheless, the applicability of the present analytical method and sensor to quantitative biological studies was restricted by the low collection efficiency of superoxide at sub-millimetric distance separating the electrode surface and the layer of cells. The spontaneous formation (without catalase involvement) of H_2O_2 in these conditions is not negligible, though it was not evaluated. Similarly, peroxynitrite formed by coupling between $NO^{\bullet}$ and $O_2^{\bullet -}$ could not be measured directly. Yet, $ONOO^-$ may re-oxidize ferricytochrome c and interfere with measurement, though this reaction is quite slow.[71] Consequently, these studies on cell colonies led to the conclusion that the detection of side-products of superoxide and nitric oxide would also be of great importance for characterizing, in real-time, the nature and origin of oxidative stress processes occurring in a cell.

Detection of Hydrogen Peroxide by Platinized Carbon Fiber Microelectrodes on Skin Fibroblasts

Among the different derivatives of superoxide anion, hydrogen peroxide is the most stable one and is, consequently, an essential marker of oxidative stress. Our group decided to focus on its detection through the development of microelectrodes, allowing its monitoring on single cells.[72] It was previously established that H_2O_2 oxidation is difficult on any type of carbon surfaces, while its oxidation on bare polished platinum surfaces induces the formation of platinum oxide and peroxide and, consequently, unstable electrochemical responses. Meanwhile, several groups working in the development of glucose sensors have shown that hydrogen peroxide produced by glucose oxidase could be easily and quantitatively detected by oxidation on black platinum or platinized electrode surfaces.[73,74] Black platinum was thus electro-deposited from a Pt(IV) solution following a 4e reduction into Pt(0) on the electrode surface:

$$PtCl_6^{2-} + 4e \rightarrow Pt^0 + 4Cl^- \qquad (13.15)$$

A stable, irreversible, voltammetric wave (2e oxidation process, $E_{1/2} = +0.25$ V vs. SSCE) for H_2O_2 could be observed on our platinized carbon fiber microelectrodes (measuring 1–30 μm in diameter). Moreover, as the platinized surface was a rough deposit, the microelectrodes possessed a huge active area, which ensured a high sensitivity for H_2O_2 detection. The limit of sensitivity determined by amperometry at $+0.55$ V vs. SSCE was a few nanomolars, the selectivity being provided by the electrode potential, as attested to by control experiments in the presence of added catalase or *o*-dianisidine.

Owing to their excellent analytical properties, these modified microelectrodes were applicable for monitoring oxidative stress on single human skin fibroblasts. These cells are the main constituents of skin dermis and, as a consequence, are the site of some of the oxidative damages induced by UV-A and -B from sunlight and other chemicals in contact with the skin. Under oxidative stress conditions, DNA alterations of fibroblasts might be amplified and cause the cell to shift toward a cancerous state.[1,75,76]

We have shown for the first time, that fibroblasts are able to produce oxidative bursts under depolarisation of the cell membrane induced by a mechanical stress,[72,77] the puncturing of the membrane with the tip of a microcapillary without inducing cell disruption or death. The shafts of the microelectrodes were insulated in order to detect electroactive species only at their very tip surface. Owing to their small size (10–15 μm in diameter), the tips could be positioned via a micromanipulator at a few tens of microns above the cell surface, i.e., in experimental conditions for which any loss of H_2O_2 through its reaction with other compounds should not occur.

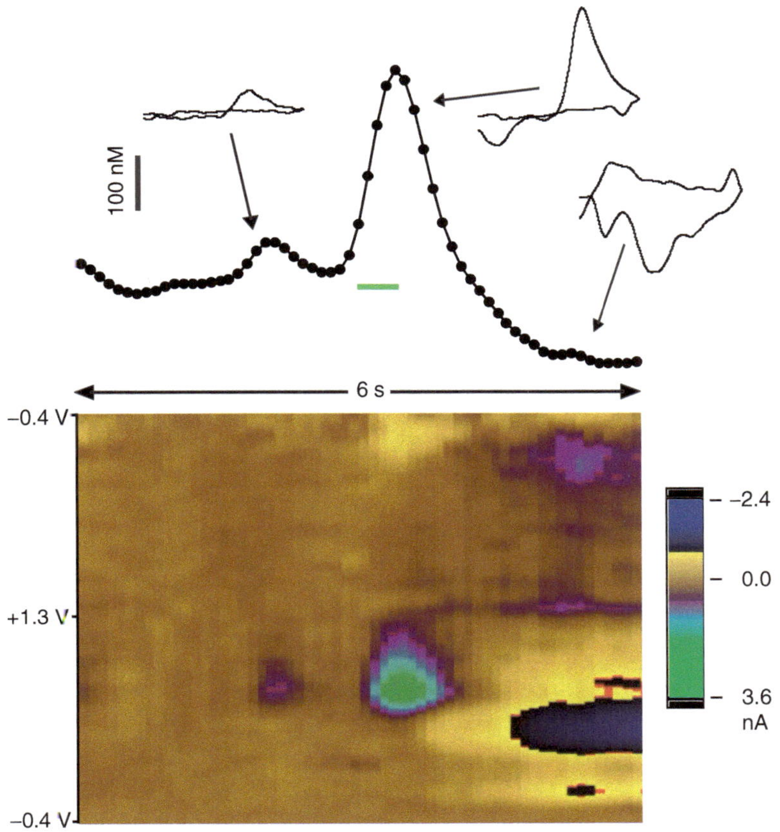

FIGURE 2.1

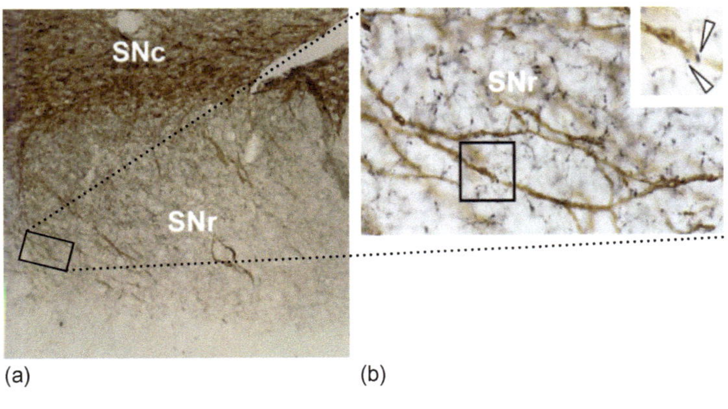

FIGURE 8.1

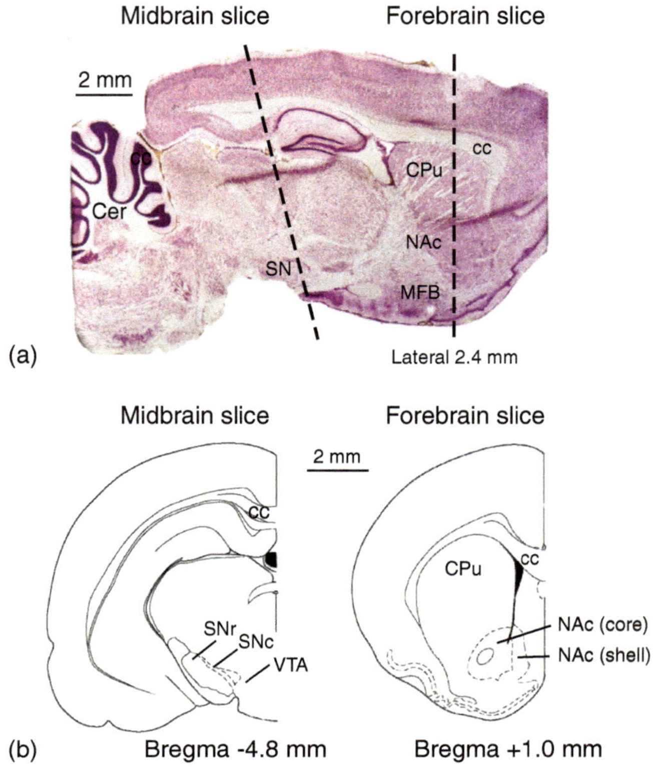
Midbrain slice
Forebrain slice
2 mm
cc
Cer
SN
CPu
cc
NAc
MFB
(a)
Lateral 2.4 mm
Midbrain slice
Forebrain slice
2 mm
cc
SNr
SNc
VTA
CPu
cc
NAc (core)
NAc (shell)
(b)
Bregma -4.8 mm
Bregma +1.0 mm

FIGURE 11.4

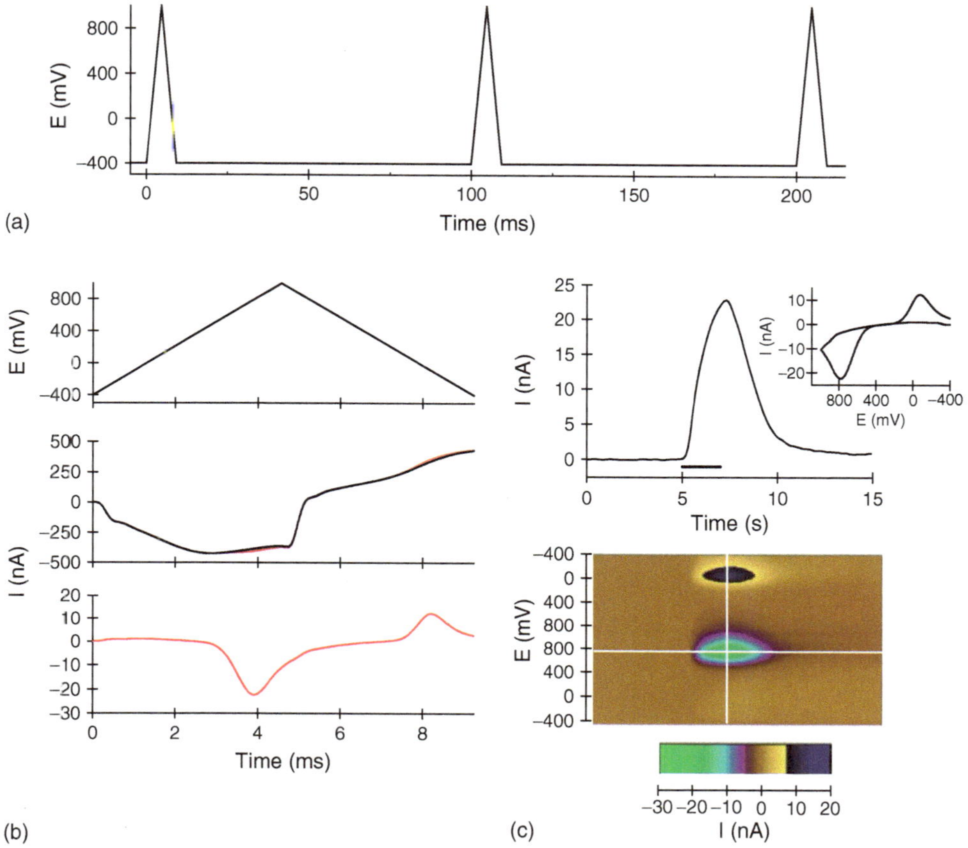

FIGURE 12.2

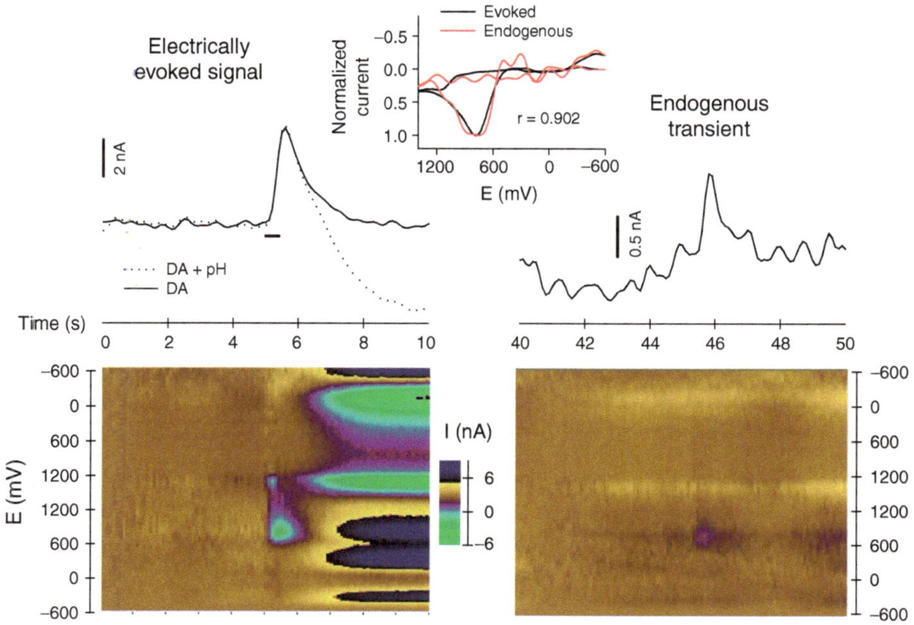

FIGURE 12.3

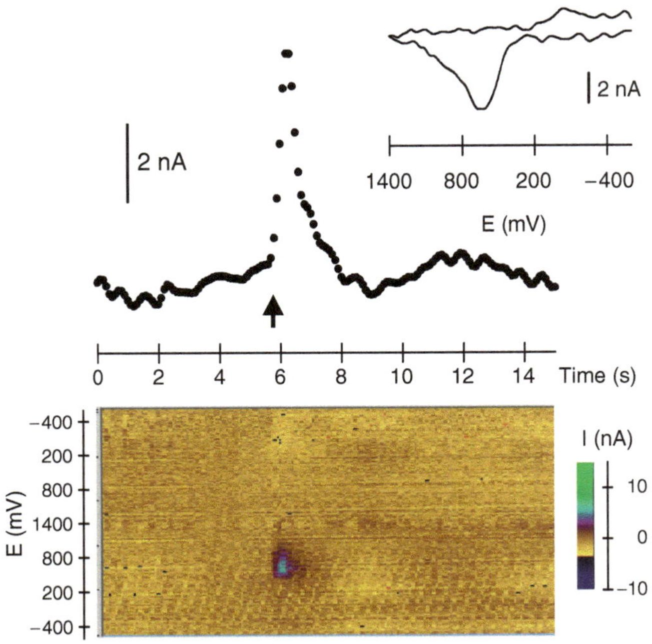

FIGURE 12.10

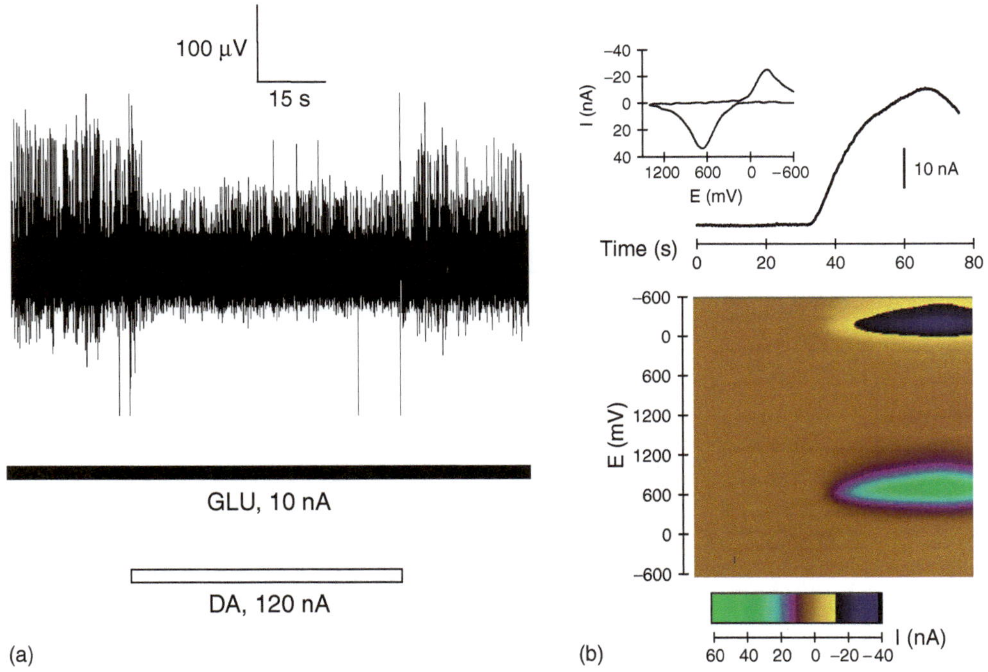

FIGURE 12.11

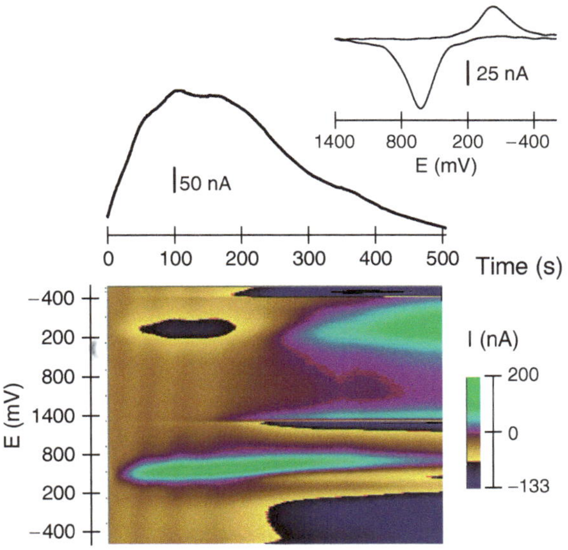

FIGURE 12.12

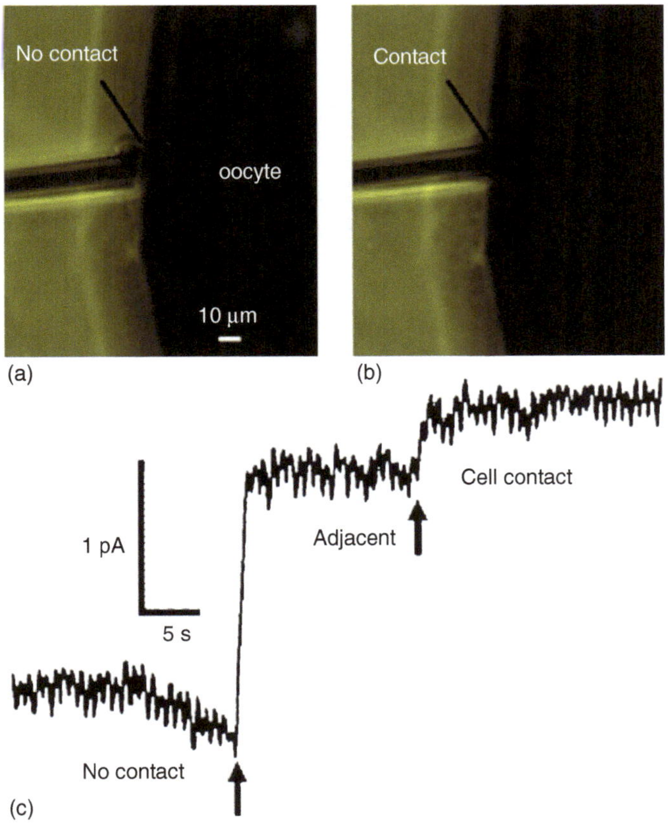

FIGURE 14.3

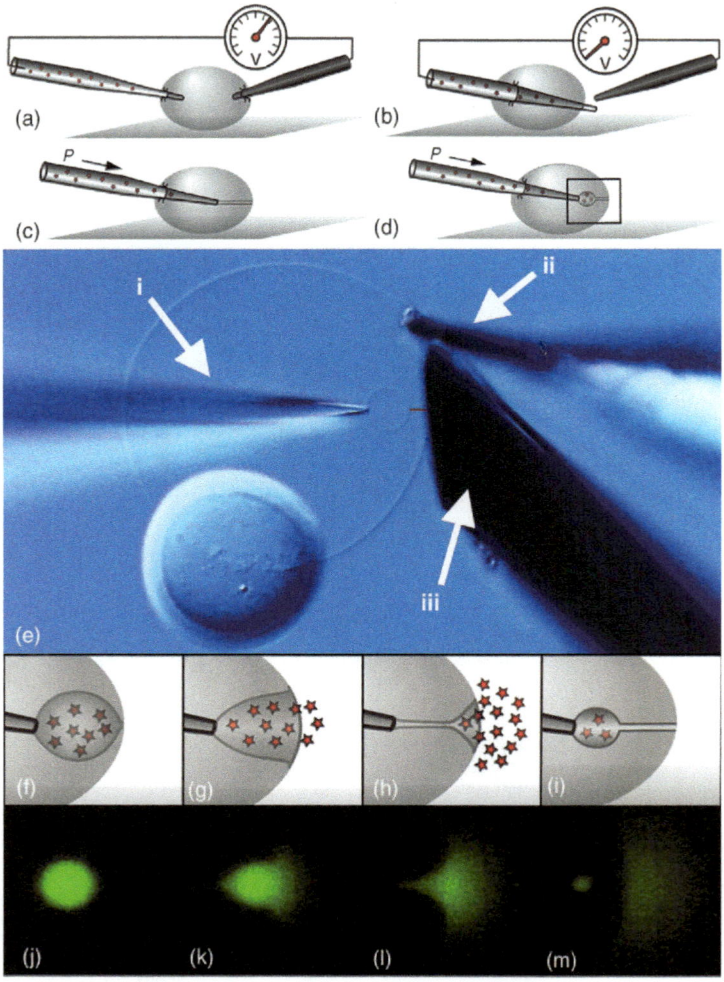

FIGURE 14.6

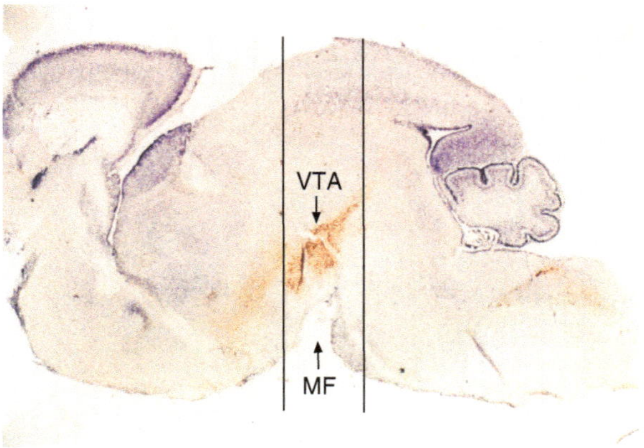

FIGURE 16.1

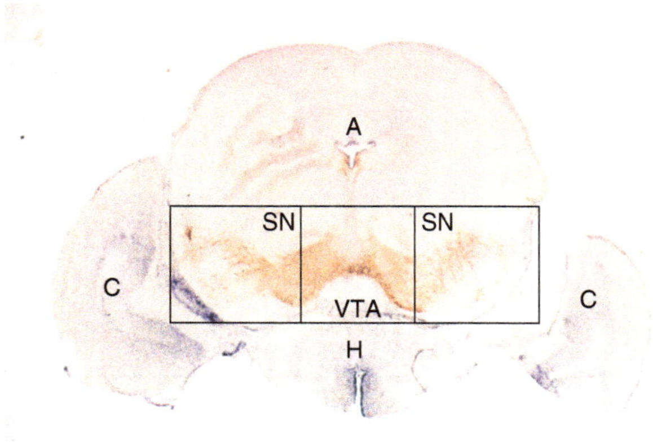

FIGURE 16.2

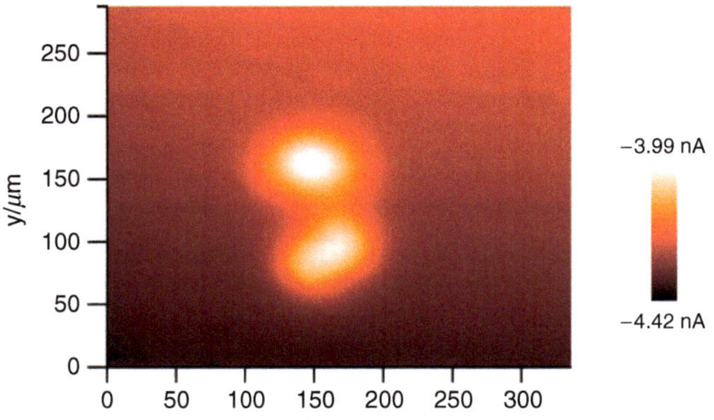

FIGURE 17.7

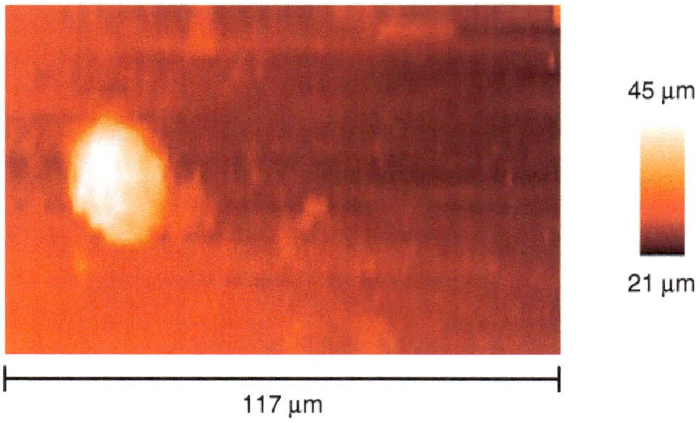

FIGURE 17.8

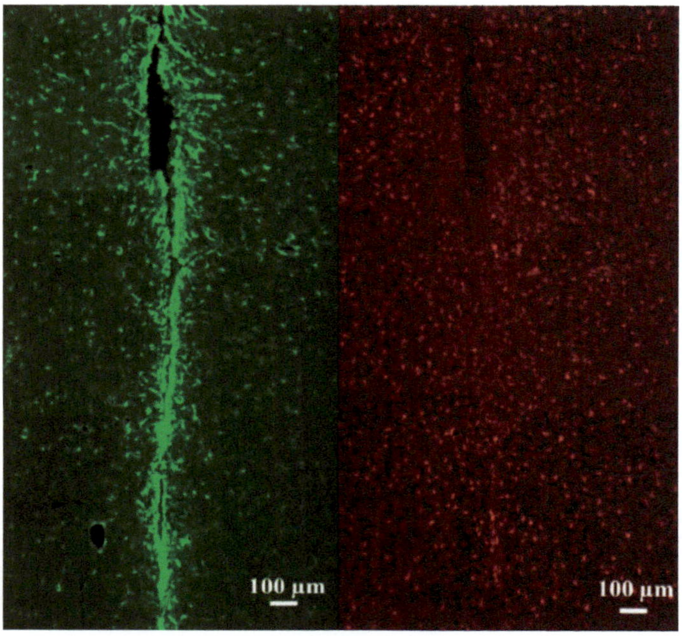

FIGURE 19.31

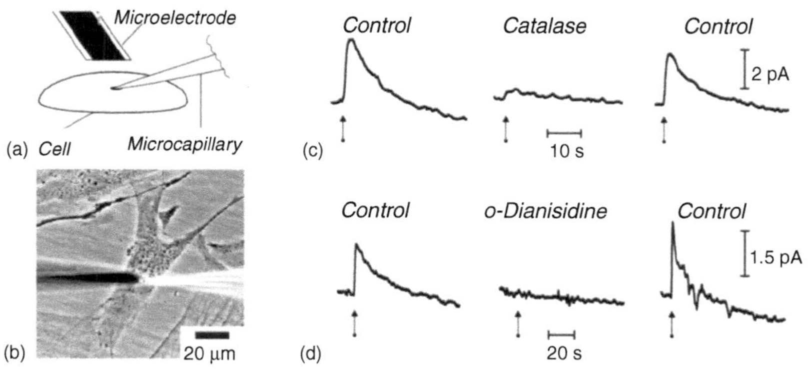

FIGURE 13.8 Detection on a platinized carbon fiber microelectrode (measuring 10–15 μm in diameter) of hydrogen peroxide release by a single human fibroblast in culture. (a) scheme and (b) view under microscope of the experimental setup. Typical responses detected on single fibroblasts after their membrane depolarization by a microcapillary and analysis of the (c) catalase or (d) *o*-dianisidine (peroxidases substrate) effect on these responses.

The variation of amperometric current detected immediately after the microcapillary puncture on the cell membrane corresponded to an active response of the cell and not to the passive release of electroactive species diffusing from the cytosol or from storage vesicles. Experiments in the presence of catalase, the natural scavenger of H_2O_2, or in the presence of a substrate for peroxidases, *o*-dianisidine, showed that hydrogen peroxide was the major component of the cell response (see Figure 13.8c and Figure 13.8d). In later studies, inhibitors of NADPH oxidases confirmed that H_2O_2 detected by the microelectrodes had an enzymatic origin[77] and that the kinetic profile of the oxidative bursts measured on fibroblasts or lymphocytes was dependent on metabolic factors governing NADPH oxidases activity.[77–79]

Nevertheless, several results led us to consider that oxidative bursts detected on fibroblasts or other cell types were not exclusively related to hydrogen peroxide production by cells. First, a residual signal was always observed in experiments with catalase, and then, the burst amplitude was increasing at amperometric potentials higher than the one of H_2O_2 plateau potential. The hypothesis of a simultaneous production of $O_2^{\bullet -}/H_2O_2$ and $NO^{\bullet}$ has been examined. An exhaustive study of the cell response has allowed a full delineation of the origin and nature of ROS and RNS produced under these conditions and established the intimate relationships and coupling between NADPH oxidases and NO synthases activities during cellular oxidative stress processes. These results will be presented in the last part of this chapter.

MICROELECTRODE DETECTION OF NITRIC OXIDE AND PEROXYNITRITE RELEASES BY LIVING CELLS

Detection of Nitric Oxide by Porphyrin Modified Microelectrodes at Single Cells from Cardiac Endothelium

Malinski's group (Oakland University, U.S.A.) was one of the first to describe an electrochemical NO sensor and its application to cells measurements of nitric oxide release. In 1992 they reported the development of a carbon fiber microelectrode, which had a surface modified by the electrodeposit of a film of nickel porphyrin.[80] We wish not to detail the preparation and characterization of this sensor as this will be documented later, in chapter 21.[81] Malinski et al. modified their electrodes to match the specific conditions of a single cell analysis on cells in culture or located on the surface of the endocardium (of a rabbit heart). The active tip of the electrochemical sensor was reduced to 2–5 μm in diameter instead of 50–60 μm, the diameter of the sensor previously used.[82–84] Secondly, the electrode was inserted into a catheter, so that the tip of the sensor was recessed 5–6 μm inside of the catheter tip, and thus, was protected in the case of experiments performed directly in the tissue. Each electrode was tested and calibrated with pure $NO^{\bullet}$ solutions of concentration ranging from $5\times10^{-}$ to 3×10^{-6} mol L^{-1}.

Experiments were performed on cells mechanically removed from the surface of the left heart ventricle of an anesthetized rabbit. Cells were plated on usual coverslips and maintained in buffer. The tip of the porphyrinic microelectrode was then positioned via a micromanipulator over the surface of a single endothelial cell; the zero distance was obtained when touching the cell membrane and henceforth a signal artifact was detected. Then, the electrode tip was placed at a 10 μm height. The cell was subsequently stimulated following the injection of a calcium ionophore solution (compound A23187, which induces transient opening of calcium channels, with a 8 μ*M* final concentration) into its environment, leading to the instant activation of NO synthases (constitutive enzymes of endothelial type). The release of $NO^{\bullet}$ detected by amperometry started in the first second following the cell stimulation (see Figure 13.9a). The signal reached its maximum in about 1 s and returned to baseline after about 10 s. Its amplitude (in concentration of $NO^{\bullet}$) was highly dependent on the distance between the sensor tip and the cell surface. Logically, the maximum value, close to 1 μ*M*, was obtained when placing the sensor at the membrane surface (see Figure 13.9b). This observation is consistent with the principles described above in the Direct Electrochemical Detection of Oxidative Release on Single Cells section of this chapter. The quantification of release from a single living cell is highly dependent on the collection efficiency of the electrode surface and of the volume of analysis defined by the distance between that sensor surface and the cell source of active species. Consequently, minimizing this distance ensures better conditions for analysis of the true nature of the biological information, though this may lead to an indirect blasting of the cell surface by the products resulting from the electrochemical detection.

This first series of studies allowed comparing the response of endothelial cells obtained from the surface of the endocardium of normal or hypertensive rats. Nitric oxide release was lower in hypertensive rats ($NO^{\bullet}$ is the major vaso-relaxing compound) in agreement with former studies reporting a weaker relaxation of smooth muscle cells from these animals compared to the wild type ones. However, the authors have shown that this lower $NO^{\bullet}$ response was very sensitive to SOD, indicating that a large part of $NO^{\bullet}$ produced by cells was not detected by the electrode because it

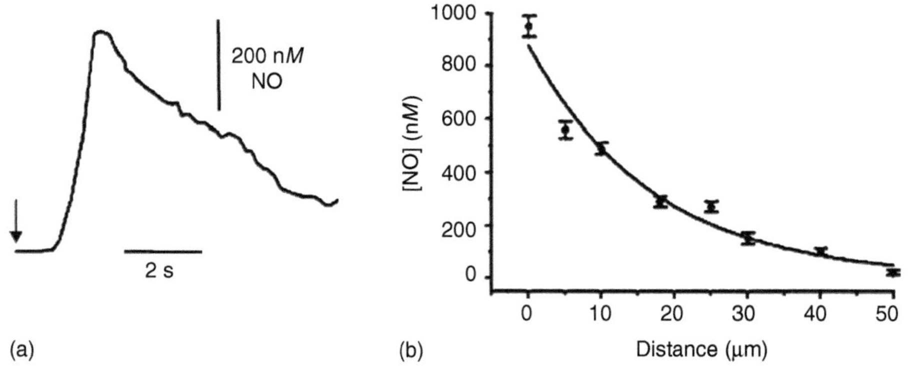

FIGURE 13.9 (a) Release of nitric oxide detected by a porphyrin modified microelectrode placed 10 μm from the surface of an isolated endothelial cell (an endocardium cell of a rabbit) stimulated by a calcium ionophore (A23187) at $t = 0$ s. (b) Exponential decay of the signal amplitude with the distance between the sensor tip and cell surface.

was consumed through its reaction with $O_2^{\bullet -}$ before diffusing to smooth muscle cells and inducing their relaxation. Consequently, these studies demonstrated that the formation of RNS occurred preferentially in the hypertensive rats, this event leading potentially to the pathophysiological consequences observed in this animal model, as well as in humans.

Detection of Peroxynitrite by Modified Microelectrodes during Ischemia of Endothelial Cells from Heart

The direct electrochemical detection on living cells of peroxynitrite anion release was first reported by two groups in 2000, each one using a different type of electrode and focusing on a different biological application. We will first emphasize studies reported by the L. T. Jin's group (ECNU, Shanghai, China), while our work will be presented in the last part of this chapter. L. T. Jin et al. had previously reported the development of microelectrodes modified by metallophtalocyanines or Schiff bases as sensors for nitric oxide and superoxide.[65,85–87] In these studies they prepared cylindrical carbon fiber or platinum microelectrodes, which had a surface modified by the deposit of tetraaminophtalocyanine manganese(III) (Mn(TAPc)), as a catalyst for peroxynitrite reduction, and subsequently modified by a membrane of the catanionic polymer poly-vinylpyridine (PVP). Briefly, cylindrical carbon fiber microelectrodes (7 μm in diameter) were chemically etched to provide electrodes of 1 μm in diameter at the tip and measuring tens of microns long. Then, their surfaces were electrochemically treated (potential cycling into H_2SO_4), dried and modified by the electrodeposition of poly-Mn(TAPc), and conducted into a solution of 5 m*M* Mn(TAPc) by voltammetric scanning (100 mV s^{-1}) of the potential between +0.5 and +1.0 V vs. Ag/AgCl. Prior to in vitro tests or cell experiments, the electrodes were modified by a layer of PVP, working as a cation repulsive barrier, as well as, as a diffusional barrier for large biological molecules, two characteristics that should significantly improve the selectivity of these microelectrodes for $ONOO^-$. PVP was simply deposited from a solution allowed to dry onto the electrode surface.

In vitro tests with pure peroxynitrite solutions have shown that this species was detectable by its reduction into nitrite and nitrogen dioxide electrocatalyzed by Mn(TAPc) at low potential (+0 V vs. Ag/AgCl). This potential value in conjunction with PVP layer properties afforded a large immunity to many interferents, including several ROS ($O_2^{\bullet -}$, H_2O_2), RNS ($NO^\bullet$, NO_2^-), dioxygen or other molecules that may be present in cell cytosol (L-Arginine, NADPH, ascorbic acid, etc.). These modified microelectrodes displayed a limit of sensitivity of 18 n*M* $ONOO^-$ when used in conjunction with differential pulse amperometry (DPA). The useful range was comprised between this lower limit up to a few hundreds of micromolars.

These electrodes were then used for intracellular measurements of $ONOO^-$ produced into myocardial cells isolated from the heart ventricle of neonatal rats. Peroxynitrite is suspected to be a major component of cardiac injury in the situation of heart ischemia followed by reperfusion, though this hypothesis was previously inferred only the basis of indirect markers of a nitrosative stress (staining of nitro-tyrosine residues). The authors chose to insert the micrometric size tip of their microelectrodes into the beating heart cell. Despite the stress it may induce to the cell (see below), they did not observe any significant change of baseline level of peroxynitrite concentration. After continuous ischemia of the cell, the signal due a priori to peroxynitrite, progressively increased to reach a plateau after 15 min that stayed at the same level of concentration (70 n*M*) for about 15 moreminutes (see Figure 13.10). Afterwards, the signal decreased until reperfusion of the cell that induced an increase of peroxynitrite concentration higher than the initial ischemic step. Control experiments made with an inhibitor of NO synthase or SOD or melatonin, the most potent pharmaceutical inhibitor of $ONOO^-$ in 2000, fully supported the conclusion that the variations of signal detected intracellularly by the modified microelectrodes were due to peroxynitrite. These results provided the first direct evidence that peroxynitrite was the actual mediator of cardiac injuries occurring during ischemia/reperfusion process.

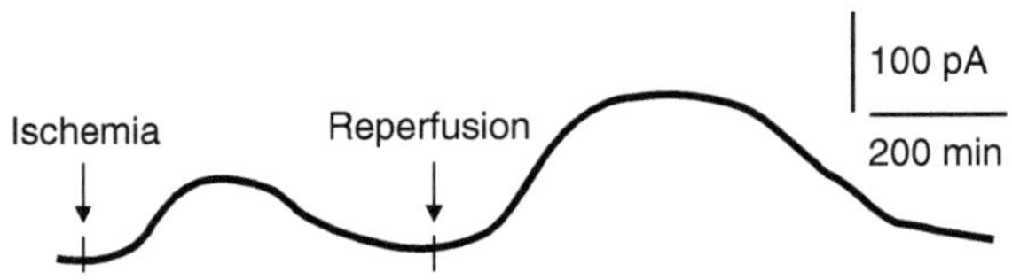

FIGURE 13.10 Time-course of peroxynitrite detected by a Mn(TAPc)/PVP modified microelectrode inside a single beating heart cell in culture after its ischemia/reperfusion.

Detection of Peroxynitrite by Platinized Microelectrodes on Single Human Fibroblasts

Beginning with the results reported in the Detection of Hydrogen Peroxide by Platinized Carbon Fiber Microelectrodes on Skin Fibroblasts section showing that platinized carbon fiber microelectrodes (10 μm in diameter) were adequate tools to measure the release of hydrogen peroxide bursts by single human skin fibroblasts, our group further demonstrated through the development of an original experimental protocol, that the oxidative bursts reflected the production and release of several important ROS and RNS, which could be identified and quantified.[46,88,89] In our second series of studies, the distance between the microelectrode tip and cell surface where it was depolarized, was varied and optimized to ensure the detection of species with short time-life that would not be able to diffuse to the electrode surface before decomposing under the conditions of Figure 13.8 ($d \approx 50$ μm). Furthermore, we have confirmed that under these new conditions, the collection efficiency of the cell response increased and was maximal (greater than 90%) when the distance was lower than 5 μm, indicating that species released by the cell in its environment were quantitatively detected.[89] Then, a large series of different experiments conducted at different amperometric oxidative potentials on about 1000 cells from the same wild type control strain, demonstrated that their response increased in amplitude with the potential and consequently, that this response included the measurement of several electroactive species (Figure 13.11a). These data enabled us to construct an equivalent steady-state voltammogram of the average cell oxidative burst contents by measuring the current versus the potential (one point at each 25 mV over the potential range between +150 and +900 mV vs. SSCE) (Figure 13.11b). The voltammogram clearly evidenced three different plateaus corresponding to at least three oxidation waves (labeled I, II and III in Figure 13.11b). However, when spikes were normalized versus their maximum current, no difference of time-profile could be observed, indicating that the different species were released following the same overall kinetic. The slowly developing kinetic (rise-time close to 1 s and total time-length of about 90 s) could not correspond to the time-course of a passive release by diffusion of ROS and RNS initially located in the cell cytosol, but rather to the kinetic of biological production of these compounds or of their mother species (compare also with other studies reported above) stimulated by the cell membrane depolarization.

The voltammogram constructed from cell experiments was then compared to the oxidative response in vitro of stable solutions of several ROS and RNS potentially derived from superoxide and nitric oxide.[46,88,89] It was determined that waves II and III correspond, in terms of their potential position and transfer kinetic to the detection of $NO^{\bullet}$ and NO_2^-, respectively. Conversely, wave I did not correspond to the detection of a single species such as H_2O_2, which had a slightly lower oxidation half-wave potential (see Figure 13.11b), or $ONOO^-$, which had a slightly higher oxidation half-wave potential. An exhaustive study of peroxynitrite anion electrochemistry[46] on platinum and the platinized surface of electrodes enabled us to show that the oxidation wave I corresponded to the mixing of current of H_2O_2 and $ONOO^-$ oxidation waves, both located in a near range of potentials. Finally, the arithmetic addition of the voltammetric response of four different species (H_2O_2, $ONOO^-$, $NO^{\bullet}$, and NO_2^-) adjusted in concentration was in perfect agreement with the voltammogram determined from the cell oxidative response (see the comparison between data

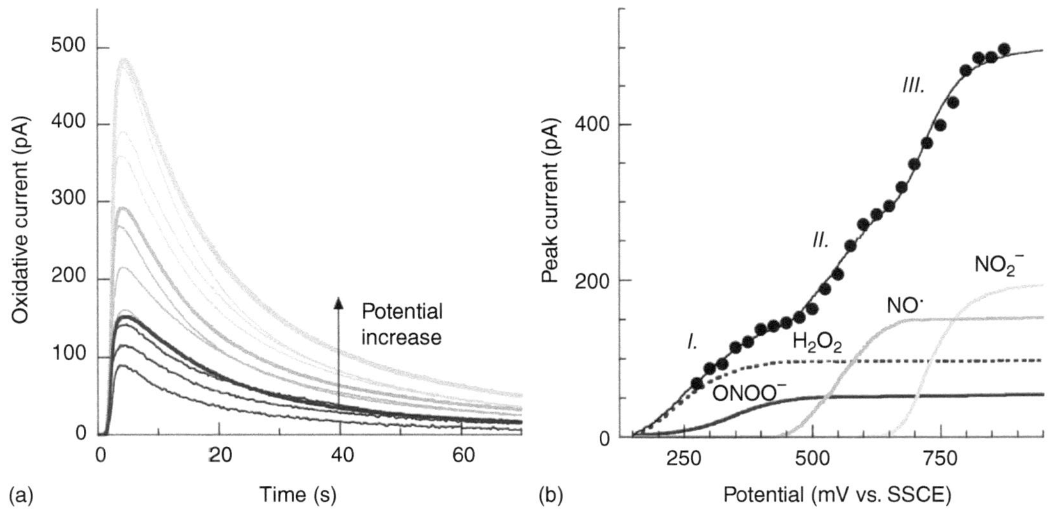

FIGURE 13.11 Determination by platinized carbon fibre microelectrodes of the species released by single human fibroblasts during oxidative burst responses induced by mechanical depolarization of the cell membrane. (a) Amperometric measurement of the cell response as a function of the potential applied at the microelectrode ($200 \leq E \leq 900$ mV vs. SSCE). (b) Evolution of the burst maximum current with the potential (filled circles). The curve was compared to the oxidation waves detected in solutions of H_2O_2, $ONOO^-$, NO_2^- and NO° at the same microelectrodes. The sum of the individual in vitro voltammograms adjusted in concentration allowed to build a global voltammogram (solid line), compared with data points from cell experiments.

points and the solid line figuring the predicted voltammogram in Figure 13.11b). This analysis has demonstrated that the oxidative bursts of fibroblasts corresponded to the emission of a cocktail of several ROS and RNS, among which the very reactive peroxynitrite was one, and that each species (though the presence of NO_3^- ions could not be detected due to their electro-inactivity in our experimental conditions) could be determined and selectively quantified by amperometric measurements at several potentials.

Further studies on cells treated with different pharmacological agents such as enzymatic inhibitors have defined that enzymatic systems of NADPH oxidase and NO synthase types were simultaneously activated following the cell stimulation by the microcapillary stress and were responsible for the production of the two primary species, $O_2^{\bullet -}$ and $NO^{\bullet}$ (Figure 13.12a).[88] Obviously, these primary species cannot be detected directly at the microelectrode surface, given that when present in the same solution they react together to form $ONOO^-$, which cannot survive long since it also rapidly evolves into NO_2^- (this later being observed into the extracellular medium by the microelectrode). However, the two types of enzymatic systems are located at different cell sites (presumably cytosol versus membrane) allowing superoxide anion to disproportionate and form hydrogen peroxide before reacting with nitric oxide. These hypotheses in conjunction with the previous characterization of a maximal collection efficiency of the species released by the fibroblasts have enabled us to convert the amperometric currents detected at the plateaus potentials of waves I, II and III into fluxes of emission (Figure 13.12b) according to Faraday's laws:

$$I_j = n_j F \times \Phi_j \tag{13.16}$$

The conversion takes into account for each species (j) the known number of electrons exchanged during its electrochemical oxidation at the electrode surface. This provides a better

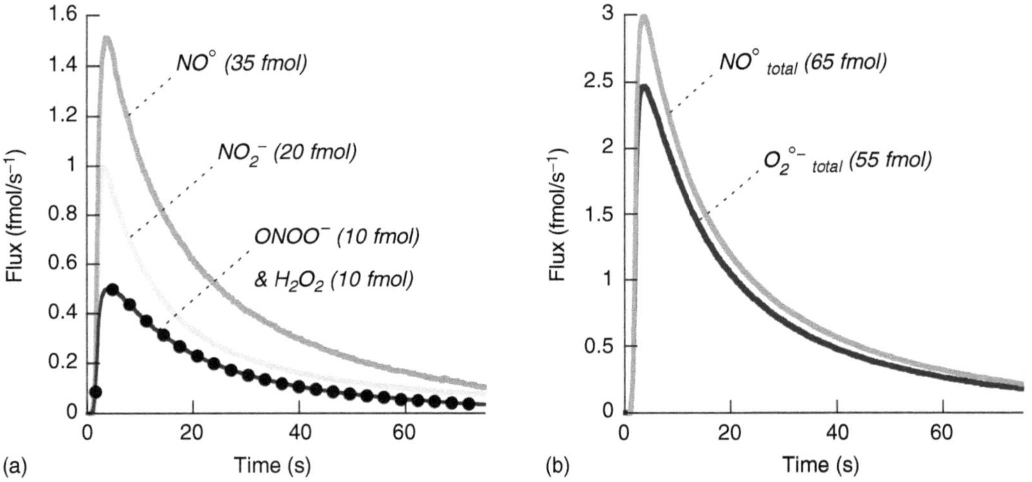

FIGURE 13.12 (a) Determination from amperometric curves in Figure 13.11a of the flux of each species (H_2O_2, $ONOO^-$, NO_2^-, and NO°) released by single human fibroblasts during oxidative burst responses. (b) Evaluation of the flux of the primary species ($O_2^{\circ-}$ and NO°) produced by a fibroblast. Total amount of each species given in parenthesis was calculated from the integration of the corresponding flux curve.

measurement than simple oxidation currents of the true exchanges between a single cell and its environment during an oxidative stress process (Figure 13.12b).

The determination of the individual fluxes of the different species released by the cells allowed us to reconstruct the fluxes of production of the primary species giving rise to these ROS and RNS. Indeed, because all species originate from $O_2^{\bullet -}$ and $NO^\bullet$, fluxes of the latter could be evaluated according to the conservation of matter Equation 13.17 and Equation 13.18:

$$(\Phi_{O_2^-})^{prod} = 2(\Phi_{H_2O_2})^{mes} + (\Phi_{ONOO^-})^{mes} + (\Phi_{NO_2^-})^{mes} \quad (13.17)$$

$$(\Phi_{NO^\circ})^{prod} = (\Phi_{NO^\circ})^{mes} + (\Phi_{ONO_2^-})^{mes} + (\Phi_{NO_2^-})^{mes} \quad (13.18)$$

Each curve was subsequently integrated and provided a measurement of the total amount of each compound produced during the oxidative burst (values in parenthesis in Figure 13.12c and Figure 13.12d). For both primary species, the quantities are in the range of tens of femtomoles and may be considered as infinitely small from a macroscopic point of view and considering actual analytical standards. Nevertheless, when considering the "living space" of a single cell, equivalent at maximum to a few times its own dimensions, these amounts are not negligible in terms of the local concentrations (by way of the kinetic effects) they produce locally. They correspond to local high, or at least efficient concentrations, of messenger, activator, or toxic molecules.

The results presented here have led ultimately to several applications for biomedical purposes, such as, studies of the oxidative stress implication in HIV infection of blood lymphocytes,[79,90] studies of the oxidative stress implication in the carcinogenesis of skin fibroblasts[78,91], or even to studies of primary biological mechanisms, such as, ROS and RNS production by macrophages during phagocytosis.

CONCLUSIONS AND PERSPECTIVES

The studies described here demonstrate that the electrochemical measurements at microelectrodes are methods of high interest for studying biological processes implicating superoxide and/or nitric oxide and their derivatives. The advantages of the electrochemical analysis over other biophysical methods include the possible direct detection, the fast response-time and selectivity of the sensors, and, above all, a single cell analysis. Clearly, when the microelectrode surface can be positioned at close distance to the cell (or the biological source), i.e., in an artificial synapse configuration, the infinitely low amplitude of the biological information can even so be detected and studied.

Future developments of the electrochemical methodologies for studies of oxidative stress processes should focus on (1) the improvement of microelectrodes selectivity for each ROS and RNS of interest; (2) multiple analysis on several cells or of several species done in parallel by several electrodes eventually integrated into microsystems; (3) intracellular measurements owing to nanometer size electrodes; (4) detection of a single molecule activity such as the one of an enzymatic system generating $O_2^{\bullet-}$ or $NO^{\bullet}$, possibly located into the cell cytosol or the cell membrane.

REFERENCES

1. Halliwell, B. and Gutteridge, J. M. C., *Free Radicals in Biology and Medecine*, 3rd ed., Oxford University Press, Oxford, 1999.
2. Davies, K. J. A. and Ursini, F., *The Oxygen Paradox*, CLEUP University Press, Padova, 1995.
3. Chance, B., In *Annual Reviews of Biophysics and Biophysical Chemistry*, Annual Reviews, Palo Alto, 20, pp.1–30, 1991.
4. Sies, H., Biochemistry of oxidative stress, *Angewandte Chemie International Edition in English*, 25, 1058–1071, 1986.
5. Landis, G. N. and Tower, J., Superoxide dismutase evolution and life span regulation, *Mechanisms of Ageing and Development*, 126 (3), 365–379, 2005.
6. Miller, A. F., Superoxide dismutases: active sites that save, but a protein that kills, *Current Opinion in Chemical Biology*, 8 (2), 162–168, 2004.
7. Kinnula, V. L. and Crapo, J. D., Superoxide dismutases in malignant cells and human tumors, *Free Radical Biology and Medicine*, 36 (6), 718–744, 2004.
8. Niviere, V. and Fontecave, M., Discovery of superoxide reductase: an historical perspective, *Journal of Biological Inorganic Chemistry*, 9 (2), 119–123, 2004.
9. Inoue, M., Sato, E. F., Nishikawa, M., Park, A. M., Kira, Y., Imada, I., and Utsumi, K., Mitochondrial generation of reactive oxygen species and its role in aerobic life, *Current Medicinal Chemistry*, 10 (23), 2495–2505, 2003.
10. Maier, C. M. and Chan, P. H., Role of superoxide dismutases in oxidative damage and neurodegenerative disorders, *Neuroscientist*, 8 (4), 323–334, 2002.
11. Zelko, I. N., Mariani, T. J., and Folz, R. J., Superoxide dismutase multigene family: a comparison of the CuZn-SOD (SOD1): Mn-SOD (SOD2): and EC-SOD (SOD3) gene structures, evolution, and expression, *Free Radical Biology and Medicine*, 33 (3), 337–349, 2002.
12. Burkitt, M. J., Chemical, biological and medical controversies surrounding the Fenton reaction, *Progress in Reaction Kinetics and Mechanism*, 28 (1), 75–103, 2003.
13. Ames, B. N. and Shigenagra, M. K., DNA damage by endogenous oxidants and mitogenesis as causes of aging and cancer, In *Molecular Biology of Free Radical Scavenging Systems*, Scandalios, J. S., Ed., Cold Spring Harbor Laboratory Press, Plainview, NY, pp. 1–22, 1992.
14. Chance, B. and Higgins, J., Peroxidase kinetics in coupled oxidation—an experimental and theoretical study, *Archives of Biochemistry and Biophysics*, 41 (2), 432–441, 1952.
15. Chance, B. and Maehly, A. C., Assay of catalases and peroxidases, *Methods in Enzymology*, 2, 764–775, 1955.
16. Sies, H., *Metabolic Compartimentation*, Academic Press, New York, 1982.
17. Christman, M. F., Morgan, R. W., Jacobson, F. S., and Ames, B. N., Positive control of a regulon for defenses against oxidative stress and some heat-shock proteins in *Salmonella-typhimurium*, *Cell*, 41, 735–762, 1985.
18. Löntz, W., Sirsjö, A., Liu, W., Lindberg, M., Rollman, O., and Törmä, H., Increased mRNA expression of manganese superoxide dismutase in psoriasis skin lesions and in cultured human keratinocytes exposed to IL-1β and TNFα, *Free Radical Biology and Medicine*, 18, 349–357, 1995.

19. Groemping, Y. and Rittinger, K., Activation and assembly of the NADPH oxidase: a structural perspective, *Biochemical Journal*, 386, 401–416, 2005.
20. Robinson, J. M., Ohira, T., and Badwey, J. A., Regulation of the NADPH-oxidase complex of phagocytic leukocytes. Recent insights from structural biology, molecular genetics, and microscopy, *Histochemistry and Cell Biology*, 122 (4), 293–304, 2004.
21. Segal, A. W., The NADPH oxidase and chronic granulomatous disease, *Molecular Medicine Today*, March, 129–135, 1996.
22. Segal, A. W. and Abo, A., The biochemical basis of the NADPH oxidase of phagocytes, *Trends in Biochemical Sciences*, 18 (2), 43–47, 1993.
23. Southan, G. J. and Szabo, C., Selective pharmacological inhibition of distinct nitric oxide synthase isoforms, *Biochemical Pharmacology*, 51, 383–394, 1996.
24. Stuehr, D. J., Mammalian nitric oxide synthases, *Biochimicaet Biophysica Acta-Bioenergetics*, 1411 (2–3), 217–230, 1999.
25. Alderton, W. K., Cooper, C. E., and Knowles, R. G., Nitric oxide synthases: structure, function and inhibition, *Biochemical Journal*, 357, 593–615, 2001.
26. Li, H. Y. and Poulos, T. L., Structure–function studies on nitric oxide synthases, *Journal of Inorganic Biochemistry*, 99 (1), 293–305, 2005.
27. Koppenol, W. H., The basic chemistry of nitrogen monoxide and peroxyntrite, *Free Radical Biology and Medicine*, 25, 385–391, 1998.
28. Koppenol, W. H., Moreno, J. J., Pryor, W. A., Ischiropoulos, H., and Beckman, J. S., Peroxynitrite, a cloaked oxidant formed by nitric oxide and superoxide, *Chemical Research in Toxicology*, 5, 834–842, 1992.
29. Nauser, T. and Koppenol, W. H., The rate constant of the reaction of superoxide with nitrogen monoxide: approaching the diffusion limit, *Journal of Physical Chemistry A*, 106 (16), 4084–4086, 2002.
30. Kissner, R., Beckman, J. S., and Koppenol, W. H., Peroxynitrite studied by stopped-flow spectroscopy, *Nitric Oxide, Pt C, Methods in Enzymology*, 303, 342–352, 1999.
31. Nauser, T., Merkofer, M., Kissner, R., and Koppenol, W. H., Gibbs energy of formation of peroxynitrite, *Chemical Research in Toxicology*, 14 (4), 348–350, 2001.
32. Meli, R., Nauser, T., Latal, P., and Koppenol, W. H., Reaction of peroxynitrite with carbon dioxide: intermediates and determination of the yield of CO_3 center dot- and NO_2 center dot, *Journal of Biological Inorganic Chemistry*, 7 (1–2), 31–36, 2002.
33. Kissner, R. and Koppenol, W. H., Product distribution of peroxynitrite decay as a function of pH, temperature, and concentration, *Journal of the American Chemical Society*, 124 (2), 234–239, 2002.
34. Kissner, R., Nauser, T., Kurz, C., and Koppenol, W. H., Peroxynitrous acid—where is the hydroxyl radical? *IUBMB Life*, 55 (10–11), 567–572, 2003.
35. Herold, S., Exner, M., and Boccini, F., The mechanism of the peroxynitrite-mediated oxidation of myoglobin in the absence and presence of carbon dioxide, *Chemical Research in Toxicology*, 16 (3), 390–402, 2003.
36. Kirsch, M., Korth, H. G., Wensing, A., Sustmann, R., and de Groot, H., Product formation and kinetic simulations in the pH range 1–14 account for a free-radical mechanism of peroxynitrite decomposition, *Archives of Biochemistry and Biophysics*, 418 (2), 133–150, 2003.
37. Pfeiffer, S., Gorren, A. C. F., Schimdt, K., Werner, E. R., Hansert, B., Bohle, D. S., and Mayer, B., Metabolic fate of peroxynitrite in aqueous solution, *Journal of Biological Chemistry*, 272 (6), 3465–3470, 1997.
38. Ford, P. C., Wink, D. A., and Stanbury, D. M., Autoxidation kinetics of aqueous nitric-oxide, *FEBS Letters*, 326 (1–3), 1–3, 1993.
39. Pogrebnaya, V. L., Usov, A. P., Baranov, A. V., Nesterenko, A. I., and Bezyazychnyi, P. I., Oxidation of nitric-oxide by oxygen in liquid-phase, *Journal of Applied Chemistry of the USSR*, 48 (5), 1004–1007, 1975.
40. Koppenol, W. H., Thermodynamics of reactions involving nitrogen–oxygen compounds, *Nitric Oxide, Pt A—Sources and Detection of NO; NO Synthase, Methods in Enzymology.*, 268, 7–12, 1996.
41. Lee, K. Y., Amatore, C., and Kochi, J. K., Electron-transfer kinetics and ternary equilibria of the NO_2^+ /NO_2/N_2O_4 system by transient electrochemistry, *Journal of Physical Chemistry*, 95 (3), 1285–1294, 1991.

42. Possel, H., Noack, H., Keilhoff, G., and Wolf, G., Life imaging of peroxynitrite in rat microglial and astroglial cells: role of superoxide and antioxidants, *Glia*, 38 (4), 339–350, 2002.
43. Gaudry-Talarmain, Y. M., Moulian, N., Meunier, F. A., Blanchard, B., Angaut-Petit, D., Faille, L., and Ducrocq, C., Nitric oxide and peroxynitrite affect differently acetylcholine release, choline acetyltransferase activity, synthesis, and compartmentation of newly formed acetylcholine in Torpedo marmorata synaptosomes, *Nitric Oxide—Biology and Chemistry*, 1 (4), 330–345, 1997.
44. Possel, H., Noack, H., Augustin, W., Keilhoff, G., and Wolf, G., 2,7-Dihydrodichlorofluorescein diacetate as a fluorescent marker for peroxynitrite formation, *FEBS Letters*, 416 (2), 175–178, 1997.
45. Wang, P. H. and Zweier, J. L., Measurement of nitric oxide and peroxynitrite generation in the postischemic heart—evidence for peroxynitrite-mediated reperfusion injury, *Journal of Biological Chemistry*, 271 (46), 29223–29230, 1996.
46. Amatore, C., Arbault, S., Bruce, D., de Oliveira, P., Erard, M., and Vuillaume, M., Characterization of the electrochemical oxidation of peroxynitrite: relevance to oxidative stress bursts measured at the single cell level, *Chemistry—A European Journal*, 7 (19), 4171–4179, 2001.
47. Kelm, M., Dahmann, R., Wink, D., and Feelisch, M., The nitric oxide/superoxide assay, *Journal of Biological Chemistry*, 272 (15), 9922–9932, 1997.
48. Pfeiffer, S., Lass, A., Schmidt, K., and Mayer, B., Protein tyrosine nitration in cytokine-activated murine macrophages—involvement of a peroxidase/nitrite pathway rather than peroxynitrite, *Journal of Biological Chemistry*, 276 (36), 34051–34058, 2001.
49. de Toledo, G. A., Fernandezchacon, R., and Fernandez, J. M., Release of secretory products during transient vesicle fusion, *Nature*, 363 (6429), 554–558, 1993.
50. Wightman, R. M., Jankowski, J. A., Kennedy, R. T., Kawagoe, K. T., Schroeder, T. J., Leszczyszyn, D. J., Near, J. A., Diliberto, E. Jr., and Viveros, O. H., Temporally resolved catecholamine spikes correspond to single vesicle release from individual chromaffin cells, *Proceedings of the National Academy of Sciences of the United States of America*, 88 (23), 10754–10758, 1991.
51. Schroeder, T. J., Borges, R., Finnegan, J. M., Pihel, K., Amatore, C., and Wightman, R. M., Temporally resolved, independent stages of individual exocytotic secretion events, *Biophysical Journal*, 70 (2), 1061–1068, 1996.
52. Wightman, R. M., Schroeder, T. J., Finnegan, J. M., Ciolkowski, E. L., and Pihel, K., Time course of release of catecholamines from individual vesicles during exocytosis at adrenal medullary cells, *Biophysical Journal*, 68 (1), 383–390, 1995.
53. Amatore, C., Arbault, S., Bonifas, I., Bouret, Y., Erard, M., and Guille, M., Dynamics of full fusion during vesicular exocytotic events: release of adrenaline by chromaffin cells, *Chemphyschem*, 4 (2), 147–154, 2003.
54. Amatore, C., Ultramicroelectrodes: their basic properties and their use in semi-artificial synapses, *Comptes Rendus de l' Académie des Sciences Paris*, 323, 757–771, 1996.
55. Bruns, D., Detection of transmitter release with carbon fiber electrodes, *Methods*, 33 (4), 312–321, 2004.
56. Schroeder, T. J., Jankowski, J. A., Kawagoe, K. T., Wightman, R. M., Lefrou, C., and Amatore, C., Analysis of diffusional broadening of vesicular packets of catecholamines released from biological cells during exocytosis, *Analytical Chemistry*, 64 (24), 3077–3083, 1992.
57. Green, M. J., Hill, H. A. O., Tew, D. G., and Walton, N. J., An opsonised electrode. The direct electrochemical detection of superoxide generated by human neutrophils, *FEBS Letters*, 170, 69–72, 1984.
58. Hill, H. A. O., Tew, D. G., and Walton, N. J., An opsonised microelecrodelectrode. Observation of the respiratory burst of single human neutrophils, *FEBS Letters*, 191, 257–263, 1985.
59. Green, M. J., Hill, H. A. O., and Tew, D. G., The rate of oxygen-consumption and superoxide anion formation by stimulated human-neutrophils—the effect of particle concentration and size, *FEBS Letters*, 216 (1), 31–34, 1987.
60. Tanaka, K., Tsuyama, T., Karatsu, Y., and Iizuka, T., Amperometric determination of superoxide anions generated from a single phagocyte with femto-ampere sensitivity, *Bioelectrochemistry and Bioenergetics*, 41, 201–203, 1996.
61. Isogai, Y., Tsuyama, T., Osada, H., Iizuka, T., and Tanaka, K., Direct measurement of oscillatory generation of superoxide anions by single phagocytes, *FEBS Letters*, 380 (3), 263–266, 1996.

62. Mesaros, S., Vankova, Z., Grunfeld, S., Mesarosova, A., and Malinski, T., Preparation and optimization of superoxide microbiosensor, *Analytica Chimica Acta*, 358 (1), 27–33, 1998.
63. Villeneuve, N., Bedioui, F., Voituriez, K., Avaro, S., and Vilaine, J. P., Electrochemical detection of nitric oxide production in perfused pig coronary artery: comparison of the performances of two electrochemical sensors, *Journal of Pharmacological and Toxicological Methods*, 40 (2), 95–100, 1998.
64. Pontie, M. and Bedioui, F., Selective and sensitive electrochemical biosensing of superoxide anion production by biological systems: a short overview of recent trends, *Analusis*, 27 (7), 564–570, 1999.
65. Xue, J. A., Xian, Y. Z., Ying, X. Y., Chen, J. S., Wang, L., and Jin, L. T., Fabrication of an ultra-microsensor for measurement of extracellular myocardial superoxide, *Analytica Chimica Acta*, 405 (1–2), 77–85, 2000.
66. Privat, C., Stepien, O., David-Dufilho, M., Brunet, A., Bedioui, F., Marche, P., and Devynck, J., Superoxide release from interleukin-1B-stimulated human vascular cells: in situ electrochemical measurement, *Free Radical Biology and Medicine*, 27, 554–559, 1999.
67. Cooper, J. M., Greenough, K. R., and McNeil, C. J., Direct electron-transfer reactions between immobilized cytochrome-*c* and modified gold electrodes, *Journal of Electroanalytical Chemistry*, 347 (1–2), 267–275, 1993.
68. Manning, P., McNeil, C. J., Cooper, J. M., and Hillhouse, E. W., Direct, real-time sensing of free radical production by activated human glioblastoma cells, *Free Radical Biology and Medicine*, 24 (7–8), 1304–1309, 1998.
69. Tammeveski, K., Tenno, T. T., Mashirin, A. A., Hillhouse, E. W., Manning, P., and McNeil, C. J., Superoxide electrode based on covalently immobilized cytochrome *c*: modelling studies, *Free Radical Biology and Medicine*, 25 (8), 973–978, 1998.
70. Tolias, C. M., McNeil, C. J., Kazlauskaite, J., and Hillhouse, E. W., Superoxide generation from constitutive nitric oxide synthase in astrocytes in vitro regulates extracellular nitric oxide availability, *Free Radical Biology and Medicine*, 26 (1–2), 99–106, 1999.
71. Thomson, L., Trujillo, M., Telleri, R., and Radi, R., Kinetics of cytochrome *c*(2+) oxidation by peroxynitrite—implications for superoxide measurements in nitric oxide-producing biological-systems, *Archives of Biochemistry and Biophysics*, 319 (2), 491–497, 1995.
72. Arbault, S., Pantano, P., Jankowski, J. A., Vuillaume, M., and Amatore, C., Monitoring an oxidative stress mechanism at a single human fibroblast, *Analytical Chemistry*, 67, 3382–3390, 1995.
73. Ikariyama, Y., Yamauchi, S., Yukiashi, T., and Ushioda, H., Surface control of platinized platinum as a transducer matrix for micro-enzyme electrodes, *Journal of Electroanalytical Chemistry*, 251, 267–274, 1988.
74. Ikariyama, Y., Yamauchi, S., Yukiashi, T., and Ushioda, H., Micro-enzyme electrode prepared on platinized platinum, *Analytical Letters*, 20, 1407–1416, 1987.
75. de Boer, J. and Hoeijmakers, J. H. J., Nucleotide excision repair and human syndromes, *Carcinogenesis*, 21 (3), 453–460, 2000.
76. Ford, J. M. and Hanawalt, P. C., Role of DNA excision repair gene defects in the etiology of cancer, *Genetic Instability and Tumorigenesis*, 47–70, 1997.
77. Arbault, S., Pantano, P., Sojic, N., Amatore, C., BestBelpomme, M., Sarasin, A., and Vuillaume, M., Activation of the NADPH oxidase in human fibroblasts by mechanical intrusion of a single cell with an ultramicroelectrode, *Carcinogenesis*, 18 (3), 569–574, 1997.
78. Arbault, S., Sojic, N., Bruce, D., Amatore, C., Sarasin, A., and Vuillaume, M., Oxidative stress in cancer prone xeroderma pigmentosum fibroblasts. Real-time and single cell monitoring of superoxide and nitric oxide production with microelectrodes, *Carcinogenesis*, 25 (4), 509–515, 2004.
79. Lachgar, A., Sojic, N., Arbault, S., Bruce, D., Sarasin, A., Amatore, C., Bizzini, B., Zagury, D., and Vuillaume, M., Amplification of the inflammatory cellular redox state by human immunodeficiency virus type 1-immunosuppressive Tat and gp160 proteins, *Journal of Virology*, 73, 1447–1452, 1999.
80. Malinski, T. and Taha, Z., Nitric oxide release from a single cell measured in situ by a porphyrinic-based microsensor, *Nature*, 358, 676–677, 1992.
81. Brovkovych, V., Stolarczyk, E., Oman, J., Tomboulian, P., and Malinski, T., Direct electrochemical measurement of nitric oxide in vascular endothelium, *Journal of Pharmaceutical and Biomedical Analysis*, 19 (1–2), 135–143, 1999.

82. Pinsky, D. J., Patton, S., Mesaros, S., Brovkovych, V., Kubaszewski, E., Grunfeld, S., and Malinski, T., Mechanical transduction of nitric oxide synthesis in the beating heart, *Circulation Research*, 81 (3), 372–379, 1997.
83. Mesaros, S., Grunfeld, S., Mesarosova, A., Bustin, D., and Malinski, T., Determination of nitric oxide saturated (stock) solution by chronoamperometry on a porphyrine microelectrode, *Analytica Chimica Acta*, 339 (3), 265–270, 1997.
84. Malinski, T., Mesaros, S., and Tomboulian, P., Nitric oxide measurement using electrochemical methods, *Nitric Oxide, Pt A—Sources and Detection of NO; NO Synthase, Methods in Enzymology*, 268, 58–69, 1996.
85. Xian, Y. Z., Zhang, W., Xue, J., Ying, X. Y., Jin, L. T., and Jin, J. Y., Measurement of nitric oxide released in the rat heart with an amperometric microsensor, *Analyst*, 125 (8), 1435–1439, 2000.
86. Mao, L. Q., Yamamoto, K., Zhou, W. L., and Jin, L. T., Electrochemical nitric oxide sensors based on electropolymerized film of M(salen) with central ions of Fe, Co, Cn, and Mn, *Electroanalysis*, 12 (1), 72–77, 2000.
87. Mao, L. Q., Tian, Y., Shi, G. Y., Liu, H. Y., Jin, L. T., Yamamoto, K., Tao, S., and Jin, J. Y., A new ultramicrosensor for nitric oxide based on electropolymerized film of nickel salen, *Analytical Letters*, 31 (12), 1991–2007, 1998.
88. Amatore, C., Arbault, S., Bruce, D., de Oliveira, P., Erard, M., Sojic, N., and Vuillaume, M., Nitrogen monoxide and oxidative stress: composition and intensity of cellular oxidative bursts cocktail. A study through artificial electrochemical synapses on single human fibroblasts, *Analusis*, 28 (6), 506–517, 2000.
89. Amatore, C., Arbault, S., Bruce, D., de Oliveira, P., Erard, M., and Vuillaume, M., Analysis of individual biochemical events based on artificial synapses using ultramicroelectrodes: cellular oxidative burst, *Faraday Discussions*, 116, 356, 2000. (see also page 116)
90. Arbault, S., Edeas, M., Legrand-Poels, S., Sojic, N., Amatore, C., Piette, J., Best-Belpomme, M., Lindenbaum, A., and Vuillaume, M., Phenylarsine oxide inhibits ex vivo HIV-1 expression, *Biomedecine and Pharmacotherapy*, 51, 430–438, 1997.
91. Raha, S. and Robinson, B. H., Mitochondria, oxygen free radicals, disease and ageing, *Trends in Biochemical Sciences*, 25 (10), 502–508, 2000.

14 Electrochemistry at the Cell Membrane/Solution Interface

Nathan Wittenberg, Marc Maxson, Daniel Eves, Ann-Sofie Cans, and Andrew G. Ewing

CONTENTS

INTRODUCTION

Electrochemistry in ultra-small environments has emerged as an increasingly important technique for fundamental studies of diffusionally restricted reactions, low sample volume analysis, and single cell neurochemistry. Development of electrochemical methods for detection of neurotransmitters began with the ground-breaking work of Adams[1] and has progressed to the point where it is now possible to detect the release of a neurotransmitter from a single vesicle, first demonstrated in the seminal work by Wightman and co-workers.[2,3] In the pioneering experiments, a carbon fiber electrode 5 μm in diameter was placed adjacent to a bovine adrenal chromaffin cell isolated in a culture dish, and the cell was stimulated to release either by chemical or mechanical means. Understanding the chemistry and structure of single cells is an area of great interest in the biological and medical sciences. Indeed, books have been written on this broad topic.[4] In neuroscience, knowledge of the chemical composition and dynamics of single nerve cells should lead to better models of the cellular neurotransmission process. The key dynamic event in neuronal communication is exocytosis. This is a process that has been investigated extensively for several decades.[5,6] The process of exocytosis can be summarized as the docking of vesicles (storage compartments) to the cell membrane and subsequent release of the contents by fusion of the vesicle and cell membranes. This process allows the conversion of an electrical signal (action potential) to a chemical signal (messenger release and receptor recognition), which is necessary for exocytotic communication between cells. Methods to observe and quantify individual exocytotic events have traditionally revolved around electron microscopy and patch-clamp capacitance measurements.[7] In 1990, Wightman and co-workers showed that they could directly monitor individual exocytotic events involving easily oxidized messengers and occurring on the millisecond time scale by use of amperometric measurements at microelectrodes.[3] This was first applied by Wightman's group to adrenal chromaffin cells,[2] and later by Neher's group.[8] This chapter will focus on the development of electrochemical and related methods to analyze exocytosis events at cells, as well as artificial cells. These methods, emphasizing amperometry at microelectrodes, will be reviewed in terms of different cell types examined and discussed for representative and specific studies of neuroscience.

Basics of Electrodes for Single Cell Measurements

Electrode Fabrication and Testing

Carbon fiber microelectrodes were developed in several laboratories in the late 1970s. Leaders among these were the Wightman[9,10] and Gonon[11,12] groups, who applied this tool to neuroscience. The method was a breakthrough for several reasons. First, the conductive carbon fiber could carry a current while maintaining sensitivity to reductants, thus increasing the working lifetime of an electrode. Second, carbon fibers as small as 5 μm were resistant to strain, allowing them to be placed snugly against cell surfaces, which allowed greater sensitivity and reproducibility. For a good discussion of the factors affecting electrode sensitivity, selectivity, and temporal response, see the paper by Cahill et al.[13] Electrodes used for studying single cells are usually prepared by aspirating a single fiber through a glass capillary. Then the capillary is pulled on a commercial puller to produce a long tapered fiber-containing pipette that is later cut at a cross sectional diameter of 8 to 10 μm on a microscope using a fine razor blade. Cut electrodes are immersed in freshly prepared high quality *epoxy* to create a seal between the glass tip and carbon fiber. Finally, no more than a few hours before experiments, electrodes are beveled to 45° on a rotary micro grinder to produce a fine-polished angled tip.[14,15] It is our experience that "bad batches" of prepared electrodes can most often be traced to old epoxy. Additionally, higher oven temperatures appear to adversely affect the curing of epoxy for electrodes.Working electrodes are generally prepared from 5 to 10 μm diameter carbon fibers. The signal-to-noise ratio improves as the electrode size approaches the size of vesicles, but larger electrodes can detect a greater number of events.[13,16] Electrode sensitivity requires that the surface be free of adsorbed molecules, such as proteins and oxidized products.[13,17] This is often achieved by frequent testing and beveling between experiments. Alternatively, electrode response can be maintained by dipping electrodes into a solution of 5.0% Nafion.[17] The Nafion coats the surface with a perfluorosolfonated anionic polymer that allows small cations to reach the surface, but prevents small anions and larger proteins from binding irreversibly to the surface.[17] In addition to selectivity for detection of cations, Nafion coating reduces fouling from proteins adsorbing to the electrode surface. Electrodes can be tested in standard solutions of the analyte using single sweep voltammetry. Typically, electrodes are back-filled with 3 *M* KCl and a silver wire inserted. The wire is connected to a current amplifier, and a AgCl reference electrode is placed into the analyte-containing solution. For catecholamines, the voltage is swept from -1.5 to $+0.5$ V to generate a voltammogram. The shape of the curve indicates various characteristics of the electrode. In general, the sharper the rise, the greater the sensitivity; the narrower the width of the curve, the better the signal-to-noise ratio.

Basic Amperometry

Electrochemical detection of biogenic amines is typically carried out in either the amperometric or voltammetric mode. Voltammetry is discussed in the next section. Amperometric experiments involve holding the electrode at high enough potentials to oxidize molecules in the solution in a diffusion-limited fashion. In the amperometric mode, changing concentration of easily oxidized species results in a current vs. time trace that changes with the amount of oxidation. By integrating the current transient, it is possible via Faraday's Law to determine the number of molecules oxidized ($N = Q/nF$, where N is the number of moles detected, Q is the integrated charge, n is the number of electrons transferred in the oxidation reaction, and F is the Faraday constant).

Amperometry has been used to study exocytosis in primary cultures,[3,10,18,19] immortalized cell lines,[20,21] brain slices,[22] and at intact neurons in vivo.[23,24] In general, candidate cell systems have been limited to those that release an oxidizable substance, usually a catecholamine-, serotonin-, or tyrosine/tryptophan-containing peptide. Immortalized cultures allow for single, isolated cells to be studied, while primary cultures offer the advantage of modeling cells that can be studied in the context of belonging to a network, continuing to receive information from adjacent cells. Since amperometry can also be used within an intact brain, the technique can potentially be applied in a setting where cell communication remains relatively undisturbed. However, difficulty in production and placement of ultrasmall electrodes limits in vivo amperometry to measurements of the extracellular "overflow" catecholamine flux, not intrasynaptic concentrations. Ultrasmall electrodes have not yet been constructed to fit into a synapse. Amperometry is well suited to measuring secretion from cells because of its ability to quantify release from vesicles on a millisecond time scale (Figure 14.1).[25] To carry out amperometry at single cells, a small electrode is placed near the cell and held at a potential where oxidation is diffusion limited. Stimulant is applied with a small pipette (Figure 14.1a) resulting in current transients for each exocytosis event. The first experiment to measure exocytosis used adrenal cells and detected the catechols norepinephrine and epinephrine.[2,3] The general oxidation reaction for catechols is shown in Figure 14.1b and a typical current–time trace is shown in Figure 14.1c. The number of molecules detected from a release event can be calculated using Faraday's law (Figure 14.1d). Experimental determination of charge allows calculation of the number of moles of molecules detected by rearranging the Faraday equation. During these experiments, the electrode is positioned near a cell, where exuded neurotransmitter can be detected (Figure 14.1a). Calculations have been carried out that estimate that when

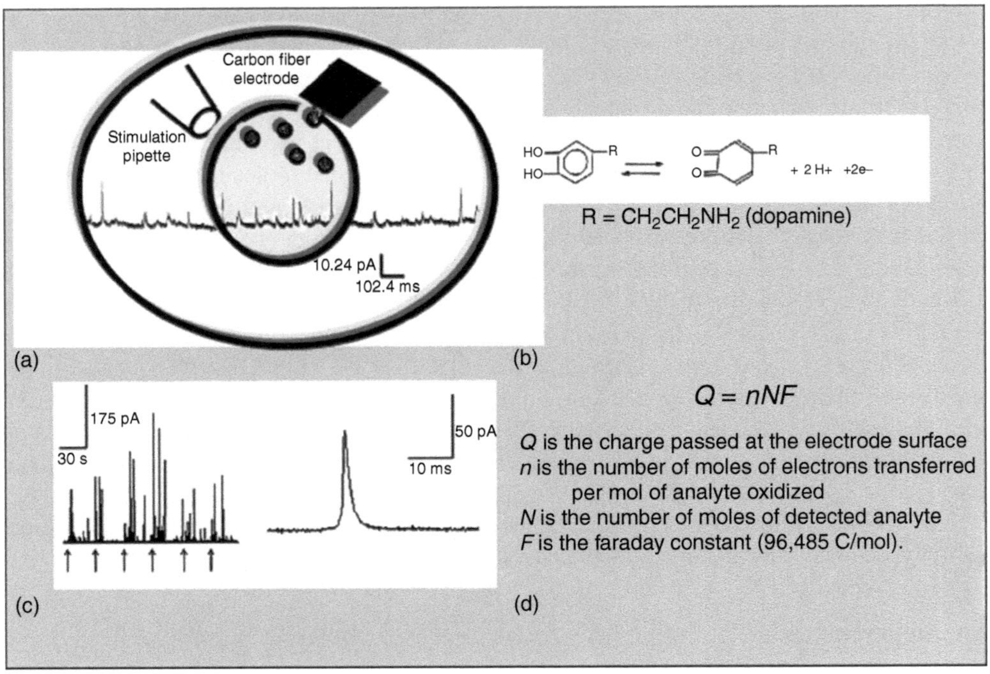

FIGURE 14.1 Basic introduction to amperometric detection of exocytosis at single cells. (a) Foreground image is the typical setup for amperometry for single cell. Exocytosis is stimulated by a pipette containing a stimulant, and the release is monitored by a carbon fiber electrode. The background image shows typical amperometric data. (b) The oxidation reaction for catecholamines. The catecholamine is oxidized to the ortho-quinone form, losing two electrons. (c) Left trace shows a series of stimulations, denoted by the arrows, and the electrochemical responses detected after each stimulation. The right hand trace shows a single amperometric spike. (d) Faraday's equation, which is used to determine the amount of material released during exocytosis. (Figure courtesy of Dr Sombers, L. A., et al. *J. Neurosci.*, 24(2), 303–309, 2004.)

the electrode is placed flush against the cell, there is a small solution filled gap of about 300 nm, between the cell and the electrode.[26,27]The characteristic shape of amperometric peaks, aids in understanding the various aspects of the exocytosis event. The peak shape is dictated by the flow of electroactive agents to the surface of the electrode. Since the electrode is held at a constant potential, 650 mV (vs. SSCE), the oxidation of the molecule generates a current that is measured by a potentiostat.[3,28] This current vs. time trace can be analyzed to provide information about exocytosis (Figure 14.1c). Specifically, the half-width of the peak (full width at half maximum) measures the duration of the exocytotic event. The rise time, or the time it takes to go from 10% to 90% of the peak height, relates to the time it takes the fusion pore to open. Thus, the basic amperometric measurement provides data on the amount of transmitter released, the duration of each event, and the opening of the fusion pore.

Fast Scan Rate Voltammetry

Fast scan cyclic voltammetry can be used to provide information about the identity of the molecule in a solution by examining characteristic oxidation–reduction peaks.[25] Cyclic voltammetry has therefore been adopted in order to measure easily oxidized substances in and at single cells. In contrast to amperometry in which the signal is proportional to the total number of molecules oxidized, in cyclic voltammetry experiments, the signal is proportional to the local concentration of electroactive molecules; it has also been shown to have nanomolar detection limits. The advent of microelectrodes has lead to the ability to scan at much higher scan rates than with conventional electrodes. High scan rates provide a means to obtain voltammograms in 10 ms or less, thus, this procedure gives time resolution necessary to monitor rapid biological events. To detect catecholamines, the voltage sweep typically begins at −0.4 V (vs. a sodium-saturated calomel reference electrode), ramps to +1.1 V and then returns back to −0.4 V. A typical scan rate is in the range of 800 V/s.[25] The rapid sweep rate creates a high background current and, therefore, background subtraction becomes very important in identifying what can often be a very small signal in a large background.

Exocytosis at Adrenal Cells

Adrenal chromaffin cells, which synthesize, store and release epinephrine and norepinephrine have been the most widely used cell model in amperometric studies to understand exocytosis. One interesting feature of these cells is a dense core at the center of the vesicles. This dense core is composed of, among other things, an anionic polypeptide called chromogranin A. The cationic messenger molecules are thought to associate with this polypeptide. During exocytosis, the vesicle fuses, beginning with the formation of a fusion pore. When the fusion pore forms, an influx of extracellular solution changes the pH of the vesicular solution. As the solution in the vesicle becomes more basic, chromogranin A releases the neurotransmitters associated in the matrix and they move into the synapse. Chromogranin A swelling has been hypothesized to force open the fusion pore during full exocytosis.[29]

To distinguish and identify epinephrine and norepinephrine release at adrenal chromaffin cells, researchers have exploited the chemical identification ability of fast scan cyclic voltammetry. Epinephrine and norepinephrine differ only by a methyl group on the amine side chain and have identical electrophores (catechol) and electrochemical formal potentials for oxidation. Hence, they are difficult to discriminate from one another. However, cyclic voltammetry can be used to discriminate these substances if one examines the re-reduction of oxidized catechol and considers that in the oxidized state these molecules undergo a cyclization reaction involving the nitrogen ending side chain to form an adrenochrome that is more difficult to reduce, resulting in a second reduction peak. The reaction rate for epinephrine is considerably faster than that for norepinephrine, hence, the adrenochrome reduction peak is larger for epinephrine. Formation of the adrenochrome from oxidized epinephrine also shifts the equilibrium potential for oxidation, providing a second means to discriminate it from norepinephrine. Figure 14.2 shows representative time traces obtained at a single adrenal cell following stimulation with nicotine.[30] In this case, the stimulated cell releases both epinephrine and norepinephrine as indicated by the voltammetry at 10 V/s. Examining several cells, 75% of the cells release primarily either norepinephrine or epinephrine, while 25% release both.

Amperometric Detection of Release via the Fusion Pore

It has been suggested that docked vesicles undergoing exocytosis transcend an intermediate state when a small fusion pore is open but cannot immediately expand (for a discourse on the forces stabilizing this fusion pore see Ref. [31]), and that this pore allows a small but measurable amount of neurotransmitter to leak out.[32] Transport of chemical messengers through the constricted pore manifests itself as a pre-spike feature or "foot" in the amperometric trace as these messengers are oxidized at the electrode. Fusion pore formation appears to be highly regulated, as numerous proteins have been implicated in this step in exocytosis.[33] Therefore, it is of interest as a mechanism by which the cell could regulate release of neurotransmitter and be involved in synaptic plasticity.

A combination of patch clamp capacitance measurements with carbon fiber amperometry was used to correlate the fusion of individual vesicles with the release of neurotransmitter.[32]

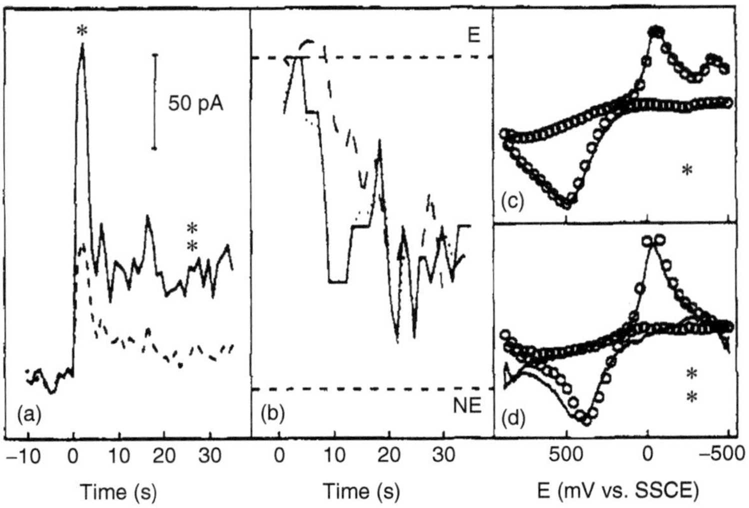

FIGURE 14.2 Cosecretion of epinephrine and norepinephrine from an adrenal cell. (a) Oxidation current (solid line), ΔE_p (dotted line), and adrenochrome current (dashed line) vs. time. (b) E_p (solid line), ΔE_p (dotted line), and i_{cyc}/i_{ox} (dashed line) vs. time. The scale for each of these parameters is adjusted so that the values fall between those for pure epinephrine (upper dashed line) and pure norepinephrine (lower dashed line). (c) Background-corrected cyclic voltammograms (solid line) taken at the asterisk in (a). Circles represent the background-corrected cyclic voltammogram of the epinephrine postcalibration. Voltammograms have been normalized to the oxidation peak currents. (d) Background-corrected cyclic voltammogram (solid line) taken 25 s after stimulation. Circles again are the postcalibration for norepinephrine. (Reproduced from Ciolkowski, E. L., Cooper, J. M., Jankowski, J. A., Jorgenson, J. W., Wightman, R. M., *J. Am. Chem. Soc.*, 114, 2815–2821, 1992. With permission.)

Alvarez de Toledo and co-workers showed that membrane capacitance appeared to transiently increase and decrease prior to the detection of amperometric events. They attributed this phenomenon to transient, reversible vesicle fusion prior to subsequent full fusion. Furthermore, they showed that the duration of the foot correlates with vesicle radius, and that the percentage of neurotransmitter released in the foot increases with decreasing vesicle size.[32] Amperometric methods were used to quantify the amount of neurotransmitter released at various points during each capacitance trace on a sub-millisecond time scale.

Insulin Release from Pancreatic Beta Cells

Although most experiments apply amperometry to catecholaminergic cells, it is not limited to them. Kennedy's group has shown that amperometry can also be used to monitor the release of peptides from cells via the same mechanisms. Highly sensitive detection (detection with a limit of 100 n*M*) of insulin (a peptide with 51 amino acids) from human pancreatic beta-cells required coating the electrode with a ruthenium oxide/cyanoruthenate film.[34] Smaller tyrosine/tryptophan-containing peptides (with 8–15 amino acids) have been detected with uncoated carbon fibers,[35–37] albeit at higher limits of detection.

It is believed that peptides are released by large dense core vesicles (LDCVs), while classical neurotransmitters are sequestered in small synaptic vesicles (SSVs) in the brain, making peptide detection potentially more feasible. Larger vesicles give rise to larger and more easily detected peaks. However, those neuropeptides, lacking oxidizable amino acid residues in their composition, would not be detected.

Exocytosis at Mast Cells

As part of the immune system, peritoneal mast cells secrete histamine and serotonin during an allergic response. Mast cells extracted from one variety of mouse, the beige mouse, have unusually large vesicles (mast cell vesicle diameters have been reported as 770 nm for normal mice and 2000 nm for beige mouse [32,38]). Beige mouse mast cells are an excellent model system for performing dual simultaneous measurements of exocytosis using amperometry and patch clamp for capacitance.[32,38] Individual vesicles are large enough to be monitored by light microscopy, and fusing of single vesicles generates discrete steps in the capacitance trace. Beige mouse mast cells have a smaller intravesicular "halo"-like lumen surrounding the core than those seen in other cell types (i.e., PC12 cells and "active" vesicles from chromaffin cells [19]) but respond to many of the same agents that affect exocytosis from chromaffin cells.[39]

Previous data have suggested that histamine and serotonin are co-released from single vesicles in mast cells; however, the evidence for this was indirect. Using electrochemically pretreated carbon fiber electrodes and a high positive potential scan limit, two oxidative waves for histamine were observed, as well as an oxidation wave for serotonin.[40] This allowed the simultaneous detection of the two analytes, leading to the conclusion that indeed both histamine and serotonin are released from the same vesicles in mast cells.

Pheochromocytoma (PC12) Cells, Undifferentiated and Differentiated

The adrenal pheochromocytoma (PC12) cell line was originally isolated from a tumor in the adrenal medulla of a rat in 1976.[41] They resemble the phenotype of sympathetic ganglion neurons upon differentiation (with nerve growth factor, NGF) and can be subcultured indefinitely. They possess slightly smaller LDCVs (75–120 nm radius compared to 170 nm[16,41,42] for chromaffin cells) and generally contain dopamine, which gives rise to spikes similar to, but smaller than, those of isolated chromaffin cells following stimulation to exocytosis.[43]

The first demonstration of zeptomole detection was shown by Chen et al. for amperometric detection of dopamine during exocytosis at undifferetiated PC12 cells.[20] The average single PC12 vesicle from this study contained 190 zmol of catecholamine. Using previous electron microscopy data [42] that estimated the average vesicular radius to be 74 nm; the concentration of catecholamine in a single vesicle was estimated to be 0.11 *M*. Chen et al. reported a limit of detection of 31 zmol with the amperometric technique, which is within the range needed to detect exocytosis at mammalian neurons. Mammalian neurons contain between 1.8 and 96.4 zmol of catecholamine in a synaptic vesicle.[44]

The minute dimensions of microelectrodes have been exploited to explore zones of exocytotic release on differentiated PC12 cells. Zerby and Ewing used 5 μm diameter carbon fiber electrodes to probe local zones of dopamine exocytosis at PC12 cells that were induced to sprout varicose neurites via exposure to NGF.[45] These experiments showed that quantal size from differentiated PC12 cell varicosities is not significantly different from undifferentiated PC12 cell quantal size, and exocytotic events were absent from the cell body of the NGF treated cells.

Release from the Cell Body of a Processed Neuron in an Animal System

Invertebrate systems offer an advantage as a model for neuronal communication in the form of larger cell dimensions than their mammalian counterparts. The Nobel prize research of Hodgkin and Huxley on membrane potential would not have been possible without the macroscopic squid axon as a research tool.[46] More recently, *Planorbis corneus* (the freshwater pond snail) has been used to study exocytosis. *Planorbis* has a giant dopamine cell from which tens of thousands of events can be elicited per stimulation. These events are derived from LDCVs and putative small synaptic vesicles (SSVs).[23,47,48] This was pioneering work to measure individual exocytosis events from an intact neuron in a living organism.

Release of Small Amounts of Catecholamine from the Leech Neuron

The Retzius cell of *Hirudo medicinalis* (the freshwater leech) has also been used to study exocytosis. This cell synthesizes, stores and releases serotonin. Bruns and Jahn detected release from two populations of vesicles (large and small) from this cell type, with the current transients from the small population averaging 3.0 fC in area (about 5000 molecules) and 595 μs wide, while the larger vesicles had an average area of 47.4 fC (about 1million molecules) and width of 3.66 ms.[49]

Release from Mammalian Cells

Neurons used to study exocytosis with amperometry have been isolated from the midbrain, hippocampus, retina, primary sensory ganglia, neostriatum, nucleus accumbens, and substantia nigra.[15,18,19,50–54] Unlike immortalized cell lines, which can be used within a few days of plating, primary neuronal cultures require 3–6 weeks of recovery after plating.[15] In addition, primary cultures are a heterogenous phenotype requiring post-experiment immunohistochemistry to confirm the presence of critical enzymes (i.e., tyrosine hydroxylase) for phenotype identification. The many brain areas, thus far examined in this way, share one characteristic: they are known to have a high density of catecholamine releasing cells. In the future, tailored genotypes of mice expressing GFP in catecholaminergic cells will allow amperometry to be applied to areas where such cells are diffuse.

Kiss-and-Run Measurements

In some cases, during exocytosis, the fusion pore is thought to open and close without a full exocytosis event occurring. This phenomenon, termed kiss-and-run has recently been observed with amperometry at mammalian cells by Sulzer's group.[55] If these events represent a significant communication pathway, then the cell's ability to modify vesicle parameters that affect the fusion pore could be a mechanism of neuronal plasticity (learning and memory).

Measurements of Cholesterol in the Cell Membrane

Although enzyme-modified electrodes are not being discussed explicitly in this review, a new application of them warrants mentioning. Devadoss et al. have developed a cholesterol oxidase modified electrode that has been used to detect cholesterol in the plasma membranes of *Xenopus* oocytes.[56] A platinum electrode modified with a lipid bilayer containing cholesterol oxidase[57] is operated in the amperometric mode and placed adjacent to an oocyte (Figure 14.3a and Figure 14.3b). As the electrode moves closer to the cell surface, the amperometric current rises (Figure 14.3c). When the electrode is placed in contact with the cell surface, cholesterol is extracted from the cell membrane, moves across the thin membrane layer, and partitions into the electrode-supported enzyme-modified lipid bilayer. Subsequent enzyme catalyzed oxidation of cholesterol by molecular oxygen produces hydrogen peroxide, which is oxidized at the platinum electrode and detected. This is an efficient and simple method to determine cholesterol levels in the plasma membrane of cells, which should find applicability in determining the role cholesterol plays in the organization of cellular membranes.[58]

OTHER TECHNIQUES

Patch Amperometry

Although electrochemical methods alone are powerful tools for the detection and analysis of single exocytotic events, they can be combined with electrophysiological methods, patch clamp for instance, to provide more detailed information about the events. This technique is known

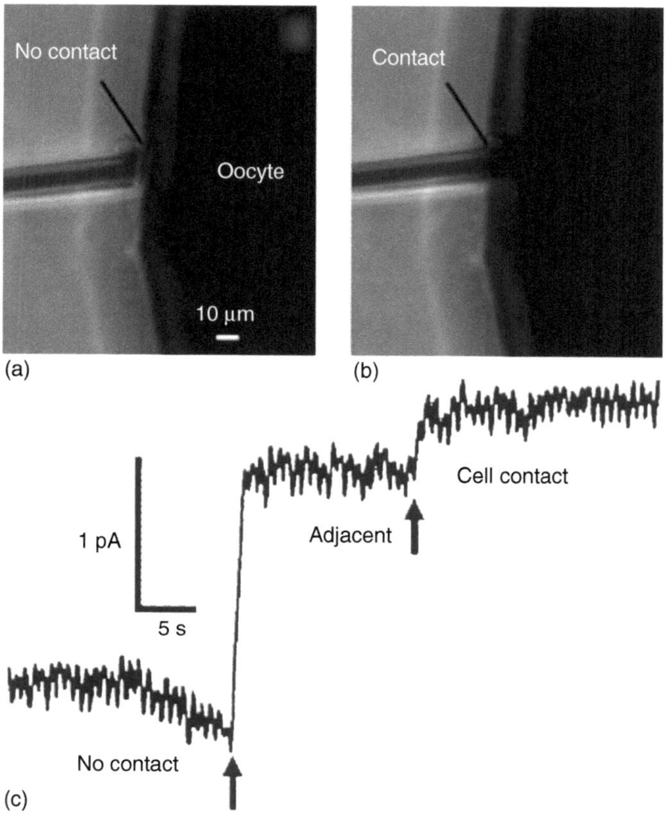

FIGURE 14.3 **(See color insert following page 272.)** Photographs and amperometric detection of cholesterol in the plasma membrane of a *Xenopus* oocyte. Photographs showing the electrode (a) positioned about 5 μm from the plasma membrane and (b) contacting the plasma membrane. (c) Amperometric data for detection of cellular cholesterol at a microelectrode (11.5 μm in diameter) modified with a lipid bilayer membrane containing cholesterol oxidase. No contact (a): baseline data; no cholesterol detection. Adjacent: data for positioning the electrode within about 1 μm of (or partially touching) the plasma membrane. Cell contact: data for contacting the oocyte plasma membrane. Arrows indicate the approximate times of changing the electrode position. The buffer was 0.1 *M* sodium phosphate, pH 6.5. The electrode potential was 800 mV vs. NHE. (Reproduced from Devadoss, A., Burgess, J. D., *J. Am. Chem. Soc.*, 126 (33), 10214–10215, 2004. With permission.)

as patch amperometry. Patch amperometry is typically accomplished by inserting a carbon fiber, which functions as the electrochemical sensor, into a glass patch clamp pipette and performing cell attached voltage clamp experiments simultaneously with electrochemical experiments. This is shown in Figure 14.4a through Figure 14.4c.

This configuration allows the simultaneous monitoring of membrane capacitance, as well as detection of the opening of individual fusion pores and the monitoring of catecholamine release kinetics from the same vesicle.[59] Membrane capacitance is directly proportional to membrane area; therefore, discrete jumps in capacitance signify fusion of individual vesicles with the plasma membrane and allow estimation of the size of the fusing vesicles (Figure 14.4d). These experiments were used with currently accepted models of the structure of the fusion pore to estimate the diameter of the pore preceding full release in chromaffin cells to be less than 3 nm. Additionally, this technique was used to determine that release of transmitter through the pore in adrenal cells is much faster than release through the pore of mast cells. In other studies on mast cells, Tabares et al. demonstrated that the size of the initial fusion pore does not determine the rate of neurotransmitter

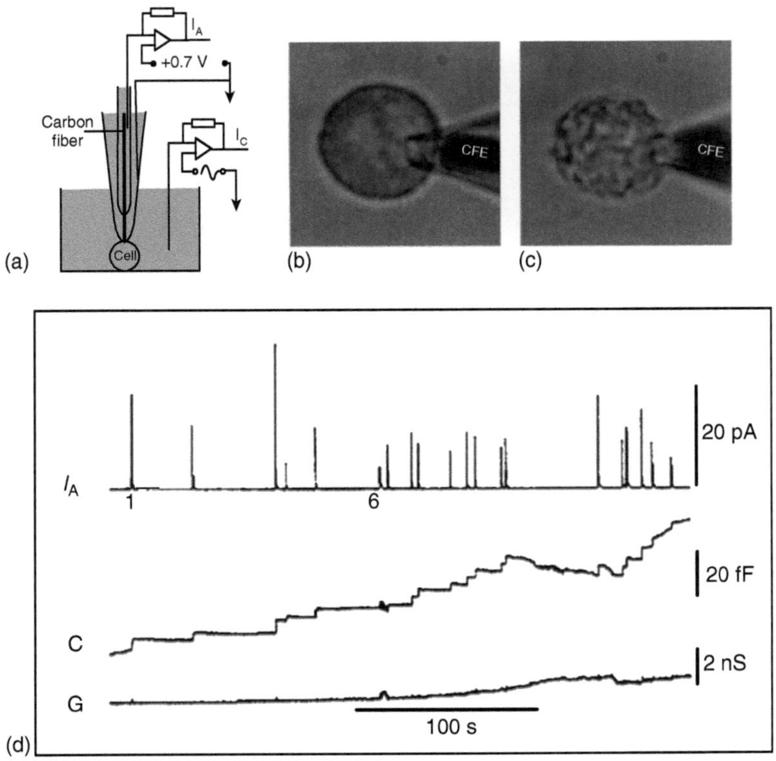

FIGURE 14.4 (a) Arrangement of a CFE inside a patch pipette. I_A, Amperometric current; I_C, sine wave current used to measure capacitance changes. (b) and (c) Chromaffin cell with attached patch pipette containing CFE at the beginning (b) and end (c) of the experiment. (d) Recording from this cell shows amperometric transients (top), associated capacitance steps (middle), and conductance trace (bottom). (Reproduced from Albillos, A., Dernick, G., Horstmann, H., Almers, W., Alvarez de Toledo, G., Lindau, M., *Nature*, 389 (6650), 509–512, 1997. With permission.)

efflux during the early or late stages of full exocytotic release.[60] This technique has also been used to elucidate the coregulation of vesicle membrane area and the amount of neurotransmitter contained in vesicles in response to pharmacological treatment.[61] Patch amperometric experiments have reinforced the conclusion put forth by Colliver et al. that neurotransmitter concentration in vesicles remains constant when vesicles are induced to grow or shrink by drug treatment.[21] While the cell attached configuration is the most prevalent configuration, patch amperometry has also been applied to excised membrane patches, thereby allowing control over the environment on both sides of the plasma membrane.[62]

Quartz Crystal Microbalance Measurements

The development of the quartz crystal microbalance (QCM) started from the predictions of Sauerbrey in the 1950s that the physical nature of a quartz crystal resonator results in a proportionality between the resonant frequency and the total mass of the crystal.[63] Therefore, a change in the crystal mass can be detected by measuring the changes in crystal resonant frequency. Further development of QCM equipment included the electrochemical quartz crystal microbalance (E-QCM), which is a widely used instrument for measurement of attached surface mass to the front side of the quartz crystal, as a working electrode.[64] This has been used for simultaneous detection of ion transfer processes that occur in adsorbed films. This has been a very useful technique in studying electrochemical deposition, etching, and electropolymerization. A QCM combined with a dissipation measuring (QCM-D) technique has been developed.[65] This approach involves simultaneous measurement of the dissipation and the resonant frequency. The QCM-D technique allows the study of surface phenomenon of adsorbed non-rigid materials in cases where the Sauerbrey equation is not entirely valid, such as living cells. Measurements of the mechanical properties of cells can also be achieved including viscosity, elasticity, density, and thickness of the adsorbing material. Cans and co-workers used the QCM-D method to monitor the mass change and rigidity of populations of excitable cells during exocytosis and endocytosis. NG108-15 and PC12 cells were grown to confluence separately on piezoelectric quartz crystals and were stimulated to exocytosis using elevated potassium solution.[66] This resulted in an increase in the frequency response corresponding to loss of mass as vesicles underwent release and a subsequent mass gain during retrieval of dense-core vesicles. The two cell lines were examined separately and used to demonstrate differences in release and retrieval for cells of different morphology, size, and number of dense-core vesicles. The amplitude and peak area in the frequency response corresponding to mass change with stimulated release was larger for PC12 cells than for NG108-15 cells, suggesting that a greater number and larger size of vesicles in PC12 cells results in a greater amount of release from these cells vs. NG108-15 cells. However, the initial rate constants for the frequency responses were similar, suggesting that the recycling of vesicles utilizes similar fusion/retrieval mechanisms in both cell types. The QCM-D measurements also revealed that vesicle retrieval had a rapid onset, masking exocytosis, and implying a rapid retrieval mechanism in the early stages of release. This demonstrates that complex, dynamic processes relating to dense-core vesicle release and retrieval can be measured simultaneously using the QCM-D technique. The QCM technique, with further development using QCM in an array format and with miniaturization of individual resonators, might offer a promising approach to study the processes of neuronal communication in neuronal networks.

Electrochemistry of Material Released from Cells in Microvials

Electrochemical detection of cellular behavior in small volume containers requires fabrication methods capable of defining micrometer size structures from a suitable substrate. Initially, Bowyer et al. sandwiched silver and platinum foil between layers of Tefzel film and glass to form an electrochemical cell with auxiliary, reference, and working band electrodes, the smallest of which measured 4 μm thick.[67] This electrochemical cell was used to investigate the differential pulse, normal pulse, and cyclic voltammetric behavior of ferrocence in aqueous solutions with volumes ranging from 2 mL to 50 nL.

More recently, however, photolithography has been employed to define and fabricate vials with a capacity as small as 1 pL.[68,69] Small vials, described by Clark et al., have been fabricated using a silicon template photolithographically, then transferring the template pattern into polystyrene using a hot press method. This results in arrays of transparent microvials. An alternative approach to producing transparent sample vials has been taken by Bratten and co-workers wherein a polyimide chamber is photolithographically defined on a transparent glass slide.[70–72] These structures differed from those of Clark and Ewing in that the reference, auxiliary, and working electrodes were formed by in situ gold deposition on the bottom surface of the container.

A third method of vial fabrication uses a combination of screen printing and laser ablation.[73,74] Briefly, carbon ink was screen printed through a patterned stencil onto a ceramic substrate. An insulating, dielectric ink was screen printed on top of the carbon layer and a Ag/AgCl ink was printed on top of the dielectric ink. A final layer of dielectric ink was printed on the outermost layer. After laser ablation, the carbon and Ag/AgCl layers eventually formed the working and reference electrodes of the device. Formation of the vials was accomplished by Kr–F excimer laser ablation at 248 nm and resulted in vial with a volume of 7.2 nL. Figure 14.5 shows scanning electron micrographs of the three types of vials described above.

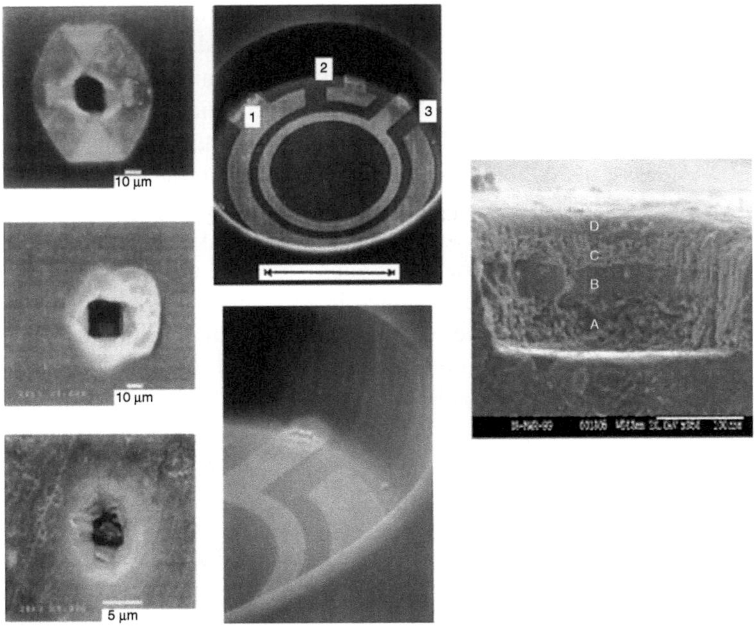

FIGURE 14.5 Scanning electron micrographs of three types of microvials. The leftmost images are polystyrene microvials fabricated with a hot press method. The center images are vials photolithographically defined polyimide wells. The rightmost image is a vial prepared by ink jet printing followed by laser ablation. (Reproduced from Clark, R. A., Heitpas, P. B., Ewing, A. G., *Anal. Chem.*, 69, (2), 259–263, 1997; Bratten, C., Cobbold, P., Cooper, J., *Anal. Chem.*, 69, (2), 253–258, 1997; Ball, J. C., Scott, D. L., Lumpp, J. K., Daunert, S., Wang, J., Bachas, L. G., *Anal. Chem.*, 72 (3), 497–501, 2000.)

There are two prominent electrochemical detection schemes when working in small volume vials. The simplest is to use micromanipulators to position the working and reference electrodes into the vial. This is the approach taken by Clark and Ewing, as well as some groups conducting biological electrochemical experiments in microvials, which will be discussed later in this section. Secondly, electrodes can be formed in situ in the chambers by traditional microfabrication and photolithography techniques[70–72,75] or by a combination of screen printing and laser ablation.[73,74] The drawback to some of these fabrication methods is that the materials that form the sides and bottom of the vials are not always the same, which could lead to differential adsorption of analytes and erroneous voltammograms.

Microvials are attractive environments for single cell experiments because of their limited volumes and the possibility of creating integrated devices with electrodes fabricated by photolithographic techniques inside the vials. Such studies include the monitoring of purine release during cardiac cell ischemia,[72] monitoring metabolic flux due to stress response,[76] and monitoring dopamine transport into cells.[77] Microvials with spatially distinct electrodes have been fabricated by Dias et al. and have been used to determine the spatial distribution of exocytosis events on the surface of adrenal chromaffin cells.[75] In a study by Troyer and Wightman, HEK-293 cells transfected to express the dopamine transporter were placed in vials ranging from 100 to 200 pL.[77] Dopamine was then injected into the vials and the rate of clearance of dopamine by the cell was monitored with fast scan rate cyclic voltammetry. Monitoring dopamine concentration change with time allows the calculation of the kinetics of the dopamine transporter, specifically the initial transport rate, V_{max}. Using the single cell microvial approach, V_{max} was determined to be 55 ± 17 amol/s-cell. In populations of cells, V_{max} was determined to 18.9 ± 1.4 amol/s-cell by rotating disk electrode voltammetry.[78] The discrepancy in V_{max} measured by competing techniques may be due to the difference in the cell-volume/solution-volume ratio. In the microvial experiments, this ratio is larger, meaning, there is less extracellular volume per cell and thus, fewer dopamine molecules are transported into the cell. Fewer transported molecules generally prevents the cytosolic concentration from reaching a high enough concentration for reverse transport to occur. However, in some cells in microvial experiments, reverse transport did occur, and was determined by a nonzero dopamine concentration plateau. Cells exhibiting this behavior were not taken into account in the calculation of V_{max}, as the calculated V_{max} from these experiments appeared lower.

ELECTROCHEMISTRY AT BILAYER MEMBRANES

Artificial Cells (Liposomes) as Models of Exocytosis

Exocytosis is a complex process that involves protein–protein and protein–lipid interactions, membrane fusion, and release of neurotransmitter. A model system that mimics this process in a protein free fashion is advantageous to study the role of the lipid bilayer in exocytosis. A totally lipidic system is useful to study the effects of altered membrane physical properties, typically accomplished by altering the components that form the membrane. Thus, an "artificial cell" composed of two liposomes with a lipid membrane nanotube connection, and one inside the other (Figure 14.6), was developed.[79]

The liposome inside a liposome configuration is analogous to a vesicle inside a secretory cell during the last stages of the release process. Microelectroporation[80] assisted insertion and careful micromanipulation of a micropipette is required for formation of the artificial cell. Briefly, a micropipette filled with the redox molecule of choice (typically catechol) is positioned next to a surface immobilized giant unilamellar vesicle (GUV) with an attached multilamellar vesicle. At the opposite side of the vesicle is the electroporation counter electrode. When a voltage pulse is applied across the GUV, the membrane is transiently destabilized, allowing insertion of the micropipette. The lipid membrane is allowed to seal around the micropipette tip before the pipette is rapidly withdrawn into the interior. This forms a nanotube connecting the tip and the liposome. When solution flows from the micropipette into the lipid nanotube, a small vesicle begins to form. When the vesicle has grown to the size where it nearly contacts the larger exterior vesicle, there is only a short nanotube separating the two liposome compartments. This geometry mimicks that of a cell undergoing the later stages of exocytosis, with a small vesicle fused to a larger vesicle and an aqueous pore connecting the vesicular contents with the extracellular space. At this point, the remaining toroidal shaped fusion pore of the artificial cell is of an energetically unfavorable lipid membrane geometry, and this results in spontaneous opening of the lipid pore and the interior vesicle emptying its content to the surrounding solution in a process that resembles exocytosis. The nanotube remains adhered to the pipette tip, allowing successive release events without pipette reinsertion. The size of the fusing vesicle can be adjusted by varying the distance from the pipette tip to the large vesicle membrane, and release from vesicles ranging from 4 μm in diameter to larger than 30 μm has been measured.

Kinetics of Membrane Distention during Artificial Cell Exocytosis

Monitoring the material released is accomplished by amperometrically detecting the redox molecule contained in the smaller interior vesicle as it is released. Typically, a carbon fiber electrode measuring 33 μm in diameter is employed for amperometric detection. The resulting current transient appears qualitatively similar to those recorded at living secretory cells, such as, mast cells, chromaffin cells, and PC12 cells (Figure 14.7a and Figure 14.7b). The obvious difference in time scale, amplitude, and area of the artificial cell current spike is due to the fact that the vesicle is larger and contains more electroactive molecules than those in living cells. However, release from the smallest vesicles measured (approximately 4 μm diameter) is similar

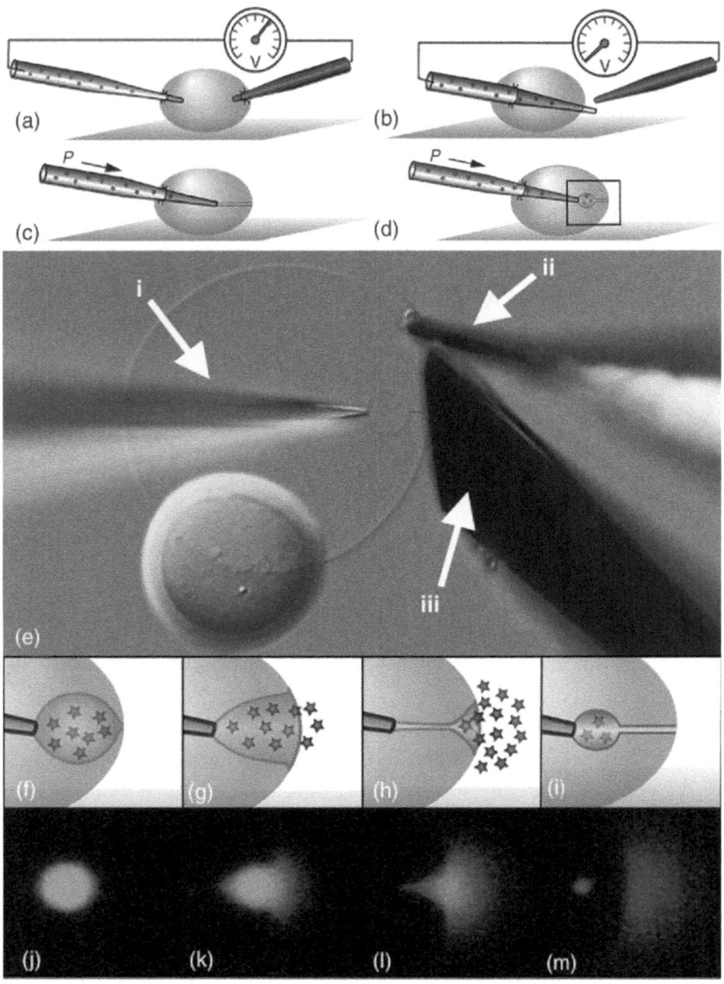

FIGURE 14.6 **(See color insert following page 272.)** Formation and release of vesicles in an artificial cell. (a) through (d) Schematics of a microinjection pipette electroinserted into the interior of a unilamellar liposome and then through the opposing wall, pulled back in to the interior, followed by spontaneous formation of a lipid nanotube and formation of a vesicle from flow out of the tip of the micropipette. (e) (i) Nomarski image of a unilamellar liposome with a multilamellar liposome attached as a reservoir of lipid, microinjection pipette, (ii) electrode for electroinsertion, and (iii) amperometric electrode measuring 30 μm in diameter, beveled to a 45° angle. A small red line depicts the location of the lipid nanotube, which is difficult to observe in the computer image using a 20× objective, illustrating a vesicle inside a liposome with a connecting nanotube inside a liposome. (f) through (i) Fluid injection at a constant flow rate results in growth of the newly formed vesicle with a simultaneous shortening of the nanotube until the final stage of exocytosis spontaneously takes place and a new vesicle is formed with the attached nanotube. (j) and (m) Fluorescence microscopy images of fluorescein-filled vesicles showing formation and final stage of exocytosis matching the events in (f) through (i). (Scale bar represents 10 μm.) (Reproduced from Cans, A. S., Wittenberg, N., Karlsson, R., Sombers, L., Karlsson, M., Orwar, O., Ewing, A., *Proc. Natl Acad. Sci. U.S.A.*, 100 (2), 400–404, 2003. With permission.)

in time course to events measured from the large vesicles of the beige mouse mast cell (which average 700 nm in diameter).[81] In Figure 14.7c, the relationship between vesicle radius and full width at half maximum (half width) is shown for experiments with artificial cells. The fit is nearly perfectly cubic, meaning that release kinetics scale linearly with vesicle volume.

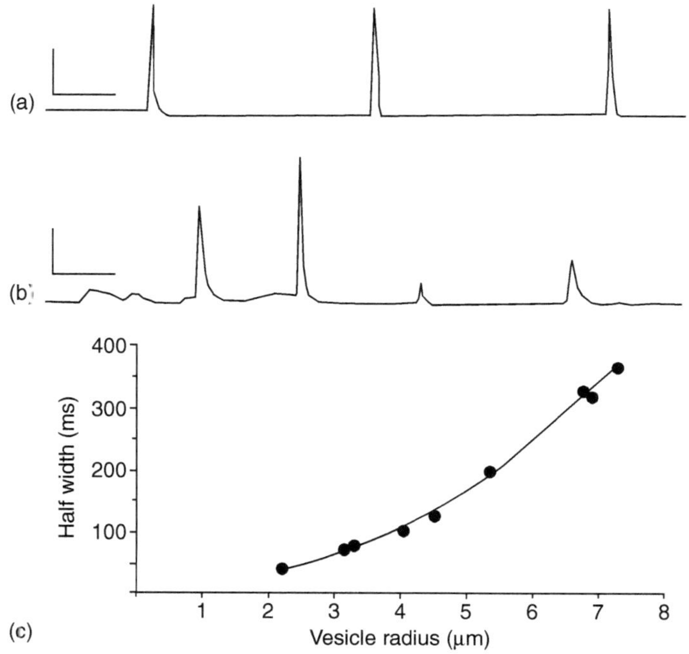

FIGURE 14.7 Amperometric monitoring of repeated exocytosis events at artificial cells and cells. (a) Amperometric detection of continuous exocytosis of three vesicles from an artificial cell. (Scale bars are 40 pA and 3000 ms.) (b) Amperometric detection of dopamine exocytosis from a PC12 cell. (Scales bar are 10 pA and 40 ms.) (c) Plot of half width vs. vesicle radius for vesicles fusing from an artificial cell, where the vesicle radius has been the only parameter varied in the experiment. (Reproduced from Cans, A. S., Wittenberg, N., Karlsson, R., Sombers, L., Karlsson, M., Orwar, O., Ewing, A., *Proc. Natl Acad. Sci. U.S.A.*, 100, (2) 400–404, 2003. With permission.)

Modeling Release via the Fusion Pore (Lipid Nanotube)

An interesting feature of this system is the transport or leakage of catechol through the nanotube prior to full release. This can be thought of as analogous to neurotransmitter leakage through the fusion pore formed before full release in living cells.[8,32] Trans-pore catechol transport is apparent on the amperometric traces as a prespike rise in current, which in live cells is termed a "foot" (Figure 14.8a and Figure 14.8b).

This transport was characterized at artificial cells as a function of vesicle size and as a function of the pressure applied to the micropipette used to inflate the vesicles (Figure 14.8c through Figure 14.8e). The pressure applied to the inflation micropipette correlates with the flow rate of catechol solution from the micropipette. The duration of the foot is strongly dependent on vesicle size for larger vesicles (size of vesicle just before release) and is also dependent on the flow rate. However, the number of molecules detected during the foot portion of the event (foot area) is not dependent on solution flow rate. The ratio of area to the duration of the foot is dependent on both vesicle size and flow rate. These electrochemical data, coupled with models of membrane and fluid dynamics, lead to a cogent explanation of leakage through the membrane nanotube (Figure 14.8f). As the vesicle is inflated, its volume grows at a constant rate, whereas the rate of surface area growth slows as the vesicle grows larger. Additional membrane required to accommodate the growing volume is drawn along the nanotube, thus inducing shear flow of the solution inside the nanotube. Consequently, shear flow has a high velocity when the vesicle is small, and slows as the surface area to volume ratio decreases. Opposing shear flow is Poiseuille flow resulting from

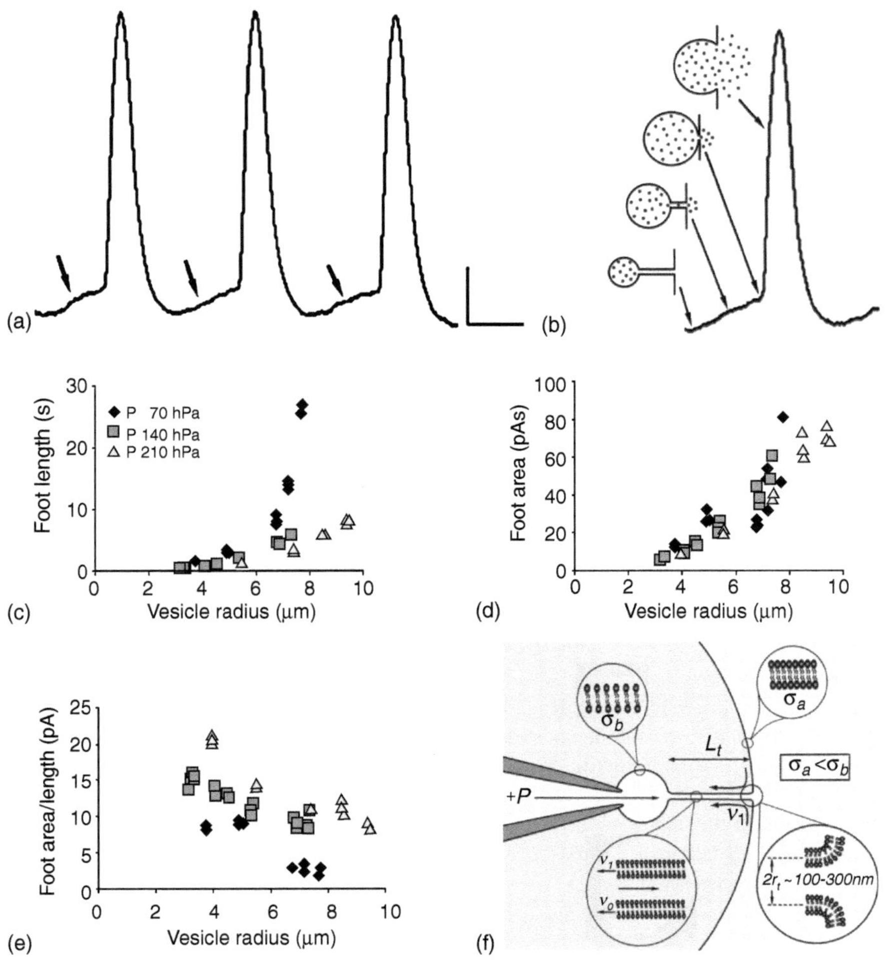

FIGURE 14.8 Amperometric monitoring of release via an artificial fusion pore. (a) Amperometric detection of release from a vesicle with a radius of 5 μm, showing prespike feet (arrows), indicating catechol transport through the lipid nanotube or fusion pore. (Scale bar is 80 pA and 500 ms.) (b) Time correlation of vesicle growth, transport of transmitter through the lipid nanotube, and the final stage of exocytosis with amperometric detection. (c) Plots of foot length, (d) foot area, and (e) the ratio of foot area over foot length, all observed with amperometry for vesicles fusing with an artificial cell, at three different pressures used to inflate the vesicles. (f) Schematic model of the factors affecting flow in the vesicle and nanotube of the artificial cell. (Reproduced from Cans, A. S., Wittenberg, N., Karlsson, R., Sombers, L., Karlsson, M., Orwar, O., Ewing, A., *Proc. Natl Acad. Sci. U.S.A.*, 100 (2), 400–404, 2003. With permission.)

the pressure difference across the nanotube. Additionally, as the vesicle grows, the nanotube connecting it to the larger vesicle shrinks, and as a result, Poiseuille flow velocity increases. When Poiseuille flow overcomes shear flow, catechol solution is transported out of the nanotube and is amperometrically detected as a prespike foot.

In this sense, transport through the nanotube is analogous to leakage through an elongated fusion pore in living cells. Intravesicular pressure and membrane transfer through the fusion pore have been suggested previously as significant driving forces for exocytosis.[33,82–84] Finally, the most significant finding from this work is that membrane mechanics alone seem to be sufficient to drive exocytosis-like behavior at biological time scales without protein intervention.

ARTIFICIAL SYNAPSES WITH LIPOSOME MODELS

In a separate set of experiments, the artificial cell model was used to develop models of coulometric efficiency as a function of electrode size and to determine the size of the space between the electrode and the membrane.[85] In this case, coulometric efficiency is defined as the ratio of the total number of molecules detected from a vesicle to the total number of molecules released by the vesicle. These studies employed two different size electrodes (5 and 33 μm in diameter) to detect release and compared the electrochemical responses as a function of electrode and vesicle size.

MODEL OF THE SYNAPSE

When a vesicle is released from the artificial cell, its size relative to the detection electrode is a strong determinant of the amount of vesicular content detected, as well as the spike amplitude. However, the measurement of the kinetics of release is not affected by the size of the electrode used for detection. To model coulometric efficiency and determine the size of the electrode membrane gap, a simple geometric representation of the system was considered (Figure 14.9). The space between the electrode and the membrane was modeled as a cylinder, and the ratio of the vesicle volume to cylinder volume determines the coulometric efficiency. If the vesicular volume is smaller than the cylinder volume, 100% oxidation is expected. For an electrode measuring 33 μm in diameter, the best fit of the data is obtained when the height of the cylinder is 300 nm, suggesting that even when the electrode is in direct contact with the membrane there is a small volume of solution trapped between the electrode and the membrane. It should be pointed out that even for somewhat smaller vesicles, a significant portion of the released molecules escape undetected. This is probably caused by solution flow out of the membrane–electrode gap due to volume limitation

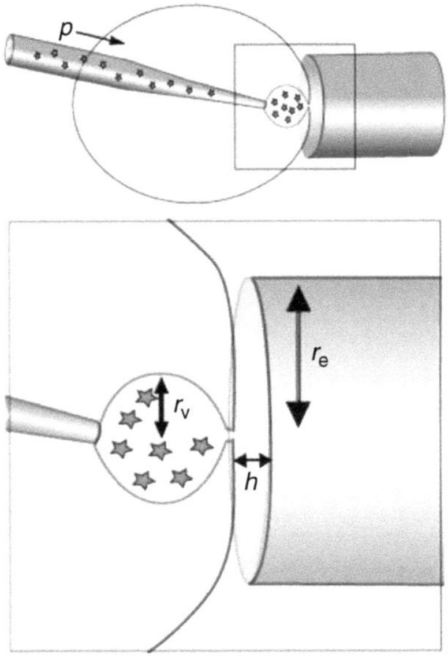

FIGURE 14.9 Simple model of coulometric efficiency for artificial exocytosis. This first-stage model assumes that the efficiency of oxidation for material released is simply the ratio of the membrane-electrode space (calculated as $\pi r_e^2 h$) over the volume of the vesicle ($4\pi r_v^3/3$). This assumes that all the catechol that is present in the membrane–electrode space after exocytosis will be oxidized, but also, that only catechol in the solution that fits in this volume will be oxidized. (Reproduced from Cans, A. S., Wittenberg, N., Eves, D., Karlsson, R., Karlsson, A., Orwar, O., Ewing, A., *Anal. Chem.*, 75 (16), 4168–4175, 2003. With permission.)

not diffusion, i.e., the volume of the gap cannot accommodate the total volume of liquid released from the vesicle.

The first model presented in this work was quite simplistic as it did not take into account the distinct stages of release that are characteristic of the artificial cell model. Therefore, a second model with distinct stages of release was proposed; each stage was represented as having a distinct geometric shape (Figure 14.10). In the first stage of release, the nanotube begins to dilate and some vesicular material escapes the vesicle as revealed by fluorescence imaging of release events (Figure 14.10b). The next stage is characterized by a membrane structure that is between a spherical vesicle and a frustum (Figure 14.10c). The final stage is after full release, when the small vesicle has become fully incorporated into the larger liposome membrane and the nanotube remains attached to the pipette tip (Figure 14.10d). During the time the vesicle is in the frustum state (calculated as twice the spike half-width for each vesicle) the charge passed is estimated from the integrated Cottrell equation.

$$Q_{\text{frustum}} = \frac{2nFAD^{1/2}C^{*}t^{1/2}}{\pi^{1/2}} \tag{14.1}$$

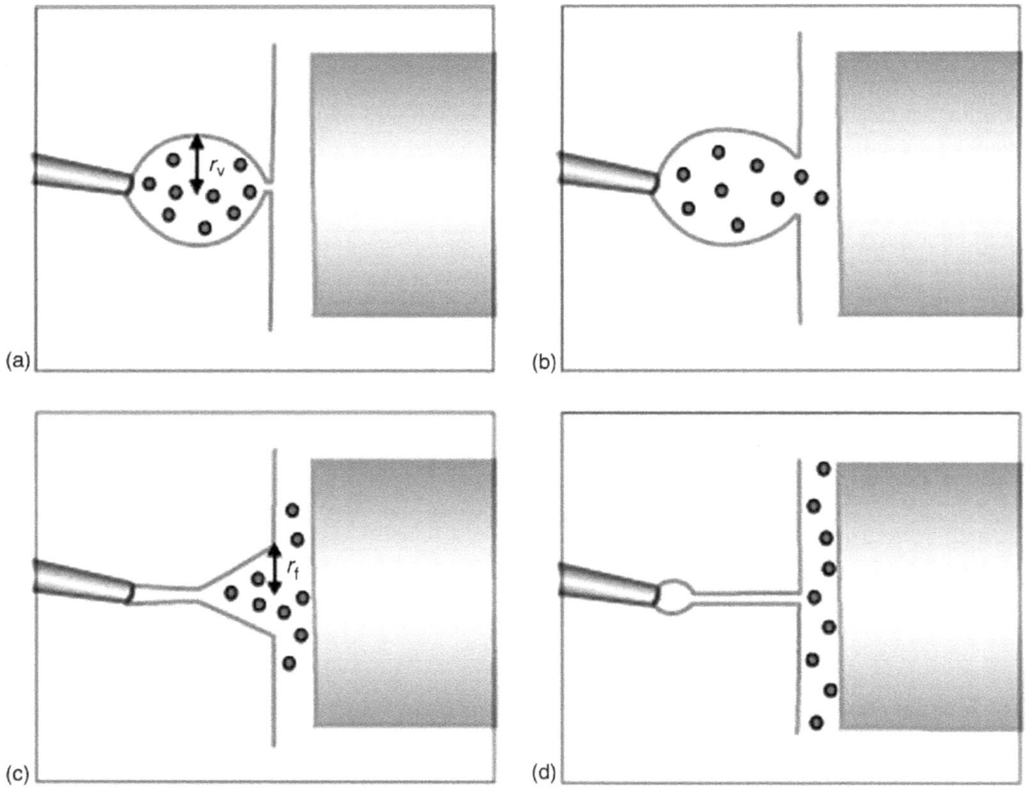

FIGURE 14.10 A more complete model of the release process during exocytosis basing on the fluorescence observations. This model assumes that as the vesicle opens, it has a transitory period where mass transport of catechol to the electrode is via diffusion from a frustum with opening r_f defining the value of r_v. Catechol diffusing to the electrode is oxidized. After exocytosis is complete, the membrane–electrode space is filled with solution from the inside of the vesicle, and catechol present continues to be oxidized as in the thin-layer representation of Figure 14.10. (Reproduced from Cans, A. S., Wittenberg, N., Eves, D., Karlsson, R., Karlsson, A., Orwar, O., Ewing, A., *Anal. Chem.*, 75 (16), 4168–4175, 2003. With permission.)

where n is the number of electrons transferred in the electrochemical reaction, F is Faraday's constant, A is the effective electrode area which is taken to be the area defined by the opening of the frustum or the electrode area (which ever is smaller), D is the diffusion coefficient (6×10^{-6} cm^2/s), C^* is the concentration of catechol in the vesicle, and t is the time the frustum is open (taken to be twice the peak half-width). During the next stage, after full exocytosis, the charge passed by oxidation of catechol in the electrode-membrane space is given by the equation

$$Q_{\text{space}} = nFC'\pi r_{e_e}^2 h \tag{14.2}$$

where r_e is the average radius of the beveled electrode, h is the distance between the membrane and the electrode, and C' is the concentration of the remaining unoxidized catechol in the membrane–electrode space, which is calculated by subtracting the number of moles oxidized during the previous (frustum) stage by the equation

$$C' = \frac{[C^*(V_v + V_{\text{flow}}] - 2AD^{1/2}C^*t^{1/2}/\pi^{1/2}}{V_v + V_{\text{flow}}} \tag{14.3}$$

where V_v is the volume of the vesicle and V_{flow} is the volume of flow out of the pipette during the release event. Flow during the event is calculated by dividing the vesicle volume by the time between events to get a volume flow rate in cm^3/s then multiplying by twice the event half-with. Combining Faraday's Law ($N = Q/nF$) and the integrated Cottrell equation, $Q = 2nFAD^{1/2}C^*t^{1/2}/\pi^{1/2}$ (Equation 14.1), the number of moles oxidized during the frustum stage is given by

$$N = 2AD^{1/2}C^*t^{1/2}/\pi^{1/2} \tag{14.4}$$

Oxidation of all the catechol in the vesicle prior to release results in the following coulometric charge:

$$Q_{\text{max}} = nFC^*(V_V + V_{\text{flow}}) \tag{14.5}$$

So, the predicted coulometric efficiency is the ratio of the sum of the charge from the two stages of opening over Q_{max}. The resulting equation that predicts percent coulometric efficiency is given as

$$\%\ \text{Coulometric Efficiency} = 0.38[(0.25Q_{\text{frustum}} + Q_{\text{space}})/Q_{\text{max}}] \times 100 \tag{14.6}$$

Only 25% of the Cottrell equation is used in the final expression to correct for the diminished diffusion in the narrowing frustum. This model correctly predicts the shape of the percent coulometric efficiency vs. vesicle radius; however, the predicted magnitude is 2.6times larger than that observed. To correct for this a factor of 0.38 has been applied to the theoretical prediction. Experimental data, predicted coulometric efficiency and the best fit of the experimental data is shown in Figure 14.11.

Quantitative Issues and Flow Effects in the Synapse

The difference in magnitude observed for the data vs. the theory for the artificial synapse model, and leading to the 0.38 correction term may arise from numerous factors. One possible explanation is that solution flow due to the distention of the vesicle membrane may lead to loss of material and the reduced oxidation efficiency. The calculation for charge passed during the frustum stage assumes mass transport by diffusion only, but it seems that the changing membrane geometry may lead to solution flow causing molecules to escape the membrane–electrode gap without being oxidized. In cell-to-cell communication, it is usually assumed that diffusion is the predominant

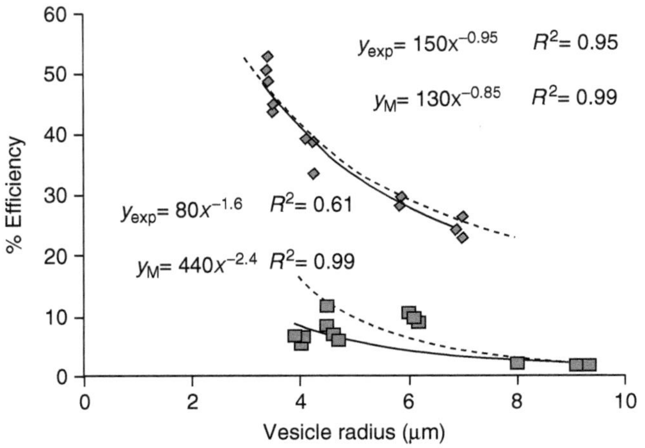

FIGURE 14.11 Coulometric efficiencies for a data set obtained with a beveled 33 μm electrode (♦) compared to data obtained with a 5 μm electrode (■) for release of catechol measured from a range of vesicle sizes. This is compared to the theoretical coulometric efficiencies for these electrode dimensions calculated with the conditions outlined in Figure 14.6 and with the model discussed in the text (dashed lines). These are compared to each set of experimental data, which are shown with best-fit equations (thick lines). The equations for all lines, and their correlation coefficients, are given with the symbols y_{exp} for experimental data sets and y_m for the modeled efficiencies, with that for the 5 μm electrode to the left, and that for the 33 μm electrode to the right. (Reproduced from Cans, A. S., Wittenberg, N., Eves, D., Karlsson, R., Karlsson, A., Orwar, O., Ewing, A., *Anal. Chem.*, 75 (16), 4168–4175, 2003. With permission.)

mass transport process, but this model shows that convective flow is likely to play an important role in mass transport over short distances.

Applying this model to exocytosis in biological systems allows the prediction of coulometric efficiency for any size electrode if the vesicle size is known. Two cell types commonly used in exocytosis experiments, adrenal chromaffin and PC12 cells, have average vesicle radii of 99 and 125 nm, respectively.[8,21] The model set forth here predicts 100% coulometric efficiency for amperometric experiments employing 5 μm electrodes at adrenal chromaffin and PC12 cells. In contrast, for cell types that release larger vesicles, such as the beige mouse mast cells, the model predicts less than 100% coulometric efficiency. In fact, for beige mouse mast cells (average vesicle radius is 1.35 μm), coulometric efficiency at a 5 μm electrode is predicted to be 47% and for the largest mast cell vesicles (vesicle radii as large as 2 μm), coulometric efficiency drops to 25%. The model predicts that even 33 μm electrodes will not quantitatively oxidize all material released from beige mouse mast cell vesicles. Most cellular experiments carried out to date have used a 5 or 10 μm diameter electrode for detection. Therefore it is doubtful, based on this model, that measurements made on cells releasing large vesicles are quantifiable with the assumption of 100% coulometric efficiency.

The data gathered with the artificial cell system shows that under certain conditions much of the released catechol will escape the membrane–electrode space undetected. Also, the model suggests that in addition to diffusion, solution flow is a significant mode of mass transport of neurotransmitter in synapses. This system can be used to model what happens in a synapse in vivo. The membrane–electrode space can be considered analogous to the synaptic cleft and the detection electrode can be thought of as the post–synaptic membrane which has receptors that "capture" the released neurotransmitter. If a synapse is considered to be a 20-nm gap between cells, and the average diameter of the post-synaptic surface is 100 nm, then 70% of the volume of a vesicle measuring 100 nm in diameter will be pumped directly to the extrasynaptic space during release.

This agrees with recent in vivo work that suggests that at dopamine and serotonin synapses, much of the neurotransmitter escapes the synapse following exocytotic release.[86–88] There are still many experimental models to examine before we will gain a thorough understanding of exocytosis. Nevertheless, the models presented here provide a means to simplify the experiments and to examine the effect of membrane mechanics and structure on exocytotic release measured with amperometry.

ELECTROCHEMICAL METHODS APPLIED TO NEUROSCIENCE

The following section of this chapter is a brief compilation of work carried out with electrochemical methods to measure release from single cells. The area has grown enormously in recent years and it would be impossible to review it comprehensively in this chapter. Hence, we have striven to discuss a few examples that provide a flavor of the kinds of work that can be carried out with this method.

STAGES OF RELEASE DURING EXOCYTOSIS ELUCIDATED WITH AMPEROMETRY

Exocytosis of dense core vesicles occurs in stages with the opening of the fusion pore, expansion to full distention, and dissociation of transmitter from the internal dense core matrix all playing important roles. All of these have been investigated with amperometry experiments. The rising portion of a current spike has been correlated to the opening of the fusion pore. Based on models that predict steeper rise times for current spikes than observed experimentally, the Amatore and Wightman groups conclude that the rise time is "due to a separate kinetic step that is temporally located between the initially formed fusion pore, where the majority of catecholamines are tightly associated with the matrix, and the final stage, where the matrix is dissociated and release of catecholamines has fully developed".[89] Using this interpretation of the rising portion of the spike, it is therefore feasible that some manipulation of the fusion pore and/or rate of catecholamine–core-dissociation may result in changes to spike rise time.

Prespike foot-like events are elicited under hypertonic conditions in chromaffin cells.[90,91] During vesicle fusion, the dense core degranulates and releases a large volume of bound neurotransmitter, producing a spike in the amperometric trace. However, under hyperosmotic (approximately 730 mOsm, compared to the normal 330 mOsm) conditions, the core appears to remain intact and only a small amount of uncomplexed neurotransmitter in the halo is released, producing flat, broad peaks about 1% of normal peak amplitude, as measured by cyclic voltammetry. These events resemble feet without associated peaks.[91] Although stimulation elicits a few small foot-like events, momentarily restoring isotonic conditions produces the massive release expected for full exocytosis from many vesicles. This is purportedly because vesicles are stalled in mid-fusion until the osmotic gradient is restored. Isotonic restoration-induced release is not accompanied by a calcium rise, supporting the idea that vesicles have already fused but cannot expel their contents. Like the patch–clamp and amperometry measurements of Alvarez de Toledo et al., the amperometric feet observed under hyperosmotic conditions appear to result from traces of neurotransmitter diffusing through the fusion pore prior to full fusion. Osmolarity is thus presented as a tool for separating initial vesicle fusion from full fusion.[91]

In a follow-up to Alvarez de Toledo's conclusion that foot duration varies inversely with vesicle size, Sombers et al. presented evidence that pharmacologically altered vesicle sizes exhibit different sized feet in PC12 cells.[92] Decreasing the amount of dopamine and the vesicular volume by treating with reserpine results in a greater portion of vesicle contents released through the constricted fusion pore, which manifests itself as a foot in the amperometric trace. Conversely, loading vesicles with L-DOPA to increase the dopamine content and their volumes, leads to a smaller fraction of total vesicular contents released through the fusion pore.[92] Evidence that a greater fraction of vesicle contents is released through a constricted fusion pore as vesicular volume decreases is consistent with recent reports that a significant portion of neurotransmitter released at synapses is through a "kiss-and-run" mechanism.[93,94] "Kiss-and-run" exocytosis involves the expulsion of neurotransmitter through a transient, constricted fusion pore without full fusion, similar to the "foot" portion of amperometric events; however, this is hypothesized to take place under physiological conditions and without full exocytosis occurring.

Theories on the Forces Involved in Opening the Fusion Pore

The rate of release of transmitter during full exocytosis is dependent on the rate of opening of the fusion pore. The dynamics of the pore have been studied by both patch–clamp[7,62,95] and amperometry.[2,8,96] Based on amperometric studies of exocytosis from adrenal chromaffin cells, a physicochemical model accurately describing the dynamics of full-fusion events from dense core vesicles has been proposed.[31,82,97] This model predicts that the vesicular matrix has a determining role in the opening of the vesicle during the final stage of exocytosis. As soon as the fusion pore opens, the concentration gradient of catecholamine cations (mostly epinephrine in chromaffin cells) provokes their outward diffusion. To maintain electroneutrality within the matrix, this is accompanied by the entry of extracellular, hydrated, monovalent cations, such as Na^+ and H^+, into the intravesicular matrix. It is usually considered that during this phase, the fusion pore has not yet expanded, but rather fluctuates between open and closed phases.[8,32] Its geometry may be imposed by a specific architecture, which is still a matter of debate, possibly implicating SNARE protein complexes, or phospholipids, or a combination of both.[98–100] However, at this step, it may be considered that the pore energy is necessarily controlled by two opposite phenomena. The first one is positive and tends to close the pore. It reflects the edge energy of the pore rim. The second one is negative and tends to open the pore. It stems from Laplace tensions of the cell and vesicle membranes.[31,82,97,101] Initially, the edge energy is larger than the surface energy so that the pore remains stable. Owing to the continuous cation exchange, which occurs as soon as the pore is opened, the matrix swells. In dense core vesicles, the matrix expansion likely increases the pressure within the vesicle and provokes a continuous increase of the vesicle membrane tension. Unless the pore closes before, the surface energy of a toroidal pore ineluctably increases until reaching the point where it compensates for the pore edge energy.[31,82,97] When this point is reached, the pore becomes unstable and enlarges irreversibly.[31,82,97,102,103]

The pore expansion initiates a massive release of catecholamines reflected by a sharp increase of current detected at the electrode immediately after the foot current. The ensuing membrane fusion ends by fully exposing the vesicular matrix to the external medium. Afterwards, the progressive depletion of catecholamines by diffusion is observed by a decrease in current. This scheme is proposed for the exocytosis of dense core vesicles that constitute the main category of vesicles within chromaffin cells. Vesicles without a dense core, as in nerve terminals,[104] or with a smaller dense core versus the total vesicle volume, as observed in PC12 cells,[21] may proceed differently in the fusion pore expansion. The main difference should lie in the physicochemical origin of the building up of the internal pressure within the vesicle and, consequently, of the surface tension of the fusion pore.

Effects of SNARE Protein Manipulation

Amperometry on PC12 cells has been used to elucidate protein interactions that regulate cellular machinery. The part of the cellular machinery that recruits vesicles to the plasma membrane is called the SNARE (SNAP receptor, where SNAP is soluble *N*-ethylmaleimide-sensitive factor-attachment protein) complex.[105] A syntaxin-binding protein, Munc18, has been shown to associate with the SNARE complex and is thought to play a role in regulating exocytosis.[106] Previous models have shown that Munc18 dissociates from syntaxin 1 prior to SNARE-complex formation and, therefore, it cannot exert any control on the later stages of exocytosis.

Based on data presented regarding a point mutation in a Munc18 homologus in *Drosophila melanogaster*,[107] Burgoyne et al. transfected PC12 cells with a mutated Munc18 protein.[105] Munc18 was believed to dissociate from syntaxin, and arginine residue 39 was identified to interact directly with syntaxin. The arginine residue was changed to a cysteine to reduce this interaction.[108] Cells transfected with the mutation in Munc18 displayed faster vesicle fusion during exocytosis. This was noted through a decrease in both amplitude and half width of the amperometric signals. The authors suggested that the reduction in amperometric signal resulted from kiss and run fusion. The conclusion drawn from this work is that Munc18 interacts with the SNARE complex and augments the later stages of exocytosis.[105] It is possible that opening of the fusion pore is kinetically hindered by the SNARE complex needed for its formation. Free Munc18 after dissociation from syntaxin might facilitate disassembly of the SNARE complex once the pore is formed, thereby leading to faster opening.

Effects of Calcium Stores on Exocytosis

Amperometry has been used to investigate the role of Ca^{2+} and ryanodine receptors in mast cells. Ryanodine receptors have been identified to modulate exocytosis by triggering a release of internal Ca^{2+} stores. Previous studies of mast cells reported that histamine release was inhibited by the presence of ryanodine, which would suggest that exocytosis was modulated by ryanodine receptors.[109] Preliminary data showed that small concentrations of ryanodine and caffeine increased intracellular Ca^{2+} levels in mast cells with no external Ca^{2+}, suggesting an increase in exocytosis. Jaffe et al. employed amperometry to investigate the role of ryanodine and caffeine in regulating serotonin release from mast cells.[111] A control group of mast cells was stimulated with a mast cell activator, 48/80, which stimulates exocytosis. These control cells were compared to mast cells treated with both a low (1 μM) and high (50 μM) dose of ryanodine. The low dose showed an increase in both amplitude and frequency when compared to control, while the higher dose greatly reduced the half width and spike area but not peak amplitude. Furthermore, addition of caffeine increased exocytosis. Caffeine (20 mM) was added to mast cells and compared to control conditions; the caffeine addition caused an increase in the number of events. Further experiments were conducted, and it was determined that caffeine increases exocytosis by increasing the release of intracellular Ca^{2+} stores. These data suggest that the combination of 1 μM ryanodine and 20 mM caffeine work by ryanodine opening ryanodine receptor channels and caffeine entering the cell through the open receptor channels, which facilitates the release of Ca^{2+} from intracellular stores. This study shows that there is a direct correlation between the ryanodine receptor and exocytosis.[110]

Regulation of Insulin Exocytosis from Pancreatic Beta Cells

One set of experiments conducted with pancreatic β cells was that of Takahashi et al. to examine both fast and slow exocytosis.[111] This reportedly takes place in two different types of vesicles: small clear vesicles, and large, dense core vesicles. Small clear vesicles have been reported to undergo fast release; where as, large dense core containing vesicles tend towards slower release. Dense core vesicles contain, among other things, the polyanionic peptide called chromogranin A. As discussed above, it is hypothesized that neurotransmitters in these vesicles are bound inside the dense core. Upon initial fusion of the vesicle to the plasma membrane, the dense core, which is in the low pH environment in the vesicle, is exposed to the more basic environment outside the cell, causing it to dissociate and release the neurotransmitter molecules. The working hypothesis is that the unraveling of the dense core takes finite time and, thus, increases the time for exocytosis.

In order to differentiate between the two types of vesicles, amperomeric spikes were compared. The average time difference between a group of fast- and a group of slow-release events was 1.5 s, an easily differentiated time difference as amperometry has submillisecond resolution. Ca^{2+} plays a large role in cell signaling for exocytosis, and in order to control the Ca^{2+} influx in this experiment, photo releasable caged Ca^{2+} was used. Slow and fast release events were also monitored using patch–clamp techniques.

Vesicle Size Changes with Messenger Amount

Adding the dopamine precursor L-DOPA to cell environments has been shown to affect quantal size in many cell systems.[19,21,61,94,112] L-DOPA is currently used to treat Parkinson's disease by apparently increasing the amount of dopamine released by neurons. Treating with L-DOPA leads to larger amounts of transmitter detected per vesicle by amperometry.[21] Reserpine, an ancient anti hypertensive drug, has the opposite effect. Reserpine blocks the vesicular monoamine transporter (VMAT) and depletes vesicles of neurotransmitter. Thus, quantal size is reduced.[21,61] L-DOPA is now frequently employed as a positive control to increase quantal size when examining novel effects on quantal size.[15,19,61,94]

Interestingly, when PC12 cells are incubated with L-DOPA to increase the dopamine level in the cells and to provide larger signals,[21,112] transmission electron microscopy experiments show that vesicle size increases. Vesicles in PC12 cells exposed to reserpine have less dopamine and are smaller.[21] This surprising result has been corroborated by Lindau and co-workers using patch amperometry on adrenal cells[113] and suggests that cells regulate vesicular volume to maintain a relatively constant neurotransmitter concentration.

SUMMARY AND FUTURE PERSPECTIVES

Electrochemistry at cells clearly produces a great deal of information pertinent to a variety of questions related to exocytosis. It has been used to study regulated exocytosis both in vivo and in isolated cultures. It has revealed fine details about exocytosis before, during, and immediately following fusion. It can reveal net changes in release as a form of toxicological assay or intracellular signal screen. Finally, quantal size manipulations have provided evidence towards the basis of possible forms of presynaptic plasticity.

The future of molecular neuroscience will depend upon a synthesis of complementary analytical techniques applied simultaneously to systems, including *in vivo,* brain slices, and cell and tissue culture preparations. Recent attempts have been made to create more complex networks of cells[114] that recreate similar cellular communication environments found in vivo. These networks lend themselves to multidimensional analytical techniques, such as, patch amperometry, fast ionophore imaging and amperometry, or amperometry with electrophysiology. Major advances are likely in the study of multicellular networks, expanding many technologies to this task, including amperometry, electrophysiology, fluorescence imaging, microfluidics, and novel techniques on the horizon, like micro arrays of quartz crystal microbalances.

ACKNOWLEDGMENTS

The authors gratefully acknowledge support from the NIH and NSF for work that is reviewed here. Additionally, the hard work of all our former co-workers that is referenced herein is gratefully acknowledged.

REFERENCES

1. Adams, R.N., Probing brain chemistry with electroanalytical techniques, *Anal. Chem.*, 48 (14), 1126A–1138A, 1976.
2. Leszczyszyn, D.J., Jankowski, J.A., Viveros, O.H., Diliberto, E.J. Jr., Near, J.A., and Wightman, R.M., Nicotinic receptor-mediated catecholamine secretion from individual chromaffin cells. Chemical evidence for exocytosis, *J. Biol. Chem.*, 265 (25), 14736–14737, 1990.
3. Wightman, R.M., Jankowski, J.A., Kennedy, R.T., Kawagoe, K.T., Schroeder, T.J., Leszczyszyn, D.J., Near, J.A., Diliberto, E.J. Jr., and Viveros, O.H., Temporally resolved catecholamine spikes correspond to single vesicle release from individual chromaffin cells, *Proc. Natl Acad. Sci. U.S.A.*, 88 (23), 10754–10758, 1991.
4. Durack, G. and Robinson, J.P., *Emerging Tools for Single-Cell Analysis: Advances in Optical Measurement Technologies Cytometric Cellular Analysis*, Wiley, New York, 2000.
5. Valtorta, F., Fesce, R., Grohovaz, F., Haimann, C., Hurlbut, W.P., Iezzi, N., Torri Tarelli, F., Villa, A., and Ceccarelli, B., Neurotransmitter release and synaptic vesicle recycling, *Neuroscience*, 35 (3), 477–489, 1990.
6. Stamford, J.A. and Justice, J.B. Jr., Probing brain chemistry, *Anal. Chem.*, 68 (11), 359A–363A, 1996.
7. Neher, E. and Marty, A., Discrete changes of cell membrane capacitance observed under conditions of enhanced secretion in bovine adrenal chromaffin cells, *Proc. Natl Acad. Sci. U.S.A.*, 79 (21), 6712–6716, 1982.
8. Chow, R.H., von Ruden, L., and Neher, E., Delay in vesicle fusion revealed by electrochemical monitoring of single secretory events in adrenal chromaffin cells, *Nature*, 356 (6364), 60–63, 1992.
9. Dayton, M.A., Ewing, A.G., and Wightman, R.M., Evaluation of amphetamine-induced in vivo electrochemical response, *Eur. J. Pharmacol.*, 75 (2-3), 141–144, 1981.
10. Ewing, A G., Wightman, R.M., and Dayton, M.A., In vivo voltammetry with electrodes that discriminate between dopamine and ascorbate, *Brain Res.*, 249 (2), 361–370, 1982.
11. Cespuglio, R., Faradji, H., Ponchon, J.L., Riou, F., Buda, M., Gonon, F., Pujol, J.F., and Jouvet, M., In vivo measurements by differential pulse voltammetry of extra-cellular 5-hydroxyindoleacetic acid in the rat brain, *J. Physiol. (Paris)*, 77 (2-3), 327–332, 1981.
12. Gonon, F., Buda, M., Cespuglio, R., Jouvet, M., and Pujol, J.F., In vivo electrochemical detection of catechols in the neostriatum of anaesthetized rats: dopamine or DOPAC? *Nature*, 286 (5776), 902–904, 1980.
13. Cahill, P.S., Walker, Q.D., Finnegan, J.M., Mickelson, G.E., Travis, E.R., and Wightman, R.M., Microelectrodes for the measurement of catecholamines in biological systems, *Anal. Chem.*, 68 (18), 3180–3186, 1996.
14. Kawagoe, K.T., Zimmerman, J.B., and Wightman, R.M., Principles of voltammetry and microelectrode surface states, *J. Neurosci. Methods*, 48 (3), 225–240, 1993.
15. Pothos, E.N., Davila, V., and Sulzer, D., Presynaptic recording of quanta from midbrain dopamine neurons and modulation of the quantal size, *J. Neurosci.*, 18 (11), 4106–4118, 1998.
16. Travis, E.R. and Wightman, R.M., Spatio-temporal resolution of exocytosis from individual cells, *Annu. Rev. Biophys. Biomol. Struct.*, 27, 77–103, 1998.
17. Gerhardt, G.A., Oke, A.F., Nagy, G., Moghaddam, B., and Adams, R.N., Nafion-coated electrodes with high selectivity for CNS electrochemistry, *Brain Res.*, 290 (2), 390–395, 1984.
18. Hochstetler, S., Puopolo, M., Gustincich, S., Raviola, E., and Wightman, R., Real-time amperometric measurements of zeptomole quantities of dopamine released from neurons, *Anal. Chem.*, 72 (3), 489–496, 2000.
19. Pothos, E.N., Mosharov, E., Liu, K.P., Setlik, W., Haburcak, M., Baldini, G., Gershon, M.D., Tamir, H., and Sulzer, D., Stimulation-dependent regulation of the pH, volume and quantal size of bovine and rodent secretory vesicles, *J. Physiol.*, 542 (Pt. 2), 453–476, 2002.
20. Chen, T.K., Luo, G., and Ewing, A.G., Amperometric monitoring of stimulated catecholamine release from rat pheochromocytoma (PC12) cells at the zeptomole level, *Anal. Chem.*, 66 (19), 3031–3035, 1994.
21. Colliver, T.L., Pyott, S.J., Achalabun, M., and Ewing, A.G., VMAT-mediated changes in quantal size and vesicular volume, *J. Neurosci.*, 20 (14), 5276–5282, 2000.

22. Schmitz, Y., Lee, C.J., Schmauss, C., Gonon, F., and Sulzer, D., Amphetamine distorts stimulation-dependent dopamine overflow: effects on D2 autoreceptors, transporters, and synaptic vesicle stores, *J. Neurosci.*, 21 (16), 5916–5924, 2001.
23. Chen, G., Gavin, P.F., Luo, G., and Ewing, A.G., Observation and quantitation of exocytosis from the cell body of a fully developed neuron in *Planorbis corneus*, *J. Neurosci.*, 15 (11), 7747–7755, 1995.
24. Schonfuss, D., Reum, T., Olshausen, P., Fischer, T., and Morgenstern, R., Modelling constant potential amperometry for investigations of dopaminergic neurotransmission kinetics in vivo, *J. Neurosci. Methods*, 112 (2), 163–172, 2001.
25. Travis, E.R. and Wightman, R.M., Spatio-temporal resolution of exocytosis from individual cells, *Annu. Rev. Biophys. Biomol. Struct.*, 27 (1), 77–103, 1998.
26. Anderson, B.B., Chen, G., Gutman, D.A., and Ewing, A.G., Demonstration of two distributions of vesicle radius in the dopamine neuron of *Planorbis corneus* from electrochemical data, *J. Neurosci. Methods*, 88 (2), 153, 1999.
27. Cans, A.S., Wittenberg, N., Eves, D., Karlsson, R., Karlsson, A., Orwar, O., and Ewing, A., Amperometric detection of exocytosis in an artificial synapse, *Anal. Chem.*, 75 (16), 4168–4175, 2003.
28. Finnegan, J.M., Pihel, K., Cahill, P.S., Huang, L., Zerby, S.E., Ewing, A.G., Kennedy, R.T., and Wightman, R.M., Vesicular quantal size measured by amperometry at chromaffin, mast, pheochromocytoma, and pancreatic cells, *J. Neurochem.*, 66 (5), 1914–1923, 1996.
29. Amatore, C., Arbault, S., Bonifas, I., Bouret, Y., Erard, M., Ewing, A.G., and Sombers, L.A. Correlation between vesicle quantal size and fusion pore release in chromaffin cell exocytosis *Biophys. J.*, 88 (6), 4411–4420, 2005.
30. Ciolkowski, E.L., Cooper, J.M., Jankowski, J.A., Jorgenson, J.W., and Wightman, R.M., Direct observation of epinephrine and norepinephrine cosecretion from individual adrenal-medullary chromaffin cells, *J. Am. Chem. Soc.*, 114, 2815–2821, 1992.
31. Amatore, C., Bouret, Y., Travis, E.R., and Wightman, R.M., Interplay between membrane dynamics, diffusion and swelling pressure governs individual vesicular exocytotic events during release of adrenaline by chromaffin cells, *Biochimie*, 82 (5), 481–496, 2000.
32. Alvarez de Toledo, G., Fernandez-Chacon, R., and Fernandez, J.M., Release of secretory products during transient vesicle fusion, *Nature*, 363 (6429), 554–558, 1993.
33. Chen, Y.A., Scales, S.J., and Scheller, R.H., Sequential SNARE assembly underlies priming and triggering of exocytosis, *Neuron*, 30 (1), 161–170, 2001.
34. Kennedy, R.T., Huang, L., Atkinson, M.A., and Dush, P., Amperometric monitoring of chemical secretions from individual pancreatic beta-cells, *Anal. Chem.*, 65 (14), 1882–1887, 1993.
35. Ishizaki, M. and Oka, Y., Amperometric recording of gonadotropin-releasing hormone release activity in the pituitary of the dwarf gourami (teleost) [correction of (teleosat)] brain-pituitary slices, *Neurosci. Lett.*, 299 (1-2), 121–124, 2001.
36. Lieste, J.R., Scheenen, W.J., Jenks, B.G., and Roubos, E.W., Single cell secretion measurements by amperometry in *Xenopus melanotropes*, *Ann. N. Y. Acad. Sci.*, 839, 546–548, 1998.
37. Paras, C.D. and Kennedy, R.T., Electrochemical detection of exocytosis at single rat melanotrophs, *Anal. Chem.*, 67 (20), 3633–3637, 1995.
38. Finnegan, J.M., Pihel, K., Cahill, P.S., Huang, L., Zerby, S.E., Ewing, A.G., Kennedy, R.T., and Wightman, R.M., Vesicular quantal size measured by amperometry at chromaffin, mast, pheochromocytoma, and pancreatic beta-cells, *J. Neurochem.*, 66 (5), 1914–1923, 1996.
39. Pihel, K., Travis, E.R., Borges, R., and Wightman, R.M., Exocytotic release from individual granules exhibits similar properties at mast and chromaffin cells, *Biophys. J.*, 71 (3), 1633–1640, 1996.
40. Pihel, K., Hsieh, S., Jorgenson, J.W., and Wightman, R.M., Electrochemical detection of histamine and 5-hydroxytryptamine at isolated mast cells, *Anal. Chem.*, 67 (24), 4514–4521, 1995.
41. Greene, L.A. and Tischler, A.S., Establishment of a noradrenergic clonal line of rat adrenal pheochromocytoma cells which respond to nerve growth factor, *Proc. Natl Acad. Sci. U.S.A.*, 73 (7), 2424–2428, 1976.
42. Schubert, D., LaCorbiere, M., Klier, F.G., and Steinbach, J.H., The modulation of neurotransmitter synthesis by steroid hormones and insulin, *Brain Res.*, 190 (1), 67–79, 1980.
43. Clark, R.A. and Ewing, A.G., Quantitative measurements of released amines from individual exocytosis events, *Mol. Neurobiol.*, 15 (1), 1–16, 1997.

44. Dahlstrom, A., Haggendal, J., and Hokfelt, T., The noradrenaline content of the varicosities of sympathetic adrenergic nerve terminals in the rat, *Acta Physiol. Scand.*, 67 (3), 289–294, 1966.
45. Zerby, S.E. and Ewing, A.G., Electrochemical monitoring of individual exocytotic events from the varicosities of differentiated PC12 cells, *Brain Res.*, 712 (1), 1–10, 1996.
46. Hodgkin, A.L. and Huxley, A.F., A quantitative description of membrane current and its application to conduction and excitation in nerve, *Bull. Math. Bio.*, 52 (1-2), 25–71, 1952 discussion 5-23.
47. Anderson, B.B., Chen, G., Gutman, D.A., and Ewing, A.G., Dopamine levels of two classes of vesicles are differentially depleted by amphetamine, *Brain Res.*, 788 (1-2), 294–301, 1998.
48. Chen, G. and Ewing, A.G., Multiple classes of catecholamine vesicles observed during exocytosis from the *Planorbis* cell body, *Brain Res.*, 701 (1-2), 167–174, 1995.
49. Bruns, D. and Jahn, R., Real-time measurement of transmitter release from single synaptic vesicles, *Nature*, 377 (6544), 62–65, 1995.
50. Abe, T., Lau, Y.Y., and Ewing, A.G., Intracelluar analysis with an immobilized-enzyme glucose electrode having a 2-μm diameter and subsecond response times, *J. Am. Chem. Soc.*, 113 (19), 7421–7423, 1991.
51. Bouchelouche, P.N., Dynamic, real time imaging of ion activities in single living cells using fluorescence video microscopy and image analysis, *Scand. J. Clin. Lab. Invest. Suppl.*, 214, 27–39, 1993.
52. Iturriaga, R., Cerpa, V., Zapata, P., and Alcayaga, J., Catecholamine release from isolated sensory neurons of cat petrosal ganglia in tissue culture, *Brain Res.*, 984 (1-2), 104–110, 2003.
53. Jaffe, E.H., Marty, A., Schulte, A., and Chow, R.H., Extrasynaptic vesicular transmitter release from the somata of substantia nigra neurons in rat midbrain slices, *J. Neurosci.*, 18 (10), 3548–3553, 1998.
54. Phillips, P.E., Hancock, P.J., and Stamford, J.A., Time window of autoreceptor-mediated inhibition of limbic and striatal dopamine release, *Synapse*, 44 (1), 15–22, 2002.
55. Staal, R.G.W., Mosharov, E.V., and Sulzer, D., Dopamine neurons release transmitter via a flickering fusion pore, *Nat. Neurosci.*, 7, 341–346, 2004.
56. Devadoss, A. and Burgess, J.D., Steady-state detection of cholesterol contained in the plasma membrane of a single cell using lipid bilayer-modified microelectrodes incorporating cholesterol oxidase, *J. Am. Chem. Soc.*, 126 (33), 10214–10215, 2004.
57. Devadoss, A. and Burgess, J.D., Detection of cholesterol through electron transfer to cholesterol oxidase in electrode-supported lipid bilayer membranes, *Langmuir*, 18 (25), 9617–9621, 2002.
58. Simons, K. and Ikonen, E., How cells handle cholesterol, *Science*, 290 (5497), 1721–1726, 2000.
59. Albillos, A., Dernick, G., Horstmann, H., Almers, W., Alvarez de Toledo, G., and Lindau, M., The exocytotic event in chromaffin cells revealed by patch amperometry, *Nature*, 389 (6650), 509–512, 1997.
60. Tabares, L., Lindau, M., and Alvarez de Toledo, G., Relationship between fusion pore opening and release during mast cell exocytosis studied with patch amperometry, *Biochem. Soc. Trans.*, 31 (Pt. 4), 837–841, 2003.
61. Gong, L.W., Hafez, I., Alvarez de Toledo, G., and Lindau, M., Secretory vesicles membrane area is regulated in tandem with quantal size in chromaffin cells, *J. Neurosci.*, 23 (21), 7917–7921, 2003.
62. Dernick, G., Alvarez de Toledo, G., and Lindau, M., Exocytosis of single chromaffin granules in cell-free inside-out membrane patches, *Nat. Cell Biol.*, 5 (4), 358–362, 2003.
63. Sauerbrey, G.Z., Verwendung von schwingquarzen zur wagung dunner schichten und zur mikrowangung, *Zeitschrift fur Physik*, 155, 206–222, 1959.
64. Finklea, H.O., Electrochemistry of organized monolayers of thiols and related molecules on electrodes, in *Electroanalytical Chemistry: A Series of Advances*, Bard, A.J., Ed., Marcel Dekker, New York, pp. 109–335, 1996.
65. Rodahl, M., Hook, F., Krozer, A., Brzezinski, P., and Kasemo, B., Quartz-crystal microbalance setup for frequency and Q-factor measurements in gaseous and liquid environments, *Rev. Sci. Instrum.*, 66, 3924–3930, 1995.
66. Cans, A.S., Hook, F., Shupliakov, O., Ewing, A.G., Eriksson, P.S., Brodin, L., and Orwar, O., Measurement of the dynamics of exocytosis and vesicle retrieval at cell populations using a quartz crystal microbalance, *Anal. Chem.*, 73 (24), 5805–5811, 2001.

67. Bowyer, W., Clark, M., and Ingram, J., Electrochemical measurements in submicroliter volumes, *Anal. Chem.*, 64 (4), 459–462, 1992.
68. Clark, R.A. and Ewing, A.G., Characterization of electrochemical responses in picoliter volumes, *Anal. Chem.*, 70 (6), 1119–1125, 1998.
69. Clark, R.A., Heitpas, P.B., and Ewing, A.G., Electrochemical analysis in picoliter vials, *Anal. Chem.*, 69 (2), 259–263, 1997.
70. Bratten, C., Cobbold, P., and Cooper, J., Micromachining sensors for electrochemical measurement in subnanoliter volumes, *Anal. Chem.*, 69 (2), 253–258, 1997.
71. Bratten, C., Cobbold, P., and Cooper, J., Enzyme assay using ultra-low volume surface micromachined sensors, *Chem. Commun.*, 4, 471–472, 1998.
72. Bratten, C., Cobbold, P., and Cooper, J., Single cell measurements of purine release using a micromachined electroanalytical sensor, *Anal. Chem.*, 70 (6), 1164–1170, 1998.
73. Ball, J., Lumpp, J., Daunert, S., and Bachas, L., Effect of fabrication factors on performance of screen-printed/laser micromachined electrochemical nanovials, *Electroanalysis*, 12 (9), 685–690, 2000.
74. Ball, J.C., Scott, D.L., Lumpp, J.K., Daunert, S., Wang, J., and Bachas, L.G., Electrochemistry in nanovials fabricated by combining screen printing and laser micromachining, *Anal. Chem.*, 72 (3), 497–501, 2000.
75. Dias, A., Dernick, G., Valero, V., Yong, M., James, C., Craighead, H., and Lindau, M., An electrochemical detector array to study cell biology on the nanoscale, *Nanotechnology*, 13 (3), 285–289, 2002.
76. Yasukawa, T., Glidle, A., Cooper, J., and Matsue, T., Electroanalysis of metabolic flux from single cells in simple picoliter-volume microsystems, *Anal. Chem.*, 74 (19), 5001–5008, 2002.
77. Troyer, K. and Wightman, R., Dopamine transport into a single cell in a picoliter vial, *Anal. Chem.*, 74 (20), 5370–5375, 2002.
78. Earles, C. and Schenk, J.O., Multisubtrate mechanism for the inward transport of dopamine by the human dopamine transporter expressed in HEK cells and its inhibition by cocaine, *Synapse*, 33 (3), 230–238, 1999.
79. Cans, A.S., Wittenberg, N., Karlsson, R., Sombers, L., Karlsson, M., Orwar, O., and Ewing, A., Artificial cells: unique insights into exocytosis using liposomes and lipid nanotubes, *Proc. Natl Acad. Sci. U.S.A.*, 100 (2), 400–404, 2003.
80. Karlsson, M., Nolkrantz, K., Davidson, M.J., Stromberg, A., Ryttsen, F., Akerman, B., and Orwar, O., Electroinjection of colloid particles and biopolymers into single unilamellar liposomes and cells for bioanalytical applications, *Anal. Chem.*, 72 (23), 5857–5862, 2000.
81. Breckenridge, L.J. and Almers, W., Final steps in exocytosis observed in a cell with giant secretory granules, *Proc. Natl Acad. Sci. U.S.A.*, 84 (7), 1945–1949, 1987.
82. Amatore, C., Bouret, Y., Travis, E.R., and Wightman, R.M., Adrenaline release by chromaffin cells: constrained swelling of the vesicle matrix leads to full fusion, *Angew. Chem. Int. Ed. Engl.*, 39 (11), 1952–1955, 2000.
83. Chizmadzhev, Y.A., Kumenko, D.A., Kuzmin, P.I., Chernomordik, L.V., Zimmerberg, J., and Cohen, F.S., Lipid flow through fusion pores connecting membranes of different tensions, *Biophys. J.*, 76 (6), 2951–2965, 1999.
84. Monck, J.R., Alvarez de Toledo, G., and Fernandez, J.M., Tension in secretory granule membranes causes extensive membrane transfer through the exocytotic fusion pore, *Proc. Natl Acad. Sci. U.S.A.*, 87 (20), 7804–7808, 1990.
85. Cans, A.S., Wittenberg, N., Eves, D., Karlsson, R., Karlsson, A., Orwar, O., and Ewing, A., Amperometric detection of exocytosis in an artificial synapse, *Anal. Chem.*, 75 (16), 4168–4175, 2003.
86. Bunin, M.A. and Wightman, R.M., Quantitative evaluation of 5-hydroxytryptamine (serotonin) neuronal release and uptake: an investigation of extrasynaptic transmission, *J. Neurosci.*, 18 (13), 4854–4860, 1998.
87. Bunin, M.A. and Wightman, R.M., Paracrine neurotransmission in the CNS: involvement of 5-HT, *Trends Neurosci.*, 22 (9), 377–382, 1999.
88. Garris, P.A. and Wightman, R.M., In vivo voltammetric measurement of evoked extracellular dopamine in the rat basolateral amygdaloid nucleus, *J. Physiol.*, 478 (Pt. 2), 239–249, 1994.

89. Schroeder, T.J., Borges, R., Finnegan, J.M., Pihel, K., Amatore, C., and Wightman, R.M., Temporally resolved, independent stages of individual exocytotic secretion events, *Biophys. J.*, 70 (2), 1061–1068, 1996.
90. Borges, R., Travis, E.R., Hochstetler, S.E., and Wightman, R.M., Effects of external osmotic pressure on vesicular secretion from bovine adrenal medullary cells, *J. Biol. Chem.*, 272 (13), 8325–8331, 1997.
91. Troyer, K.P , Mundorf, M.L., Fries, H.E., and Wightman, R.M., Separating vesicle fusion and exocytosis in hypertonic conditions, *Ann. N. Y. Acad. Sci.*, 971, 251–253, 2002.
92. Sombers, L.A., Hanchar, H.J., Colliver, T.L., Wittenberg, N., Cans, A., Arbault, S., Amatore, C., and Ewing, A.G., The effects of vesicular volume on secretion through the fusion pore in exocytotic release from PC12 cells, *J. Neurosci.*, 24 (2), 303–309, 2004.
93. Ales, E., Tabares, L., Poyato, J.M., Valero, V., Lindau, M., and Alvarez de Toledo, G., High calcium concentrations shift the mode of exocytosis to the kiss-and-run mechanism, *Nat. Cell Biol.*, 1 (1), 40–44, 1999.
94. Staal, R.G., Mosharov, E.V., and Sulzer, D., Dopamine neurons release transmitter via a flickering fusion pore, *Nat. Neurosci.*, 7 (4), 341–346, 2004.
95. Haller, M., Heinemann, C., Chow, R.H., Heidelberger, R., and Neher, E., Comparison of secretory responses as measured by membrane capacitance and by amperometry, *Biophys. J.*, 74 (4), 2100–2113, 1998.
96. Leszczyszyn, D.J., Jankowski, J.A., Viveros, O.H., Diliberto, E.J. Jr., Near, J.A., and Wightman, R.M., Secretion of catecholamines from individual adrenal medullary chromaffin cells, *J. Neurochem.*, 56 (6), 1855–1863, 1991.
97. Amatore, C., Arbault, S., Bonifas, I., Bouret, Y., Erard, M., and Guille, M., Dynamics of full fusion during vesicular exocytotic events: release of adrenaline by chromaffin cells, *Chemphyschem*, 4 (2), 147–154, 2003.
98. Almers, W., Fusion needs more than SNAREs, *Nature*, 409 (6820), 567–568, 2001.
99. Brunger, A.T., Structural insights into the molecular mechanism of Ca(2+)-dependent exocytosis, *Curr. Opin. Neurobiol.*, 10 (3), 293–302, 2000.
100. Zimmerberg, J., How can proteolipids be central players in membrane fusion? *Trends Cell Biol.*, 11 (6), 233–235, 2001.
101. Chizmadzhev, Y.A., Kuzmin, P.I., Kumenko, D.A., Zimmerberg, J., and Cohen, F.S., Dynamics of fusion pores connecting membranes of different tensions, *Biophys. J.*, 78 (5), 2241–2256, 2000.
102. Sandre, O., Moreaux, L., and Brochard-Wyart, F., Dynamics of transient pores in stretched vesicles, *Proc. Natl Acad. Sci. U.S.A.*, 96 (19), 10591–10596, 1999.
103. Taupin, C., Dvolaitzky, M., and Sauterey, C., Osmotic pressure induced pores in phospholipid vesicles, *Biochemistry*, 14 (21), 4771–4775, 1975.
104. Li, J.Y., Edelmann, L., Jahn, R., and Dahlstrom, A., Axonal transport and distribution of synaptobrevin I and II in the rat peripheral nervous system, *J. Neurosci.*, 16 (1), 137–147, 1996.
105. Burgoyne, R.D., Fisher, R.J., Graham, M.E., Haynes, L.P., and Morgan, A., Control of membrane fusion dynamics during regulated exocytosis, *Biochem. Soc. Trans.*, 29, 467–472, 2001.
106. Graham, M.E., Washbourne, P., Wilson, M.C., and Burgoyne, R.D., Molecular analysis of SNAP-25 function in exocytosis, *Ann. N. Y. Acad. Sci.*, 971 (1), 210–221, 2002.
107. Wu, M.N., Littleton, J.T., Bhat, M.A., Prokop, A., and Bellen, H.J., ROP, the *Drosophila* Sec1 homolog, interacts with syntaxin and regulates neurotransmitter release in a dosage-dependent manner, *EMBO J.*, 17 (1), 127–139, 1998.
108. Misura, K.M.S., Scheller, R.H., and Weis, W.I., Three-dimensional structure of the neuronal-Sec1-syntaxin 1a complex, *Nature*, 404 (6776), 355, 2000.
109. Takei, M., Ueno, M., and Endo, K., Effect of ryanodine on histamine release from rat peritoneal mast cells induced by anti-IgE, *J. Pharm. Pharmacol.*, 44, 523–525, 1992.
110. Jaffe, E.H., Bolanos, P., Galvis, G., and Caputo, C., Ryanodine receptors in peritoneal mast cells: possible role in the modulation of exocytotic activity, *Pflugers Arch.*, 447 (4), 377–386, 2004.
111. Takahashi, N., Kadowaki, T., Yazaki, Y., Miyashita, Y., and Kasai, H., Multiple exocytotic pathways in pancreatic beta cells, *J. Cell Biol.*, 138 (1), 55–64, 1997.

112. Kozminski, K.D., Gutman, D.A., Davila, V., Sulzer, D., and Ewing, A.G., Voltammetric and pharmacological characterization of dopamine release from single exocytotic events at rat pheochromocytoma (PC12) cells, *Anal. Chem.*, 70 (15), 3123–3130, 1998.
113. Gong, L.W., Hafez, I., Alvarez de Toledo, G., and Lindau, M., Secretory vesicles membrane area is regulated in tandem with quantal size in chromaffin cells, *J. Neurosci.*, 23 (21), 7917–7921, 2003.
114. Misgeld, U., Zeilhofer, H.U., and Swandulla, D., Synaptic modulation of oscillatory activity of hypothalamic neuronal networks in vitro, *Cell. Mol. Neurobiol.*, 18 (1), 29–43, 1998.

15 The Patch Amperometry Technique: Design of a Method to Study Exocytosis of Single Vesicles

Gregor Dernick, Guillermo Alvarez de Toledo, and Manfred Lindau

CONTENTS

INTRODUCTION: CAPACITANCE MEASUREMENTS OF MEMBRANE FUSION WITH AMPEROMETRIC DETECTION OF RELEASED MOLECULES

Exocytosis of vesicles as small as 60 nm in diameter can be detected by cell membrane admittance measurements with the patch clamp technique in the cell attached configuration [2–4], and by amperometry with a carbon fiber electrode (CFE) [5]. The admittance measurement provides the membrane capacitance that reveals changes in surface area due to the incorporation of the vesicular membrane into the plasma membrane. Each fusing vesicle causes a stepwise increase in capacitance [2]. The admittance measurement also provides the membrane conductance and the fusion pore conductance during an exocytotic event [4]. With amperometry, released molecules that are readily oxidizable are electrochemically detected at the surface of the CFE. This technique is able to resolve release events down to a few ten-thousands of molecules [6].

In this article we describe a technique that combines high-resolution patch capacitance measurements with amperometry by placing the amperometric detector inside the patch pipette. We call this patch amperometry. It allows the simultaneous detection of membrane fusion and release of transmitter molecules originating from vesicles as small as 100 nm in diameter. In chromaffin cells it was shown that with patch amperometry, the "foot", which sometimes precedes amperometric spike signals [7], is due to the narrow fusion pore restricting the flux of molecules in the early stages of exocytosis [8,9]. Using this technique, it was shown that even during short flickers, the entire contents of a chromaffin granule can be secreted, and that flickers are more likely to occur at high intra-pipette Ca^{2+}-concentrations [10]. Patch amperometry was also used to investigate the regulation of quantal size and granule size in chromaffin cells [11,12]. Another recent application of patch amperometry was the measurement of the cytosolic pool of catecholamines [13].

More than 10 years ago, the first experiments combining whole-cell capacitance measurements with carbon fiber amperometry were carried out in order to reveal the correlation between fusion and release in mast cells and chromaffin cells [14–16]. The granules in mast cells are very large (greater than 0.5 μm), with an associated single granule capacitance greater than 7 fF, which is resolvable in the whole-cell configuration. One limitation of whole-cell capacitance measurements with carbon fiber amperometry is that the carbon fiber electrode does not capture exocytotic events from the entire cell surface, whereas the capacitance measurement does. As a result, for some capacitance steps, the released molecules are only partially detected or not detected at all. However, in 1993, using this approach, Alvarez de Toledo et al. [14] were the first to show a proportionality of the fusion pore conductance with the flux of molecules through that pore. They also showed that fusion of a granule can be transient ("flicker" or "kiss-and-run") and that release can occur during transient fusion through a fusion pore of less than 1 nS.

Chromaffin granules are much smaller than mast cell granules. Thus, in whole-cell capacitance measurements from chromaffin cells, only the largest granules could be detected as capacitance steps greater than 5 fF associated with amperometric spikes [15–17] due to a noise level of 2–3 fF. From morphological studies, it is known that most chromaffin granules are much smaller, with diameters in the range of 60–450 nm (Fenwick et al. 1978), which would correspond to capacitance steps of 0.1–5.7 fF. Accordingly, whole-cell capacitance measurements capture only the largest granules as discrete steps. However, when the capacitance traces were aligned according to the amperometric spikes and averaged, an "average" step size of 2.7 fF was obtained. Whole-cell capacitance measurements did not allow determination of the fusion pore conductance in chromaffin granules.

The overall time course of amperometric spikes is still not fully understood. If release occurred instantaneously, the spike shape would be governed exclusively by diffusion from the site of release to the site of detection. In this case, the signal shape can be well described analytically [7,18]. The experimentally measured time course of amperometric spikes, on the other hand, indicates that release may not be instantaneous, but may follow a certain kinetics [19,20]. Also, from fusion pore and matrix dynamics obtained by deconvolution of amperometric recordings, it was suggested that the shape of the amperometric signals is governed by release of catecholamines from the intragranular matrix [21]. Recent experiments employing microfabricated electrochemical detector arrays suggest that reversible binding of catecholamines to the cell surface also affects the amperometric spike shape [22].

With patch amperometry, the different features of amperometric spikes can now be directly correlated with the fusion pore restriction [8,9]. Furthermore, the cell attached configuration limits the area of observed cell membrane and measures the signals in the confined space between cell, pipette wall, and the detector. Hence, there is much less variation in signal width in a given experiment, which allows the diffusion-controlled contributions to the signal shape to be distinguished from the non-diffusion related components.

PATCH AMPEROMETRY

Patch amperometry combines the patch clamp technique with the amperometry technique by including a CFE inside the patch pipette. Several alterations to the stand-alone techniques were necessary to use this new configuration. The combination of the two techniques required the design of a novel pipette holder. To accommodate the carbon fiber electrode inside the confined space of the patch pipette, the properties of both the patch pipettes and the carbon fiber electrodes needed to be modified. Here we describe the details of these modifications as well as the analysis of the signals that are recorded with patch amperometry.

Patch Amperometry Requires Reversed Patch Clamp Electrode Configuration

The regular patch clamp electrode configuration is shown in Figure 15.1a. The electrode, which is connected to the patch clamp amplifier, is located inside the patch pipette. The reference electrode, which is connected to ground, is placed in the bath. For conventional amperometry (Figure 15.1b), the CFE is connected to the input of the amplifier, and is held at a potential of 650–800 mV versus the Ag|AgCl reference electrode in the bath. The two techniques are combined in patch amperometry by placing the CFE inside the patch pipette (Figure 15.1c). If both the patch clamp and the amperometric electrode were located inside the pipette and the reference electrode for both measurements located in the bath, then the CFE would be separated from the reference (ground) electrode by the high giga-seal resistance, which would make the recordings unstable and would cause excessive noise. As a result, the electrode configuration for the patch clamp experiments has to be reversed. When the electrode connected to the patch clamp amplifier is located in the bath, then the CFE and the reference electrode can both be placed in the same compartment, the patch pipette. Hence, the high potential for amperometry is not applied across the cell membrane. Although the configuration of the patch clamp measurements is reversed, it still measures the current passing through the cell between the patch pipette and the electrode in the bath. Since the bath is not connected to ground, it is important to reduce its stray capacitance by reducing the surface area of the recording chamber and by using a long working distance objective (see below). Noise reduction through proper shielding of the setup and avoiding ground loops need to receive particular attention, too.

Equipment and Setup

Data Acquisition

Figure 15.2 shows a schematic view of the signal processing devices and their connections to each other for patch amperometry. In general, a data acquisition system is required that can record five traces for at least 30 min at a sampling rate of ≥ 1 kHz. We typically used 1 kHz.

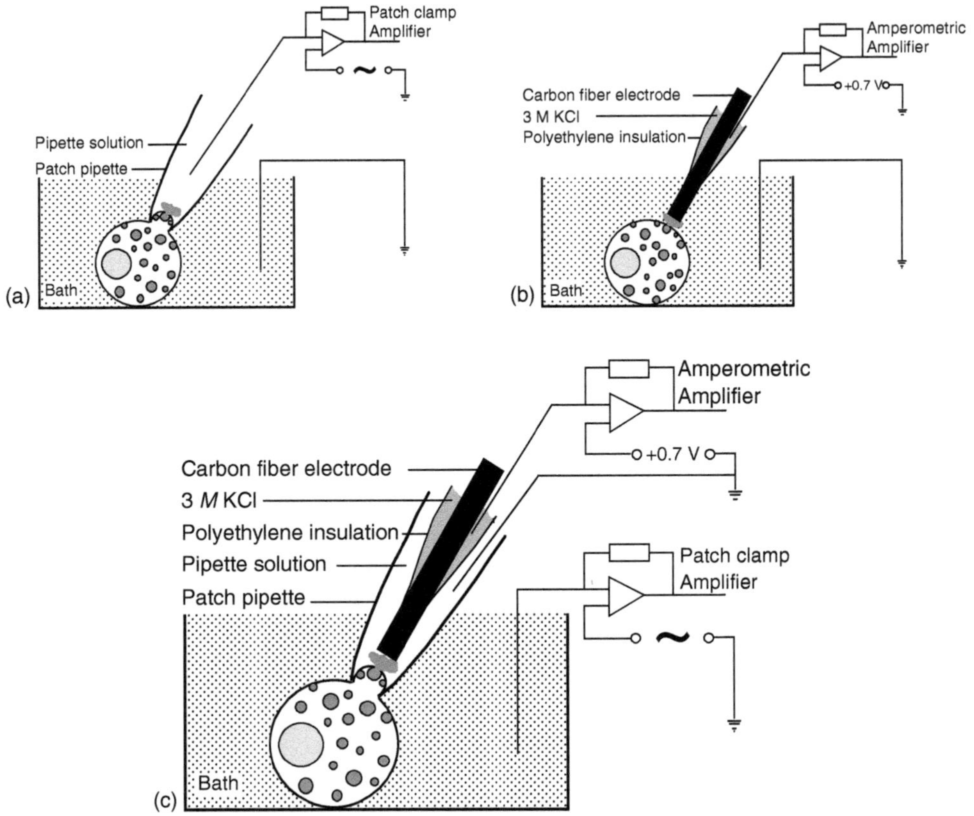

FIGURE 15.1 Electrode configurations.(a) Conventional cell-attached patch clamp configuration. (b) Conventional carbon fiber amperometry configuration. (c) Patch amperometry configuration, where the patch clamp electrode configuration is reversed in order to place the reference electrode for amperometry in the same compartment as the CFE. The connections from the respective solutions to the amplifiers or to ground is made by Ag|AgCl electrodes.

Experiments were controlled and data acquired with an Apple Macintosh G4 or G5 computer running MacOS9 by procedures written in Igor 4 (WaveMetrics Inc., Lake Oswego, OR, U.S.A). The signals were digitized by a 16-bit analog-digital converter (ADC). We currently use the PCI-MIO-16XE-50 DAQ board with a shielded cable and the BNC adapter BNC-2090 (National Instruments Inc., Austin, TX, U.S.A). This data acquisition board has eight differential inputs, two 12-bit digital-analog converter (DAC) outputs and eight selectable digital inputs/outputs.

Signals were passed through RG58C/U type coaxial cables with 50 Ω impedance (Newark Electronics, Chicago, IL, U.S.A.). One DAC output supplied the voltage clamp holding potential and pulses for seal formation, the other a trigger signal to synchronize the oscilloscopes with the pulses. One digital output gave a TTL signal to switch the adder and the image acquisition. For the use in our labs we created a set of Igor Procedures for acquisition (PA_Acquire) and for analysis (PA_Viewer). These are available online [1] and may be modified for the development of the reader's own customized software.

Capacitance Measurements

In the following, we describe the setup for two different patch clamp amplifiers, the HEKA EPC-7 (Heka Electronics, Darmstadt, Germany) and the Axopatch 200B (Axon Instruments/Molecular Devices, Sunnyvale, CA, U.S.A.).

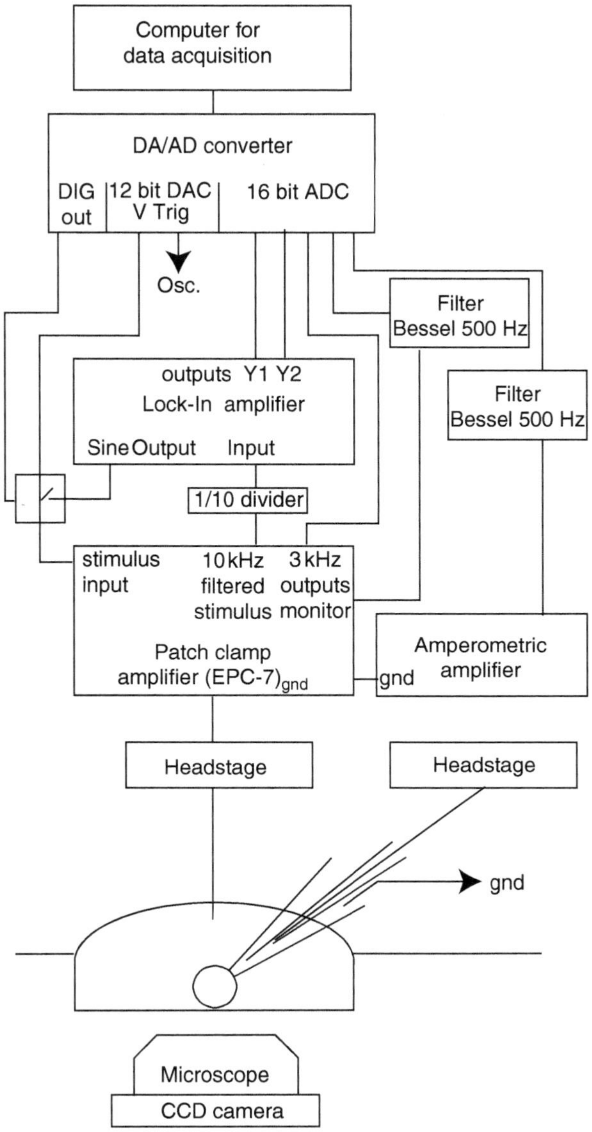

FIGURE 15.2 Patch amperometry set-up. For details and description of signal pathways see section "Equipment and Set-up". (From Dernick, G. et al., *Nat. Methods*, 2, 699–708, 2005.)

For patch capacitance measurements, we used a lock-in amplifier (SR 830; Stanford Research Instruments, Sunnyvale, CA, U.S.A.), whose settings were controlled from within Igor via a RS-232 serial port of the data acquisition computer. The sine wave from the lock-in amplifier (50 mV r.m.s, 20 kHz) was added to the holding potential from the DAQ board in a home-built switchable adder. This device adds two analog inputs, where a transistor–transistor logic (TTL) signal determines if only the first input goes through or if the second input is added to the first. A circuit diagram of the adder is available online [1].

The output of the adder was fed into the stimulus input of the patch clamp amplifier. For recording purposes, the stimulus monitor was low pass filtered at 500 Hz to remove the sine wave and was acquired by the ADC.

The EPC-7 was set to voltage-clamp mode, C-slow compensation to the 10-pF range; the C-slow compensation value was set to 0.2 pF (minimum setting of dial), the G-series compensation to 0.2 μS. The series resistance compensation was switched off. The compensation is chosen for a patch with a typical capacitance of 200–300 fF and an access resistance of about 5 MΩ, which results in a time constant of about 1.5 μs [23]. For high-resolution measurements, a gain of 50 mV/pA or higher must be used. In this range, the amplifier uses the high resistance feedback resistor in the head stage, which allows for low-noise recordings. During the setup of a recording and seal formation, a stimulus scaling of 0.01 is used; during the recording itself, it is set to 0.1. The input filter T_R is set to 2 μs. To avoid amplifier clipping, the pipette current was filtered by the built-in 10 kHz filter (filter 1). Although a 20 kHz sine wave command voltage is used, the signal-to-noise ratio is not significantly compromised because the 10 kHz filter reduces both signal and noise around 20 kHz by the same factor [23]. This output is displayed on an oscilloscope. Additionally, the pipette current was filtered with the built-in 3 kHz filter (filter 2) from the patch clamp amplifier and acquired by the ADC. The 3 kHz filter suppresses the 20 kHz sine wave such that the net patch current can be monitored.

The Axopatch 200B is used in an analogous way. It has two command signal inputs that are added internally, one being selected or deselected by a front panel switch. With this amplifier, the signal adder is not necessary. However, the sine wave must then be switched on and off manually. The scaling factors differ for the front panel- and back panel-switched inputs. If the front panel-switched input is used, then the scaling factor is 0.02 and the lock-in output voltage must be set to 2.5 V to obtain a 50 mV r.m.s. pipet voltage signal. For patch capacitance measurements, the "whole cell $\beta = 1$" mode can be used. We were not able to achieve low noise recordings in the capacitive feedback "patch" mode. The gain setting is not critical since all gain settings use the same feedback resistor. We still use relatively high gain to minimize noise contributions from other sources, including ADC quantization noise. For the patch compensation, the SERIES RESISTANCE is set to 5 MΩ and the WHOLE CELL CAP to the minimum, which is according to the manual about 0.2 pF. The pipette compensation dials of the Axopatch 200B (termed C-slow) may not be used for the patch compensation because the time constant of C-slow is too slow. However, they may be used freely to compensate pipette and bath capacitances and to minimize the sine wave during a recording.

We have, so far, mainly used the two amplifiers described above. We have done initial tests with the EPC-10 dual amplifier (HEKA), which is suitable because filter 1 can be set independently from filter 2; this is important to avoid clipping without over-filtering the sine wave current. In its double-amplifier version it may serve for low noise recording of both the patch clamp capacitance changes and the amperometric signals (see below).

The output of either patch clamp amplifier is scaled down by a factor of 10 before input to the lock-in amplifier by a 10 kΩ:100 kΩ voltage divider. The downscaling is necessary to avoid saturation of the lock-in amplifier because the output range of the EPC-7 is ± 10 V, whereas the maximum input range of the lock-in amplifier is only (± 1 V). The lock-in amplifier settings can be found in Table 15.1.

At the correct phase setting, the lock-in amplifier computes the changes in the real and the imaginary part of the patch current, which is recorded by the ADC as Y1 and Y2, respectively. Finding the correct setting of the phase is important, only then is the lock-in amplifier output Y1 proportional to the membrane conductance, and the output Y2 proportional to the membrane capacitance [23].

The correct phase for the lock-in amplifier can be found in various ways [23]. One way is to vary slightly the patch capacitance compensation (C-slow in EPC-7; WHOLE CELL CAP in Axopatch 200B), which can be done with open headstage while the sine wave is applied to the stimulus input. The amplitude of the sine wave in the pipette current is minimized to zero with the pipette compensation dials (C-fast and τ-fast in the EPC-7, C-fast and C-slow dials in the Axopatch 200B) observing it on the oscilloscope. Then, the patch capacitance compensation

TABLE 15.1
SR830 Lock-in Amplifier Settings

Variable	Setting
Time constant	1 ms, 24 dB
Signal input	Input:A, couple:AC, ground:FLOAT
Sensitivity	1 V
Reserve	LOW NOISE
Filters	LINE & 2×LINE
Channel 1 (Y1)	Output:X, OFFSET:Off, RATIO:Off, Expand:Off
Channel 2 (Y2)	Output:Y, OFFSET:Off, RATIO:Off, Expand:Off
Interface	RS232, 8, 9600, none
Reference	Phase +120°[a], Freq:20 kHz, Ampl.:0.5 V[b], Harm:#1, Trig:SINE

[a] The phase setting needs to be determined as described below.
[b] For Axopatch 200B: 2.5 V.

Source: Dernick, G. et al. *Nat. Methods*, 2, 699–708, 2005.

(C-slow in EPC-7; WHOLE CELL CAP in Axopatch 200B) is slightly varied observing the two lock-in outputs Y1 and Y2. The phase is adjusted until changes in patch capacitance compensation appear only in the Y2 channel and the Y1 channel remains unchanged. Here, an increase of the compensation is a capacitance decrease and must therefore produce a decrease of Y2. The found phase setting applies only to the used feedback resistor in the headstage and to the used filter settings. As mentioned above, the EPC-7 has two feedback resistors in the headstage. One for gain settings below 20 mV/pA and another for settings above 50 mV/pA. The Axopatch 200B has only one feedback resistor in the headstage in the "whole cell $\beta = 1$" mode, so the gain may be changed without changing the phase. In our tests the capacitive feedback in the Axopatch 200B headstage ("patch" mode) did not provide low noise in patch amperometry recordings.

Other methods to find the correct phase include the application of suction pulses to a patched cell followed by adjustments of the phase until only changes in Y2 are observed [23], and the utilization of a capacitance dither switch on the patch clamp amplifier. Maladjustments of the phase may be corrected off-line, after the recording. This is only possible if the amplifier was not saturated (clipping). Unfortunately, the clipping indicator LED's on the amplifiers may fail to indicate slight amplifier saturation at the extrema of the sine wave current. Thus, it is imperative to display the pipette current on an oscilloscope and to observe the sine wave amplitude during all operations.

The application of a defined capacitance compensation change, such as, 200 fF, with a capacitance dither may also be used to calibrate the recording. However, in this case such a large change may saturate the amplifier. It would then be necessary to use 0.01 stimulus scaling, effectively generating a calibration signal ten times smaller. In the EPC-7, for convenience, a switchable 20 Ω resistor was inserted in series to the C-slow potentiometer which provides a 20 fF capacitance compensation change for calibration. In the Axopatch 200B, the built-in 100 fF capacitance dither may be used, but a low enough gain must be selected to avoid clipping. Instructions to modify the dither circuit of the Axopatch 200B for a 10 or 20 fF capacitance compensation change can be obtained from Axon Instruments upon request. However, the small resistors used may result in an imprecise calibration. Note that the resistor for calibration is added to the compensation in the EPC-7, while it is subtracted (switched off) in the Axopatch 200B. The orientation of the calibration pulses wil thus be downwards for th EPC-7 but upwards in the Axopatch 200B.

After the recording, the calibration pulses were used to convert units of the Y1 and Y2 traces from raw counts into pS and fF, respectively. This is described in the data analysis section of this article.

Amperomety

The VA-10 (NPI Electronics, Tamm, Germany) or a patch clamp amplifier may be used as an amperometric amplifier. Also, a simple home-built amplifier can be used; a circuit diagram may be obtained from the Alvarez de Toledo laboratory. As mentioned before, initial tests indicate that the EPC-10 dual amplifier (HEKA Electronics) is a practical solution for patch amperometry because it may serve as both an amperometric and a patch clamp amplifier. Capacitance measurements with a 20 kHz sine wave have been implemented in the new software for this instrument. This instrument will be tested in the future to determine its noise performance in patch amperometry recordings.

When using the VA-10 or the EPC-10, a special version of the head stage must be ordered, in which the BNC shield is grounded. This is available upon request from the manufacturers. The modification is necessary because in patch amperometry, the reference electrode for amperometry is connected to the shield of the BNC. The standard head stages where the BNC shield is connected to V_{ref} will not work for patch amperometry.

The amplifier typically applies a potential of +700 mV to the CFE relative to the reference electrode. The amperometric current is low pass filtered at 500 Hz and acquired by the ADC. Since both the patch clamp amplifier and the amperometric amplifier use the same reference electrode that is connected to the ground of the amperometric amplifier headstage, it is essential to connect the signal-ground connectors of the two amplifiers to each other, but to make no connections to the ground pin of the patch damp headstages to avoid ground loops.

Microscopy

The experiment was observed through an inverted microscope (Axiovert 100; Zeiss, Oberkochen, Germany), using an objective with long working distance (Zeiss 20×0.5 Plan Neofluar) to minimize stray capacitance. Objectives with short working distance, and in particular immersion objectives, may not be used because the stray capacitance between such an objective and the bath (which is not grounded in patch amperometry) cannot be compensated for.

As for other patch clamp applications, the microscope was mounted on a vibration-isolated table (Physik Instrumente, Waldbronn, Germany) hooked up to house pressurized air. Coarse positioning of the self-built microscope stage and the electrode holder arm was done with tower-mounted linear micrometers (Spindler&Hoyer, Göttingen, Germany, now LINOS Photonics). For the fine positioning of the patch pipette, a hydraulic high-resolution micromanipulator (Narishige, Tokyo, Japan) was used.

Images were taken with a CCD camera (SSC M-370, Sony, Japan), which was mounted on the microscope. Any USB or FireWire camera may be used for this purpose. Images were taken automatically during acquisition of electrophysiological data with a PixelBuffer frame grabber (Perceptics, Knoxville, TN, U.S.A) in an additional Apple Macintosh computer. As explained earlier, during data acquisition, a TTL signal given from the DAQ board switches on the sine wave from the lock-in amplifier. The same signal was fed into the trigger input of the frame grabber. A script in IPLab (Scanalytics Inc., Fairfax, VA, U.S.A.) took pictures at 60 s intervals whenever the trigger was enabled. The images were time stamped for off-line synchronization with the recorded data.

Design of the Electrode Holders

The pipette holder for patch amperometry had to combine several different features in order to measure fusion and release from single chromaffin granules simultaneously in the cell-attached configuration. The design was guided by the two techniques, patch clamp and amperometry, which were combined. The patch clamp part of the holder had to accommodate a silver wire for the electrical connection as well as the patch pipette. It needed to be airtight, so that suction could be applied to it via a small tube to obtain a pipette-membrane seal [24]. For the patch amperometry configuration, the CFE as well as the reference electrode had to be inserted into the pipette. As already mentioned, the electrode configuration for the patch clamp technique was reversed such that the reference electrode located inside the pipette was connected to ground. Therefore, the BNC shield of the pipette holder had to be connected to ground on the amplifier headstage. In addition, the CFE needed to be movable with respect to the patch pipette, or the pipette relative to the CFE, to adjust the variations in length of both the patch pipettes and CFEs. The CFE has to be positioned as close as possible to the patch to minimize diffusional broadening of the amperometric signals [25].

Two different types of electrode holders were developed: one with a fixed CFE and a manually adjustable pipette, the other with a fixed pipette and a CFE that was moved by a motor inside the electrode holder.

Manually Adjustable Electrode Holder

The manually adjustable electrode holder is shown in Figure 15.3. The CFE was attached to the housing; the patch pipette was clamped to the cap, whose position could be changed with respect to the housing by means of a fine-pitched thread. The insertion of CFE and reference electrode into the patch pipette was done under a binocular dissecting microscope, and coarse adjustment of the CFE to the tip of the pipette was made, so that the distance between pipette tip and CFE tip was roughly 50–100 μm. The holder was then mounted onto the headstage of the amperometric amplifier in the patch clamp setup and the pipette tip was observed through the inverted microscope, which was used for the experiments. The final adjustment bringing the pipette tip close to the CFE tip was performed by hand, right before seal formation, when the pipette was in the bath. A distance of a few micrometers was desired. Since touching the pipette holder (like touching any other part on the

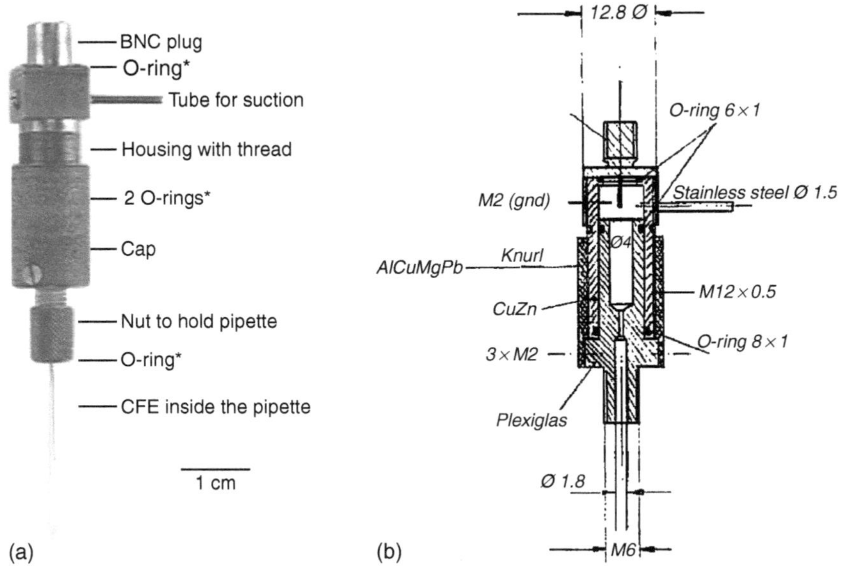

FIGURE 15.3 Manually adjustable electrode holder. (a) Photograph of electrode holder. The lower part can be adjusted against the upper part by the fine-pitched thread. The CFE is mounted fixed and the pipette is moved in respect to it. The asterisk denotes parts inside the holder not visible on this picture. (b) Technical drawing. (Modified from Dernick, G. et al., *Nat. Methods*, 2, 699–708, 2005.)

optical table) caused large movements of the pipette tip, no further adjustments could be made after a seal was formed without losing the seal. Recordings with the manually adjustable holder were, therefore, done with the CFE at a fixed position that was preset before seal formation. Nevertheless, slight fluctuations in distance did sometimes occur. Presumably, due to the release of tension that had built up in the polyethylene tube of the CFE during assembly and adjustments.

The connection to the amperometric amplifier was made with a BNC connector (Figure 15.3, top). A silver wire (0.25 mm in diameter; Goodfellow, Cambridge, England) was soldered to the inner pin of the BNC plug and chlorinated. The wire both guided the CFE and made electrical contact with the carbon fiber via the 3M KCl solution inside the CFE. A short piece of polyethylene tubing attached the CFE mechanically to the BNC plug. The reference electrode (chlorinated silver wire, 0.25 mm in diameter) was soldered to a small screw in the housing of the holder and was in this way connected to ground. The cap was screwed onto the housing with a fine pitched thread. O-rings in the housing and the cap of the holder ensured that the junction was airtight and provided additional stability to prevent the tilting of the cap against the housing. The pipette was clamped onto the holder in a manner similar to conventional patch pipette holders. A step in the insertion hole provided a stop position for the pipette. An O-ring in the small screwcap tightened and held the pipette in place. Suction was applied to the entire lumen of the holder through the metal tube close to the BNC plug.

Motorized Electrode Holder

Although successful recordings were made with the manually adjustable holder, it has some drawbacks. The holder has to be removed from the head stage every time a new pipette is mounted, and the adjustments under the inverted microscope must to be done very carefully to avoid breaking the newly mounted pipette. The most serious drawback of the manually adjustable holder is that the distance between patch and CFE cannot be changed after the seal is formed. In some cases the CFE was too far away, resulting in very small and broad amperometric signals that were hard to analyze. In other cases, the CFE was positioned too close to the pipette tip preventing seal formation when upon applying suction, the patch touched the CFE before a giga-seal was achieved. The motorized electrode holder improved convenience of handling, control and variation of CFE-patch distance during an experiment and revealed information about the release mechanism of catecholamines after the fusion pore expanded completely.

An electrode holder that allows adjustment of the CFE-patch distance during an experiment required a stationary pipette to keep the seal intact. It had to be robust and airtight like the manually adjustable holder. Inside, a small device was needed to move the CFE, while the electrical connection between the CFE and the Ag|AgCl wire via the KCl-Solution was maintained. These considerations yielded the design of a new electrode holder, which accommodated a small motor to move the carbon fiber (Nanomotor, Klocke, Nanotechnik, Aachen, Germany). This type of piezo-driven linear motor has a wide spectrum of applications in micro-positioning and micromanipulation, ranging from scanning tunneling microscopes to micro-assembly of electronics and telecommunication devices [26]. The motor was mounted inside the holder by the manufacturer. The first model of the motorized holder is shown in Figure 15.4a and b. Improvements in airtightness and sturdiness were achieved for the second model, shown in Figure 15.4c and d.

As the manually adjustable holder, the motorized holder's connection to the amperometric amplifier was made by a BNC plug. A teflon-coated silver wire (0.125 mm in diameter; World Precision Instruments, Sarasota, Florida, U.S.A.) soldered to the inner pin of the BNC plug connects to the CFE. The teflon coat serves three purposes: (1) it insulates the wire against the housing when the wire is ducted through the movable inner sheath of the nanomotor; (2) since the CFE is moved against the wire, the teflon coat minimizes friction; and (3) it prevents the 3 *M* KCl solution in the CFE from ascending to the motor where the solution could cause short circuits and crystals from dried KCl solution could damage the motor. The CFE was attached mechanically to the inner

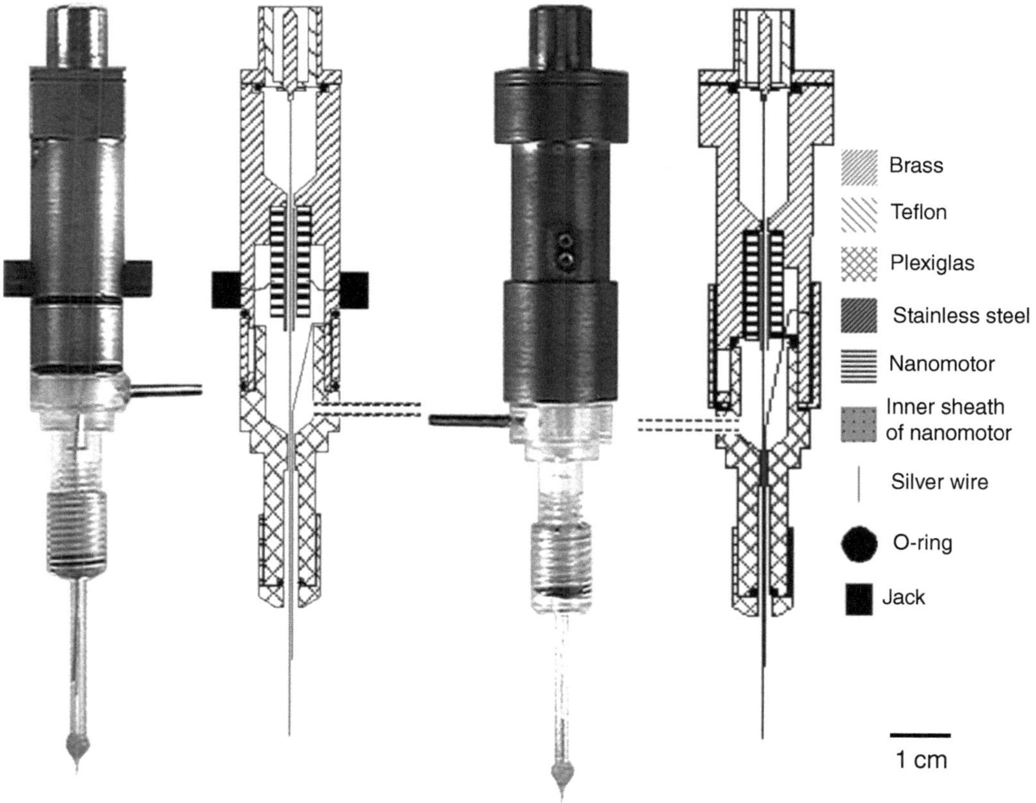

FIGURE 15.4 Motorized electrode holders. (a) Photograph of the first model of the motorized electrode holder. (b) Schematic sketch of the first model of the motorized electrode holder. (c) Photograph of the second, improved model. (d) Sketch of the second, improved model. For better clarity, the pipette and the CFE are omitted in the sketches. The CFE is attached to the sheath of the nanomotor, which is the only moving part. (Modified from Dernick, G. et al., *Nat. Methods*, 2, 699–708, 2005.)

moving sheath of the nanomotor via a tightly fitting, short (approximately 5 mm in length) piece of silicon tubing. In the first model (Figure 15.4 a and b), the motor received power via two jacks, one on each side of the holder. A hole in the housing of the holder, which is required for mounting the motor, was closed by a tightly fitting brass ring that slides over the hole. This ring also holds a small pin to connect the reference electrode (silver wire, 0.25 mm in diameter) to the grounded holder housing. The holder is further sealed with two O-rings. The perspex cap resembles that of commercially available electrode holders and was screwed onto the brass housing. The features of this part are similar to those of the manually adjustable holder described above.

In the improved second model of the motorized holder (Figure 15.4c and d), the perspex cap screws against the rim above the mounting hole, thus improving the air-tightness and stability. Because of this modification, the reference electrode pin had to be moved beyond this rim. The pin was soldered into the hole in the housing to make it airtight. The two contact jacks were placed on the same side of the holder making it sleeker and fit better into the setup.

The controller for the nanomotor was built according to a circuit design provided by the manufacturer. The motor moves by means of a high frequency control signal, which is switched on briefly to move the CFE. It is switched off to stop CFE movement and during recording. The controller should be placed inside the faraday cage of the setup to minimize noise pick-up.

Patch Pipettes

Patch pipettes were pulled from borosilicate glass (outer diameter of 2.0 mm, inner diameter of 1.4 mm, length of 85 mm, ends fire polished; Hilgenberg GmbH, Malsfeld, Germany) with a programmable horizontal puller (P-97; Sutter Instruments, Novato, CA, U.S.A). The puller was set up with a box filament, which had a box size of 3×3 mm and a width of 3 mm. The settings for the puller are given in the following in the arbitrary units which are used to program the device. The *air pressure* was set to 500, the *air before pull* and the *air after pull* were 5 s each. The *ramp test* for these capillaries was 655. A single line program looped four times and had the parameters as follows: *heat* 615, *pull* 0, *velocity* 50, and *time* 200. Pipettes ideally had a tip diameter of 2–3 μm, an opening angle of at least 20°, and a shank length of 5 ± 0.5 mm. A large opening angle was necessary to bring the tip of the CFE close to the pipette tip. For larger angles the velocity was sometimes reduced. However, a compromise had to be found between a large enough opening angle and a small enough tip diameter. After inspection under the microscope, only those pipettes which fulfilled the requirements were used. Pipette tips were dipped in melted wax to reduce stray capacitance (Sticky Wax; Kerr Inc., Orange, CA, U.S.A.). The wax coating procedure is quicker than using sylgard, and formation of seals appeared to be facilitated with wax-coated pipettes. Right before use, pipettes were fire polished with a microforge (microscope: Axiovert 25, Zeiss, Oberkochen, Germany; microforge: CPM-2; ALA Scientific Instruments Inc., Westbury, NY, U.S.A.). The pipettes were placed near a heated wire. As a result, the wax in the tip melted and flowed out of the pipette tip. Subsequently the tip narrowed due to the hot filament (polishing). Pipettes were backfilled with pipette solution, which was filtered through a 0.22 μm syringe filter and applied with a hypodermic needle. The pipette solution was connected to ground via the Ag|AgCl reference electrode in the holder. Typical pipette resistance was approximately 1 MΩ.

Carbon Fiber Electrode Fabrication

The procedure followed the method described by Chow and V. Rüden [18], but was modified so that the CFE fit inside the patch pipette close to the tip. Briefly, the entire electrode was made from polyethylene tubing, no glass capillary was used, and the big and the small drop-like beads formed during fabrication were manipulated to minimize their size. The differences between a conventionally fabricated CFE and one produced with our method are shown in Figure 15.5.

Carbon Fiber Fabrication Setup

Fabrication of CFEs was performed under a dissecting microscope (Stemi 2000, Zeiss, Oberkochen, Germany). A Pt/Ir wire (0.4 mm in diameter) measuring about 50 mm in length served as a heating filament. The wire was bent in the middle into a loop with a diameter of about 6 mm. The ends were held by an insulated clamp and connected to a variable power supply (12 V, greater than 3 A) via a foot switch. The wire loop was positioned under the dissecting microscope in the middle of the field of view. The 0.4 mm wire was chosen as a trade off between stability of the loop geometry and required heating current.

Pulling CFEs requires some practice. The beginner may train by pulling just the PE tubing without the inserted carbon fiber until appropriate heat settings are found and the desired shape, as described below, can be pulled consistently. Heat settings were chosen, such that the PE-tubing in the loop softened within 10–15 s and melted within 20–25 s. Upon melting, the PE tubing became clearly transparent. It may be helpful to set up a programmable CFE puller as described (Supplementary Note of [1]).

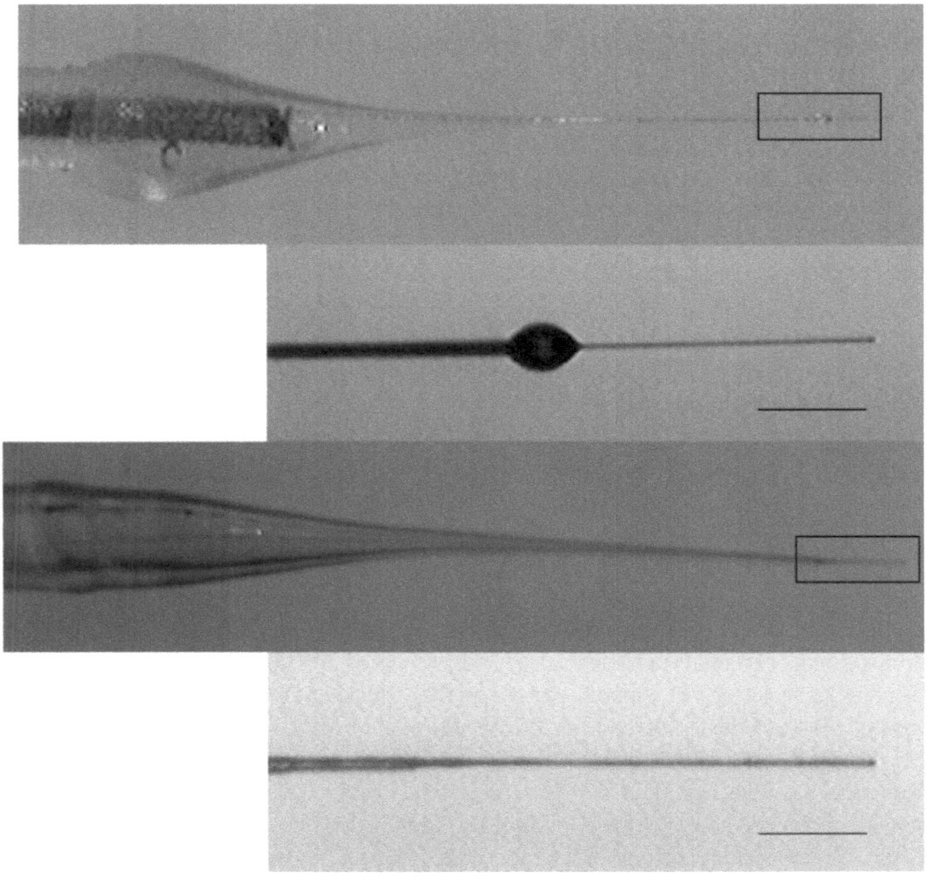

FIGURE 15.5 Carbon fiber electrodes.(a) Conventionally fabricated CFE. Due to symmetrical pulling, a big PE-bead is present on the left. Inside the big bead the silver wire (0.25 mm in diameter) can be seen. (b) Magnification of the box in (a) with the small bead. The right part of the carbon fiber seems uncoated but is insulated with a polyethylene layer of about 0.1 μm. (c) The CFE for patch amperometry prepared by asymmetrical pulling does not have the big bead. (d) Magnification of the box in (c) shows that the tip of the CFE lacks the small bead. The material of which the small bead consisted was melted and distributed over the thin part of the fiber so that it was covered with a slightly thicker layer of polyethylene than the CFE in (b). Scale bars in (b) and (d) are in 100 μm. Such a CFE is ideal to be inserted into the patch pipette, and its tip will easily be placed at the pipette tip. (From Dernick, G. et al., *Nat. Methods*, 2, 699–708, 2005.)

Carbon Fiber Pulling

To handle the rigid but brittle carbon fibers without breaking them, the tips of two pairs of watchmakers forceps (#5; TechniTools, Switzerland) were coated with 0.5 cm of polyethylene tubing (0.80 mm outer diameter, 0.40 mm inner diameter; Sims-Portex Ltd. Hythe, Kent, England). The same type of polyethylene (PE) tubing was used to fabricate the CFEs. A piece 10 cm in length was dipped in methanol and allowed to fill completely with the solvent by capillary action. This reduced the static attraction between the plastic and the carbon fiber during the next (cannulation) step. A single carbon fiber (Thornell T650/42, 5 μm in diameter; Amoco Co., Atlanta, GA, U.S.A.) of at least 5 cm in length was taken from a strand of carbon fibers with the watchmakers forceps and its tip was inserted into the methanol-filled PE-tubing under a dissecting microscope using bright illumination. The carbon fiber was then pushed into the PE-tubing completely. The solvent was wicked away by tapping the tubing onto a piece of tissue paper, followed by drying for 30 min at 50°C. It should be noted that production of the 5 μm carbon fibers was discontinued by the manufacturer but they are still available from the Lindau laboratory upon request.

To pull the electrode, the PE-tubing with the inserted carbon fiber is held in the center of the filament loop with the thumb and index finger of both hands, on both sides of the filament. The wrists of both hands should be stabilized by resting them on the table. About 2 cm of the tubing should stick out of the filament loop on one side and about 8 cm on the other. The pull is performed asymmetrically to obtain the desired shape of electrode. The shorter part is discarded after pulling and only the longer end is used as a CFE. The asymmetrical pull is done by first switching on the heating filament, and when the PE starts to soften, both ends are moved in the same direction with the long end being pulled away from the loop with twice the velocity as the short end is moved towards the loop. The longitudinal axis of the tubing remains in the center of the loop during the pull. In this way, the large bead on the long side of the PE-tubing is avoided and the shape shown in Figure 15.5c should be obtained. As the two parts are pulled apart, the part in the filament narrows and the PE in that area becomes transparent as it melts. Care must be taken that the polyethylene forms a tightly sealed coating around the carbon fiber, not just a narrow tube. The narrow section is then cut with a fresh scalpel blade on a glass slide. The PE layer on the tip of the carbon fiber should now be about 0.1 mm thick. The tip is then moved towards the heated filament again. Upon melting, the PE will retract from the tip, forming a small, drop-like bead. When the PE has retracted by about 0.5 mm from the tip of the carbon fiber the heat is turned off, and within a second, the filament is touched with the small bead, thus wiping it off on the warm filament by pulling the electrode away in a quick move. The small bead on the tip is thus removed and the tip of the electrode coated with a slightly thicker layer of PE, which also reduces the CFE current noise. Using a fresh scalpel blade, the end of the electrode is then cut to the desired length to fit the appropriate holder and pipette length. The CFE is backfilled with 3 *M* KCl utilizing a patch pipette filling needle (World Precision Instruments, Sarasota, FL, U.S.A.). In the manual holder we stick the CFE onto the 0.25 mm diameter chlorinated silver wire (Goodfellow, Cambridge, U.K.) that connects the CFE to the BNC pin. In the motorized holder a teflon coated silver wire measuring 0.125 mm in diameter (World Precision Instruments) is used, with 2 cm of teflon coating scraped off and the blank part of silver wire chlorinated.

Testing the CFE in the Patch Amperometry Configuration

CFEs were tested prior to use. When CFEs were immersed in Ringer saline with the reference electrode, the amperometric current ideally decayed to a few pA within a minute or two, and the current noise level was significantly below 1 pA r.m.s. These properties were determined by observing the amperometric current on an oscilloscope. CFEs that did not fulfill these criteria were discarded. If a CFE shows rapid current spikes in this configuration, it may help to rechlorinate the silver wires thoroughly and to fill the pipette and CFE such that only the chlorinated parts of the wires are immersed into the respective solutions.

It is important, that the CFE is insulated all the way to the tip and that its surface is working properly. This can be tested in the patch amperometry configuration: With the pipette immersed in the bath solution, but no cell attached to it, the CFE is positioned close to the pipette tip. Positive pressure is applied to the pipette, and a drop of dopamine solution is added to the bath (e.g., 10 μl of 1 *M* dopamine to 100 μl of bath solution). When the pressure is released from the pipette, dopamine diffuses into the pipette tip and a large oxidation current (I_A) appears. When positive pressure is applied again, I_A quickly decreases to background current level. The CFE is then moved further away from the tip. Now, when pressure is released, it takes longer for the dopamine to diffuse into the small orifice of the pipette, so the onset of the signal is delayed. CFE's were usually tested in this way prior to the experiments. If a CFE was used for more than a few days its functionality was tested again in this way.

CFE's can be stored and used for a several weeks. If a CFE is not in use it is advisable to withdraw the KCl solution from the inside to avoid crystallization, which may damage the CFE.

RECORDING CHAMBER DESIGN

Because of the reversed electrode configuration, the bath is not held at ground potential as usual, but at the potential applied by the patch clamp amplifier. If the bath surface is large, then the stray capacitance between the bath on one side and the microscope stage and objective on the other side cannot be compensated for with the patch clamp amplifier compensation circuit. A smaller bath surface also reduces noise. We found that a bath camber with a diameter of 10 mm and an amount of bath solution of 100–150 μL gave good results.

Recording chambers were prepared by drilling a 10 mm hole into a petri dish that fit into the respective patch clamp rig. A coverslip measuring 12 mm in diameter or more, was glued to the bottom of the dish with Sylgard. Sylgard is non-toxic and broken coverslips are easily exchanged. Only the inner 10 mm diameter well formed by the hole with the coverslip bottom serves as recording chamber.

A small coverslip with adherent cells (e.g., chromaffin cells) is placed in this inner well and 100–150 μL of bath solution is added. Care must be taken not to spill solution over the edge of the inner well to avoid large and fluctuating bath capacitance. The small volume requires that bath solution and cells are exchanged at least every hour because the bath solution's evaporation leads to unacceptable changes in concentration.

We performed experiments mostly with chromaffin cells. Bath and pipette solutions were typically based on the following compositions, with alterations according to experimental conditions:

- Bath solution (for chromaffin cells):140 mM NaCl, 5 mM KCl, 5 mM $CaCl_2$, 1 mM $MgCl_2$, 10 mM HEPES–NaOH, 10 mM glucose (pH 7.3).
- Pipet solution (for chromaffin cells): 50 mM NaCl, 100 mM tetraethylammonium ion(TEA)-Cl, 5 mM KCl, 10 mM $CaCl_2$, 1 mM $MgCl_2$, 10 mM HEPES–NaOH (pH 7.3).
- The osmolality of the solutions was checked such that bath and pipette solution had matching osmolalities not exceeding 300 mosmol.

PATCHING OF CHROMAFFIN CELLS

Experiments began with mounting a CFE and testing its quality as described above. During a series of experiments, the fiber typically remained attached to the electrode holder unless it broke or became noisy or leaky. A coverslip with cells from the incubator was washed once with bath solution and placed in the recording chamber. Under the microscope, a suitable cell for an experiment was chosen. A patch pipette was fire polished and filled with pipette solution as described above, and mounted on the electrode holder.

The manual holder always needed to be removed from the setup for mounting the patch pipette and for coarse adjustment of the CFE, close to the tip of the pipette, under the dissecting microscope. The fine adjustment of the carbon fiber to the tip of the pipette was done on the manual holder with the pipette just barely immersed in the bath and by observing it through the inverted microscope. Great care must be taken not to move the holder too much while turning the thumbscrew of the holder, which could break the pipette tip by touching the bath chamber. This was rather tedious because the pipette tip very often comes out of focus during the adjustment.

Adjustment of the CFE in the motorized holder was much easier. With some practice and good lighting, the pipette could be mounted on the motorized holder while the holder remained on the head stage in the setup. Some caution had to be taken not to break the CFE. While observing the pipette tip immersed in the bath through the inverted microscope, the CFE was then adjusted inside the pipette using the nanomotor, first at the highest speed, i.e., with the highest frequency selected at the motor control unit. We always operated the controller at maximal voltage amplitude and varied only the frequency to change speed. As soon as the CFE came into the field of view, the frequency of the control signal was reduced at the motor controller. With the smallest possible frequency (but still with maximum amplitude) the CFE was slowly positioned approximately 15 μm back from the pipette tip.

Seal formation followed standard patch clamp procedures as described elsewhere [27]. A test pulse was applied and the pipette current monitored on an oscilloscope with a gain of 2 mV/pA and a stimulus scaling of 0.01 on the EPC-7 (similarly, a low gain is used in the Axopatch 200B; the built-in test pulse capability can be used for pipette and seal resistance determination). If during application of suction to the pipette for seal formation the patch moved too far into the pipette near the CFE, then the CFE was retracted with the motorized holder. When working with the manual holder, a CFE too close to the tip sometimes prevented seal formation when the patch was sucked too far into the pipette. Sometimes increasing of the holding potential facilitated seal formation. Note that because the electrode configuration is reversed, the polarity of the applied voltage is reversed.

After seal formation, the gain was increased to 50 mV/pA for low noise, and the stimulus scaling set to 0.1. A gain as low as 10 mV/pA could be used with the Axopatch 200B if a high-resolution data acquisition system was used. With the test pulses still switched on, fast transients of the pipette current were minimized with the pipette compensation dials (C-fast and τ-fast dials on the EPC-7; dials in the "C-slow" area on the Axopatch 200B). Then the test pulses were switched off, the sine wave was switched on, and the amplitude of the sine wave in the pipette current minimized again using the respective pipette compensation dials.

Capacitance Calibration

After starting the data acquisition, the recording was calibrated with the capacitance dither to determine the amplitude factor at the chosen amplifier and filter settings. For calibration, we typically applied three capacitance calibration pulses for easy identification during analysis by transiently changing the capacitance compensation by 20 fF. Larger changes may saturate the amplifier and a reduced stimulus scaling would have needed to be used. Chromaffin cell exocytosis was stimulated during seal formation, such that the exocytotic events typically occurred at the beginning of a recording. It was consequently advisable to perform all the adjustments between seal formation and start of the data acquisition in less than 30 s.

During the recording, the amplitude of the sine wave in the pipette current had to be observed on the oscilloscope to prevent clipping of the amplfiers. The clipping indicator LED's were not reliable when a 20 kHz sine wave was used. The amplitude of the patch clamp amplifier output should never exceed plus or minus 10 V to avoid saturation of the lock-in amplifier. When the sine wave of the patch clamp amplifier output exceeded 5 V (100 pA at 50 mV/pA gain) we re-adjusted the pipette capacitance compensation to minimize the sine wave current amplitude. Occasionally checking of the seal resistance was appropriate, particularly when recordings became very unstable. For seal check, the sine wave was switched off momentarily and the test pulses switched on. During such a seal check capacitance recordings were invalid because of the absence of the sine wave.

The end of a recording is either marked by loosing the seal or by the cell going to the whole cell configuration. In the latter case, the cytosolic catecholamines diffused out of the cell and caused a rather slow but fairly big spike in the oxidation current [13]. Sometimes a cell went whole cell and then resealed to the cell-attached configuration. In such cases, some chromaffin granules may have been freely floating outside the patch in the pipette tip. When they touched the CFE, they gave very fast amperometric signals that were not accompanied by capacitance steps.

Capacitance steps without amperometric signals may reflect exocytosis of granules that do not contain catecholamines [11]. However, if all events in a patch lack amperometric signals, then the CFE may not function properly and should be checked. In a given experiment it may be helpful to break the patch at the end and check the CFE by eliciting spikes from bursting granules sucked out of the cell.

ANALYSIS OF EXOCYTOTIC EVENTS

VESICLE CAPACITANCE AND FUSION PORE CONDUCTANCE

The practical approach and the underlying theoretical treatment of capacitance measurements of individual chromaffin vesicles is described in detail elsewhere [4,28,38,39]. Here we discuss the steps in analysis of patch amperometry recordings. We wrote a set of procedures for data analysis that are available online [1] and may serve as a guide to the development of custom data analysis procedures.

After a recording, the five traces were loaded into Igor, where the "fifo" file format from the acquisition was converted into "waves". Next, the traces V (stimulus voltage monitor), I (filtered pipette current) and A (ampermetric current) were converted into mV and pA, respectively. This required entering the gain settings that were used for the amplifiers during the recording. In the next step, the admittance traces were inspected. The phase was initially set by visual inspection of the calibration pulses. The calibration pulses should only occur in the imaginary part of the pipette current (Im-trace, Y2). If a projection was seen in the conductance trace (Re-trace, Y1), then the phase was adjusted according to the following relation:

$$\begin{pmatrix} Y_G \\ Y_C \end{pmatrix} = \begin{pmatrix} \cos(\Delta\varphi) & -\sin(\Delta\varphi) \\ \sin(\Delta\varphi) & \cos(\Delta\varphi) \end{pmatrix} \begin{pmatrix} Y_1 \\ Y_2 \end{pmatrix} \tag{15.1}$$

until the calibration pulses did not show any projection into the real part of (Re-trace, Y1) anymore.

This relation was converted into an Igor procedure as a function as follows:

```
Function Phaseshift()
Variable Y1, Y2
Wave Y1_ = Y1_, Y2_ = Y2_
nvar Ph_ch = Ph_ch
   Variable wavecount = 0
   Variable endpoint = numpnts(Y1_)
   Variable cp = cos(Ph_ch*2*Pi/360)
   Variable sp = sin(Ph_ch*2*Pi/360)
   Do
         Y1 = (Y1_[wavecount]*cp)+(Y2_[wavecount]*sp)
         Y2 = (Y2_[wavecount]*cp)-(Y1_[wavecount]*sp)
         Y1_[wavecount] = Y1
         Y2_[wavecount] = Y2
         wavecount+ = 1
         while (wavecount<endpoint)
End
```

This function applied phase changes to the entire length of the Y1 and Y2 traces. Phase changes were made absolute (not relative) by first resetting the phase to the originally recorded traces. After the phase adjustment, the calibration pulses were measured and the Y1 and Y2 traces converted into fF (1 femto Farad $= 10^{-15}$ F) and nS (1 nano Siemens $= 10^{-9}$ S $= 10^{9}$ Ω). The real and imaginary traces were displayed at the same "magnification": The axis range for the imaginary trace was set by the user, and the axis range for the real trace calculated using the relation:

$$\mathrm{Re(nS)} = \omega\ \mathrm{Im}\ 10^{-6}\mathrm{(fF)} \tag{15.2}$$

If the phase is set correctly, the real and imaginary part of the pipette current can be measured at the two outputs of the lock-in amplifier. In this ideal case:

$$\text{Re} = \frac{(\omega C_V)^2/G_p}{1 + (\omega C_V/G_p)^2} \quad (15.3)$$

$$\text{Im} = \frac{\omega C_V}{1 + (\omega C_V/G_P)^2} \quad (15.4)$$

The true vesicle capacitance and the true conductance of the fusion pore can be calculated as [4,28]:

$$C_V = [(\text{Re}^2 + \text{Im}^2)/\text{Im}]/\omega \quad (15.5)$$

$$G_P = (\text{Re}^2 + \text{Im}^2)/\text{Re} \quad (15.6)$$

The expected trace appearance, as well as the parameter calculations, for these traces is illustrated for an idealized exocytotic event in Figure 15.6. Here we assume that the imaginary part of the patch admittance (Figure 15.6a, solid line) increases linearly until it reaches the level of the full vesicle capacitance (ωC_V). The real part of the patch admittance was calculated from the imaginary part by substituting Equation 15.5 and Equation 15.6 into Equation 15.3 and is shown in the same panel as a dashed line. As the imaginary part increases, the real part increases steeply, becomes shallower, and reaches its maximum when $G_P = \omega C_V$. At this point, the real and imaginary part are equal to $\omega C_V/2$ (Re = Im = 1/2 ωC_V). As the imaginary part increases further towards the actual vesicle capacitance (ωC_V), the real part returns to baseline. From such lock-in amplifier output traces, the time course of C_V and G_P are calculated using Equation 15.5 and Equation 15.6, as shown in Figure 15.6b. C_V increases stepwise at fusion pore opening. While C_V is constant, the fusion pore conductance G_P increases in a time-dependent manner and eventually becomes immeasurably large.

Figure 15.6c and d show the traces one would obtain if the fusion pore conductance increased linearly to $10 \times \omega C_V$ and then increased rapidly to infinity (full flattening in the membrane). From

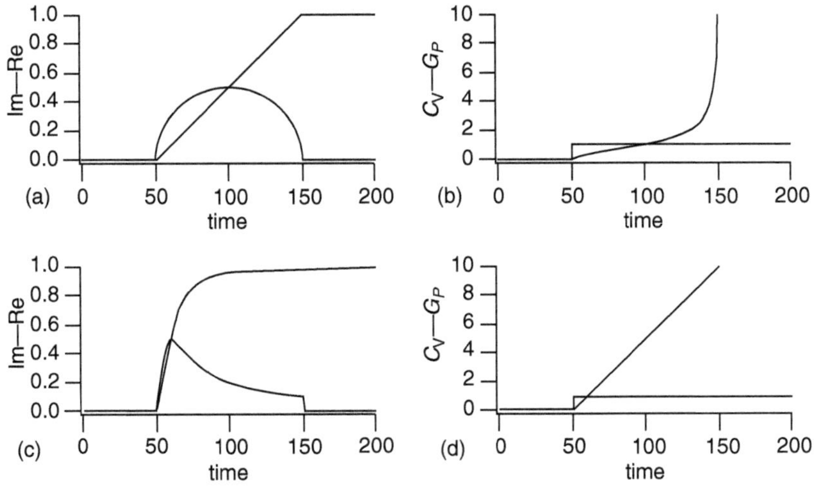

FIGURE 15.6 Theoretical traces for two cases of fusion pore expansion. Linear increase of imaginary part (a),(b); linear increase of fusion pore conductance (c),(d). Abscissae: units of ωCV; ordinate: arbitrary units of time. For details see text.

the two traces for G_P and C_V (Figure 15.6d), the real and the imaginary parts were calculated as they were expected to be measured at the outputs of the lock-in for such a pore (Figure 15.6c). In actual recordings, the time course of G_P is usually more complex but a flat C_V is a good indication that the fusion pore analysis is reasonable.

The actual calculation of recorded traces requires correct setting of base lines in the real and the imaginary traces in the section preceding an event. If that is not possible, one can also set a base line for only the imaginary trace, use C_V obtained from the imaginary part at late times, and calculate the fusion pore conductance and the expected real part. The expected and measured real part should not show significant differences.

The vesicle capacitance can be converted into membrane area by using an appropriate conversion factor (9 fF/μm^2 for chromaffin granules) [8]. Assuming spherical shape, this can also be transformed into vesicle volume.

The fusion pore may be characterized by various parameters. We found it helpful to use the *initial fusion pore conductance*, *initial expansion rate*, and *fusion pore lifetime* [9]. The initial opening is marked by the first increase in Re and Im trace above the noise. Interestingly, this *initial fusion pore conductance* is not randomly distributed; it shows a distribution with peak indicating the most likely fusion pore conductance for a particular cell type. The initial fusion pore conductance represents a sudden transition from a closed to an open state of the fusion pore, similar to that of ion channels. However, the distribution of initial fusion pore conductances is wider than for most ion channels. We quantified the *initial fusion pore expansion rate* following fusion pore opening using the slope of $G_P(t)$ over the next 25 ms [9]. As fusion pore lifetime, we defined the time between fusion pore opening and the time where G_P exceeded 1 nS. This value may be compared with the foot duration compared in conventional amperometric recordings.

The calculation of fusion pore dimensions from the conductance is somewhat uncertain. Although the thickness of the pore can be assumed to be about two lipid bilayers (10 nm), the specific conductance of the solution in the pore is not precisely known. Usually the conductance of physiological saline is assumed for fusion pore calculations [4,9,29,30], however, the pore will be filled with a mixture of extracellular and intravesicular solution. Chromaffin granules contain about 1 *M* catecholamines, which are mostly monovalent cations. Mast cell granules contain the divalent cation histamine at a concentration of hundreds of m*M*. In contrast, the ion content of neutrophil azurophil granules and esosinophil granules may be much lower. This may explain why the fusion pore conductance in mast cells [29] and chromaffin cells [9] is somewhat higher than in neutrophils [4] and eosinophils [30].

Fusion pores may open transiently (for review see [31]). In several studies, amperomtrically observed changes in quantal size were attributed to transient vs. full fusion pore openings or to changes in the vesicular transmitter content. Patch amporemetry can directly distinguish between these possibilities [12].

Release of Molecules

The amperometric spikes that are recorded upon release of molecules from single vesicles were analyzed by several groups already and the methods have been discussed elsewhere [18,32–34]. Frequently used parameters are the presence or absence of a foot signal, foot duration, foot amplitude and foot charge; the maximum amplitude of a spike and its width at half the maximum (half width or t_{half}); and the integrated current. The latter yields the charge in coulombs. Assuming that z electrons are transferred during the electrochemical oxidation of catecholamines, the number of released molecules can be calculated from the charge with the following relation:

$$Q = \int I \, \mathrm{d}t = \frac{z\mathrm{FM}}{N_\mathrm{A}} = eM \qquad (15.7)$$

Here, Q represents the total charge involved in the redox reaction, which is obtained by integration of the current (I) transient; M represents the number of molecules reacted; z is the number of electrons transferred per oxidized molecule; F is Faraday's constant, 96485 C/mol; N_A is Avogadro's number, 6.023×10^{23}; and e is the elementary charge 1.602×10^{-19} C. For catecholamines, $z = 2$ is generally accepted, and thus the number of molecules per vesicle can be estimated.

Quantal Size, Vesicle Size, and the Fusion Pore

From the capacitance step the vesicle surface area can be determined as described above. This can be converted into diameter or volume of the vesicle. The associated amperometric spike provides the charge from which the number of molecules can be calculated. From these two parameters the concentration of individual vesicles can be determined. Typical catecholamine concentrations in vesicles are on the order of 1 *M* [8]. It was shown that the drugs L-Dopa and reserpine, which modulate vesicle content, also modulate vesicle membrane area, keeping the concentration virtually constant [12]. However, exocytosis of empty vesicles that do not contain catecholamines has also been reported [11].

Correlating the foot signal with fusion pore conductance [8,9,14,35], it was shown that the fusion pore conductance controls the flow of molecules through the fusion pore. Patch amperometry may be useful to investigate the effect of fusion pore mutants that may differentially affect fusion pore conductance and selectivity as determined from the correlation between fusion pore conductance and flux of catecholamines.

SUMMARY AND DISCUSSION

We developed patch amperometry, a new technique that combines high-resolution cell-attached capacitance measurements with simultaneous amperometric detection by placing the amperometric electrode inside the patch pipette. For this purpose we designed a new electrode holder, which accommodates the connections for two electrodes and has the capability to change the position of the CFE tip versus the pipette tip. The method makes it possible to determine vesicular concentrations, to distinguish transient and full fusion events, and to analyze individual fusion pore openings and the role of the fusion pore in limiting release. While it is now clear how the size of the fusion pore determines release from chromafin granules during the amperometric foot signal, it is not yet well understood what determines the shape of the spike after the fusion pore expansion. After the fusion pore dilates, the molecules are released rapidly, giving rise to the amperometric spike. It is likely that in this phase release is not hindered by the fusion pore restriction any longer. The vesicular contents should be able to diffuse freely to the detector. However, when a CFE is placed very close to the cell surface, spikes are broader than expected from diffusion theory if the release giving rise to the amperometric spike were instantaneous [18–20,36,37]. Using patch amperometry it was shown that in mast cells the upstroke of the amperometric spike is not limited by the fusion pore [35]. It is widely believed that the time course of the amperometric spike reflects dissociation of bound molecules from the matrix followed by diffusion to the CFE. However, recent amperometric measurements from chromaffin cells using four-electrode electrochemical detector arrays have suggested that diffusion near the membrane is very slow, presumably due to reversible binding of catecholamines to the cell surface [22]. Whether a larger fusion pore in chromaffin cells is also participating in limiting release during the amperometric spike is still to be answered.

ACKNOWLEDGMENTS

We thank M. Montesinos, R. Borges, R. Staal, and D. Sluzer for their contributions, in particular in the use of the Axopatch 200B for patch amperometry. This work was supported by grants form NIH

(RO1 NS38200-01A2) and the Nanobiotechnology Center (an STC program of NSF, Agreement No. ECS-9876771) to M. L., and a grant from the Ministerio de Educación y Cultura, Spain to G.A.d.T.

REFERENCES

1. Dernick, G. *et al.*, Patch amperometry: High resolution measurements of single vesicle fusion and release, *Nat. Methods*, 2, 699–708, 2005.
2. Neher, E. and Marty, A., Discrete changes of cell membrane capacitance observed under conditions of enhanced secretion in bovine adrenal chromaffin cells, *Proc. Natl Acad. Sci. U.S.A.*, 79, 6712–6716, 1982.
3. Lollike, K., Borregaard, N., and Lindau, M., Capacitance flickers and "pseudoflickers" of small granules, measured in the cell attached configuration, *J. Biophys.*, 75, 53–59, 1998.
4. Lollike, K., Borregaard, N., and Lindau, M., The exocytotic fusion pore of small granules has a conductance similar to an ion channel, *J. Cell Biol.*, 129, 99–104, 1995.
5. Leszczyszyn, D. J. *et al.*, Nicotinic receptor-mediated catecholamine secretion from individual chromaffine cells, *J. Biol. Chem.*, 265, 14736–14737, 1990.
6. Bruns, D. and Jahn, R., Real-time measurement of transmitter release from single synaptic vesicles, *Nature*, 377, 62–65, 1995.
7. Chow, R. H., Rüden, L. v., and Neher, E., Delay in vesicle fusion revealed by electrochemical monitoring of single secretory events in adrenal chromaffin cells, *Nature*, 356, 60–63, 1992.
8. Albillos, A. *et al.*, The exocytotic event in chromaffin cells revealed by patch amperometry, *Nature*, 389, 509–512, 1997.
9. Dernick, G., Alvarez De Toledo, G., and Lindau, M., Exocytosis of single chromaffin granules in cell-free inside-out membrane patches, *Nat. Cell Biol.*, 5, 358–362, 2003.
10. Alés, E. *et al.*, High calcium concentrations shift the mode of exocytosis to the kiss-and-run mechanism, *Nat. Cell Biol.*, 1, 40–44, 1999.
11. Tabares, L. *et al.*, Exocytosis of catecholamine-containing and catecholamine-free granules in chromaffin cells, *J. Biol. Chem.*, 276, 39974–39979, 2001.
12. Gong, L. W., Alvarez De Toledo, G., and Lindau, M., Secretory vesicles membrane area is regulated in tandem with quantal size in chromaffin cells, *J. Neurosci.*, 23 (21), 7917–7921, 2003.
13. Mosharov, E. V., Intracellular patch electrochemistry: Regulation of cytosolic catecholamines in chromaffin cells, *J. Neurosci.*, 23, 5835–5845, 2003.
14. Alvarez de Toledo, G., Fernández-Chacón, R., and Fernandez, J. M., Release of secretory products during transient vesicle fusion, *Nature*, 363, 554–558, 1993.
15. Chow, R. H. *et al.*, Mechanisms determining the time course of secretion in neuroendocrine cells, *Neuron*, 16, 369–376, 1996.
16. Oberhauser, A. F., Robinson, I. M., and Fernandez, J. M., Simultaneous capacitance and amperometric measurements of exocytosis: A comparison, *Biophys. J.*, 71, 1131–1139, 1996.
17. Robinson, I. M. *et al.*, Colocalization of calcium-entry and exocytotic release sites in adrenal chromaffin cells, *Proc. Natl Acad. Sci. U.S.A.*, 92, 2474–2478, 1995.
18. Chow, R. H. and Rüden, L. v., In *Electochemical Detection of Secretion from Single Cells, in Single Channel Recording*, Sakmann, B. and Neher, E., Eds., Plenum Press, New York, 1995.
19. Wightman, R. M. *et al.*, Time course of release of catecholamines from individual vesicles during exocytosis at adrenal medullary cells, *Biophys. J.*, 68, 383–390, 1995.
20. Schroeder, T. J. *et al.*, Temporally resolved, independent stages of individual exocytotic secretion events, *Biophys. J.*, 70 (2), 1061–1068, 1996.
21. Amatore, C. *et al.*, Interplay between membrane dynamics, diffusion and swelling pressure governs individual vesicular exocytotic events during release of adrenaline by chromaffin cells, *Biochimie*, 82 (5), 481–496, 2000.
22. Hafez, I. *et al.*, Electrochemical imaging of fusion pore openings by electrochemical detector arrays, *Proc. Natl Acad. Sci. U.S.A.*, 102, 13879–13884, 2005.
23. Debus, K. and Lindau, M., Resolution of patch capacitance recordings and of fusion pore conductances in small vesicles, *Biophys. J.*, 78, 2983–2997, 2000.

24. Hamill, O. P. *et al.*, Improved patch-clamp technique for high-resolution current recording from cells and cell-free membrane patches, *Pflügers Arch. Eur. J. Physiol.*, 391, 85–100, 1981.
25. Schroeder, T. J. *et al.*, Analysis of diffusional broadening of vesicular packets of catecholamines released from biological cells during exocytosis, *Anal. Chem.*, 64, 3077–3083, 1992.
26. Klocke, V., Atomic precision and millimeter range, *F & M*, 104 (4), 274, 1996.
27. Penner, R., In *A Practical Guide to Patch Clamping, in Single Channel Recording*, Sakmann, B. and Neher, E., Eds., Plenum Press, New York, 1995.
28. Lindau, M., Time-resolved capacitance measurements: Monitoring exocytosis in single cells, *Q. Rev. Biophys.*, 24, 75–101, 1991.
29. Breckenridge, L. J. and Almers, W., Currents through the fusion pore that forms during exocytosis of a secretory vesicle, *Nature*, 328, 814–817, 1987.
30. Hartmann, J. and Lindau, M., A novel Ca^{2+}-dependent step in exocytosis subsequent to vesicle fusion, *Fed. Eur. Biochem. Soc. Lett.*, 363, 217–220, 1995.
31. Lindau, M. and Alvarez de Toledo, G., The fusion pore, *Biochim. Biophys. Acta*, 1641 (2–3), 167–173, 2003.
32. Gomez, J. F. *et al.*, New approaches for analysis of amperometrical recordings, *Ann. N. Y. Acad. Sci.*, 971, 647–654, 2002.
33. Travis, E. R. and Wightman, M. R., Spatio-temporal resolution of exocytosis from individual cells, *Ann. Rev. Biophys. Biomol. Struct.*, 27, 77–103, 1998.
34. Mosharov, E. V. and Sulzer, D., Analysis of exocytotic events recorded by amperometry, *Nat. Methods*, 2 (9), 651–658, 2005.
35. Tabares, L., Lindau, M., and Alvarez De Toledo, G., Relationship between fusion pore opening and release during mast cell exocytosis studied with patch amperometry, *Biochem. Soc. Trans.*, 31 (4), 837–841, 2003.
36. Jankowski, J. A., Finnegan, J. M., and Wightman, R. M., Extracellular ionic composition alters kinetics of vesicular release of catecholamines and quantal size during exocytosis at adrenal medullary cells, *J. Neurochem.*, 63, 1739–1747, 1994.
37. Walker, A., Glavinovic, M. I., and Trifaró, J. M., Time course of release of content of single vesicles in bovine chromaffin cells, *Pflügers Arch. Eur. J. Physiol.*, 431, 729–735, 1996.
38. Lindan, M. and Neher, E., Patch-clamp techniques for time-resolved capacitance measurements in single cells, *Pflugers Arch.*, 411 (2), 137–146, 1988.
39. Gillis, K. D., In *Techniques for membrane capacitance measured, in single channel recording*, Sakmann, B. and Neher, E., Eds., Plenum Press, New York, 1995.

16 Amperometric Detection of Dopamine Exocytosis from Synaptic Terminals

Roland G. W. Staal, Stephen Rayport, and David Sulzer

CONTENTS

INTRODUCTION TO DOPAMINE NEUROTRANSMISSION

Dopamine (DA) was discovered as the precursor of the neurotransmitter norepinephrine in 1957 (Montagu 1957). Carlsson demonstrated that DA levels in the striatum were much higher than in the rest of the brain despite its low level of norepinephrine, suggesting that DA was not merely a norepinephrine precursor but was itself a neurotransmitter (1959) (Carlsson 1959). Subsequently, DA's role as a neurotransmitter has been confirmed and populations of DA neurons have been identified in various regions of the central nervous system (olfactory bulb, substantia nigra, ventral tegmental area, and the retina) and the periphery (enteric nervous system of the gut). These neurons are involved in modulating intestinal motility, the sense of smell, light sensitivity, and control of fine motor movements. They are also integral to higher brain functions such as salience, euphoria, pleasure and appetitive and consummatory aspects of reward. Their demise or dysfunction plays vital roles in neuropathologies such as Parkinson's disease, schizophrenia and depression.

Thus, understanding how DA neurons function normally in neurotransmission and in pathological processes is essential for elucidating their roles in behavior and disease.

Electrochemical Detection of DA

The ability to detect DA and its metabolites electrochemically (Kissinger, Hart, and Adams 1973) greatly advanced the study of dopaminergic transmission. Microdialysis provided the first studies of DA levels in the brain in response to pharmacological treatments and during behavioral tests. Limitations of microdialysis include the relatively slow time resolution (on the order of several minutes) and the size of the probes, which lead to local tissue damage (Borland et al. 2005). The development of the carbon fiber electrode (CFE) enabled the detection of oxidizable transmitters with much faster time resolution and at the cellular and quantal level (transmitter released by a single vesicle) (Gonon et al. 1978). Initial studies performed by Wightman and colleagues on adrenal chromaffin cells which contain large dense core vesicles (LDCVs) (Leszczyszyn et al. 1990; Wightman et al. 1991) were followed by other studies utilizing carbon fiber amperometry to detect oxidizable transmitter release from both LDCVs and small synaptic vesicles (SSVs) from various cell types (Alvarez de Toledo, Fernandez-Chacon, and Fernandez 1993; Chen, Luo, and Ewing 1994; Bruns and Jahn 1995; Chen and Ewing 1996; Zhou, Misler, and Chow 1996; Pothos, Davila, and Sulzer 1998). The techniques outlined here are the first to provide for measurement of neurotransmitter release by SSV exocytosis from a central nervous system synaptic terminal.

Quantal DA Release

A major challenge in the study of DA neurotransmission has been the development of a system (cell culture, slice or other) in which it is possible to record quantal DA release. Most studies of quantal transmitter release have been performed on neuroendocrine cells such as adrenal chromaffin cells. Stimulation of these cells yields many readily detectable release events, although the release occurs from LDCVs in the cell body (Wightman et al. 1991), not from SSVs on axons as in neurons. Brain slices are widely used to study DA overflow, but are not widely used for amperometric recordings of quantal DA release because it is virtually impossible to record directly from varicosities due to steric hindrance by glia or other cells and structures. There is a single report of quantal recordings in the slice by Jaffe, Marty, and Schulte (1998), in which the surfaces of cell bodies were "cleaned" by a pressure stream, providing evidence for quantal release from substantia nigra cell bodies. This approach has proven technically challenging, and it is impossible to differentiate between release from cell bodies versus release from axonal terminals that contact these cell bodies, such as serotonergic inputs from the raphe. The laboratory of John Dani also has had some success in recording relatively small spontaneous DA release events in the slice (2005), and Mark Wightman's group has produced analogous recordings in vivo (2004). These events, however, are composed of multiple overlapping quanta that cannot be resolved.

Another challenge exists in developing a preparation to study the exocytosis of SSVs from synaptic terminals due to their much smaller quantal size and rapid kinetics. We have shown that with available hardware, the amperometric signal resulting from the oxidation of transmitter release from a SSV is lost in the baseline noise of the current due to diffusion at distances greater than 300 nm (Sulzer and Pothos 2000). Therefore, the CFE must essentially touch the release site in order to minimize the filtering effect of diffusion on the amperometric signal. In addition, the background noise must be kept to a minimum and the sampling rate must be fast enough to obtain a reliable report of the event. Finally, the release site may also be blocked by the presence of a postsynaptic cell or by glial cells that cover the terminal. Despite this, there are a sufficient number of exposed release sites to make studies of quantal DA release possible, although patience is required to locate exposed sites. While it remains unproven, we suspect that SSVs may release from multiple sites at or near a varicosity, and this in part increases the chances of successful quantal recordings (Gonon et al. 2000; Sulzer and Pothos 2000).

Culture systems have several advantages for the monitoring of quantal DA release from SSVs. First, the DA neurons are intact and can survive for long periods during which they can be treated with drugs, transfected or otherwise manipulated. Second, ascorbate concentrations in the media are very low, and projections from other neurons with oxidizable transmitters are rare. Third, the absence of brain waves and the reduced electrical activity greatly reduce the background noise. The only system that has to date consistently provided recordings of quantal DA release from SSVs is the postnatally-derived ventral midbrain culture system. Amperometric studies of DA release from this preparation have already yielded important insights into the regulation of pre-synaptic quantal size and modes of transmitter release in DA neurons (Pothos, Davila, and Sulzer 1998; Pothos et al. 2000; Staal, Mosharov, and Sulzer 2004).

Source of DA Neurons: Selection of Animals and Brain Regions

Several factors need to be taken into consideration when determining the source of DA neurons. The choice of SN, VTA or both (ventral midbrain, VM) depends mostly on the neuropathology or neurological system to be modeled (Parkinson's disease, SN; schizophrenia and depression, VTA; etc.) as well as the type of experiments (multi-dish toxicity studies, VM). For toxicity studies, the source of DA neurons is often less important than producing an adequate yield of DA neurons. If a high percentage of DA neurons within a dish are needed, only the VTA is dissected. This may be necessary for assays where measured parameters need to be maximized to reach detectable levels (DA levels, TH activity etc). If a high yield (many cultures) is desired, the entire VM is used. This would be appropriate for studies with many treatment groups and an easily measured parameter (such as number of cells/dish). Since we are trying to increase our success rate of finding active release sites, we usually opt for VTA cultures which have a very high density of DA neurons and DAergic axons and varicosities versus SN or VM cultures. Approximately 1 rat cortex yields enough glia to plate ~125 dishes, and one rat VTA yields enough DA neurons to plate 3–4 dishes. There is no distinct boundary between the lateral VTA and the medial SN, however; Figure 16.1 and Figure 16.2 show the approximate location of the incisions we make when

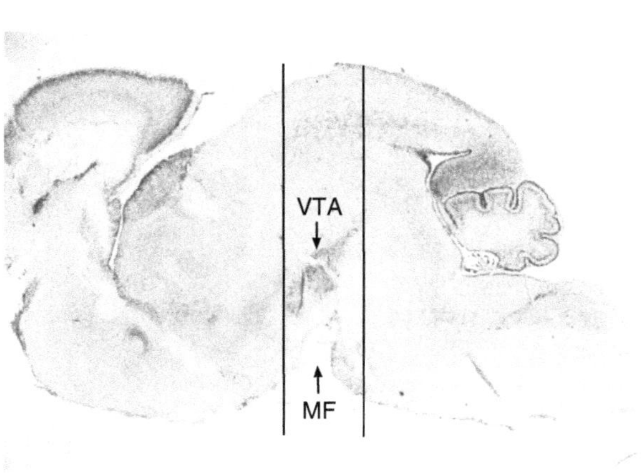

FIGURE 16.1 **(See color insert following page 272.)** Sagittal section of P1 mouse brain stained for TH using horseradish peroxidase and diaminobenzidine. Black lines indicate the incisions made during the dissection. (MF, midbrain flexure; VTA, ventral tegmental area.)

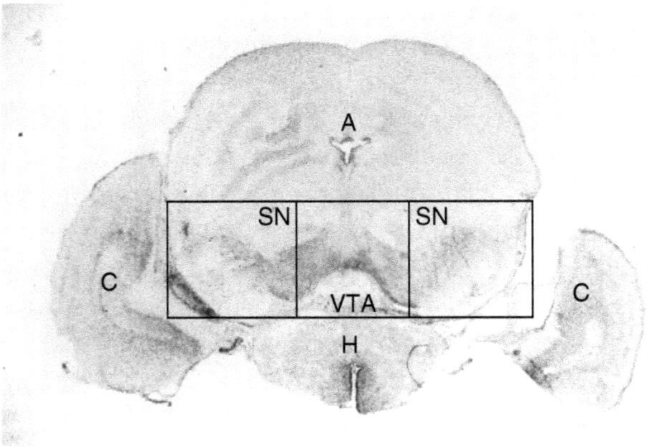

FIGURE 16.2 **(See color insert following page 272.)** Coronal section of P1 mouse brain stained for TH using horseradish peroxidase and diaminobenzidine. Black lines indicate the incisions made during the dissection (A, aqueduct; C, cortex; H, hypothalamus; MF, midbrain flexure; SN, substantia nigra; VTA, ventral tegmental area.)

dissecting these nuclei. In those studies where a comparison between VTA and SN is required, we dispose of the border regions between the nuclei (Burke, Antonelli, and Sulzer 1998).

The choice of animal species and strain is also critical for successful experiments. For example, amperometric recordings from cultured DA neurons from rats yield readily detectable amperometric spikes, whereas those from mice are much smaller, making their detection difficult if not impossible (unpublished observations). This may be due to lower levels of VMAT2 in striatal SSVs in mice (Staal et al. 2000). One drawback to using rats, however, is that it is currently impossible to visually differentiate between varicosities from DA neurons and other neurons. In our experiments to date (Pothos, Davila, and Sulzer 1998; Staal, Mosharov, and Sulzer 2004), axonal varicosities were located without knowing whether the varicosity was DAergic or not. Electrodes were placed onto the varicosity and secretagogue was applied and only the presence of amperometric events confirmed the varicosity as DAergic. Thus, recordings may be attempted at dozens of sites before finding an active DAergic release site. One potential solution is to perform recordings on cultured DA neurons derived from strains of mice expressing fluorescent proteins under appropriate promoters, such as the TH or DAT promoter or expressing proteins tagged with fluorescent proteins (GFP-hDAT, Staal et al, unpublished) (Matsushita et al. 2002; Zhang et al. 2004; Zhuang et al. 2005).

METHODS

Various methods for preparing dissociated cultures of DA neurons are available. Most preparations produce cultures of embryonically-derived neurons of which approximately 1% are dopaminergic. The postnatally derived cultures developed by Rayport and Sulzer (1992) have several advantages, the most striking being that the fraction of DA neurons is usually at least 20% for VM and as high as 70% for VTA. Another advantage of postnatally derived neurons is that they display higher levels of the DA uptake transporter (Staal, unpublished observations).

Under optimal conditions, these cultures can survive up to 6 months, although most survive about 2 months. We suspect that the death is mostly due to increased osmolarity of the medium, and could probably be controlled by partially exchanging the older media with fresh conditioned media, testing the osmolarity of the medium over time with an osmometer and compensating with the addition of water or using a dish cover that allows air transfer but not evaporation. The latter approach has recently been introduced by Steve Potter (2001), but has not been attempted to date for the postnatal DA neurons.

We have cultured DA neurons on various types of surfaces and dishes, including multi-well plates, although we generally prefer to use glass coverslips that are glued to petri dishes using SYLGARD (SYLGARD® 184 Elastimer kit, Dow Corning Corporation, Midland, MI), so that the neurons produce small, dense cultures. This system maintains the cultures in a healthy state for weeks to months without requiring feeding with new media (for up to several weeks). The glass allows optimal microscopy with DIC/Nomarski imaging and fluorescent imaging. The central glass well allows most of the neurons to be approached with electrodes (except those within ~50 μm of the edge of the well).

Preparation of Glass Coverslips

1. The utmost care should be taken to always maintain a sterile environment and sterile technique.
2. Drop 100 coverslips [15×15 mm coverslips (deckglöser), Assistant, Germany purchased through Carolina Biological Supply Co., Burlington, NC] one at a time into a 200 ml beaker containing 100 ml of 95% ethanol (leave for at least 1 min).
3. Prepare a solution of poly-D-lysine (PDL) in water (H_2O). Always use ultra-pure water such as tissue culture water or water purified by reverse osmosis to a resistance of ≥18 Ω. The final concentration should be 40 μg/ml, and 250 ml are adequate for 100 coverslips. Place the PDL solution in a glass tray (~4 cm H×100 cm^2) so that the PDL solution is at least 2 cm deep.
4. Pour off excess ethanol and place wet cover slips on to a paper or glass container.
5. Using forceps, pass individual coverslips through the flame of a Bunsen burner once.
6. Cool the coverslip by gently waving it in the air in the air for approximately 10 s.
7. Place the coverslip into the PDL solution (cracking or sizzling sounds mean it did not cool enough). Repeat for all coverslips.
8. Let the slips sit for 1 h, then pour off the PDL and allow the slips to dry.
9. Punch 10 mm holes into 50×9 mm Petri culture dishes (or whichever dishes fit the desired dish holder for the physiology set up) using a bench punch (Roper and Whitney No. XX, Rockford, IL).
10. Wearing gloves, prepare about 5 ml of SYLGARD in a large weighing boat by combining 10 parts resin with 1 part catalyst (by weight) and stirring extensively with a glass stirrer. Pour contents into new weighing boat to avoid use of resin without catalyst.
11. Dip an empty 15 ml conical culture tube or glass scintillation vial upside-down into the SYLGARD, using it like a rubber stamp to apply a circle of SYLGARD to the underside of the dish, around the hole. Use caution when applying the SYLGARD as too much will result in excess SYLGARD covering the culturing surface of the coverslip whereas inadequate SYLGARD will result in a leaky well.
12. Place (drop) the coverslips on the circle of SYLGARD taking care to prevent bubbles.
13. Allow SYLGARD to spread (about 15 min).
14. Cure the SYLGARD, place dishes, upside-down, at 37°C overnight, or at room temperature for 2–3 days (cured SYLGARD is no longer sticky to the touch). Sterlize dishes under UV light for 2 h.
15. Note that these glass coverslips must be coated with laminin before culturing (see glia dissection protocol). Once the dishes are cured they can be stored in zip-lock bags at 4°C until needed.

Preparation of Pipette "Tech-Tips" for Triturating Cells

1. Obtain a box of 1000 μl pipette tips (no-filters) and flame all tips in order to seal off the hole.
2. Using either a 21G 1½ (Large), 22G 1½ (Medium) or 26G 1½ (Small) needle, poke a hole as close to the tip as possible passing straight through, in one side and out the other. This will result in two holes on opposite sides of the tip. The purpose of piercing through both sides is to prevent cells from being caught in the pipette during trituration.

SYLGARD Circles

SYLGARD circles are used as a dissection platform enabling one to pin down the brain using forceps without damaging the tips of the forceps or blunting the scalpel blade. SYLGARD that is left over from the preparation of dishes can be used to make SYLGARD circles.

1. Weigh out SYLGARD, mix the components well and pour into wells of a 12 well plate to a depth of approximately 3–4 mm.
2. Cure in oven at 60°C for 30 min or at room temperature for 2–3 days.
3. Pull circles out of wells using razor and individually package each circle in aluminum foil and autoclave.

Glia (Using Rat Pups)

1. Add 100 μl laminin (10 μg/ml) per round well for a minimum of 1 h at room temperature before plating cells and leave in flow bench (or leave overnight at 4°C) . Aspirate off laminin and wash each well with 200 μl MEM. Aspirate off all remaining MEM and place into 37°C incubator (CO_2 5%) to warm up before plating cells.
2. Obtain one or two P1–P3 mouse or rat pups (animals which are 1–3 days old).
3. Pre-heat the water bath to 37°C (Mistral Multi Stirrer, Lab-Line, Melrose Park IL; Neslab EX-7, Thermo Electron Corp, Austin, TX).
4. Prepare the papain dissociation solution (Table 16.1B). Each vial should hold tissue from a maximum of 4–6 animals.
5. Sterilize the top of each Nunc specimen vial (25 ml) with ethanol. Punch two holes in the top with an 18 gauge needle, using a separate needle to bore holes. In one hole place a new needle with a 0.22 μm filter attached. Leave the other hole empty for ventilation.
6. Add an HCl-cleaned, autoclaved micro-stir bar to the vial. Fill each tube with 5–10 ml of freshly made papain solution (sterile filtered with 0.22 μm filter). Place in temperature-controlled 37°C water bath with magnetic stirrers.
7. Arrange dissection tools, microscope, dissection light, transfer pipettes, sterile Petri dish with SYLGARD square, beaker, and 70% ethanol in a 15 ml tube in the sterile hood.
8. Prepare two buckets of ice to chill the PBS and for the pups.
9. Cut aluminum foil squares (about 10×10 cm).
10. Perfuse dissociation tube (papain tube) with a steady flow of humidified carbogen (95% oxygen, 5% CO_2) delivered through the filter/needle (flow rate adjusted to about 1 bubble per second).
11. When the papain has reached a pH of 7.2–7.4 in about ½–1 h (indicated by a red color; purple indicates that the papain solution is too alkaline and orange that it is too acidic), anesthetize pups and begin dissection.
12. Anesthetize pups with an intraperitoneal injection of ketamine (0.01 and 0.05 ml of 100 mg/ml for mouse and rat pups respectively).

13. When pups begin to show sedation, put them on ice until unresponsive and hypothermic.
14. Rinse aluminum foil square, scissors and pup with 70% ethanol.
15. Decapitate pup, allowing head to fall onto aluminum foil square and move to laminar flow hood. Using (toothed) forceps hold head by the eyes (using left hand if you are right handed). Using the other hand and a curved or angled forceps, pinch the scalp just behind the eyes and pull back, tearing it off. Next, use a micro-scissor to cut around the circumference of the skull, and gently peel the skull off with a forceps, being careful in case any tissue still connects it to the rest of the skull. Gently remove brain (using a small scoop shaped scalpel) and place into Petri dish with one SYLGARD circle stuck onto the bottom and filled ice cold sterile PBS (use enough PBS so that the brain lying on the SYLGARD circle is submerged).
16. Dissect out the cortex taking care to remove the hippocampus. Cut into 1 mm^3 chunks.
17. Place chunks in papain (Table 16.1A), without kynurenate (1 cortex/vial). Chunks should be digested until the tissue looks a bit fluffy like cotton candy (10–30 min, so it can be easily triturated). In the meantime, warm up 45 ml sterile M10C-G for rinsing and trituration (Table 16.2).
18. From this point forward, everything should be done in a sterile laminar hood with sterile technique.
19. Remove the chunks from enzyme vial using a transfer pipette and place in 15 ml conical test tube.
20. Rinse the chunks using transfer pipettes 3 times with 2 ml M10C-G, allowing them to settle each time, and discard supernatant.
21. Using 1 ml warm M10C-G triturate up to a maximum of 10 times (3–5 times should be sufficient) using a regular 1 ml pipette tip.
22. Remove supernatant to a separate tube, and replace the volume with 1 ml fresh M10C-G.
23. Repeat trituration, first using a large, then a medium, then a small bore "tech-tip" until the chunks are dissociated.
24. Pellet the glial cells at 1000×g for 5 min.
25. Resuspend the pellet in 8 ml M10C-G, triturate 5 times to mix, and then count the cells.
26. Dilute the cell suspension to a plating density of 1,000,000–1,500,000 cells/ml and plate about 80,000 cells per well.
27. Plate by dripping the cell suspension from about 10 cm above the dish to help kill neurons.
28. Tap the sides of the trays in order to evenly distribute the cell suspension.
29. Allow cells to settle for 1½–2 h in incubator. Chill the M10C-G in a 4°C refrigerator or on ice.
30. For monolayers: add 2 ml of cold MEM into each well, moving the pipette around in a circular motion to wash each well evenly.
31. Aspirate off MEM and repeat the wash. Aspirate off the MEM and feed with 2 ml of cold M10C-G (Table 16.3Table 16.4).
32. Add FDU (Table 16.5) to the dishes when glia are ~70% confluent (about 3–5 days). If there are still neurons present, remove the medium and cold-feed with 2.5 ml M10C-G before adding FDU. If neurons still persist, place cells at 4°C for 1 h.
33. Astrocytes dishes should be at least one week old and at most 4 weeks old at the time the neurons are plated. Feed with cold M10C-G at one month if not already used.

DA Neurons

1. The day before culturing neurons, wash glial cultures twice with 2 ml of cold MEM. Replace MEM with 2 ml of SF1C (Rosenberg 1988). (Table 16.7, Table 16.8) Alternatively, neurobasal medium A/B27 can also be used (Table 16.9). While A/B27 is easier to make,

we have had slightly better success with SF1C in toxicity assays (unpublished observations).

2. Place sterile rings inside the dish but to the side of the well, not around the well, using forceps with sterile tips (Thomas Scientific, catalog no. 6705R12). The rings will later be placed around the wells to ensure that when the neurons are plated they settle into the well and do not float or wash away into the rest of the dish. Rings should be cleaned by washing with 0.1 *M* HCl for at least 1 h, and then rinsed at least 10 times with distilled water before autoclaving. Any remaining acid will kill neurons the next time the rings are used!
3. Leave dishes with SF1C in the incubator overnight.
4. Pre-heat the water bath to 37°C.
5. Make the papain dissociation solution (Table 16.1B). Each vial should hold tissue from a maximum of 4–6 animals.
6. Sterilize the top of each Nunc specimen vial with ethanol. Make two holes in the top with an 18 gauge needle, using a separate needle to bore holes. In one hole place a new needle with a 0.22 μm filter attached. Leave the other hole empty for ventilation.
7. Add an HCl-cleaned, autoclaved micro-stir bar to the vial. Fill each tube with 5–10 ml for neuronal cultures (2 ml for neuronal cultures from individual pups) of freshly made papain solution (sterile filtered with 0.22 μm filter). Place in temperature-controlled 37°C water bath with magnetic stirrers.
8. Arrange dissection tools, microscope, dissection light, a bunch of transfer pipettes, sterile Petri dish with SYLGARD square, beaker, and 70% ethanol in a 15 ml tube in the sterile hood.
9. Get two buckets of ice to chill the PBS and for pups.
10. Cut aluminum foil squares (about 10 cm^2).
11. Obtain as many P1–P3 rat or mouse pups as needed (animals which are one to three days old).
12. Perfuse dissociation tube (papain tube) with a steady flow of humidified carbogen delivered through the filter/needle (flow rate adjusted to about 1 bubble per second).
13. When the papain has reached a pH of 7.2–7.4 in about ½–1 h (see Glia step 11), anesthetize pups and begin dissection.
14. Anesthetize pups with an intraperitoneal injection of ketamine (0.01 ml and 0.05 ml of 100 mg/ml for mouse and rat pups).
15. Gently remove brain into ice cold PBS-filled Petri dish as described for glia. With anterior of brain facing left on the SYLGARD circle (dissection will be described for right handed people) use a fine forceps in the left hand to pin the two hemispheres of the brain to the SYLGARD. Remove any remaining meninges.
16. Place the brain, ventral side down, anterior facing left onto the SYLGARD. Using forceps, pin the brain to the SYLGARD to immobilize. Make the initial cut through the entire brain caudal to the midbrain flexure (right line, Figure 16.1), and the second cut rostral to the flexure, including the caudal aspect of the hypothalamus. Lay the resulting slice flat on the SYLGARD with the ventral side facing right, and the dorsal side to the left and the caudal aspect up. Remove any remaining cortex. Place the forceps through the aqueduct and push into the SYLGARD to immobilize the slice. Cut the ventral edge of the slice along the top of the hypothalamus as indicated by the line in Figure 16.2. Next cut approximately halfway between the new ventral edge of the slice and the aqueduct. Cut into 1 mm^3 segments and place all three segments into the vial with papain solution for ventral midbrain cultures. For VTA cultures only use the two outer segments and for SN only use the inner segment.
17. Set the magnetic stirrer on low. Chunks should gently swirl around vial.
18. Digest for approximately 10–20 min or until chunks have fallen apart and look fluffy, like cotton candy. Periodically examine the tubes to see if segments are sticking to the stir bars; if so, tap vials to knock the tissue loose.
19. Place tubes containing SF1C for rinsing and triturating in the incubator until ready to triturate, so the pH stays between 7.2 and 7.4. The media must be at the correct pH (indicated by the red

color) or the neurons will die.

20. Use forceps with sterile pipette tips at the end (so they can be changed between dishes) to slide the rings from the side of the dish over the well so the ring acts as a wall extending the height of the well. The rings should be removed 2 h after plating.
21. From this point forward, everything should be done in a sterile laminar hood with sterile technique.
22. Remove chunks from enzyme vial using transfer pipette and place in 15 ml conical test tube. Take great care to prevent bubble formation because bubbles are harmful to cells. If bubbles occur, aspirate them off.
23. Rinse chunks 3 times using about 2 ml pre-warmed SF1C. After adding SF1C for each wash, flick the tube so that the tissue swirls around and is washed. Allow the tissue to settle each time and discard as much of the supernatant as possible without exposing the tissue to air.
24. Triturate in 1 ml of room temperature SF1C (for tissue from 4 to 6 pups or 0.5 ml for individual pups) with large-bore "tech-tips" up to 10 times, preferably fewer (when done properly the medium becomes slightly cloudy with dissociated cells). Let the tube stand until the cells settle, remove the supernatant to a new tube and replace with equal amount of fresh SF1C.
25. Repeat the trituration process up to three times with a medium-, then a small-bore "tech-tip." The chunks should now be completely dissociated into individual cells.
26. Centrifuge cells at 1000×g for 5 min. Re-suspend the pellet in 0.2–0.5 ml (for individual and pooled mouse cultures) or 1 ml (for cultures from 4 rat pups) of SF1C media, mix and count (see below). Dilute the cell suspension until 75–120 μl of media contain 80,000 cells. Add the appropriate volume to the well (with the slide ring around it) of the SF1C-containing dishes so that each well contains 80,000 cells (use a circular motion to spread the cells around). Again, the rings should ensure that the neurons settle into the well and do not float away or get washed away when moving the dishes.
27. After 1½–2 h, cells should have attached to the well. Remove rings from each well using forceps with sterile pipette tips at the end, again, changing tips between dishes.
28. 1½–2 h after plating, add 100 μl GDNF (200 ng/ml, frozen aliquots) to all dishes (containing 2 ml SF1C media), for a final concentration of 10 ng/ml. We have shown that GDNF enhances quantal size (Pothos, Davila, and Sulzer 1997).
29. Allow cells to settle in a 37°C incubator overnight (5% CO_2).
30. The next day add 20 μl of FDU (diluted from stock) to inhibit the growth of non-neural cells (use 1000× stock).

TABLE 16.1A
Papain Solution for Glia from 1 to 2 Cortices (Prepare Just prior to Culturing)

Ingredient	Amount	Final Concentration
1. Cysteine water (Table 16.2)	7.87 ml	1 mM cysteine
2. Papain	Varies	20 units/ml
3. H&B concentrate (Table 16.3)	2 ml	
4. Carbogen	95% O_2+5% CO_2	
5. 0.5% Phenol red	20 μl	0.001%

Note: Dissolve papain in cysteine water first by mixing well. Then add remaining ingredients. Otherwise, papain will not dissolve and will be removed by filtering. Check the H&B concentrate for crystals and if crystals are present placed the solution in the warm water bath for approximately 5 min or until crystals disappear.

TABLE 16.1B
Papain Solution for Neurons (Prepare Just prior to Culturing)

Ingredient	Amount	Final Concentration
1. Cysteine water (Table 16.2)	15.75 ml	1 mM cysteine
2. Papain	Varies	20 units/ml
3. H&B concentrate (Table 16.3)	4 ml	
4. Carbogen	95% O_2+5% CO_2	
6. 0.5% Phenol red	40 μl	0.001%
5. HCL 5 *N*	15 μl	
7. Kynurenate 0.5 *M*	20 μl	0.5 mM

Note: Dissolve papain in cysteine water first by mixing well. Then add remaining ingredients. Otherwise, papain will not dissolve and will be removed by filtering. Check the H&B concentrate for crystals and if crystals are present placed the solution in the warm water bath for approximately 5 min or until crystals disappear.

TABLE 16.2
Cysteine Water

	Components		
Ingredient	Stock Concentration (mM)	ml	Final Concentration (mM)
$CaCl_2$	500	0.60	1.9
Cysteine	19.8	10	1.25
Tissue culture H_2O		146.9	

157.5 ml of 1.25×; sterile filter into bottle and keep refrigerated. Make stocks of 0.5 *M* $CaCl_2$ and 0.02 *M* cysteine in water. Mix 10 ml cysteine stock with 0.6 ml of 0.5 *M* $CaCl_2$.

TABLE 16.3
H&B Concentrate (5×)

Ingredient	Stock Concentration (M)	Final Concentration (mM)
NaCl	4	116
KCl	1	5.4
$NaHCO_3$	1	26
$NaH_2PO_2 \cdot H_2O$	1	2
$MgSO_4$	1	1
EDTA (5%)	0.17	0.5
Glucose (45%)	2.5	25
Tissue culture H_2O		61.84 ml

For H&B concentrate (From Huettner, J. E. and Baughman, R. W., *J. Neurosci.*, 6, 3044–3061, 1986) combine amounts indicated in Table 16.8. Sterile filter (0.22 μm) and divide into 4 ml aliquots and store at −20°C. Add aliquot to cysteine water after adding papain.

TABLE 16.4
M10C-G

Component	Amount
1. MEM	180 ml
2. Calf serum (not heat-inactivated)	20 ml
3. Glucose 45%	1.5 ml
4. Pen-Strep[a]	0.24 ml
5. Insulin (25 mg/ml of 0.02 *M* HCl)	40 μl
6. Glutamine (200 mM)	0.5 ml

[a] Penicillin-Streptomycin solution (100×, 10,000 units/ml penicillin and 10 mg/ml streptomycin in 0.9% NaCl; from Sigma). Filter using a 0.22 μm filter unit, label, and store at 4°C until needed.

TABLE 16.5
5-Fluorodeoxyuridine (FDU) Solution

Ingredient	Final Concentration
Uridine	16.5 mg/ml
5-FDU	6.7 mg/ml
Tissue culture water	15 ml

Make 1000× stock. Add 15 ml uridine solution to 100 mg bottle of FDU to make 6.7 mg/ml solution FDU. Divide into 200 μl aliquots and freeze. Add 1.8 ml of MEM into the 200 μl aliquot of FDU and vortex. Add 20 μl of diluted FDU solution to each dish containing 2 ml of media.

TABLE 16.6
SF1C

Ingredient	Amount	Notes
1. BSA	0.50 g	Fraction V
2. MEM liquid	94.0 ml	
3. DME liquid	80.0 ml	
4. F-12 nutrient liquid	20.0 ml	
5. Glucose 45% liquid	1.5 ml	Solution
6. Glutamine 200 mM	0.50 ml	Aliquotted and frozen
7. DiPorzio concentrate Table 16.7 and Table 16.8	2.0 ml	See below
8. Calf serum	2.0 ml	Heat-inactivated
9. Liquid catalase	0.10 ml	Solution
10. Kynurenic acid 0.5 *M*	200 μl	0.5 *M* kynurenic acid is 946 mg in 10 ml. Dissolve in 1 *N* NaOH
11. HCl 5 *N*	50 μl	

Place on a shaker until BSA goes into solution. Filter using a 0.22 μm filter unit, label, and store at 4°C until needed. Place on a shaker until BSA goes into solution. Filter using a 0.22 μm filter unit, label, and store at 4°C until needed.

TABLE 16.7
Stock Solutions for DiPorzio Concentrate

	Need:		
Additive	Solvent	Tube	Aliquot Concentration
Insulin	20 mM HCl	Plastic	25 mg/ml
Transferrin	Hank's		100 mg/ml
SOD[a]	Hank's		5 mg/ml
Putrescence	Hank's		20 mg/ml
Na_2SeO_3	Hank's		10.4 μg/ml
T3[b]	10 mM NaOH		0.2 mg/ml
Progesterone	100% Ethanol	Glass	1.25 mg/ml
Cortisol	100% Ethanol	Glass	2.0 mg/ml

[a] Superoxide dismutase.

[b] 3,3',5-triiodo-L-thyronine (Na salt) or Liothyronine; Stock solutions should be kept frozen at −80°C.

Source: From di Porzio, U., Daguet, M. C., Glowinski, J., and Prochiantz, A., *Nature*, 288, 370–373, 1980.

TABLE 16.8
DiPorzio Concentrate (10 ml, 100×)

			Final Concentration in SF1C Buffer		
Additive	Amount (ml)	Final Concentration in Stock			ml for Two Batches
1. Progesterone	0.05	6.25 μg/ml	62.5 ng/ml	200 nM	0.1
2. Cortisol	0.02	4 μg/ml	40 ng/ml	125 nM	0.04
3. Hank's BSS	6.21				12.42
4. Insulin	1	2.5 mg/ml	25 μg/ml		2
5. Na_2SeO_3	0.5	0.52 μg/ml	5.2 ng/ml	30 nM	1
6. T3[a]	0.1	2 μg/ml	20 ng/ml	30 nM	0.2
7. Super oxide dismutase	1	0.5 mg/ml	5 μg/ml		2
8. Putrescence	0.12	0.24 mg/ml	2.4 μg/ml	15 μM	0.24
9. Transferrin	1	10 mg/ml	100 μg/ml		2

[a] 3,3′,5-triiodo-L-thyronine (Na salt) or Liothyronine; Stocks of single substances for making DiPorzio concentrate are listed in Table 16.7. To make the DiPorzio concentrate, add the progesterone and the cortisol to a sterile 15 ml polypropylene test tube (translucent). Use a Pasteur pipette attached to vacuum to speed up the evaporation of the ethanol. Add the subsequent aliquots in order. After the addition of the insulin which makes the solution cloudy, add 20 μl of 1 *M* NaOH to neutralize the pH. Solution should go from yellow to pink and immediately clear. Also, after the addition of transferrin, immediately add 20 μl of 1 *N* NaOH, to neutralize the pH and prevent the formation of precipitates. Divide into 5 aliquots and store at −20°C.

TABLE 16.9
Neurobasal A/B27 Medium

Ingredient	Concentration	Amount	Final Concentration
1. Neurobasal A		193 ml	
2. B27		4 ml	
3. Calf serum		2 ml	
4. Glutamine	200 mM	400 μl	0.4 mM
5. HCl	5 *N*	20 μl	0.5 mM
6. Kynurenic acid	0.5 *M*	20 μl	0.05 mM
7. CNQX	50 mM	40 μl	0.01 mM
8. FDU	6.7 mg/ml	200 μl	
9. β-Mercapto-ethanol	139 mM	36 μl	25 μM

Pour about 150 ml Neurobasal A into 250 ml receiver bottle, add the rest of the ingredients, then add more Neurobasal A to make a final total volume of 200 ml. Filter using 0.22 μm filter unit and refrigerate.

31. SF1C media can be changed when necessary (change of color) by aspirating half the media and replacing an equal volume of fresh media (pre-warmed in incubator!).
32. Disturb cultures as little as possible.

Media and Reagents

Table 16.1 through Table 16.9

Carbon Fiber Amprometry

Electrode Fabrication

1. A carbon fiber bundle (individual fibers are 5 μm in diameter, Type T650-42, grade 12K, finish 309, Thornel® carbon fibers, Cytec Industries, West Paterson, NJ) is cut into approximately 10 cm sections (the length of the glass capillary tube) and placed into the middle of a folded piece of paper so that the bundle protrudes about 1 cm. This is placed onto another sheet of white paper (so that the black carbon fibers are more easily seen). The protruding fibers may be gently rolled between thumb and forefinger or otherwise manipulated with the other hand until the carbon fiber cable is frayed. While one hand holds down the folded piece of paper containing the carbon fiber bundle, the index finger of the other hand pins down an individual carbon fiber. The folded paper holding the cable of fibers is pulled back, leaving behind a single carbon fiber.
2. A glass capillary filament tube (12 mm×0.68 mm, 10 cm long, A–M Systems, Inc Everett, WA) which has been kept in 70% ethanol is attached to a vacuum socket which dries the capillary tube. Pinning the carbon fiber down with an index finger at one end, the other end of the carbon fiber is sucked into the capillary tube which is moved towards the base of the fiber/index finger until the entire fiber is in the tube and tube is touching the index finger pinning down the fiber. Disconnect the tube from the vacuum before releasing the carbon fiber! Then, trim the protruding ends of the carbon fiber with a razor or scalpel blade.
3. The carbon fiber-containing capillary tube is then pulled into two equal parts using a Sutter Flaming/Brown Micropipette Puller (P-97; Novato, CA). Usually the carbon fiber remains intact, so it is necessary to cut the fiber using a scissor before removing electrodes from the

puller. Otherwise it will pull out of one side. Settings on the micropipette puller vary, depending on the installation of the filament as well as the filament geometry. The pipette should be pulled so that the tip is gently tapered from the original capillary diameter to the point where the carbon fiber protrudes (approximately 5 mm). A short length (approximately 0.1–0.3 mm) of the carbon fiber should be tightly encased by the glass pulled tightly around the carbon fiber. In order to reduce noise, the pipette tips are quickly dipped into molten stickly wax (Grobet, Carlstadt, NJ), held for about 5 seconds to allow the wax to seal any remaining gaps or cracks between the pipette and carbon fiber and slowly removed to allow the excess wax to run off the pipette and carbon fiber, thus preventing beading up of the wax along the fiber. Pipettes are then beveled (marked side up) at approximately 45° (the same angle that will be used for recording when the electrode is in the holder). Electrodes may be purchased commercially, although we have found to date that commerically produced electrodes are too nosiy to record from neurons.

Testing and Selection of Electrodes

Electrodes are backfilled with K^+ acetate (4 *N*). Silver wires for the pipette holder and ground electrodes (0.25 mm diameter, Warner Instrument Corp., Hamden, CT) are coated with chloride using 1 *N* HCl and a 9 V battery. Electrodes are tested by running a voltammogram (−200–800 mV at 200 mV/s) with the tip of the CFE immersed in 100 μM DA in physiological saline (in mM:128 NaCl, 2 KCl, 10 HEPES, 1 Na_2HPO_4, 2 $MgCl_2$, 1.2 $CaCl_2$, 10 D-glucose; ~300 mOsm and pH 7.4) using a Ensman potentiostat (Ensman instruments, Bloomington IN). Electrodes with a sigmoid shaped voltammogram are then mounted on the headstage of an Axopatch 200 B amplifier (Axon Instruments, Foster City, CA). A voltage of +700 mV (filtered with a built in 4-pole 10 kHz Bessel filter) is applied with the tip of the CFE and the reference electrode immersed in physiological saline. The root mean square of the background noise should be less than 1.1 pA for neuronal recordings.

Recording

Neurons cultured for at least 3 weeks are used for amperometric recordings as it takes this long for VMAT2 and DAT to reach maximum expression levels (Staal unpublished observations). We have rarely had success obtaining quantal DA release before 2 weeks. Neuronal cultures are typically incubated for at least 1 h with 100 μM L-DOPA prior to changing to physiological saline (listed above), which enhances quantal size (Pothos, Davila, and Sulzer 1998). 100 μM L-DOPA may be added to the saline while recording. This will raise the baseline current but will not increase the noise. Using micromanipulators (Newport micromanipulator MX300R, Irvine, CA), electrodes are positioned above the desired recording site and slowly lowered until minimal depression of the structure (cell body or varicosity) is seen. Carbon fibers are held at +700 mV and the resulting current is filtered using a low pass 4-pole, 10 kHz Bessel filter built into an Axopatch 200B amplifier. The current is digitized and sampled at 100 kHz (PCI-6052E, National Instruments, Austin, TX). Start recording as soon as possible followed by application of secretagogue (usually after 10 s). Data is recorded using IGOR PRO (WaveMetrics, Lake Oswego, OR) and saved as an IGOR binary file for further analysis.

We generally use potent secretagogues, such as high potassium (in mM: 52 NaCl, 80 KCl, 10 HEPES, 1 Na_2HPO_4, 2 $MgCl_2$, 1.2 $CaCl_2$) or α-latrotoxin (a generous gift from A. Petrenko, New York University, New York, NY) and potassium (80 mM K^+ solution and 20 nM α-LTX). Secretagogues are applied by local perfusion through a glass micropipette (Picospritzer, General Valve, Fairfield, NJ) for 3–10 s at 10–15 psi and ~5–15 μm from the recording site depending on response. Data should be acquired for several minutes as events often appear at long intervals and sometimes after a significant delay following secretagogue application.

Since amperometric spikes are so small and fast they are often difficult to distinguish from action potentials or a spike due to a noisy electrode. While electrodes with any current spikes should be discarded, an occasional noise-related spike may occur even with good electrodes. However, non-amperometric spikes often consist of a single point (when recording at 100 kHz), are typically symmetrical (like an isosceles triangle) or have a component right before or after the spike that drops below the baseline (a negative current component). Another method to verify that the amperometric events are not mere artifacts depends on finding a site that continuously releases DA. First, the electrode can be raised off the recording site. All spikes should cease, confirming that they are not due to current leaks in a bad electrode. Second, the holding potential can be slowly reduced from +700 to 0 mV and after 30 s or a min, brought back to +700 mV. All amperometric events should disappear as the potential goes to zero and reappear as it approaches the oxidation potential of DA. Only spikes due to action potentials will persist while the voltage is 0 mV.

On rare occasions, varicosities with a small fraction of very large amperometric events have been observed. The quantal size of these spikes resembled large vesicles as utilized by serotonergic neurons (Bruns and Jahn 1995). Due to their infrequency in DA neurons, they could also be eliminated statistically as they were more than 5 standard deviations larger than the average quantal size (Staal, Mosharov, and Sulzer 2004). These events could also correspond to exocytosis of "small" dense core vesicles (~100 nm diameter) that we have noted in electron micrographs, which typically represent less than 1% of the vesicles in synaptic terminals (Pothos et al. 1998).

Analysis

The detection and analysis of amperometric peaks has been reviewed by Mosharov and Sulzer (in presss), and our analysis program, written for Igor Pro (4.07 or later), can be downloaded from our laboratory web site:

http://cumc.columbia.edu/dept/neurology/sulzer/download.html

REFERENCES

Alvarez de Toledo, G., Fernandez-Chacon, R., and Fernandez, J. M., Release of secretory products during transient vesicle fusion, *Nature*, 363 (6429), 554–558, 1993.

Borland, L. M., Shi, G., Yang, H., and Michael, A. C., Voltammetric study of extracellular dopamine near microdialysis probes acutely implanted in the striatum of the anesthetized rat, *J. Neurosci. Methods*, 146 (2), 149–158, 2005.

Bruns, D. and Jahn, R., Real-time measurement of transmitter release from single synaptic vesicles, *Nature*, 377, 62–65, 1995.

Burke, R. E., Antonelli, M., and Sulzer, D., Glial cell line-derived neurotrophic growth factor inhibits apoptotic death of postnatal substantia nigra dopamine neurons in primary culture, *J. Neurochem.*, 71 (2), 517–525, 1998.

Carlsson, A., The occurrence, distribution and physiological role of catecholamines in the nervous system, *Pharmacol. Rev.*, 11 (2, Part 2), 490–493, 1959.

Chen, G. and Ewing, A. G., Multiple classes of catecholamine vesicles observed during exocytosis from the *Planorbis* cell body, *Brain Res.*, 701, 167–174, 1996.

Chen, T. K., Luo, G., and Ewing, A. G., Amperometric monitoring of stimulated catecholamine release from rat pheochromocytoma (PC12) cells at the zeptomole level, *Anal. Chem.*, 66, 3031–3035, 1994.

di Porzio, U., Daguet, M.-C., Glowinski, J., and Prochiantz, A., Effect of striatal cells on in vitro maturation of mesencephalic dopaminergic neurones grown in serum-free conditions, *Nature*, 288, 370–373, 1980.

Gonon, F., Cespuglio, R., Ponchon, J. L., Buda, M., Jouvet, M., Adams, R. N., and Pujol, J. F., In vivo continuous electrochemical determination of dopamine release in rat neostriatum, *C. R. Acad. Sci. Hebd. Seances Acad. Sci. D*, 286 (16), 1203–1206, 1978.

Gonon, F., Burie, J. B., Jaber, M., Benoit-Marand, M., Dumartin, B., and Bloch, B., Geometry and kinetics of dopaminergic transmission in the rat striatum and in mice lacking the dopamine transporter, *Prog. Brain Res.*, 125 (8), 291–302, 2000.

Huettner, J. E. and Baughman, R. W., Primary culture of identified neurons from the visual cortex of postnatal rats, *J. Neurosci.*, 6, 3044–3061, 1986.

Jaffe, E. H., Marty, A., Schulte, A., and Chow, R. H., Extrasynaptic vesicular transmitter release from the somata of substantia nigra neurons in rat midbrain slices, *J. Neurosci.*, 18 (10), 3548–3553, 1998.

Kissinger, P. T., Hart, J. B., and Adams, R. N., Voltammetry in brain tissue–a new neurophysiological measurement, *Brain Res.*, 55 (1), 209–213, 1973.

Leszczyszyn, D. J., Jankowski, J. A., Viveros, O. H., Diliberto, E. J. Jr., Near, J. A., and Wightman, R. M., Nicotinic receptor-mediated catecholamine secretion from individual chromaffin cells. Chemical evidence for exocytosis, *J. Biol. Chem.*, 265 (25), 14736–14737, 1990.

Matsushita, N., Okada, H., Yasoshima, Y., Takahashi, K., Kiuchi, K., and Kobayashi, K., Dynamics of tyrosine hydroxylase promoter activity during midbrain dopaminergic neuron development, *J. Neurochem.*, 82 (2), 295–304, 2002.

Montagu, K. A., Catechol compounds in rat tissues and in brains of different animals, *Nature*, 180 (4579), 244–245, 1957.

Mosharov, E. and Sulzer. D., Analysis of exocytotic events recorded by amperometry. *Nat. Methods*, in press.

Pothos, E. N., Davila, V., and Sulzer, D., Glial-derived neurotrophic factor increases quantal size in midbrain dopamine neurons, *Soc. Neurosci. Abstr.*, 23, 1173, 1997.

Pothos, E. N., Davila, V., and Sulzer, D., Presynaptic recording of quanta from midbrain dopamine neurons and modulation of the quantal size, *J. Neurosci.*, 18 (11), 4106–4118, 1998.

Pothos, E. N., Przedborski, S., Davila, V., Schmitz, Y., and Sulzer, D., D2-Like dopamine autoreceptor activation reduces quantal size in PC12 cells, *J. Neurosci.*, 18 (15), 5575–5585, 1998.

Pothos, E. N., Larsen, K. E., Krantz, D. E., Liu, Y., Haycock, J. W., Setlik, W., Gershon, M. D., Edwards, R. H., and Sulzer, D., Synaptic vesicle transporter expression regulates vesicle phenotype and quantal size, *J. Neurosci.*, 20 (19), 7297–7306, 2000.

Potter, S. M. and DeMarse, T. B., A new approach to neural cell culture for long-term studies, *J. Neurosci. Methods*, 110 (1–2), 17–24, 2001.

Rayport, S., Sulzer, D., Shi, W. X., Sawasdikosol, S., Monaco, J., Batson, D., and Rajendran, G., Identified postnatal mesolimbic dopamine neurons in culture: morphology and electrophysiology, *J. Neurosci.*, 12 (11), 4264–4280, 1992.

Roitman, M. F., Stuber, G. D., Phillips, P. E., Wightman, R. M., and Carelli, R. M., Dopamine operates as a subsecond modulator of food seeking, *J. Neurosci.*, 24 (6), 1265–1271, 2004.

Rosenberg, P. A., Catecholamine toxicity in cerebral cortex in dissociated cell culture, *J. Neurosci.*, 8 (8), 2887–2894, 1988.

Staal, R. G., Hogan, K. A., Liang, C. L., German, D. C., and Sonsalla, P. K., In vitro studies of striatal vesicles containing the vesicular monoamine transporter (VMAT2): rat versus mouse differences in sequestration of 1-methyl-4-phenylpyridinium, *J. Pharmacol. Exp. Ther.*, 293 (2), 329–335, 2000.

Staal, R. G., Mosharov, E. V., and Sulzer, D., Dopamine neurons release transmitter via a flickering fusion pore, *Nat. Neurosci.*, 7, 341–346, 2004.

Sulzer, D. and Pothos, E. N., Presynaptic mechanisms that regulate quantal size, *Rev. Neurosci.*, 11, 159–212, 2000.

Wightman, R. M., Jankowski, J. A., Kennedy, R. T., Kawagoe, K. T., Schroeder, T. J., Leszczyszyn, D. J., Near, J. A., Diliberto, E. J. Jr., and Viveros, O. H., Temporally resolved catecholamine spikes correspond to single vesicle release from individual chromaffin cells, *Proc. Natl Acad. Sci. U.S.A.*, 88, 10754–10758, 1991.

Zhang, D. Q., Stone, J. F., Zhou, T., Ohta, H., and McMahon, D. G., Characterization of genetically labeled catecholamine neurons in the mouse retina, *Neuroreport*, 15 (11), 1761–1765, 2004.

Zhou, Z., Misler, S., and Chow, R. H., Rapid fluctuations in transmitter release from single vesicles in bovine adrenal chromaffin cells, *Biophys. J.*, 70 (3), 1543–1552, 1996.

Zhou, F. M., Liang, Y., Salas, R., Zhang, L., De Biasi, M., and Dani. J. A., Co-release of dopamine and serotonin from striatal dopamine terminals, *Neuron*, 1(46), 65–74, 2005.

Zhuang, X., Masson, J., Gingrich, J. A., Rayport, S., and Hen, R., Targeted gene expression in dopamine and serotonin neurons of the mouse brain, *J. Neurosci. Methods*, 143 (1), 27–32, 2005.

17 Scanning Electrochemical Microscopy as a Tool in Neuroscience

Albert Schulte and Wolfgang Schuhmann

CONTENTS

INTRODUCTION

During evolution, humans and other vertebrates developed a dense network of highly specialized secretory cells, which were eventually fine-tuned to effectively synthesize, store, and release small, diffusible chemical messenger molecules. These transmitters work for intercellular communication and the associated control of physical and cognitive activities such as fine motor coordination, body movement and balance, sensing, thinking, learning, and establishing memory.

In fact, chemical signal transduction all through the body is governed by a vast number of neurotransmitter-releasing nerve cells or neurons from the central and peripheral nervous system including the brain and spinal cord and many hormone-releasing secretory cells inside the glands of the endocrine system, all acting together in a well-synchronized fashion. With some exceptions such as the gaseous messenger nitric oxide (NO), neurotransmitters from nerve cell terminals at synapses and hormones from neuroendocrine cells are released via the exocytosis of bubble-like membranous storage vesicles that are formed inside the cytoplasm by the Golgi apparatus.

As part of their life cycle, the initially empty vesicles fill with a transmitter that is characteristic for a given type of cells and later move and dock via firm membrane protein interactions to the inner side of the plasma membrane. The resulting complex between the two membranes travels through a sequence of maturation steps before, in rapid response to a sudden raise in the intracellular Ca^{2+} concentration as the stimulus, full-membrane fusion and exocytosis occur. Initially, only a narrow aqueous channel or "fusion pore" forms that connects the interior of a fused vesicle with the outside of the cell. However, only fractions of seconds later the fusion pore usually expands and the vesicle collapses, resulting in the complete release of its transmitter contents into the extracellular fluid. Once released, neurotransmitters move through the narrow synaptic cleft to reach and interact rapidly with receptors on the nearby postsynaptic membrane, whereas hormones usually are brought to more distant target cells using the blood stream as a carrier.

Knowledge about the mechanisms and molecular dynamics of chemical cell-to-cell signaling in particular, about the process of vesicular transmitter release and the modulation of exocytosis by endogenous and exogenous factors, not only is fundamental to unveiling the complexity of brain function and motivated behavior, but also to understanding the onset and progression of neurodegenerative diseases such as Alzheimer's disease and Parkinson's syndrome.

Fortunately, many isolated and cultured neuronal and endocrine cells keep their ability to transduce and convey the relevant biochemical and -physical signals and to release their secretory products in response to proper stimulation. Accordingly, individual secretory cells became an excellent scaled-down model system for exploring the process of neurotransmitter and hormone release in controlled manner and steering clear of the difficulties of in-vivo measurements with sensors implanted e.g., directly in brain tissue. Living cells, on the other hand, are tiny objects and their responses to signal transduction and metabolism are diminutive and often occur rapidly. Sophisticated methodologies with an adequate sensitivity, selectivity, and an excellent spatio-temporal resolution are needed to observe neurochemical events on the level of single cells.

Scientists have developed innovative optical assays combining confocal, multi-photon or evanescent-wave microscopy with the application of fluorescence-labeled cellular constituents for looking inside cells and imaging the function of their molecular machinery, in particular, the kinetics and directions of protein dynamics.[1–3] Structural information against which fluorescent images can be interpreted typically comes from high-resolution scanning electron microscopy.[4] The patch-clamp technique Invented by Neher and Sakmann for detecting tiny currents flowing through single ion channels in cell membranes, matured into the most prominent and powerful tool in single-cell electrophysiology.[5] The capability to look at single ion channel activity led to major advancements in the conception of cell excitability and knowledge about the dependency of the function of ion channels on specific metabolic or pharmacological factors. Moreover, patch-clamp capacitance measurements proved to be an excellent approach for monitoring in real time the fusion of intracellular secretory vesicles with the plasma membrane, which is important for instance when studying exocytosis.[6–8]

Common secretory products of nerve and endocrine cells are catecholamine and indolamine derivatives (e.g. dopamine, epinephrine, norepinephrine, and serotonin), histamine, insulin, and NO. These compounds can be oxidized at a suitable electrode surface at an appropriate applied potential and thus they are electrochemically detectable. As early as in the 1970s, R. Adams and coworkers established through their pioneering work, voltammetry at implanted microelectrodes. The result was a powerful electroanalytical approach for in vivo measurements of bulk catecholamine release in animal brains.[9–11] Carbon microelectrodes showed improved characteristics over noble metal electrodes for neurochemical studies due to their lower residual background currents and their better resistance against electrode fouling causing improved long-term sensitivity.

Advancements in instrumentation such as the development of low-noise electrochemical amplifiers with current sensitivities in the low pA range and improved micropositioning systems became available. The development of disk-shaped carbon-fiber microelectrodes (CFMEs) with overall tip diameters in the μm range and millisecond response times allowed then to proceed to localized voltammetry at single cells. Scientists obtained the first successful electrochemical single-cell secretion measurements on chromaffin cells from the adrenal medulla by carefully placing the tip of properly polarized disk-shaped CFMEs closest as possible to the cell membrane and monitoring the amperometric tip current as a function of time.[12,13]

Stimulation of approached chromaffin cells resulted in the appearance of a series of anodic current spikes each of them related to the oxidation of catecholamine molecules leaking out of a single secretory vesicle that underwent exocytosis. Motivated by the capability to watch the rapid process of single-vesicle exocytosis with an exquisite sensitivity in real time, researchers have used carbon-fiber amperometry to study single-cell transmitter release from a number of different neuronal and neuroendocrine cells.[14–24] The analysis of individual current transients in amperometric recordings (e.g., the rise time, amplitude, charge, and half width) provides the number of transmitter molecules released per vesicle and features of the time course of release, both valuable information for elucidating the mechanism and kinetics of exocytosis. If not known, the chemical nature of the released transmitter and its concentration can be assessed by operating the CFME next to a secretory cell in the fast-scan cyclic voltammetry instead of the amperometry mode. A number of recently published comprehensive review articles[25–33] contain detailed information about the analysis, advantages, and limitations of constant-potential carbon-fiber amperometry and fast-scan cyclic voltammetry for the detection of chemical secretion from single cells as well as selected applications.

As expected, scientists have applied different variants of high-resolution scanning probe microscopes as biological imaging, sensing, or manipulation devices. In addition, they have opened new opportunities for probing the properties of individual cells under near-physiological conditions. Atomic force microscopy (AFM) found, for instance, not only widespread application for topographical imaging of cellular structures with up to nanometer lateral resolution[34,35] but also for evaluating adhesion forces between biological (macro) molecules and cell membranes[36] or the elasticity of cells.[37] The fine, nanometer-sized tip apertures of pulled fiber-optic wave guides operated as optical scanning probes and the special properties of near-field illumination made scanning near-field optical microscopy (SNOM) a promising technique for resolving cell structures and molecular dynamics well beyond the resolution of conventional light microscopes.[38–40] Even though offering a little lower spatial resolution than SNOM and AFM, scanning ion conductance microscopy (SICM)[41–43] became another attractive tool for imaging cell topography essentially because of its potential for functional mapping of cell volume changes and movements[44] and of ion fluxes through channels in the membrane surface of living cells.

In the late 1980s, the development of scanning electrochemical microscopy (SECM) brought the small tips of amperometric ultramicroelectrodes into play as electrochemical scanning probes (SECM tips).[45,46] Scanning electrochemical microscopy imaging involves monitoring the current of SECM tips that are scanned close across a surface as a function of lateral tip position. In contrast to other scanning probe techniques, SECM is not only able to reveal the topography of samples in solution, but it may also detect and visualize local variations in the (electro) chemical reactivity of the investigated sample. Originally, researchers developed SECM merely for applications in surface science; however, the potential of the method for biological applications was recognized soon. Scanning electrochemical microscopy-based bioelectroanalysis developed into a powerful approach for imaging chemical gradients in the diffusion layer surrounding cells and tissue,[47,48] the catalytic activity of immobilized enzymes[49,50] or antibodies.[51] Oxygen permeability of cartilage,[52] photosynthetic electron transport in guard cells,[53] cell respiration,[54] calcium release during bone resorption,[55,56] hybridization events on DNA microarrays[57,58] or differences in redox activities of cancer cells[59] are further biological phenomena that were investigated at high spatial resolution by SECM.

Scientists have proposed potentiometric SECM with ion-selective microelectrodes as scanning probes aimed at the determination of the local activity of a number of ions including protons, chloride, sodium, potassium, and calcium.[60–65] All of these ions are of physiological importance and, although not shown to date, assessing their local concentrations in vicinity of individual living cells with scanned H^+-, Na^+-, K^+-, Ca^{2+}- or Cl^--selective tips could be attractive. However, this review focuses on the latest efforts in developing amperometric SECM into a high-resolution electrochemical imaging tool applicable for research and development in the field of neuroscience, with special emphasis given to the local amperometric detection of chemical secretion from single secretory cells. In order to help understand the method, this chapter will first introduce the relevant features of the amperometric microelectrodes and the basic operational principles and technical concept of SECM including the amperometric feedback and the generator collector mode of imaging.

Taking into account the fragile nature of soft living cells, this chapter will discuss the limitations of the ordinary constant-height mode of SECM and present approaches for establishing a constant-distance mode. Only by using SECM in constant-distance mode, the SECM tip is forced to carefully follow the contours of the cell throughout scanning to avoid damaging the tip and the investigated biological sample. Finally, this section will present selected examples of single-cell SECM studies, which highlight the potential impact of this modern electrochemical assay in neurochemistry.

OPERATIONAL PRINCIPLES OF THE AMPEROMETRIC FEEDBACK AND THE GENERATOR/COLLECTOR MODE OF SECM

The key component of SECM in the amperometric feedback and the generator/collector mode is the tip of a mobile disk-shaped carbon or noble metal microelectrode, which is connected to a low-noise potentiostat for sensitive current measurements and made accurately movable through the attachment to computer-controlled high-precision *x*-, *y*-, and *z*-micropositioning devices. Accompanied by fixed reference and counter electrodes, the SECM tip completes an electrochemical cell that is filled with supporting electrolyte containing as well a redox active compound (the "mediator"). Applying, for example, reducible redox mediators such as $[Ru(NH_3)_6]^{3+}$, $[Fe(CN)_6]^{3-}$ or even dissolved oxygen and tip potentials notably more negative than the standard potential of the chosen redox couple, a diffusion limited cathodic tip current is generated and used as specific signal for SECM imaging.

With the SECM tip located in the bulk of the electrolyte far above the sample surface, hemispherical diffusion of the mediator toward the electroactive disk of the microelectrode is controlling the tip current and a steady-state diffusion-limited current value is observed in cyclic voltammograms (Figure 17.1) or in constant-potential amperometry.

However, when one maneuvers the SECM tip closer to a sample, the vicinity and nature of the approached surface will actually start to affect the tip response. For instance, electrochemical recycling of tip-consumed mediator molecules can take place at a neighboring electrochemically active surface. A considerable increase in tip current ($I > I_{lim}$, positive feedback, Figure 17.2a) is observed if the kinetics of the electron-transfer reaction at the sample is faster than the diffusion of the redox mediator in the narrow gap between tip electrode and surface. In contrast, proximity to an electrochemically inactive surface physically obstructs diffusion of species toward the tip leading to a decrease in tip current ($I < I_{lim}$, negative feedback, Figure 17.2b).

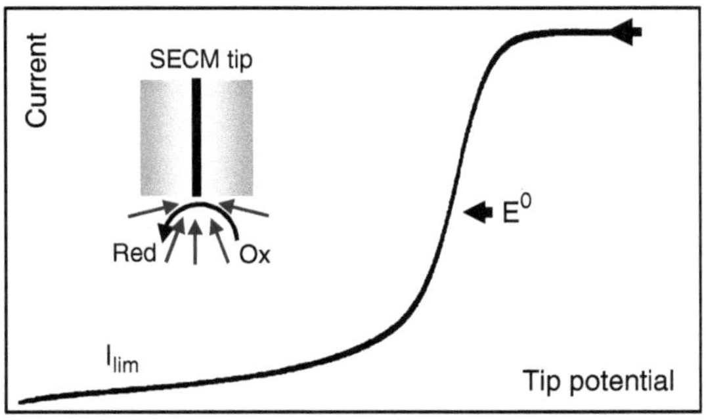

FIGURE 17.1 Voltammetric behavior of disk-shaped microelectrodes (scanning electrochemical microscopy (SECM) tips). Cyclic voltammetry at disk-shaped carbon or noble metal disk-shaped microelectrodes in solutions containing a supporting electrolyte and an e.g., reducible redox active compound, leads to sigmoidal *I*/*E* curves with a steady-state, diffusion-limited current, I_{lim}, measured at electrode potentials significantly beyond the standard redox potential of the dissolved redox mediator.

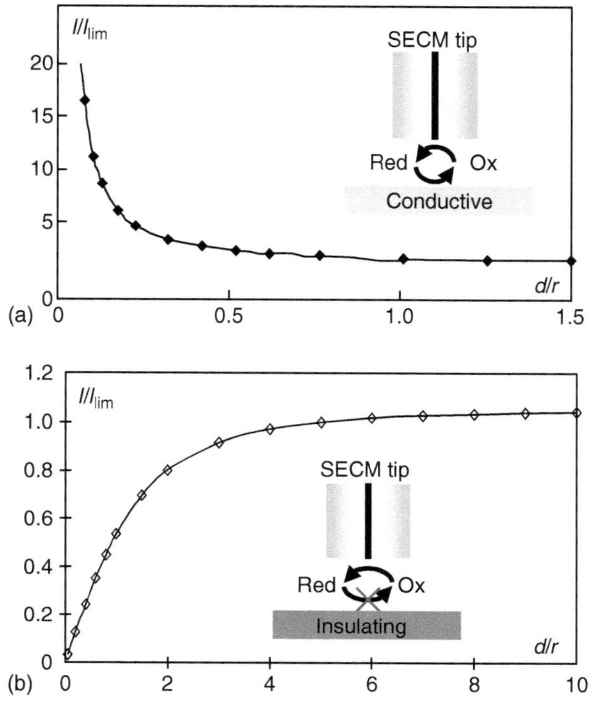

FIGURE 17.2 *Z*-approach curves showing the positive and negative feedback above electrochemically active and passive surfaces. In close proximity to a conducting surface, tip-consumed redox species can be recycled to its original oxidation state. As the tip-to-sample distance, *d*, decreases (*d* here normalized to *r*, the radius of the microelectrode) recycling becomes more effective and the tip-current (normalized to I_{lim}) increases (positive feedback, (a)). Close to an insulator, free diffusion of redox active species toward the microdisk electrode is physically hindered, an effect leading to a drop in the tip-current as compared to the value in bulk (negative feedback, (b)).

Positive and negative feedback effects appear gradually with decreasing tip-to-sample distance, *d*, however, the associated up- or downward deflections in typical approach curves (I/I_{lim} vs. d/r; $r =$ radius of the active disk of the microelectrode) become steeper the closer the tip gets to the sample. For imaging in the feedback mode, approach curves have to be recorded and used to place the SECM tip at an appropriate working distance within the "electrochemical" near field, which starts at distances of about two times the tip electrode diameter. Image acquisition is then achieved by moving the SECM tip laterally at predetermined constant *z*-height along the sample surface and simultaneously monitoring the tip current as a function of *x* and *y* location.

As schematically shown in Figure 17.3, increases or decreases in the tip current obtained in line scans at homogenously insulating or conducting surfaces can be understand as up and down in surface topography while current variations gained at flat surfaces with neighboring conductive and insulating regions represent the local changes in electrochemical activity of the sample surface. In samples with variations in both interfacial electrochemical activity and changes in surface topography, the tip response reflects a convolution of contributions from alterations in electrochemical activity and topography, and, in fact, additional information about the morphology of the specimen is obligatory for an interpretation of the constant-height SECM images.

One can effectively detect and visualize the release of redox-active species liberated from localized spots on the surface via a properly positioned SECM tip capable of oxidizing or reducing the molecules that are ejected from the active site. Clearly, this strategy, the so-called substrate

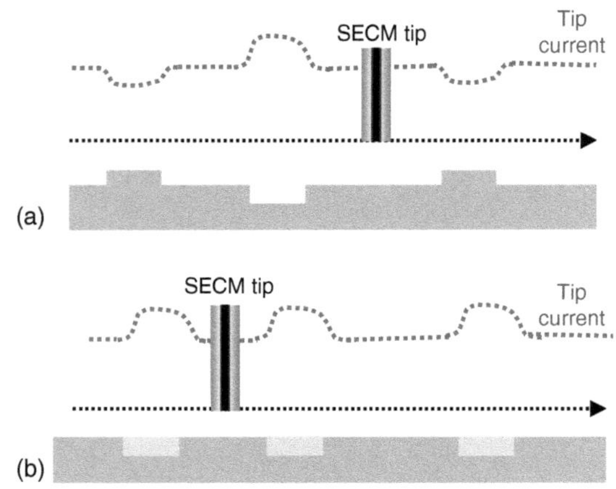

FIGURE 17.3 Principles of constant-height scanning electrochemical microscopy (SECM) feedback imaging. In the feedback mode of operation, SECM imaging takes advantage of an amperometric tip current that originates from the redox conversion of a redox mediator, which is modulated by variations in the tip-to-sample distance and by the local electrochemical activity of the sample surface. On totally insulating or homogeneously active surface variations in the tip current, reflect variations in the sample topography (a). On the other hand, on reasonably flat surfaces, variations in the tip current are indicative for variations in the local electrochemical activity of the sample (b).

generator/tip collector mode of SECM is not only appealing when aiming on the local detection of pitting corrosion, diffusion through micropores of semi permeable polymer membranes or the confined activity of enzyme microstructures, but is also ideal for spotting the dynamic chemical release of biologically important species from single secretory cells (see Figure 17.4).

Traditionally, SECM imaging is carried out with the tip scanned at constant height above the sample surface. However, due to the convolution of topography and electrochemical activity

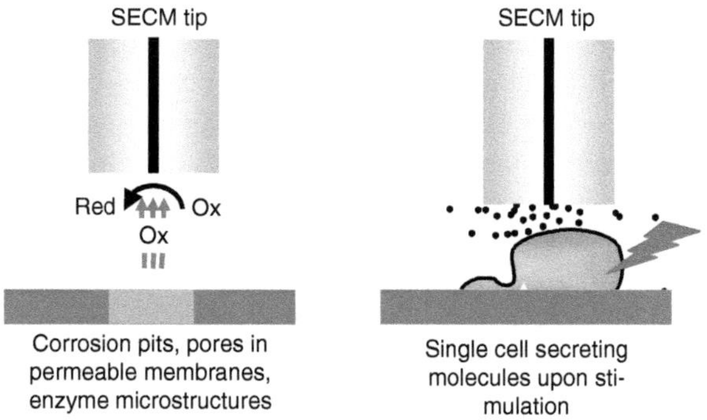

FIGURE 17.4 Generation/collection mode of scanning electrochemical microscopy (SECM). A tip current originates from an electrochemical-induced conversion of a redox species that is generated at locally confined small release sites at the sample surface. Typical examples of phenomena that can be studied in the generation/collection mode of operation are: the release of metal ions from locally corroding metals or alloys, the diffusion of redox species through microscopic pores in semipermeable membranes and their emergence in the vicinity of the pore opening, and the generation of redox active species close to enzyme microstructures via localized enzyme–substrate interaction (left); the release of readily oxidizable or reducible chemical messenger molecules such as hormones and neurotransmitters from single secretory cells upon appropriate stimulation (right).

pointed out before constant-height mode SECM has significant drawbacks when applied for the investigation of samples with three-dimensional objects on the surfaces or a high relief as compared to the radii of the applied SECM tips. Tip crash and tip/sample damage may occur in case the tip is positioned in too close proximity to the surface. The following sections will discuss constant-height mode SECM measurements as well as constant-distance mode measurements at individual cells.

CONSTANT-HEIGHT MODE SECM MEASUREMENT AT INDIVIDUAL CELLS

For electrochemical single-cell measurements, the SECM or any positioning unit that allows for the positioning of a microelectrode in close proximity to a single cell has to be placed on the stage of an inverted optical microscope in order to combine an independent inspection of the biological object together and the microelectrode tip. In addition, by this pre-positioning the tip relative to the object becomes possible. For an efficient detection of chemical release from a secretory cell the tip of the voltammetric microsensor has to be close to the release sites in the cell membrane. This is especially true for vesicle exocytosis events with a very limited amount of neurotransmitter molecules being released to the extracellular medium within a very short time.

Early theoretical calculations revealed that a separation between the collecting electrode and the exocytosis site of larger than about 1 μm already lead to a significant diffusional loss of the released compounds due to their escape from the gap between the disk-shaped sensor and the cell membrane.[66] Also in the case of stimulated release of NO, which diffuses through the cell membrane for longer times after intracellular formation the exact knowledge about the distance between the detecting sensor tip and the releasing cell is indispensable for a quantitative treatment of the obtained results.

Interestingly, scientists introduced rather early the detection of single vesicle exocytosis using a manual positioning of a carbon fiber disk electrode in close proximity to the surface of an individual cell followed by concepts as developed for patch-clamp measurements. In these "conventional" amperometric[37,67–73] or fast-scan cyclic voltammetry[74–76] measurements of single-vesicle exocytosis, the tip of a disk-shaped CFME is carefully moved with manually operated micromanipulators to a cell of choice until a small deformation of the cell can be seen in the optical microscope invoked by gently pressing the microelectrode against the cell's membrane. The microelectrode is then retracted a little to ensure proximity and thus good collection efficiency. This approach heavily stimulated research in brain physiology and understanding of cell-to-cell communication and made accurately positioned microelectrodes an important tool in brain research. However, one potential drawback of this approach was seen in the direct physical contact between the microelectrode tip and the cell membrane, which may lead to a (local) damage of the cell and/or a contamination of the carbon microdisk electrode with cell components. In addition, some compounds are released already upon application of a shear stress to the cell and hence the conventional manual positioning may already lead to shear stress stimulated exocytosis during the microelectrode positioning.

Based on the above, a major question, which had to be solved, was the accurate positioning of a microdisk electrode in known and extremely close proximity to an individual cell without touching the cell. As shown above, with the feedback mode of SECM the positioning of the microelectrode tip relative to the sample surface is supported by the distance-dependent current response of the amperometric SECM tip in combination with a computer-controlled positioning. Because living cells as well as the glass cover slips carrying them are behaving as electrochemically inactive surface, only the negative feedback mode of SECM is relevant for cell measurements. The effective working distance for imaging in the negative feedback mode is about the diameter of the electroactive disk of the SECM tip[77,78] with some dependence from the ratio between the diameter of the active electrode disk and the insulating glass sheath (Figure 17.5). As a consequence, the smaller the electroactive disk electrode of the SECM tip the smaller is the working distance and the more difficult are constant height scans over surfaces with pronounced three-dimensional topography.

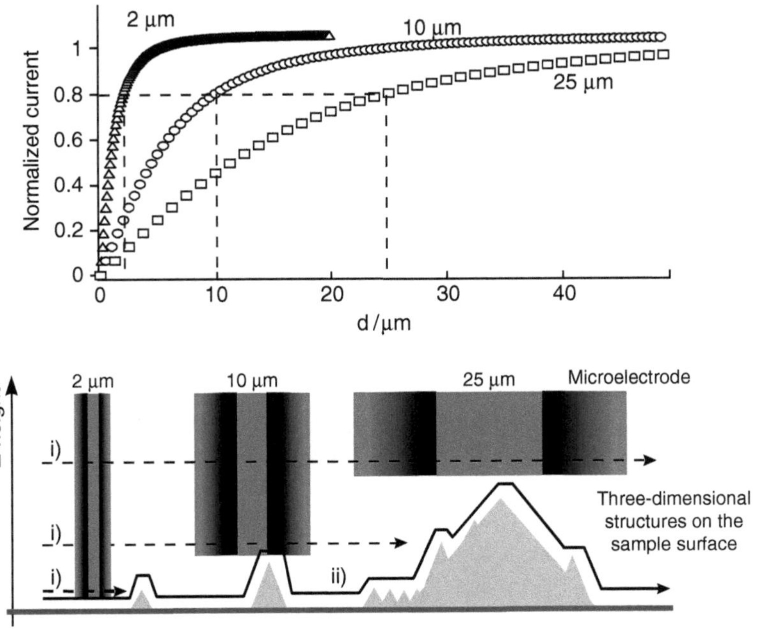

FIGURE 17.5 Negative feedback approach curves in dependence from the active diameter of the scanning electrochemical microscopy (SECM) tip. The working distance (i) is significantly decreased with decreasing electrode size. If the changes in topography are in the same order as the electrode size, the electrode has to follow the sample topography to avoid tip-crash (ii).

Topographical imaging of adherently growing living cells can be achieved using feedback mode SECM using a free-diffusing redox-active compound which is converted at the SECM tip in a diffusion controlled manner at an appropriate electrode potential. In fact, if the SECM approach curve using negative feedback mode is performed at a x, y-location without a cell, the tip has to be retracted to a distance, which is at least as big as the expected three-dimensional topography of the sample. If the electrode is very small, one may even loose the feedback related nearfield distance established previously over the glass cover slip. Two-dimensional scans at this predetermined constant height will lead to enhanced negative feedback at areas where the tip is moving across cells and thus cells become visualized as areas of lower currents in plots of the electrochemical tip response vs. x, y tip position (Figure 17.6).

This procedure is useful for imaging the morphology of individual cells, however, potentially overlaid with information about cell metabolism. Researchers have successfully used a set of membrane-impermeable, hydrophilic, and membrane-permeable, hydrophobic redox mediators along with specific scanning protocols to probe and image the redox activity of individual metastatic and non-metastatic human breast cells.[79] By means of biocompatibility tests, a collection of redox active compounds was selected which should be most suitable for biological SECM experiments on cultured cells.[80] The best mediators were employed in SECM experiments on cultured PC12 pheochromocytoma cells, a model cell type for studying neuronal development and function. One could visualize the cell bodies and branching neurites of differentiated PC12 cells by means of constant height feedback SECM and changes in cell height induced by an exposure of PC12 cells to hypotonic or hypertonic solutions were detectable in real time.

Scientist exploited SECM in the constant-height mode for the local detection of variations in the O_2 partial pressure at single living cells in the late 1990s of the last century.[54,81–83] A disk-shaped Pt microelectrode was kept at a constant potential sufficiently cathodic to drive

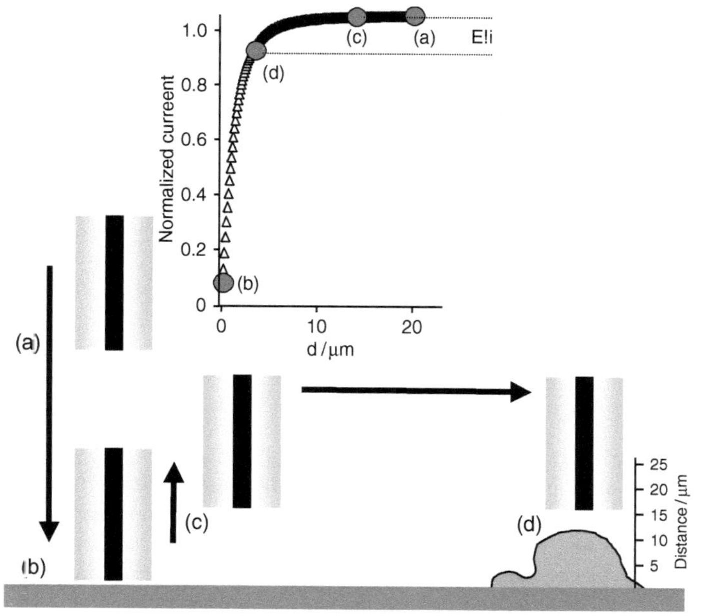

FIGURE 17.6 Tip movements from the bulk (a) toward the glass cover slip (b) allows to record the feedback approach curve which represents a measure for the absolute distance once the electrode diameter and the ratio between the diameter of the electrode and the insulating glass sheath is known. Retracting the electrode to a distance, which is bigger than the expected dimensions of the cell (c) nearly re-established bulk current. Scanning over the cell at distance (d) will lead to a current decrease due to diffusional blocking of the redox mediator and solution and hence to a decrease in the current (Δi) representing the cell topography.

diffusion-limited reduction of dissolved O_2 and scanned in a horizontal plane over an individual cell or a population of adherently growing cells while simultaneously recording the amperometric O_2 reduction current as a function of the SECM tip location. As a representative example, Figure 17.7 illustrates the constant height feedback SECM image of adherent Retzius cells that were dissected from a leech ganglion and maintained in cell culture before SECM imaging. In order to visualize the isolated neuron, dissolved oxygen as usually present in physiological buffer was used as redox mediator and a glass-insulated Pt disk microelectrode of 25 μm diameter as SECM tips. At a constant potential of −600 mV vs. Ag/AgCl the leech neuron clearly became visible in the *x*/*y* plot as area of reduced tip (oxygen) current. In fact, the measured current at the SECM tip is a convolution of potentially decreased O_2 concentrations due to the respiratory activity of the cell and the topography modulating the diffusional access of molecular oxygen to the SECM tip surface.

One might partly overcome this problem by using very small electrodes (typically 3–5 μm diameter), which are retracted to a distance at which one could at least reduce the impact of the cell topography on the observed current variations. The approach was employed for local oxygen measurements at individual guard cells from plants,[53] cells from cultivated cell lines,[81,84] or single protoplasts from algae,[85,86] and more recently for investigations at single bovine embryos[87] and individual neuronal cells.[88] Furthermore, SECM-based local detection of respiration-induced changes of O_2 concentrations were performed in combination with specially designed micropatterns of cultivated cells on glass,[89] or silicon[90] substrates to develop sensitive cell proliferation assays[91] or screening the mode of action of anticancer drugs.[92,93]

As mentioned above, the interpretation of the results of single-cell oxygen measurements is often difficult taking into account the permeability of cells for molecular oxygen and the continuous consumption of molecular oxygen at the SECM tip leading to a hemispherical depletion of O_2 in

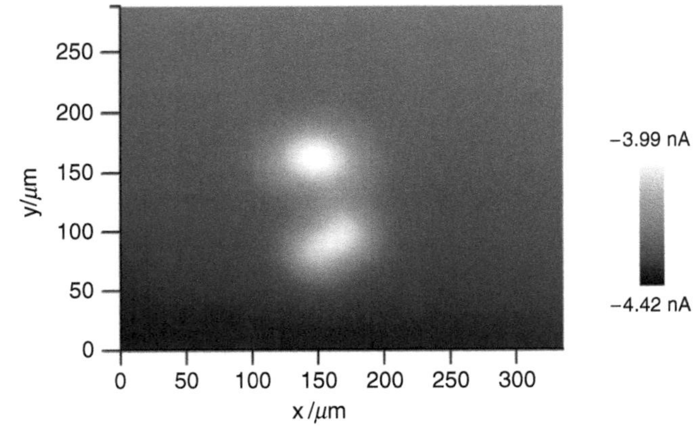

FIGURE 17.7 **(See color insert following page 272.)** Constant-height scanning electrochemical microscopy (SECM) feedback image of a pair of Retzius cells from leech ganglion. The SECM image of the leech neuron was recorded in the negative feedback mode in physiological buffer solution using a 10 μm diameter, glass-insulated Pt disk microelectrode polarized to −600 mV vs. Ag/AgCl pseudo-reference electrode for diffusion limited reduction of molecular oxygen at the tip electrode. Scientists scanned the tip electrode at predetermined (constant) z-height over the cell while monitoring the tip current as a function of lateral position. The leech neuron is visible in the *x*/*y* plot as an area of reduced oxygen reduction current.

its vicinity. Thus, it seems to be indispensable to develop potential pulse detection schemes avoiding the competition between the positioned SECM tip and the cell for O_2 or to follow the cell topography in a defined and constant distance. Alternatively, SECM negative feedback curves can be recorded with the SECM tip directly over a cell in absence or presence of a drug. The difference in the current-distance curve should then represent the changes in cell metabolism assuming constant cell morphology.

A further gaseous compound that plays an important role in the nervous system, is involved in vasodilatation, the control of cerebral blood flow, modulation of synaptic plasticity, memory formation, and the initiation of host defense against tumor cells and infectious pathogens is NO. Because of its diverse biological functions and the involvement of abnormalities in physiological NO levels in, e.g., hypertension, diabetes, ischemia, and atherosclerosis, the local detection of NO release from single cells or cell populations became a topic of increasing interest. Aiming at the elucidation of NO signaling pathways and the basic mechanisms underlying NO release, NO selective amperometric microsensors were developed[94–96] and applied in physiological systems.[97] Generally, electrodes chemically modified with NO oxidizing catalysts such as metalloporphyrins and phthalocyanines in combination with permselective and negatively charged membranes for exclusion of negatively charged potentially interfering compounds were proposed were proposed. Already in 1992, it was shown that porphyrin-modified CFMEs are suitable for NO determination on the level of a single cell.[15] However, although NO sensors became commercially available, quantification of NO released from a single cell or a cell population remains difficult due to the complex diffusion profile of NO in vicinity of the releasing cells. On one hand, a NO diffusion profile is established that depends on the rate of the enzymatic NO formation at the NO synthase by converting arginine to citrulline, reactions within the diffusion zone with molecular oxygen and other reactive oxygen species, its trapping by radical scavengers. On the other hand, the NO sensor locally decreases the NO concentration, which leads to a diffusional flux of NO toward the NO sensor surface. Due to the still-limited performance of available electrocatalysts, the sensitivity of the NO sensors requires often rather large sensor areas of several μm^2. Therefore, the NO conversion at the electrode cannot be neglected with respect to the rate of NO formation inside a single cell and hence an exact knowledge about the distance of the NO sensor and the NO release site is indispensable.

In a recent attempt to improve the accuracy and reproducibility of the positioning of NO microsensors relative to NO releasing (endothelial) cells, scientists designed, and fabricated dual disk-shaped microelectrodes to support an SECM feedback mode z-approach and precise adjustment of the cell-to-sensor separation[98,99] One of the two electrodes, actually a bare 10 μm diameter Pt disk electrode, was operated as conventional amperometric SECM tip and thus could be used via feedback-mode z-approach curves as "travel guide" for the slightly larger second electrode, which was a porphyrin-modified 50 μm diameter Pt disk. With a set of stimulated NO release measurements made at different distances between the disk surface of NO microsensors and endothelial cells it could be clearly demonstrated that the magnitude of the acquired NO signals was as expected strongly dependent on the position of the NO sensor relative to the release sites. In view of that, the SECM-based strategy of employing dual disk, bifunctional probe tips for the reproducible and accurate positioning of NO-specific sensors at well-defined distances from NO releasing cells is a significant advancement for the local detection of NO release. This approach has been extended to the simultaneous detection of the release of NO and glutamate from adherently growing cell populations[100] and could even be integrated into a microtiter-plate compatible electrochemical robotic system for cell assays with increased throughput.[101] However, this type of electrodes is not suitable for lateral scanning in the x, y plane and hence other techniques have to be combined with the conventional SECM modes for keeping the distance between SECM tip and sample constant.

CONSTANT-DISTANCE MODE SECM AT INDIVIDUAL CELLS AND CELL POPULATIONS

Monitoring simultaneously the morphology and the chemical activity of dynamic living cells in a non-invasive manner is the most challenging goal of biological SECM. For accomplishing proper tip positioning constant-distance mode SECM was developed, which enables to continuously control the tip-to-sample distance via a computer-controlled feedback circuitry and guides an SECM tip at constant spacing across the sample surface during scanning. Several strategies have been implemented for providing a distance-dependent input signal that is required for the closed-loop feedback of the distance control including shearforce-based feedback mechanisms with optical[102,103] piezoelectric,[104] or tuning fork based[105,107] detection of shearforces between the vibrating tip and the sample surface. In addition, the maintenance of constant tip currents throughout line-scans[106–108] and a tip impedance-based feedback system have been proposed.[108,109] In any case, the established constant-distance mode forces the tip electrode to follow the specimens shape und thus provides visualization of the sample topography simultaneously with the local detection of electrochemical activity or released compounds. Tip-sample collisions can be avoided and precise positioning of SECM tips at preferred spots and adjusted distances is facilitated. The different schemes for establishing a distance-sensitive feedback signal have their specific assets and drawbacks. Tuning fork and piezoelectric shear force detection systems have been shown well feasible for SECM imaging of individual living cells allowing for positioning of the SECM tip in a constant distance of 100–300 nm (Figure 17.8).

The constant-current mode, however, works with exogenous redox mediators in physiological buffer solution, compounds that may be harmful to cultured cells or change their behavior. Advantageous from this point of view are systems using the tip impedance or optically translated shearforces for the establishment of a distance control being applicable in absence of a possibly toxic mediator even in pure cell growth media. It is additionally useful that both positioning modes allow for synchronized detection of a faradaic and topographic signal due to the fact that the high-frequency tip impedance can be separated from the low-frequency voltammetric tip response and shear forces are current-independent in any case. Thus, efforts were put forward for establishing constant-distance mode biological SECM as a specialized tool for the synchronized topographical

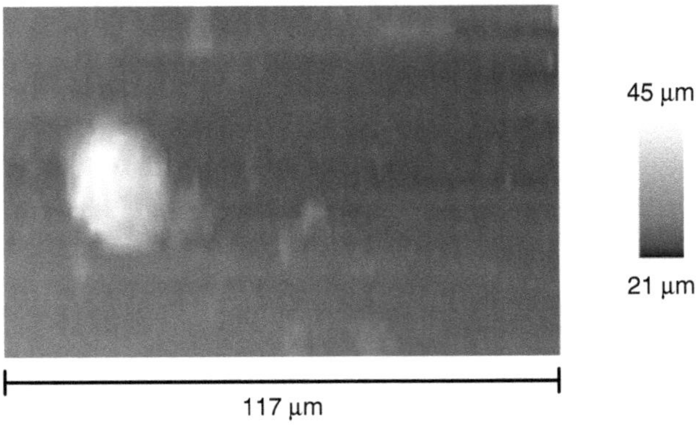

FIGURE 17.8 **(See color insert following page 272.)** Constant-distance topography image of a PC12 cell. A vibrating carbon fiber tip was scanned over the surface using a shearforce-based constant distance mode with optical read-out for repositioning of the scanning electrochemical microscopy (SECM) tip by keeping the damping of the resonance amplitude of the vibrating tip constant.

and chemical imaging of single cells. The setups that were realized are similar in their basic design and composed of the constant-distance SECM unit located on the stage of an inverted optical microscope (Figure 17.9).

As already mentioned earlier, researchers have used different strategies to provide the Bio-SECM with the input signal for the computer-controlled closed-loop feedback system that continuously detects the tip-to-sample separation and maintains it constant throughout lateral scanning. One possibility is to take advantage of optical and non-optical schemes for the detection of hydrodynamic shearforces occurring between liquid/solid interfaces and specially-designed, flexible SECM tips that are brought to vibration at resonance.[102–104] The distance control benefits from a shearforce-induced dampening of tip vibration in extreme proximity to surface. The integrated computer-controlled feedback loop of the device continually compares actual measured oscillation amplitudes and/or phase shifts with a user-defined set point by means of lock-in amplification and responds to alter damping caused by distance variations with tip repositioning. Thus, a constant level of damping is maintained and non-contact scanning at constant distance of about 100 nm to a few hundred nm is guaranteed. It has to be pointed out that vibration of the SECM tip in its resonance frequency as described in[102–104] has advantages over tuning-fork systems due to a softer interaction between tip and sample at an increased tip-to-sample distance.

Alternatively, scientists applied pure electrochemical feedback signals for distance control, namely the tip impedance at high frequencies and the amperometric tip current that occurs at proper tip potentials in the presence of a suitable exogenous redox mediator.[108] With shearforce, impedance and current-based constant-distance mode, conventional-sized carbon fiber microelectrodes (Ø 5–10 μm) and carbon microelectrodes of reduced diameter (Ø 1–2 μm) could be scanned in constant distance across cover slips carrying individual cells, thus revealing the topography image of the cell.

In order to detect single exocytosis events upon stimulation of a secretory cell, the cell is first visualized using non-contact constant-distance topography imaging (see Figure 17.8). The SECM tip is then brought right beside the cell, and using constant-distance scanning the tip is moved exactly above the cell. Scanning is stopped and by addition of a suitable stimulant by means of a positioned microcapillary, compound release is invoked (Figure 17.10).

Obviously, the knowledge about the exact topography of the cells made a precise positioning of SECM tips in close proximity of the membrane of secretory cells possible and allowed the local detection of transmitter release. In addition to the detection of neurotransmitter release from

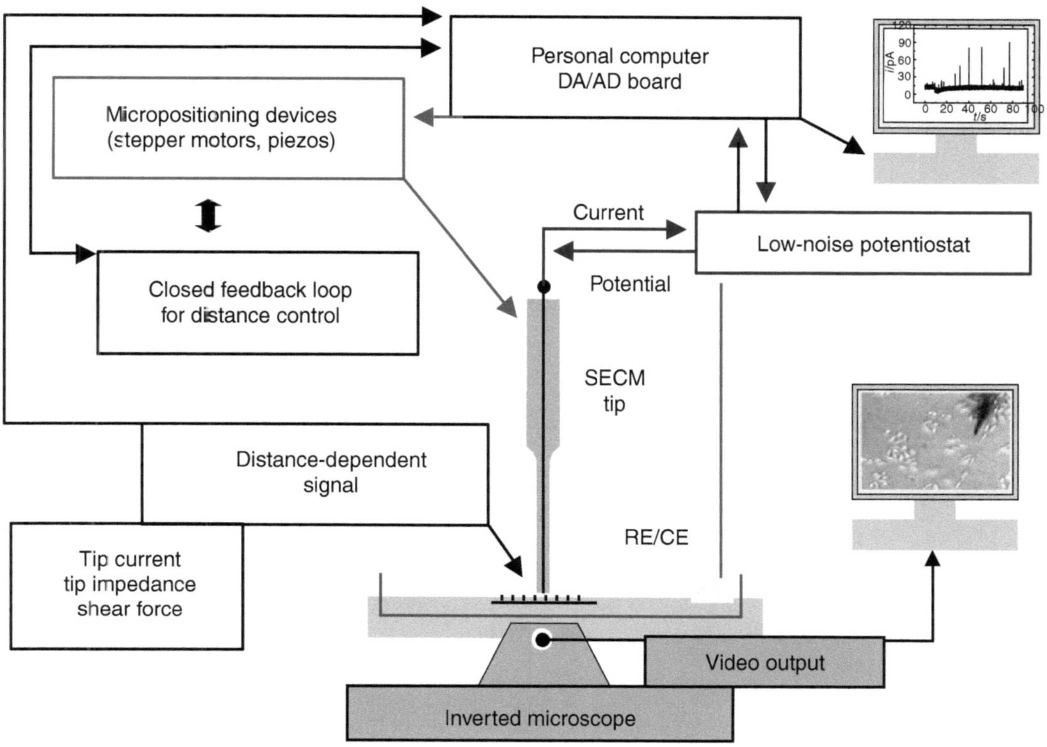

FIGURE 17.9 Schematic representation of a biological scanning electrochemical microscopy (SECM) set-up designed for constant-distance mode topographical and chemical imaging on living cells. The main functional components of the BioSECM are a SECM tip, a micropositioning device, an inverted microscope with optional video output, a distance control unit, a low-noise potentiostat, and a PC connected to an AD/DA board. Not shown is a micropipette-based injection system used for local applications of stimulatory agents.

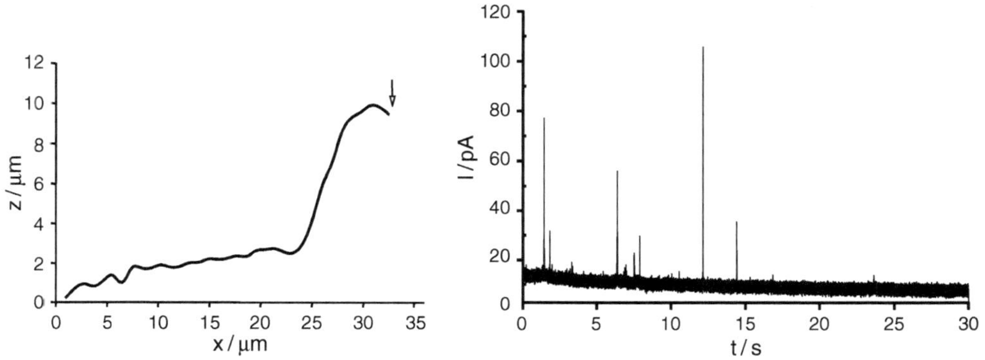

FIGURE 17.10 Left: Shearforce-based constant-distance mode scanning electrochemical microscopy (SECM) scan for positioning the SECM tip exactly above a PC12 cell. Right: Current trace after depolarization of the cell membrane by means of local application of high K^+ concentrations using a micropipette. The current spikes represent individual exocytotic events of noradrenalin release from the PC12 cell.

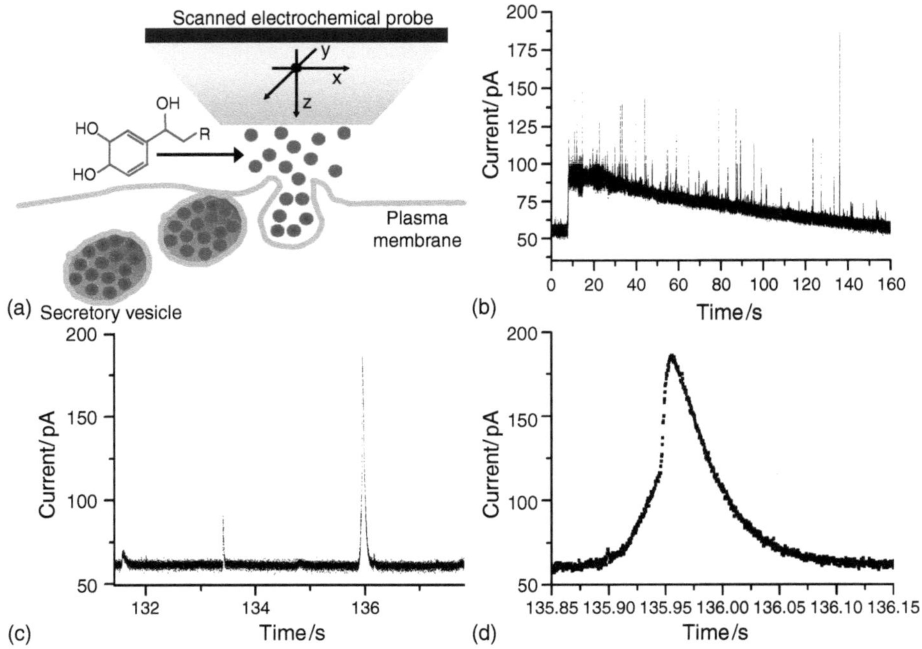

FIGURE 17.11 Local amperometric detection of single-vesicle adrenaline release from an individual bovine chromaffin cell using constant-distance mode Bio-SECM. A 5 μm diameter carbon-fiber disk electrode was polarized to +800 mV vs. Ag/AgCl and positioned within a fraction of 1 μm to the membrane of a bovine chromaffin cell using the topographical information available from constant-distance mode line scans across the cell. In response to stimulation with 100 m*M* KCl solution exocytosis takes place (a) which is detected in recordings of the amperometric tip current through the appearance of multiple current spikes (b), each of them representing adrenaline release from an individual storage vesicle. The traces in (c) and (d) are illustrating parts of the curve in (b) at increasing temporal resolution.

PC12[108] cells (Figure 17.10), catecholamine release from bovine chromaffin cells[110,111] could be successfully demonstrated using unetched and etched carbon fiber electrodes (Figure 17.11).

Recently, constant-distance positioning over single cells could be extended to catalyst-modified carbon fiber electrodes suitable for the selective determination of NO.[112] A Ni-phthalocyanine film was electrochemically deposited on the surface of a platinized carbon fiber and positioned using shearforce-based constant-distance positioning under simultaneous optical control through the inverted microscope. The NO microsensor was positioned exactly above an adherently growing human umbilical vein endothelial cell (HUVEC) using a constant-distance SECM line scan. NO release was then stimulated by local application of bradykinin leading to a current increase at a constant applied potential of 750 mV vs. Ag/AgCl (Figure 17.12).

CONCLUSION AND FUTURE ASPECTS

Due to significant technical advancements, SECM matured into a powerful and valuable high-resolution imaging instrument for applications at the cutting-edge of life science. Biological SECM, especially in the constant distance mode of scanning, has been proven well suitable for visualizing simultaneously the topography and activity of single living cells. Imaging the exact contours of neuronal and secretory cells concomitantly with the local detection of single-vesicle exocytosis, visualization of cell metabolism and release of NO at accurately positioned SECM tips have been successfully demonstrated.

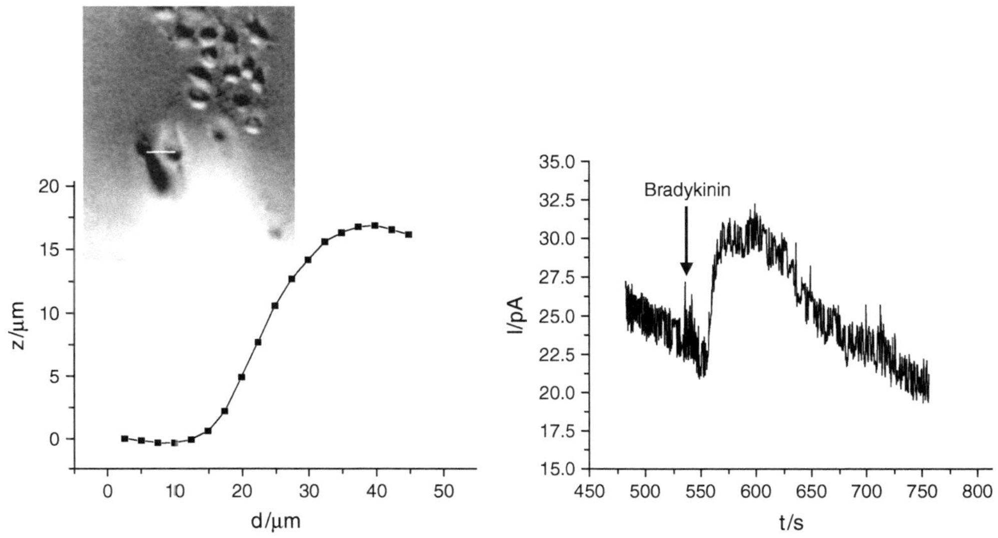

FIGURE 17.12 Left: Shearforce-based constant-distance mode scanning electrochemical microscopy (SECM) scan for positioning the Ni-phthalocyanine-modified SECM tip exactly above a single human umbilical vein endothelial cell (HUVEC) cell. The insert shows the situation before scanning with the shadow of the nitric oxide (NO) sensor to the left of the selected cell. Right: Current trace after stimulation of the cell by means of local application of bradykinin using a micropipette. The increase in current is due to the oxidation of NO at a constant potential of +750 mV vs. Ag/AgCl.

The ability of Bio-SECM to monitor at cellular level dynamic morphological and functional processes is of special importance for the elucidation of underlying mechanisms of cell-to-cell communication. Taking Bio-SECM to the next level of sophistication clearly requires the development, implementation and routine operation of SECM tips with sizes down to the 100 nm range, an improvement of the sensitivity and selectivity of the scanned tips with respect to the target analytes, and an advancement of the software controlling the action of the instrument including the tip-to-sample distance control, data acquisition and data analysis. Achieving these goals in principle should allow approaching more complex cellular systems with Bio-SECM and investigate in detail the influence of the chemical environment of neuronal cells and morphology changes during neurodevelopment, synaptogenesis and neurodegeneration.

REFERENCES

1. Dailey, M., Marrs, G., Satz, J., and Waite, M., Concepts in imaging and microscopy. Exploring biological structure and function with confocal microscopy, *Biol. Bull.*, 197, 115, 1999.
2. Stephens, D. J. and Allan, V. J., Light microscopy techniques for live cell imaging, *Science*, 300, 82, 2003.
3. Zipfel, W. R., Williams, R. M., and Webb, W. W., Nonlinear magic: Multiphoton microscopy in the biosciences, *Nat. Biotechnol.*, 21, 1369, 2003.
4. Koster, A. J. and Klumpermann, J., Electron microscopy in cell biology: Integrating structure and function, *Nat. Rev. Mol. Cell Biol.*, 4, SS6, 2003.
5. Sakmann, B. and Neher, E., *Single Channel Recording*, 2nd ed., Plenum Press, New York, 1995.
6. Lindau, M. and Neher, E., Patch-clamp techniques for time-resolved capacitance measurements in single cells, *Pflugers Arch.*, 411, 137, 1988.
7. Gillis, K. D., Techniques for membrane capacitance measurements, In *Single Channel Recording*, Sakmann, B. and Neher, E., Eds. 2nd ed., Plenum Press, New York, 1995 chap. 7.

8. Heidelberger, R., Electrophysiolgical approaches to the study of neuronal exocytosis and synaptic vesicle dynamics, *Rev. Physiol. Biochem. Pharmacol.*, 143, 1, 2001.
9. Kissinger, P. T., Hart, J. B., and Adams, R. N., Voltammetry in brain tissue—a new neurophysiological measurement, *Brain Res.*, 55, 209, 1973.
10. Adams, R. N., Probing brain chemistry with electroanalytical techniques, *Anal. Chem.*, 48, 1126A, 1976.
11. Adams, R. N., In vivo electrochemical measurements in the CNS, *Prog. Neurobiol.*, 35, 297, 1990.
12. Duchen, M. R., Millar, J., and Biscoe, T. J., Voltammetric measurement of catecholamine release from isolated rat chromaffin cells, *J. Physiol.*, 426, 5P, 1990.
13. Leszczyszyn, D. J., Jankowski, J. A., Viveros, O. H., Diliberto, E. J. Jr., Near, J. A., and Wightman, R. M., Nicotinic receptor-mediated catecholamine secretion from individual chromaffin cells, *J. Biol. Chem.*, 265, 14736, 1990.
14. Chow, R. H., von Rüden, L., and Neher, E., Delay in vesicle fusion revealed by electrochemical monitoring of single secretory events in adrenal chromaffin cells, *Nature*, 356, 60, 1992.
15. Malinski, T. and Taha, Z. H., Nitric oxide release from a single cell measured in situ with a porphyrinic-based microsensor, *Nature*, 358, 676, 1992.
16. Alvarez de Toledo, G., Fernandez-Chacon, R., and Fernandez, J. M., Release of secretory products during transient vesicle fusion, *Nature*, 363, 554, 1993.
17. Duchen, M. R., Voltammetric detection of quantal secretory events from rat peritoneal mast cells, *J. Physiol.*, 467, 2p, 1993.
18. Kennedy, R. T., Huang, L., Atkinson, M. A., and Dush, P., Amperometric monitoring of chemical secretion from individual pancreatic β-cells, *Anal. Chem.*, 65, 1882, 1993.
19. Urena, J., Fernández-Chachón, R., Benot, A. R., Alvarez De Toledo, G., and López-Barneo, J., Hypoxia induces voltag-dependent Ca2+ entry and quantal dopamine secretion in carotid body glomus cells, *Proc. Natl Acad. Sci. U.S.A.*, 91, 10208, 1994.
20. Bruns, D. and Jahn, R., Real-time measurement of transmitter release from single synaptic vesicles, *Nature*, 377, 62, 1995.
21. Jaffe, E. H., Marty, A., Schulte, A., and Chow, R. H., Extrasynaptic vesicular transmitter release from the somata of substantia nigra neurons in rat midbrain slices, *J. Neurosci.*, 18, 3548, 1998.
22. Pothos, E. N., Davila, V., and Sulzer, D., Presynaptic recording of quanta from midbrain dopamine neurons and modulation of the quantal size, *J. Neurosci.*, 18, 106, 1998.
23. Tran, V. S., Serotonin secretion by human carcinoid BON cells, *Ann. N.Y. Acad. Sci.*, 1014, 179, 2004.
24. Staal, R. G. W., Mosharov, E. V., and Sulzer, D., Dopamine neurons release transmitter via a flickering fusion pore, *Nat. Neurosci.*, 7, 341, 2004.
25. Wightman, R. M., Finnegan, J. M., and Pihel, K., Monitoring catecholamines at single cells, *Trends Anal. Chem.*, 14, 154, 1995.
26. Chen, G. Y. and Ewing, A. G., Chemical analysis of single cells and exocytosis, *Crit. Rev. Neurobiol.*, 11, 59, 1997.
27. Angleson, J. K. and Betz, W. J., Monitoring secretion in real time: Capacitance, amperometry and fluorescence compared, *Trends Neurosci.*, 20, 281, 1997.
28. Travis, E. R. and Wightman, R. M., Spatio-temporal resolution of exocytosis from individual cells, *Annu. Rev. Biophys. Biomol. Struct.*, 27, 77, 1998.
29. Henry, J. P., Darchen, F., and Cribier, S., Physical techniques for the study of exocytosis in isolated cells, *Biochemie*, 80, 371, 1998.
30. Bruns, D., Detection of transmitter release with carbon fiber electrodes, *Methods*, 33, 312, 2004.
31. Westerink, R. H. S., Exocytosis: Using amperometry to study presynaptic mechanism of neurotoxicity, *Neurotoxicology*, 25, 461, 2004.
32. Mosharov, E. V. and Sulzer, D., Analysis of exocytotic events recorded by amperometry, *Nat. Methods*, 2, 651, 2005.
33. Dernick, G., Dernick, G., Gong, L. W., Tabares, L., de Toledo, G. A., and Lindau, M., Patch amperometry: high-resolution measurements of single-vesicle fusion and release, *Nat. Methods*, 2, 699, 2005.
34. Dufrene, Y. F., Atomic force microscopy, a powerful tool in microbiology, *J. Bacteriol.*, 184, 5205, 2002.

35. Hörber, J. K. and Miles, M. J., Scanning probe evolution in biology, *Science*, 302, 1002, 2003.
36. Benoit, M. and Gaub, H. E., Measuring cell adhesion forces with the atomic force microscope at the molecular level, *Cells Tissues Organs*, 172, 174, 2002.
37. Arnoldi, M., Fritz, M., Bauerlein, E., Radmacher, M., Sackmann, E., and Boulbitch, A., Bacterial turgor pressure can be measured by atomic force microscopy, *Phys. Rev. E*, 61, 1034, 2000.
38. Hecht, B., Sick, B., Wild, U. P., Deckert, V., Zenobi, R., Martin, O. J. F., and Pohl, D. W., Scanning near-field optical microscopy with aperture probes: Fundamentals and applications, *J. Chem. Phys.*, 112, 7761, 2000.
39. Edidin, M., Near-field scanning optical microscopy, a siren to biology, *Traffic*, 2, 797, 2001.
40. de Lange, F., Cambi, A., Huijbens, R., de Bakker, B., Rensen, W., Garcia-parajo, N, M., Van Hulst, N., and Figdor, C. G., Cell biology beyond the diffraction limit: near-field scanning optical microscopy, *J. Cell Sci.*, 114, 4513, 2001.
41. Korchev, Y. E., Milovanovic, M., Bashford, C. L., Bennett, D. C., Sviderskaya, E. V., Vodyanoy, I., and Lab, M. J., Specialized scanning ion-conductance microscope for imaging living cells, *J. Microsci.*, 188, 17, 1997.
42. Ying, L., Bruckbauer, A., Zhou, D., Gorelik, J., Shevchuk, A., Lab, M., Korchev, Y., and Klenerman, D., The scanned nanopipette: a new tool for high-resolution bioimaging and controlled deposition of biomolecules, *Phys. Chem. Chem. Phys.*, 7, 2859, 2005.
43. Mann, S. A., Hoffmann, G., Hengstenberg, A., Schuhmann, W., and Dietzel, I. D., Pulse-mode scanning ion-conductance microscopy—a method to investigate cultured hippocampal cells, *J. Neurosci. Methods*, 116, 113, 2002.
44. Happel, P., Hoffmann, G., Mann, S. A., and Dietzel, I. D., Monitoring cell movements and volume changes with pulse-mode scanning ion-conductance microscopy, *J. Microsci.*, 212, 144, 2003.
45. Engstrom, R. C., Weber, M., Wunder, D. J., Burgess, R., and Winquist, S., Measurements within the diffusion layer using a microelectrode probe, *Anal. Chem.*, 58, 844, 1986.
46. Bard, A. J., Fan, F. R. F., Kwak, J., and Lev, O., Scanning electrochemical microscopy. Introduction and principles, *Anal. Chem.*, 61, 132, 1989.
47. Horrocks, B. R. and Wittstock, G., Biological systems, In *Scanning Electrochemical Microscopy*, Bard, A. J. and Mirkin, M. V., Eds., Marcel Dekker, New York, 2001 chap. 11.
48. Barker, A. L., Gonsalves, M., Macpherson, J. V., Slevin, C. J., and Unwin, P. R., Scanning electrochemical microscopy: beyond the solid/liquid interface, *Anal. Chim. Acta*, 385, 223, 1999.
49. Kranz, C., Wittstock, G., Wohlschläger, H., and Schuhmann, W., Imaging of microstructured biochemically active surfaces by means of scanning electrochemical microscopy, *Electrochim. Acta*, 42, 3105, 1997.
50. Gàspàr, S., Mosbach, M., Wallman, L., Laurell, T., Csöregi, E., and Schuhmann, W., A novel method for the design and study of enzyme microstructures formed by means of a flow-through microdispenser, *Anal. Chem.*, 73, 4254, 2001.
51. Wittstock, G., Yu, K. J., Halsall, H. B., Ridgway, T. H., and Heineman, W. R., Imaging of immobilized antibody layers with scanning electrochemical microscopy, *Anal. Chem.*, 67, 3578, 1995.
52. Gonsalves, M., Barker, A. L., Macpherson, J. V., Unwin, P. R., O'Hare, D., and Winlove, C. P., Scanning electrochemical microscopy as a local probe of oxygen permeability in cartilage, *Biophys. J.*, 78, 1578, 2000.
53. Tsionsky, M., Cardon, Z. G., Bard, A. J., and Jackson, A. J. R. B., Photosynthetic electron transport in single guard cells as measured by scanning electrochemical microscopy, *Plant Physiol.*, 113, 895, 1997.
54. Yasukawa, T., Kaya, T., and Matsue, T., Imaging of photosynthetic and respiratory of a single algal protoplast by scanning electrochemical microscopy, *Chem. Lett.*, 9, 975, 1999.
55. Berger, C. E. M., Horrocks, B. R., and Datta, H. K., Application of ion-selective microelectrodes for the detection of calcium release during bone resorption, *Electrochim. Acta*, 44, 2677, 1999.
56. Berger, C. E. M., Rathod, H., Gillespie, J. I., Horrocks, B. R., and Datta, H. K., Scanning electrochemical microscopy at the surface of bone-resorbing osteoclasts: Evidence for steady-state disposal and intracellular functional compartmentalization of calcium, *J. Bone Miner. Res.*, 16, 2092, 2001.
57. Turcu, F., Schulte, A., Hartwich, G., and Schuhmann, W., Label-free electrochemical recognition of DNA hybridization by means of modulation of the feedback current in SECM, *Angew. Chem. Int. Ed. Engl.*, 43, 3482, 2004.

58. Turcu, F., Schulte, A., Hartwich, G., and Schuhmann, W., Imaging immobilised ssDNA and detecting DNA hybridisation by means of the repelling mode of scanning electrochemical microscopy (SECM), *Biosens. Bioelectron.*, 20, 925, 2004.
59. Feng, W. J., Rotenberg, S. A., and Mirkin, M. V., Scanning electrochemical microscopy of living cells. 5. Imaging of fields of normal and metastatic human breast cells, *Anal. Chem.*, 75, 4148, 2003.
60. Denuault, G., Nagy, G., and Toth, K., Potentiometric probes, In *Scanning Electrochemical Microscopy*, Bard, A. J. and Mirkin, M. V., Eds., Marcel Dekker, Inc., New York, 2001 chap. 10.
61. Denuault, G., Frank, M. H. T., and Peter, L. M., Scanning electrochemical microscopy—potentiometric probing of ion fluxes, *Faraday Discuss. Chem. Soc.*, 94, 23, 1992.
62. Horrocks, B. R., Mirkin, M. V., Pierce, D. T., Bard, A. J., Nagy, G., and Toth, K., Scanning electrochemical microscopy. 19. Ion-selective potentiometric microsocopy, *Anal. Chem.*, 65, 1213, 1993.
63. Wei, C., Bard, A. J., Nagy, G., and Toth, K., Scanning electrochemical microscopy. 28. Ion-selective neutral carrier-based microelectrode potentiometry, *Anal. Chem.*, 67, 1346, 1995.
64. Etienne, M., Schulte, A., Mann, S., Jordan, G., Dietzel, L. D., and Schuhmann, W., Constant-distance mode scanning potentiometry: Visualization of calcium carbonate dissolution in aequous solution, *Anal. Chem.*, 76, 3682, 2004.
65. Schröck, K., Schulte, A., and Schuhmann, W., Visualisation of the reaction zones between two miscible solutions with potentiometric and amperometric microsensors, *Electroanalysis*, 17, 489, 2005.
66. Chow, R. H. and von Rüden, L., Electrochemical detection of secretion from single cells, In *Single Channel Recording*, Sakmann, B. and Neher, E., Eds. 2nd ed., Plenum Press, New York, 1995, chap. 11.
67. Cahill, P. S., Walker, Q. D., Finnegan, J. M., Mickelson, G. E., Travis, E. R., and Wightman, R. M., Microelectrodes for the measurement of catecholamines in biological systems, *Anal. Chem.*, 68, 3180, 1996.
68. Wightman, R. M., Jankowski, J. A., Kennedy, R. T., Kawagoe, K. T., Schroeder, T. J., Leszczyszyn, D. J., Near, J. A., Diliberto, J. E., and Viveros, O., Temporally resolved catecholamine spikes correspond to single vesicle release from individual chromaffin cells, *Proc. Natl Acad. Sci. U.S.A.*, 88, 10754, 1991.
69. Schroeder, T. J., Jankowski, J. A., Kawagoe, K. T., Wightman, R. M., Lefrou, C., and Amatore, C., Analysis of diffusional broadening of vesicular packets of catecholamines released from biological cells during exocytosis, *Anal. Chem.*, 64, 3077, 1992.
70. Schroeder, T. J., Borges, R., Finnegan, J. M., Pihel, K., Amatore, C., and Wightman, R. M., Temporally resolved, independent stages of individual exocytotic secretion events, *Biophys. J.*, 70, 1061, 1996.
71. Finnegan, J. M., Pihel, K., Cahill, P. S., Huang, L., Zerby, S. E., Ewing, A. G., Kennedy, R. T., and Wightman, R. M., Vesicular quantal size measured by amperometry at chromaffin, mast, pheochromocytoma, and pancreatic beta-cells, *J. Neurochem.*, 66, 1914, 1996.
72. Hochstetler, S. E., Puopolo, M., Gustincich, S., Raviola, E., and Wightman, R. M., Real-time amperometric measurements of zeptomole quantities of dopamine released from neurons, *Anal. Chem.*, 72, 489, 2000.
73. Troyer, K. P. and Wightman, R. M., Temporal separation of vesicle release from vesicle fusion during exocytosis, *J. Biol. Chem.*, 277, 29101, 2002.
74. Michael, D. J., Joseph, J. D., Kilpatrick, M. R., and Travis, E. R., Improving data acquisition for fast scan cyclic voltammetry, *Anal. Chem.*, 71, 3941, 1999.
75. Bath, B. D., Martin, H. B., Wightman, M. R., and Anderson, M. R., Dopamine adsorption at surface modified carbon-fiber electrodes, *Langmuir*, 17, 7032, 2001.
76. Troyer, K. P. and Wightman, R. M., Temporal separation of vesicle release from vesicle fusion during exocytosis, *J. Biol. Chem.*, 277, 29101, 2002.
77. Kwak, J. and Bard, A. J., Scanning electrochemical microscopy: theory of the feedback mode, *Anal. Chem.*, 61, 1221, 1989.
78. Bard, A. J., Fan, F. R. F., and Mirkin, M. V., Scanning electrochemical microscopy, In *Electroanalytical Chemistry*, Bard, A. J., Ed., Vol. 18, Marcel Dekker, New York, p. 244, 1994.

79. Liu, B., Rotenberg, S. A., and Mirkin, M. V., Scanning electrochemical microscopy of living cells: Different redox activities of nonmetastatic and metastatic human breast cells, *Proc. Natl Acad. Sci. U.S.A.*, 97, 9855, 2000.
80. Liebetrau, J. M., Miller, H. M., and Baur, J. E., Scanning electrochemical microscopy of model neurons: Imaging and real-time detection of morphological changes, *Anal. Chem.*, 75, 563, 2003.
81. Yasukawa, T., Kondo, Y., Uchida, I., and Matsue, T., Imaging of cellular activity of single cultured cells by scanning electrochemical microscopy, *Chem. Lett.*, 767, 1998.
82. Yasukawa, T., Kaya, T., and Matsue, T., Dual imaging of topography and photosynthetic activity of a single protoplast by scanning electrochemical microscopy, *Anal. Chem.*, 71, 4637, 1999.
83. Yasukawa, T., Kaya, T., and Matsue, T., Characterization and imaging of single cells with scanning electrochemical microscopy, *Electroanalysis*, 12, 653, 2000.
84. Kaya, T., Torisawa, Y. S., Oyamatsu, D., Nishizawa, M., and Matsue, T., Monitoring the cellular activity of a cultured single cell by scanning electrochemical microscopy (SECM). A comparison with fluorescence viability monitoring., *Biosens. Bioelectron.*, 18, 1379, 2003.
85. Yasukawa, T., Kaya, T., and Matsue, T., Imaging of photosynthetic and respiratory activities of a single algal protoplast by scanning electrochemical microscopy, *Chem. Lett.*, 975, 1999.
86. Yasukawa, T., Uchida, I., and Matsue, T., Microamperometric measurements of photosynthetic activity in a single algal protoplast, *Biophys. J.*, 76, 1129, 1999.
87. Shiku, H., Shiraishi, T., Ohya, H., Matsue, T., Abe, H., Hoshi, H., and Kobayashi, M., Oxygen consumption of single bovine embryos probed by scanning electrochemical microscopy, *Anal. Chem.*, 73, 3751, 2001.
88. Takii, Y., Takoh, K., Nishizawa, M., and Matsue, T., Characterization of local respiratory activity of PC12 neuronal cell by scanning electrochemical microscopy, *Electrochim. Acta*, 48, 3381, 2003.
89. Nishizawa, M., Takoh, K., and Matsue, T., Micropatterning of HeLa cells on glass substrates and evaluation of respiratory activity using microelectrodes, *Langmuir*, 18, 3645, 2003.
90. Kaya, T., Nagamine, K., Oyamatsu, D., Nishizawa, M., and Matsue, T., A microbial chip for glucose sensing studied with scanning electrochemical microscopy (SECM), *Electrochemistry*, 71, 436, 2003.
91. Torisawa, Y. S., Kaya, T., Takii, Y., Oyamatsu, D., Nishizawa, M., and Matsue, T., Scanning electrochemical microscopy-based drug sensitivity test for a cell culture integrated in silicon microstructures, *Anal. Chem.*, 75, 2154, 2003.
92. Torisawa, Y., Shiku, H., Yasukawa, T., Nishizawa, M., and Matsue, T., Multi-channel 3-D cell culture device integrated on a silicon chip for anticancer drug sensitivity test, *Biomaterials*, 26, 2165, 2004.
93. Torisawa, Y. S., Shiku, H., Kasai, S., Nishizawa, M., and Matsue, T., Proliferation assay on a silicon chip applicable for tumors extirpated from mammalians, *Int. J. Cancer*, 109, 302, 2004.
94. Pontie, M. and Bedioui, F., Design of electrochemical microsensors to monitor nitric oxide production in biological systems: A global compilation, *Analusis*, 28, 465, 2000.
95. Bedioui, F. and Villeneuve, N., Electrochemical nitric oxide sensors for biological samples—Principle, selected examples and applications [Review], *Electroanalysis*, 15, 5, 2003.
96. Caro, C. A., Zagal, J. H., and Bedioui, F., Electrocatalytic activity of substituted metallophthalocyanines adsorbed on vitreous carbon electrode for nitric oxide oxidation, *J. Electrochem. Soc.*, 150, e95, 2003.
97. Taha, Z. H., Nitric oxide measurements in biological samples, *Talanta*, 61, 3, 2003.
98. Isik, S., Etienne, M., Oni, J., Blöchl, A., Reiter, S., and Schuhmann, W., Dual microelectrodes for distance control and detection of nitric oxide from endothelial cells by means of scanning electrochemical microscope, *Anal. Chem.*, 76, 6389, 2004.
99. Pailleret, A., Oni, J., Reiter, S., Isik, S., Etienne, M., Bedioui, F., and Schuhmann, W., In situ formation and scanning electrochemical microscopy assisted positioning of NO-sensors above human umbilical vein endothelial cells for the detection of nitric oxide release, *Electrochem. Commun.*, 5, 847, 2003.
100. Isik, S., Castillo, J., Blöchl, A., Csöregi, E., and Schuhmann, W., Simultaneous detection of L-glutamate and nitric oxide from adherently growing cells at known distance using disk shaped dual electrodes, *Bioelectrochem*, doi: 10.1016/j.bioelechem.2006.03.037.

101. Borgmann, S., Radtke, I., Erichsen, T., Blöchl, A., Heumann, R., and Schuhmann,W., Electrochemical high-content screening of nitric oxide release from endothelial cells, *Chembiochem*, 7, 662, 2006.
102. Ludwig, M., Kranz, C., Schuhmann, W., and Gaub, H. E., Topography feedback mechanism for the scanning electrochemical microscope based on hydrodynamic forces between tip and sample, *Rev. Sci. Instrum.*, 66, 2857, 1995.
103. Hengstenberg, A., Kranz, C., and Schuhmann, W., Facilitated tip-positioning and applications of non-electrode tips in scanning electrochemical microscopy using a shear force based constant-distance mode, *Chem. Eur. J.*, 6, 1547, 2000.
104. Ballesteros Katemann, B., Schulte, A., and Schuhmann, W., Constant-distance mode scanning electrochemical microscopy (SECM)—Part I: adaptation of a non-optical shear-force-based positioning mode for SECM tips, *Chem. Eur. J.*, 9, 2025, 2003.
105. James, P. J., Garfias-Mesias, L. F., Moyer, P. J., and Smyrl, W. H., Scanning electrochemical microscopy with simultaneous independent topography, *J. Electrochem. Soc.*, 145, L64, 1998.
106. Wipf, D. O., Bard, A. J., and Tallman, D. E., Scanning electrochemical microscopy. 21. Constant-current imaging with an autoswitching controller, *Anal. Chem.*, 65, 1373, 1993.
107. Lee, Y., Ding, Z. E., and Bard, A. J., Combined scanning electrochemical/optical microscopy with shear force and current feedback, *Anal. Chem.*, 74, 3634, 2002.
108. Kurulugama, R. T., Wipf, D. O., Takacs, S. A., Pongmayteegul, S., Garris, P. A., and Baur, J. E., Scanning electrochemical microscopy of model neurons: constant distance imaging, *Anal. Chem.*, 77, 1111, 2005.
109. Alpuche-Aviles, M. A. and Wipf, D. O., Impedance feedback control for scanning electrochemical microscopy, *Anal. Chem.*, 73, 4873, 2001.
110. Pitta Bauermann, L., Schuhmann, W., and Schulte, A., An advanced biological scanning electrochemical microscope (Bio-SECM) for studying individual living cells, *Phys. Chem. Chem. Phys.*, 6, 4003, 2004.
111. Fletcher, H., Mini probes for looking at living cells, *Chem. World*, 1 (9), 18, 2004.
112. Isik, S., *PhD thesis*, Ruhr-Universität Bochum, 2006.

18 Principles, Development and Applications of Self-Referencing Electrochemical Microelectrodes to the Determination of Fluxes at Cell Membranes

Peter J. S. Smith, Richard H. Sanger, and Mark A. Messerli

CONTENTS

INTRODUCTION

The complexity of the neural microenvironment leads to two inevitable conclusions—in vivo studies are difficult and reduced preparations are simplistic. However, studying the brain environment and cellular function from both angles provides information that clearly advances our understanding of cellular function, signal processing, and diseased states. Implanted sensors in whole brain, both electrical and electrochemical, have contributed to our knowledge of Central Nervous System (CNS) organization and function. Recent moves to merge techniques of on-line capillary electrophoresis and detection of multiple chemicals simultaneously are exciting advances in understanding the CNS microenvironment.[1] Clearly, the chemical composition of this environment is important in not just sustaining the cells of the nervous system, but also in providing and controlling information transfer between neurons, glia and others. In vivo, however, studies are usually restricted to populations of cells and specific mechanisms are difficult to examine on a cell-by-cell basis. Reduced preparations and single cell studies offer insights into mechanisms, even if integration has been largely lost.

Tremendous advances in the approach to single cell studies has occurred in the past two decades. Electrophysiology has been dominated by the powerful derivatives of the original patch clamp technique, addressing the biophysics of single channels or whole cells.[2] Imaging has advanced our understanding of architectural dynamics and intracellular ion activities.[3,4] Indeed, the success of the latter displaced the use of intracellular, chemically-selective electrodes where voltage changes and competing ions make the technique difficult and the interpretation complex. Despite these limitations, electrochemical sensors have continued to demonstrate strength and versatility when applied to the external boundary layers of single cells and tissues. For example, Travis and Wightman[5] demonstrated a correlation between vesicle fusion and cell capacitance for dopamine release. In addition, scientists used electrochemistry of 5-HT and histamine to study quantal corelease.[6] Both of these examples require the placement of a carbon fiber, amperometric microelectrode, in close apposition to the plasma membrane where strong electrochemical signals can be derived over low background. These cases are special in that a discrete cellular event carries a clear electrochemically detectable, phasic, signal into the boundary layer. However, all cells modify the diffusive boundary layers because of physiology, for example transporter activities or respiration. These signals can be weak and often imposed on top of significant background levels of the same chemical–oxygen and calcium flux detection are examples. Problems in measuring changes of such chemical signals are further compounded by the instability of any electrode design. How can these limitations be overcome to make boundary layer measurements with high temporal and spatial fidelity?

This chapter proposes two approaches to making measurements in the boundary layer. Firstly, the gradient can be constrained by a restrictive space, effectively "amplifying" and defining the signal to be detected. Secondly, detection can be coupled directly to the chemical gradients radiating into or out of the cell. Two elegant examples of the first approach are available. Poitry et al. sought to measure the oxygen consumption and nucleotide levels, under closely matching conditions, for single cells.[7] To acquire the oxygen levels, they captured single rod photoreceptorsin glass micropipettes such that they lodged approximately 10 μm from the tip. Next, they inserted a Whalen-style oxygen electrode[8] through the pipette tip and measured oxygen consumption in a static configuration. Kang and Hilgemann[9] recently published results from a similar approach, but for the study of the Na^+/Ca^{2+} exchanger in a cardiac muscle macropatch. Here an ion-selective electrode (ISE) was advanced into the body of a pipette supporting the macropatch. As the pipette geometry could be precisely modeled, Kang and Hilgemann were able to derive the stochiometry of this electroneutral transporter.[9]

Coupling directly to the diffusive boundary layer, by alternating the placement of a chemically selective microelectrode in the proximity of active tissues or cells, was first reported by Gow et al.[10] These investigators monitored pH changes proximal to a growing fungal hypha. The process of microelectrode placement and modulation was automated by Jaffe and Levy and applied to biological preparations by Kühtreiber and Jaffe.[11,12] In a more recent study, Kang et al. made measurements of the transmembrane potassium transport by "self-referencing" through movement of the macropatch relative to a K^+ ISE.[13] The development of this "self-referencing" technique, defined below, has occurred primarily at the BioCurrents Research Center, funded as a national resource by the National Institutes of Health (NIH:NCRR) since 1996, but also in other laboratories.[14] Smith et al.[15] coined the term "self-referencing" as it applies to this technique.

In the context of the electrochemical technique reviewed here, self-referencing refers to a modulation approach. The method belongs to a family of techniques employing temporal and time dependent, positional modulation of microprobes to enhance signal detection. Variants can be seen in scanning electrochemical microscopy, atomic force microscopy, and scanning reference electrode techniques used in both biological and materials science.[16–19] In the application described here, the term implies that a single sensor is modulated such that signal values obtained at the near pole (closest to the tissue or cell) can be compared to the values at the far pole (a known distance away). As will be discussed below, this delivers significant advantages in signal analysis with background, drift, and noise reduction—the electrode signal is effectively compared to itself at two locations. A second and independent reference electrode is also included to complete the circuit.

When positioned in a chemical gradient surrounding a single cell, or tissue surface, this type of electrochemistry generates a differential output that can be converted to a flux value through calibration and Fick's First Law. Using this approach with potentiometric electrodes, differential signals of low μvolts can be observed and with amperometric electrodes values of femtoamps are measured. The technique can be coupled with electrooptical and positional methods.[20,21] Overall, self-referencing of an electrochemical sensor offers the advantage of low signal detection, observation of non-electrogenic transport, use of ion-selective and redox based reactions, and enzyme-assisted flux detection, all at the single cell level.

SELF-REFERENCING: PRINCIPLES

Cells modify the immediate chemical composition of the medium surrounding them. This is an inevitable consequence of physiologically driven molecular movements, both active and passive. For example, as a cell respires oxygen will be taken up from the medium and a shallow depletion

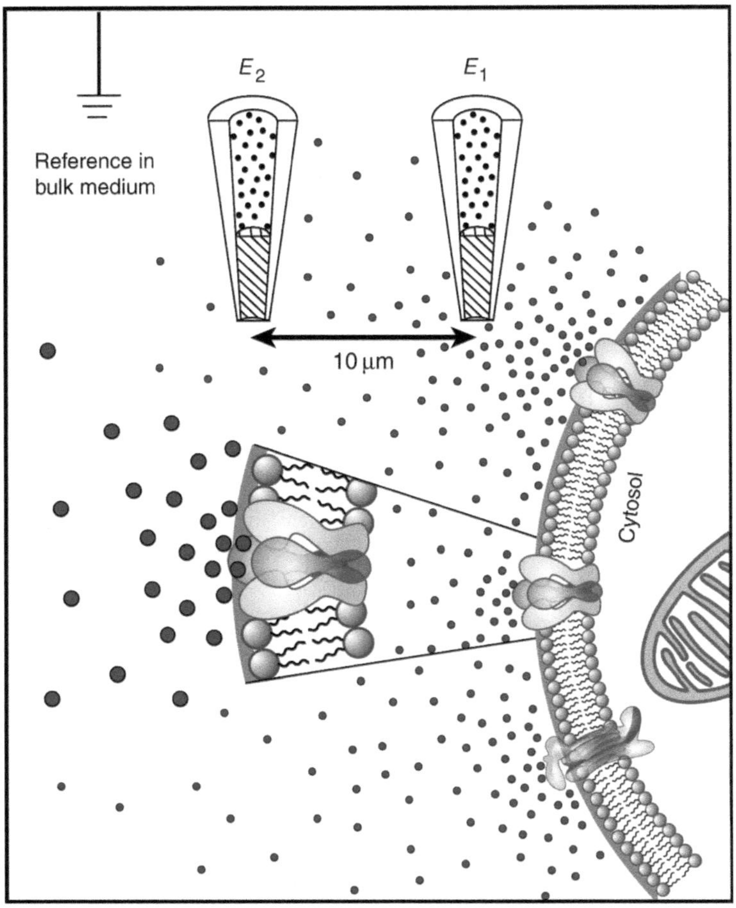

FIGURE 18.1 Self-referencing electrodes operate in the diffusive boundary layers surrounding cells and tissues. Within these layers, analytes form gradients depending on cellular activity, such as channel openings, as illustrated, or metabolism. This figure illustrates an ion-selective microelectrode located at both the near pole and far pole within an ion gradient. The voltages recorded, E_1 and E_2, will depend on the activity of the ion. Comparing these two values generates a differential value minimizing the impact of non-relevant voltages generated by background levels of the analyte, junction potentials and competing analyte levels. The same principle applies to amperometric microelectrodes.

gradient will exist around the plasma membrane, particularly in locations where mitochondria are localized.[22] Similarly the opening of a channel, or the activity of a pump, will locally modify the chemical concentrations/activities in a regional manner (Figure 18.1). These gradients reveal aspects of cell physiology in both normal and diseased states, making it of interest to couple to the gradients and thus gain insight into cell function. To measure a gradient, one needs two points of reference, a known distance apart, recording an identifiable analyte, with a known diffusion coefficient.

In a static configuration, drift, sensitivity, and chemical selectivity, or the lack of it, can have complicated implications for the use of these sensors, whether potentiometric or amperometric. In a self-referencing, non-invasive, mode of operation, many of these conventional problems can have a minimal impact, although other aspects influence the quantification of data recovered. The purpose of this paper is to highlight both the strengths and weaknesses of the self-referencing methodology.

The basic operational configuration of a self-referencing electrode is shown in Figure 18.2, where a calcium ISE is illustrated near a pipette source of calcium forming a defined gradient in the bulk solution (for details of this procedure and comparisons to modeled gradients see Smith et al.[15]). The ISE is moved in a square wave between two points in the gradient, usually 10 μm apart. This distance can be varied. The frequency of movement is usually 0.3 Hz. At that frequency, the electrode comes to rest for approximately 1 s at each pole after translation. As with distance, frequency can be independently controlled. The point nearest the source is referred to as the near-pole and the other as the far-pole. The convention is followed that if the electrode detects a higher concentration at the near-pole the flux is given a positive value indicating efflux. Signals are sampled 1000 times a second with the first 30% collected during and after translation being discarded. A final differential value is stored along with the correlated DC value, which can be compared to an original calibration slope.

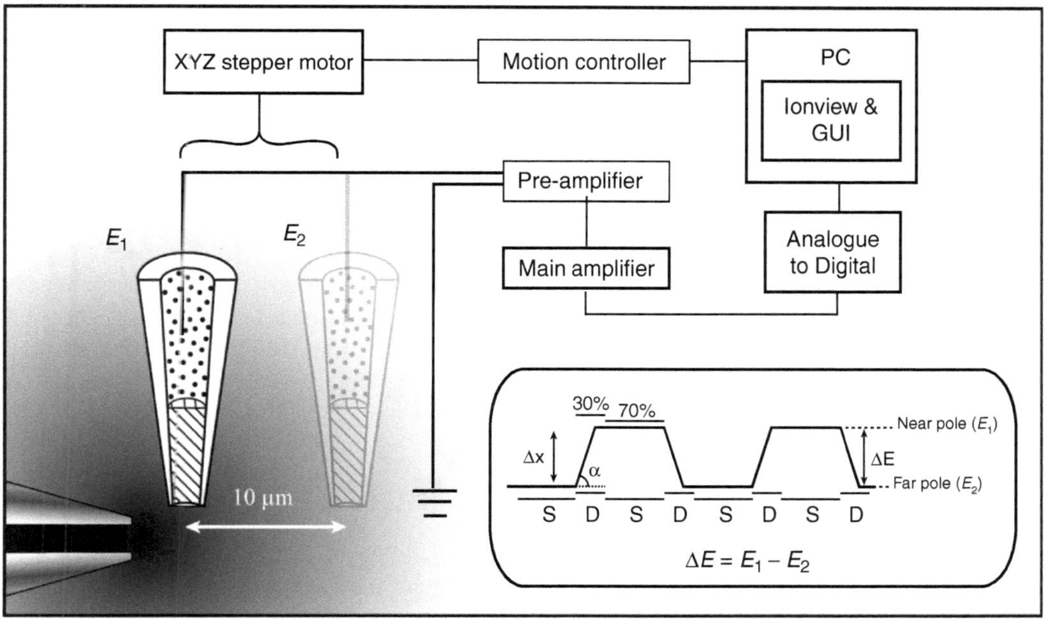

FIGURE 18.2 A schematic illustration of the self-referencing set-up. In this case a electrode is shown near a source of the analyte, moving in a "square" wave step between two positions 10 μm apart. Data is collected at 1000 points per second with the first 30% collected during and after translation being discarded (D). Movement is frequently set for a 10 μm displacement (Δ×) at a frequency of 0.3 Hz, thus data is collected at each pole for approximately 1 s or 70% of the cycle time (S). Comparing, in this case, voltages collected between the two extremes of movement, gives rise to a differential value (ΔE). A description of the electronics and software components are given in "Equipment and Software".

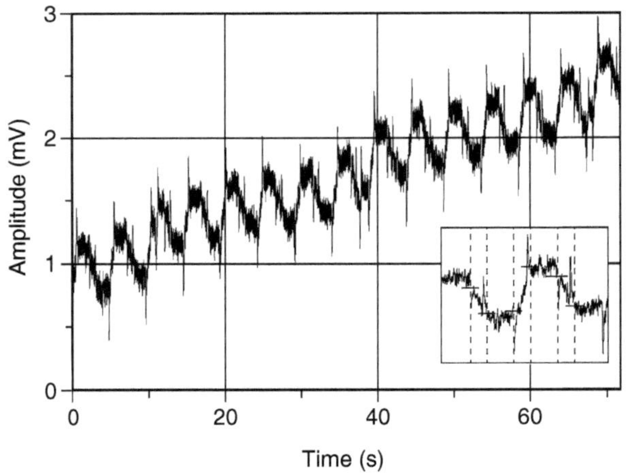

FIGURE 18.3 Raw data collected from a calcium ISE moving in a steep gradient generated by a 100 mM source of Calcium Chloride. A clear stepping of the data can be seen as the ISE moves between the two poles of translation. Significant drift in the base line is also evident. The insert shows the time scale for two cycles. The broken lines divide the signal into the discard and saved sections—see Figure 18.2. Biological signals will normally be in the microvolt range and therefore buried in the noise.

The key to the sensitivity of this approach is that drift and background are relatively common to both positions of translation. Thus, the signal seen by the amplifiers when a calcium selective electrode is placed in a steep ion gradient, from an artificial source, may be similar to the result illustrated in Figure 18.3. No signal processing has taken place. A clear drift is seen in the signal from both poles of translation. Extracting a differential voltage or current, by subtracting the far pole signal from the near pole, minimizes the impact of drift and subtracts out background signals as well as contaminating voltages—such as junction potentials.

Although originally conceived for calcium flux detection,[12,15,23] uses of ISEs for the detection of potassium and hydrogen[24] as well as chloride ion flux[25,26] have been popular extensions. Amperometric-based detection of nitric oxide (NO), oxygen, glucose and ascorbate, have proven powerful additional applications (references below and additional measurable analytes). No other approach permits repeated non-invasive observations on the same intact cell over hours or days, recording flux over periods of seconds, with a spatial resolution of a few microns. Figure 18.4 illustrates a Ca^{2+} ISE in the near pole of measurement next to an INS-1 cell. In this sense, the technique is unique, and complements other methods available to observe the movements of analytes across membranes or within cells.

SELF-REFERENCING: POTENTIOMETRIC ION-SELECTIVE ELECTRODES (ISE)

Several recent studies have contributed to our understanding of the factors that influence ISE performance. Bakker et al. provide a review.[27] A key point they discuss is the importance of the internal electrolyte on detection limits, a matter dealt with below. However, it is not the purpose of this paper to review methods of electrochemical detection so readers are referred to other reviews.[27,28] Here the researchers focus on the use of these designs in a self-referencing mode.

The ISEs are based on an ion-selective solvent or liquid membrane, immobilized in a silanized micropipette with tip diameters of 1–4 μm. For an ideal electrode the measured voltage, E, is related to the activity of the ion by the Nernst equation:

$$E = E_0 + S \log a_i \tag{18.1}$$

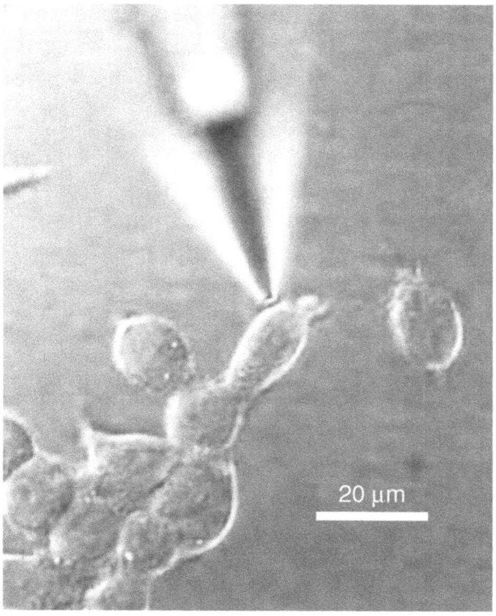

FIGURE 18.4 A real microelectrode near a real cell. This figure illustrates a Ca^{2+} ISE located at the near pole position for measurement from a cultured INS-1 cell. As the electrode is silanized, and the cocktail lipophilic, stability of electrode placement and movement is critical. The electrode must not touch the cell. Control of motion is discussed in "Equipment and Software".

where E_0 is an offset potential, S is the Nernstian slope

$$\frac{2.3RT}{z_i F} \tag{18.2}$$

and a_i is the activity of the primary ion. R, T, and F have their usual meaning. The offset potential is composed of the boundary potentials and liquid junction potentials that exist across any circuit made up of a reference and measuring electrode. Through calibration of each ISE a value for the slope of the line, describing the voltage output and the change in ionic activity is collected. Since ionic activity is directly proportional to ion concentration, via the activity coefficient, and the changes that occur to this coefficient due to changes in ionic strength are usually negligible during self-referencing, we will use concentration in place of activity.

Minimizing drift and noise for a potentiometric sensor operating in a self-referencing mode (termed self referencing ion selective: SERIS[15]) can be explained by introducing a formulation of the Nernst equation. The voltage difference at the two poles of excursion within an ionic gradient is related to the concentrations at the two poles by the following equation:

$$\begin{aligned} E_1 - E_2 &= (E_0 + S \log C_i)_1 - (E_0 + S \log C_i)_2 \\ \Delta E &= (S \log C_i)_1 - (S \log C_i)_2 \\ \Delta E &= \log C_{i(1)}^{S_1} - \log C_{i(2)}^{S_2} \\ \Delta E &= \log \left(\frac{C_{i(1)}^{S_1}}{C_{i(2)}^{S_2}} \right) \end{aligned} \tag{18.3}$$

E_1, $C_{i(1)}$, and S_1 are the measured voltage, primary ion concentration, and slope of the voltage-log plot for the near pole of excursion. Subscript 2 labels the same parameters for the far pole of excursion. By calculating the difference in potentials over short periods of time the impact of slow drift, due to the constant potential differences throughout the system, are reduced. This places the emphasis of the potential difference measurement on the concentration difference of the ion.

Equation 18.3 can be simplified to determine the relationship between the ionic concentrations at the two points of excursion:

$$C_{i(1)} = C_{i(2)}^{S_2/S_1} \cdot 10^{\Delta E/S_1} \tag{18.4}$$

The slopes at two points of the calibration curve are included to address measurement of the primary ion under different circumstances; (1) differential concentration of the primary ion in the absence of an interfering ion; (2) differential concentration of the primary ion in a constant concentration of an interfering ion and; (3) differential concentration of a primary ion coexisting with a gradient of an interfering ion.

For an ideal electrode, used in the absence of interfering ions, the slope is constant over a wide range of the primary ion concentration and close to Nernstian. Under these conditions Equation 18.4 simplifies to:

$$C_{i(1)} = C_{i(2)} \cdot 10^{\Delta E/S} \tag{18.5}$$

Under many circumstances, the average concentration of the ion at the far pole, position 2, is not that different from the average concentration of the ion in the bulk solution. Therefore, the difference in ion concentration between the two points of excursion can be described as follows:

$$\Delta C = C_{i(1)} - C_{i(2)} = C_{\text{bath}} 10^{\Delta E/S} - C_{\text{bath}} \tag{18.6}$$

A primary assumption here is that the concentration difference measured between the two excursion points is linear. This is only true if the excursion distance is small compared to the extent that the gradient extends out into the bulk solution. For small cells an excursion of 10 μm will most likely sample over a distance in which the concentration difference is not linear and could, therefore, lead to an underestimate of the flux. An incorrect estimation of flux could also occur during a two-point measurement in an intense extended gradient, where the concentration of the ion at the far pole is substantially different from the background concentration of that ion. The solution to these issues is the use of a three-point measurement to; (1) more carefully map the concentration gradient with a third point to either ensure a linear relationship or determine a more accurate nonlinear relationship and; (2) to determine the concentrations in the gradient relative to the background concentration of the ion in the bath. For slow drift conditions, the electrode can simply be moved from position 3 (outside of the gradient) to position 2 (slightly into the gradient) to position 1 (deeper into the gradient). In general, regular measurements at all three positions should be attempted.

Measurement of a concentration difference for a primary ion in the presence of a constant concentration of an interfering ion presents us with a more difficult situation. Messerli et al.[14] have dealt with this in detail.Their results will be summarized here. Two variables in Equation 18.4 now need to be determined, specifically the concentration of the ion in the near pole $C_{i(1)}$ and the new slope of the calibration curve in the near pole, S_1. Therefore, a relationship between the change in voltage and the change in slope needs to be established in order to measure two points in a concentration gradient of a primary ion in the presence of a constant concentration of an interfering ion. The Nicolsky–Eisenman equation provides a means for predicting the change in slope due to the change in voltage in the presence of an interfering ion, assuming the electrode responds to the interfering ion in a Nernstian manner.

Rearranging the Nicolsky–Eisenman equation to get primary ion concentration in terms of voltage, we can determine the relationship of the slope at any point along the curve in terms of voltage:

$$\begin{aligned} E &= S_N \cdot \log\left(C_i + K_{ij} \cdot C_j^{z_i/z_j}\right) \\ C_i &= 10^{E/S_N} - K_{ij} \cdot C_j^{z_i/z_j} \\ \log(C_i) &= \log\left(10^{E/S_N} - K_{ij} \cdot C_j^{z_i/z_j}\right) \\ \frac{1}{S} &= \frac{d\log(C_i)}{dE} = \frac{10^{E/S_N}}{S_N(10^{E/S_N} - K_{ij} \cdot C_j^{z_i/z_j})} \end{aligned} \tag{18.7}$$

S_N is $2.3RT/z_iF$ and S is the slope along the curve. Z is the relevant valence and i is the measured ion and j the competing. In order to achieve this sort of measurement we need to know the starting position (voltage point) along the theoretical curve. One can calculate the starting voltage point on the theoretical curve by determining the slope in the bath empirically and solving for E in Equation 18.7.

$$E = S_N \log\left(\frac{S_N \cdot K_{ij} \cdot C_j^{z_i/z_j}}{S_N - S}\right) \tag{18.8}$$

The unknown S in the denominator is determined by calculating the slope between two different concentrations of the primary ion in the working solution. More points can be used to validate the starting position along the curve. From this point on the theoretical curve, a small change in measured voltage, ΔE, either up or down the curve, can be used to determine both the new slope at the near pole from Equation 18.7, and the concentration in the near pole from Equation 18.4. The differential concentration can now be determined as the concentration in the far pole and near pole are known.

The final case to be considered is measurement of a gradient of a primary ion coexisting with a gradient of an interfering ion. If a relatively ideal ISE can be used to measure the interfering ion then the situation discussed above with Equation 18.7 can be used to solve for the primary ion flux. However, if the ISE used to measure the interfering ion can also sense the primary ion then no simple electrochemical method to determine the absolute flux of either ion exists. Selective pharmacological block of targeted transporters can go some way to solving this problem by removing ion specific flux values. However, this is dependent on prior knowledge of the system.

TYPES OF POTENTIOMETRIC ION-SELECTIVE MICROELECTRODES

Many more ISE designs, based on liquid membranes, exist than have been used in a self-referencing mode. Lists are available in Umezawa et al.[29] for inorganic cations, and Bühlmann et al.[30] with 30 available commercially from Sigma/Fluka. Calbiochem is another source. Fluka published data sheets for the ion selective solvents they offer.[31] As this publication is no longer generally available Table 18.1 gives the details for the ion-selective solvents that have been used in a self-referencing mode and unless stated otherwise this material comes from Fluka.[31] These data are also included to highlight the fact that all sensors are selective not specific. This critical point will be discussed below.

Not dealt with in Table 18.1 are interferents frequently used to block or activate molecular transport. This is a complex issue with no simple solution; examples are given for calcium detection[15] and for chloride detection.[25,32] Detecting a chloride flux over the actions of blockers or

TABLE 18.1
Selectivities, Concentration-Dependent Slopes and Response Times for the LIXs Used in a Self-Referencing Mode

Calcium

Fluka calcium ionophore I-Cocktail A. It has the following selectivity values determined by the fixed interference method:

$\log K^{Pot}_{CaNa}$	−5.5
$\log K^{Pot}_{CaK}$	−5.4
$\log K^{Pot}_{CaMg}$	−4.9

Electrode function: Slope of the linear regression is 28.1+/−1.8 mV (10^{-2} to 10^{-7} $CaCl_2$ determined in calcium buffered solutions at a constant background of 125 mM K^+). With a tip diameter between 1–2 μm the expected resistance is in the range of 2×10^{10} Ω. Data from Fluka (1996). In our applications with larger tips (2–4 μm) and short column lengths (30 μm) we record a resistance of $2\text{–}3\times10^{9}$ Ω. Response times evaluated at the BRC give a mean value of 48 ms to 90% response (see below)

Chloride

Fluka chloride ionophore 1-Cocktail A. It has the following selectivity factors as determined by the separate solution method

$\log K^{Pot}_{ClHCO_3}$	−1.5 [31]
	−0.9 [32]
$\log K^{Pot}_{ClAcetate}$	−1.3 [31]
	−1.2 [32]
$\log K^{Pot}_{ClGlutamate}$	−3.2 [32]
$\log K^{Pot}_{ClGlutamate}$	−3.0 [32]
$\log K^{Pot}_{ClCitrate}$	−1.6 [32]
$\log K^{Pot}_{ClThiosulfate}$	−0.8 [32]
$\log K^{Pot}_{ClSO_4}$	−2.6 [31]
$\log K^{Pot}_{ClSCN}$	3.4 [31]
$\log K^{Pot}_{ClSalicylate}$	3.0 [31]

Electrode function: Slope of the linear regression is 57.5+/−0.5 mV (0.001–0.5 *M* NaCl on 0.01 *M* Tris/H_2SO_4, pH 7.4). The resistance for an approximately 1 μm tip will be in the region of 7×10^{10} Ω. These data are taken from [31]. The 90% response time is given as 1.8 s [32]

Hydrogen

Fluka hydrogen ionophore 1-Cocktail B. It has the following selectivity factors determined by the fixed interference method:

$\log K^{Pot}_{HNa}$	−10.4
$\log K^{Pot}_{HK}$	−9.8
$\log K^{Pot}_{HCa}$	<-11.1

Electrode function: Slope of the linear regression is 58.0+/−0.4 mV pH range 5.5–12.0. The resistance for an approximately 1 μm tip will be in the region of 1×10^{11} Ω. These data are taken from [31]

Response times evaluated at the BRC give a mean value of 88 ms to 90%.

Potassium

This cocktail has the following selectivity values based on the separate solution method:

$\log K^{Pot}_{KLi}$	−4.2
$\log K^{Pot}_{KNa}$	−3.9
$\log K^{Pot}_{KMg}$	−4.6
$\log K^{Pot}_{KCa}$	−4.9
$\log K^{Pot}_{KAcetylcholine}$	−4.9

Electrode function: Slope of the linear regression is 58.8+/−1.2 mV (20°C:10^{-4} to 10^{-1}). The detection limit of this ionophore is given as $\log a_K = -5.0$ against 140 mM Na^+. These data are from [31]. The response time to 95% is 41–77 ms depending on the change in concentration [14]

(continued)

TABLE 18.1 *Continued*

Cadmium	
These data are taken from Pineros et al. (1996). Reference should be made to this paper for ISE preparation and testing.	
Selectivity based on the separate solutions method:	
$\log K^{Pot}_{CdZn}$	−2.9
$\log K^{Pot}_{CsPb}$	−4.6
$\log K^{Pot}_{CdCu}$	−4.8
$\log K^{Pot}_{CdMn}$	−5.2
$\log K^{Pot}_{CdFe}$	−8.6
$\log K^{Pot}_{CdNi}$	−10.4
$\log K^{Pot}_{CdCa}$	−10.8
$\log K^{Pot}_{CdMg}$	−12.2
$\log K^{Pot}_{CdNH_4}$	−6.2
$\log K^{Pot}_{CsNa}$	−7.1
$\log K^{Pot}_{CdK}$	−7.9

interferents is particularly interesting and has led to erroneous data getting into the literature.[14,33,47] The BioCurrents Research Center is compiling a web site database highlighting known problems (www.BioCurrents.edu) but any investigator is advised to conduct tests using their own system and solutions.

An interesting design for the detection of Cu fluxes has also been published.[34] As in Gow et al. self-referencing movement was manually controlled.[10]

Choosing an Electrolyte

Before the front filling of the ISE with a column of ion-selective solvent, the micropipette is back filled with an electrolyte. The back filling solution is used to provide electrical contact between the back surface of the ion-selective solvent and the voltage-sensing node, via a Ag/AgCl wire. Usually a 100 mM solution of the ion being measured is used as the backfill to ensure that the ionic concentration between the back of the ion-selective solvent and the backfilling solution does not change significantly. However, it is now apparent that regulating this parameter can significantly alter the performance of the ISE.

Bakker and Meyerhoff review the phenomenon that the ion of interest from the back filling solution can easily move from the solution through the membrane into the bulk.[28] This will raise the local ion concentration effectively "blinding" the electrode to small activity changes at the tip. Clearly, there is a need to re-examine the use of more dilute solutions, and/or the use of ion buffers, in the electrolyte. Sokalski et al. show that the addition of buffer to the inner solution yielded picomolar detection levels for Pb^{2+}.[35] The influence of the electrolyte composition has also been demonstrated for K^+ detection using valinomycin.[36] Additionally, the backfilling solution should be matched osmotically to the bath solution. Osmotic gradients across the ion-selective solvent can lead to its movement into or out of the electrode tip, either causing the electrode to fail or an increase in noise (unpublished observation—Mark Messerli).

SELF-REFERENCING: AMPEROMETRIC MICROELECTRODES

Researchers have used amperometric electrodes in the self-referencing mode, taking advantage of the features discussed above for the potentiometric electrodes. To date all of the amperometric microelectrodes have been used in a constant potential mode. The electrode is clamped at a specific holding voltage to oxidize or reduce the analyte. This type of design and self-referencing application is termed self referencing potentiometric: SERP. Electrodes are built so that measurement of the analyte is limited by diffusion of the analyte itself, attempting to ensure a current dependent on the analyte concentration and not the concentration of a secondary factor or byproduct. The current generated by this redox reaction is, therefore, proportional to the concentration of the analyte. In the special case of enzyme-assisted electrodes (SERE) using glucose oxidase to measure a concentration difference of glucose, a by-product of the enzyme-substrate reaction, H_2O_2, is measured.[37]

For an ideal electrode under constant holding potential the measured current is proportional to the concentration of analyte. During self-referencing the differential current can be used to give a measurement of the differential analyte concentration:

$$I_1 - I_2 = (I_0 + S \cdot C)_1 - (I_0 + S \cdot C)_2$$

$$I_1 - I_2 = S(C_1 - C_2) \tag{18.9}$$

$$\Delta C = \frac{\Delta I}{S}$$

I_1 and C_1 are the current and concentration of the analyte in the near pole of excursion. The subscript 2 has been used to identify the same parameters at the far pole of excursion. S is the slope of the calibration curve. Unlike ISEs, knowing the background concentration of the analyte is not necessary to determine the differential concentration due to the linearity of the current–concentration relationship.

Types of Amperometric Microelectrodes

The amperometric microelectrodes are solid-state, with a core of platinum, gold, or carbon as the electrolytic surface. The BRC has developed and published the application of Amperometric, and enzyme-assisted, self-referencing electrodes for the detection of oxygen, NO, hydrogen peroxide, ascorbate, glucose and glutamate.[37–43] Mancuso et al. independently published this approach for oxygen detection.[44] Oxygen electrodes produced by depositing thin layers of gold on optical fibers are also possible.[20]

Calculation of Flux

To quantify the analyte movement in the bulk the differential concentration measurement is converted to flux. Not only does flux give a value proportional to transporter activity but it can also be used to calculate the total amount of analyte uptake or release by integrating the flux over space and time. Flux takes into account the diffusion coefficient of the analyte being measured, the distance over which the differential concentration measurement is acquired, the surface geometry of the source and the distance from the source. For cases where the measuring electrode is relatively close to a large source of the analyte and the differential concentration is measured over a small distance Δx within the gradient, the source can be modeled as a planar source so the flux (J) is:

$$J = -D\frac{\Delta C}{\Delta x} \tag{18.10}$$

where D is the diffusion coefficient of the measured analyte. By this model, the flux measured at some distance from the source is the same as the flux at the surface of the source. What constitutes a planar source? Kochian et al. found that the planar flux calculation is adequate for H^+ fluxes determined between a 30 μm distance, within 100 μm of a 1 mm diameter plant root.[45] However, for smaller cells or tissues, the geometry of the source must be taken into account.[46,47]

In order to determine flux at the cell surface for known surface geometries it is useful to calculate analyte flow i.e., the quantity of substance (Q) moving per unit time.[46] Flow is the same for all concentric regions from the source surface. Flux at the source surface is the flow divided by the surface area of the source. Therefore, radially emanating flow from a cylindrical surface is:

$$\text{Flow} = \frac{Q}{t} = \frac{2\pi D}{\ln(b/a)}(\Delta C) \qquad (18.11)$$

where D is the diffusion coefficient of the analyte and a and b are the radial distances between the center of the cylinder and the electrode tip at the near and far poles, respectively. Analyte flux at the surface of the cylinder is then determined by dividing by its surface area $2\pi r^2$. A caveat to this approach is the assumption that the flow is equal at all points around the cylinder and along the shaft of the cylinder. An alternative is to calculate flux per unit length by dividing by $2\pi r$.[46]

The flow from a spherical source is:

$$\text{Flow} = \frac{Q}{t} = 4\pi D\frac{ab}{b-a}(\Delta C) \qquad (18.12)$$

Flux at the cell surface can then be determined by dividing by the sphere surface area $4\pi r^2$. These equations have been adapted from Crank.[48]

A similar approach has been used in modeling a cluster of Hamster Insulinoma Tumor cells as an oblate spheroid to calculate the uptake of glucose.[37] The calculated glucose consumption per unit volume was 61.7 ± 9.5 fmol $\text{nl}^{-1}\,\text{s}^{-1}$. This value is in close agreement with the results (48 fmol $\text{nl}^{-1}\,\text{s}^{-1}$) using an absorption method and a mouse tumor cell line—EMT6/Ro.[49]

Some researchers have chosen to forgo flux calculations as they are only interested in changes of the concentration measurement for cells and tissues under different conditions.[50,51] Given that the cells being used have relatively similar surface geometry, and that the electrodes are positioned at a similar position with respect to the cell surface, this simpler method is adequate.

Setting the Frequency and Distance of Translation

Unfortunately there is no simple rule concerning the selection of the correct translational frequency and excursion distance. These are influenced by the specific conditions of cells and electrodes. Put simply, an investigator should ensure that both poles of translation are within the gradient—preferably over a "linear" portion—and that the frequency is adjusted to allow a reasonable sampling period but one that avoids the impact of drift. The electrode must also have time to fully respond before data is collected. Messerli et al. deal with these matters in more depth.[14] Within the BRC default starting parameters of 0.3 Hz over a 10 μm translation are usually used.

ELECTRODE CONSTRUCTION

Below are described the basic methods for making both potentiometric ion-selective solvent based electrodes and amperometric solid-state electrodes. With the correct apparatus, making ISEs is simple, applications being limited primarily by suitable ion selective cocktails (see above). Preparation of solid-state electrodes, at the dimensions and stability needed for self-referencing, is a different matter. The production of both metal and carbon core electrodes is tedious, with calibrations (see below) only indicating which electrodes *may* be useable in a self-referencing mode. Whereas, for example, an oxygen electrode calibrated by both dipping in solutions with different pO_2, or by cyclic voltammetry at different pO_2, can show a good DC response, the same electrode may prove too unstable or noisy for self-referencing applications. The basic electrode designs are

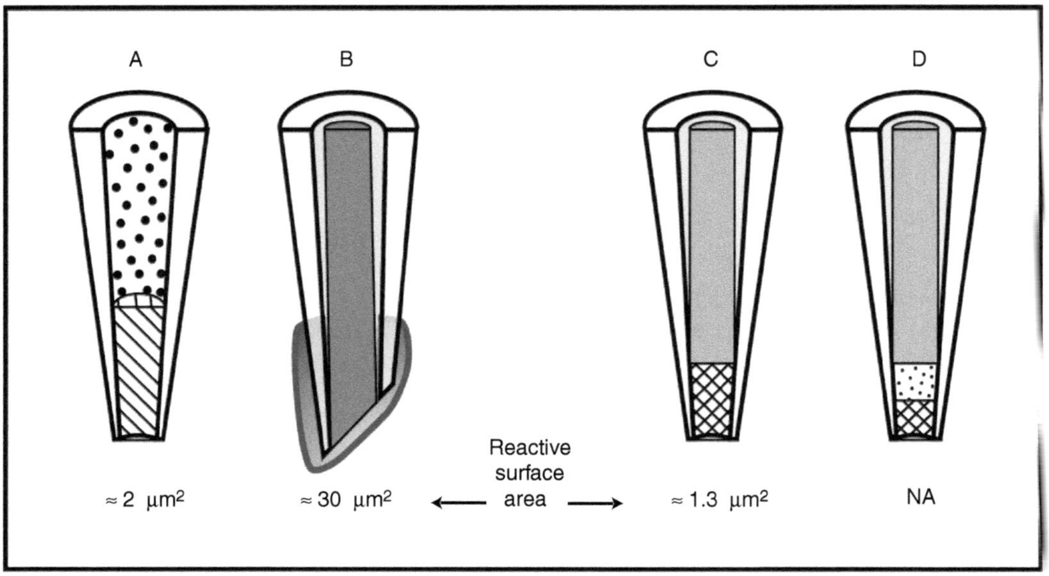

FIGURE 18.5 Diagrammatic representation of the potentiometric (A) and amperometric (B–D) electrode designs. In A, the electrode is made up of a pulled pipette backfilled with electrolyte and front filled with a short column of ion-selective solvent. Figure 18.4 shows a real example for the 30 μm column of Ca^{2+} selective solvent. The electrode design shown in B is a carbon fiber type as used for ascorbate and NO. The beveled tip has been coated with membranes to impart some degree of selectivity—see "Electrode Construction." The electrode types illustrated in C and D have a platinum core etched back to form a recess and closed with a cellulose acetate membrane. C is a design used for O2 detection. D has the recess pre-filled with platinum spheres and an enzyme before capping. The latter design has been used for glucose detection. Electrode reactive surfaces are also given. These are approximate as tip diameters can vary.

shown in Figure 18.5, along with the reactive surface areas. Note that, simplistically, tip diameters do not represent reactive surface areas; it is the exposed area of the electrode that counts.

Electrode Construction: Potentiometric

The manufacturing of micron-sized ISEs used in biological systems has changed little in two decades and a detailed description of the process, as applied to self-referencing ISEs, is available.[15] Briefly, this method involves pulling borosilicate glass capillaries, without fibers, to the appropriate taper and tip size, 1–4 μm, using a Sutter P-97. 1.5 mm outer diameter, thin walled glass is normally used (World Precision Instruments TW150-4). The glass should be handled with non-powdered latex gloves. The pipettes are made in batches of around 50 and dried at >200°C for at least 2 h. Hydrophobicity is produced by vapor silanization using *N,N*-dimethyltrimethylsilylamine (Sigma–Aldrich/Fluka cat # 41716). 50 μl is injected under the lip of glass beaker placed over the electrodes, standing vertically in a custom built rack (15 min: BioCurrents Research Center (BRC), Marine Biological Laboratory (MBL), Woods Hole, MA). Silanized pipettes can be stored for several days at 200°C or at room temperature in a container with desiccant.

Electrodes are made by backfilling the pipette with the appropriate electrolyte under slight positive pressure and then tip filling with the appropriate column length of ion-selective solvent from a micropipette reservoir (Table 18.2). The electrolyte column is kept short-≈5 mm. This process is observed under the microscope and the two pipettes are aligned before applying pressure to backfill. It is particularly important that after backfilling, the tip is brought to the reservoir quickly or else the tip blocks.[15] Researchers choose the column length primarily for stability of

TABLE 18.2
Characteristics and Response Times Determined for a Variety of Electrochemical Microsensors under Different Conditions. All Tip Diameters and Reactive Areas Are Approximate

Electrode	Column Length	Tip Size	Medium	Response Times (t_{95}% ms) for Concentration Ranges			
				10–1 mM	1–0.1 mM	0.1–1 mM	1–10 mM
K^+	100 μm	1 μm	1	195±59	376±86	165±41	114±26
	1000 μm	1 μm	1	369±91	516±86	247±54	191±46
	100 μm	2–3 μm	1	41±3	77±4	53±4	44±3
	100 μm	2–3 μm	2	64±9	225±6	91±8	69±15
				pH 6–7	pH 7–8	pH 8–7	pH 7–6
H^+	30 μm	1 μm	3	209±12	220±15	214±27	202±13
	300 μm	1 μm	3	251±21	269±14	245±11	244±17
	30 μm	2–3 μm	3	135±23	130±32	126±30	124±35
				10–1 mM	1–0.1 mM	0.1–1 mM	1–10 mM
Ca^{2+}	30 μm		4	58±9	81±10	48±7	53±10

Electrode		Surface Area	Medium	Response Times (t_{90}% ms) to Different Partial Pressures.	
				Air—N_2 saturated	$N_{saturated}$—Air
O_2	n.a.	2 μm^2	4	17.95±5.26	17.03±5.75
				Saline—NO bubbled	NO bubbled—Saline
NO	n a.	30–40 μm^2	4	49.63±13.21	61.00±14.33

Medium 1. 100 mM HEPES (pH 7.0) with 0.1, 1.0 or 10 mM KCl Medium 2. Medium 1 with 450 mM NaCl Medium 3. 100 mM MES (pH 6), 100 mM HEPES (pH 7,8) set with KOH Medium 4. In mM, 120 NaCl, 5 KCl, 2 $CaCl_2$, 2 $MgCl_2$ 10 HEPES (pH 7.4).

the solvent in the pipette tip. For the examples given in Table 18.2, the H^+ and Ca^{2+} electrodes are made with a 30-μm column of solvent while the K^+ electrode is made with a 100–150 μm column length. This provides a more stable potassium electrode. Shorter lengths and the K^+ ion selective solvent is lost. The reason for these stability differences is probably the physical properties of the solvents used in the ion-selective cocktail.

The solvent used for H^+(Sigma–Aldrich/Fluka cat #95293-1EA) and Ca^{2+}(Sigma–Aldrich/-Fluka cat # 21048-1EA) cocktails is *o*-nitrophenyl-*n*-octyl ether, which is 3.9 times more viscous, and nearly 3 orders of magnitude less soluble in water than 2,3-dimethyl-nitrobenzene.[52] The latter is used as the K^- -selective solvent (Sigma–Aldrich/ Fluka cat # 60398-1EA-F). There is certainly flexibility in the column length of the electrodes. ISEs are made fresh for each experiment.

A Ag/AgCl wire connects the electrolyte to the preamplifier via a standard electrode holder, available from several companies (for example Warner Instrument Corp: QSW-B15P). Clean silver wire is plated with chloride by passing 9 V, through the cathode via a 1 KΩ resistor, while immersed in a 1 *N* HCl solution. The process takes approximately 30 s. An Ag/AgCl reference electrode (MEH3815: World Precision Instr. Sarasota FL), connected via a 3 *M* KCl/5% agar bridge placed in the bulk solution, completes the circuit. The bridge is made with sodium acetate for experiments where K^+ ISEs are used.

ISEs should be calibrated in different concentrations of the targeted analyte made up in the physiological solution to be used in the experiment. A good practice is to use three concentrations, each an order of magnitude apart, that bracket the expected concentration in the saline to be used.

For example, calcium might be used at 0.1 mM, 1.0 mM, and 10 mM. For H^+ sensors, unless there is a reason to expect problems from an unusual saline, calibration in commercially available buffer solutions are adequate provided a good pH value is determined on immersion in the saline at the start of the experiment. Under these conditions, the output from any ion selective solvent-based ISE should be close to Nernstian (see above). Extreme deviations indicate a failed electrode or the presence of a competing analyte in the saline to be used. Adding pharmacological compounds to these solutions is a simple and straightforward way to check for that type of interference. Garber et al. and Messerli et al. give good examples of this.[14,32] Electrodes should be calibrated before and after each experiment.

An additional calibration step is to compare the self-referenced output to a semi-empirical, modeled, gradient. Smith et al. explain this for the ISEs.[15]

Electrode Construction: Amperometric

A number of solid-state amperometric microelectrode designs have been used in the self-referencing mode (Figure 18.5) and will be discussed below. A Ag/AgCl reference electrode, connected via a 3 *M* KCl/5% agar bridge placed in the bulk solution, completes the circuit. Reactive surface areas for the different designs are shown in Figure 18.5.

Making an Oxygen Electrode

Platinum wire, 25 μm in diameter, (1.5 cm length: Johnson Matthey, Ward Hill, MA) is straightened by rubbing with cardboard against a flat surface. These wires are then glued into 27 gauge hypodermic tubes (5 cm long: Small Parts, Inc., Miami Lakes, FL) with silver epoxy (Epoxy Technology, Inc., Billerica, MA) so that the Pt wire extends $\approx$1 cm from the tip. The wire is then electrochemically etched in an aqueous solution of 3 *M* KCN and 1 *M* NaOH by applying square waves of 4.0 V amplitude and a 4 ms period (pulse generator model 114 Tektronix, Inc., Portland, OR) until the tips are reduced to a fine point of $<$1 μm. The etched Pt wires are rinsed with water and isopropyl alcohol and inserted into borosilicate glass capillaries (1B150, WPI) pulled to outer tip diameters of $\approx$3μm using a Sutter P-97, so that the wire protrudes slightly from the pipette tip. The electrodes are dipped into epoxy (Epon 828 with *m*-phenylenediamine curing agent; Miller-Stephenson, Danbury, CT) and heated to 90 ± 6°C on a hot plate. The electrodes dry overnight. At this stage the epoxy has sealed the Pt to the glass. A thin layer of epoxy probably coats the Pt protruding from the glass. Briefly dipping into acetone cleans the Pt. The epoxy is then cured by baking at 100°C for 2 h and 150°C for 2 h. The exposed Pt electrodes are etched again (same conditions as above) to form a recessed electrode with a cavity 2–3 μm deep. This etching takes no more than 5 min and can be monitored by examining the electrode under a microscope every 60 s. Finally, the electrodes are coated by dipping in a solution of 10% cellulose acetate (30 kDa: w/v in tetrahydrofuran, THF) for 60 s and drying for 10 min. In a more recent paper THF is replaced by acetone.[37,53,54]

For oxygen detection, the electrodes are polarized to -0.6 V. Electrodes should be calibrated before and after each experiment by making measurements in nitrogen-bubbled (2–5 pA dark current) and normal air saturated salines (70 pA: 210 μmol.l^{-1}). When polarized incrementally from 0 to -1.0, the output in air-saturated saline should demonstrate a plateau between -0.6 and -0.8 V at $\approx$70 pA. Oxygen electrodes have a rapid response time (see below) and are normally used at a modulation frequency of 0.5–0.3 Hz over 10–20 μm. These parameters are, however, case specific. Land et al. present a measured versus a semi-empirically derived gradient.[38]

Making a Hydrogen Peroxide Electrode

H_2O_2 microelectrodes are prepared as described above for oxygen sensors. The final tip diameter of the electrode is about 3 μm. The microelectrode is polarized to $+0.6$ V against a Ag/AgCl reference electrode with a sensitivity of 0.85 ± 0.12 pA/μM (mean ± SD, $n = 4$). Although the electrode can potentially detect other reactive oxygen intermediates beside H_2O_2 (such as NO and O_2^-), in the applications cited, H_2O_2 is the major component of the concentration gradient, considering its longer half-life and the composition of the media (< 100 μM L-arginine).[40,41]

Making an Ascorbate Electrode

This involves sealing a 5 μm carbon fiber (Amoco, Greenville, SC) in a glass microcapillary. Heating and pulling the capillary draws it down to make contact with the fiber. The fiber is stabilized and sealed in the pulled glass pipette using epoxylite (Epoxylite Corp., Westerville, OH). After the epoxylite is cured in an oven at 110°C for 5–10 h, the electrode is backfilled with a graphite/epoxy paste (PX-grade Graphpoxy, Dylon Industries, Cleveland, OH). A copper wire can then be inserted to make electrical contact with the carbon fiber through the paste. The paste is cured (110°C for 5–10 h), and the excess carbon fiber chopped with a scalpel and beveled to 30° using a Narishige Model EG-44. This generates a comparatively large reactive surface area of ≈30 μm². The electrodes are then electrochemically treated, based on previous studies. A 70 Hz triangle wave that cycles between +3.0 V and +1.5 V (7 s) is followed by a constant potential of +1.5V (7 s), then held at 0.0 V for 30 min. The electrochemical treatment decreases the polarization voltage required for oxidation of ascorbic acid down to the 0.15–0.3 V range and delivers a reasonable selectivity for ascorbate under the conditions described by Pepperell et al.[42]

Experimentally the ascorbate microelectrodes have an applied voltage of 0.3 V, conferring good selectivity for ascorbic acid, and is oscillated over a distance of 30 μm at a frequency of 0.3 Hz.[42]

Making a Nitric Oxide Electrode

The construction of the electrode body is the same as for the ascorbate electrode described above. To impart selectivity for the oxidation of NO, the carbon fibers are treated with Nafion and o-phenylenediamine (OPD).[55] Nafion is a polysulfonated Teflon carrying an intrinsic negative charge that repels electrochemically active anions (e.g., nitrate, nitrite, and ascorbate). OPD is an electrochemically active material imparting a degree of selectivity for NO. The exposed carbon fiber surface is first coated with Nafion (5% in aliphatic alcohols, Aldrich cat #27,470-4) and dried at 110°C for 5–10 min. After two additional Nafion coatings, the fiber is plated using a 5 mM OPD plating solution containing 0.1 mM ascorbic acid in 100 mM PBS (pH 7.4). The OPD, prepared fresh for each use, is plated on the fiber at a constant +0.9 V until a stable current is obtained.[39]

The modified carbon fiber electrodes have final tip diameters of 7–8 μm, but after beveling have a reactive surface area of approximately 30–40 μm², and are operated at +0.90 V vs a Ag/AgCl return electrode, which completes the circuit in solution via a 3 *M* KCl/5% agar bridge. While the sensitivity of the NO electrodes for sodium nitrite is 1.43 ±0.16 pA/mM, the sensitivity for NO is 730 ±0.11 pA/mM.[39] Therefore, the NO electrodes are ≈500 times more sensitive to NO than nitrite.

NO electrodes can be calibrated with a standard 2 mM NO solution prepared by bubbling HEPES buffered saline with argon gas for 30 min and then compressed NO for 40 min in a fume hood. The saturation concentration of NO in aqueous media at 22°C is 2 mM (see Porterfield et al. 2001). Porterfield et al. present a comparison of the modeled versus measured NO gradient generated using the NO donor *S*-nitroso-*N*-acetylpenicillamine (SNAP). The modeling equation includes a correction for the ½ life of NO in an aqueous solution.[39] The electrode is operated at 0.3 Hz over a translational distance of 10 μm.[39,56]

Making a Glucose (Enzyme-Assisted) Electrode

Glucose microsensors are fabricated in the first instance as described above for oxygen electrodes. The exposed Pt wires are etched to form a cavity ≈5 μm deep. Once etched, Pt particles are electrochemically plated on the recessed Pt wire at −0.2 V vs Ag/AgCl in a solution of 10 mM hydrogen hexachloroplatinate (IV) hydrate (Cat # 20,608-3: Aldrich, Milwaukee, Wisc.) for 3–5 min. The amount of charge required for platinization is $1.2 \pm 0.3 \times 10^{-5}$ coulombs (mean ± SD, $n = 26$). Glucose oxidase (GOx: EC 1.1.3.4, type X-S from *Aspergillus niger*: Sigma, St. Louis, MO) is then loaded onto the platinum particles by immersing the tip of the electrodes in 1% (w/v) GOx solution for 10 min. The enzyme is immobilized with 10% glutaraldehyde vapor for 1 min. The immobilized enzyme layer is then coated with 10%, 30 kD, cellulose acetate (w/v in acetone) by dipping (10 s) and drying (60 s), repeated 4 times. The final tip diameter is ~8 μm.[37]

Calibration of the glucose electrode, polarized to +0.6 V, is against 2, 5 and 10 mM glucose in 50 mM/l HEPES and 150 mM NaCl at pH 7.4 and 37°C. Experimentally the glucose electrode is operated at a frequency of 0.2 Hz over a 10 μm excursion. The speed of response to 90% is 0.46 s.[37]

Making a Glutamate Electrode

Microelectrodes are fabricated in a manner similar to oxygen sensors, except that 8-μm carbon fiber, 12-μm gold wire, and 10- and 25-μm platinum wire were tested as a reactive surface. Electrodes were dip-coated with Os-gel-HRP redox polymer (BAS, W. Lafayette, IN) and allowed to dry for 10 min. Electrodes were then dip-coated in 50 units/ml glutamate oxidase from *Streptomyces* sp. (Cat. # G-0400, Sigma) and allowed to dry for 10–30 min. Glutamate electrodes were polarized to either 0 or +100 mV against a Ag/AgCl reference electrode, connected via a 3 M/l KCl/5% agar bridge placed in the bulk solution, to complete the circuit. The electrode was operated at a frequency of 0.1 Hz over a translational distance of 25 μm. The average response of platinum-based electrodes was 0.45 pA/μM of glutamate, whereas gold-based electrodes produce a weaker signal of 0.21 pA/μM, and carbon gave the weakest signal, 0.1 pA/μM. This electrode has not yet been used on a biological preparation.[43]

RESPONSE TIMES

Sensor response times in rapidly changing flowing media frequently reveal values to 90% of the full response in the 10–100 ms range (see below and Table 18.2). These numbers are similar to some reports in the literature[57] but contradict others where response times in seconds have been reported.[58,59] As cited by Bakker et al.[27]—"Responses of ISEs can be so fast that the electronic equipment may become limiting ....".[52] Similarly, millisecond response times for amperometric electrodes have been predicted by Schneiderman and Goldstick.[60] An increase in temperature can further decrease the response times of various ion selective sensor designs, an effect attributed to alteration in the diffusion coefficient or relative ionic mobilities.[57]

Response time tests are conducted by changing the analyte activity in the bulk medium surrounding the microelectrode. To change solutions, work at the BRC uses an automated, parallel, flow-through system (Warner Instruments Corp.). This device allows three parallel streams of test solution to be passed over the sensor, with switching taking place in less than 10 ms for the complete movement. The microelectrode will see a faster event as it crosses the interface between the two streams. A manifold allows rapid changing of each stream to other solutions with minimal dead space. Using this system solutions of known composition, with or without potential interferents/toxins being present, can be passed over the reactive surface of the electrodes. Under these experimental conditions, the unstirred boundary layer at the bulk surface of the electrode will be minimal. This layer determines much of the limitation in response times for neutral carrier ISEs.[27,61] If electrodes are held stationary in a non-flowing analyte, this layer will be thicker.

Response Times: ISEs

Messerli et al. present data from response time testing on several ISEs (see Table 18.2).[14] All fall in the millisecond range. Micron-sized, ISEs have high resistances; 1–20 GΩ.[61] Amplifier input impedances of 10^5–10^6 GΩ are typically used in conjunction with these electrodes to minimize current drawn and allow sampling of the ion concentration without significantly changing the local ion concentration. If the amplifier is ideal, what determines the speed of response for an ISE? Diffusion through the stagnant aqueous layer at the surface of the electrode defines the response time of the electrode when the membrane is equilibrated with the ion to which the electrode responds (Bakker et al. 1997). To put this in perspective consider the following theoretical analysis. If a system is diffusion limited, how quickly will an analyte diffuse from the edge to the center of the microelectrode? For a K^+ ISE, with a tip diameter of 4 μm, it would require only 0.3 ms ($t = L^2/6D_{K+}$ for diffusion in 3 dimensions, $D_{K+} = 2.0 \times 10^{-5}$ cm/s and L is the distance traveled). If the electrode were 1 mm in diameter, this diffusion to the center takes 21 s. All electrodes used with self-referencing have tip diameters of less than 10 μm and diffusion coefficients in the range of 10^{-4}–10^{-6} cm²/s. Under these conditions the electronics will be the primary factor limiting the response time of the measurement under relatively ideal conditions.

Response Times: Amperometric

Tests using similar protocols to those described above also demonstrate relatively rapid response times for the amperometric electrodes (Table 18.2). Figure 18.6 gives an example for step change in the partial pressure of oxygen flowing past an O_2 electrode. The time to a 90% response is 17.95 ± 5.26 ms averaged for 4 electrodes. For a NO microelectrode (Figure 18.7) the values are 49.63 ± 13.21 ms averaged for 3 electrodes. Even the more complex glucose enzyme-assisted electrode reaches 90% response in ≈500 ms (0.46 ± 0.07 s).[37]

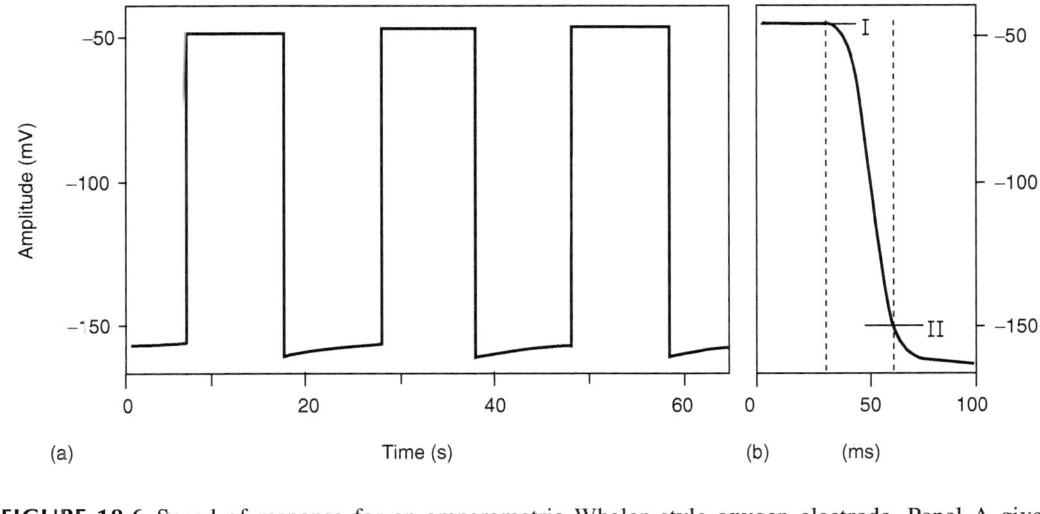

FIGURE 18.6 Speed of response for an amperometric Whelan-style oxygen electrode. Panel A gives examples of three exchanges between air equilibrated (more negative) and nitrogen bubbled. Panel B expands one exchange from low pO_2 to air-saturated pO_2. Position I marks the beginning of the exchange to a new value and position II the arrival at 90% of the response. For this example this occurs in approximately 25 ms. These tests were conducted in a standard animal saline with 10 mM HEPES buffer-pH 7 saline.

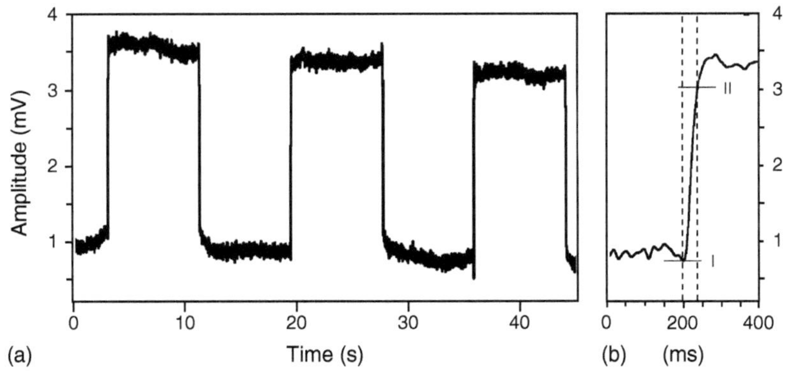

FIGURE 18.7 Speed of response for an amperometric nitric oxide electrode. Panel A gives examples of three exchanges between control and NO bubbled physiological saline. Panel B expands one exchange from control to NO bubbled. Position I marks the beginning of the exchange to a new value and position II the arrival at 90% of the response. For this example this occurs in approximately 50 ms. These tests were conducted in a standard animal saline with 10 mM HEPES buffer-pH 7 saline.

RESPONSE TIMES: INTERFERENTS

As the self-referencing method requires that the electrodes remain stationary at the two poles for finite periods, any change to the response time is a source of concern. Such changes do occur. A 2–3 μm K^+ -selective electrodes has a normal response time to 95% ($t_{95\%}$) of 41–77 ms depending on the change in concentration.[14] This ISE possesses $10^{3.9}$ selectivity for K^+ over Na^+, implying that a lot of Na^+ must be present to interfere with K^+ detection. However, animal salines can have significantly high levels of sodium. For example, in the presence of 450 mM Na^+(seawater levels) the response time increases nearly 3-fold.[14] Further, the Nernstian response decreases by 8%.[14] Both of these features are classic indications of an interferent.[27,62] However, given that the data collected during the first 30% of the cycle time is discarded, the slower response time need not affect the data collected.

SPATIAL RESOLUTION

Micron-sized electrodes provide good spatial resolution. For potentiometric electrodes the surface area of the ion-selective solvent in contact with the bulk medium defines this. The depth of sensing from the surface of a potentiometric electrode is considered negligible compared to the diameter of the ion-selective solvent interface, usually 1–4 μm. For example, if an electrode is being used to measure ionic concentration at a point near the surface of the cell, a 1–4 μm diameter electrode placed with the surface of the tip orthogonal to the cell surface will report the average concentration over the 1–4 μm tip diameter. Amperometric sensors used in constant potential mode, on the other hand, possess a sensing field that extends out into the medium away from the electrolytic surface. Take for example the O_2-sensing electrode. The electrode generates current by reducing O_2 at its sensing surface. Reduction of O_2 consumes O_2 in the region immediately in front of the tip creating an O_2 depleted zone. These O_2 depleted fields have been modeled for both macro and micro-electrodes with recessed and/or coated tips.[60,63] Carefully constructed electrodes, with recessed electrolytic surfaces and/or coatings, will limit the extent of this O_2 depleted field keeping the depletion zone restricted close to the tip. A well-constructed electrode can have virtually the entire

O_2 depleted zone located within the recess at its tip. Therefore, the choice of design and construction of amperometric electrodes will have the greatest influence on defining the sensing zone.

PROBLEMS AND PITFALLS

POTENTIOMETRIC: SELECTIVITY

It cannot be overemphasized that all the electrochemical techniques used in a self-referencing mode to date are selective not specific. There is always a finite chance that an interferent can be detected under experimental conditions designed for the primary analyte. As discussed above the differential approach inherent to self-referencing a microelectrode will greatly reduce the probability of simple background interferents being detected. However, unexpected problems can happen. Chloride detection using an ISE built with the Fluka Cl^- selective liquid ion exchanger cocktail A, is a good example. This ISE should more accurately be described as an anion sensor, even detecting a H^+ flux via sensitivity to the unprotonated anionic form of a Good buffer.[14,47] This ISE is also sensitive to a pharmacological agent used to block chloride transport, 4,4′-diisothiocyanatostilbene-2, 2′-disulphonic acid (DIDS).[47] Measurements of Cl^- with the more commonly used Cl^- ionophore I, cocktail A (Fluka Chemical Co) can sense or be poisoned by chloride channel blockers such as 5-nitro-2-(3-phenylpropyl-amino)-benzoic acid (NPPB), indanyloxyacetic acid 94 (IAA-94), and DIDS.[25,32]

Exposing an ISE to excitation light during imaging causes an additional area of interference. Hydrogen flux detection, while imaging the cytosolic pH sensor BCECF, is not possible with global illumination. In this case, the microelectrode couples to the excitation wavelength altering the Nernstian values and sensitivity (unpublished observation—Anthony Molina). Caged calcium has been released while using the amperometric oxygen microelectrode without adverse effect.[64]

AMPEROMETRIC: SELECTIVITY

Amperometric electrodes are also susceptible to the detection of multiple molecules. For example, NO oxidation at a carbon fiber begins to plateau between +0.8 and +0.9.[65] Neurotransmitters, ascorbic acid, nitrite and H_2O_2 can also be oxidized at this high voltage. A charged coating on the carbon fiber surface can eliminate passage of charged compounds.[55,65] However, H_2O_2, an uncharged molecule, will be oxidized as it can pass through the coatings. Addition of a layer of catalase to the charged coating layers, similar to the enzyme layers used by Csöregi et al.[66] can eliminate H_2O_2 before it reaches the electrolytic surface.

MIXING

As self-referencing involves movement there is an inevitable risk of disrupting the gradient. This disruption can be minimal if a sufficient amount of time is allowed to pass for the gradient to be reestablished after sensor movement. The slow movement of the self-referencing microelectrodes, usually 40 μm/s, does not cause turbulence that can be detected visually. Also, the Reynolds numbers calculated for the speed of movement of either the small tip, 1–4 μm diameter, or even up the shaft where the diameter reaches 50 μm, are in the range of 10^{-4} to 10^{-3}, respectively, indicating that only laminar flow occurs around the tip of the electrode. However, in the worst-case scenario, if gradient disruption does occur how long must the microelectrode remain stationary for the gradient to be reestablished? If we consider the case of reestablishing the potassium gradient, after a 10 μm move between the near and far pole, we find that it will take about 8.3 ms for K^+ to diffuse this distance. This indicates that an ionic gradient would be rebuilt in a few tens of milliseconds. As long as turbulent flow does not occur, most small analyte gradients will be rebuilt quickly.

POSITIONAL ARTIFACTS

Positional artifacts can be broken into two types. The first concerns access or egress of analytes to or from the microelectrode tip. Specific cases are discussed below. The second potential artifact comes from a change in the geometrical relationship between the electrode and the target cell or tissue. By definition, the amplitude of a differential signal will depend on the position from the source and the recording position within the gradient. If this is not controlled then apparent flux changes will occur that actually represent a changing electrode to target distance. Although the motion controller design developed by the BRC shows exceptional stability,[67] cells can move. This has constrained most studies to recordings made perpendicular to the visible edge of a cell or tissue, thus stabilizing distance by visual observation and software based manual adjustment.

Recently a new approach has been developed for the amperometric microelectrode that allows positional control through coupled electrochemical techniques.[21] In this case, an AC signal of 10 mV at 100 kHz is superimposed around the −0.6 V working potential of a self-referencing oxygen electrode. The monitored amplitude of this signal is dependent on the access resistance and, therefore, sensitive to the distance from an object. There is no interference between the two voltages and oxygen flux changes from a single INS-1 cell can be recorded while correcting for nanometer changes in electrode to target distance.[21]

POSITIONAL ARTIFACTS: POTENTIOMETRIC

The positioning of ISEs near surfaces can generate artifacts. Of particular relevance to the self-referencing technique is the possibility of ions emanating from the tip of the sensor and accumulating between the electrode tip and the target cell or tissue. The source of this potential artifact is a combination of current leakage from the electronics and zero-current (electroneutral) ion exchange of the primary ion for an interfering ion. This phenomenon has been shown for K^+ and Ca^{2+} -selective solvents.[36,57] The artifacts are most apparent in solutions of a low background primary ion. Reducing the concentration of the primary ion in the backfilling solution has been one compensation method.[36] Additionally, an applied current can control the flux across the membrane.[57] Pergel et al. used a current to compensate for the zero current phenomenon with Pb^{2+} detection.[68] None of these solutions have as yet been applied to self-referencing microelectrodes.

POSITIONAL ARTIFACTS: AMPEROMETRIC

In amperometric applications, the analyte is actually being consumed on the electrode surface. Therefore, an analyte depletion zone can form. An ideal electrode is built such that the analyte-depleted region is located primarily within the electrode itself. However, it can quickly be envisioned that even a small depletion of the analyte in the bulk can give rise to a measurement artifact. In the extreme case, placing the electrode up against an insulator will prevent any analyte from reaching the surface of the electrode resulting in no change in current for a change in analyte. Backing the electrode away from the insulator, such that analyte can now diffuse, albeit in a restricted manner, will still give rise to a value below the real level. Moving back further, such that diffusion of analyte is not impeded, will result in maximal current for a change in analyte concentration. This phenomenon has been observed with the H_2O_2 product of the glucose electrode reaction building up between the electrode and a cluster of Hamster Insulinoma Tumor cells.[54] In that specific case, the artifact was removed by the addition of catalase to the medium.

VOLTAGE FIELDS

In low conductivity medium, cellular currents can generate substantial voltage gradients next to the cells, coexisting with the chemical gradients of the analytes being measured. The potentiometric electrodes may sense these voltage gradients. Therefore, the voltage difference measured by the potentiometric electrode will be the sum of the voltage differences, due to the analyte concentration difference and the voltage difference, based on the current density and medium conductivity. In an animal system, however, the effect of voltage fields will be minimal due to the conductivity of the medium. Earlier studies using the vibrating voltage electrode confirm that these voltages are in the nV range and will not affect significantly the ISEs.[18]

Static surface charges exist within the glycocalyx of cells because of molecular and ionic charge distribution. These charges will project their electrostatic effects into the bulk medium in accordance with the Debye length, $1/\mathcal{K}$. In the case of a mammalian Ringer, this is 7.8 Å (0.78 nm).[69] All self-referencing microelectrode studies have been conducted at significantly greater distances.

USE OF BUFFERS

There are a number of circumstances that cause the calculated flux to be smaller than measured. Buffering is clearly one of them. In the example discussed here, extracellular buffers can collapse gradients of free analytes. This has been addressed for the collapsing of H^+ gradients by H^+ buffers.[33,70,71] The analyte can diffuse from the surface of the cell in either its free state or bound to the buffer. ISEs only measure the free concentration of the analyte. The actual H^+ flux from a source is the sum of the measured free H^+ flux and the unmeasured H^+ flux moving as H^+ bound to buffer. The standard equations described above allow us to obtain the flux of H^+ ions through an unbuffered solution. It is, however, still possible to calculate a value, x_i, to represent the ratio of H^+ flux bound to buffer and H^+ unbound, a method recently applied to visual horizontal cells.[70–72]

The ratio can be obtained using the equation:

$$x_i = \frac{D_B}{D_{H^+}} \cdot [B] \cdot \frac{Ka}{(Ka + [H^+])^2} \tag{18.13}$$

where D_B is the diffusion coefficient for protonated HEPES buffer ($0.62 \times 10^{-5}\ cm^2 \cdot s^{-1}$); D_{H+} is the diffusion coefficient for H^+ ions ($9.22 \times 10^{-5} \cdot cm^2\ s^{-1}$); and [B] is the buffer concentration in $mol \cdot cm^{-3}$, which in experiments on the horizontal cells was $2.0 \times 10^{-6}\ mol \cdot cm^{-3}$ (that is, equivalent to 2.0×10^{-3} M), and the value for Ka (the dissociation constant for the buffer) is $3.16 \times 10^{-11}\ mol \cdot cm^{-3}$.[72] Once the ratio is obtained, the total flux of H^+ can be calculated using the equation:

$$J_{Htotal} = J_{Hmeasured} + J_{Hbuffered} = J_{Hmeasured} \cdot (1 + x_i) \tag{18.14}$$

That is, the total hydrogen ion flux from a cell is equal to the sum of the hydrogen ions detected by the sensor plus the hydrogen ions bound to buffer, which is not detectable, by the sensor. In order to minimize the effect of buffer on H^+ efflux a buffer with a Ka below the $[H^+]$ should be chosen, such that most of the buffer will be in its protonated form reducing the effective buffering capacity. The case of multiple buffers can also be considered.[14,70]

In the absence of intentionally added H^+ buffers, water and dissolved carbonates can have a significant buffering effect at neutral to alkaline pH.[70] Water, at a concentration of about 56 M with a Ka of 10^{-16} begins to impact the correction factor above pH 6. The correction factor due to water acting as a H^+ buffer contributes to the total flux by less than 1% when the pH$\leq$6. Bicarbonate in solution, due to atmospheric CO_2, will also have more significant impact at neutral to alkaline pH. Bicarbonate will increase in concentration at more alkaline pH, and can add to the H^+ flux by about 2 and 21% at pH 7 and 8 respectively.[14]

EQUIPMENT AND SOFTWARE

The equipment and software described below are designed and assembled at the BioCurrents Research Center (Woods Hole, MA). All the development and microelectrode testing described above has been done using these designs. The basic components required to do self-referencing are the pre-amplifiers designed for potentiometric and amperometric detection, the main amplifier, a suitable PC with an analog to digital board, running the graphic user interface (GUI: IonView: Figure 18.2). This software also runs the motion controller interfacing with the 3-axis manipulator.

PREAMPLIFIERS

Voltage Head Stage Potentiometric

For potentiometric measurement the active element of the voltage head stage is an AD549 operational amplifier (op-amp: Analog Devices, Norwood, Ma.) configured as a voltage follower. This allows the microelectrode voltage to be sensed while drawing little current.

Current Head Stage-Amperometric

For amperometric measurements the components of the head stage are reconfigured to be a voltage clamp. The microelectrode is held at a command potential and the current that flows through the microelectrode is collected. An AD549 op-amp is configured as a current to voltage converter, having a 1 GΩ sense resistor. The command voltage is applied to the positive input of the op-amp and the microelectrode is attached to the negative input. An AD620 instrumentation amplifier is used to subtract the command voltage from the output of the op-amp leaving a voltage that is proportional to the microelectrodes current flow, times the value of the sense resistor, or 1 V per nA.

AMPLIFIER

Amplification of the position dependant portion of the microelectrode signal must be sufficient for digitizing with reasonable resolution. The analog signal from the electrode must be amplified enough to make the sub-μV changes in voltage resolvable by the analog to digital converter without exceeding the dynamic range of the converter. Currently in use is a 16-bit data acquisition system (Datatranslation, Marlboro, MA) with a dynamic range of ± 10 V. This gives 65536 possible numbers to represent a voltage in the ± 10 V range, or 305 μV per least significant bit change. To resolve μV changes, a gain of 1000 times is applied, giving 305 nV per step or least significant bit change. The problem here is that the microelectrodes background voltage is frequently high, up to 100 mV. Since the dynamic range is ± 10 V the signal will have to be offset closer to zero volts before the gain is applied. Many methods can be used to do this. Two are currently in use at the BioCurrents Research Center. In the first a resistor and capacitor (RC), having a time constant of 10 s, generates a value that is subtracted from the non-filtered signal before the gain is applied, thus constructing a high pass filter. This allows signals faster than the time constant to be amplified while the impact of DC levels, and signals slower than the time constant, are minimized. The second method in use is a sample and hold approach. Here the computer, either by timed control or manual control (using a software button), sends a digital signal to a sample and hold component in the amplifier, latching the real time analog voltage to be the new reference or offset value of the live signal. Gain can be applied. In the above example the gain needed, and therefore the maximum offset that can be tolerated without exceeding the dynamic range of the analog to digital converter, is determined by the resolution of the converter (number of bits). If a high enough resolution converter becomes available, no nulling of the offset would be necessary.

Motion Control

Microelectrode movement is accomplished by a robotic manipulator and driving electronics. The system designed and used within the BRC comprises 0.9-degree stepper motors (Oriental Motors, Torrance, CA) fitted to Newport 433 and 432 translation stages (Newport Corporation, Irvine CA). The motors drive a 100 pitch lead screw via a zero backlash flexible bellows coupling (Servometer, Cedar Grove, NJ). A motor, in turn, is driven by a 128 micro-step drive. Micro stepping increases the resolution of motion and smoothness of operation. The software increments or decrements the micro-step count to the motor and the shaft angle follows. Driving the motors with linear potential, rather than using the popular chopper method, reduces electrical interference but at the expense of motor torque. The system is capable up to several hundred microns per second at sub micron resolution and repeatability.[67]

Software

IonView is the name given to the BRC windows program used for data acquisition and motion control. It has a standard graphic user interface (GUI). The current version of IonView provides simple access to data acquisition parameters-choice of RC subtract versus sample and hold, distance and speed of translation, and the choice of signal averaging. Data are processed and presented on screen, in real-time. As the polarity of the signal recorded is entirely dependent on which pole is subtracted from which, by convention one pole of translation is designated the +ve pole, as selected by the angle of movement, and is always the near pole of translation. A separate drop down window handles electrode positioning.

APPLICATIONS

As the purpose of this paper is to introduce advances in a method, the many applications of self-referencing to biological systems will not be reviewed. A collection of references to publications from the BRC, and others, can be found at the BRC web site (www.mbl.edu/BioCurrents). However, selected studies and publications, relevant to the neurosciences, are briefly introduced below.

Neural Repair and Disease

Microglia

These reactive cells of both vertebrate and invertebrate nervous systems have been the subjects of three studies using self-referencing microelectrodes. Chromogranin A (CGA) is a recently identified endogenous component of neurodegenerative plaques in Alzheimer's and Parkinson's diseases. CGA stimulates microglial secretion of NO and tumor necrosis factor-α, resulting in neuronal and microglial apoptosis. Using an electrochemical H_2O_2 microelectrode, it was demonstrated that CGA produces a rapid oxidative burst in primary microglia in culture.[41]

Using both self-referencing K^+ and H^+ ISEs the presence of an K^+/H^+ -ATPase in the microglia, with a K_D in the physiological range of the $[K^+]_0$, has been reported.[50] The pump was ouabain insensitive but modulated by Omeprazole and Sch28080. In this study, comparisons were also made between current values measured locally with an ISE and those calculated from whole cell patch clamp as related to the inward rectifying potassium channel. The ISEs followed ~0.5 of the whole cell current. Better signal processing methods developed since this study, and taking into account the problem of the electronics speed-of-response (discussed above), should now result in better correspondence to the total cell current.

In an invertebrate system, the leech, damage to the interganglionic connectives is hypothesized to stimulate the activity of eNOS, with the resultant production of NO causing the accumulation of migratory microglia at the lesion site. The production of the NO was measured directly from the lesion site using the electrode design discussed above.[5]

Neural Tube Defects

One of the numerous complications of diabetes is the occurrence of embryonic congenital malformations, notably neural tube defects. Li et al. correlated this to the expression of Pax3 and tested the hypothesis that expression and embryopathy is related to a hypoxic state.[73] Hypoxia arises through diffusion limited oxygen delivery caused by excessive glucose metabolism. In this study, the self-referencing oxygen microelectrode was used to demonstrate that embryonic oxygen consumption was suppressed after exposure to high maternal glucose levels. Along with other experiments, it was concluded that a depletion of oxygen in the embryo gives rise to impaired embryonic gene expression and defective development through induced oxidative stress.[73]

Sensory Neurobiology

Hair Cells

The increase in K^+ flux through hair cells upon stimulation—whether by sound in the cochlea or acceleration in the vestibular labyrinth—is known to increase the perilymph $[K^+]$ on the basolateral side. Self-referencing K^+ ISEs, in conjunction with several other techniques, including micro-Ussing chambers, have followed these fluxes from both the strial marginal cells and the vestibular dark cells.[74–77] In both tissues, the potassium channel responsible for the flux is the slowly activated K^+ channel (K_{sK}). The laboratories involved in this research have also made extensive use of the older vibrating voltage probe. This device has not been covered in this review. Interested parties are directed to Lee and Marcus for applications to the current studies in the ear and to Nuccitelli for an applications overview.[78,18]

Hair cells of both the auditory and vestibular systems use calcium to control aspects of performance and tuning.[79] Self-referencing Ca^{2+} ISEs have been used to follow the role of the plasma membrane Ca^{2+}/H^+ ATPase (PMCA) in the regulation of this flux. A substantial calcium efflux could be measured, suggesting, in conjunction with pump density, that the PMCA activity should generate a substantial membrane current as the bundles expel cytosolic Ca^{2+}.[79]

Vision

Horizontal cells isolated from the retina of the skate have provided a model system in vision research. Self-referencing hydrogen-selective ISEs have been applied to examine the relationship between glutamate and regional changes in extracellular pH.[72] The results suggested that glutamate modulation of H^+ flux arises from calcium entry into the cells and the subsequent activity of the plasma membrane Ca^{2+}/H^+ ATPase. The data from this reduced preparation did not support the action of hydrogen ions as part of an inhibitory feedback loop but further experiments are underway (R. Paul Malchow, unpublished).

Olfaction

Olfaction in estuarine animals, moving through different salinities, poses unique problems. To sample for chemical cues the sensillae must be exposed to changing osmotic and ionic conditions. Gleeson et al. used ISEs selective for K^+ and Ca^{2+} to question whether at low salinities the integrity of the olfactory dendrites of the Blue Crab are sustained by a diffusion generated ionic microenvironment within the aesthetascs.[80,81] Both ISEs confirmed this hypothesis with a clear loss of these ions from the hymolymph at low salinities. Further, the flux rates fell well within the range estimated from $^{22}Na^+$ flux data from previous studies.[81]

ADDITIONAL ELECTRICALLY ACTIVE CELLS

Neurons

Isolated neurons have been the subject of several self-referencing studies although limited by technical aspects of the technique. With the availability of positional control, as discussed above, and faster temporal resolution, one can anticipate more work on isolated neurons or networks, in addition to those mentioned above. Knox et al. in an early study of an isolated neuron, observed calcium influx and activation of a cation current, coupled to intracellular Ca^{2+} release, in peptidergic neurons of *Aplysia californica*.[82] Ca^{2+} released from a thapsigargin-sensitive store activated a non-selective cation current that may sustain depolarization of the somata. Using the same cell, Duthie et al. measured calcium fluxes in oxidatively challenged neurons.[83] Preliminary data on oxygen consumption, at the level of a single neuron, has been shown and correlated to membrane potential under voltage clamp conditions.[84]

Muscle

Using self-referencing approaches on muscle are complicated by contraction. Contraction, particularly rapid contraction, will disturb the chemical gradient the microelectrodes are coupling to. However, under some conditions the technique can be successfully used, for example in muscle from the Sea Cucumber.[85] In this case Ca^{2+} flux changes were correlated with neurotransmitter application. Devlin also recorded Ca^{2+} flux modulation, to FMRFamide and 5-HT application, from the muscular trabeculae of the gastropod ventricle.[86,87]

A mammalian muscular system studied with self-referencing is within the cerebral arteries. Membrane depolarization from -65 to -35mV is associated with myogenic contraction. Depolarization is graded to transmural pressure and sufficient to act on the open probability of the dihydropyridine-sensitive Ca^{2+} channel, augmenting the calcium contraction mechanism.[25] Using a self-referencing Cl^- ISE arterial Cl^- efflux can be measured with a clear temperature dependent pressure component that correlated closely to myogenic contractility.[25] This study concludes that Cl^- efflux, through Cl^- channels, contributes to the depolarization of the muscle cell. An additional attribute of this study is the realization that the ISE is affected by Cl^- channel blockers, such as DIDS, NPPB, and IAA-94. Pharmacological experiments are conducted using tamoxifen.

Amperometric detection has also been applied to L6 myotubes. Individuals suffering from type 2 diabetes mellitus have reduced expression of genes for key proteins in oxidative metabolism and mitochondrial function in muscle.[88] Measurements of glucose oxidation from myotubes are unaltered by moderate overexpression of mitochondrial uncoupling protein 3 (UCP3) but markedly increased with the chemical uncoupler dinitrophenol (DNP). Glucose oxidation is measured using U-^{14}C Glucose and scintillation counting. This result is closely matched to the pattern of oxygen uptake monitored using self-referencing technologies. No difference in UCP3 expressing cells and control myotubes is found but a significant increase in oxygen uptake was seen with DNP treatment.[88]

Pancreatic β-Cells

Pancreatic β-cells are the electrically active component of the Islets of Langerhans and secrete insulin in response to depolarization through voltage dependent Ca^{2+} channels. Part of the control pathway is the closure of the K_{ATP} channel regulated by changes in the ATP/ADP ratio as this relates to increased metabolism and circulating glucose levels. This predicts an oscillation in single-cell oxygen consumption at similar frequencies to insulin vesicle fusion. Porterfield et al. studying INS cells in culture, show this at the level of a single cell.[89] Damon Osbourn and Emma Heart have confirmed this observation on primary β-cell cultures and correlated activity with Ca^{2+} signaling (unpublished). Using INS cells, a method has been developed to simultaneous monitor oxygen flux with sub-micron spatial resolution of the microelectrode.[21]

Additional Pumps and Channels

Apoptosis and Volume Changes

Apoptosis occurs in most cell types and can be triggered by a variety of physiological and pathological stimuli. It is a key mechanism in structuring tissues during development. This phenomenon has been studied using a mouse zygote (the condition prior to the first cleavage) and a self-referencing K^+ ISE.[51,90] Cell shrinkage is an incipient hallmark of apoptosis and relates to changes in potassium flux that decrease the intracellular K^+ concentration, regulating apoptotic progression. This flux is measured at the single zygote level with apoptosis induced by H_2O_2 treatment.[90] Using a collection of pharmacological blockers for various K^+ channels, the mechanism underlying this change in K^+ flux is identified as the two-pore domain K^+ channel (K_{2P}).[51]

Ion fluxes, moving as a result of hyposmotic challenge and cell swelling, have been studied using single human epithelial tsA201a cells.[32] A role for an anion flux is critical in the induction of a recovery mechanism that ultimately involves the activation of voltage dependent potassium channels, a reduction in cytoplasmic osmolarity, and the initiation of flow of osmotically obligated water out of the cell. A clear chloride flux was followed using a self-referencing Cl^- ISE and was correlated to cell volume changes and membrane currents. This paper is of particular interest for the assessment of interferents to what would be better described as an anion selective ISE.

Acidification and Alkalinization

Self-referencing microelectrodes, although designed for applications to single or small groups of cells, can equally be applied to semi-intact epithelial structures. A series of papers studying the acidification of the vas deferens and epididymus of the rat is a good example.[91–96] The lumen of the mammalian vas deferens and epididymus is held at a pH of around 6.6, an important factor in the immobilization and maturation of sperm. Using a self-referencing H^+ ISE Breton et al. scanned the surface of the vas deferens locating hydrogen transport hotspots inhibited by Bafilomycin A1—a highly selective blocker for the vacuolar type proton ATPase (V-Type pump), a non-phosphorylating H^+ pump.[91] Subsequent studies have shown the involvement of chloride independent bicarbonate transport and an inhibition by cadmium.[92,95] Monitoring the localized H^+ flux has also allowed the study of the vesicle fusion event and a role for the cytoskeleton in trafficking and recycling of the V-Type pumps.[93,96] These studies have been conducted in conjunction with molecular and imaging techniques.

Using another epithelial structure, larval mosquito midgut, processes driving the extreme alkalinization of the anterior segment (pH $\approx$ 11) have been studied with both H^+ and Cl^- self-referencing ISEs.[97,98] Proton and chloride regional fluxes were mapped down the length of the midgut and pharmacological blockers used to determine their inter-relationship. Application of DIDS produced consistent results when measuring either H^+ or Cl^- flux. However, at 10^{-4} M DIDS caused a small reversal of the Cl^- flux, a phenomenon not seen with methazolamide treatment.[98] This difference may be attributed to the interference of DIDS with the electrode performance as discussed above and by Messerli et al.[47] The authors conclude that an H^+ V-Type ATPase energizes an anion exchange across the apical membrane. The electrophoretic Cl^-/HCO_3^- exchanger and carbonic anhydrase are seen as critical components in the alkalinization of the anterior midgut.[98] This study also presents an alternative approach to buffer correction when the diffusion coefficient of the buffer is unknown.[98]

CONCLUSIONS

Electrochemical microelectrodes, whether stationary or self-referencing, offer unique advantages in studying the chemical composition of the cellular boundary layer. When operated in a self-referencing modulation mode these electrodes, both potentiometric and amperometric, follow the dynamics of chemical movement with high spatial and temporal fidelity. This format minimizes the impact of electrode and circuit drift, cuts out electrical boundary potentials, and chemical background signals. Being non-invasive, the approach further permits the tracking of changes from different locations over periods of seconds to days. Self-referencing microelectrodes, therefore, offer a powerful additional tool to an understanding of neuronal, glial, and cellular dynamics generally.

ACKNOWLEDGMENTS

The authors would like to thank Robert Lewis for performing the response time studies, Tamara Clark for her graphics work, and our numerous collaborators, mentioned and unmentioned, for contributing so significantly to the further development of the techniques available at the BioCurrents Research Center. We also gratefully acknowledge the NIH: NCRR for funding the BRC as a national research resource (P41 RR001395 to PJSS).

REFERENCES

1. Shou, M., Smith, A. D., Shackman, J. G., Peris, J., and Kennedy, R. T., In vivo monitoring of amino acids by microdialysis sampling with on-line derivatization by naphthalene-2, 3-dicarboxyaldehyde and rapid micellar electrokinetic capillary chromatography, *J. Neurosci. Methods*, 38, 189–197, 2004.
2. Sakmann, B. and Neher, E., *Single-Channel Recording*, 2nd ed., Plenum Press, New York, London, 1995.
3. Inoué, S. and Spring, K. R., *Video Microscopy: The Fundamentals*, Plenum Press, New York, 1997.
4. Haugland, R. P., *The Handbook—A Guide to Fluorescent Probes and Labeling Technologies*, 10th ed., Invitrogen, Carlsbad, California, CA, 2005.
5. Travis, E. R. and Wightman, R. M., Spatio-temporal resolution of exocytosis from individual cells, *Annu. Rev. Biophys. Biomol. Struct.*, 27, 77–103, 1998.
6. Pihel, K., Hsieh, S., Jorgenson, J. W., and Wightman, R. M., Quantal corelease of histamine and 5-Hydroxytryptamine from mast cells and the effects of prior incubation, *Biochemistry*, 37, 1046–1052, 1998.
7. Poitry, S., Tsacopoulos, M., Fein, A., and Cornwall, M. C., Kinetics of oxygen consumption and light-induced changes of nucleotides in solitary rod photoreceptors, *J. Gen. Physiol.*, 108, 75–87, 1996.
8. Whalen, W. J., Riley, J., and Nair, P., A microelectrode for measuring intracellular PO_2, *J. Appl. Physiol.*, 23, 798–801, 1967.
9. Kang, T. M. and Hilgemann, D. W., Multiple transport modes of the cardiac Na+/Ca2+exchanger, *Nature*, 427, 544–548, 2004.
10. Gow, N. A. R., Kropf, D. L., and Harold, F. M., Growing hyphae of *Achlya bisexualis* generate a longitudinal pH gradient in the surrounding medium, *J. Gen. Microbiol.*, 130, 2967–2974, 1984.
11. Jaffe, L. F. and Levy, S., Calcium gradients measured with a vibrating calcium-selective electrode, *Proc. IEEE/EMBS Conf.*, 9, 779–781, 1987.
12. Kühtreiber, W. M. and Jaffe, L. F., Detection of extracellular calcium gradients with a calcium-specific vibrating electrode, *J. Cell Biol.*, 110, 1565–1573, 1990.
13. Kang, T. M., Markin, V. S., and Hilgemann, D. W., Ion fluxes in giant excised cardiac membrane patches detected and quantified with ion-selective microelectrodes, *J. Gen. Physiol.*, 121, 325–347, 2003.
14. Messerli, M. A., Robinson, K. R., and Smith, P. J. S., Electrochemical sensor applications to the study of molecular physiology and analyte flux in plants. In *Plant Electrophysiology Theory and Methods*, Springer, New York, 2006.

15. Smith, P. J. S., Hammar, K., Porterfield, D. M., Sanger, R. H., and Trimarchi, J. R., Self-referencing, non-invasive, ion selective electrode for single cell detection of trans-plasma membrane calcium flux, *Micros. Res. Tech.*, 46, 398–417, 1999.
16. Wipf, D. O. and Bard, A. J., Scanning electrochemical microscopy. 15. Improvements in imaging via tip-position modulation and lock-in detection, *Anal. Chem.*, 64, 1362–1367, 1992.
17. Siedlecki, C. and Marchant, R. E., Atomic force microscopy for characterization of the biomaterial interface, *Biomaterials*, 19, 441–454, 1998.
18. Nuccitelli, R., Vibrating probe technique for studies of ion transport, In *Noninvasive Techniques in Cell Biology*, Wiley-Liss, New York, pp. 273–310, 1990.
19. Kinlen, P. J., Menon, V., and Ding, Y., A mechanistic investigation of polyaniline corrosion protection using the scanning reference electrode technique, *J. Electrochemical Soc.*, 146, 3690–3695, 1999.
20. Smith, P. J. S., Haydon, P. G., Hengstenberg, A., and Jung, S. K., Analysis of cellular boundary layers and their modulation by plasma membrane transporters: Application of electrochemical microsensors, *Electrochim. Acta*, 47, 283–292, 2001.
21. Osbourn, D., Sanger, R. H., and Smith., P. J. S., Determination of single cell oxygen consumption with impedance feedback for control of sample-probe separation, *Anal. Chem.*, 77(21), 6999-7004
22. Malchow, R. P., Land, S. C., Patel, S., and Smith, P. J. S., Consumption of oxygen by isolated skate retinal photoreceptors measured using a self-referencing oxygen-selective microelectrode, *Biol. Bull.*, 193, 231–232, 1997.
23. Smith, P. J. S., Sanger, R. H., and Jaffe, L. F., The vibrating Ca^{2+} electrode: A new technique for detecting plasma membrane regions of Ca^{2+} influx and efflux, In *Methods in Cell Biology. A Practical Guide to the Study of* Ca^{2+} *in Living Cells*, Wilson, L., Matsudaira, P., Series Eds., Nuccitelli, R., Ed., Vol. 40, Academic Press, San Diego, pp. 115–134, 1994.
24. Smith, P. J. S. and Trimarchi, J. R., Noninvasive measurement of hydrogen and potassium ion flux from single cells and epithelial structures, *Am. J. Physiol.*, 280, C1–C11, 2001.
25. Doughty, J. M. and Langton, P. D., Measurement of chloride flux associated with the myogenic response in rat cerebral arteries, *J. Physiol.* 534, 753–761, 2001.
26. Land, S. C. and Collett, A., Detection of Cl^- flux in the apical microenvironment of cultured foetal distal lung epithelial cells, *J. Exp. Biol.*, 204, 785–795, 2001.
27. Bakker, E., Bühlmann, P., and Pretsch, E., Carrier-based ion-selective electrodes and bulk optodes. 1. General characteristics, *Chem. Rev.*, 97, 3083–3132, 1997.
28. Bakker, E. and Meyerhoff, M. E., Ionophore-based membrane electrodes: New analytical concepts and non-classical response mechanisms, *Anal. Chim. Acta*, 416, 121–137, 2000.
29. Umezawa, Y., Bühlmann, P., Umezawa, K., Tohda, K., and Amemiya, S., Potentiometric selectivity coefficients of ion-selective electrodes: Part 1. Inorganic cations, *Pure Appl. Chem.*, 72, 1851–2082, 2000.
30. Bühlmann, P., Pretsch, E., and Bakker, E., Carrier-based ion-selective electrodes and bulk optodes. 2. Ionophores for potentiometric and optical sensors, *Chem. Rev.*, 98, 1593–1687, 1998.
31. Fluka, *Selectophore, Ionophores for Ion-Selective Electrodes and Optrodes*, Fluka Chemie A.G., Ronkonkoma, New York, 1996.
32. Garber, S. S., Messerli, M. A., Hubert, M., Lewis, R., Hammar, K., Indyk, E., and Smith, P. J. S., Monitoring Cl^- movement in single cells exposed to hypotonic solution, *J. Memb. Biol.*, 203, 101–110, 2005.
33. Messerli, M. A., Danuser, G., and Robinson, K. R., Pulsatile fluxes of H^+, K^+, and Ca^{2+} lag growth pulses of *Lilium longiflorum* pollen tubes, *J. Cell Sci.*, 112, 1497–1509, 1999.
34. Papeschi, G., Mancuso, S., and Marras, A. M., Electrochemical behavior of a Cu/CuSe microelectrode and its application in detecting temporal and spatial localization of copper(II) fluxes along *Olea europaea* roots, *J. Solid State Electrochem.*, 4, 325–329, 2000.
35. Sokalski, T., Ceresa, A., Zwickl, T., and Pretsch, E., Large improvement of the lower detection limit of ion-selective polymer membrane electrodes, *J. Am. Chem. Soc.*, 119, 11347–11348, 1997.
36. Mathison, S. and Bakker, E., Effect of transmembrane electrolyte diffusion on the detection limit of carrier-based potentiometric ion sensors, *Anal. Chem.*, 70, 303–309, 1998.
37. Jung, S. K., Trimarchi, J. R., Sanger, R. H., and Smith, P. J. S., Development and application of a self-referencing glucose microsensor for the measurement of glucose consumption by pancreatic β-cells, *Anal. Chem.*, 73, 3759–3767, 2001.

38. Land, S. C., Porterfield, D. M., Sanger, R. H., and Smith, P. J. S., The self-referencing oxygen-selective microelectrode: Detection of transmembrane oxygen flux from single cells, *J. Exp. Biol.*, 202, 211–218, 1999.
39. Porterfield, D. M., Laskin, J. D., Jung, S.-K., Malchow, R. P., Billack, B., Smith, P. J. S., and Heck, D. E., Proteins and lipids define the diffusional field of nitric oxide. Measurement of nitric oxide fluxes from macrophages using a self-referencing electrode, *Am. J. Physiol.*, 281, L904–L912, 2001.
40. Twig, G., Jung, S. K., Messerli, M., Smith, P. J. S., and Shirihai, O., Real-time detection of reactive oxygen intermediates from single microglial cells, *Biol. Bull.*, 201, 261–262, 2001.
41. Twig, G., Graf, S. A., Messerli, M. A., Smith, P. J. S., Yoo, S. H., and Shirihai, O. S., Synergistic amplification of beta-amyloid and interferon-gamma-induced microglial neurotoxic response by the senile plaque component chromogranin A, *Am. J. Physiol.*, 288, C169–C175, 2005.
42. Pepperell, J. R., Porterfield, D. M., Keefe, D. L., Behrman, H. R., and Smith, P. J. S., Control of ascorbic acid efflux in rat luteal cells: Role of intracellular calcium and oxygen radicals, *Am. J. Physiol.*, 285, C642–C651, 2003.
43. Bogorff, D. J., Messerli, M. A., Malchow, R. P., and Smith, P. J., Development and characterization of a self-referencing glutamate-selective micro biosensor, *Biol. Bull.*, 205, 207–208, 2003.
44. Mancuso, S., Paeschi, G., and Marras, A. M., A polarographic, oxygen-selective, vibrating-microelectrode system for the spatial and temporal characterization of transmembrane oxygen fluxes in plants, *Planta*, 211, 384–389, 2000.
45. Kochian, L. V., Shaff, J. E., Kühtreiber, W. M., Jaffe, L. F., and Lucas, W. J., Use of an extracellular, ion-selective, vibrating microelectrode system for the quantification of K^+, H^+, and Ca^{2+} fluxes in maize roots and maize suspension cells, *Planta*, 188, 601–610, 1992.
46. Henriksen, G. H., Raman, D. R., Walker, L. P., and Spanswick, R. M., Measurement of net fluxes of ammonium and nitrate at the surface of barley roots using ion-selective microelectrodes, *Plant Physiol.*, 99, 734–747, 1992.
47. Messerli, M. A., Smith, P. J. S., Lewis, R. C., and Robinson, K. R., Chloride fluxes in lily pollen tubes: A critical reevaluation, *Plant J.*, 40, 799–812, 2004.
48. Crank, J., *The Mathematics of Diffusion*, Oxford University Press, London, 1967.
49. Freyer, J. P and Sutherland, R. M., A reduction in the in situ rates of oxygen and glucose consumption of cells in EMT6/Ro spheroids during growth, *J. Cell. Physiol.*, 124, 516–524, 1985.
50. Shirihai, O., Smith, P. J. S., Hammar, K., and Dagan, D., H^+ and K^+ gradient generated by microglia H/K ATPase, *Glia*, 23, 339–348, 1998.
51. Trimarchi, J. R., Liu, L., Smith, P. J. S., and Keefe, D. L., Apoptosis recruits two-pore domain potassium channels used for homeostatic volume regulation, *Am. J. Physiol.*, 282, C588–C594, 2002.
52. Ammann, D., *Ion-Selective Micro-Electrodes Principles, Design and Application*, Springer-Verlag, Germany, 1986.
53. Jung, S. K., Kauri, L. M., Qian, W. J., and Kennedy, R. T., Correlated oscillations in glucose consumption, oxygen consumption, and intracellular free Ca^{2+} in single islets of Langerhans, *J. Biol. Chem.*, 275, 6642–6650, 1999.
54. Jung, S. K., Hammar, K., and Smith, P. J. S., Development of self-referencing oxygen microsensor and its application to single HIT cells, *Biol. Bull.*, 199, 197–198, 2000.
55. Friedemann, M. N., Robinson, S. W., and Gerhardt, G. A., *O*-phenylendediamine-modified carbon fiber electrodes for the detection of nitric oxide, *Anal. Chem.*, 1023, 421–425, 1996.
56. Kumar, S. M., Porterfield, D. M., Muller, K. J., Smith, P. J., and Sahley, C. L., Nerve injury induces a rapid efflux of Nitric Oxide (NO) detected with a novel NO microsensor, *J. Neurosci.*, 21, 215–220, 2001.
57. Lindner, E., Gyurcsányi, R. E., and Buck, R. P., Tailored transport through ion-selective membranes for improved detection limits and selectivity coefficients, *Electroanalysis*, 11, 695–702, 1999.
58. Mückenhoff, K., Schreiber, S., De Santis, A., Okada, Y., and Scheid, P., Ion-sensitive microelectrode system with short response time, *J. Neurosci. Methods*, 51, 147–153, 1994.
59. Voipio, J., Pasternak, M., and MacLeod, K., Ion-sensitive microelectrodes, In *Microelectrode Techniques. The Plymouth Workshop Handbook*, Ogden, D., Ed., The Company of Biologists, Ltd., Cambridge, MA, pp. 275–316, 1994.
60. Schneiderman, G. and Goldstick, T. K., Oxygen fields induced by recessed and needle oxygen microelectrodes in homogenous media, *Adv. Exp. Med. Biol.*, 75, 9–16, 1976.

61. Morf, W. E., Lindner, E., and Simon, W., Theoretical treatment of the dynamic response of ion-selective membrane electrodes, *Anal. Chem.*, 47, 1596–1601, 1975.
62. Fleet, B., Ryan, T. H., and Brand, M. J. D., Investigations of the factors affecting the response time of a calcium selective liquid membrane electrode, *Anal. Chem.*, 46, 12–15, 1974.
63. Schneiderman, G. and Goldstick, T. K., Oxygen electrode design criteria and performance characteristics: recessed cathode, *J. Appl. Physiol.*, 45, 145–154, 1978.
64. Dumollard, R., Hammar, K., Porterfield, M., Smith, P. J., Cibert, C., Rouviere, C., and Sardet, C., Mitochondrial respiration and Ca^{2+} waves are linked during fertilization and meiosis completion, *Development*, 130, 683–692, 2003.
65. Zhang, X., Kislyak, Y., Lin, J., Dickson, A., Coradosa, L., Broderickand, M., and Fein, H., Nanometer size electrode for nitric oxide and *S*-nitrosothiols measurement, *Electrochem. Commun.*, 4, 11–16, 2002.
66. Csöregi, E., QuinnD, C. P., Schmidtke, W., Lindquist, S. E., Pishko, M. V., Ye, L., Katakis, I., Hubbell, J. A., and Heller, A., Design, characterization, and one-point in vivo calibration of a subcutaneously implanted glucose electrode, *Anal. Chem.*, 66, 3131–3138, 1994.
67. Danuser, G., Photogrammetric calibration of a stereo light microscope, *J. Microsc.*, 193, 62–83, 1999.
68. Pergel, E., Gyurcsanyi, R. E., Toth, K., and Lindner, E., Picomolar detection limits with current-polarized Pb^{2+} ion-selective membranes, *Anal. Chem.*, 73, 4249–4253, 2001.
69. Hille, B., *Ion channels and Excitable Membranes*, 3rd ed., Chap. 10, 342, 2001.
70. Arif, I. and Newman, I. A., Proton efflux from oat coleoptile cells and exchange with wall calcium after IAA or fusicoccin treatment, *Planta*, 189, 377–383, 1993.
71. Demarest, J. R. and Morgan, J. L. M., Effect of pH buffers on proton secretion from gastric oxyntic cells measured with vibrating ion-selective microelectrodes, *Biol. Bull.*, 189, 219–220, 1995.
72. Molina, A. J., Verzi, M. P., Birnbaum, A. D., Yamoah, E. N., Hammar, K., Smith, P. J. S., and Malchow, R. P., Neurotransmitter modulation of extracellular H^+ fluxes from isolated retinal horizontal cells, *J. Physiol.*, 560, 639–657, 2004.
73. Li, R., Chase, M., Jung, S. R., Smith, P. J. S., and Loeken, M. R., Hypoxic stress in diabetic pregnancy contributes to impaired embryo gene expression and defective development by inducing oxidative stress, *Am. J. Physiol.*, 289, E591–E599, 2005.
74. Wangemann, P., Liu, J., and Marcus, D. C., Ion transport mechanisms responsible for K^+ secretion and the transepithelial voltage across marginal cells of stria vacularis in vitro, *Hearing Res.*, 84, 19–29, 1995.
75. Marcus, D. C., Sunose, H., Liu, J., Bennett, T., Shen, Z., Scofield, M. A., and Ryan, A. F., Protein kinase C mediates P_{2U} purinergic receptor inhibition of K^+ channel in apical membrane of strial marginal cells, *Hearing Res.*, 115, 82–92, 1998.
76. Marcus, D. C. and Shipley, A. M., Potassium secretion by vestibular dark cell epithelium demonstrated by vibrating probe, *Biophys. J.*, 66, 1939–1942, 1994.
77. Shen, Z., Liu, J., Marcus, D. C., Shiga, N., and Wangemann, P., DIDS increases K^+ secretion through an IsK channel in apical membrane of vestibular dark cell epithelium of gerbil, *J. Membr. Biol.*, 146, 283–291, 1995.
78. Lee, J. H. and Marcus, D. C., Endolymphatic sodium homeostasis by Reissner's membrane, *Neuroscience*, 119, 3–8, 2003.
79. Yamoah, E. N., Lumpkin, E. H., Dumont, R. A., Smith, P. J. S., Hudspeth, A. J., and Gillespie, P. G., Plasma-membrane Ca^{2+} -ATPase ensures low Ca^{2+} concentration in hair-cell stereocilia, *Neurosci. J.*, 18, 610–624, 1998.
80. Gleeson, R. A., Hammar, K., and Smith, P. J. S., Sustaining olfaction at low salinities: Mapping ion flux associated with the olfactory sensilla of the blue crab *Callinectes sapidus*, *J. Exp. Biol.*, 203, 3145–3152, 2002.
81. Gleeson, R. A., McDowell, L. M., Aldrich, H. C., Hammar, K., and Smith, P. J. S., Sustaining olfaction at low salinities: evidence for a paracellular route of ion movement from the hemolymph to the sensillar lymph in the olfactory sensillar of the blue crab, *Callinectes sapidus*, *Cell Tissue Res.*, 301, 423–431, 2000.
82. Knox, R. J., Kao, L. S., Jonas, E., Connor, J., Smith, P. J. S., and Kaczmarek, L. K., Calcium influx and activation of a cation current are coupled to an intracellular Ca^{2+} mobilization in peptidergic neurons, *J. Physiol.*, 494, 627–663, 1996.

83. Duthie, G. G., Shipley, A., and Smith, P. J. S., Use of a vibrating electrode to measure changes in calcium fluxes across the cell membranes of oxidatively challenged Aplysia nerve cells, *Free Rad. Res.*, 20, 307–313, 1994.
84. Hammar, K., Malchow, R. P., and Smith, P. J. S., Electrochemical measurement of O_2 consumption in single excitable cells, *Biophys. J.*, 84, 457A, 2003.
85. Devlin, C. L., A vibrating Ca^{2+} -selective electrode measures Ca^{2+} flux induced by the neuropeptide FMRFamide in a gastropod ventricle, *Comp. Biochem. Physiol.*, 116A, 93–100, 1997.
86. Devlin, C. L., 5-Hydroxytryptamine stimulates Ca^{2+} flux in the ventricular muscle of mollusk (Busycon canaliculatum) during cardioexcitation, *Biol. Bull.*, 200, 344–350, 2001.
87. Devlin, C. L. and Smith, P. J. S., A non-invasive vibrating calcium-selective electrode measures acetylcholine-induced calcium flux across the sarcolemma of a smooth muscle, *J. Comp. Physiol.*, 166, 270–277, 1996.
88. MacLellan, J D., Gowing, A., Gerrits, M., Smith, P. J. S., Sivitz, W., Wheeler, M. B., and Harper, M. H., Increased uncoupling protein 3 stimulates fatty acid, but not glucose oxidation, and decreases reactive oxygen species in muscle cells, *Diabetes*, 54(8), 2343-2350, 2005.
89. Porterfield, D. M., Corkey, R. F., Sanger, R. H., Tornheim, K., Smith, P. J. S., and Corkey, B. E., Oxygen consumption oscillates in single clonal pancreatic β-cells (HIT), *Diabetes*, 49, 1511–1516, 2000.
90. Trimarchi, J R., Liu, L., Smith, P. J. S., and Keefe, D. L., Non-invasive measurement of potassium efflux as an early indicator of cell death in mouse embryos, *Biol. Reprod.*, 63, 851–857, 2000.
91. Breton, S., Hammar, K., Smith, P. J. S., and Brown, D., Acidification of the male reproductive tract by bafilomycin-sensitive H^+ ATPase, *Nat. Med.*, 2, 470–472, 1996.
92. Breton, S., Hammar, K., Smith, P. J. S., and Brown, D., Proton secretion in the male reproductive tract: Involvement of chloride-independent bicarbonate transport, *Am. J. Phyiol.*, 275, C1134–C1142, 1998.
93. Breton, S., Nsuma, N. N., Galli, T., Smith, P. J. S., and Brown, D., Tetanus toxin-mediated cleavage of cellubrevin impairs proton secretion in the male reproductive system, *Am. J. Physiol.*, 278, F717–F725, 2000.
94. Brown, D., Smith, P. J. S., and Breton, S., Role of V-ATPase rich cells in acidification of the male reproductive tract, *J. Exp. Biol.*, 200, 257–262, 1997.
95. Herak-Kramberger, C. M., Sabolic, I., Blanusa, M., Smith, P. J. S., Brown, D., and Breton, S., Cadmium inhibits vacuolar H^+ ATPase-mediated acidification in the rat epididymus, *Biol. Reprod.*, 63, 599–606, 2000.
96. Beaulieu, V., Da Silva, N., Pastor-Soler, N., Brown, C. R., Smith, P. J. S., Brown, D., and Breton, S., Modulation of the actin cytoskeleton via gelsolin regulates vacuolar H^+ ATPase (V-ATPase) recycling, *J. Biol. Chem.*, 280, 8452–8463, 2005.
97. Boudko, D., Moroz, L., Linser, P., Trimarchi, P. J., Smith, P., and Harvey, W., In situ analysis of pH gradients in mosquito larvae using non-invasive, self-referencing, pH-sensitive microelectrodes, *J. Exp. Biol.*, 204, 691–699, 2001.
98. Boudko, D. Y., Moroz, L. L., Harvey, W. R., and Linser, P. J., Alkalinization by chloride/bicarbonate pathway in larval mosquito midgut, *Proc. Natl Acad. Sci.*, 98, 15354–15359, 2001.

19 Second-by-Second Measures of L-Glutamate and Other Neurotransmitters Using Enzyme-Based Microelectrode Arrays

Kevin N. Hascup, Erin C. Rutherford, Jorge E. Quintero, B. Keith Day, Justin R. Nickell, Francois Pomerleau, Peter Huettl, Jason J. Burmeister, and Greg A. Gerhardt

CONTENTS

INTRODUCTION

L-Glutamate is the major excitatory neurotransmitter in the mammalian central nervous system (CNS) and is implicated in a number of brain disorders including Parkinson's disease (PD), cognitive disturbances, epilepsy, schizophrenia, attention deficit hyperactivity disorder (ADHD) and drug abuse. While microdialysis methods have been used extensively over the last decade to investigate minute-by-minute measures of L-glutamate, the rapid time dynamics of L-glutamate signaling in the CNS has warranted a technique to measure L-glutamate release on a second-by-second basis. A major goal of the research is to develop a recording technology for recording second-by-second measurements of L-glutamate and other neurotransmitters—specifically a mass-fabricated microelectrode technology that could be (1) mass produced such that other laboratories could utilize the same recording technology and (2) configured for "self-referencing" recordings, which allows for second-by-second cross-checking of the selectivity of the microelectrode measures and improved signal-to-noise of the recording methods. The present chapter documents current capabilities of measuring L-glutamate and several other neurotransmitters on a second-by-second basis using mass-fabricated microelectrode arrays formed on ceramic. While not a comprehensive assessment of the technology, this chapter contains a large amount of information regarding the fabrication, use, and potential pitfalls of this technology. The reader should refer to numerous articles[1–6] for additional details regarding measuring neurotransmitters in the CNS.

PRINCIPLES OF IN VIVO ELECTROCHEMISTRY

In vivo electrochemistry utilizes microelectrodes that can be implanted into the mammalian CNS, thus providing a means to record chemical signaling of neurons. The microelectrode's recording surface, or working electrode (normally an inert metal such as platinum (Pt) or carbon), can oxidize or reduce compounds of interest. A potential is applied versus a reference electrode, normally a Ag/AgCl reference electrode that is in ionic contact with the working microelectrode. Low noise headstages, which include potentiostats, apply this potential and are computer-controlled with multiple inputs to simultaneously record from several microelectrodes or several recording surfaces on a single microelectrode array. If the potential at the microelectrode surface is sufficient, then molecules directly at the recording surface of the working electrode are either oxidized or reduced depending upon their intrinsic electrochemical properties. Oxidized molecules give up one or more electrons to the recording surface while reduced molecules receive electrons from the recording surface. The currents generated from those Faradaic reactions are linear with respect to concentration of the electroactive molecule(s) in the tissue surrounding the microelectrode. This basic principle allows for in vitro calibration methods for in vivo studies. (For a more complete overview, please see Gerhardt and Burmeister's chapter in the *Encyclopedia of Analytical Chemistry*.)

One of the simplest electrochemical techniques is amperometry, which involves the measurement of current at a constant fixed potential. The current can be monitored continuously, thus, events can be measured as quickly as ≤ 1 ms[6]. Because the voltage is applied continuously to the working electrode, the non-Faradaic background current recorded from the microelectrode is low allowing for sensitive measurements of electrochemically active molecules.[6,7]

Several neurochemicals reside in the extracellular space of the CNS that are electrochemically active on Pt recording surfaces at low or high oxidation potentials including ascorbic acid (AA), dopamine (DA), norepinephrine (NE), serotonin (5-HT), 5-hydroxyindoleacetic acid (5-HIAA), homovanillic acid (HVA), nitric oxide, uric acid, and 3,4-dihydroxyphenylacetic acid (DOPAC). Additionally, enzymes are applied to the recording surface to produce an electrochemically active reporter molecule (usually hydrogen peroxide) to allow measurements of molecules that are not inherently electrochemically active. When a potential of $+0.7$ V versus a Ag/AgCl reference is applied to the Pt recording sites, the newly formed peroxide oxidizes, and the resulting change in current from the transfer of electrons to the Pt recording surface is detected. While O_2 is required as a cofactor for oxidase enzymes, the oxidation of the reporter molecule, hydrogen peroxide, generates a portion of the O_2 for continued reactions. This, however, does not allow measures using oxidase enzymes in zero O_2 environments.

An important and often poorly considered aspect of microelectrodes is that analytes from only a small area (microns) of tissue surrounding the recording surface are detected. This aspect is useful for studying small or layered structures in the brain or the spinal cord. Additionally, microelectrodes with multiple recordings surfaces can be geometrically arranged to measure analyte concentrations from two or more distinct brain regions. Since microelectrodes coupled with amperometric techniques can measure analytes of interest on rapid time scales (1–1000 ms), uptake and release kinetics of neurotransmitters are easily studied.

ENZYME-BASED MULTISITE MICROELECTRODE ARRAYS

FABRICATION

Enzyme-based multisite microelectrode arrays are mass fabricated using photolithographic methods.[8–11] An advantage of photolithography allows routine production of reproducible recording surfaces as small as 5–10 μm. In addition to this advantage, multiple microelectrodes are patterned onto a single fabrication substrate (usually 2.5 cm$\times$2.5 cm) allowing for increased fabrication number at a decreased cost. Finally, photolithographic methods are used to manufacture numerous microelectrode designs with multiple recordings sites in well-defined, highly reproducible geometrical configurations.

The multisite microelectrodes used in the research are currently constructed in conjunction with Thin Films Technology, Inc. (Buellton, California). The fabrication process is explained thoroughly in previously published papers.[1,2,11–13] Initially, microelectrode photographic masks are designed on a computer aided design (CAD) program where arrays of 4–16 recording sites are arranged on templates. A 2.5$\times$2.5 cm$\times$125 m thick ceramic wafer (alumina, Al_2O_3, Coors Ceramic, Coors Superstrate 996) serves as a common substrate for the microelectrode arrays. Ceramic reduces the cross-talk from adjacent connecting lines. Additionally, ceramic is strong and rigid, which aids in precise stereotaxic placement into tissues without flexing or breaking. The ceramic substrate may be polished or lapped down to achieve microelectrodes as thin as 37.5 μm.[12]

Following cleaning, photoresist is spun onto the ceramic wafer. Collimated light passing through the photomask exposes the photoresist, thus transferring the microelectrode images onto the wafer. Spaces for the recording sites, connecting lines and bonding pads are not exposed to light. Solvents are used to remove unexposed photoresist from the wafer. Many 1-cm long micro-electrodes (34–64) can be patterned onto the wafer simultaneously to facilitate production and to increase the number of microelectrodes that are made from the ceramic substrate (Figure 19.1). Patterning also can be performed on the reverse side of the ceramic wafer to increase recording site density.

Next, an adhesion layer of Titanium (Ti) is used to allow Noble metals (Pt or Ir) to adhere to the ceramic substrate. Titanium (500 Å thick) is sputtered onto the developed photoresist covered ceramic wafer. Following the adhesion layer, the active recording metal layer, Pt for most of the work, is sputtered onto the substrate ($\sim$1.5 μm thick). Solvents are used to remove the developed photoresist and unwanted metals leaving the recording pads, connecting lines, and connecting pads.

Once the Pt recording sites, connecting lines, and bonding pads are in place, the connecting lines are coated with an insulator using another photolithographic step. The connecting lines act as wires to connect the bonding pads to the recording sites and must be insulated from aqueous environments. To accomplish this, the microelectrodes are coated again with photoresist.

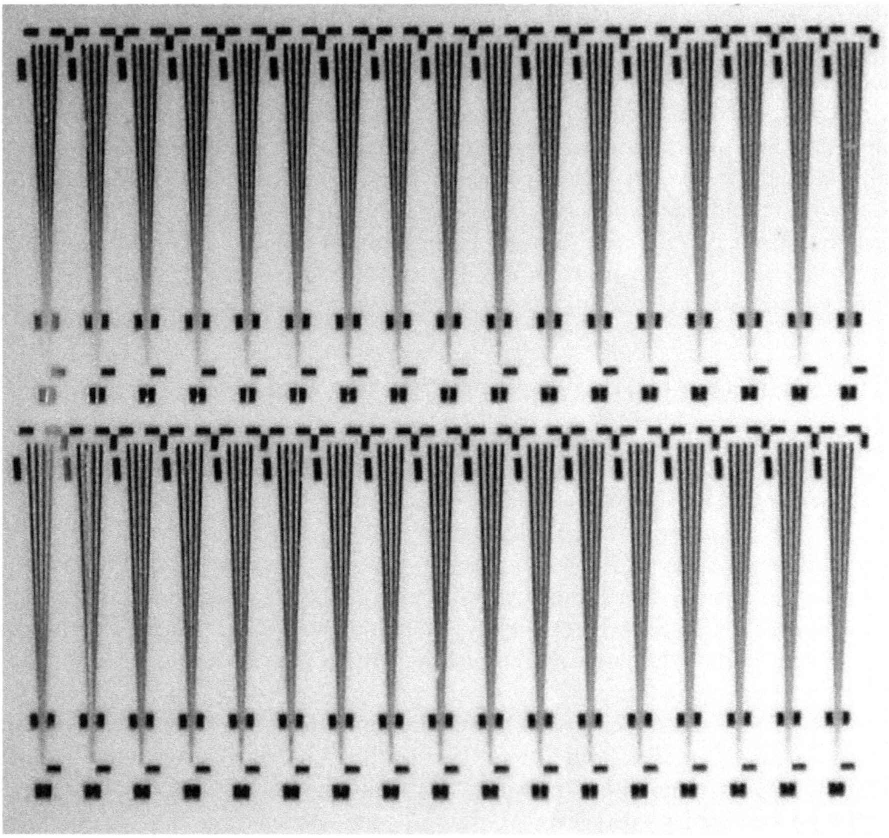

FIGURE 19.1 Photograph of a ceramic wafer containing 34 microelectrodes with Pt bonding pads, connecting lines and recording sites (not visible) are patterned onto the 2.5 cm × 2.5 cm, 125 μm thick substrate.

A second photomask is used to define the areas where polyimide is to be placed. Polyimide acts as an insulator to define the recording site active area as well as the bonding pads. It also reduces cross talk between the connecting lines. After the photoresist is developed, polyimide is spun onto the wafer (2–4 μm thick). Once the insulating layer is applied, only the recording sites and bonding pads are exposed. The photoresist and excess polyimide are removed and the remaining polyimide is cured at 200°C. The complete fabrication process is depicted in Figure 19.2.

After the formation of the microelectrode on the ceramic wafer, a diamond saw or laser is used to form or "cut out" the individual microelectrodes. A major advantage of the diamond saw is that it produces highly polished edges for reduced tissue damage during implantation. The ceramic wafer with Pt recording sites is attached to a printed circuit board (PCB) holder for handling and connection to recording equipment. To connect the ceramic wafer to the PCB holder, each bonding pad is wire bonded to an individual Pt recording site on the ceramic wafer. The tips are epoxied onto the paddle for stability and to insulate the wire bonds. Cutting and assembly is performed currently in conjunction with Hybrid Circuits, Inc. (Sunnyvale, CA). The PCB holder and fully constructed multisite microelectrode are shown in Figure 19.3a and b, respectively. Figure 19.3c is a version of the microelectrode for awake animal recordings (see the section in this chapter on Awake, Freely-Moving Rats and Mice for further details).

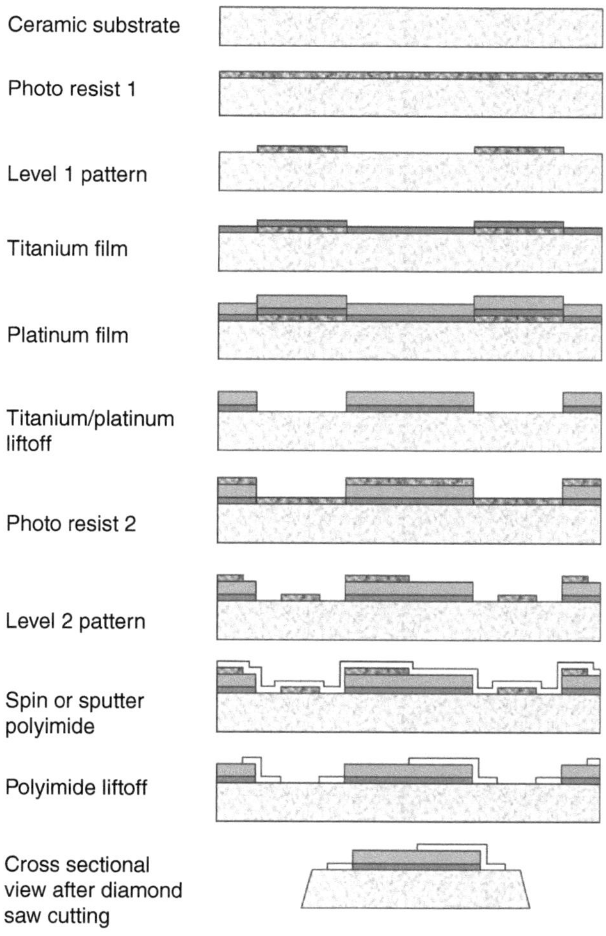

FIGURE 19.2 Schematic of the fabrication sequence for the ceramic wafers with Pt recording sites.

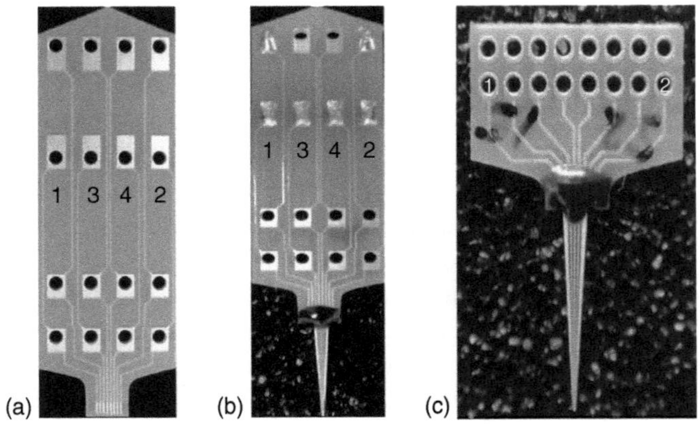

FIGURE 19.3 Photograph of the microelectrode PCB with and without attached ceramic microelectrode tip. (a) Close-up view of the PCB. Numbers beneath the pinholes indicate the specific Pt recording site connection. (b) Fully assembled microelectrode array. Black epoxy is used to insulate the wire bonding from the ceramic tip to the PCB as well as to provide stability. Pins have been soldered into place for microelectrode attachment into a dual inline pin (DIP) socket. Pins placed in the upper left and right of the PCB provides added stability. (c) Fully assembled microelectrode array for recordings in freely moving rats or mice.

Multisite Microelectrode Array Designs

Due to the flexibility of the fabrication process, the microelectrodes can be manufactured with different recording site geometric configurations. For this research, the most commonly used configurations have four Pt recording sites. However, microelectrodes with eight recording sites are in use and those for sixteen recording sites are being developed as discussed in later sections. Figure 19.4 shows two popular four site configurations designated as the R1 (a row of four in-line sites, Figure 19.4a) and S2 (two side-by-side site pairs, Figure 19.4b). Individual microelectrodes are selected based on the type of recordings and brain regions of interest. The R1s provide a larger recording distance that is useful for large brain regions or layered structures, while the S2s provide dual detection in smaller brain structures. Furthermore, the advantage of multiple recording sites allows for self-referencing techniques (see the section on Amperometric Recordings Utilizing

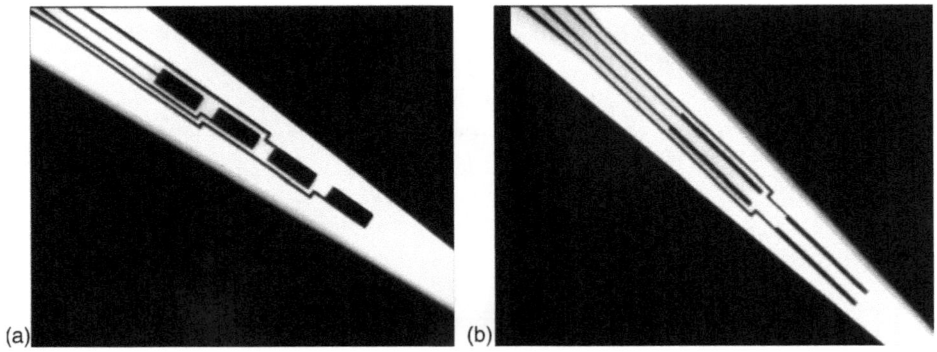

FIGURE 19.4 Photomicrographs of two ceramic-based multisite microelectrode designs. Each has four Pt recording sites arranged in different geometrical configurations. (a) Photomicrograph of an R1 microelectrode with a row of four Pt recording sites. Each site measures 50×150 μm with 50 μm spacing between sites. (b) A photomicrograph of a S2 microelectrode with two pairs of side-by-side Pt recording sites. Each site measures 333×15 μm with a 30 μm between side-by-side sites and 100 μm separation among paired sites.

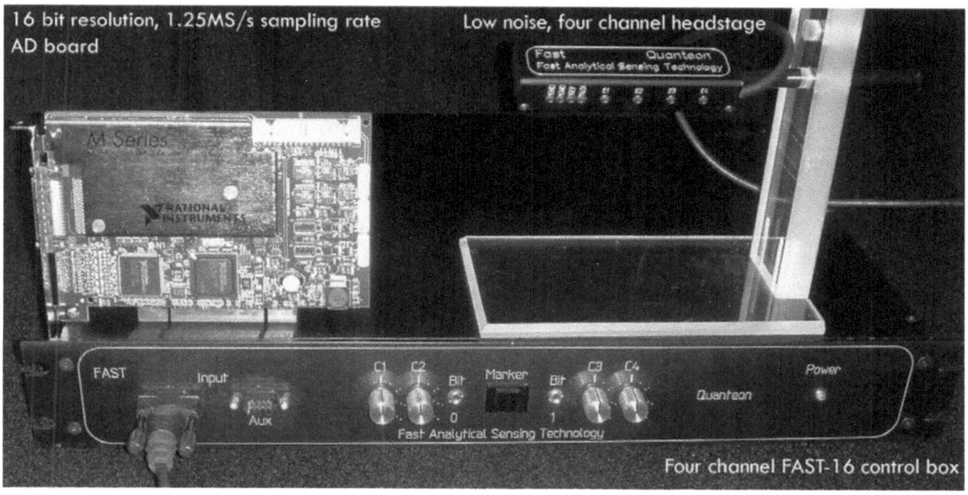

FIGURE 19.5 Photograph of the control box (bottom), headstage (right) and A/D board (left) encompassing the Fast Analytical Sensing Technology (FAST-16). The A/D board has a sampling rate of 1.25 million samples per second (MS/s). This value refers to the speed at which the digitizer converts the input signal to digital values. The FAST-16 system measures 17.5″×12″×3″, making it ideal for bench top placement or for rack mounting. (Photograph courtesy of Quanteon L.L.C.)

Self-Referencing) and/or detection of multiple analytes. Additional designs are presented in Gerhardt and Burmeister's work in the *Encyclopedia of Sensors.*

Headstage/Recording System

A headstage/potentiostat creates a potential difference between the microelectrode array and a reference electrode to carryout electrochemical recordings. The recording system used also amplifies the electrochemical signal and is part of the Fast Analytical Sensing Technology (FAST-16) system (Quanteon L.L.C, Nicholasville, KY). Figure 19.5 shows a photograph of the current FAST-16 recording system. In conjunction with other hardware elements, including the control box and a 16 bit A/D card in the computer, the FAST-16 system amplifies the electrical current generated by oxidation or reduction reactions and digitizes this current. The Windows™ based FAST-16 system software records, creates and displays second-by-second (1 Hz), or faster (1–40 Hz), data files (nominal 1–10 Hz). The software is written for eight simultaneous channel recordings, which can be increased as the number of Pt recording sites increases in future generations of microelectrodes. Recorded files are exported to other Windows™ based applications, such as Excel™, for easier data processing; or processed directly through the data analysis software package on the FAST-16 system software. The low noise design and software oversampling allow the FAST-16 recording system to be used for bench top experimentation without the need of a Faraday cage. However, greater performance can be achieved with the use of a Faraday cage or appropriate shielding for most applications

MICROELECTRODE PREPARATION

Many CNS neurotransmitters are not inherently electrochemically active and therefore must be converted to a reporter molecule by cross-linking enzymes to our microelectrode surface. Below are the general procedures involved in preparing a microelectrode for use in vitro or in vivo.

Cleaning Procedures

Microelectrodes received from Hybrid Circuits, Inc., undergo a cleaning procedure to remove any particulate matter or residue from the Pt recording sites that may have been deposited during the manufacturing process. New microelectrodes are placed into a stirred solution of Fisherbrand Citrisolv™ (Fisher Scientific, Catalog #22-143-975) for 5 min followed by a 5-min rinse in ddH_2O. After the chemical cleaning procedure, a thin film can develop on the Pt recording sites, so a Kimwipe® (Kimberly-Clark Professional) is swiped delicately across the tip of the microelectrodes to remove this film and any excess ddH_2O. Then microelectrodes are allowed to dry for 15 min in an oven at 105°C–115°C prior to coating.

Typically, enzyme-based microelectrodes are reused for more than one experiment, provided they meet calibration criteria. Over prolonged use, applied enzyme layers (described below) may degrade analyte detection. A slightly different cleaning procedure is employed for microelectrodes that have been previously coated with an enzyme layer. This enzyme layer must first be removed prior to re-use. The microelectrode tip is placed into a stirred solution of ddH_2O at 80°C to soften the protein matrix on the Pt recording sites. Following 30 min of soaking, the microelectrodes are cleaned as described in the preceding paragraph and tested for response to hydrogen peroxide to ensure good sensitivity of the microelectrode recording surfaces.

Exclusion Layer Coatings

Exclusion layer coatings are applied after successful cleaning of the microelectrode and are used to alter the recording properties of the microelectrode arrays. Exclusion layers are applied to block or minimize undesirable electrochemically active compounds found in high concentrations in the CNS such as AA (250–500 μM) or DOPAC. By blocking these undesired electrochemically active molecules from the Pt recording sites, the microelectrode has better selectivity for the analyte of interest. The planar geometry of the multisite microelectrodes often affects how materials adhere to the recording surfaces. It should be noted that exclusion and enzyme layer coatings can inhibit compounds from diffusing to the microelectrode surface by creating a diffusion barrier, thus limiting the response time of the microelectrode. For this reason, coating parametrics are performed to determine optimal coating procedures for both exclusion layer and enzyme layer coatings.

Nafion Exclusion Layer

Nafion® is an anionic Teflon® derivative and is one of the most widely used methods for improving the selectivity of voltammetric recordings in CNS tissue. Historically, Nafion is used on the surface of carbon fiber microelectrodes for the detection of catecholamines.[14–19] The negatively charged sulfonic acid groups substituted into the polymer matrix of Nafion repel anions from the Carbon or Pt recording surfaces. If these anions are not blocked, they cause high background signals that interfere with reliable recordings. At the same time, the negatively charged sulfonic acid groups on Nafion concentrate cationic species such as DA, NE and 5-HT. With the development of the multisite microelectrodes, Nafion is a logical choice for an exclusion layer and is used reliably in our studies of L-glutamate.[1–5,12] However, because Nafion does attract catecholamines to the Pt recording surface, which are oxidizable on Pt, other exclusion layers are used in CNS regions with high monoamine levels.

Nafion (5%, Sigma-Aldrich, Catalog #27,470-4) is aliquoted into a 300 μl centrifuge tube. The microelectrode tip is lowered halfway into the aliquot of Nafion and rotated in a circle five times lasting approximately 1 s per rotation. Next the tip is pulled straight out of the Nafion solution. Finally, microelectrodes are placed into an oven at 165°C–175°C for 4 min to allow the Nafion layer to cure on the tip. In general, lower temperature curing produces thicker films compared to higher temperature curing.[20] Also, longer drying times at higher temperatures produce microelectrodes with lower responses to both AA and to analytes such as L-glutamate.

Microelectrodes coated with excessive Nafion show little or no detectable response to electrochemically active molecules. Microelectrodes coated with not enough Nafion record high levels of background AA once inserted into the brain. After the curing procedure, microelectrodes are allowed to cool to room temperature for 30 min before enzyme coating (described in the section on Enzyme Coating Layers). An important concept to remember is that once a prepared Nafion coated microelectrode is placed into an aqueous solution (either a beaker or the brain), then the tip must remain wet or at the very least, removed from solution for only a limited time. When Nafion dries after soaking, it can crack allowing areas where interferents pass through and are oxidized on the Pt recording surfaces. If this occurs, the electrode is no longer selective for the analyte being measured.

1,3-Phenylenediamine

The organic molecule 1,3-phenylenediamine (mPD) is another chemical used to create an exclusion layer for Pt recording electrodes.[21,22] A potential is applied to a solution of mPD, thus causing mPD to electropolymerize onto the Pt recording surfaces. Historically, mPD or its derivate 1,2-phenylenediamine (*o*-phenylenediamine), is used on carbon fiber microelectrodes for selectivity and to prevent electrode fouling—a process where oxidizable molecules such as 5-HT can adhere to the carbon fiber surface thus preventing detection. Electropolymerized mPD selectivity is achieved by forming a size exclusion layer that prevents larger molecules such as AA, DA, and DOPAC from reaching the recording surface. Smaller molecules, such as nitric oxide and peroxide, are still able to pass through the matrix.[23] Since peroxide is a reporter molecule for oxidase enzymes, it makes mPD an ideal exclusion layer for our enzyme-based multisite microelectrodes.[21,24]

The electropolymerizing procedure is completely different from Nafion coating but has four distinct advantages: (1) The matrix formed by mPD blocks DA, NE, and 5-HT from reaching the Pt recording sites. This helps make the microelectrode more selective for measuring the analyte of interest; (2) The mPD is electropolymerized onto the microelectrode after the microelectrode is coated with an enzyme layer. The mPD electropolymerizes through the enzyme layer to form the matrix. Since mPD is electropolymerized after enzyme coating, additional coatings of mPD can be applied to the microelectrode surface without having to clean the microelectrode. If the matrix degrades during an experiment, the microelectrode can be replated, recalibrated, and re-used to finish the experiment; (3) Because a potential must be applied, Pt recording sites can also be selectively coated with mPD. Removing the applied potential between a Pt recording site and the headstage prevents mPD from being electropolymerized onto that site; and (4) Once mPD has been successfully electropolymerized onto the Pt recording sites, the solution does not have to remain wet. In fact, our laboratory has found that exposure to aqueous solutions slowly degrades the matrix over time. The procedure for electropolymerizing mPD to the Pt recording sites is outlined below.

Our laboratory has found that 1,3-phenylenediamine dihydrochloride, 99% (Sigma-Aldrich, Catalog #235903-25g) works best for electroplating and matrix formation on our Pt recording sites. First, a solution of 5 mM mPD is prepared in a degassed solution of 0.05 M phosphate buffered saline (PBS). Degassing is accomplished by bubbling nitrogen gas through the 0.05 M PBS solution for 20 min to remove oxygen before dissolving mPD. Once the 5 mM mPD is dissolved in the degassed 0.05 M PBS, the mPD solution is stored in a brown glass bottle to prevent oxidation from light. If the solution turns yellow, it has oxidized and will no longer electropolymerize. Even with these storage methods, the 5 mM mPD solution can oxidize after 3 h so it is recommended using the solution immediately after preparation. When the mPD solution is prepared, the microelectrode is connected to the FAST-16 recording system and approximately 40 ml of the 5 mM mPD is poured into a 50 ml beaker. The microelectrode tip is lowered halfway into the solution along with a glass, Ag/AgCl reference electrode (Bioanalytical Systems, Inc. RE-5B, Catalog #MF-2079). A constant potential of +0.5 V versus a Ag/AgCl reference electrode is applied for 15 min to allow the mPD to electropolymerize onto the Pt recording sites. At 15 min, the microelectrode tip is removed from the 5 mM mPD solution, rinsed with ddH_2O, and stored at room temperature for twenty-four hours prior to calibration.

ENZYME LAYER COATINGS

Oxidase Enzymes

Enzymes provide a means to convert a molecule that is not inherently electroactive and thus not measurable, into a reporter molecule such as peroxide that is oxidized at the Pt recording surfaces of the microelectrodes. The current measured from the oxidation of peroxide generated during the enzymatic breakdown is directly proportional to the analyte concentration. Table 19.1 provides a list of available oxidase enzymes and their potential uses. Some compounds require multiple enzymes to convert them to a reporter molecule, such as acetylcholine and γ-aminobutyric acid (GABA). Oxidase enzymes that our laboratory has used to measure neurochemicals are highlighted in Table 19.1 and include L-glutamate oxidase,[1–5] choline oxidase,[25,26] L-lactate oxidase,[27] and L-glucose oxidase. Furthermore, acetylcholine esterase is used to measure acetylcholine (in conjunction with choline oxidase) and catalase is used to prevent peroxide detection.

As previously mentioned, the O_2 dependence of oxidase enzyme-coated microelectrodes is also a concern. It is widely known that O_2 is required by the enzymes to measure the analyte, but a portion is returned to the tissue by the oxidation of peroxide to O_2 and H^+ at the microelectrode surfaces. Often oxidase enzyme-coated microelectrodes are relatively free from O_2 dependence over a useful range in vivo,[25] but certain applications, such as those during recordings of stroke episodes, may require correction by use of an O_2 sensing microelectrode.

Immobilization of enzymes to the Pt recording surface helps to stabilize the enzymes and makes them active for longer periods of time. Our laboratory has immobilized several of the oxidase enzymes onto the Pt recording surfaces. While the general procedure remains the same for cross-linking the enzymes to the Pt recording sites, relative ratios of the enzyme to protein mixtures may vary. Our laboratory focuses extensively on L-glutamate oxidase for the detection of L-glutamate dynamics in the mammalian CNS whose immobilization procedure is outlined in detail below.

L-Glutamate Oxidase Coating Procedure

Properly cleaned microelectrodes are ready for enzyme immobilization. A stock solution of L-glutamate oxidase (Seikagaku America, Catalog #100645-1) is prepared by adding 50 μl of ddH_2O to the lyophilized, purified enzyme in a vial to make a final concentration of 1 U/μl. All proteins and enzymes are brought to room temperature and 0.10 g of bovine serum albumin (BSA) Fraction V, 99% (Sigma-Aldrich, Catalog #A-3059) is dissolved in a 1.5 ml microcentrifuge tube containing 985 μl ddH_2O by manual agitation. Once dissolved, 5 μl of glutaraldehyde solution, Grade I, 25% (Sigma-Aldrich Catalog #G-6257) is added to the BSA mixture and manually mixed by inversion five times. The solution is then set aside for 5 min until it turns a faint yellow color. Glutaraldehyde crosslinks proteins to the microelectrode surface when cured. The BSA serves as a matrix to protect the oxidase enzyme activity during immobilization. Next, 9 μl of the BSA/glutaraldehyde mixture is removed and added to a 300 μl microcentrifuge tube and to this 1 μl of the L-glutamate stock solution (1 U/μl) is added and mixed by pipette agitation. This 10 μl solution has a final concentration of 1% BSA, 0.125% glutaraldehyde, and approximately 1% L-glutamate oxidase. An extensive series of parametric studies were carried out to determine this optimized stoichiometry for enzyme immobilization.

All enzymes are currently applied to the Pt recording sites by hand. Solutions are drawn up into a 25 μl Gastight® Hamilton microsyringe (Hamilton Co., Catalog #80065) and slowly dispensed to form a small droplet of solution at the tip of the microsyringe. Using a dissecting microscope (and a steady set of hands), the droplet is applied briefly to the Pt recording sites. The solution quickly

TABLE 19.1
Available Oxidase Enzymes with Their Substrates, Products, and Neurochemicals

Enzyme	Substrate	Product	Use
Acetylcholinesterase	**Acetylcholine**	**Choline, Acetic Acid**	**Measure Choline with Choline oxidase**
Alcohol oxidase	Alcohol, O_2	H_2O_2, Aldehyde	Measure Alcohols
D-amino acid oxidase	D-amino acids, O_2	H_2O_2, NH_4^+, 2-oxo acid	Measure D-amino acids
L-amino acid oxidase	L-amino acids, O_2	H_2O_2, NH_4^+, 2-oxo acid,	Measure L-amino acids
Ascorbate oxidase	**Ascorbate, O_2**	**Dehydroascorbate**	**Remove ascorbate**
Aspartate oxidase	Aspartate, O_2	H_2O_2, NH_4^+, Oxaloacetate	Measure aspartate
Catalase	**H_2O_2**	**O_2**	**Remove H_2O_2**
Cholesterol esterase	Cholesteryl oleate, or other Cholesterol ester, Taurocholate	Cholesterol, Oleic acid, or other corresponding ketone	Measure cholesterol ester with cholesterol oxidase
Cholesterol oxidase	Cholesterol, O_2	H_2O_2, 4-cholesten-3-one	Measure cholesterol
Choline oxidase	**Choline, O_2**	**H_2O_2, Betaine**	**Measure choline**
Galactose oxidase	D-galactose, O_2	H_2O_2, D-galacto-hexodialose	Measure galactose
Glucose oxidase	**Glucose, O_2**	**H_2O_2, Gluconic Acid**	**Measure glucose**
L-glutamate oxidase	**L-glutamate, O_2**	**H_2O_2, α-ketoglutarate**	**Measure L-glutamate**
GABase	**GABA, α-ketoglutarate**	**NADPH, H^+**	**Measure GABA with L-glutamate oxidase**
Glutaminase	Glutamine	NH_4^+, L-glutamate	Measure Glutamine with L-glutamate oxidase
Glycerol kinase	Glycerol, ATP	Glycerol-3-phosphate, ADP	Measure glycerol with glycerol-3-phosphate oxidase
Glycerol-3-phosphate oxidase	Glycerol-3-phosphate, O_2	H_2O_2, Dihydroxyacetone phosphate	Measure glycerol
Hexokinase	ATP, Glucose	Glucose-6-phosphate, ADP	Measure ATP with L-glutamate oxidase
Horseradish peroxidase	H_2O_2	H_2O, O_2	Use as mediator for H_2O_2 detection
Lactate oxidase	**Lactate, O_2**	**H_2O_2, Pyruvate**	**Measure lactate**
Lysine oxidase	Lysine, O_2	H_2O_2, NH_4^+, 6-amino-2-oxohexanoic acid	Measure lysine
Pyruvate oxidase	Pyruvate, O_2, Phosphate	H_2O_2, CO_2, acetyl phosphate	Measure pyruvate
Sarcosine oxidase	Sarcosine, O_2	H_2O_2, Glycine, Formaldehyde	Measure sarcosine
Xanthine oxidase	**Xanthine, O_2**	**H_2O_2, Uric acid**	**Measure xanthine**

Note: Oxidase enzymes that have been successfully immobilized onto our Pt recording surfaces and are shown in bold.

dries, leaving behind a thin, translucent layer of enzyme that is visible underneath a dissecting microscope. Two additional coats of enzyme are applied in the same manner with 1-min dry times in between each coat. This procedure is complete once a visible film remains following coating. The advantage of using multisite microelectrodes is that a pair of recording sites is coated with the enzyme mixture and the adjacent sites are coated with the inactive protein matrix (BSA and glutaraldehyde) in the same manner as applying the enzyme mixture. The laboratory refers to this technique as "self-referencing",[1] and this approach has many advantages, which are discussed later in this chapter.

Once the enzyme and/or inactive protein matrix is applied, microelectrodes are stored at room temperature. *Enzyme-coated microelectrodes must cure at room temperature for 48–72 h prior to calibration and experimentation.* Complete curing increases enzyme layer adhesion to the microelectrode surface. This process provides better sensitivity to L-glutamate as well as increases the shelf-life of the microelectrodes. Our laboratory recommends using the microelectrodes within three weeks after coating. The enzyme/protein layers on "uncured microelectrodes" can dissolve when put into solution. Optimal curing times may vary depending on temperature, humidity, and the type of enzyme. Appendix A provides a step-by-step procedure for cleaning and preparing L-glutamate microelectrodes.

Since photolithographic procedures can place adjacent Pt recording sites within tens of microns from one another, coating individual Pt sites with enzyme layers by hand is often difficult if not an impossible task. For this reason, our laboratory is developing a microcoater that can precisely control the application of microliter amounts of enzyme solutions. This instrument can precisely coat individual recording sites while significantly decreasing the variability caused by hand coating procedures. Additional details of this new technology will be published elsewhere once work is completed.

CALIBRATION

Our laboratory has observed that the micro-fabricated microelectrode arrays have highly reproducible Pt recording surfaces. However, manufacturing procedures only investigate the geometric surface area of the microelectrodes. Since each Pt recording site on a microelectrode can respond differently to peroxide and L-glutamate, they must be calibrated in vitro prior to experimentation to determine standard curves. Thus, calibrations are used to equate a change in current from the oxidation of peroxide to a proportional change in analyte concentration from the L-glutamate generating the peroxide. A known concentration of analyte is added to 40 ml of 0.05 M PBS solution at 37°C to generate a current in picoamperes (pA) that is measured by the FAST-16 system. The FAST-16 system software records the current for each addition of analyte, creates a calibration curve for each Pt recording site, and stores the slope of this calibration. Also, known interferents such as AA are added during the calibration to test the selectivity of the recording sites to the analyte of interest versus interferents. When the experiment is being performed, the calibration data are recalled and used to determine the concentration of analyte being measured from the change in current during experiments in vivo.

Calibration Preparation

Nafion coated microelectrode tips are soaked in a solution of 0.05 M PBS at 37°C for at least one hour prior to calibration. These microelectrodes can be placed in the PBS solution overnight for calibration the next morning. The soaking time allows for better diffusion of analytes through the Nafion layer as well as activation of the enzyme layer. The mPD-coated microelectrodes, on the other hand, do not require soaking. The time period that the microelectrode tip sits in the calibration solution is sufficient for proper enzyme activation without additional soaking.

All stock solutions for L-glutamate calibrations are prepared in ddH_2O and listed below:

1. 20 mM AA (Ascorbic acid powder, Fisher Scientific, Catalog #A62)
2. 20 mM L-glutamic acid sodium salt hydrate, minimum 99% (Sigma-Aldrich, Catalog #G1626)
3. 2 mM DA (3-Hydroxytyramine hydrochloride, Sigma-Aldrich, Catalog #H-8502)
4. 8.8 mM hydrogen peroxide (Rite Aide, 3% Hydrogen Peroxide Topical Solution)

Solutions are stored in a refrigerator at 4°C. AA and hydrogen peroxide are made fresh daily since they oxidize in solution. L-glutamate is made fresh on a weekly basis. DA is kept for a month providing the stock solution contains 0.01 M perchloric acid.

Calibration Procedures

Using the FAST-16 recording system, a combination headstage gain of 2 nA/V and secondary gain of 10X generates a final 200 pA/V gain for all calibrations. Forty milliliters of 0.05 M PBS is measured out with a graduated cylinder and added to a 50 ml beaker. This beaker is placed in a recirculating water bath (Quanteon, L.L.C.) set at 37°C resting upon a battery operated, portable magnetic stir plate (Barnant Co., Model #700-0153). The calibration temperature is maintained at 37°C to activate L-glutamate oxidase for physiologically relevant temperatures. Our laboratory uses a battery operated stir plate to reduce potential AC (60 Hz) current that can affect the recordings. A 10×3 mm^2 stir bar is added to the 0.05 M PBS and the solution is slowly stirred to prevent forming a vortex in the solution. A glass Ag/AgCl reference electrode is placed in the buffer solution. Finally, the microelectrode tip is lowered halfway into the buffer solution. Figure 19.6a shows a photograph of the calibration system.

After initial system settings are completed on the FAST-16 software, the FAST-16 recording system applies a potential of +0.7 V versus the Ag/AgCl reference to the microelectrode recording surfaces. Once calibration has begun, the current is allowed to reach a stable baseline (approximately 10–15 min, but may take longer). When a stable baseline is achieved, the user marks the baseline. (When marking, the user selects a time point for an event, such as baseline, interferent, or analyte addition, which is recorded by the FAST-16 software. A series of data points before the mark are averaged and used in the calibration analysis.) Next, the user adds 500 μl of 20 mM AA (interferent) for a final beaker concentration of 250 μM. When a new stable baseline is reached the user marks the interferent. Next, three 40 μl additions of 20 mM L-glutamate are added for final buffer concentrations of 20, 40, and 60 μM L-glutamate. Analyte marks are recorded after each addition to create the calibration curve. After 3 additions of analyte, 40 μl of 2 mM DA (2 μM, final beaker concentration) is added to the solution as a test substance and marked. Finally, 40 μl of 8.8 mM (8.8 μM, final beaker concentration) hydrogen peroxide is added to the solution to confirm microelectrode sensitivity, and the addition is marked. The test substances are used to check selectivity and can be used to normalize the enzyme coated and inactive protein matrix sites for self-referencing recordings. The additions of test substances do not factor into calculating the standard curve for the analyte of interest. All chemicals used in vivo should be tested in vitro to ensure that they are not electrochemically active on the Pt recording surface. It is important to know whether a chemical is electrochemically active in vitro prior to its local application near the Pt recording sites during experimentation. This ensures that in vitro or in vivo analyte concentration changes are from the analyte of interest rather than due to local application of drugs. Upon completion of the calibration, the program is stopped, the data are saved, and the microelectrode is removed from the buffer solution and stored appropriately. Nafion coated microelectrode tips are placed in 0.05 M PBS while mPD electroplated microelectrodes are kept dry. Graphs of typical self-referencing L-glutamate calibrations with either Nafion or mPD applied as exclusion layers are shown in Figure 19.6b and Figure 19.6c, respectively.

Microelectrode Calibration Criteria

During calibration, the FAST-16 recording software automatically calculates selectivity ratios for L-glutamate over AA as well as the slope (microelectrode sensitivity for L-glutamate), limit of detection (LOD), and linearity (R^2). What do all these numbers mean, and how does one know if they have a satisfactory L-glutamate microelectrode? A poor R^2 is seen rarely with good microelectrodes. Since the microelectrode array fabrication procedure is highly reproducible, our

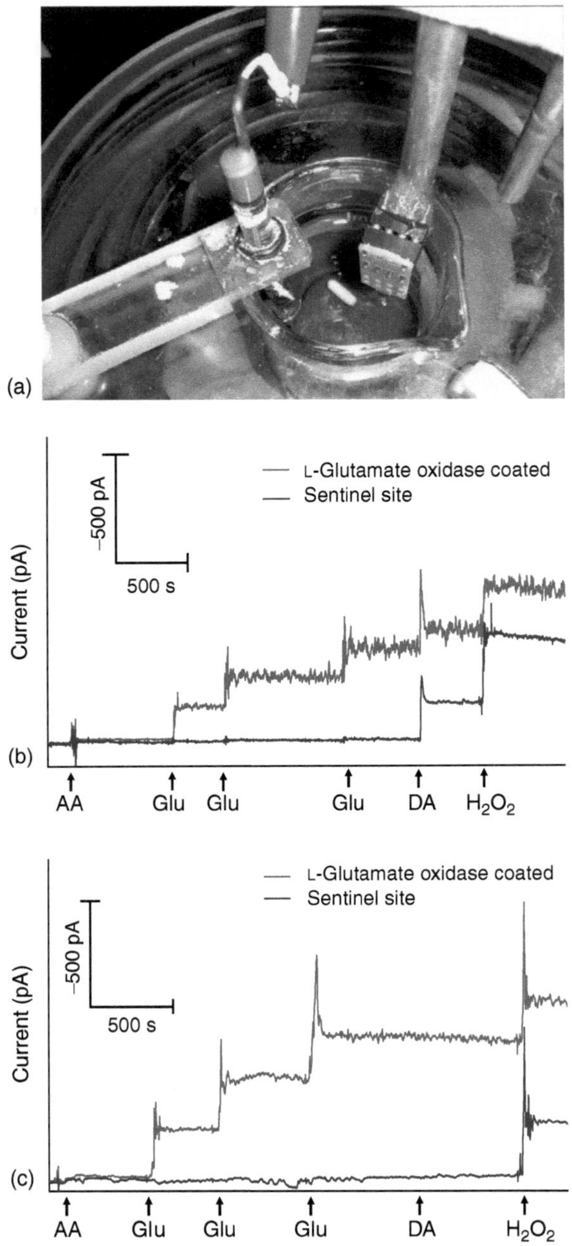

FIGURE 19.6 (a) Photograph of a microelectrode immersed in the calibration chamber. The calibration solution is maintained at 37°C with a heated waterbath. (b) Calibration of a Nafion coated L-glutamate (Glu) self-referencing microelectrode. During calibration, solutions are added to determine each recording site's sensitivity to Glu, selectivity for Glu over AA, and responsiveness to other electroactive molecules (e.g. DA and H_2O_2). The enzyme coated Glu recording site responds well to Glu and some electroactive molecules. The enzyme-free sentinel sites does not respond to Glu but shows similar responses to the same electroactive molecules. Final concentrations of each addition: 250 μM AA, 20, 40, and 60 μM Glu, 2 μM DA, and 8.8 μM H_2O_2. (c) Calibration of an mPD coated Glu self-referencing microelectrode. All calibration concentration are the same. Note how mPD excludes the DA response on both the L-glutamate oxidase and sentinel recording sites.

research has found that L-glutamate responses are extremely linear and should result in linear regression curve fits with $R^2 \geq 0.99$.

Slope or sensitivity of the microelectrode refers to how well it can measure the change in L-glutamate. The number is used to equate a change in current to a change in L-glutamate concentration. The slope also is used to calculate LODs, which are the most important criteria for determining if a microelectrode is satisfactory for use. Here LODs are used to select microelectrodes because slopes can be misleading. LOD refers to the limit of detection for the microelectrodes defined as the analyte concentration that yields an electrode response that is equivalent to three times the background noise of the recording system. This is the lowest detectable change in analyte concentration that cannot be attributed to noise. The LODs for the study's microelectrodes range from 0.2 to 1.0 μM, depending on the recording enzyme composition and stability of the sites. The user must select microelectrodes with LODs that are lower than the response they expect to observe. Normally it is recommended using microelectrodes with LODs ≤ 1 μM.

Selectivity refers to a ratio of the microelectrodes sensitivity for L-glutamate over interferents such as AA. It is calculated by dividing the L-glutamate slope by the AA slope. A microelectrode with a selectivity of 100:1 means that a 100 μM concentration increase of AA results in an apparent 1 μM concentration increase in L-glutamate. With a selectivity of 100:1, the user knows that the microelectrode is 99% effective at blocking AA. Selectivity ratios of 100:1 or greater are ideal, but selectivity ratios of 20:1 are used and are still extremely effective because this ratio is not a linear correlation. In other words, a microelectrode with a selectivity of 50:1 is blocking approximately 98% of the AA while a microelectrode with a selectivity of 20:1 is still blocking approximately 95% of the AA signal.

Glass Micropipettes for Local Drug Delivery

After calibrations are completed, a glass micropipette often is attached to the PCB holder so the tip lies among the Pt recording sites. Through this glass micropipette, drug solutions are pressure ejected for local pharmacological manipulations of the CNS. Our laboratory uses a vertical pipette puller (David Kopf Instruments, Model 700C, Catalog #730) to prepare single barrel glass (1 mm o.d., 0.58 mm i.d., A-M Systems, Inc., Catalog #601500) micropipettes. Using a dissecting microscope, the tip of the micropipette is bumped against a glass stir rod to create an internal diameter of 12–15 μm. Next, modeling clay is attached to the PCB holder. The glass micropipette is attached to the modeling clay and used to hold the glass micropipette steady while it is positioned between the recording sites. Once the glass micropipette is properly positioned, it is held more firmly in place by applying melted Sticky Wax (Kerr Corporation, Catalog #00625) onto the PCB between the ceramic wafer and the modeling clay. The drying of the Sticky Wax may reposition the glass micropipette relative to its placement amongst and above the Pt recording sites. If necessary, the tip of a spatula can be heated to help remodel the Sticky Wax and thus reposition the glass micropipette. Figure 19.7a shows a glass micropipette attached to the PCB with modeling clay and Sticky Wax. Figure 19.7b and Figure 19.7c show the tip of a glass micropipette positioned within the recording sites for an R1 and S2 microelectrode, respectively. The tip of the glass micropipette must not touch the ceramic surface. Instead it should be positioned between 50 and 100 μm above the Pt recording sites (as shown in Figure 19.7d). This distance allows an optimum amount of CNS tissue between the glass micropipette and microelectrode when it is lowered into the brain. Additionally, if the microelectrode is too close to the Pt recording sites, pressure ejection of fluid causes a dilution effect in vivo (an actual decrease in analyte concentration due to the introduction of exogenous fluid in the area of the Pt recording site). If positioned too far, the rapid uptake of L-glutamate into glia limits its diffusion to the Pt recording sites on the microelectrode surface. To properly position the glass micropipette on the microelectrode, the microelectrode must be viewed at two angles underneath a dissecting microscope. Our lab has designed a special holder for this application consisting of two glass microscope slides glued

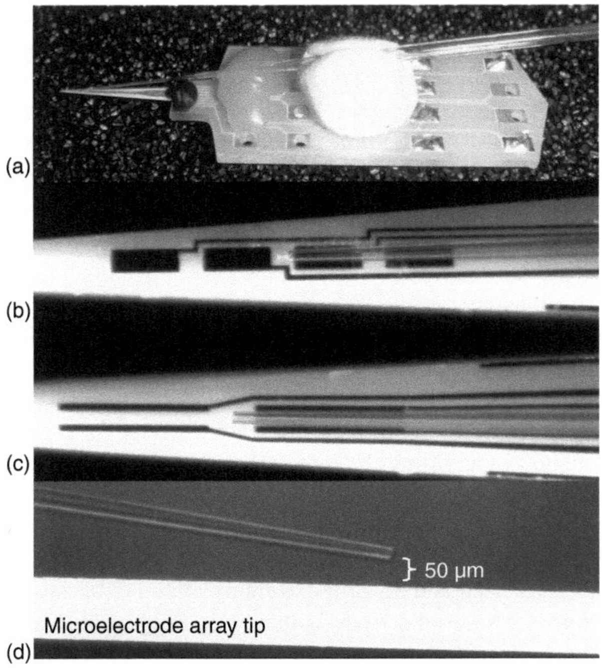

FIGURE 19.7 Photographs of a glass micropipette attached to a multisite microelectrode. (a) Photograph of the glass micropipette held in place with modeling clay and Sticky Wax. (b) Photomicrograph of a glass micropipette tip positioned between sites 2 and 3 on an R1 microelectrode. (c) Photomicrograph of a glass micropipette tip positioned among the four Pt recording sites on an S2 microelectrode. (d) Side view photomicrograph of the glass micropipette tip held approximately 50 μm above the microelectrode array tip.

perpendicular to one another with an eight pin DIP socket glued onto one of the flat surfaces (Figure 19.8). Additionally, double and triple barrel glass micropipettes (World Precision Instruments Catalog #2B150F-6, and #3B120F-6, respectively) can be used to locally apply two or three different drug solutions into the same brain region. These are harder to utilize because the adjoining

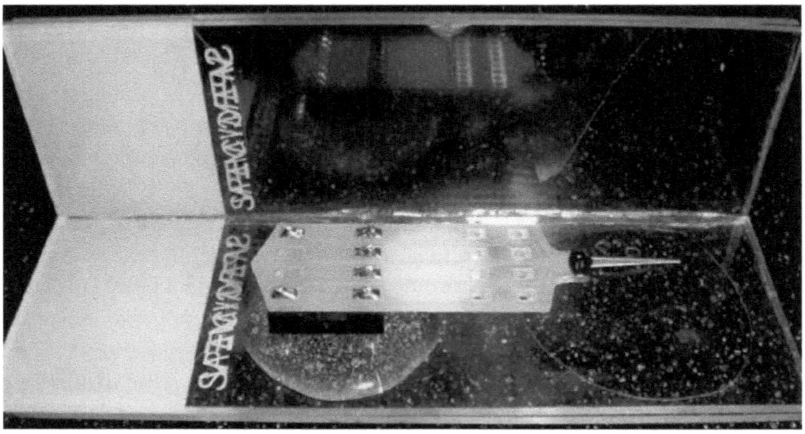

FIGURE 19.8 Photograph of the microelectrode holder used for attaching glass micropipettes to the microelectrode arrays. The microscope slides fit any standard dissection microscope and allows for easier placement and viewing of the micropipette at both angles.

micropipettes prevent tubing from easily sliding over the openings of the glass micropipette. Instead, smaller tubing (Small Parts, Inc., PTFE Tube Regular Wall, 28-gauge, Catalog #SWTT-28) is wedged into the open end of each micropipette, which attaches to the tubing from a pressure ejection system such as a Picospritzer® III (Parker Hannifan, Corp., General Valve Operation).

Attaching the glass micropipette is probably one of the most difficult tasks of an experiment and takes practice to properly achieve. Practicing is also ideal for success of experiments because positioning the glass micropipette on a microelectrode must occur within a short time period for optimum microelectrode performance.

The glass micropipette is filled with solution using a 30-gauge, 4-in. long hypodermic needle (Popper and Sons, Inc., Standard Female Luer Hub, Catalog #7400). Glass micropipettes always are filled prior to implantation. First, drug solutions are drawn into a 1 ml tuberculin tip syringe (Beckton-Dickinson, and Co., Catalog #309602), which is then attached to the intake end of a μStar 0.22 μm sterile filter (Costar Corp, Catalog #8110). The other end of the filter attaches to a 4-in., 30-gauge, hypodermic needle. The solution is dispensed through the filter and ultimately through the hypodermic needle. The needle is inserted into the glass micropipette and filled from the pulled tip up to the open end making sure not to create air bubbles within the glass micropipette. Air bubbles prevent effective delivery of drugs by pressure ejection and make it difficult to determine how much fluid is dispensed. Solutions are normally pressure ejected prior to microelectrode implantation into the brain to determine that there is a smooth fluid flow out of the glass micropipette. Solutions are removed from a glass micropipette by inserting the 30-gauge filling needle with an attached tuberculin syringe and drawing up the solution.

Ag/AgCl Reference Electrode

While larger glass Ag/AgCl reference electrodes are suitable for mPD electropolymerization and microelectrode calibration, this reference electrode is too large for many biological applications. For this reason, a smaller Ag/AgCl reference electrode must be made at the start of every experiment. Miniature Ag/AgCl reference electrodes are prepared by first stripping 0.25 in. of the Teflon coating from each end of a silver wire (0.008 in. bare, 0.011 in. coated; A-M Systems, Inc., Catalog #786500). One of the stripped ends is soldered to a wire crimp pin (Mill-Max Mfg. Corp., Part #3603) for connection to the FAST-16 headstage. The other end is placed into a solution of 1 M HCl saturated with NaCl. A Pt wire acts as the counter electrode and is also placed into the solution. The laboratory routinely uses a 9V DC adapter fitted with alligator clips on both the positive and negative leads. However, DC adapters ranging from 1.5 to 9 V can be used. The negative lead is attached to the Pt counter electrode while the positive lead is connected to the Ag wire. With both wires in solution, the applied potential attracts Cl^- on the wire to form AgCl thus making the Ag/AgCl reference. The plating potential is applied for approximately 10–15 min. When properly connected, bubbles are seen around the Pt counter electrode. These miniature Ag/AgCl references are soaked in 3 M NaCl prior to use in vivo.

Signal Analysis

These are the basic steps necessary for successfully measuring L-glutamate in the CNS using our multisite microelectrode arrays. The steps for setting up different experiments using this technology will be further discussed later in this chapter. Before there is an explanation of the experimental protocol, there must be an examination of the features of a typical, rapid L-glutamate signal and how it is analyzed.

The Picospritzer III or related hardware connects to the FAST-16 recording system, so when solutions are pressure ejected through the glass micropipette, or experimental events occur, the software records these events. The FAST-16 recording system automatically saves amperometric or chronoamperometric data, time, and pressure ejections marks (external events) for all recording

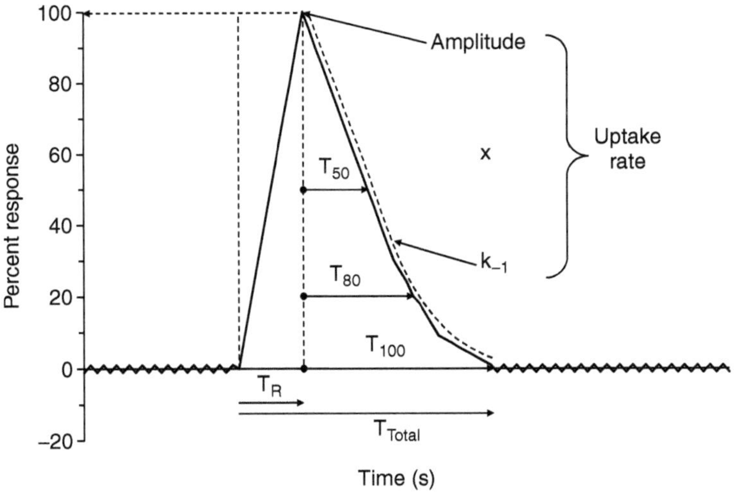

FIGURE 19.9 A schematic diagram of an L-glutamate signal is shown to illustrate how different signal parameters are measured and calculated. See text for details. Typically L-glutamate or other neurotransmitter signals are plotted as concentration versus time. The *y*-axis has been changed to percent response to clarify the definitions of certain signal parameters.

sites for a specified time period. Figure 19.9 shows a typical L-glutamate signal with some of the analysis terms. A stereomicroscope fitted with a reticule monitors the change in solution in the glass micropipette after pressure ejection. Pressure ejection of 70 mM KCl (70 mM KCl, 79 mM NaCl, 2.5 mM $CaCl_2$, pH 7.4) leads to a depolarization event and release of L-glutamate into the extracellular space, causing a robust and reproducible rise in the L-glutamate signal compared to baseline. The maximum change in concentration compared to baseline is referred to as maximum amplitude or amplitude of the L-glutamate signal. This value is normally recorded in μM units. Next, the maximum amplitude of the signal is divided by the amount of stimulus (i.e. volume ejected from the glass micropipette) necessary to elicit the depolarization event. The stimulus ejected is visualized via a stereomicroscope[28] mounted on a universal boom stand (Meiji EMZ Zoom, Labtek, Inc.) fitted with a reticule. Volume displacement is calibrated according to the pipette inner diameter. (Note: the single barrel glass micropipettes used have a solution volume of approximately 250 nl/mm of glass, but this varies with the inner diameter/total volume of the glass micropipette. For this reason, one should use the type of glass micropipettes from A-M Systems, Inc., Catalog #601500 as discussed in the section on glass micropipettes for local drug delivery.) The volumes of stimulus ejected are recorded in nanoliters, so this calculation is referred to as amplitude per nanoliter (μM/nl) and gives an interpretation of the excitability of the neurons. Finally, the time period for the signal to reach the maximum amplitude from baseline is recorded (in s) and referred to as rise time (T_r).

Signal decay refers to the point of maximum amplitude of the signal to its return to baseline. The clearance of L-glutamate is a result of the rapid uptake of L-glutamate into glial and neuronal transporters.[29] L-Glutamate uptake follows an apparent first order decay rate. The decay of the L-glutamate signal is fitted to the slope of the linear regression of the natural log transformation of the data over time and this is referred to as k_{-1} and has units of s^{-1}. When k_{-1} and the maximum amplitude of the signal are multiplied, the uptake rate, or micromolar of L-glutamate removed from the extracellular space per second (μM/s), is obtained. Finally, the decay of the signal can be examined at different percentages of decay. These often include T_{50}, T_{80}, and T_{100}, and follow the general format of T_D (s), where T refers to the time from maximum amplitude for the signal to decay by D percent. Typically T_{50} and T_{80} decay times are used.

AMPEROMETRIC RECORDINGS UTILIZING SELF-REFERENCING MICROELECTRODES

IMPORTANCE OF SELF-REFERENCING MICROELECTRODE ARRAYS

An advantage of photolithographic fabrication of microelectrode designs is the precise control of the layout of the microelectrodes allowing for multiple recording sites geometrically configured in a small area. Multiple recording sites on a single microelectrode can be used to remove chemical interferents that may contribute to a portion of the analyte signal.[1] Our laboratory refers to this principle as self-referencing, and it is commonly used in the data analysis. As mentioned in the coating section, the L-glutamate oxidase is applied to a pair of Pt recording sites while the chemically inactive protein matrix (BSA and glutaraldehyde) is applied to the adjacent pair of recording sites referred to as "sentinel sites". A schematic of the coating procedure is shown in Figure 19.10. The L-glutamate oxidase coated Pt recording sites detect extracellular L-glutamate concentrations (the analyte) in addition to molecules that the inactive protein matrix coated recording sites can detect (interferents). These additional molecules detected can include interferents such as AA and DOPAC as well as other neurotransmitters such as DA, NE, 5-HT. Figure 19.6b and Figure 19.6c show a typical L-glutamate self-referencing calibration where the L-glutamate oxidase coated sites respond to L-glutamate while the inactive protein matrix coated sites do not.

The purpose of the inactive protein matrix on the self-referencing, or sentinel sites, is to make the response times of the sentinel sites similar to the L-glutamate oxidase recording sites. As mentioned earlier, protein coats can alter the diffusional properties of molecules to the Pt recording

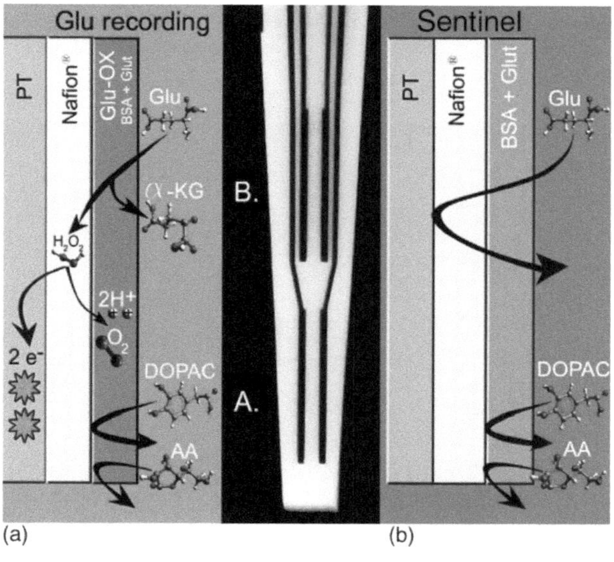

FIGURE 19.10 Schematic and photomicrograph of an S2, Nafion coated, self-referencing, L-glutamate selective multisite microelectrode. (a) Schematic showing the bottom pair of Pt recording sites coated with L-glutamate oxidase (Glu-ox), BSA, and glutaraldehyde (Glut). L-glutamate that comes into contact with the Glu-ox is converted to α-ketoglutarate and peroxide. The peroxide readily diffuses through the Nafion layer that blocks anionic substances such as AA and DOPAC. When the peroxide formed comes into contact with the Pt recording sites at a potential of +0.7 V versus a Ag/AgCl reference electrode, the peroxide is oxidized and the resulting two electrons are transferred to the Pt recording site and the change in current is measured by the FAST-16 system. (b) Schematic showing the top pair of Pt recording sites coated with the inactive protein matrix of BSA and Glut, also referred to as the sentinel sites. Since no Glu-ox is coated onto this pair of recording sites, L-glutamate is not converted to a reporter molecule and thus is not measured on the Pt recording sites. Once again, the Nafion on these sites repels AA and DOPAC.

surfaces. Without this inactive protein matrix coat, the sentinel sites would respond faster to interferents than the L-glutamate oxidase sites. Protein layers on both pairs of Pt recordings sites are necessary to minimize differences between the temporal recording properties of the different microelectrode surfaces. Ideally, the sentinel vs. analyte recording sites should be positioned side-by-side. However, the current, manual enzyme coating procedures preclude this arrangement.

Experimentally, self-referencing recordings are beneficial for providing qualitative as well as quantitative assessment of the analyte of interest. During an experiment, the FAST-16 recording software allows the user to simultaneously view changes in L-glutamate concentration for all four recording sites. If a peak is seen on the L-glutamate oxidase coated sites, but not the inactive protein matrix coated sites, this provides a good qualitative assessment that L-glutamate is being measured. Figure 19.11 shows a screen capture of an L-glutamate self-referencing microelectrode during an in vivo experiment.

Quantitatively, the self-referencing method is even more powerful. Since some of the response on the L-glutamate oxidase coated sites is attributable to interferents the sentinel sites are used to record the interferents contributing to the L-glutamate response. If the recording sites' measure of Faradaic responses to DA and H_2O_2 are close ($<10\%$), the raw current data collected for a sentinel recording site is subtracted from the raw current obtained from the L-glutamate oxidase coated sites. This new raw current, once divided by the slope of the corresponding L-glutamate oxidase recording site, provides the adjusted concentration of L-glutamate that is free of interferents. If the recording sites' calibration non-Faradaic currents and Faradaic response to DA and H_2O_2 are not close

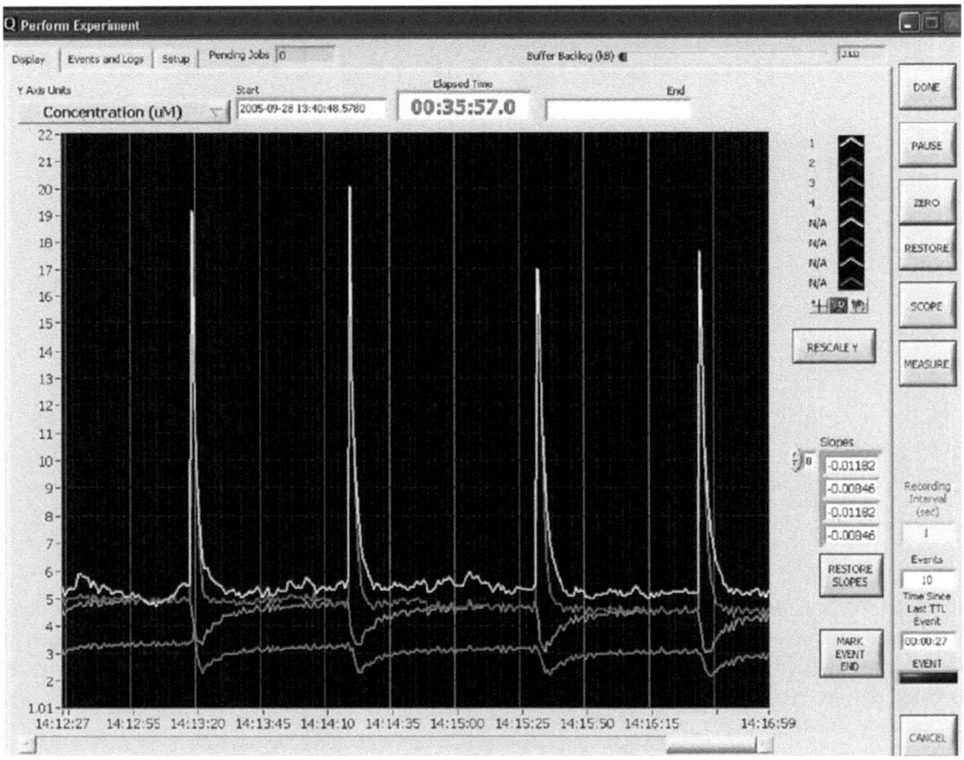

FIGURE 19.11 Screen capture of the FAST-16 system with a self-referencing L-glutamate microelectrode measuring potassium-evoked L-glutamate release in the motor cortex of a rhesus monkey. Each line corresponds to a different Pt recording site. Sites 1 and 2 are coated with L-glutamate oxidase and show a rapid, robust increase in extracellular L-glutamate compared to the two sentinel recording sites (Sites 3 and 4).

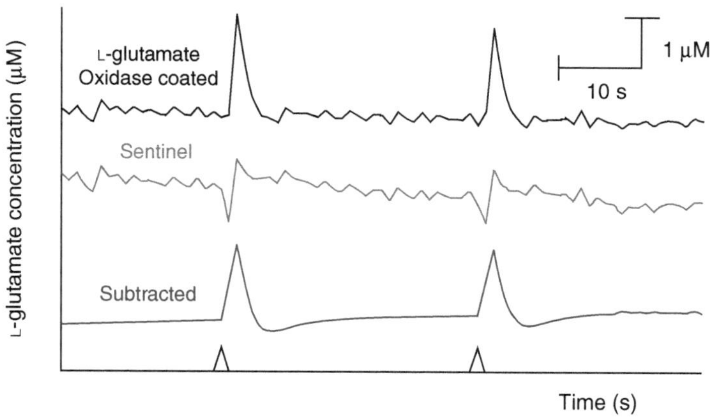

FIGURE 19.12 Pressure ejection of 70 mM KCl (black arrowheads) causes release of L-glutamate in the striatum of a C57BL/6 mouse. Non-Faradaic current decreases the signal-to-noise ratio making it difficult to distinguish between current fluctuations and low L-glutamate responses. The noise occurs on both the L-glutamate oxidase coated sites and the sentinel recordings sites. When the raw current from the sentinel recording site is subtracted from the raw current on the L-glutamate oxidase coated sites, the noise is removed creating a higher signal-to-noise ratio for L-glutamate measurements.

($>$10%), then a normalization procedure is used to correct for recording site coating inconsistencies by factoring in the DA and/or the peroxide responses during calibration.[1]

Besides being able to remove interferents that contribute to an analyte signal, self-referencing recordings can remove periodic or random noise.[2] This is an obvious advantage because smaller changes in current (lower detection limits) can be achieved. Figure 19.12 illustrates removal of noise using self-referencing recordings. The noisy upper trace is a response from an L-glutamate oxidase coated recording site, while the lower trace is the resulting sentinel site response. When the sentinel site raw current is subtracted from the L-glutamate raw current and divided by the slope of the L-glutamate oxidase coated site, the signal-to-noise ratio of the peaks is greatly improved. This dual-channel noise subtraction approach is widely used in other methodologies such as electrophysiology and spectroscopy and can now be used with the precise microelectrode arrays.

Phasic Release of Neurotransmitters

Depolarization of neurons using increased KCl as a stimulus is a common method for in vivo analysis of neurotransmitter release.[3,4,24–26,30–33] While a stimulus does depolarize neurons, nerve fibers, and nerve terminals it does not selectively release neurotransmitters. On the contrary, the solution causes release of a multitude of neurotransmitters. These additional neurotransmitters released into the extracellular space have the potential to contribute to a portion of the analyte response. Luckily, few neurotransmitters are electrochemically active (DA, NE, and 5-HT) without having to be converted to a reporter molecule. DA and 5-HT usually exist in much higher levels than NE in the CNS and therefore are of greater concern for contributing to a portion of the KCl-evoked L-glutamate response. Using the self-referencing technique, contributions from DA, NE, and 5-HT to the L-glutamate response are easily removed.

Interestingly, the present studies support that while release of L-glutamate and DA are relatively fast and can occur simultaneously with applications of KCl, DA has a much slower uptake process compared to L-glutamate.[2,3] This slower uptake gives DA a very distinguishable response signal compared to L-glutamate in vivo. The decay of the DA signal may last minutes while L-glutamate decay takes only seconds. This DA response coincides with the L-glutamate response on the L-glutamate oxidase coated sites, and gives a very distinguishable response that is referred to as

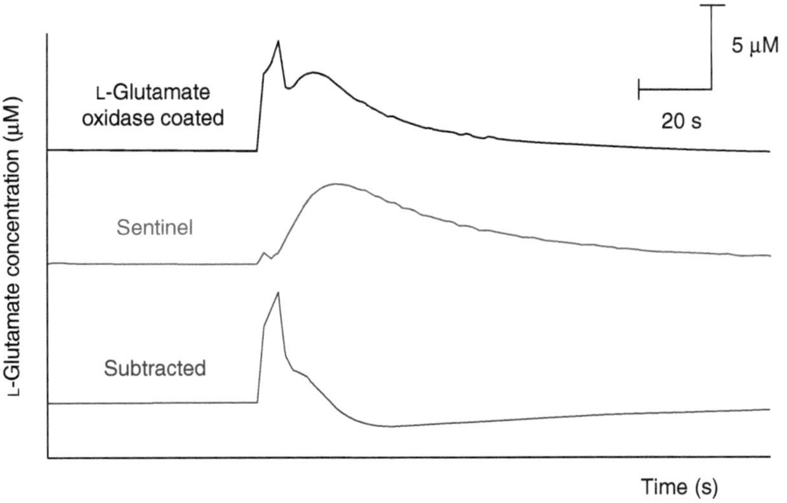

FIGURE 19.13 Stimulus evoked L-glutamate release sometimes results in measuring both L-glutamate (the fast spike) and DA (the slow dome) on the L-glutamate oxidase coated Pt recording sites. The sentinel Pt recording sites only measure the DA contribution to the response. When the raw current from the L-glutamate oxidase coated site is subtracted by the raw current on the sentinel site, only the DA signal is removed leaving primarily the L-glutamate response.

a spike-dome response. The spike corresponds to the quick release and re-uptake of L-glutamate. However, as this L-glutamate response begins to decay, DA release and diffusion continues and its much slower uptake creates a dome response. Essentially, the DA response contributes to a portion of the L-glutamate response making analysis difficult.

As mentioned before, Nafion coated microelectrodes can actually attract DA, NE, and 5-HT to the Pt recording sites and mPD coatings may not always block the DA, NE, and 5-HT responses. Therefore, self-referencing recordings are crucial for removing interferent responses from the analyte response. Since interferent responses occur on the sentinel and L-glutamate oxidase coated sites, the raw current from the sentinel site can be subtracted from the L-glutamate oxidase coated sites. This generates the true L-glutamate response without interferents contributing to the signal. Figure 19.13 shows a typical spike-dome response in a mouse striatum from an L-glutamate oxidase coated site, the DA response on the sentinel site, and the subtracted L-glutamate response.

Tonic or Basal Release of Neurotransmitters

The photolithographic fabrication methods used allow the multiple recording sites to be patterned in close proximity to one another. This provides a novel opportunity to measure basal levels of neurotransmitters. Basal measures refer to the low levels of neurotransmitters constantly found in the extracellular space that are important for CNS homeostasis. High in vivo concentrations of interferents, such as AA may contribute to a portion of the L-glutamate basal responses. Once again, self-referencing is utilized to subtract off the portion of the interferent signal that contributes to the L-glutamate signal thus providing basal L-glutamate measurements.

Furthermore, basal L-glutamate regulation is studied using pharmacological methods. Local applications of Tetrodotoxin (TTX), a sodium ion channel blocker, or DL-threo-beta-benzyloxyaspartate (TBOA), an excitatory amino acid uptake inhibitor, are utilized to study their effects on basal L-glutamate concentrations. The laboratory has demonstrated successfully that local application of TTX decreases basal L-glutamate while TBOA increases basal L-glutamate on the

L-glutamate oxidase coated microelectrodes.[5] Sentinel sites do not show changes in their basal recordings after administration of either drug. In this scenario, a self-referencing microelectrode qualitatively validates the effects of pharmacological manipulations. This assessment is made directly after local application of the drug and is easily seen using the FAST-16 recording system. Once again, self-referencing can quantitatively determine the change in basal L-glutamate since the "normalized" raw currents from the sentinel sites are subtracted from the raw currents of the L-glutamate oxidase coated sites to obtain changes in basal L-glutamate. This is a unique recording feature of the microelectrode arrays that cannot be carried out using single microelectrode recordings.

The principle of self-referencing holds great promise for the future of in vivo electrochemical recordings. Self-referencing involves a control recording site introduced in close proximity to the analyte recording sites. This technique allows the user to qualitatively assess the validity of an analyte signal during an experiment. Additionally, the ability to subtract off interferents and noise during off-line data analysis allows the user to obtain accurate and detect smaller concentrations of analytes

APPLICATION TO BIOLOGICAL SYSTEMS

IN VITRO SYSTEMS

In vitro cell culture techniques provide scientists with simpler, more controllable conditions for studying isolated systems. The multisite microelectrodes are small enough for basic cell culture studies as well as brain slice measures. Below the experimental procedures and techniques used for both are discussed.

Cell Culture Techniques

The culturing of cells does not change and standard culture dishes are used with the enzyme-based multisite microelectrodes. In fact the microelectrodes are capable of recording from a variety of cell culture well sizes.

Culture plates are placed on top of a heated water pad (Gaymar Industries, Inc Model #TP12E) that holds the temperature constant at 37°C and insulated with towels to help prevent heat loss from the air. An electrode manipulator on a stereotaxic frame (Kopf Instruments, Model #960) is positioned next to the culture dish, and the microelectrode connector with attached microelectrode is secured to the electrode manipulator. The glass micropipette attached to the microelectrode is filled with solution as previously described. The tubing from the Picospritzer III is laced through the electrode manipulator and secured onto the glass micropipette. Lacing the tubing through the electrode manipulator prevents breakage of the glass micropipette by an accidental tug on the tubing. The solution flow through the glass micropipette is tested with several pressure ejections before lowering into a well. If an air bubble is found, the solution is removed, and the glass micropipette is refilled and tested again. Finally, the microelectrode is lowered into a culture well with the aid of a stereomicroscope to ensure that the Pt recording sites are immersed in the culture media, without breaking the tip on the bottom of the plate. The coated tip of the Ag/AgCl reference electrode is placed into the same well and held in place with tape. Figure 19.14 shows an in vitro cell culture recording set-up used in the laboratory.

The laboratory has used cell culture techniques with enzyme-based multisite microelectrodes to examine L-glutamate uptake in astrocyte cell cultures. By pressure ejecting a solution of 5 mM L-glutamate into the culture media, one can determine the rate of uptake of the exogenously supplied L-glutamate into astrocytes. Furthermore, by making serial dilutions of the astrocyte cultures, one can examine the efficiency of L-glutamate uptake. These studies hold great promise for a better understanding of L-glutamate uptake.

(a)

(b)

FIGURE 19.14 Photograph of an in vitro cell culture recording set-up. (a) The microelectrode and connector are attached to an electrode manipulator on a stereotaxic frame (not shown) and positioned over a culture dish. The reference wire is attached to the culture dish to keep it in the appropriate well. (b) Close-up view of the microelectrode tip inserted into one of the culture wells. The miniature Ag/AgCl reference electrode is positioned in the same well.

Brain Slice Techniques

Our laboratory uses the R1 microelectrode for brain slice experiments. The slice is only thick enough for 1 Pt recording site to be inserted, and the R1 configuration is best for this recording method. Because only one Pt recording site is inserted, a slightly different coating method is employed. The bottom and top sites are coated with L-glutamate oxidase while the two middle sites are left uncoated and used qualitatively for self-referencing recordings. If a micropipette is attached, the pipette is centered over the bottom recording site since this site is inserted into the brain slice.

Brain slices are harvested using standard procedures.[23,24] The artificial cerebral spinal fluid (aCSF) used consists of the following (mM): NaCl (124); KCl (5); $MgCl_2$ (1.5); $CaCl_2$ (2.5); NaH_2PO_4 (1.4); D-glucose (10); $NaHCO_3$ (26). aCSF is bubbled continuously with 95% O_2/5% CO_2. After the brain is removed, it is placed in ice cold oxygenated aCSF, blocked, and then, using a Vibratome® Series 1000 Sectioning System (Technical Products International, Inc., Product #054018), the brain is sectioned to form coronal slices that are 350–400 μm in thickness. Brain slices containing brain regions of interest are immersed in a holding chamber filled with bubbled room temperature aCSF and allowed to acclimate for at least 1h prior to recording.

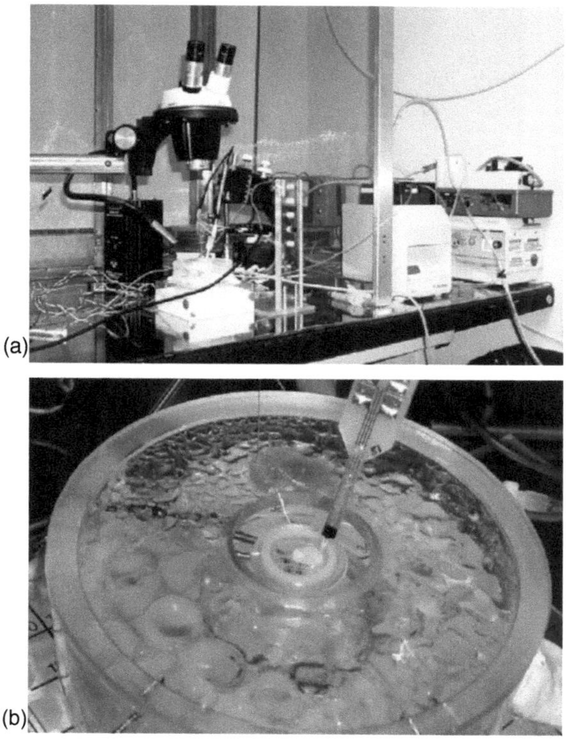

FIGURE 19.15 (a) Recording system for measuring L-glutamate in brain slices. (b) An immersion-style chamber helps maintain brain slices during L-glutamate measures. The slice is perfused from below and only a reference electrode and the microelectrode tip are placed in the bath. Note: a section of rat neocortex is in the chamber.

An immersion-style perfusion chamber maintains the brain slices during the recordings at a temperature of 33°C–35°C (Figure 19.15). Pre-warmed solution is perfused at a rate of 1.5–2.0 ml/min through the chamber with the aid of a peristaltic pump. Prior to placing the slice in the chamber, the coronal slices are cut along the midline and only one hemisphere is immersed in the chamber. After securing the microelectrode-connector-headstage assembly to a micromanipulator, the microelectrode is lowered at a 60° angle into the slice. Once in the slice, a stereomicroscope is used to visualize the lowering of the microelectrode such that site 1 (the site closest to the microelectrode tip) is completely in the slice (Figure 19.16a and b). Slices are allowed to acclimate and the recordings are allowed to reach a steady, level baseline prior to studies.

Delivery of drugs is provided through superfusion of the slices or local application from glass micropipettes. L-glutamate release is often evoked by stimulation through one of two means: (1) bath application of 40 mM KCl (40 mM KCl, 109 mM NaCl, 2.5 mM $CaCl_2$, pH 7.4) or (2) direct, local application of 70 mM KCl (70 mM KCl, 79 mM NaCl, 2.5 mM $CaCl_2$, pH 7.4) via pressure ejection from glass micropipettes. Bath application of 40 mM KCl (aCSF where the KCl concentration is raised to 40 mM with an equivalent decrease in NaCl concentration) for 2 min elicits an increase in L-glutamate concentration (Figure 19.17). A 30 min recovery period is carried out before a second KCl stimulation is repeated. A comparison between the ratio of the amplitudes of the second response to that of the first response in control slices versus slices treated with drugs (e.g., transporter or presynaptic receptor modulators) serves as a useful technique for assessing the effects of these treatments.

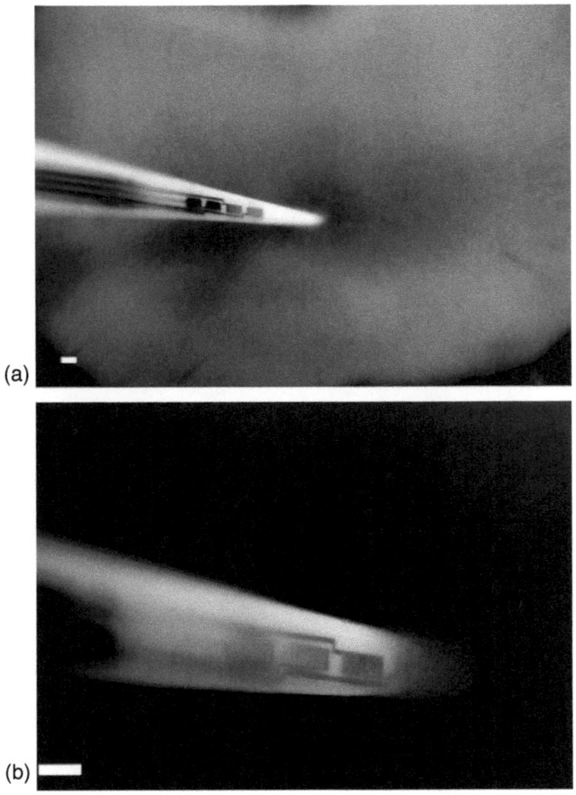

FIGURE 19.16 Microelectrode array in a brain slice. (a) Photograph of a microelectrode above a brain slice. The microelectrode is lowered at a 60° angle to provide adequate contact with the tissue. (b) An L-glutamate oxidase coated site is placed into the brain slice. Scale bars correspond to 150 μm.

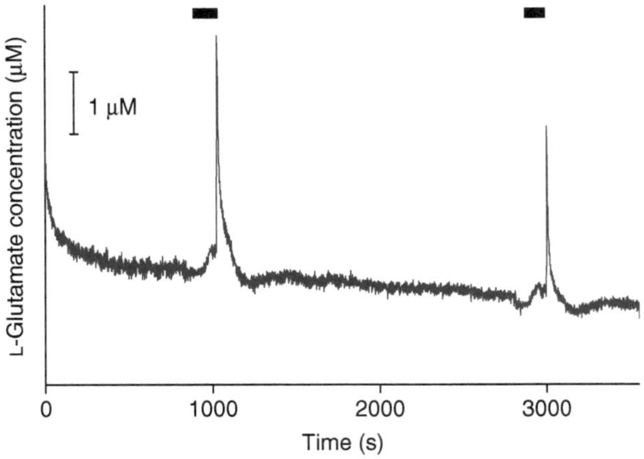

FIGURE 19.17 Representative L-glutamate signals seen following perfusion of the brain slice with 40 mM KCl (black bars). The KCl solution depolarizes cells and evokes the release of extracellular L-glutamate.

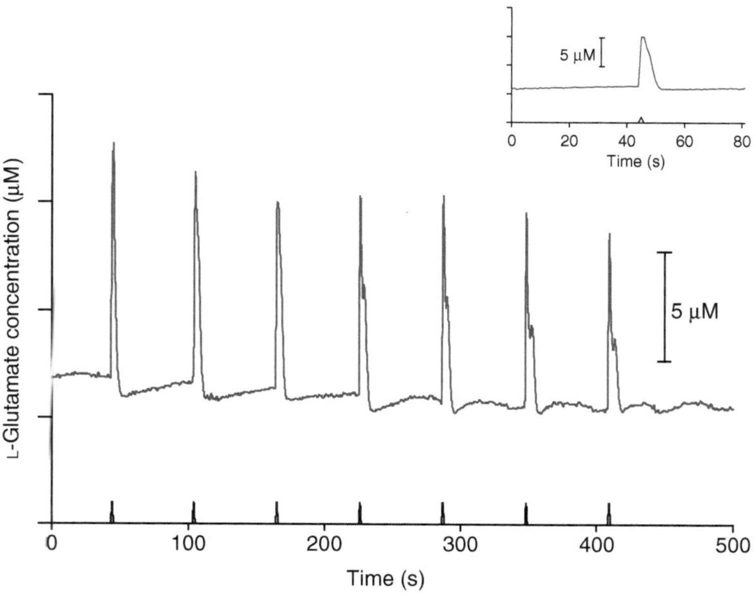

FIGURE 19.18 Pressure ejection induced L-glutamate release in the rat frontal cortex. Repeated ejections of 70 mM KCl (25–50 nl) produced rapid L-glutamate responses in the frontal cortex. Inset, L-glutamate release after a single pressure ejection of potassium solution.

A second method employed to stimulate release is pressure ejection of excess KCl directly into the brain slices. A glass micropipette filled with Isotonic 70 mM KCl (pH 7.4) is attached to the microelectrode as described above and pressure ejections of 70 mM KCl are used to evoke L-glutamate responses in brain areas (Figure 19.18). Both of these methods provide reliable, reproducible means of measuring L-glutamate in brain slices. The ability to measure the effects of known and highly controlled concentrations of drugs on L-glutamate signaling in the living brain slice provides a helpful compliment to L-glutamate measurements in vivo.

In Vivo Anesthetized Animal Recordings

The ability to perform in vivo recordings in anesthetized animals with enzyme-based multisite microelectrodes is a hallmark of this technology. The laboratory has recorded successfully L-glutamate (as well as other neurotransmitters including L-lactate, acetylcholine and choline) from the brains of rats, mice, and monkeys as well as L-glutamate in the rat spinal cord. Outlined below is the procedure for preparing an animal for in vivo brain recordings. Spinal cord recordings follow a similar procedure except a laminectomy is performed rather than a craniotomy.

Anesthetized Rats and Mice

Once the microelectrode is calibrated, rodents are anesthetized with urethane (Sigma, U-2500) (1.25 g/kg, i.p.) or another suitable anesthetic such as chloral hydrate. Our laboratory began employing urethane as an anesthetic during DA studies using carbon fiber microelectrodes. Urethane, unlike other anesthetics, does not greatly affect the clearance of DA from the extracellular space.[35] Furthermore, little is known regarding the mechanism of action of urethane. Urethane is also a useful anesthetic for long-term surgeries. Other anesthetics that are used for these types of experiments, such as ketamine, are documented to affect glutamatergic function.[36–38] As the recording technology has evolved, and the understanding of the glutamatergic system has increased, the laboratory has noticed that urethane does decrease basal levels of L-glutamate (unpublished data).

Once fully anesthetized, the rodent is placed in a stereotaxic frame (Kopf Instruments) A Deltaphase® Isothermal Pad (Braintree Scientific, Inc. Model #39DP) is placed between the metal frame and the animal to maintain its body temperature at 37°C. In the case of mice, these heating pads are bulky so a heated water pad (Gaymar Industries, Inc., TP3E) connected to a water bath (Gaymar Industries, Inc, T/Pump, TP500) is used to maintain body temperature. An incision is made on the midline of the scalp and the skin is reflected. Once the skull is exposed, a Dremel® (local hardware retailer) with bit size 107 (rats) or 105 (mice) is used to remove a portion of the skull over the recording sites. With forceps, the overlying dura is pulled laterally to expose the surface of the brain. The dura is removed gently as it is strong enough to break the microelectrode tip if it is not reflected prior to microelectrode insertion. Finally, a small hole is drilled in a remote location from the recording site for a Ag/AgCl reference electrode placement. The coated tip of the Ag/AgCl reference electrode is inserted into the brain and held in place using Durelon® Carboxylate dental cement (CMA/microdialysis Catalog #030 0003801).

It is recommended that a micromanipulator (Narishige Scientific Instrument Lab, RO-10) that attaches to the microelectrode manipulator on the stereotaxic frame is used to lower or raise the microelectrode by small increments. The microelectrode attached to the connector is fitted onto the micromanipulator by means of a Lexan® microelectrode holder designed by our lab. Once fitted to the stereotaxic frame, the glass micropipette attached to the microelectrode is filled with solution. Next, the tubing from the Picospritzer III is laced through the electrode manipulator and secured onto the glass micropipette. The flow of the solution through the glass micropipette is tested with several pressure ejections before implantation. Once properly configured, the microelectrode is lowered into the brain region of interest. The calibration parameters are recalled on the FAST-16 recording software and an experimental recording is initiated. The microelectrode equilibrates until a steady baseline is reached (usually 15–20 min) at which time the experimental study begins. Figure 19.19a and b show photographs of the microelectrode inserted into the striatum of a Fischer 344 (F344) rat and C57BL/6 mouse, respectively.

This laboratory has examined the effects of aging on the glutamatergic system using the F344 rat model of aging.[4] Using the microelectrodes the differences in L-glutamate release and re-uptake in the striatum of anesthetized six, eighteen and twenty-four-month-old F344 rats have been determined. Nickell et al.[4] found significant changes in L-glutamate release and re-uptake in young versus aged rats. Figure 19.20 shows representative tracings for 70 mM KCl-evoked

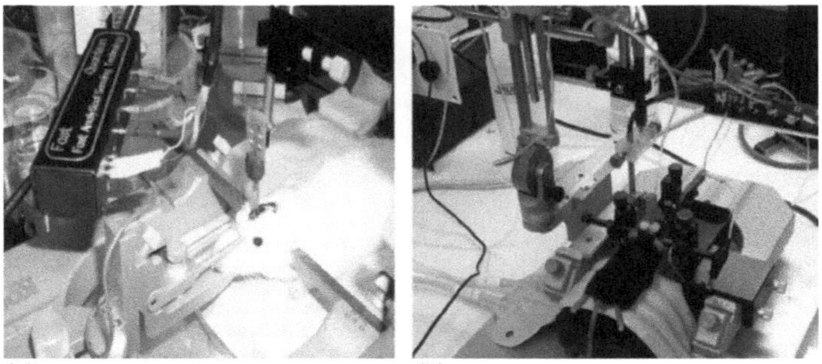

(a) (b)

FIGURE 19.19 Photographs of the anesthetized rodent recording apparatus. (a) A close-up view of a F344 rat placed in a stereotaxic frame with a microelectrode array inserted into the striatum. (b) A far view of a C57BL/6 mouse placed in a mouse stereotaxic frame designed to fit a rat stereotaxic frame. The microdrive is attached to the electrode manipulator and by means of a Lexan plastic holder, which raises and lowers the microelectrode. Cotton balls soaked with 0.9% physiological saline (pH ~ 7.4) are placed over the exposed areas of brain.

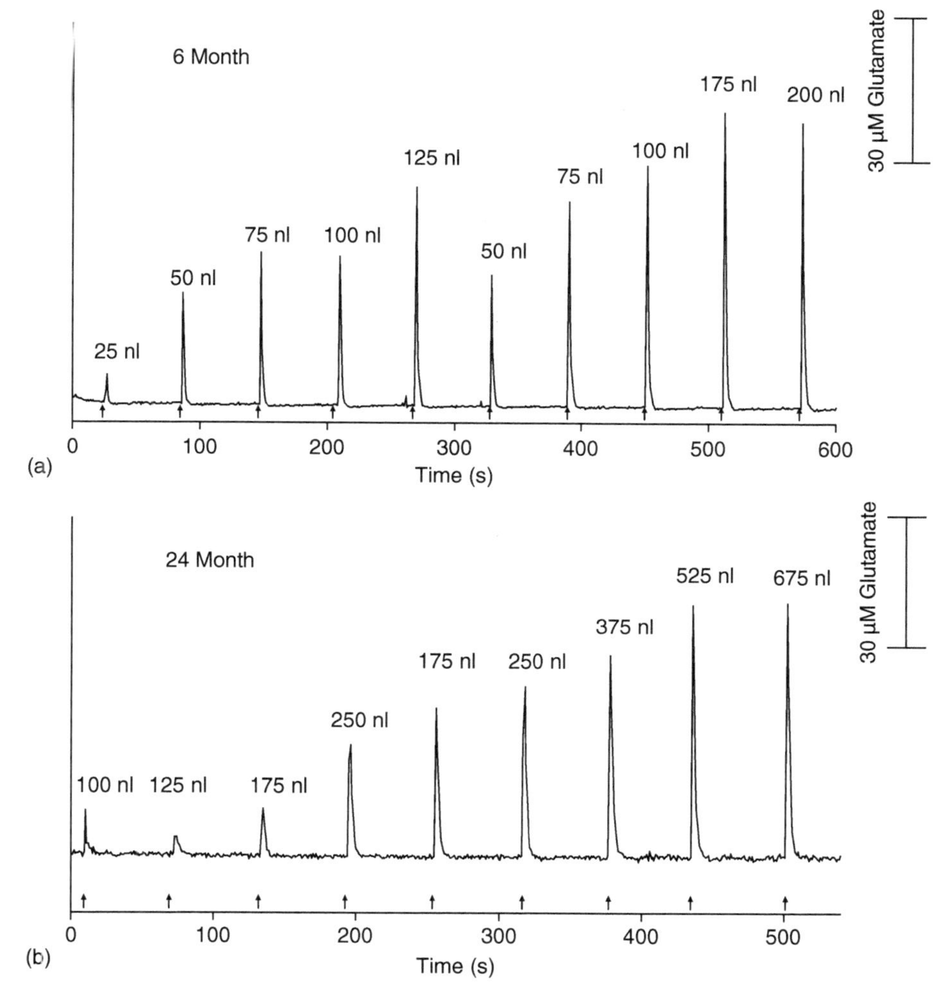

FIGURE 19.20 Tracings of 70 mM potassium-evoked L-glutamate release in the young, 6 month (a), and aged, 24-month (b), F344 rats. Both age groups produced robust, reproducible L-glutamate responses. However, larger responses in the six-month rats were obtained with less stimulus compared to the 24 month rats. Volumes of 70 mM potassium solutions applied by pressure ejection (denoted by the arrows) are listed above each peak.

L-glutamate release in six and twenty-four-month old F344 rats. Six-month-old F344 rats released more L-glutamate with less stimulus compared to the aged animals (note scale bars). Furthermore, the uptake of L-glutamate was significantly decreased in the twenty-four-month-old F344 animals compared to the six and eighteen-month-old F344 animals as shown in Figure 19.21.

Anesthetized Rhesus Monkeys

Much of the basic science that is conducted in rodents is assumed to be applicable to human conditions; however, important biological species differences may exist that limit the relevance of experimental findings.[39] Species that are more closely related to humans, such as rhesus monkeys, represent a logical step along the pathway toward applying the technology in the clinical setting. Additionally, differences exist between working with rodents under acute recording conditions in the laboratory setting and chronically investigating monkeys in the veterinary operating

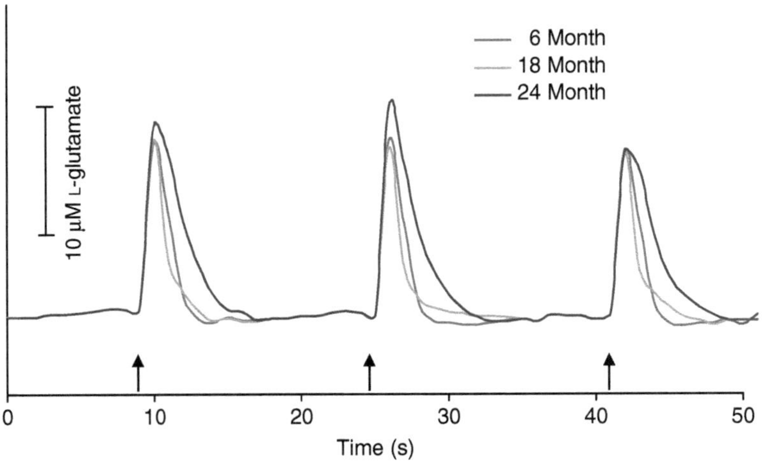

FIGURE 19.21 Representative traces of 70 mM potassium-evoked L-glutamate signals in 6, 18, and 24-month F344 rats. Note that the 24-month F344 rats had a significantly longer rate of L-glutamate clearance compared to the 6-month and 18-month F344 rats.

room. The tasks of general animal handling, anesthesia, surgery, and recovery fall on teams of experts.

The primary concern centered on creating a mobile FAST-16 recording system that is easily moved between settings and can be closely positioned to the anesthetized monkey in the operating room without compromising the sterile field. It was sought to make the system self-contained, to move all components as a single unit, and to require minimal setup and additional equipment. As shown in Figure 19.22, the FAST-16 recording system was placed onto a standard push cart that is easily moved into and around the operating room.

Similar to anesthetized rodent studies, microelectrodes are coated and calibrated as previously described. However, due to the expensive costs of these monkeys and the limited time allotted for these procedures, four microelectrodes are prepared for these experiments. Based on calibration data, the four microelectrode arrays are ranked by LOD, slope, and selectivity. The top two ranking microelectrode arrays are assigned for use with KCl induced L-glutamate release and local applications of exogenous L-glutamate, respectively. The other two microelectrodes are backups in case of accidental breakage or other unforeseen problems.

Rhesus monkey surgical procedures are handled by qualified technicians and outlined below. Following initial sedation with ketamine hydrochloride (~20 mg/kg; i.m.) plus atropine sulfate (~0.04 mg/kg; i.m.), each rhesus monkey is intubated via the orotracheal method, and intravenous lines are secured. Then, the animal is anesthetized with isoflurane (1%–3%) and placed in a magnetic resonance image-compatible stereotaxic apparatus (Kopf Instrument, Model #1430M) in a ventral–lateral position. The animal is maintained on a heated blanket and has cardiac and respiratory parameters monitored during the procedure. Coordinates for microelectrode array implantation are determined by magnetic resonance imaging prior to the surgery. After being shaved, the scalp area is cleaned with antiseptic. Starting at the incision site and moving circumferentially to the periphery, the shaved area is gently scrubbed with sterile 4″×4″ sponges soaked in chlorhexidine diacetate surgical scrub followed by 70% isopropyl alcohol. After the alcohol dries, Betadine® prep is applied, and the animal is covered with sterile drapes. Then, a 5–7 cm incision is made through the scalp, and the skin and muscles are reflected to expose the skull. Small (2 mm diameter) holes are drilled in the skull directly over the targeted areas using a dental drill and a rounded drill bit, and the overlying meninges are removed to expose the surface of the brain.

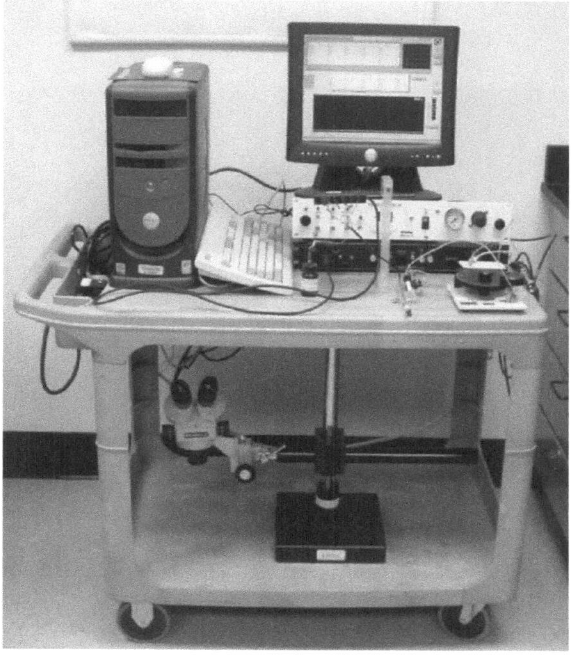

FIGURE 19.22 Photograph of the mobile FAST-16 recording system used for anesthetized monkey recordings.

A lateral pocket is made in the fascia at the posterior extent of the incision using blunt dissection. A 70% isopropyl alcohol-cleaned glass-bodied Ag/AgCl reference electrode (Bioanalytical Systems, Inc., RE-5b) is inserted into the pocket. Sterile saline irrigation is used to maintain ionic contact between the reference electrode and the exposed brain.

Microelectrode/micropipette assemblies are slowly lowered into each brain region using a stereotaxic holder as previously described. Figure 19.23 shows photographs of the anesthetized rhesus monkey set-up in the operating room. After completion of the recording session, bone wax is placed in the drill holes, and the scalp incision is sutured over the exposed areas as per normal procedures. The animal then is given an analgesic (buprenorphine, 0.01 mg/kg, i.m.) and prophylactic antibiotics (combination Penicillin G Benzathine and Penicillin G Procaine, 20,000 U/kg, s.c.). Temperature, heart rate, respiratory rate, and femoral pulse are monitored until the animal recovers from anesthesia. Full recovery is defined as self-sustained balance and posture.

Our laboratory has conducted several recordings in the anesthetized monkey. We have found that stimulus-evoked release of L-glutamate in the motor cortex and prefrontal cortex is similar to that seen in anesthetized rodents. Figure 19.24 shows potassium-evoked L-glutamate release in the cortex of young and aged anesthetized rhesus monkeys.

Awake, Freely Moving Rats and Mice

Our laboratory is transitioning from anesthetized animal recordings to recordings in awake freely moving animals. Recordings of L-glutamate and other neurotransmitters in awake freely moving animals have several distinct advantages. First, this approach allows for monitoring L-glutamate neurotransmission without the effects of anesthesia. Anesthetics are known to alter basal and phasic

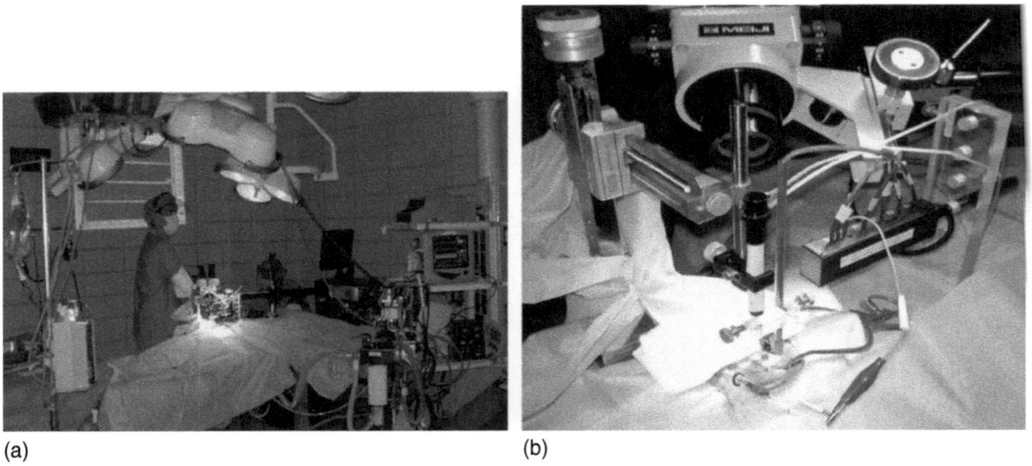

FIGURE 19.23 Photographs of L-glutamate recordings in anesthetized rhesus monkeys. (a) A photograph of the operating room. Rhesus monkeys are anesthetized and prepared for electrochemical recordings under sterile field conditions. The mobile FAST-16 is positioned in the room at the head of the operating table. (b) The microelectrode/micropipette assembly is lowered stereotaxically into cortical regions based on coordinates derived individually from each monkey's MRI scan. The four-channel headstage plugged into the FAST-16 recording system is placed near the recording area and connected to the microelectrode array using a Lexan microelectrode holder. A stereomicroscope on a boom stand, focused on the meniscus within the micropipette, is used to monitor the volumes of solutions that are locally applied into the brain.

L-glutamate release dynamics.[37,40,41] Second, freely moving animal recordings can be coupled to behavioral studies to investigate changes in L-glutamate that are dynamically related to behaviors. Third, the microelectrodes can reliably record L-glutamate for over two weeks in freely moving animals. This means that multiple measures can be made over multiple days instead of only one day as in anesthetized and microdialysis studies.

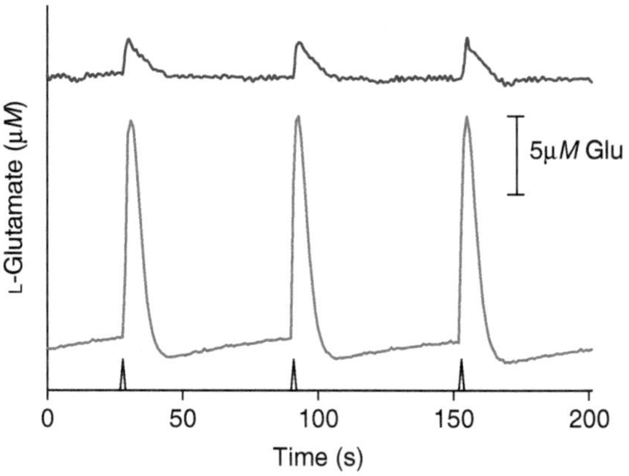

FIGURE 19.24 Rapid potassium induced release of L-glutamate in the motor cortex of young (top trace) and aged (bottom trace) rhesus monkeys. Potassium ejections at one minute intervals (black arrows) in the monkey motor cortex evoked fast L-glutamate release and uptake. Smaller amplitude signals were observed in the motor cortex of the young rhesus monkey (responses to 100 nl potassium volumes) compared to the aged monkey (responses to 250 nl potassium volumes).

Implantation and Recording Procedure

First, the microelectrode used for in vitro or anesthetized in vivo animal recordings is modified to make it smaller and lighter weight for use in chronic implantations. Refer to Rutherford et al. for a detailed description of freely moving microelectrode preparation and implantation (manuscript in preparation).[42] The PCB holder of the enzyme-based microelectrodes is scored just beneath the first set of pinholes on the PCB (Figure 19.3c), without splitting the microelectrode into two pieces. The microelectrode is handled easily as a single unit, but the scoring allows for the top part of the PCB to be easily broken off at a later time. Since scoring produces dust that can cling to the Pt recording sites, the microelectrodes are then cleaned as previously described. Once cleaned, the scored microelectrode is coated with Nafion (it has been found this exclusion layer works longer for chronic implantations compared to mPD) and enzyme solution.

Next, microelectrode connecting wires are prepared by first stripping both ends of 30 AWG varnished copper wire (Radio Shack). One end of the copper connecting wire is soldered to a pinhole on the row closest to the microelectrode tip. The other end is attached to a gold-plated socket amphenol (Ginder Scientific Part #220-S02). Four copper wires are soldered to the PCB, each corresponding to one of the Pt recording sites. Once all wires are soldered, the microelectrode is coated with 5-min epoxy, which protects the microelectrode connection from potential water or other liquid damage during the remaining portion of the procedure. After the epoxy dries for approximately 30 min, the remaining portion of the PCB is split in two using pliers to bend the PCB at the score mark. Now, the modified microelectrode is ready for enzyme coating and allowed to cure for 48 h before final modifications.

Once the enzyme layer has properly cured to the microelectrode surface, the socket amphenol attached ends of the reference and microelectrode wires are inserted into a miniature connector (Ginder Scientific, 9-pin ABS Plug, Part #GS09PLG-220). Pliers are used to push the socket amphenols into the connector so that the socket amphenols are completely encased. A miniature Ag/AgCl reference electrode is prepared and inserted into the connector. The copper wires are wrapped around the connector and the microelectrode tip is positioned parallel to the connector. Correct positioning of the microelectrode tip is essential for accurate stereotaxic placement into the brain. Five-minute epoxy is used to secure the microelectrode array and wires, making sure that all wires and solder points are covered with epoxy. The openings on the end of the connector closest to the microelectrode tip are also covered with epoxy to ensure that moisture does not penetrate the connector and disrupt the recorded signal.

Similar to anesthetized studies, drugs are locally applied close to the Pt recording sites for pharmacological studies in freely moving animals. Instead of a glass micropipette, a 26-gauge stainless steel guide cannula (Plastics One Inc., C315G Cannula Guide 26GA 38172, PO: 052604AA, SO: 41123-1) is centered among the recording sites with sticky wax, Figure 19.25. A stainless steel dummy cannula (Plastics One Inc., C315DC Cannula dummy, PO: 121409, SO: 45203-3), slightly longer than the guide, is inserted into the guide cannula and screwed into place. The dummy cannula tip is positioned among the four recording sites of the microelectrode at a distance of approximately 100 μm above the microelectrode tip. The stainless steel dummy cannula remains in the guide at all times, except during ejection when it is replaced with a 33 gauge internal acute cannula (Plastics One Inc., C315IA Cannula Internal Acute, PO: 121409, SO: 45203-1) through which solutions are locally applied. This cannula is the same length as the dummy cannula so it is positioned in the same location. The dummy cannula helps prevent particulate matter from clogging the guide cannula.

The recording headstage is slightly different from those used for in vitro and in vivo anesthetized animal recordings. This headstage screws directly onto the rat microelectrode assembly connecting the chronically implanted microelectrode to the FAST-16 recording system. The headstage consists of a miniature connector with five connector pins (one connecting each of the four channels and one for the Ag/AgCl reference electrode). The connector pins lead to the four channel

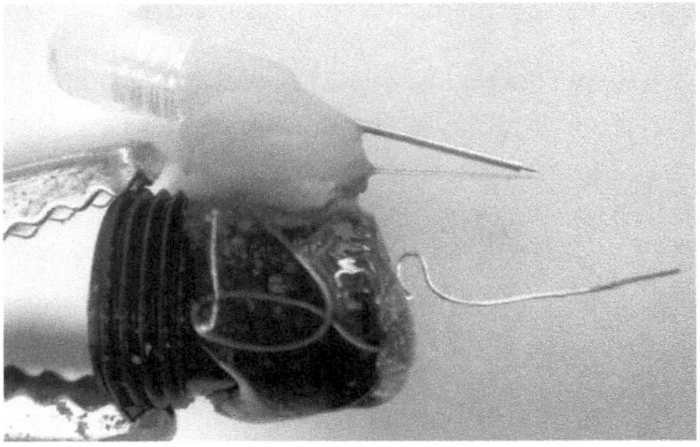

FIGURE 19.25 Photograph of the modified microelectrode with attached guide cannula used for freely moving recordings in rats and mice. The attached guide cannula is centered among the 4 recording sites and held approximately 100 μm above the ceramic surface with modeling clay and Sticky Wax.

mini-amplifier, which is positioned as close as possible to the animal in order to minimize noise artifacts as shown in Figure 19.26. Shielded connecting wire, leads to an electrical swivel (Airflyte Electronics Co., P/N 1001460-012) at the top of the recording chamber. The headstage is connected at the top center of the recording apparatus to the electrical swivel (commutator) that contains twelve electrical contacts as seen in Figure 19.27a. The commutator allows the animal to move freely to all areas of the behavior chamber.

For chronic implantation surgeries, animals are anesthetized with sodium pentobarbital solution (50 mg/ml), administered in one or two intraperitoneal (i.p.) doses, and placed in a stereotaxic apparatus. Animal body temperature is maintained with an isothermal heating pad at 37°C. The animals' eyes are coated with artificial tears (www.medi-vet.com, Catalog #11970) to help maintain fluids and prevent infection. All surgeries are performed in a Vertical Laminar Flow Workstation (Microzone Corp., VLF-2-4). Prior to incision, excess fur on the head is shaved and the skin directly on top of the animal's head (between the ears and from just behind the eyes to the neck) is wiped with an iodine solution to keep the incision area clean and prevent infection. The skin on top of the head is reflected, making as small an incision as necessary. The rat undergoes a craniotomy to remove a 2 mm×2 mm portion of the skull for rat microelectrode implantation.

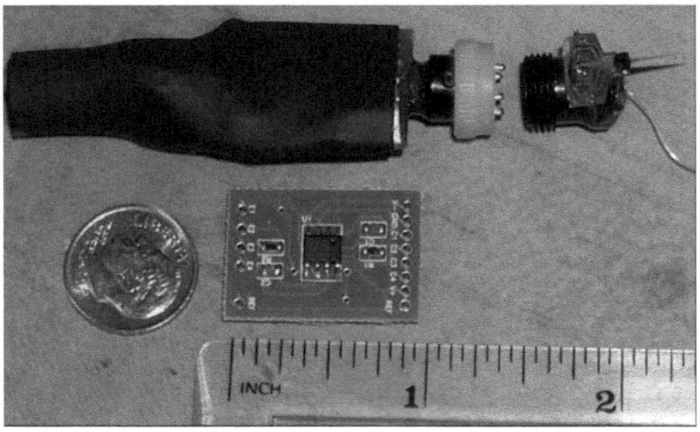

FIGURE 19.26 Photographs of the freely moving rat and mouse connector. The circuit board at the center of the photograph is the smaller, lighter headstage. This is then wired to the tether and shielded (upper left). Finally, the fully assembled tether and modified microelectrode are connected together for chronic recordings in freely moving rodents. A dime and ruler are placed in the photograph to show the size of these recording devices.

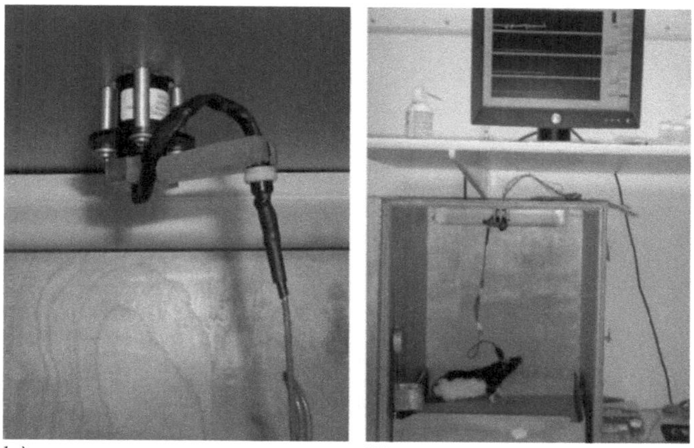
(a) (b)

FIGURE 19.27 Photographs of an awake rat connected to the commutator and FAST-16 recording system. (a) Close-up of the commutator with the attached tether for freely moving recordings. (b) A freely moving Long Evans rat is connected to the FAST-16 recording system.

Additionally, four small holes (<0.5 mm) are drilled in the skull with Dremel engraving cutting bit (#105) for placement of the reference electrode and three stainless steel skull screws (Small Parts Inc., Part #MPX-080-02-M) that help hold the assembly in place. Similar to the anesthetized studies, a Ag/AgCl reference electrode is implanted remotely from the recording site. The assembly is secured with approximately four layers of Ortho-Jet Powder acrylic resin (Lang Dental Manufacturing, Co., Inc., Reference No. 1330) mixed with Jet Acrylic Liquid (Lang Dental Manufacturing Co., Inc., Reference #1406), which the skull screws help anchor in place. The dental acrylic covers as much of the microelectrode PCB holder as possible, without adhering to the threads of the cannula or the microelectrode connector. The dental acrylic has a smooth texture so as not to promote the rodent to scratch the implanted microelectrode. Figure 19.28a is a photograph of a Long Evans rat with an implanted microelectrode held in place with dental acrylic.

Following surgery, rats are placed on a heating pad to help maintain body temperature until the animal recovers from the anesthesia. Three subcutaneous injections of 1 ml each of Ringer's Solution, Mammal (Fisher Scientific, Catalog #S77939) are administered subcutaneously immediately following surgery. Animals are given three days to recover prior to the start of a recording session.

Also our laboratory is adapting this technology for use in the freely moving mouse. The smaller size and thinner skull of the mouse makes implantation more difficult than in the rat. Although the basic microelectrode preparation remains similar, the implantation is slightly different. Mice are anesthetized with a diluted dose of the sodium pentobarbital (5 mg/ml in 0.9% physiological saline, pH ~ 7.4). Once again, a 2 mm × 2 mm craniotomy is performed for microelectrode implantation. Next, three holes are drilled in the skull, two for the skull screws placed diagonal from one another and the third for a Ag/AgCl reference placed diagonal from the craniotomy. Because the skull of the mouse is considerably thinner than the rat, the skull screws are cut in half with wire cutters prior to implantation. Again, the assembly is held in place with the Hygenic cold cure denture resin and Jet acrylic liquid. Because the brain of the mouse does not have the same volume of the rat, the microelectrode tip is not lowered to the same depths in rats.[43,44] The assembly rests higher on the head of the mouse than the rat; therefore, additional layers of dental resin are needed to secure the assembly. Figure 19.28b shows a photograph of a C57BL/6 mouse chronically implanted with an L-glutamate microelectrode. Despite the current size of the microelectrode, the mouse is not hindered by the additional weight.

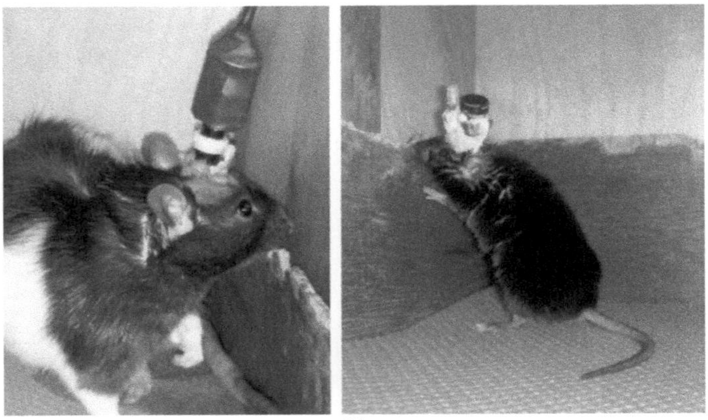

FIGURE 19.28 Photographs of a Longs Evan rat (a) and a c57BL/6 mouse (b) with implanted microelectrode arrays held in place with dental acrylic. Note that the animal in (a) is connected to the miniature recording headstage seen in Figure 19.26.

On experiment days, the animal is allowed to explore the behavior chamber for 15–20 min before being connected to the FAST-16 recording system. Usually, the first day of recordings involves locally applying 5 mM L-glutamate (in 0.9% physiological saline, pH ~ 7.4) to determine the microelectrode response before behavioral studies are conducted. Locally applied solutions are filtered and then delivered into the internal guide cannula with the aid of 28-gauge tubing (Small Parts Inc.) fitted over a Hamilton 10 μl microsyringe to provide accurate volume ejections of locally-applied exogenous 5 mM L-glutamate. The microsyringe and tubing are taped to the tether so it can freely rotate with the rat. The dummy cannula is removed, and the internal cannula is inserted in its place. The animals should not be restrained to insert the cannula as restraining the rat alters behavioral studies by stressing the rodent.

The freely moving rodent recording system allows the effects of behavior on L-glutamate neurotransmission to be examined. In particular, different types of stressors to elicit L-glutamate signals have been used successfully. Rats exposed to a 5 min tail pinch (a clothes pin around the base of the tail) exhibit novel L-glutamate responses from the striatum and the prefrontal cortex. Figure 19.29 shows a representative tracing of an L-glutamate signal from a 5 min tail pinch in the

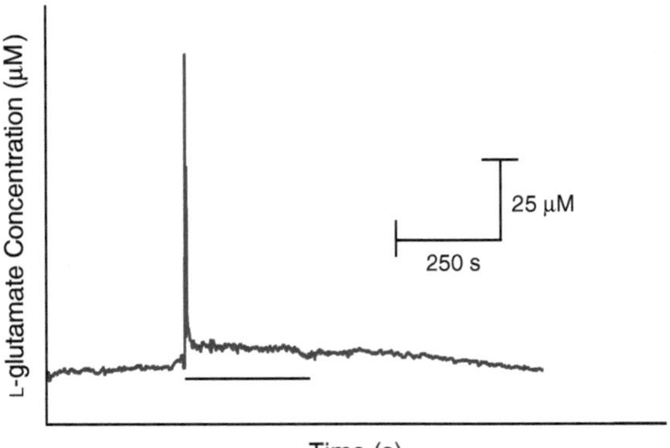

FIGURE 19.29 Representative tracing of an L-glutamate response elicited by a five minute tail pinch stress in the freely moving F344 rat. The duration of the tail pinch is denoted by the solid black line under the trace.

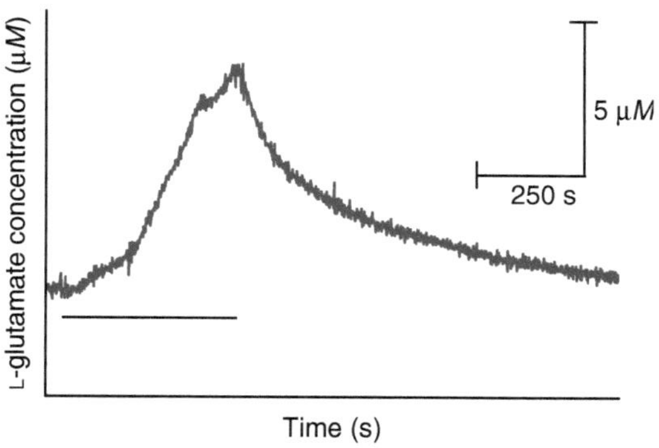

FIGURE 19.30 Representative tracings from a typical L-glutamate signal recorded in the striatum of a freely moving c57BL/6 mouse from a fox urine stressor. The duration of the fox urine exposure is indicated by the black line under the trace.

striatum of a F344 rat. A typical tail pinch produces a bi-modal response in the striatum of a F344 rat—a fast L-glutamate spike with an elevated plateau. The spike occurs immediately at the start of the tail pinch while the plateau lasts for the duration of the tail pinch stressor, followed by a slow return to baseline. The clothespin does not hurt the rodent, but the rodent normally gnaws and/or flips around in the recording chamber in an attempt to remove the clothes pin.

Additionally, predatory stressors[45,46] have been used to evoke L-glutamate release in both rats and mice. Fox urine applied to a cotton ball and introduced into the recording chamber for 5 min produces increases in L-glutamate release in the mouse striatum. Figure 19.30 shows a typical L-glutamate signal from a 5 min fox urine exposure in the right striatum of a c57BL/6 mouse. The L-glutamate concentration steadily increases throughout the duration of the odor stressor. Introduction of the fox urine soaked cotton ball also causes the rodent to move to a far corner of the recording chamber at the same time as an increase in L-glutamate is measured. When the odor stressor is removed from the behavior chamber, the L-glutamate levels decrease until the response returns to baseline (approximately 15 min later).

Chronic Implantation Histopathology

Finally, the chronically implanted L-glutamate microelectrodes have been seen to produce minimal to no damage to the surrounding brain tissue in rats. Rats were chronically implanted with microelectrodes for either two, four or eight weeks. At the end of these time points, the rats were perfused; and the brains removed, sectioned, and stained for both astrocytes (GFAP) and microglia (Iba1). Staining showed a slight increase in astrocytes immediately surrounding the implant site and minimal activated microglia (rounded microglia) surrounding the implant site for all three time periods compared to the contralateral control hemisphere. Figure 19.31 shows photomicrographs of the GFAP and Iba1 stained brain sections. Our lab is continuing to study tissue reactivity in longer term implants and to better understand the effects of extended (i.e., six month) microelectrode implants. However, there is encouragement to be found by the lack of scarring from these devices, thus, what factors (ceramic material, shape, etc) contribute to the limited tissue changes will continued to be studied.

FUTURE DIRECTIONS

Eight Site Microelectrode Arrays

Most of the applications discussed center around the use of enzyme-based microelectrodes with four Pt recording surfaces. Photolithography allows the design of microelectrodes with even more

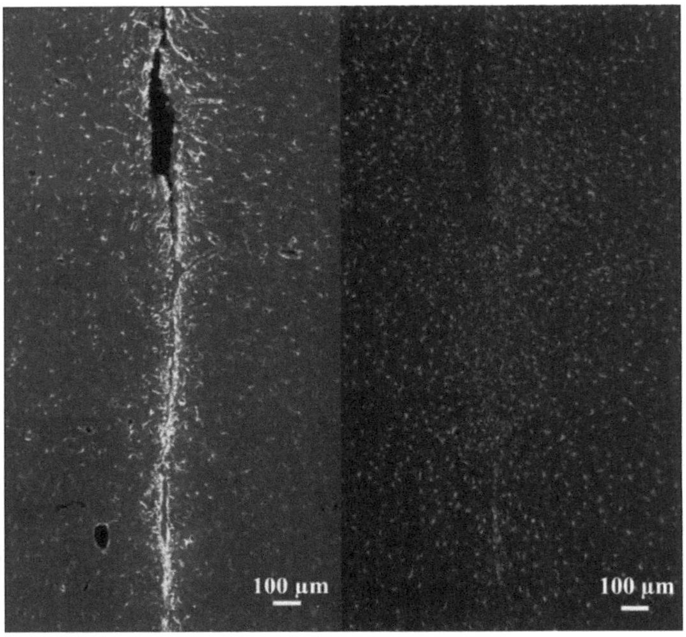

FIGURE 19.31 **(See color insert following page 272.)** Immunohistochemical staining of a microelectrode recording tract of a Long Evans rat 8 weeks post implantation. The brain was sectioned and stained for GFAP (left, green) and Iba1 (right, red). The microelectrode tract mark is clearly visible in both sections.

Pt recording sites. Now there has been the fabrication of microelectrodes that contain eight Pt recording sites, and those with sixteen Pt recording sites are being developed. Furthermore, a more flexible multisite microelectrode for recording from deep brain structures in monkeys and humans is being tested. These new electrode designs can further the understanding of L-glutamate and other neurotransmitters in the CNS.

The eight site microelectrodes coupled with the new FAST-16 recording system can simultaneously record amperometric data from eight sites on a second-by-second basis. Figure 19.32 shows photomicrographs of three current eight site microelectrode designs. There are two advantages of the increased number of Pt recording sites: (1) multiple analytes can be measured from a single microelectrode in the same brain region; and (2) multisite microelectrodes can be designed specifically to simultaneously record from several brain regions.

The long term purpose of designing these multisite microelectrodes is to simultaneously record multiple neurotransmitters in the same brain region. The eight site microelectrodes allow the user to

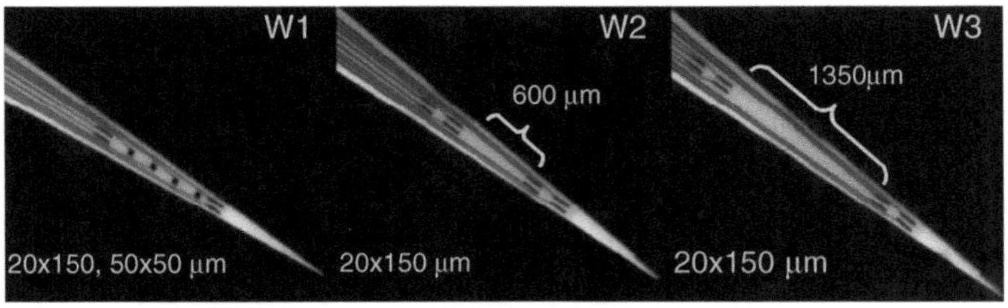

FIGURE 19.32 Photomicrographs of three, eight site microelectrode arrays. Pt recording site dimensions are labeled at the bottom left of each photomicrograph while the dimensions between groups of recording sites are labeled between the hash marks.

coat a single microelectrode as if it were two microelectrodes. For example, on the W2 and W3 designs the bottom set of Pt recording sites are configured for L-glutamate with self-referencing measures. The top set of Pt recording sites is coated with a different enzyme/protein matrix also configured for self-referencing recordings. This allows for simultaneous recording of both L-glutamate and other neurotransmitters and observing any possible interactions during a depolarization or behavioral event. With future designs that incorporate sixteen Pt recording sites on the same microelectrode, four or more analytes can be simultaneously measured. Furthermore, with the development of a microcoater to facilitate more precise enzyme coating, side-by-side pairs of Pt recording sites can be configured for self-referencing recordings. This allows measurements of four different analytes on an eight site microelectrode array and eight different analytes on a 16 site microelectrode.

The second advantage of eight Pt recording sites per array is that the Pt recording sites can be geometrically positioned to encompass two separate anatomical brain structures. For example, 600 μm spacing between sets of Pt recordings sites on the W2 enzyme-based microelectrode is the average distance between the rat CA1 and CA3 cell regions in the rat hippocampus. The bottom set of Pt recordings sites rests in the CA3 regions while the top set of Pt recording sites can reside in the CA1 region. By configuring both sets of Pt recordings sites for L-glutamate/self-referencing recordings, neurons in the CA3 region can be depolarized and the subsequent L-glutamate signaling to the CA1 region can be recorded.

USING MICROELECTRODE MICRODRIVES FOR CHRONIC IMPLANTATIONS

A microdrive (Figure 19.33) developed at Wake Forest University (courtesy of Dr. Sam Deadwyler) allows one to replace and lower microelectrodes for freely moving animal studies. Freely

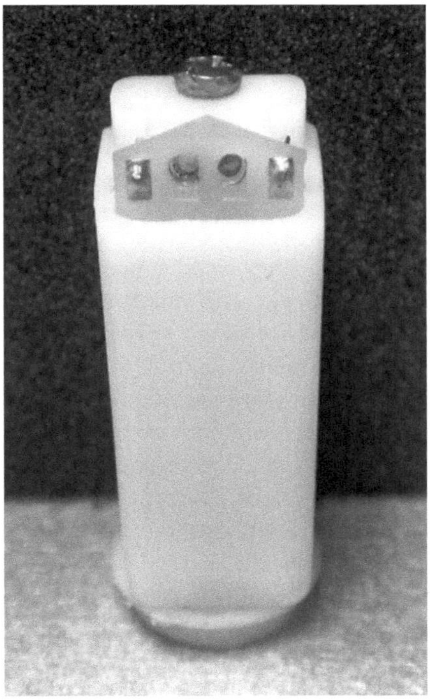

FIGURE 19.33 Photograph of a microelectrode inserted into a microdrive. The microdrive is used to reposition a microelectrode array once chronically implanted in a freely moving rodent. The microdrive unit is attached to the base plate, which is attached to the rodent cortex.

moving rodent craniotomies are performed as described earlier in the chapter. However, here a base plate is attached to the skull. The microelectrode fits into the microdrive; this assembly attaches to the base plate. Once connected to the base plate, the microelectrode is lowered to the desired depth. This microdrive allows one to measure multiple brain regions within the same animal by raising or lowering the microelectrode as desired. Additionally, if a microelectrode loses sensitivity for L-glutamate, the microelectrode can be replaced allowing for longer recording capability in a single animal. This has applicability for studying L-glutamate dynamics throughout the lifespan of aging or disease animals.

Telemetry

The tether attaching the freely moving rodents to the recording hardware hinders the ability to perform certain type of behaviors, such as learning tasks in a radial arm maze. Due to this limitation, our laboratory is developing a wireless transmitter for multisite recordings from freely moving rodents as well as primates. This wireless transmitter includes a smaller lighter-weight version of the current headstage. We envision being able to chronically implant a multisite microelectrode as previously discussed. The wireless transmitter with headstage connects to the microelectrode and is strapped to the animals back with a harness. The distance the animal moves no longer is limited by the length of the tether, but rather by the range of the wireless transmitter (currently 20 ft.). Additionally, the headstage now is closer to the multisite microelectrode, thus improving the signal-to-noise ratio recording by the microelectrodes.

Despite the fact that much of this technology is currently available, it is not yet practical for neurobiological studies. Battery life is a major limiting issue. The multisite microelectrodes are capable of recording L-glutamate in the CNS for over two weeks, and current battery life falls well short of this. In order to replace the battery the animal must be sedated every few days in order to prevent undue harm or stress. The continued work on this project has made inroads into resolving these issues and progressing towards the ultimate goal of eliminating the need for restraining cable connections.

Deep Brain Structure Enzyme-Based Microelectrodes

A version of the enzyme-based multisite microelectrodes has been designed for recording from deep brain structures such as the hippocampus and temporal lobe in monkeys as well as humans. This new microelectrode called the Spencer-Gerhardt electrode is modeled after the Spencer electrode (AD-TECH®) currently used for monitoring electrophysiological signals in humans. The advantage of this design increases the flexibility of the microelectrode so it can be positioned and attached to the skull with little concern for tissue damage as the brain moves. This version of the microelectrode is being utilized in the nonhuman primate testing before being used in human clinical trials. Future designs will incorporate both components of the two electrodes so that a single electrode can simultaneously monitor electrophysiological recordings of multiple single neurons and L-glutamate dynamics. This application has important benefits for helping neurosurgeons excise brain tissues in patients with epilepsy.

ACKNOWLEDGMENTS

Special thanks to Ingrid Strömberg for her help with the sectioning and staining of brains from chronically implanted rats. Special thanks to AD-TECH Medical (David Putz) and Dr. Dennis Spencer (Yale University) for their help with designing new microelectrode arrays for nonhuman primate and human studies. In addition, special thanks to Drs. Sam Deadwyler and Robert Hampson

(Wake Forest University) for the design of the microdrive system for recordings in awake animals. Finally, we thank Andrew Allen for her dedication to these on-going projects as well as Heather Davis for her help with the photographs.

None of this work would be possible without generous support from the following agencies:

DARPA: DAMD17-99-1-9480
NIDA: 5 R01 DA017186
NIA: 2 P01 AG013494
NINDS: 1P50 NS39787
NSF: DBI-0352848

APPENDIX A

Preparing Glutamate Microelectrodes

1. *Cleaning (first time use)*:

   5 min in Citrisolv (constant stirring)
   5–10 min in filtered (0.22 μm) DI water (constant stirring)
   Dry at low temp (105–115°C) oven (~15–20 min)
   (When reusing electrodes, stir 30 min in 80°C filtered DI water and continue as above)

2. *Nafion coating*: *(This step is skipped if electropolymerizing with mPD)*

   Use fresh Nafion weekly
   (Store in fridge ~ 1ml aliquots in glass container)
   Use Nafion at room temp; Insert tip (more than half-way), swirl 5 times, pull out
   Keep level (horizontal) and bake 4 min @ 165–175°C
   (This time will vary according to humidity levels in your area i.e. more humid = more time)
   ***Do not heat microelectrodes above 175°C.

3. *Enzyme coating*:

   BSA (Sigma A-3059), Glutaraldehyde (Sigma G-5882), Glutamate oxidase (Glu-ox) (Seikagaku America, 100645)
   (Dissolve Glu-Ox @ 1U/μl in filtered DI water)
   3.1 Bring all chemicals to room temp before use
   3.2 Weigh 0.01 g BSA in 1.5 ml microcentrifuge tube
   - Add 985 μl of filtered DI water, DO NOT vortex vigorously
   - Add 5 μl of glutaraldehyde
   (Final concentrations: BSA; 1%, glut; 0.125%)
   3.3 Transfer 9 μl of BSA/Glut into a 500 μl microcentrifuge tube
   - Add 1 μl of Glu-Ox
   (Final concentration: 2%)

3.4 Apply enzyme solution using a 10 μl Hamilton syringe with a dome tip needle.
   - Draw up ~1 μl of solution
   - Create a bead (drop) of solution at the tip of the syringe
   - Under a dissecting stereoscope, apply bead on surface of the electrode, wait 2–3 s, pull tip away
   - Let dry for a minute, typically 3 applications are enough

3.5 Set aside in a clean dry place
   - Room temp for at least 48–72 h—Time required for full cross-linking of the protein to the surface of the electrode.

4. *mPD electropolymerization 1,3 phenylinediamine*, dihydrochloride 99% (Sigma-Aldrich, Cat #235903)

   4.1 Degas 100 ml of 0.05 M PBS with nitrogen for 20 min
   4.2 Dissolve 0.0905 g mPD in 100 ml 0.05 M PBS (5 mM)
   4.3 Add ~40 ml of 5 mM mPD to a 50 ml beaker
   - Lower microelectrode tip into 5 mM mPD solution
   - Place glass Ag/AgCl reference electrode into beaker
   - Apply a potential of +0.5 V vs. Ag/AgCl reference for 15 min
   - Remove microelectrode from 5 mM mPD and rinse tip with filtered, DI H_2O.
   - Cure at room temperature for 24 h prior to calibration.

REFERENCES

1. Burmeister, J. J. and Gerhardt, G. A., Self-referencing ceramic-based multisite microelectrode for the detection and elimination of interferences from the measurement of L-glutamate and other analytes, *Anal. Chem.*, 73, 1037, 2001.
2. Burmeister, J. J., Pomerleau, F., Palmer, M., Day, B. K., Huettl, P., and Gerhardt, G. A., Improved ceramic-based multisite microelectrode for rapid measurements of L-glutamate in the CNS, *J. Neuro. Methods*, 119, 163, 2002.
3. Pomerleau, F., Day, B. K., Huettl, P., Burmeister, J. J., and Gerhardt, G. A., Real time in vivo measurements of L-glutamate in the rat central nervous system using multisite ceramic-based microelectrode arrays, *Ann. N.Y. Acad. Sci.*, 1003, 454, 2003.
4. Nickell, J., Pomerleau, F., Allen, J., and Gerhardt, G. A., Age-related changes in the dynamics of potassium-evoked L-glutamate release in the striatum of Fischer 344 rats, *J. Neural. Transm.*, 112, 87, 2005.
5. Day, B. K., Pomerleau, J., Burmeister, J. J., Huettl, P. F., and Gerhardt. G. A., Microelectrode array studies of basal and potassium-evoked release of L-glutamate in the anesthetized rat brain, *J. Neurochem.*, 96(6), 1626, 2006.
6. Gerhardt, G. A. and Burmeister, J. J., Voltametry in vivo for chemical analysis of the nervous system, In *Encyclopedia of Analytical Chemistry*, Meyers, R. A., Ed., Wiley, Chichester, p. 710, 2000.
7. Cahill, P. S., Walker, Q. D., Finnegan, J. M., Mickelson, G. E., Travis, E. R., and Wightman, R. M., Microelectrodes for the measurement of catecholomines in biological systems, *Anal. Chem.*, 68, 1380, 1996.
8. Texas Engineering Extension Service, Ed., *Photolithography*, Texas Engineering Extension Service, Bryant, TX, 2001.
9. Bargon, J. Ed., *Methods and Materials in Microelectronic Technology* (*IBM Research and Symposia Series*). Plenum Pub. Corp., New York, Stenvenson Ranch, CA 1984.
10. Texas Engineering Extension Service, Ed., *Wire Bond*, Texas Engineering Extension Service, Bryant, TX, 1998.

11. Gerhardt, G. A. and Burmeister, J. J., Neurochemical arrays, In *Encyclopedia of Sensors*, Grimes, C. A., Ed., American Scientific Publishers New York Stenvenson Ranch, CA., 2005.
12. Burmeister, J. J. and Gerhardt, G. A., Ceramic-based multisite microelectrode arrays for in vivo electrochemical recordings of glutamate and other neurochemicals, *Trends Anal. Chem.*, 22, 48, 2003.
13. Burmeister, J. J., Moxon, K., and Gerhardt, G. A., Ceramic-based multisite microelectrodes for electrochemical recordings, *Anal. Chem.*, 72, 187, 2000.
14. Gerhardt, G. A., Rapid chronocoulometric measurements of norepinephrine overflow and clearance in CNS tissues, *Neuromethods-Voltammetric Methods Brain Syst.*, 27, 117, 1995.
15. May, L. J. and Wightman, R. M., Heterogenity of stimulated dopamine overflow within rat striatum as observed with in vivo voltametry, *Brain Res.*, 487, 311, 1989.
16. Kawagoe, K. T., Zimmerman, J. B., and Wightman, R. M., Principles of voltametry and microelectrodes surface states, *J. Neurosci. Methods*, 48, 225, 1993.
17. Rice, M. E. and Nicholson, C., Measurement of nanomolar dopamine diffusion using low-noise perfluorinated ionomer coated carbon fiber microelectrodes and high-speed cyclic voltametry, *Anal. Chem.*, 61, 1085, 1989.
18. van Horne, C., Bement, B., Hoffer, B. J., and Gerhardt, G. A., Multichannel semiconductor-based electrodes for in vivo electrochemical and electrophysiological studies of rat CNS, *Neurosci. Lett.*, 120, 249, 1990.
19. Suaud-Chagny, M. F., Brun, P., Buda, M., and Gonan, F., Fast in vivo monitoring of electrically evoked dopamine release by differential pulse amperometry with untreated carbon fiber electrodes, *J. Neurosci. Methods*, 45, 183, 1992.
20. Gerhardt, G. A. and Hoffman, A. F., Effects of recording media compostion on the responses of nafion-coated carbon fiber microelectrodes measured using high-speed chronoamperometry, *J. Neurosci. Methods*, 109, 13, 2001.
21. Lowry, J. P., Miele, M., O'Neill, R. D., Boutelle, M. G., and Fillenz, M., An amperometric glucose-oxidase/poly(o-phenylenediamine) biosensor for monitoring brain extracellular glucose: in vivo characterization in the striatum of freely behaving rats, *J. Neurosci. Methods*, 79, 65, 1998.
22. McMahon, C. P. and O'Neill, R. D., Polymer-enzyme composite biosensor with high glutamate sensitivity and low oxygen dependence, *Anal. Chem.*, 77, 1196, 2005.
23. Friedemann, M. N., Robinson, S. W., and Gerhardt, G. A., o-Phenylenediamine-modified carbon fiber electrodes for the detection of nitric oxide, *Anal. Chem.*, 68, 2621, 1996.
24. Ryan, M. R., Lowery, J. P., and O'Neill, R. D., Biosensor for neutrotransmitter L-glutamic acid designed for efficient use of L-glutamate oxidase and effective rejection of interference, *Analyst*, 122, 1419, 1997.
25. Burmeister, J. J., Palmer, M., and Gerhardt, G. A., Ceramic-based multisite microelectrode array for rapid choline measures in brain tissue, *Anal. Chem. Acta*, 481, 65, 2003.
26. Parikh, P., Pomerleau, F., Huettll, P., Gerhardt, G. A., Sarter, M., and Bruno, J. P., Rapid assessment of in vivo cholinergic transmission by amperometric detection of extracellular choline levels, *Eur. J. Neurosci.*, 20, 1545, 2004.
27. Burmeister, J. J., Palmer, M., and Gerhardt, G. A., L-lactate measures in brain tissue with ceramic-based multisite microelectrodes, *Biosens. Bioelectron.*, 20, 1772, 2005.
28. Friedemann, M. N. and Gerhardt, G. A., Regional effects of aging on dopaminergic function in the Fischer-344 rat, *Neurobiol. Aging*, 13, 325, 1992.
29. Danbolt, N. C., Glutamate uptake, *Prog. Neurobiol.*, 65, 1, 2001.
30. Hoffman, A. F., Lupica, C. R., and Gerhardt, G. A., Dopamine transporter activity in the sustanstia nigra and striatum assessed by high-speed chronoamperometric recordings in brain slices, *J. Pharmacol. Exp. Ther.*, 287, 487, 1998.
31. Molchanova, S., Koobi, P., Oja, S. S., and Saransaari, P., Interstitial concentrations of amino acids in the rat striatum during global forebrain ischemia and potassium-evoked spreading depression, *Neurochem. Res.*, 29, 1519, 2004.
32. Su, M. T., Dunwhittle, T. V., Mynlieff, M., and Gerhardt, G. A., Electrochemical characterization of stimulated norepinephrine overflow in locus coeruleus-hippocampus double brain grafts grown in oculo, *Neurosci. Lett.*, 110, 186, 1990.

33. Hoffman, A. F. and Gerhardt, G. A., Differences in pharmacological properties of dopamine release of the substantia nigra and striatum: an in vivo electrochemical study, *JPET*, 289, 455, 1999.
34. Madison, D. V. and Edson, E. B., Preparation of hippocampal brain slices, Crawley, J. *et al.*, Eds. In *Current Protocols of Neuroscience*, pp. 6.4.1–6.4.7, Wiley, New York, 1997.
35. Sabeti, J., Gerhardt, G. A., and Zahniser, N. R., Chloral hydrate and ethanol but not urethane, alter the clearance of exogenous dopamine recorded by chronoamperometry in striatum of unrestrained rats, *Neurosci. Lett.*, 343, 9, 2003.
36. Anis, N. A., Berry, S. C., Burton, N. R., and Lodge, D., The dissociative anasthetics, ketamine, phencyclidine, selectively reduce excitation of central mammalian neurones by N-methyl-aspartate, *Br. J. Pharm.*, 79, 565, 1983.
37. MacDonald, J. F., Miljkovic, Z., and Pennefather, P., Use-dependent block of excitatory amino acids currents in cultured neurones by ketamine, *J. Neurophysiol.*, 58, 251, 1987.
38. Rozza, A., Masoero, E., Favalli, L., Lanza, E., Govoni, S., Rizzo, V., and Montalbetti, L., Influence of different anaesthetics on extracellular amino acids in rat brain, *J. Neurosci. Methods*, 101, 165, 2000.
39. Roth, G. S., Ingram, D. K., and Lane, M. A., Caloric restrictions in primates and relevance to humans, *Ann. N.Y. Acad. Sci.*, 928, 305, 2001.
40. Liachenko, S., Tang, P., Somogyi, G. T., and Xu, Y., Comparison of anaesthetic and non-anaesthetic effects on depolarization-evoked glutamate and GABA release from mouse cerebrocortical slices, *Br. J. Pharm.*, 123, 1274, 1998.
41. Liachenko, S., Tang, P., Somogyi, G. T. and Xu, Y., Concentration-dependent isoflurane effects on depolarization-evoked glutamate and GABA outflows from mouse brain slices, *Br. J. Pharm.*, 127, 131, 1999.
42. Rutherford, E. C., Pomerleau, F., Strömberg, I., and Gerhardt G. A., Chronic second-by-second measures of L-glutamate in the CNS of rats using enzyme-based microelcetrode arrays, (in preparation).
43. Paxionos, G. and Watson, C., *The Rat Brain in Stereotaxic Coordinates*, 4th ed., Academic Press, New York, 1998.
44. Franklin, K. B. J. and Paxinos, G., *The Mouse Brain in Stereotaxic Coordinates*, Academic Press, New York, 1997.
45. Sullivan, R. M. and Gratton, A., Relationships between stress-induced increases in medial prefrontal cortical dopamine and plasma corticosterone levels in rats: role of cerebral laterality, *Neuroscience*, 83, 81, 1998.
46. Stevenson, C. W., Sullivan, R. M., and Gratton, A., Effects of basolateral amygdala dopamine depletion on the nucleus accumbens and medial prefontal cortical dopamine response to stress, *Neuroscience*, 166, 285, 2003.

20 Telemetry for Biosensor Systems

David A. Johnson and George S. Wilson

CONTENTS

INTRODUCTION

This chapter will provide an overview of radio frequency (RF) telemetry systems by examining the design requirements of two contrasting systems: a wireless system designed to support Fast Scan Cyclic Voltammetry (FSCV) and a wireless system designed to support selective biosensors. Traditionally, RF design and analog circuit design were very complicated endeavors. Component interactions and non-ideal behaviors made such designs as much art as science. Today however, there are a wide variety of analog integrated circuits with near ideal behavior, and self-contained RF modules with built-in standardized protocols are available. In short, it is now very possible for a neuroscientist, with little or no knowledge of telemetry, to design and construct robust wireless acquisition systems.

In addition, there are a wide variety of commercial implants available for identification and physiological monitoring in animals. Currently available wireless implants can measure activity, core temperature, blood pressure, heart rate, blood flow, pH, biopotentials Electrocardiography (ECG), Electroencephalography (EEG), Electromyography (EMG) and respiratory rate.[1] Major suppliers include Data Sciences International (www.datasci.com), Biomedic Data Systems (www.bmds.com), and Respironics (www.minimitter.com). Most of these systems use some form of inductive power and telemetry, very low power transmitters or both. That is, these devices normally use proprietary low power analog transmitters that can only communicate over short distances, and in some cases the power required to operate the device is derived from RF energy transmitted from the receiver (Reader) to the device. A wide and impressive variety of Application Specific Integrated Circuits (ASIC) has also been developed. Mohensi et al. provide an excellent overview of current systems.[2]

The goal of this chapter is simply to provide an overview based on familiar examples in the hope that this will help the reader to make informed decisions, whether building or buying telemetry based systems. To that end, we will consider the design of a digital wireless FSCV system for the measurement of electroactive species such as dopamine, and a biosensor based digital wireless system.

FAST SCAN CYCLIC VOLTAMMETRY

FSCV has become the standard tool for neuroscientists to study the kinetics of neurotransmitter release and uptake in the rat brain.[3] In vivo cyclic voltammetry is a powerful technique that allows the direct measurement and identification of compounds in the brain. Typically, FSCV for neurotransmitter applications uses scan rates around 300 V/s repeated 10 times per second. FSCV, in combination with very small carbon fiber electrodes is used for a number of reasons. Microelectrodes (6–30 μm diameter) provide superior spatial resolution from a group of synapses, however use of microelectrodes requires very fast scan rates due to the diffusion characteristics and kinetics of the electrochemical reaction.[4,5] FSCV provides a unique "fingerprint" for specific electroactive species, which serves to increase selectivity from potential interferences. The small electrode size creates a very small diffusion layer, thus leading to enhanced spatial and temporal resolution. In addition, the amount of toxic electrogenerated species will be quite small, preserving the integrity of the measured environment.[6] One limitation of FSCV is high background. As a result of the double layer capacitance, a large background current is always observed. Background currents can be easily identified in vitro. Wightman's group has described methods by which average background current can be acquired in vivo. In general, this involves acquiring multiple scans while biological activity is minimal and averaging them.[7]

FSCV is widely used for the in vivo measurement of dopamine.[8–11] While other techniques, such as microdialysis, can be used for this application, FSCV has a number of advantages, as outlined above. In addition to superior spatial resolution, FSCV also has better temporal resolution. Speed is essential because neurotransmitter (such as dopamine) activity tends to occur in phasic bursts that only last for 1–2 s.[11] Microdialysis probes have also been shown to produce damage that alters measurements near the probe.[12]

FSCV SYSTEM REQUIREMENTS

Typical FSCV system requirements are:

- Three electrodes (1 working, 1 reference, and 1 counter)
- User-controlled sweep repeat rate
- User-controlled scan rate (0–1000 V/s)
- User-controlled voltage range of -1.0 V to 2.0 V

- User-controlled Low Pass Filters
- Total samples per sweep of 1000
- At least 12-bit analog output resolution
- At least 14-bit analog input resolution ($\geq$ 100 kHz sampling rate).

From this list the telemetry data rate requirements can be estimated. Assuming a repetition rate of 10 scans per second and a total sweep time of less than 30 ms, 1000 14-bit data points will be acquired every 100 ms, and there will be a 70 ms window in which to transmit these data between scans. Given processing overhead it is prudent to assume that only 50 ms will be available for transmission. Further assuming a 2-byte per sample transmission, the minimum data rate required is 16 (bits) * 1000 (Samples/Sweep)/0.05 (seconds/Transmission) $=$ 320 kbits/s or 40 kBytes/s. This estimate does not include compression or packetization/balancing overhead. FSCV also requires a significant amount of set-up (scan rate, scan type, upper and lower bias, repeat rate, gain, stimulus). This indicates the need for a 2-way telemetry scheme. That is, a scheme in which data can be both transmitted and received.

IN-VIVO ELECTROCHEMICAL SENSORS

The pioneering work of Adams[13,14] and Wightman[15] on in vivo electrochemistry has demonstrated the utility of microsensors in the study of important issues in neurophysiology and neuropharmacology. These devices are generally confined to species that have well-behaved direct electrochemistry such as dopamine.

By definition, an electrochemical biosensor is "a self-contained integrated device, which is capable of providing specific quantitative or semi-quantitative analytical information using a biological recognition element (biochemical receptor) which is retained in direct spatial contact with an electrochemical transduction element."[16] In general, enzymes are used to selectively convert the substrate of interest into an equivalent amount of a readily detected electroactive species. Microbiosensors offer high selectivity, spatial, and temporal resolution, and considerable progress has been made in their design and application.[17,18] A recent review provides an excellent overview of the current state-of-the-art.[19] Biosensors have been developed for the detection of wide range of analytes including glutamate, choline and acetylcholine, lactate, pyruvate, and of course, glucose. Figure 20.1 shows an illustration of a glutamate biosensor available from Pinnacle Technology, Inc. Lawrence, KS.

The electronics support system required for biosensors is quite simple. For amperometric biosensors, a standard potentiostat circuit is used to maintain a fixed voltage while measuring a very small signal current. Biosensors are selective by design, and their temporal response is typically longer than 50 ms. This makes the design of a telemetry system much simpler than it is for FSCV.

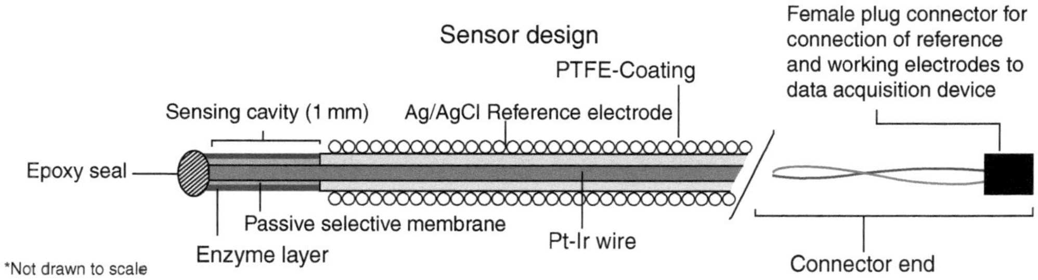

FIGURE 20.1 Glutamate biosensor overview (Pinnacle technology).

Biosensor System Requirements

These specifications will change with the type of biosensor targeted, but a typical set of biosensor system requirements is:

- 2 or 3 electrodes
- Static voltage (cannot be removed)
- User controlled bias range of (0–1.3 V)
- Gain of 20 million V/A
- Sample rate of 1 sample/second
- At least 12 bits of analog output resolution
- At least 16 bits of analog input resolution.

The telemetry and power requirements in this case are much lower than those required for FSCV. Even allowing a factor of two safety margin for various overhead parameters and a couple of channels, the required data rate is 2 (safety) * 2 (channels) * 1 (sample/seconds) * 16 (bits/sample) = 64 bits/s or 8 Bytes/s. The number of changeable parameters in this type of system is also low. That is, it may be enough to allow bias to be changed via a wired interface. In this case, a one-way (transmit only) telemetry system may be employed.

GENERAL DESIGN CONSIDERATIONS

Wireless systems offer a number of advantages, the primary two being freedom of movement for the animal under examination and elimination of all cable artifacts. There are, of course, associated drawbacks. In particular, these systems require that all functional blocks including power, telemetry, data acquisition, and conditioning, are mounted on or in the animal.

As a rule, any system developed should relieve stress for both the animal under test and the researcher carrying out the experiment. To this end, it is very important for system designers to work with end users to optimize the mechanical aspects of the system (reduced surgery time, simple sensor placement, simple battery replacement/recharging/recalibration), and software (simple, intuitive graphical user interfaces, reliable telemetry, reliable, and robust data storage).

Another obvious, but very important consideration for external systems designed for use with rats or mice, is that both are very small. Animal testing at Northwestern University has shown that mice will tolerate a head-mounted unit of up to 1 cm^3 without modifying their running wheel behavior, as long as the weight does not exceed 0.5 g (E. Naylor, personal communication). Others report upper weight limits for head mounted devices as 25 g for rats and 5 g for mice.[20] This presents a significant problem, and for this reason, most wireless systems produced for mice are internal implants. In the discussion that follows, we will not make any distinction between external systems (head mount, back mount) and implants since the operating principles are very similar. For rats there is an additional concern. Rats are quite aggressive in the removal of any external systems, so any external monitoring system developed for rats needs to be well protected.

For a wireless system, the designer has the choice of transmitting the data acquired in an analog form, or of digitizing on board before transmitting. Frequency modulated analog transmission can be relatively noise free, but the resolution of such systems is constrained by the resolution of the frequency modulation electronics on the transmitter, and the demodulator on the receiver. As was previously mentioned, there are a number of ASICs that have been developed for this purpose, but in general, it is preferable to digitize at the point of measurement if possible. In this case, the resolution and accuracy of the system are under direct control of the sensor electronics, and once the signal is digitized, no resolution is lost in the telemetry regardless of type.

Electronics

Batteries and other power sources will be discussed in a subsequent section, but it is important to note that when calculating battery life, the average current consumed is the key parameter—not the peak current. It will often be desirable to use a high data rate link in order to minimize the amount of time that the transmitter is on. Typically, the RF transmitter will consume more power than any other part of the system, so minimizing the amount of time that the telemetry system is on minimizes power consumption. FCC licensing is also simplified if the duty cycle (on time/total time) is less than 1%. In some situations, a fast, high-power telemetry system, on average, may require less power than a slow, low-power telemetry system.

A digital wireless data acquisition system will normally contain a small, low-power controller of some type, as well as analog to digital converters (A/D), digital to analog converters (D/A), sensor conditioning, power and telemetry. Battery voltage is typically limited by the voltage requirements of the controller (usually 2.7 V), and headroom for the amplifiers.

Assuming that the system being developed will digitize the data onboard, then the A/D type and range must also be carefully evaluated. For many low frequency and high-resolution applications sigma-delta A/Ds are very attractive. A/Ds of this type basically trade bit-resolution for sample speed. That is, they can be extremely accurate if the sampling rate is low. They are capable of measuring low-level signals directly and they normally implement an internal filter that can remove, or reduce the size of required anti-aliasing filters.

In many cases, this type of system can be designed to run directly off a voltage reference. References come in a variety of forms, but they are basically just Zener diodes held in breakdown with some form of buffered output. Voltage references provide a very stable, known voltage that ensures that the A/D bit measurement can be accurately translated to voltage. A linear voltage regulator may also be required. If needed, a regulator should be employed that works when the input voltage is very close to the desired voltage (low drop out), and for highest efficiency, the systems should be operated as close to the drop-out point as possible. Switching regulators should be avoided for most designs of this type. Switching regulators effectively operate by changing a DC input to an AC intermediary and then back to a DC output at the same or different voltage. These devices feature low cost, small size and low parts count, but they typically are not suitable for low noise and low power applications.

POWER SOURCES

New power sources are currently attracting considerable attention. These include advanced battery designs, microfuel cells, biofuel cells, and energy harvesters that convert light or vibration to energy. In the future, one or more of these solutions may become valuable for wireless design, but at present, they either target incompatible primary markets such as hybrid vehicles or cell phones, they are not commercially available at a reasonable price or they do not produce enough power to be useful in a data streaming system. For most research applications, it is preferable to use a power solution that is readily available and inexpensive. For this reason, we will focus on primary (not rechargeable) and secondary (rechargeable) batteries. Note: milliamp-hour (mAh) energy specifications are often optimistic, and normally involve discharge to a voltage level that may be significantly less than the minimum operating voltage of the system.

There are a large number of battery chemistries to choose from, however, if one of the system design parameters is small size, then the options are very limited. Battery availability is generally driven by the laptop, cell phone, hearing aid, and game markets. For many animal and human telemetry designs the main concerns are size, weight, capacity, discharge profile, and pulse current capability. In general, size and weight are a function of energy density. That is, how much energy can a given battery chemistry store in a given weight and volume. However, be cautious of data sheet claims. Practical energy density, including enclosure materials, electrolytes, etc., are often significantly lower than the theoretical energy density that is calculated based solely on the properties of the active materials. In addition, the chemical reaction in a primary cell will slow down at low temperatures degrading their ability to supply energy. At high temperatures, self-discharge can become a problem. For animal research size and weight are normally the most important parameters.

Primary batteries can be subdivided into five main classes: zinc/carbon/ammonium chloride (dry cell), zinc-air, silver oxide, alkaline, and lithium. Of these, the silver oxide and lithium chemistries are most useful for small wireless systems. Silver oxide button cells offer reasonable peak currents and a flat discharge curve. Lithium generally supports lower discharge currents, however lithium offers light-weight, high energy density, high open circuit voltage, low self-discharge, a flat discharge curve and a variety of small form factors. Three primary lithium chemistries are of specific interest, lithium carbon monofluoride ($LiCF_x$), lithium manganese dioxide ($LiMnO_2$), and lithium thionyl chloride ($LiSOCl_2$). Zn-Air energy density and storage life are also excellent and this chemistry may be a good choice for some systems. Table 20.1 gives an overview of the general characteristics of several important primary battery types. Pulse capability describes the ability of a battery to deliver pulse currents. This is important in RF systems since relatively high pulse currents may be required when transmitting and receiving data. Flat discharge describes the discharge characteristics of a battery under load. A flat discharge profile is desirable in most cases since the battery will maintain a steady voltage throughout the bulk of its discharge, and then the voltage will fall quickly. This makes it easier to design systems that use more of the available energy in the battery.

Rechargeable batteries generally offer low energy densities, and they require support electronics to implement the correct charging protocols. In the case of rechargeable lithium cells, special protection circuitry may also be required to ensure user safety. The cell voltage of rechargeable batteries also tends to be lower. However, for many applications, the convenience of recharging by direct, inductive or optical means may make a rechargeable chemistry preferable. For very small footprint devices, useful chemistries include: nickel cadmium (NiCd), nickel metal hydride (NiMH), and lithium-ion. Memory effects in Ni chemistries due to crystal formation are not normally a problem with modern battery designs. A wide range of lithium-ion and lithium-ion polymer (solid electrolyte) chemistries are available. The construction of rechargeable lithium cells from different manufacturers can be quite different. This makes it difficult to summarize the behavior. Table 20.2 shows the general characteristics of two of the most common materials.

TABLE 20.1
Primary Battery Comparison by Type

Type	Cell Voltage	Size	Energy Density	Flat Discharge	Pulse	Self-Discharge	Comments
Alkaline	1.5	Various large	Moderate	No	Excellent	Fair	Inexpensive, ubiquitous
$LiCF_x$	3.0	Various	High	Yes	Poor	Excellent	Long term low current
$LiMnO_2$	3.0	Various	High	Yes	Good	Excellent	High pulse
$LiSOCl_2$	3.6	Various	Very high	Yes	Good	Excellent	Highest pulse extended temp
Silver Oxide	1.5	Small	High	Yes	Excellent	Good	Low temp, high pulse
Zn-Air	1.45	Various	Very high	Yes	Good	Excellent	Must have access to air

TABLE 20.2
Secondary Battery Comparison by Type

Type	Cell Voltage	Energy Density	Flat Discharge	Pulse	Self-Discharge	Recharge Cycles[a]	Comments
NiCd[b]	1.2	Moderate	No	Good	Poor	>2000	Best recharge behavior
NiMH	1.2	Good	No	Good	Poor	>200	Good general characteristics and energy density
Lithium[b]	1.5–4.0	High	Yes	Good	Good	>500	Many different chemistries available

[a] Full discharge.
[b] Cadmium and Cobalt (some Lithium chemistries) are toxic and require proper disposal.

It is also worth noting that it is possible to form thin film lithium-ion batteries that can be molded into the mechanical structure of a device.

Specialty suppliers such as Wilson-Greatbatch (www.greatbatch.com) and Quallion (www.quallion.com) offer a wide variety of solutions targeting medical implants, and some of these can be useful for wireless designs. For higher power and high pulse applications, Tadiran (www.tadiran.com) offers a wide range of lithium solutions. Another option that should be considered is supercapacitors. Supercapacitors offer the possibility of very fast charging, high discharge currents, light-weight, and low cost. As more small format options become available, supercapacitors may come to play a greater role in wireless designs.

RF TELEMETRY

This section will focus on devices that operate in the Industrial Scientific and Medical (ISM) band of the RF spectrum. The ISM band includes 315–433 MHz, 902–928 MHz, 2.402–2.480 GHz, 5.725–5.875 GHz, and 24.0–24.25 GHz. In general, the maximum data rate is lower at lower frequencies. Telemetry in these bands is unlicensed in the U.S., but specific FCC regulations apply to the various bands. European regulations differ from those in the U.S., so careful attention must be paid to the various regulatory requirements if a device is intended to operate internationally. One drawback of the unlicensed ISM band is that it suffers from the "Tragedy of the Commons."[21] That is, use within these bands is minimally restricted and in some cases abused, so freedom from interference can never be guaranteed. A current example is the performance degradation and sometimes complete failure of wireless datalinks from laptop computers, wireless cell phones, etc., all vying for bandwidth in the very crowded 2.4 GHz band. Depending on the application, the Medical Implant Communications Service MICS (402–405 MHz), and the Wireless Medical Telemetry Services (608–614 MHz) may also be of interest.

True discrete RF circuit design requires considerable expertise. In most cases, it is preferable to buy and implement a module than to implement any solution from scratch. If you decide to pursue a discrete solution, follow the manufacturer's circuit and layout guidelines *exactly*. The following sections will discuss some of the currently available RF module and protocol options.

PROTOCOL INDEPENDENT TRANSMITTERS AND RECEIVERS

The simplest telemetry systems require nothing more than an inductor/capacitor (LC) circuit and an amplifier. These systems are very easy to construct, but they are relatively inefficient and subject to drift based on proximity to ground planes, temperature, etc. Transmitters that are much more stable can be constructed from crystals, phase locked loops (PLLs) or surface acoustic wave (SAW) elements. A variety of vendors (RF Monolithics www.rfm.com, Linx Technologies www.linxtechnologies.com, etc.) offer prefabricated and precertified transmit and receive modules that require little more than a data input/output, and antenna connection to set up a simple wireless link. Transceiver modules are also readily available. The data rates of these modules tend to be low ($<$ 200 kbits/s), but for simple systems, they can be very useful. One drawback to solutions of this type is that they are normally fixed to one frequency. This can create problems when multiple devices are in range of one another. For low duty cycle transmitters, simple time domain multiplexing (TDM) and error correction schemes can be used to overcome these limitations.

ZIGBEE (IEEE 802.15.4)

The ZigBee (IEEE 802.15.4) protocol offers reasonable data rates (250 kbit/s at 2.4 GHz), low power, small size, and mesh capable operations. ZigBee devices are specifically targeted to very low power, battery operated applications where the device itself is expected to be off 99.9% of the time. ZigBee is also self-assembling. Up to 65,000 devices can auto-assemble in a mesh with a single coordinating node. ZigBee is also "self-healing" in that it can detour data around malfunctioning nodes. ZigBee can operate in three bands for international compliance (868 MHz—1 channel, 915 MHz—10 Channels, 1.4 GHz—16 channels). Major module vendors include, Crossbow (www.xbow.com), Ember (www.ember.com) and Maxstream (www.maxstream.com). ZigBee modules typically require approximately 25–50 mA at 3 V when operating, but the current consumption in shutdown is very low (<25 uA). The power requirements are dependent on the desired range.

Bluetooth

In its base form, Bluetooth provides a two-way data link at up to 721 kbits/s and 3 voice channels for distances up to 100 m. The version 2.0+ EDR specification extends the data rate to 3 Mbit/s. Bluetooth is a robust specification operating in the 2.4 GHz ISM band. Bluetooth specifies three separate power ranges depending on the range required: Class 1 (100 m), Class 2 (10 m) and Class 3 (10 cm). Bluetooth normally requires that a user manually pair devices. The primary markets for Bluetooth devices are in local device-to-device communications at moderate data rates—primarily wireless headsets for phones and audio equipment. A class 2 Bluetooth device typically requires 25–50 mA when operating and 1–2 mA when in idle mode. The power requirements are dependent on desired range. Bluetooth modules are available from a wide range of suppliers.

WiFi (IEEE 802.11)

Of the various incarnations of 802.11, 802.11b at 11 Mbit/s is most commonly used in embedded systems, but 802.11 g at 54 Mbit/s can be easily implemented. WiFi targets wireless computer networking and as such it is typically overkill for sensor systems. However, there are applications that require the robust infrastructure and high data rates offered by this protocol. Unfortunately, in this case, speed requires power. WiFi modules will normally consume about 1 watt when operating. WiFi modules for embedded systems are becoming commonplace (www.lantronix.com, www.digi.com).

Ultra Wideband

Ultra Wideband (UWB) is a very interesting new protocol that targets very high speed 100–480 Mbit/s, short-range (10–20 m) video and high-speed data applications. The basic technology uses 500 MHz channels in a 7 GHz range (3.1–10.6 GHz). By using very wide bands, the power at each frequency can be very low, so this technology is not likely to interfere with other devices. The wide bands also preclude interference from other technologies that may be using the same bands. The power required by these devices is also very low, and the module size is small, so in the future UWB should be very useful for a wide range of data streaming applications. At present, consumer devices such as wireless USB 2.0 and Bluetooth links over UWB are just being introduced.

Radio Frequency Identification

Radio Frequency Identification (RFID) is being developed to replace UPC bar codes. RFID devices may be passive or active. Passive devices derive the power from the electric or magnetic field generated by a reader, and respond to the reader by detuning a tuned circuit. From the reader's point of view, less energy is coupled to the RFID device when it is detuned and this can be interpreted as binary data. The typical range of passive RFID tags is about 15 ft. (dependent on frequency). Data rates are very dependent on the carrier frequency and type of data modulation used. Amplitude shift key (ASK), frequency shift key (FSK), and phase shift key (PSK), are common approaches. For example, in an FSK scheme, the tuned circuit may be tuned and detuned for 4 and 4 carrier cycles to represent a binary 0, and 5 and 5 carrier cycles to represent a binary 1 with resulting data rates in the range of 30 kbits/s. Active RFID devices typically contain a battery that drives a transmitter in response to an RF query. These devices can operate over much longer distances. A common example is the devices used for automatic toll collection on many highways. RFID devices operate in a wide range of bands (125–135 kHz, 13.56 MHz, 868–928 MHz, 2.45 GHz, 5.8 GHz). RFID devices can be, and have been, applied to sensor applications, however, there are two main problems. The first is that the driving markets for RFID do not require data acquisition of any kind. They just return a unique identification in response to a query. For this reason, the bulk of the devices that are designed specifically for RFID have no acquisition capability. The second problem is that many of the techniques relevant to neuroscience require very accurate measurement of low-level signals. The RF interrogation and power coupling from an RFID reader generates a substantial amount of electrical noise which can severely degrade the signal to noise ratio if the system is not very carefully designed. The amount of power that can be coupled in is also very limited. Another problem unique to biosensors is that the applied potential on the sensor must be maintained. This can be achieved with RFID only if the on-board capacitance is kept sufficiently high.

Interference

In a normal operating environment, many different devices may be sharing the same RF band, and non-intentional radiators such as motors and microwave ovens can contribute a significant amount of RF interference in a band. This is particularly true in the ISM bands. Transmission in these bands is governed by the Federal Communication Commission's FR47 Part 15 rules. These rules place significant restrictions on power and modulation format. The simplest RF systems use only a single carrier frequency. Very low power transmitters can be located in close proximity as long as the transmitter range is small in comparison to the unit separation. In many cases, higher power single frequency systems can also be closely spaced if the duty cycle of the transmitter and receiver is very low and each unit contains a unique digital identification. More advanced systems such as ZigBee, Bluetooth and WiFi use spread spectrum techniques to overcome these limitations. As an example, Bluetooth implements frequency hopping spread spectrum modulation (FHSS) changing between 79 different randomly chosen frequencies in the 2.402–2.48 GHz range 1600 times each second. ZigBee and WiFi use a related technique called direct sequence spread spectrum in which the data are transmitted over a broad range of frequencies. In this way, a single frequency interference source cannot disrupt communications although broad frequency non-intentional noise sources can, and as an added benefit, the signals are generally "noiselike" and therefore, they are difficult to intercept. Hybrid modulation formats such as distributed frequency spread spectrum (DFSS) attempt to combine the benefits of the FHSS and DSSS.

Mesh Networks

Although not necessary for most neuroscience applications, it is worth noting that ZigBee, Bluetooth, and WiFi can all be operated in a mesh network architecture. With this structure, each unit communicates with its nearest neighbor until the data are finally transferred to/from a central controller. Systems of this type are robust since the loss of one node in the network will not affect the overall signal flow; they are simple to implement, they can use very little power since the transmitter only needs to be powerful enough to communicate with its closest neighbors, and they can work well in difficult RF environments since multi-path effects are greatly reduced.

Antenna Design

Regardless of telemetry type, an antenna will be needed. The antenna must be small enough to meet the requirements of a particular application, but it must also be matched to the RF system. That is, in the case of a transmitter, power flowing from the transmitter should couple to the antenna without reflection. In general, matching an antenna to a particular RF system will require consultation with someone who has the necessary tools and expertise to evaluate the design. In typical small medical systems everything in the system contributes to the RF circuit. This can include the particular size and type of battery being used, thickness of substrate, available ground plane, proximity of active components, and proximity of other ground planes (i.e., the animal itself), etc. That said, it is possible to get a good solution by trial and error. Most vendors of RF systems will provide on-line antenna design guides that are an excellent starting point. Beyond that, most small antenna designs will look capacitive, so in the case of a transmitter, a reasonable approach is to place a range of inductors in series with the transmitter and monitor the signal strength at a known receiver. When the inductor that provides the highest signal strength or transmit distance is found, the design is probably not perfect, but it may be adequate. Also, note that all small antennas will be directional to some extent. If all orientations are possible in an experimental situation, it is important to specify the range of a newly designed RF system in its worst possible orientation.

Any antenna designed for animal use should also be comfortable for the animal. The frequencies and levels of RF energy under discussion are both non-ionizing and non-thermal (no local heating). There is still some debate about possible health issues related to cell phones, but the power output of cell phones is much higher than any of the RF systems under discussion. In short, there is no evidence that exposure to digital RF energy at these levels will have any adverse effects on the animals under test.

IR Telemetry

Infrared (IR) telemetry should also be considered for any external design. IR telemetry systems can be very small, require no antenna, and although they require a considerable amount of power to operate, high data rates are possible, and as mentioned previously, this means that a capacitor can be used to create a reservoir of charge to supply power for a short period of time. The average power draw can be very small. Standards set by the Infrared Data Association specify 16 Mbit/s transmission rates currently with 100 Mbit/s standards planned. The primary drawback for IR telemetry is that it is line of sight. It is possible to use multiple receivers to eliminate this problem to some extent, but in general, IR is not appropriate for small animals that may sleep curled up, or under bedding.

DESIGN EXAMPLES

FSCV

With size and power consumption in mind, it is likely that the computational power available on a wireless FSCV system will be quite small. However, computationally simple Huffman coding techniques in conjunction with derivative computation have been shown to deliver $<50\%$ compression on EEG signals.[22] When all of these requirements are taken into account, the end result is that using available technology, a wireless FSCV system will require a 2-way data link capable of at least 200,000 bits per second (bit/s) and the overall system will consume approximately 0.1 watts of power.

Of the RF solutions discussed, only Bluetooth and UWB are appropriate. At this time, there are many more Bluetooth embedded solutions available, so this would be the most reasonable choice.

The power consumption of this system will be considerable and the duty cycle will be high, so a relatively large power source will be required. Of those discussed, two $LiSOCl_2$ half AA batteries at 3.6 V may be ideal. With a capacity of approximately 1.5 Ah, this could be used to power the system for a few days of continuous use. Secondary lithium cells would also be appropriate, but for most research use, the cost of primary cells will be insignificant, and the added battery lifetime will be valuable.

Selective Amperometric Biosensors

As described the data rate and power requirements of amperometric biosensor telemetry systems is quite low. This offers a range of RF and power solutions that will work well. ZigBee is a very reasonable choice since the duty cycle of this type of system will be very low. That is, data is transmitted in very short bursts with a long period of inactivity between transmissions. ZigBee offers two-way secure communications at very low power levels, and the ability to place numerous devices in close proximity with no interference, however, there are also some disadvantages to this choice: ZigBee module sizes are relatively large, and ZigBee requires a significant amount of power while transmitting and receiving. Typical lithium coin cells cannot deliver a pulse of more than approximately 1 mA without experiencing a severe voltage drop. Other small button cells such as silver oxide and zinc air can deliver the required pulse currents without a large capacitive assist, but the cell voltages are low, so multiple batteries may be required to implement a solution.

As an alternative, a simple transmission only solution using any of the available 315, 433 or 915 MHz (868 MHz for Europe) modules may be implemented. The advantage to this type of solution is that lithium coin cells with a capacitive assist can be used to implement a very small, light solution. There are several disadvantages. The first is that this type of solution uses a fixed frequency and therefore is more susceptible to interference from other systems and noise sources. It is also not possible to communicate commands to the device. That is, it is not possible to request a gain or bias change without some form of secondary wired interface. This type of solution will also have lower data rates than a ZigBee based solution. The result is that although the transmitter draws less power when on, the transmitter may be on for a much longer time. A final decision requires a full energy analysis for the specific application and a full evaluation of user requirements.

CONCLUSIONS

The purpose of this chapter was to present an overview of the design of wireless systems that is appropriate for in vivo measurements. The general topic is very broad, and many details have been omitted for the sake of brevity. The gist of the argument is that a great deal of industrial research has targeted the design of small, plug-and-play telemetry modules, and with these and some general knowledge of electronics, any neuroscientist can evaluate and design wireless data acquisition systems.

REFERENCES

1. Kramer, K. and Kinter, L., Evaluation and application of radio telemetry in small laboratory animals, *Physiol. Genomics*, 13, 197–205, 2003.
2. Mohensi, P., Najafi, K., Eliades, S., and Wang, X., Wireless multichannel biopotential recording using an integrated FM telemetry circuit, *IEEE Trans. Neural Syst. Rehabil. Eng.*, 13, 263–271, 2005.
3. Robinson, D., Venton, B. J., Heien, M. L. A. V., and Wightman, R. M., Detecting subsecond dopamine release with fast-scan cyclic voltammetry in vivo, *Clin. Chem.*, 10, 1763–1773, 2003.
4. Bard, A. J. and Faulkner, L. R., *Electrochemical Methods; Fundamentals and Applications*, Wiley, New York, 2001.

5. Brett, C. M. A. and Brett, A. M. O., *Electrochemistry Principles, Methods, and Applications*, Oxford University Press, Oxford, 2000.
6. Zimmerman, J. B. and Wightman, R. M., Simultaneous electrochemical measurements of oxygen and dopamine, *Anal. Chem.*, 63, 24–28, 1991.
7. Michael, D., Joseph, J., Kilpatrick, M., Travis, E., and Wightman, R. M., Improving data acquisition for fast-scan cyclic voltammetry, *Anal. Chem.*, 71, 3941–3947, 1999.
8. Jones, S. R., Garris, P. A., Kilts, C. D., and Wightman, R. M., Comparison of dopamine uptake in the basolateral amygdaloid nucleus, caudate-putamen, and nucleus accumbens of the rat, *J. Neurochem.*, 64, 2581–2589, 1995.
9. Cragg, S. J., Hille, C. J., and Greenfield, S. A., Dopamine release and uptake dynamics within nonhuman primate striatum in vitro, *J. Neurosci.*, 20, 8209–8217, 2000.
10. Schmitz, Y., Lee, C. J., Schmauss, C., Gonon, F., and Sulzer, D., Amphetamine distorts stimulation-dependent dopamine overflow: effects on D2 autoreceptors, transporters, and synaptic vesicle stores, *J. Neurosci.*, 21, 5916–5924, 2001.
11. Robinson, D. L., Heien, M. L. A. V., and Wightman, R. M., Frequency of dopamine concentration transients increases in dorsal and ventral striatum of male rats during introduction of conspecifics, *J. Neurosci.*, 22, 10477–10486, 2002.
12. Borland, L. M., Shi, G., Yang, H., and Michael, A. C., Voltammetric study of extracellular dopamine near microdialysis probes acutely implanted in the striatum of the anesthetized rat, *J. Neurosci. Methods*, 146, 149–158, 2005.
13. Kissinger, P. T., Hart, J. B., and Adams, R. N., Voltammetry in brain tissue: a new neurophysiological measurement, *Brain Res.*, 55, 209–213, 1973.
14. Adams, R. N., In-vivo electrochemical measurements in the CNS, *Prog. Neurobiol.*, 35, 297–311, 1990.
15. Wightman, R. M., Jankowski, J. A., Kennedy, R. T., Kawagoe, K. T., Schroeder, T. J., Leszczyszyn, D. J., Near, J. A., Diliberto, E. J., and Viveros, O. H., Temporally-resolved catecholamine spikes correspond to single vesicle release from individual chromaffin cells, *Proc. Natl Acad. Sci. U.S.A.*, 88, 10754–10758, 1991.
16. Thévenot, D., Toth, K., Durst, R., and Wilson, G., Electrochemical biosensors: recommended definitions and classification, *Pure Appl. Chem.*, 71, 2333–2348, 1999.
17. O'Neill, R. D., Microvoltammetric techniques and sensors for monitoring neurochemical dynamics in-vivo: a review, *Analyst*, 119, 767–779, 1994.
18. Pantano, P. and Kuhr, W. G., Enzyme-modified microelectrodes for in-vivo neurochemical experiments, *Electroanalysis*, 7, 405–416, 1995.
19. Wilson, G. S. and Gifford, R., Biosensors for real-time in vivo measurements, *Biosens. Bioelectron.*, 20, 2388–2403, 2004.
20. Fee, M. and Leonardo, A., Miniature motorized microdrive and commutator system for chronic recording in small animals, *J. Neurosci. Methods*, 112, 83–94, 2001.
21. Hardin, G., The tragedy of the commons, *Science*, 162, 1243–1248, 1968.
22. Antoniol, G. and Tonella, P., EEG data compression techniques, *IEEE Trans. Biomed. Eng.*, 2, 105–114, 1997.

21 The Principles, Development and Application of Microelectrodes for the In Vivo Determination of Nitric Oxide

Michael J. Serpe and Xueji Zhang

CONTENTS

INTRODUCTION

Nitric oxide (NO) is a hydrophobic, highly labile free radical that is catalytically produced in biological systems from the reduction of L-arginine by nitric oxide synthase (NOS) to form L-citrulline, which produces NO in the process. In biological systems NO has long been known to play various roles in physiology, pathology and pharmacology.[1] In 1987, NO was identified as being responsible for the physiological actions of endothelium-derived relaxing factor (EDRF).[2] Since that discovery, NO has been shown to be involved in numerous biological processes such as: vasodilatation and molecular messaging,[2] penile erection,[3] neurotransmission,[4,5] inhibition of platelet aggregation,[6] blood pressure regulation,[7] immune response,[8] and as a mediator in a wide range of both anti-tumor and anti-microbial activities.[9,10] In addition, NO has been implicated in a number of diseases including: diabetes,[11] Parkinson's and Alzheimer's diseases.[12] The importance of NO was confirmed in 1992 when *Science Magazine* declared NO the "Molecule of the Year" and in 1998, when F. Furchgott, Louis J. Ignarro, and Ferid Murad were awarded the Nobel Prize in Physiology and Medicine for unraveling the complex nature of this simple molecule. Despite the obvious importance of NO in so many biological processes, less than 10% of the thousands of scientific publications over the last decade dedicated to the field of NO research involved its direct measurement.

As stated above, NO plays a significant role in a variety of biological processes where its spatial and temporal concentration is of extreme importance. However, the measurement of NO is quite difficult due to its short half-life ($\sim$5 s) and high reactivity with other biological components such as: superoxide, oxygen, thiols and others. To date, several techniques have been developed for the measurement of NO including: chemiluminescence,[13,14] Griess method,[15] paramagnetic resonance spectrometry,[16] paramagnetic resonance imaging, spectrophotometry,[17] and bioassay.[18] Each of these techniques has certain benefits associated with it, but suffer from poor sensitivity and the need for complex and often expensive experimental apparatus. In addition, the above NO sensing techniques are limited when it comes to continuous monitoring of NO concentration in real-time and most importantly in vivo. To date, electrochemical (amperometric) detection of NO is the only available technique sensitive enough to detect relevant concentrations of NO in real-time and in vivo, and suffers minimally from potential interfering species such as: nitrite, nitrate, dopamine, ascorbate, and L-arginine. Also, because electrodes can be made on the micro and nano-scale, these techniques also have the benefit of being able to measure NO concentrations in living systems without any significant effects from electrode insertion.

The first amperometric NO electrode used for direct measurement was described in 1990.[19] In 1992, World Precision Instruments Inc. (WPI) developed the first commercial NO sensor system called the ISO-NO. Over subsequent years a range of highly specialized and sensitive NO electrodes have been developed offering detection limits for NO ranging from below 1 nM up to 100 μM.[20] Most recently, a unique range of high sensitivity NO sensors based on a membrane coated activated carbon microelectrode, with diameters ranging from 200 μm down to 100 nm, have been developed by this lab. These electrodes exhibit superior performance during NO measurement and feature a detection limit of less than 0.5 nM NO.

NO MICROELECTRODES

Among the several electrochemical techniques that have been shown to be useful for the measurement of NO, the most popular is amperometry. This technique uses the model set forth by Clark and Lyons in 1962 for continuous gas monitoring during cardiovascular surgery.[21] Generally, this technique involves applying a fixed (poise) voltage potential to a working electrode, vs. a reference electrode, and monitoring the very low redox current produced (e.g., pA's) by the oxidation of NO. This technique has proven to be very useful for NO detection due to its fast response time, which is less than a few seconds, and its high sensitivity. As a result it is possible to monitor changes in NO concentration on biologically relevant time scales and concentrations, which are typically in the nM range. A multitude of other electrochemical techniques have been used to detect NO including: differential pulse voltammetry (DPV), differential normal pulse voltammetry (DNPV), linear scanning voltammetry (LSV), square wave voltammetry (SWV) and fast scan voltammetry (FSV). These methods typically employ a classical 3-electrode configuration consisting of a working electrode, reference electrode, and a counter electrode. Scanning techniques, with the exception of fast scanning voltammetry, require approximately ten seconds for the voltammogram to be recorded, which precludes its use in most NO research applications. Moreover, since scanning voltammetry-based NO instrumentation is not commercially available, NO researchers typically prefer to use the 2-electrode amperometric technique. Since the amperometric method is so widely used the discussion below will focus on these techniques.

CLARK TYPE NO MICROELECTRODES

The first described electrochemical NO sensor was based on a classical Clark electrode design, where NO was directly oxidized on the working electrode surface.[22] The NO sensor was composed of a fine platinum wire and a separate silver wire, which were then inserted into a glass micropipette. The micropipette was then filled with 30 mM NaCl and 0.3 mM HCl and sealed at the tip with a chloroprene rubber membrane. The platinum (working) electrode was positioned close to the surface of the membrane. The silver wire was then used as the reference/counter electrode. Although such electrodes could be used to measure NO, their inherent low sensitivity, narrow linear concentration measurement range, and fragility rendered them unsuitable for most research applications. In 1992, utilizing the Clark type design, WPI produced the first commercial electrochemical NO sensor (ISO-NOP) for use with their NO detection meter (ISO-NO). The ISO-NOP sensor consisted of a platinum wire disk working electrode and an Ag/AgCl reference electrode. Both electrodes were encased within a protective Faraday-shielded stainless steel sleeve. The tip of sleeve was covered with a NO-selective membrane and the sleeve itself contained an electrolyte. The rugged design of this sensor made it extremely convenient in many research applications and the sensor became widely used and established in numerous NO measurement research applications. The basic design of this type of NO sensor is illustrated in Figure 21.1.[23]

In principle the ISO-NOP sensor works as follows. The sensor is immersed in a solution containing NO and a positive potential of ~860 mV (vs. Ag/AgCl reference electrode) is applied. NO diffuses across the gas permeable/NO-selective membrane and is oxidized at the working electrode surface producing a redox current. This oxidation proceeds via an electrochemical reaction followed by a chemical reaction. The electrochemical reaction is a one-electron transfer from the NO molecule to the electrode, resulting in the formation of the nitrosonium cation:

$$NO - e^- \rightarrow NO^+$$

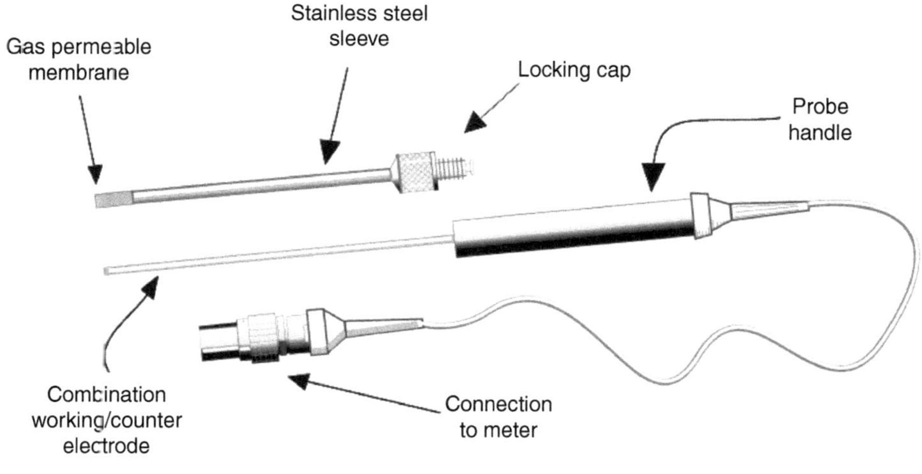

FIGURE 21.1 Illustration of WPI's ISO-NOP NO sensor. (Reprinted from Zhang, X. J., *Front. Biosci.*, 9, 3434, 2004. With permission.)

NO^+ is a relatively strong Lewis acid and in the presence of OH^-, it is converted into nitrite (NO_2^-):

Hence:

$$NO^+ + OH^- \rightarrow HNO_2 \rightarrow H^+ + NO_2^-$$

Nitrite can then be further oxidized into nitrate.

The amount of NO oxidized is thus proportional to the current flow between the working and reference electrodes, which is measured by an NO meter.

The amount of redox current that is typically generated by the oxidation of NO in biological systems is extremely small, typically on the order of 1–10 pA. As a result of these extremely small currents, the design of an amperometric-based electrode NO detection system requires very sensitive electronics and ultra low noise amplification circuitry. These measurement limitations were overcome by this lab with the development of the low noise and isolated circuit electronic devices ISO-NO and ISO-NO Mark II NO meters. These meters employed a unique electrically isolated low noise circuit that permitted measurement of redox currents as small as 0.1 pA. The design of the instrument also allowed measurement of NO to be performed without the need for special electrical screening, such as a Faraday cage. Recently, the world's only fully integrated multiple channel electrochemical NO/free radical detection system, the Apollo-4000 was developed at WPI. The detection system is an optically isolated multiple-channel NO/free radical analyzer designed specially for the detection of NO and other free radicals such as oxygen, hydrogen peroxide, hydrogen sulfide, and superoxide.[19] Although the Clark type electrode has enjoyed such great success for NO detection, there have been other significant advances for the detection of NO that deserve mention.

Metalloporphyrin and Metallophthalocyanine Modified NO Microelectrodes

Surface modified NO sensors incorporate an electrode surface that has been modified or treated in some way so as to increase the selectivity of the sensor for NO and promote catalytic oxidation of NO. An early example of such a sensor was presented by Malinski and Taha in 1992.[24] In this publication a ~500 nm diameter carbon fiber electrode was coated with tetrakis (3-methoxy-4-hydroxyphenyl) porphyrin, via oxidative polymerization, and Nafion. This electrode was shown to have a detection limit of ~10 nM for NO and great selectivity against common interferences. However, recently it has been shown that this electrode suffers severe interference from H_2O_2.[25] Following this publication, Schuhmann showed that pyrrole functionalized porphyrins, containing metals such as Ni, Pd and Mn, can be immobilized on carbon microelectrode surfaces via oxidative polymerization and may be used for NO detection.[26,27] Other researchers have shown that carbon fibers coated with a variety of porphyrins such as iron porphyrin,[28] and cobalt porphyrin[24,29–32] are also effective for NO detection. Although metal porphyrin coated electrodes were successfully used to some extent for various applications;[33–35] subsequent studies have shown that carbon fibers modified with unmetallated porphyrins, as well as bare carbon fibers, can detect NO with similar sensitivity.[36,37] The sensitivity and selectivity of porphyrinic NO sensors vary significantly from electrode to electrode and depend not only on the potential at which NO oxidizes, but also on the effects of axial ligation to the central metal in the porphyrin, modification/treatment procedure and other experimental variations. Furthermore, because the surface of the electrode remained in direct contact with the measurement medium a variety of biological species were shown to interfere (i.e., give false responses) with the measurement of NO. Adding a Nafion layer to the porphyrin-coated fibers could minimize these interferences. Other practical problems such as easy porphyrin removal and degradation have limited their usefulness for most applications.[38] Phthalocyanines, with a similar structure to porphyrins, containing metals such as Fe, Ni and Co have also been used to modify electrode surfaces for NO sensing.[39,40] Phthalocyanine modified electrodes have comparable detection limits and selectivity to porphyrin modified electrodes with the added benefit of being more stable to degradation.

COMBINATION NO MICROELECTRODES

During the mid to late 1990s a new range of combination NO sensors with tip diameters between 7 and 200 μm were developed by this lab.[41] These sensors combined a carbon fiber working electrode with a separate integrated Ag/AgCl reference electrode. The resulting combination sensor was then coated with a proprietary gas permeable/NO selective membrane. A high performance Faraday-shielded layer was then added to the sensor outer to minimize susceptibility to environmental noise. This electrode was then operated exactly as outlined above for the Clark type NO sensors. The use of these proprietary diffusion membranes and the novel design allows for NO measurement in small volumes and confined spaces with great selectivity against a wide range of interferences such as ascorbic acid, nitrite and dopamine. This sensor design was elaborated upon by creating L-shaped sensors designed specifically for tissue bath studies.[19] The above design was further elaborated upon by our lab by creating flexible, virtually unbreakable NO sensors designed specifically for use in measuring NO concentrations in arteries and microvessels. This electrode combines a Pt/Ir wire with a separate integrated Ag/AgCl reference electrode.[42] The resulting combination sensor was again coated with a proprietary gas permeable/NO selective membrane and a high performance Faraday-shielded layer and operated as described above. Recently, we developed a novel combination NO nanosensor, which had a tip diameter of just 100 nm.[43] The design of this sensor can be seen in Figure 21.2. This sensor was constructed using a 7 μm carbon fiber that was etched with an Ar ion beam to result in fibers with diameters in the 100 nm range. These 100 nm fibers were then used to construct combination electrodes, as described above. These sensors are capable of making NO measurements on the cellular level. Later in 2000, our group developed a unique "microchip" combination sensor.[44] This sensor is quite unique in its design but its operation principle is the same as for all the other combination electrodes presented in this section. Briefly, a 5000 Å thick carbon film was deposited on a Si wafer by RF sputtering. This carbon layer was then covered with a resist layer. 2 μm diameter carbon electrodes were then exposed by a dry etching process in a 50×50 array on the wafer surface, where each carbon electrode is 20 μm away from its nearest neighbor. An Ag/AgCl reference electrode and a Pt counter electrode were added to the wafer surface. The surface was then modified with WPI's proprietary gas permeable, NO selective membrane and ready for use. The resulting sensor (ISO-NOP-MC) exhibited an extremely low NO detection limit on the order of 300 pM, great selectivity against ascorbic acid, nitrite and dopamine, and a superior response time.

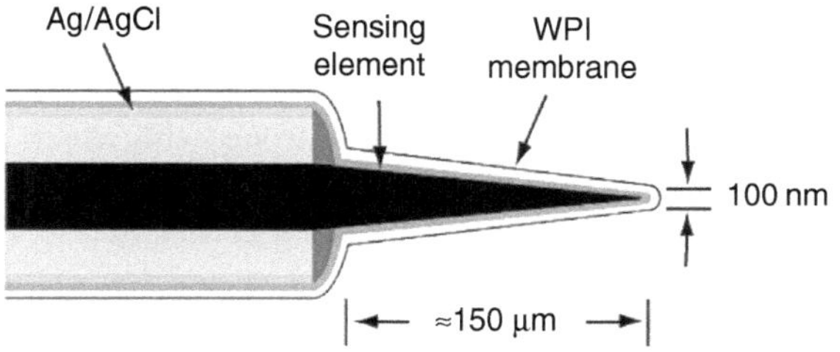

FIGURE 21.2 Illustration of WPI's ISO-NOPNM 100 nm combination NO nanosensor. (Reprinted from Zhang, X. J. et al., *Electrochem. Commun.*, 4, 11, 2002. With permission.)

Other NO Microelectrodes

Various other types of carbon fiber NO sensors, that utilize a variety of different coatings, have been described. Coatings used for these sensors include: conducting and non-conducting polymers,[45–49] multiple membranes,[50,51] ruthenium,[52] iridium and palladium,[53] heated-denatured cytochrome *c*,[54] nafion-CoII-1,10-phenanthroline,[55] ferrioxamine,[56] a microcoaxial microelectrode was reported for in vivo NO measurement,[57] siloxane polymer,[58] nafion and cellulose,[59] Hb/phosphatidylcholine films,[60] hemoglobin–DNA film,[61] ionic polymers and α-cyclodextrin.[62] Recently, Mizutani's group reported the coating of a 10 μm Pt disc electrode with a cross-linked sol–gel Langmuir–Blodgett film of siloxane polymer to render the electrode permselective for NO.[49] Schoenfisch's group recently reported the addition of a sol–gel film to a Pt electrode for NO detection.[63] Meyerhoff's group described an improved planar amperometric NO sensor based on a platinized anode[64,65] and its application for measurement of NO release from NO donors. Scheler and co-workers explored using a myoglobin–clay modified electrode for NO detection.[66] Kamei and co-workers fabricated a NO sensing device for drug screening using a polyelectrolyte film.[67] Indium hexacyanoferrate film-modified electrodes were used for NO detection by Casero and co-workers.[68] Schuhmann's group[69] developed a device for both in situ formation and scanning electrochemical microscopy assisted positioning of NO sensors above human umbilical vein endothelial cells for the detection of nitric-oxide release.

Unfortunately, despite the novelty of the above approaches, none of the sensors have stood the test of time, mostly due to various practical difficulties and/or poor sensitivity/selectivity. Furthermore, the lack of any published data describing the use of these sensors in any biological research applications limits any conclusion that can be made on their individual performance.

CALIBRATION OF NO MICROELECTRODES

Routine calibration of an NO sensor is essential in order to ensure accurate experimental results. One of three calibration techniques is generally used, depending on the sensor type, and will be described in the following section. Each of these methods has already been the subject of several reviews,[20,70–72] and will therefore, only be summarized here. NO sensors are typically sensitive to temperature and thus, calibration is usually best performed at the temperature at which the measurements will be made.

Calibration Based on Chemical Generation of NO

This method of calibration generates known concentrations of NO based on the reaction of nitrite with iodide in acid according to the following equation:

$$2KNO_2 + 2KI + 2H_2SO_4 \rightarrow 2NO + I_2 + 2H_2O + 2K_2SO_4$$

The NO generated from the reaction is then used to calibrate the sensor. Since the conversion of NO_2^- to NO is stoichiometric (and KI and H_2SO_4 are present in excess) the final concentration of NO generated is equal to the concentration of KNO_2 in the solution. Hence, the concentration of NO can be easily calculated by simple dilution factors. Experiments have demonstrated that NO generated from this reaction will persist sufficiently long enough to calibrate an NO sensor. However, since this technique involves the use of a strong acid, which can damage the delicate selective membrane of most NO microsensors, it is only suitable for use with Clark type stainless steel encased NO sensors such as the ISO-NOP. Figure 21.3 illustrates the amperometric response of a 2.0 mm ISO-NOP sensor following exposure to increasing concentrations of NO. The sensor responds rapidly to NO and reaches steady state current within a few seconds. The data generated from Figure 21.3 is then used to construct a final calibration curve (Figure 21.3, inset). The calibration curve illustrates the good linearity that exists between NO concentration and the current produced by its oxidation.

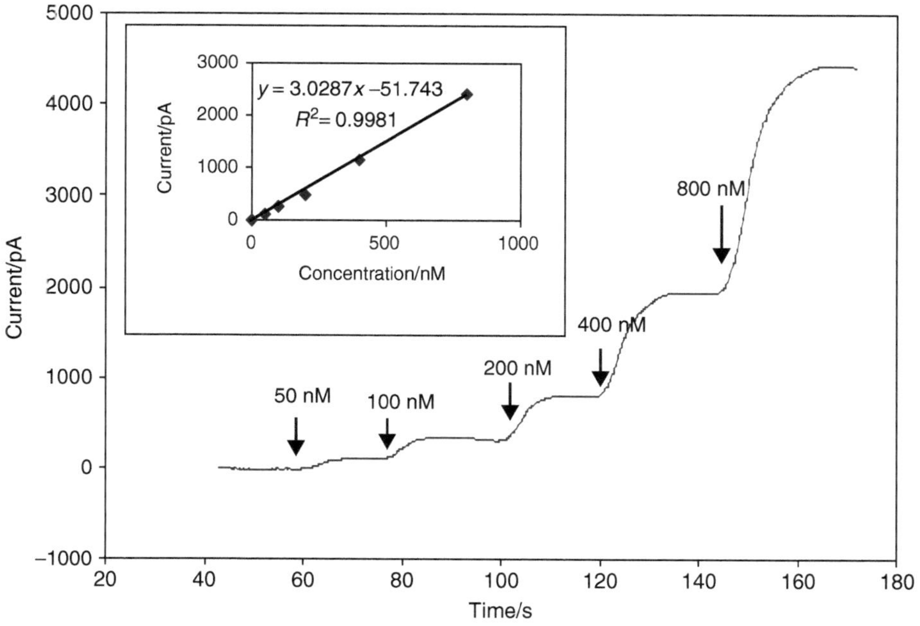

FIGURE 21.3 Response of WPI's ISO-NOP NO sensor to increasing concentration of chemically generated NO, inset shows the resulting calibration curve. (Reprinted from Zhang, X. J., *Front. Biosci.*, 9, 3434, 2004. With permission.)

Calibration Using an NO Standard Solution

This technique involves the production of an NO stock solution using a supply of compressed NO gas. One advantage of this method is that it allows an NO sensor to be calibrated in a similar environment in which the experimental measurements will be made. However, the major drawback is that it requires a source of compressed NO gas, and since NO gas is toxic, the whole procedure must be performed in a fume hood. This method can be summarized as follows. A vacutainer is first filled with 10 mL deionized water and agitated ultrasonically for 10 min. Purified argon is then passed through an alkaline pyrogallol solution (5% w/v) to scavenge any traces of oxygen before being purged through a deionized water solution for 30 min. NO stock solution is prepared by bubbling compressed NO gas through the argon-treated water for 30 min. The NO gas is first purified by passing it through 5% pyrogallol solution in saturated potassium hydroxide (to remove oxygen) and then 10% (w/v) potassium hydroxide to remove all other nitrogen oxides. The resultant concentration of saturated NO in the water is 2 mM at 22°C.[73] This can be confirmed further by a photometric method based on the conversion of oxyhemoglobin to methemoglobin in the presence of NO.[74] NO standard solutions can then be freshly prepared by serial dilution of saturated NO solution with oxygen-free deionized water prior to each experiment.

Calibration Based on Decomposition of SNAP

For this method *S*-nitroso-*N*-acetyl-D, L-penicillamine (SNAP, FW = 220.3) is decomposed to NO in solution in the presence of a Cu(I) catalyst.[75] The resultant NO generated can then be used to calibrate the sensor. The reaction proceeds in accordance to the following reaction:

$$2RSNO \rightarrow 2NO + RS-SR$$

The stoichiometry of the reaction dictates that the final generated NO concentration will be equal to the concentration of SNAP in the solution. The method can be summarized as follows. Saturated cuprous chloride solution is first prepared by adding 150 mg CuCl to 500 ml distilled water. This solution is then deoxygenated by purging with pure nitrogen or argon gas for 15 min. The final, saturated CuCl solution will have a concentration of approximately 2.4 mM at room temperature. The solution is light sensitive and must therefore, be kept in the dark prior to use.

The SNAP solution is then prepared separately as follows. EDTA (5 mg) is dissolved in 250 mL of pure water (HPLC grade, Sigma) and then adjusted to pH 9.0 using 0.1 M NaOH. The solution is then deoxygenated using the method described above. 5.6 mg of SNAP is then added to the solution to result in a SNAP concentration of ~0.1 mM. SNAP solution is also extremely sensitive to light and temperature and must be stored refrigerated and in the dark until required. Under these conditions, and in the presence of the cheating reagent (EDTA), the decomposition of SNAP occurs extremely slowly. This allows the solution to be used to calibrate NO probes throughout the day. In practice, calibration is performed by placing an NO sensor into a vial containing a measured amount of the CuCl solution and known volumes of the SNAP stock solution are then injected into the vial and the final concentration of NO can be calculated using dilution factors.

The concentration of SNAP in the stock solution is calculated as follows:

$$[C] = [AW/(MV)] \times 1000$$

where C, concentration of SNAP (μM); A, purity of SNAP; W, weight of SNAP in milligrams (mg); M, MW of SNAP (mg/mmole); and V, volume of the solution in liters (L).

If SNAP purity, for example, is 98.5% then the concentration of SNAP is calculated as:

$$[C] = [98.5\% \times 5.6/(220.3 \times 0.25)] \times 1000 = 100.1\ \mu M$$

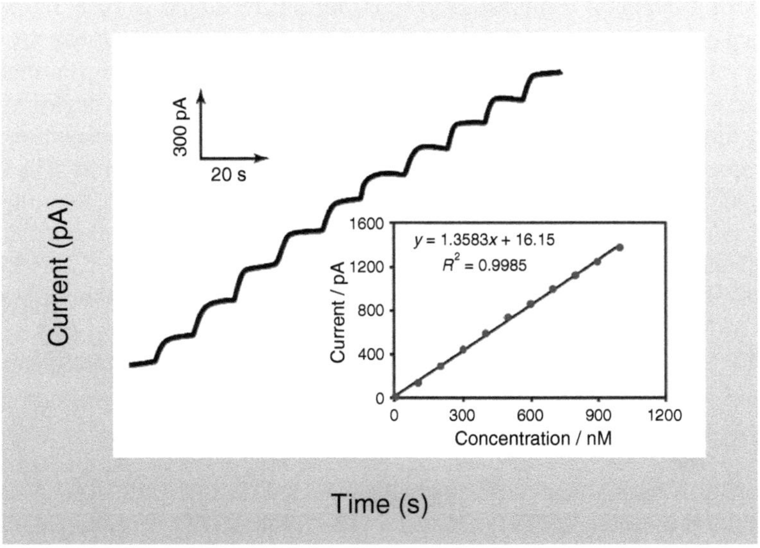

FIGURE 21.4 Response of WPI's ISO-NOP007 NO sensor to increasing concentration of NO produced by introduction of SNAP to a solution of CuCl. (Reprinted from Zhang, X. J. et al., *Electroanalysis*, 12, 425, 2000. With permission.)

Figure 21.4 shows a typical calibration curve generated using an NO microsensor and the SNAP method described.

CHARACTERIZATION OF NO MICROELECTRODES

NO microsensors can be characterized in terms of sensitivity, detection limit, selectivity, response time, stability, linear range, lifetime, reproducibility and biocompatibility. Sensor stability is important especially when measuring low NO concentrations. For example, when measuring low NO concentrations it must be the case that the noise is lower than the anticipated current change upon NO addition. The linear range of an NO sensor is important to know before performing any measurement. For example, the concentration of NO that is being measured must be in the linear range of the sensor in order for the measurement to be accurate. Typical commercially available NO sensors from WPI have a large linear range of 1 nM to 10 μM. Sensor lifetime and reproducibility is important to know when using a sensor and can be determined by frequent calibration, as described above. Biocompatibility is also extremely important when making NO measurements in vivo. Frequently, NO sensors are impaled into living tissue so the reactivity of the tissue toward the NO selective membrane must be minimal and cause a minimal amount of irritation in the tissue. The NO selective membranes used by this lab fulfill the requirements of being unreactive toward biomolecules and do not contribute to tissue inflammation. In most applications detection limit, sensitivity, selectivity and response time are usually the most important requirements, and will be described in further detail.

SENSITIVITY AND DETECTION LIMIT

The sensitivity of an NO sensor depends largely on the reactive surface area of the sensor and the electrode materials used in the design and can range from 0.03 to 100 pA/nM NO. Generally speaking, this sensitivity is directly proportional to the electrode size and surface status where an electrode with a small surface area will generally have a lower sensitivity compared to one with a larger surface area. Although a sensor's sensitivity is clearly important, its detection limit is often more important to the investigator. High sensitivity of a sensor does not necessarily equate to a low detection limit. For example, a highly sensitive NO sensor may have a high background noise level, which at a high NO concentration may not be a problem. However, at lower NO concentrations, measurement can be hindered by excessive noise. Accordingly, in evaluating the performance of an NO sensor, the ultimate detection limit is usually more critical than the sensitivity. Fortunately, most commercial NO sensors can detect NO at levels of 1 nM or less and are therefore well suited for the majority of research applications.

SELECTIVITY

An NO sensor is practically useless unless it is immune to interference from other species likely to be present in the measurement environment. Selectivity is usually controlled by both the voltage applied between the working and reference electrode (poise voltage) and the selective membrane used to coat the sensor. Many species present in a biological matrix are easily oxidized at the poise voltage employed to detect NO (i.e., +860 mV vs. Ag/AgCl). For example, monoamines such as dopamine (DA), 5-hydroxytryptamine (5-HT), and norepinephrine (NE), as well as their primary metabolites, can be oxidized at 0.3 V (and higher) vs. Ag/AgCl. Ascorbic acid can be oxidized at 0.4 V (and higher). A Clark type NO sensor (e.g., ISO-NOP) is covered with a gas permeable membrane, hence the selectivity of such sensors in biological samples is extremely good. With other types of NO sensors selectivity is usually achieved by coating the sensor surface with Nafion and other gas permeable membranes. Nafion is widely used to eliminate interference caused by anions, such as ascorbate and nitrite, during measurement of catecholamine species. When used for

NO detection the negatively charged Nafion layer can stabilize NO^+ formed upon the oxidation of NO and prevent a complicated pattern of reactions that could lead to the formation of nitrite and nitrate. However, the main drawback with Nafion is that it does not eliminate interference from cationic molecules such as dopamine, serotonin, epinephrine and other catecholamines. Consequently, selectivity of the conventional Nafion coated NO sensors is very poor. Nafion coated NO sensors also exhibit other undesirable characteristics including: unstable background current, continuous drift in the base line, and extended polarization requirements. These problems significantly limit the use of Nafion coated carbon fiber electrodes for measurement of NO. During the late 1990s this lab developed a unique multi-layered proprietary membrane configuration. NO sensors coated with this membrane exhibited increased selectivity and sensitivity for NO, and moreover were shown to be immune from interference caused from a wide range of potentially interfering species.[41]

Response Time

Response and recovery times of NO sensors are extremely important for their use in vivo. Theoretically, since the rate of mass transport at a microelectrode is very high it should have a response time on the order of μsec, however the addition of NO selective membranes to the electrode surface decreases this response time significantly. With a membrane present the response time is now dependent on the diffusion rate of NO across the membrane, which is highly dependent on the nature of the membrane as well as its thickness. The response time not only depends on the electrode being used, but also on the electronics being used to read out the current. For example, since the current being read out is typically on the order of pA's an electronic filter is usually applied, which also slows the system response. Since the half life of NO is ~5 s, a sensor response on the order of 3–4 s will work fine.

SELECTED NO MICROELECTRODE APPLICATIONS

Several hundred research papers have been published over the last decade describing the amperometric detection of NO in biological systems. This is mainly due to the fact that these type of electrochemical sensors are the only way NO can be measured in vivo in biological systems. Because there has been such an explosion in the development of NO microsensors these measurements can now be made in a variety of biological tissues and organs, as well as on the cellular level without significant damage to the system. This section will point out several examples where NO microsensors were used to determine its biological effects.[76]

Determinations of NO in a variety of biological systems have been made. For example, measurement of NO has been made in eyes,[77–79] gastrointestinal tracts,[80,81] brain tissue,[45,48,82–85] kidney and kidney tubule fluids,[86–91] rat and guinea pig isolated and intact hearts,[92,93] rat spinal cords,[94] human monocyte cells,[95] human endothelial cells,[96] mitochondria,[97,98] rat penis corpus cavernosum,[99] granulocytes,[100] invertebrate ganglia and immunocytes,[101] choroidal endothelial cells,[102] cancer cells,[103,104] peripheral blood,[105] human blood,[106] human leukocytes,[107] platelets,[108–110] ears,[111,112] plants,[113–116] and pteropod mollusks.[117]

Levine and co-workers first reported on the real-time profiling of kidney tubular fluid nitric-oxide concentration in vivo.[87,89] In the 2001 publication, a modified version of a combination NO electrode (WPI, ISONOP007) was successfully used to measure NO concentration profiles along the length of a single nephron of a rat kidney tubular segment. Since it was shown that the electrode is sensitive to NO in the rat tubule, it was used to detect NO concentration differences in rat kidney tubules before and after 5/6 nephrectomy. The results clearly showed that the NO concentration was much higher in nephrectomized rats vs. unnephrectomized rats.

In a recent publication, investigators used a specially customized ISO-NOP sensor to monitor, in real-time, NO production in the stomach and esophagus of human patients.[80] In this method a patient first swallowed two NO electrodes (see Figure 21.5), which were then withdrawn slowly at 1 cm increments every 2 min. The investigators were then able to establish a profile of NO concentration in the upper gastrointestinal tract.

Simonsen's group has performed some elegant work over the years on NO release characteristics from rat superior mesenteric arteries. Initially, Simonsen's group simultaneously monitored artery relaxation and NO concentration in the artery using a NO microsensor in response to various drugs.[118] NO concentration was monitored via an ISONOP30 electrode, purchased from WPI, that was inserted into the artery lumen using a micromanipulator. The results of this work are shown in Figure 21.6. The figure shows the force (upper traces) and NO concentration (lower traces) in an endothelium intact (+E) segment of the rat superior mesenteric artery and the same segment after mechanical endothelial cell removal (−E). As can be seen from the traces, if endothelial cells are present the artery is capable of relaxation due to the endothelial cells releasing NO in response to acetylcholine, but if the endothelial cells are removed the artery is insensitive to acetylcholine injection, relaxing upon the introduction of the NO releasing molecule SNAP. In a subsequent publication, Simonsen's group again monitored artery relaxation and NO concentration in the rat superior mesenteric artery to monitor its hyporeactivity to various endotoxins.[119] In this study an ISONOP30 electrode (WPI) was inserted into the lumen of the artery and the NO concentration, as well as artery relaxation, was measured as a function of endotoxin introduction to determine what effects lipopolysaccharide had on the artery function. The results showed that

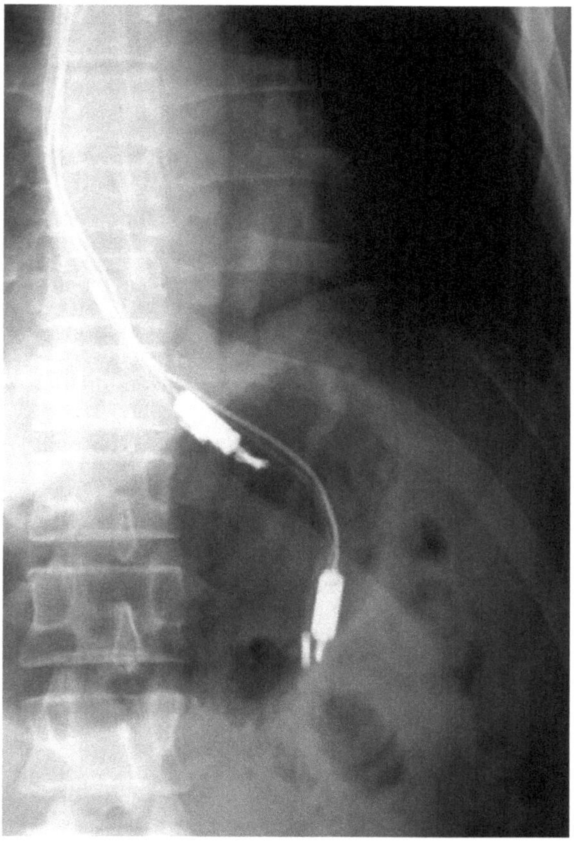

FIGURE 21.5 Abdominal x-ray showing the apparatus consisting of two nitric-oxide sensors, a 4-channel pH catheter, and a Teflon nasogastric tube. (Reprinted from Iijima, K. et al., *Gastroenterology*, 122, 1248, 2002. With permission.)

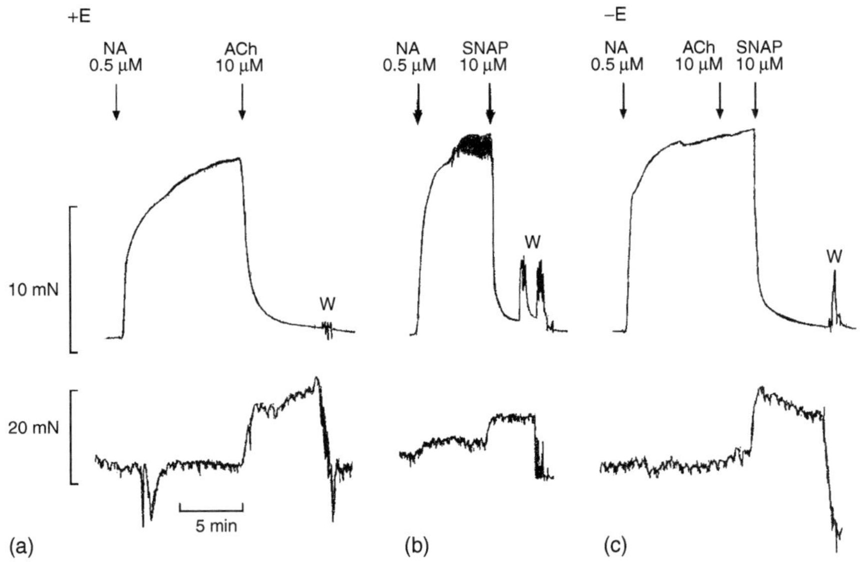

FIGURE 21.6 Simultaneous measurements of force (upper traces) and NO concentration (lower traces) in an endothelium intact (+E) segment of rat superior mesenteric artery contracted with 0.5 μM noradrenaline (NA) and relaxed with either 10 μM acetylcholine (ACh) (a), or 10 μM SNAP (b). Panel c shows a similar measurement in the rat superior mesenteric artery after mechanical endothelial cell removal. As can be seen in c, ACh addition does not cause NO production from the artery but shows a NO increase upon SNAP addition causing artery relaxation. W, washout. (Reprinted from Simonsen, U. et al., *J. Physiol. London*, 516, 271, 1999. With permission.)

lipopolysaccharide resulted in induction of iNOS and SOD associated with endotoxin, and NO concentration increased only in response to L-arginine. Simonsen's group has also experimented with measuring NO concentration in the artery of a hypertensive rat[120] and in isolated human small arteries.[121]

Schuhmann's group has recently published some very interesting results describing the measurement of NO release from human umbilical vein endothelial cells (HUVEC) using a unique array of microelectrodes.[122] Figure 21.7 (top) shows an SEM image of the microelectrode array. Following microelectrode array construction the surface of the electrodes were modified with nickel tetrasulfonate phthalocyanine tetrasodium salt using electrochemically induced deposition. Following this deposition, the HUVEC cells were allowed to grow in the interstitial spaces between the electrodes (Figure 21.7, bottom), and NO release from the cells was monitored as a function of growth and stimulation from bradykinin. This work is unique because the cells are actually being grown on the Si_3N_4 insulating layer so the cells are not being affected by the microelectrode working potential, which typically causes cell death, allowing for NO concentration to be monitored.

A recent publication by Millar has shown that NO concentrations can be measured in bovine eyes trabecular meshwork in situ using a ISONOP200 electrode from WPI.[77] For this study, the tip of the 200 μm diameter electrode was inserted into the region of the trabecular meshwork and NO monitored as a function of epinephrine concentration. This study found that NO generation increased as a function of epinephrine addition resulting in a reduction of intraocular pressure. This finding is important for shedding light on treatments for patients with primary open-angle glaucoma. Researchers such as Akeo and Amaki have used a Pt/Ir electrodes, coated with NO selective membranes, for the determination of what effects L-DOPA has on eyes.[78,79]

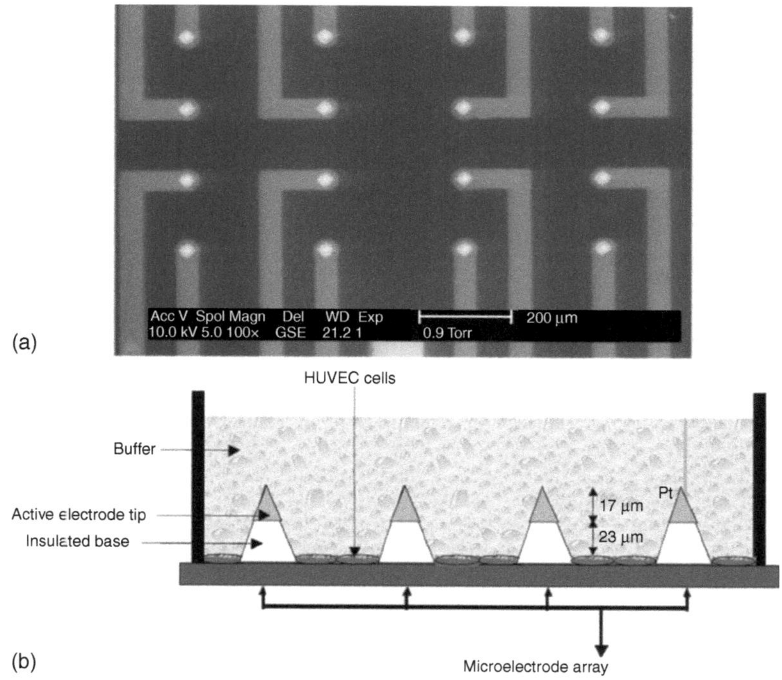

FIGURE 21.7 (a) A representative SEM image of the microelectrode array and (b) a schematic representation of the experimental setup. (Reprinted from Isik, S. et al., *Biosens. Bioelectron.*, 20, 1566, 2005. With permission.)

Because NO is a diffusible messenger molecule in the brain, the measurement of NO concentration in vitro is quite important to understand its action. In order to accomplish this general goal, Barbosa's group recently developed a 8 μm diameter carbon fiber electrode coated with Nafion and *o*-phenylenediamine.[48] Nafion was added to the electrode surface by dipping the electrode into a 5% solution followed by drying for 10 min at 170°C, followed by *o*-phenylenediamine electropolymerization on the electrode by holding the potential at +0.9 V vs. Ag/AgCl reference electrode. These electrodes were then inserted into the CA1 region of the hippocampus, from a Wistar rat, and the NO concentration was monitored as a function of L-glutamate and *N*-methyl-D-aspartate introduction. The results can be seen in Figure 21.8. The main finding of this study was that these electrodes were sufficiently sensitive to monitor brain NO concentrations with a sensitivity of 954 ± 271 pA/μM and a limit of detection of 6 ± 2 nM. These measurements can also be made with minimal interference from common interferences such as: ascorbate, nitrite, and H_2O_2. These electrodes can be useful for unraveling pathways for memory and learning processes in the hippocampus.

The Mas group recently published results on the measurement of NO release from the corpus cavernosum of the penis and its relation to penile erection.[99] For this study a 30 μm diameter carbon fiber was coated with nickel tetrakis (3-methoxy-4-hydroxyphenyl) porphyrin, via electrodeposition using DPV, and Nafion. This electrode was then inserted into the cavernous bodies of a rat penis. The results showed an increase in NO concentration, and intracavernous pressure, upon cavernous nerve stimulation and subsequently decreased upon introduction of NO synthase isoenzymes. This study was important to facilitate further measurements of NO concentrations in vivo in the penis.

NO measurements have also been made in the ears of a guinea pig. The first of these studies was performed by Nuttall and co-workers in 2002.[111] For this investigation a ISONOP30 carbon fiber

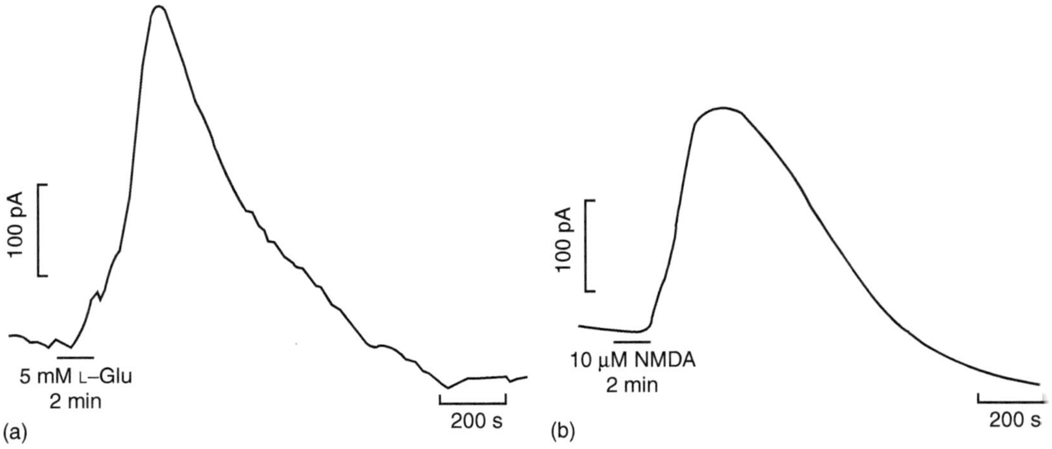

FIGURE 21.8 NO production from the CA1 region of the hippocampal slice following addition of 5 mM L-glutamate (a) and 10 μM *N*-methyl-D-aspartate (b). (Reprinted from Ferreira, N. R. et al., *Anal. Chim. Acta*, 535, 1, 2005. With permission.)

electrode from WPI was inserted into the perilymph of the basal turn of the guinea pig ear to measure changes in NO concentration as a result of noise stimulation. This study showed that guinea pigs exposed to broadband noise for 3 h/day at 120 dBA, for 3 days, exhibited an increase in NO concentration in the perilymph. This result is important in understanding the role that NO plays in hearing loss. A subsequent publication by Nuttall and co-workers used the ISONOP200 from WPI to measure NO concentration in the spiral modiolar artery (SMA) of a guinea pig.[112] This study is important in gaining a better understanding of how NO potentially regulates cochlear blood flow to set a benchmark for pharmacological and pathological evaluations. To perform this study a 3 mm section of the (SMA) was added to a bath solution and the complete tip of the 200 μm electrode was inserted into the bath, parallel to the SMA, to measure basal NO concentration as well as drug induced NO release, and how that relates to cochlear blood flow regulation. The key findings of this study were that the SMA continuously released NO, and that a blockage of this release by L-NAME causes a decrease in NO production and a vasoconstriction. These finding are important in understanding and interpreting future findings.

An interesting study was performed by Kashiwagi and co-workers studying the role that NO plays in tumor vessel morphogenesis and maturation.[103] For this study, B16 tumor cells were injected into mice, and tumor tissue removed from the mice when it reached ~8 mm in diameter. The tip of a Nafion polymer coated Au microelectrode was subsequently inserted into the tumor to monitor NO production. The results of this study showed that NO mediates mural cell coverage and vessel branching/longitudinal extension, but does not play a part in the growth of tumor blood vessels. The investigators also used a NOS inhibitor to show that NOS from endothelial cells in tumors is the primary source of NO and mediates tumor growth.

Kishi and co-workers used a commercial NO-selective microelectrode to monitor the effect that exercise has on platelet-derived NO.[108] This study used 23 healthy, male, non-smokers that underwent treadmill exercise. Blood samples were taken from the subjects before and directly following exercise, and blood platelets were isolated. The study showed that NO concentration and platelet levels were increased following exercise. This increase in NO concentration is thought to play a role in the prevention of exercise-induced platelet activation in humans.

Kellogg and co-workers recently reported on the measurement of NO under the human skin in response to heat stress. For this study a flexible 200 μm NO microelectrode from WPI was inserted into the cutaneous interstitial space of the forearm of nine human patients to measure NO

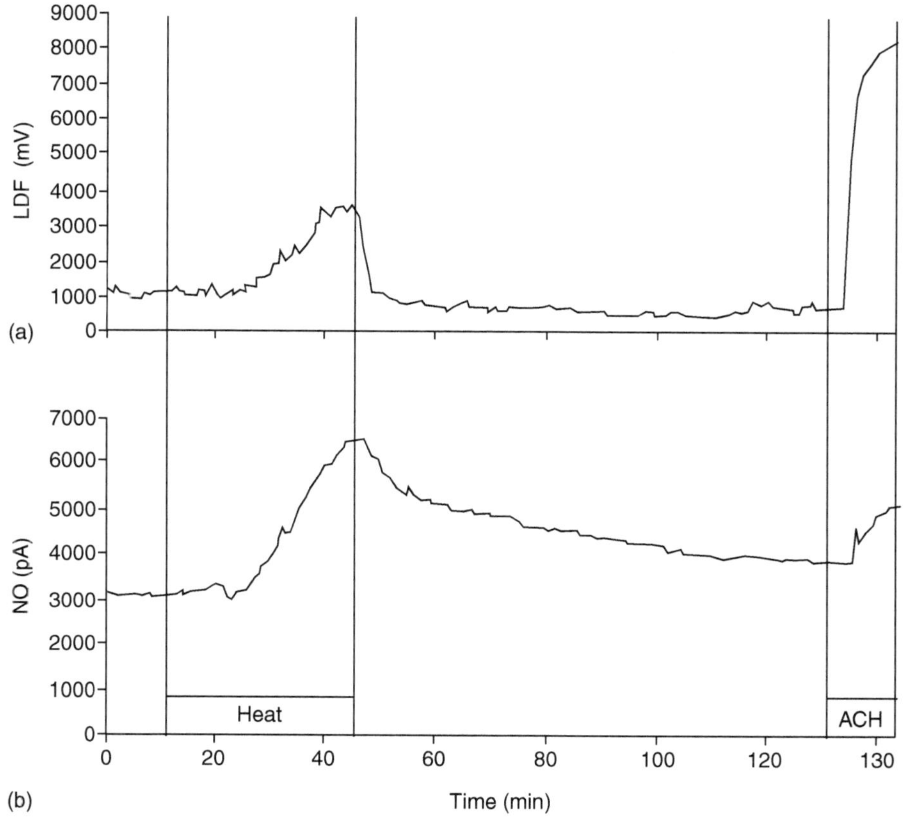

FIGURE 21.9 Laser-doppler flowmetry (a) and NO measurement (b) from one subject. Upon heating the subject to 39°C, at ~10 min, NO production and skin blood flow increased, which instantly returned to normal upon cooling the subject at ~45 min. After heat stress and cooling, ACh was administered by intradermal microdialysis to confirm the ability of the microelectrode to measure NO concentrations. (Reprinted from Kellogg, D. L. et al., *J. Appl. Physiol.*, 94, 1971, 2003. With permission.)

concentration while the subjects were at low (34°C) and high (39°C) temperatures. Laser-Doppler flowmetry (LDF) was used to monitor skin blood flow (SkBF).[123] This publication demonstrated that NO concentration, as well as SkBF increased in the cutaneous interstitial space during heat stress in humans. Figure 21.9 shows the key results of these experiments. As the temperature was increased at ~10 min, the data show that the blood flow (top plot) and NO concentration (bottom plot) increased as a function of temperature. Also, as the temperature was decreased again, at ~45 min, the blood flow and NO concentration returned to its original value. At ~130 min, while the subjects were at low temperature, the investigators injected acetylcholine to show that NO was indeed being detected in the subjects. The same group used a similar experimental design to monitor NO concentration, as well as SkBF, during reactive hyperemia under the human skin.[124]

Thom and co-workers published results on the stimulation of perivascular NO synthesis by oxygen.[125] To perform this study a 200 μm diameter electrode (ISONOP200) was placed between the aorta and vena cava of anesthetized rats and mice (rodents) and then the rodents were placed inside a hyperbaric chamber. Inside the hyperbaric chamber the partial pressure of O_2 was regulated/ changed as NO concentration was monitored. Figure 21.10 shows that NO concentration increased as a function of O_2 partial pressure. This experiment is important in understanding how NO synthesis, by NOS, is altered and regulated by O_2.

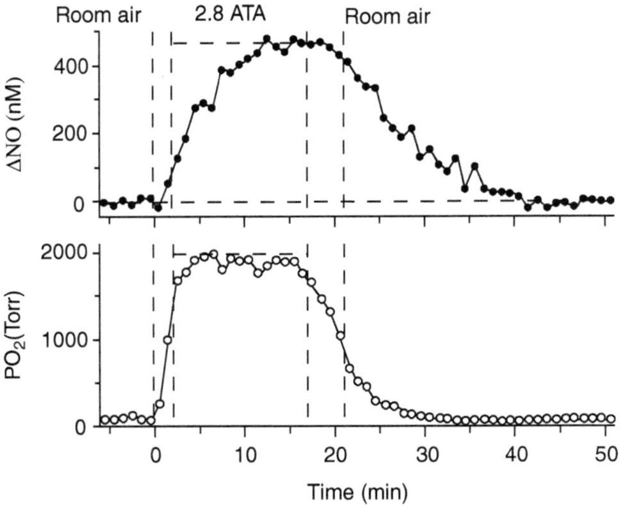

FIGURE 21.10 NO concentration (top) and O_2 concentration (bottom) as a function of O_2 pressure. As can be seen, NO and O_2 concentrations increase significantly when a rat is exposed to pressurized O_2 atmospheres (2.8 atmospheres absolute, ATA). (Reprinted from Thom, S. R. et al., *Am. J. Physiol.-Heart Circul. Physiol.*, 284, H1230, 2003. With permission.)

Studies have also been conducted on the NO release from plants and the effects of a plant diet on artherosclerosis and endothelial cell dysfunction. Visioli and co-workers studied the effects that a diet of wild artichoke and thyme had on the release of NO from porcine aortic endothelial cells and cerebral cell membranes.[116] For this study a rat brain was homogenized and cell membranes

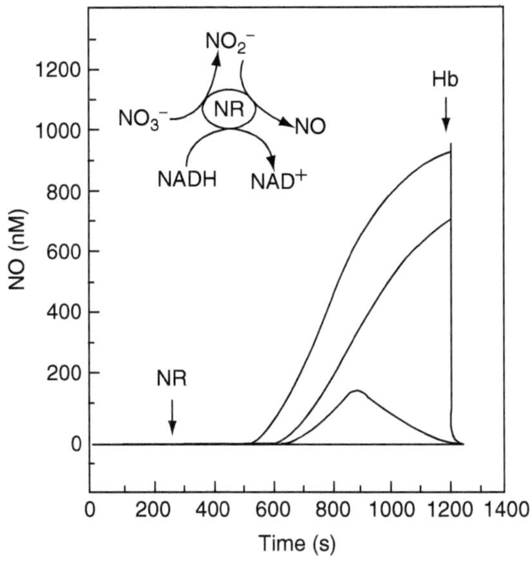

FIGURE 21.11 NO production from the addition of nitrate reductase (NR = 15 mU/mL) to a solution containing 50 μM sodium nitrate and various concentrations of NADH. The curves from top to bottom were obtained in solutions containing 100, 50, 40, and 0 μM NADH. Hemoglobin (Hb) was introduced to quench the production of NO. (Reprinted from Yamasaki, H. and Sakihama, Y., *FEBS Lett.*, 468, 89, 2000. With permission.)

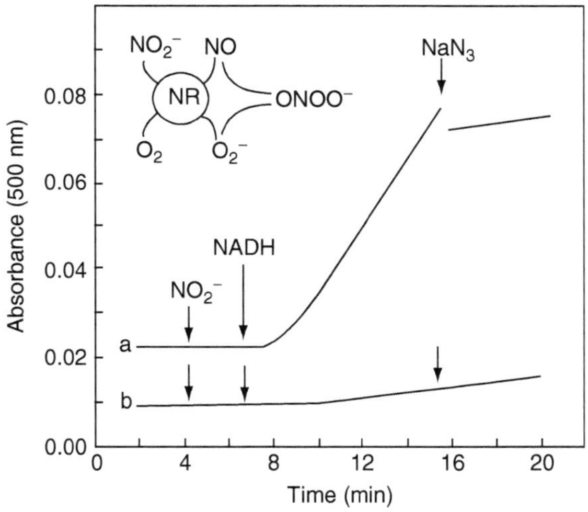

FIGURE 21.12 Absorbance measurement of 2′,7′-dichlorodihydrofluorescein (DCDHF), a peroxynitrite sensitive dye, as a function of nitrite (1 mM) and NADH (1 mM) introduction to a solution containing 100 μM DCDHF and 30 mU/mL NR under ambient (a) and nitrogen saturated conditions (b). As can be seen, the absorbance of DCDHF increases upon nitrite and NADH introduction only under an oxygen atmosphere, indicative of peroxynitrite production. (Reprinted from Yamasaki, H. and Sakihama, Y., *FEBS Lett.*, 468, 89, 2000. With permission.)

isolated by ultracentrifugation and NO release was monitored using a 2 mm NO sensor (ISONOP). The cell membranes were then exposed to wild artichoke and thyme extracts. This study showed that NO release was significantly increased following wild artichoke and thyme extract addition. These results suggest that eating a diet rich in phenolic compounds, such as wild artichoke and thyme, contributes to maintenance of a healthy cardiovascular system. Yamasaki and co-workers used an ISONOP from WPI to study the ability of plant nitrate reductase to produce NO.[115] Figure 21.11 and Figure 21.12 show that plant nitrate reductase produces NO in vitro as well as its toxic derivative peroxynitrite.

CONCLUDING REMARKS

The use of the electrochemical NO microsensor provides an elegant and convenient way to detect NO in real time and in biological samples. Currently they provide the only means by which to measure NO continuously, accurately and directly within living tissue without significant damage. The increasing acceptance of such sensors and their diversity of use in many NO research applications will help to further the current understanding of the various clinical roles of this interesting and ubiquitous molecule. Continual improvements being made to NO microsensor design and technology will facilitate these studies in the foreseeable future.

ACKNOWLEDGMENTS

This research was supported by NIH grants (1 R43 GM62077-01, 2R44 GM62077-02, 5R GM62077-3) to XZ and a WPI R&D priority research funds.

REFERENCES

1. Moncada, S., Palmer, R. M. J., and Higgs, E. A., Nitric-oxide—physiology, pathophysiology, and pharmacology, *Pharmacol. Rev.*, 43, 109, 1991.
2. Ignarro, L. J., Buga, G. M., Wood, K. S., Byrns, R. E., and Chaudhuri, G., Endothelium-derived relaxing factor produced and released from artery and vein is nitric-oxide, *Proc. Natl Acad. Sci. USA*, 84, 9265, 1987.
3. Ignarro, L. J., Nitric-oxide—a novel signal transduction mechanism for transcellular communication, *Hypertension*, 16, 477, 1990.
4. Bredt, D. S., Hwang, P. M., Glatt, C. E., Lowenstein, C., Reed, R. R., and Snyder, S. H., Cloned and expressed nitric-oxide synthase structurally resembles cytochrome-P-450 reductas, *Nature*, 351, 714, 1991.
5. Feldman, P. L., Griffith, O. W., and Stuehr, D. J., The surprising life of nitric-oxide, *Chem. Eng News*, 71, 26, 1993.
6. Radomski, M. W., Palmer, R. M. J., and Moncada, S., An L-arginine nitric-oxide pathway present in human platelets regulates aggregation, *Proc. Natl Acad. Sci. USA*, 87, 5193, 1990.
7. Moncada, S., Radomski, M. W., and Palmer, R. M. J., Endothelium-derived relaxing factor—identification as nitric-oxide and role in the control of vascular tone and platelet-function, *Biochem. Pharmacol.*, 37, 2495, 1988.
8. Akaike, T., Yoshida, M., Miyamoto, Y., Sato, K., Kohno, M., Sasamoto, K., Miyazaki, K., Ueda, S., and Maeda, H., Antagonistic action of imidazolineoxyl *n*-oxides against endothelium-derived relaxing factor: NO through a radical reaction, *Biochemistry*, 32, 827, 1993.
9. Hibbs, J. B., Taintor, R. R., Vavrin, Z., and Rachlin, E. M., Nitric-oxide—a cyto-toxic activated macrophage effector molecule, *Biochem. Biophys. Res. Commun.*, 157, 87, 1988.
10. Anbar, M., Hyperthermia of the cancerous breast—analysis of mechanism, *Cancer Lett.*, 84, 23, 1994.
11. Schmidt, H., Warner, T. D., Ishii, K., Sheng, H., and Murad, F., Insulin-secretion from pancreatic B-cells caused by L-arginine derived nitrogen-oxides, *Science*, 255, 721, 1992.
12. Moncada, S., Palmer, R. M. J., and Higgs, E. A., Biosynthesis of nitric-oxide from L-arginine—a pathway for the regulation of cell-function and communication, *Biochem. Pharmacol.*, 38, 1709, 1989.
13. Beckman, J. S. and Congert, K. A., Direct measurement of dilute nitric oxide in solution with an ozone chemiluminescent detector, *Methods*, 7, 35, 1995.
14. Robinson, J. K., Bollinger, M. J., and Birks, J. W., Luminol/H2O2 chemiluminescence detector for the analysis of nitric oxide in exhaled breath, *Anal. Chem.*, 71, 5131, 1999.
15. Green, L. C., Wagner, D. A., Glogowski, J., Skipper, P. L., Wishnok, J. S., and Tannenbaum, S. R., Analysis of nitrate, nitrite, and N-15-labeled nitrate in biological-fluids, *Anal. Biochem.*, 126, 131, 1982.
16. Wennmalm, A., Lanne, B., and Petersson, A. S., Detection of endothelial-derived relaxing factor in human plasma in the basal state and following ischemiausing electron-paramagnetic resonance spectrometry, *Anal. Biochem.*, 187, 359, 1990.
17. Bredt, D. S. and Snyder, S. H., Nitric-oxide mediates glutamate-linked enhancement of cGMP levels in the cerebellum, *Proc. Natl Acad. Sci. USA*, 86, 9030, 1989.
18. Wallace, J. L. and Woodman, R. C., Detection of nitric oxide by bioassay, *Methods*, 7, 55, 1995.
19. *World Precision Instruments Catalog*, 42–49, 2005.
20. Zhang, X. and Broderick, M., Amperometric detection of nitric oxide, *Mod. Aspects Immunobiol*, 1, 160, 2000.
21. Clark, L. C. and Lyons, C., Electrode systems for continuous monitoring in cardiovascular surgery, *Ann. N. Y. Acad. Sci.*, 102, 29, 1962.
22. Shibuki, K., An electrochemical microprobe for detecting nitric-oxide release in brain-tissue, *Neurosci. Res.*, 9, 69, 1990.
23. Zhang, X. J., Real time and in vivo monitoring of nitric-oxide by electrocehmical sensors—from dream to reality, *Front. Biosci.*, 9, 3434, 2004.
24. Malinski, T. and Taha, Z., Nitric-oxide release from a single cell measured in situ by a porphyrinic-based microsensor, *Nature*, 358, 676, 1992.

25. Nagase, S., Ohkoshi, N., Ueda, A., Aoyagi, K., and Koyama, A., Hydrogen peroxide interferes with detection of nitric oxide by an electrochemical method, *Clin. Chem.*, 43, 1246, 1997.
26. Diab, N. and Schuhmann, W., Electropolymerized manganese porphyrin/polypyrrole films as catalytic surfaces for the oxidation of nitric oxide, *Electrochim. Acta*, 47, 265, 2001.
27. Diab, N., Oni, J., Schulte, A., Radtke, I., Blochl, A., and Schuhmann, W., Pyrrole functionalised metalloporphyrins as electrocatalysts for the oxidation of nitric-oxide, *Talanta*, 61, 43, 2003.
28. Hayon, J., Ozer, D., Rishpon, J., and Bettelheim, A., Spectroscopic and electrochemical response to nitrogen monoxide of a cationic iron porphyrin immobilized in nafion-coated electrodes or membranes, *J. Chem. Soc. Chem. Commun.*, 619, 1994.
29. Kashevskii, A. V., Lei, J., Safronov, A. Y., and Ikeda, O., Electrocatalytic properties of meso-tetraphenylporphyrin cobalt for nitric oxide oxidation in methanolic solution and in Nafion((R)) film, *J. Electroanal. Chem.*, 531, 71, 2002.
30. Brunet, A., Privat, C., Stepien, O., David-Dufilho, M., Devynck, J., and Devynck, M. A., Advantages and limits of the electrochemical method using Nafion and Ni-porphyrin-coated microelectrode to monitor NO release from cultured vascular cells, *Analysis*, 28, 469, 2000.
31. Malinski, T., Taha, Z., Grunfeld, S., Burewicz, A., Tomboulian, P., and Kiechle, F., Measurements of nitric-oxide in biological-materials using a porphyrinic microsensor, *Anal. Chim. Acta*, 279, 135, 1993.
32. Bedioui, F., Trevin, S., and Devynck, J., Chemically modified microelectrodes designed for the electrochemical determination of nitric oxide in biological systems, *Electroanalysis*, 8, 1085, 1996.
33. Malinski, T., Bailey, F., Zhang, Z. G., and Chopp, M., Nitric-oxide measured by a porphyrinic microsensor in rat-brain after transient middle cerebral-artery occlusion, *J. Cereb. Blood Flow Metab.*, 13, 355, 1993.
34. Malinski, T., Radomski, M. W., Taha, Z., and Moncada, S., Direct electrochemical measurement of nitric-oxide released from human platelets, *Biochem. Biophys. Res. Commun.*, 194, 960, 1993.
35. Zhang, Z. G., Chopp, M., Bailey, F., and Malinski, T., Nitric-oxide changes in the rat-brain after transient middle cerebral-artery occlusion, *J. Neurol. Sci.*, 128, 22, 1995.
36. Lantoine, F., Trevin, S., Bedioui, F., and Devynck, J., Selective and sensitive electrochemical measurement of nitric-oxide in aqueous-solution—discussion and new results, *J. Electroanal. Chem.*, 392, 85, 1995.
37. Bedioui, F., Trevin, S., and Devynck, J., The use of gold electrodes in the electrochemical detection of nitric-oxide in aqueous-solution, *J. Electroanal. Chem.*, 377, 295, 1994.
38. Yokoyama, H., Mori, N., Kasai, N., Matsue, T., Uchida, I., Kobayashi, N., Tsuchihashi, N., Yoshimura, T., Hiramatsu, M., and Niwa, S. I., Direct and continuous monitoring of intrahippocampal nitric oxide (NO) by an NO sensor in freely moving rat after *N*-methyl-D-aspartic acid injection, *Denki Kagaku*, 63, 1167, 1995.
39. Kim, K. I., Chung, H. T., Oh, G. S., Bae, H. O., Kim, S. H., and Chun, H. J., Integrated gold-disk microelectrode modified with iron(II)-phthalocyanine for nitric oxide detection in macrophages, *Microchem. J.*, 80, 219, 2005.
40. Vilakazi, S. L. and Nyokong, T., Voltammetric determination of nitric oxide on cobalt phthalocyanine modified microelectrodes, *J. Electroanal. Chem.*, 512, 56, 2001.
41. Zhang, X. J., Cardosa, L., Broderick, M., Fein, H., and Lin, J., An integrated nitric oxide sensor based on carbon fiber coated with selective membranes, *Electroanalysis*, 12, 1113, 2000.
42. Dickson, A., Lin, J., Sun, J., Broderick, M., Harry, F. A., and Zhang, X. J., Construction and characterization of a new flexible and nonbreakable nitric oxide microsensor, *Electroanalysis*, 16, 640, 2004.
43. Zhang, X. J., Kislyak, Y., Lin, H., Dickson, A., Cardosa, L., Broderick, M., and Fein, H., Nanometer size electrode for nitric oxide and *S*-nitrosothiols measurement, *Electrochem. Commun.*, 4, 11, 2002.
44. Zhang, X. J., Lin, J., Cardoso, L., Broderick, M., and Darley-Usmar, V., A novel microchip nitric oxide sensor with sub-nM detection limit, *Electroanalysis*, 14, 697, 2002.
45. Fabre, B., Burlet, S., Cespuglio, R., and Bidan, G., Voltammetric detection of NO in the rat brain with an electronic conducting polymer and Nafion(R) bilayer-coated carbon fiber electrode, *J. Electroanal. Chem.*, 426, 75, 1997.
46. Friedemann, M. N., Robinson, S. W., and Gerhardt, G. A., *o*-Phenylenediamine-modified carbon fiber electrodes for the detection of nitric oxide, *Anal. Chem.*, 68, 2621, 1996.

47. Park, J. K., Tran, P. H., Chao, J. K. T., Ghodadra, R., Rangarajan, R., and Thakor, N. V., In vivo nitric oxide sensor using non-conducting polymer-modified carbon fiber, *Biosens. Bioelectron.*, 13, 1187, 1998.
48. Ferreira, N. R., Ledo, A., Frade, J. G., Gerhardt, G. A., Laranjinha, J., and Barbosa, R., Electrochemical measurement of endogenously produced nitric oxide in brain slices using Nafion/*o*-phenylenediamine modified carbon fiber microelectrodes, *Anal. Chim. Acta*, 535, 1, 2005.
49. Kato, D., Kunitake, M., Nishizawa, M., Matsue, T., and Mizutani, F., Amperometric nitric oxide microsensor using two-dimensional cross-linked Langmuir–Blodgett films of polysiloxane copolymer, *Sens. Actuator B-Chem.*, 108, 384, 2005.
50. Ichimori, K., Ishida, H., Fukahori, M., Nakazawa, H., and Murakami, E., Practical nitric-oxide measurement employing a nitric-oxide-selective electrode, *Rev. Sci. Instrum.*, 65, 2714, 1994.
51. Pontie, M., Bedioui, F., and Devynck, J., New composite modified carbon microfibers for sensitive and selective determination of physiologically relevant concentrations of nitric oxide in solution, *Electroanalysis*, 11, 845, 1999.
52. Allen, B. W., Piantadosi, C. A., and Coury, L. A., Electrode materials for nitric-oxide detection, *Nitric Oxide-Biol. Chem.*, 4, 75, 2000.
53. Xian, Y. Z., Sun, W. L., Xue, J. A., Luo, M., and Jin, L. T., Iridium oxide and palladium modified nitric oxide microsensor, *Anal. Chim. Acta*, 381, 191, 1999.
54. Haruyama, T., Shiino, S., Yanagida, Y., Kobatake, E., and Aizawa, M., Two types of electrochemical nitric-oxide (NO) sensing systems with heat-denatured cyt *c* and radical scavenger PTIO, *Biosens. Bioelectron.*, 13, 763, 1998.
55. He, X. C. and Mo, J. Y., Electrocatalytic oxidation of NO at electrode modified with Nafion-Co-II-1,10-phenanthroline film and its application to NO detection, *Analyst*, 125, 793, 2000.
56. Smith, S. R. and Thorp, H. H., Application of the electrocatalytic reduction of nitric oxide mediated by ferrioxamine B to the determination of nitric oxide concentrations in solution, *Inorg. Chim. Acta*, 273, 316, 1998.
57. Kitamura, Y., Uzawa, T., Oka, K., Komai, Y., Ogawa, H., Takizawa, N., Kobayashi, H., and Tanishita, K., Microcoaxial electrode for in vivo nitric oxide measurement, *Anal. Chem.*, 72, 2957, 2000.
58. Mizutani, F., Yabuki, S., Sawaguchi, T., Hirata, Y., Sato, Y., and Iijima, S., Use of a siloxane polymer for the preparation of amperometric sensors: O-2 and NO sensors and enzyme sensors, *Sens. Actuator B-Chem.*, 76, 489, 2001.
59. Katrlik, J. and Zalesakova, P., Nitric oxide determination by amperometric carbon fiber microelectrode, *Bioelectrochemistry*, 56, 73, 2002.
60. Fan, C. H., Pang, J. T., Shen, P. P., Li, G. X., and Zhu, D. X., Nitric-oxide biosensors based on Hb/phosphatidylcholine films, *Anal. Sci.*, 18, 129, 2002.
61. Fan, C. H., Li, G. X., Zhu, J. Q., and Zhu, D. X., A reagentless nitric-oxide biosensor based on hemoglobin–DNA films, *Anal. Chim. Acta*, 423, 95, 2000.
62. Kitajima, A., Teranishi, T., and Miyake, M., Detection of nitric-oxide on carbon electrode modified with ionic polymers and alpha-cyclodextrin, *Electrochemistry*, 69, 16, 2001.
63. Shin, J. H., Weinman, S. W., and Schoenfisch, M. H., Sol–gel derived amperometric nitric oxide microsensor, *Anal. Chem.*, 77, 3494, 2005.
64. Lee, Y., Oh, B. K., and Meyerhoff, M. E., Improved planar amperometric nitric-oxide sensor based on platinized platinum anode. 1. Experimental results and theory when applied for monitoring NO release from diazeniumdiolate-doped polymeric films, *Anal. Chem.*, 76, 536, 2004.
65. Lee, Y., Yang, J., Rudich, S. M., Schreiner, R. J., and Meyerhoff, M. E., Improved planar amperometric nitric oxide sensor based on platinized platinum anode. 2. Direct real-time measurement of NO generated from porcine kidney slices in the presence of L-arginine, L-arginine polymers, and protamine, *Anal. Chem.*, 76, 545, 2004.
66. Kroning, S., Scheller, F. W., Wollenberger, U., and Lisdat, F., Myoglobin–clay electrode for nitric-oxide (NO) detection in solution, *Electroanalysis*, 16, 253, 2004.
67. Kamei, K., Haruyama, T., Mie, M., Yanagida, Y., Aizawa, M., and Kobatake, E., The construction of endothelial cellular biosensing system for the control of blood pressure drugs, *Biosens. Bioelectron.*, 19, 1121, 2004.

68. Casero, E., Pariente, F., and Lorenzo, E., Electrocatalytic oxidation of nitric oxide at indium hexacyanoferrate film-modified electrodes, *Anal. Bioanal. Chem.*, 375, 294, 2003.
69. Pailleret, A., Oni, J., Reiter, S., Isik, S., Etienne, M., Bedioui, F., and Schuhmann, W., In situ formation and scanning electrochemical microscopy assisted positioning of NO-sensors above human umbilical vein endothelial cells for the detection of nitric oxide release, *Electrochem. Commun.*, 5, 847, 2003.
70. Magazine, H. I., Detection of endothelial cell-derived nitric-oxide: Current trends and future directions, *Adv. Neuroimmunol.*, 5, 479, 1995.
71. Kiechle, F. L. and Malinski, T., Nitric-oxide—biochemistry, pathophysiology, and detection, *Am. J. Clin. Pathol.*, 100, 567, 1993.
72. Archer, S., Measurement of nitric-oxide in biological models, *FASEB J.*, 7, 349, 1993.
73. *Handbook of Chemistry and Physics*, 76th ed., CRC Press, Boca Raton, FL, 1995.
74. Kelm, M. and Schrader, J., Control of coronary vascular tone by nitric-oxide, *Circ. Res.*, 66, 1561, 1990.
75. Zhang, X. J., Cardosa, L., Broderick, M., Fein, H., and Davies, I. R., Novel calibration method for nitric oxide microsensors by stoichiometrical generation of nitric-oxide from SNAP, *Electroanalysis*, 12, 425, 2000.
76. Tristani-Firouzi, M., DeMaster, E. G., Quast, B. J., Nelson, D. P., and Archer, S. L., Utility of a nitric-oxide electrode for monitoring the administration of nitric oxide in biologic systems, *J. Lab. Clin. Med.*, 131, 281, 1998.
77. Millar, J. C., Real-time direct measurement of nitric-oxide in bovine perfused eye trabecular meshwork using a Clark-type electrode, *J. Ocular Pharmacol. Ther.*, 19, 299, 2003.
78. Akeo, K., Amaki, S., Suzuki, T., and Hiramitsu, T., Melanin granules prevent the cytotoxic effects of L-DOPA on retinal pigment epithelial cells in vitro by regulation of NO and superoxide radicals, *Pigment Cell Res.*, 13, 80, 2000.
79. Amaki, S. K., Oguchi, Y., Ogata, T., Suzuki, T., Akeo, K., and Hiramitsu, T., L-DOPA produced nitric-oxide in the vitreous and caused greater vasodilation in the choroid and the ciliary body of melanotic rats than in those of amelanotic rats, *Pigment Cell Res.*, 14, 256, 2001.
80. Iijima, K., Henry, E., Moriya, A., Wirz, A., Kelman, A. W., and McColl, K. E. L., Dietary nitrate generates potentially mutagenic concentrations of nitric oxide at the gastroesophageal junction, *Gastroenterology*, 122, 1248, 2002.
81. Stefano, G. B., Zhu, W., Cadet, P., Bilfinger, T. V., and Mantione, K., Morphine enhances nitric-oxide release in the mammalian gastrointestinal tract via the mu 3 opiate receptor subtype: A hormonal role for endogenous morphine, *J. Physiol. Pharmacol.*, 55, 279, 2004.
82. Cherian, L., Goodman, J. C., and Robertson, C. S., Brain nitric-oxide changes after controlled cortical impact injury in rats, *J. Neurophysiol.*, 83, 2171, 2000.
83. Buerk, D. G., Ances, B. M., Greenberg, J. H., and Detre, J. A., Temporal dynamics of brain tissue nitric oxide during functional forepaw stimulation in rats, *Neuroimage*, 18, 1, 2003.
84. Rocchitta, G., Migheli, R., Mura, M. P., Esposito, G., Desole, M. S., Miele, E., Miele, M., and Serra, P. A., Signalling pathways in the nitric-oxide donor-induced dopamine release in the striatum of freely moving rats: Evidence that exogenous nitric-oxide promotes Ca^{2+} entry through store-operated channels, *Brain Res.*, 1023, 243, 2004.
85. Meulemans, A., A brain nitric-oxide synthase study in the rat: Production of a nitroso-corn pound NA and absence of nitric oxide synthesis, *Neurosci. Lett.*, 321, 115, 2002.
86. Saito, M. and Miyagawa, I., Real-time monitoring of nitric oxide in ischemia-reperfusion rat kidney, *Urol. Res.*, 28, 141, 2000.
87. Levine, D. Z., Iacovitti, M., Burns, K. D., and Zhang, X. J., Real-time profiling of kidney tubular fluid nitric-oxide concentrations in vivo, *Am. J. Physiol.-Renal Physiol.*, 281, F189, 2001.
88. Levine, D. Z. and Iacovitti, M., Real time microelectrode measurement of nitric oxide in kidney tubular fluid in vivo, *Sensors*, 3, 31, 2003.
89. Levine, D. Z., Burns, K. D., Jaffey, J., and Iacovitti, M., Short-term modulation of distal tubule fluid nitric-oxide in vivo by loop NaCl reabsorption, *Kidney Int.*, 65, 184, 2004.
90. Thorup, C., Kornfeld, M., Winaver, J. M., Goligorsky, M. S., and Moore, L. C., Angiotensin-II stimulates nitric-oxide release in isolated perfused renal resistance arteries, *Pflugers Arch.*, 435, 432, 1998.

91. Arregui, B., Lopez, B., Salom, M. G., Valero, F., Navarro, C., and Fenoy, F. J., Acute renal hemodynamic effects of dimanganese decacarbonyl and cobalt protoporphyrin, *Kidney Int.*, 65, 564, 2004.
92. Fujita, S., Roerig, D. L., Bosnjak, Z. J., and Stowe, D. F., Effects of vasodilators and perfusion pressure on coronary flow and simultaneous release of nitric-oxide from guinea pig isolated hearts, *Cardiovasc. Res.*, 38, 655, 1998.
93. Novalija, E., Fujita, S., Kampine, J. P., and Stowe, D. F., Sevoflurane mimics ischemic preconditioning effects on coronary flow and nitric-oxide release in isolated hearts, *Anesthesiology*, 91, 701, 1999.
94. Schulte, D. and Millar, J., The effects of high- and low-intensity percutaneous stimulation on nitric-oxide levels and spike activity in the superficial laminae of the spinal cord, *Pain*, 103, 139, 2003.
95. Stefano, G. B., Prevot, V., Beauvillain, J. C., Fimiani, C., Welters, I., Cadet, P., Breton, C., Pestel, J., Salzet, M., and Bilfinger, T. V., Estradiol coupling to human monocyte nitric-oxide release is dependent on intracellular calcium transients: Evidence for an estrogen surface receptor, *J. Immunol.*, 163, 3758, 1999.
96. Stefano, G. B., Prevot, V., Beauvillain, J. C., Cadet, P., Fimiani, C., Welters, I., Fricchione, G. L., Breton, C., Lassalle, P., Salzet, M., and Bilfinger, T. V., Cell-surface estrogen receptors mediate calcium-dependent nitric-oxide release in human endothelia, *Circulation*, 101, 1594, 2000.
97. Beltran, B., Quintero, M., Garcia-Zaragoza, E., O'Connor, E., Esplugues, J. V., and Moncada, S., Inhibition of mitochondrial respiration by endogenous nitric oxide: A critical step in Fas signaling, *Proc. Natl Acad. Sci. USA*, 99, 8892, 2002.
98. Shiva, S., Brookes, P. S., Patel, R. P., Anderson, P. G., and Darley-Usmar, V. M., Nitric-oxide partitioning into mitochondrial membranes and the control of respiration at cytochrome *c* oxidase, *Proc. Natl Acad. Sci. USA*, 98, 7212, 2001.
99. Mas, M., Escrig, A., and Gonzalez-Mora, J. L., In vivo electrochemical measurement of nitric-oxide in corpus cavernosum penis, *J. Neurosci. Methods*, 119, 143, 2002.
100. Kedziora-Kornatowska, K. Z., Luciak, M., Blaszczyk, J., and Pawlak, W., Effect of aminoguanidine on the generation of superoxide anion and nitric-oxide by peripheral blood granulocytes of rats with streptozotocin-induced diabetes, *Clin. Chim. Acta*, 278, 45, 1998.
101. Stefano, G. B., Cadet, P., Breton, C., Goumon, Y., Prevot, V., Dessaint, J. P., Beauvillain, J. C., Roumier, A. S., Welters, I., and Salzet, M., Estradiol-stimulated nitric-oxide release in human granulocytes is dependent on intracellular calcium transients: Evidence of a cell surface estrogen receptor, *Blood*, 95, 3951, 2000.
102. Uhlmann, S., Friedrichs, U., Eichler, W., Hoffmann, S., and Wiedemann, P., Direct measurement of VEGF-induced nitric-oxide production by choroidal endothelial cells, *Microvasc. Res.*, 62, 179, 2001.
103. Kashiwagi, S., Izumi, Y., Gohongi, T., Demou, Z. N., Xu, L., Huang, P. L., Buerk, D. G., Munn, L. L., Jain, R. K., and Fukumura, D., NO mediates mural cell recruitment and vessel morphogenesis in murine melanomas and tissue-engineered blood vessels, *J. Clin. Invest.*, 115, 1816, 2005.
104. Tsatmali, M., Graham, A., Szatkowski, D., Ancans, J., Manning, P., McNeil, C. J., Graham, A. M., and Thody, A. J., Alpha-melanocyte-stimulating hormone modulates nitric-oxide production in melanocytes, *J. Invest. Dermatol.*, 114, 520, 2000.
105. Rysz, J., Luciak, M., Kedziora, J., Blaszczyk, J., and Sibinska, E., Nitric-oxide release in the peripheral blood during hemodialysis, *Kidney Int.*, 51, 294, 1997.
106. Rievaj, M., Lietava, J., and Bustin, D., Electrochemical determination of nitric-oxide in blood samples, *Chem. Pap.-Chem. Zvesti.*, 58, 306, 2004.
107. Larfars, G., Lantoine, F., Devynck, M. A., and Gyllenhammar, H., Electrochemical detection of nitric-oxide production in human polymorphonuclear neutrophil leukocytes, *Scand. J. Clin. Lab. Invest.*, 59, 361, 1999.
108. Kasuya, N., Kishi, Y., Sakita, S., Numano, F., and Isobe, M., Acute vigorous exercised primes enhanced NO release in human platelets, *Atherosclerosis*, 161, 225, 2002.
109. De La Cruz, J. P., Gonzalez-Correa, J. A., Guerrero, A., Marquez, E., Martos, F., and de la Cuesta, F. S., Differences in the effects of extended-release aspirin and plain-formulated aspirin on prostanoids and nitric-oxide in healthy volunteers, *Fundam. Clin. Pharmacol.*, 17, 363, 2003.
110. Freedman, J. E., Loscalzo, J., Barnard, M. R., Alpert, C., Keaney, J. F., and Michelson, A. D., Nitric-oxide released from activated platelets inhibits platelet recruitment, *J. Clin. Invest.*, 100, 350, 1997.

111. Shi, X. R., Ren, T. Y., and Nuttall, A. L., The electrochemical and fluorescence detection of nitric-oxide in the cochlea and its increase following loud sound, *Hear. Res.*, 164, 49, 2002.
112. Jiang, Z. G., Shi, X. R., Zhao, H., Si, J. Q., and Nuttall, A. L., Basal nitric-oxide production contributes to membrane potential and vasotone regulation of guinea pig in vitro spiral modiolar artery, *Hear. Res.*, 189, 92, 2004.
113. Sakihama, Y., Nakamura, S., and Yamasaki, H., Nitric-oxide production mediated by nitrate reductase in the green alga *Chlamydomonas reinhardtii*: An alternative NO production pathway in photosynthetic organisms, *Plant Cell Physiol.*, 43, 290, 2002.
114. Yamasaki, H., Sakihama, Y., and Takahashi, S., An alternative pathway for nitric oxide production in plants: New features of an old enzyme, *Trends Plant Sci.*, 4, 128, 1999.
115. Yamasaki, H. and Sakihama, Y., Simultaneous production of nitric-oxide and peroxynitrite by plant nitrate reductase: In vitro evidence for the NR-dependent formation of active nitrogen species, *FEBS Lett.*, 468, 89, 2000.
116. Grande, S., Bogani, P., De Saizieu, A., Schueler, G., Galli, C., and Visioli, F., Vasomodulating potential of Mediterranean wild plant extracts, *J. Agric. Food Chem.*, 52, 5021, 2004.
117. Moroz, L. L., Norekian, T. P., Pirtle, T. J., Robertson, K. J., and Satterlie, R. A., Distribution of NADPH-diaphorase reactivity and effects of nitric-oxide on feeding and locomotory circuitry in the pteropod mollusc, *Clione limacina*, *J. Comp. Neurol.*, 427, 274, 2000.
118. Simonsen, U., Wadsworth, R. M., Buus, N. H., and Mulvany, M. J., In vitro simultaneous measurements of relaxation and nitric oxide concentration in rat superior mesenteric artery, *J. Physiol.-Lond.*, 516, 271, 1999.
119. Hernanz, R., Alonso, M. J., Zibrandtsen, H., Alvarez, Y., Salaices, M., and Simonsen, U., Measurements of nitric-oxide concentration and hyporeactivity in rat superior mesenteric artery exposed to endotoxin, *Cardiovasc. Res.*, 62, 202, 2004.
120. Stankevicius, E., Martinez, A. C., Mulvany, M. J., and Simonsen, U., Blunted acetylcholine relaxation and nitric-oxide release in arteries from renal hypertensive rats, *J. Hypertens.*, 20, 1571, 2002.
121. Buus, N. H., Simonsen, U., Pilegaard, H. K., and Mulvaney, N. J., Nitric-oxide, prostanoid and non-NO, non-prostanoid involvement in acetylcholine relaxation of isolated human small arteries, *Br. J. Pharmacol.*, 129, 184, 2000.
122. Isik, S., Berdondini, L., Oni, J., Blochl, A., Koudelka-Hep, M., and Schuhmann, W., Cell-compatible array of three-dimensional tip electrodes for the detection of nitric-oxide release, *Biosens. Bioelectron.*, 20, 1566, 2005.
123. Kellogg, D. L., Zhao, J. L., Friel, C., and Roman, L. J., Nitric-oxide concentration increases in the cutaneous interstitial space during heat stress in humans, *J. Appl. Physiol.*, 94, 1971, 2003.
124. Zhao, J. L., Pergola, P. E., Roman, L. J., and Kellogg, D. L., Bioactive nitric-oxide concentration does not increase during reactive hyperemia in human skin, *J. Appl. Physiol.*, 96, 628, 2004.
125. Thom, S. R., Fisher, D., Zhang, J., Bhopale, V. M., Ohnishi, S. T., Kotake, Y., Ohnishi, T., and Buerk, D. G., Stimulation of perivascular nitric-oxide synthesis by oxygen, *Am. J. Physiol. Heart Circ. Physiol.*, 284, H1230, 2003.

22 In Vivo Fast-Scan Cyclic Voltammetry of Dopamine near Microdialysis Probes

Hua Yang and Adrian C. Michael

CONTENTS

INTRODUCTION

As several of the previous chapters in this volume have explained, the various electrochemical techniques for the detection of catecholamine and indolamine neurotransmitters are finding their way into an increasing number of application areas. Electrochemical techniques offer high speed capabilities, which make them particularly useful for monitoring dynamic processes such as the clearance of neurotransmitters from extracellular fluids. This has provided for the detailed analysis of the clearance kinetics of dopamine, norepinephrine, and serotonin transporters in a number of preparations (Chapter 3 through Chapter 8). High speed electrochemical recording has enabled detailed investigations of exocytotic events at the single cell level (Chapter 14 through Chapter 16). The recent ability to record spontaneous dopamine transients in vivo is providing new insights into the functional properties of dopaminergic systems (Chapter 2).

Despite the recent advances in the capabilities of the electrochemical techniques, where in vivo measurements in living animals are concerned, many laboratories continue to employ microdialysis sampling. However, studies have suggested that the tissue trauma associated with the implantation of microdialysis probes represents a shortcoming of the technique.[1–3] If the tissue surrounding the probe is in a traumatized state, it is necessary to ask how similar the properties of the traumatized tissue are to normal, uninjured tissue. A strategy that we have pursued to address this question, in the case of dopamine measurements, is to implant carbon fiber voltammetric electrodes at various distances from microdialysis probes and compare the results. In several papers, we have demonstrated that voltammetric results vary dramatically as a function of distance from the probe, suggesting that the dopamine system in the traumatized tissue near the probe is very different from that in normal, uninjured tissue.

Given that several histological studies have pointed to probe-induced tissue trauma,[4–6] it is perhaps surprising that more attention has not been paid to the effects of that trauma on the outcome of microdialysis measurements. The origin of this can be linked to the absence of procedures for the in vivo calibration of microdialysis probes. Without calibration, it is difficult to know exactly how the concentration of dopamine (as measured in a microdialysate sample) is related to the actual in vivo concentration of dopamine in the tissue surrounding the probe. For this reason, it is not particularly obvious that the dopamine is derived from traumatized tissue. It is also true, however, that procedures for the in vivo calibration of electrochemical techniques do not exist. Electrodes are calibrated on the lab bench either before or after they are used for in vivo measurements but not during the actual in vivo measurements themselves. The approach we have adopted to examining the effects of trauma on the dopamine system is to compare responses recorded simultaneously at similar electrodes at different recording locations. So, it is the comparison between similar electrodes that provides the insights into the effects of trauma, not the absolute value of the dopamine concentrations that we are measuring.

The rationale of using voltammetry as a tool for investigating the effect of trauma is based on the difference in size of the microdialysis probes and the carbon fiber microelectrodes. The probes have diameters of 225 μm or more and some studies have noticed disrupted tissue as far as 1.4 mm from the probe implantation site.[6] The carbon fiber electrodes we use for measuring dopamine have a diameter of 7 μm and the majority of the tissue disruption they cause is confined to within 3 μm of the electrode.[7] On a volume:volume basis, the probes are at least 10,000 times larger than the microelectrodes:

$$\frac{\pi r_{\text{probe}}^2 \ell_{\text{probe}}}{\pi r_{\text{elect}}^2 \ell_{\text{elect}}} = \frac{\pi(0.110\ \text{mm})^2(4\ \text{mm})}{\pi(0.0035\ \text{mm})^2(0.4\ \text{mm})} \cong 10{,}000$$

So, the concept of using a carbon fiber microelectrode as a tool for probing the trauma layer surrounding a microdialysis probe seems legitimate.

The suggestion that in vivo calibration of microdialysis probes is not routinely performed might appear surprising considering that a number of papers on the topic of quantitative microdialysis exist.[4,8–11] However, findings from voltammetry next to microdialysis probes suggest that the originally proposed approach to quantitative microdialysis has a problem. The original approach involved the determination of the quantity called the concentration of no-net-flux, which was assumed to be equal to the concentration in the fluid surrounding the probe. But, the no-net-flux concentration is only equal to the external concentration if the microdialysis extraction fraction and relative recovery are the same during the measurement. Voltammetry near microdialysis probes shows that these parameters are not the same when dopamine is measured in the striatum of chloral-hydrate anesthetized rats.[2,3]

In this chapter, we present some findings that extend our previous studies of voltammetry near microdialysis probes. A variable that is considered important in microdialysis protocols is the time after implantation before results are collected. Our previous work has used a 2 h wait after implantation, which is very different from the 24 h wait that is conventional. Because these experiments are performed in anesthetized animals, a 24 h wait time is difficult. But, we have examined a 16 h wait time. Also, our previous work has placed a great deal of emphasis on the ability of dopamine uptake to prevent dopamine from reaching microdialysis probes by diffusion. This is based on observations of the effect of the dopamine uptake inhibitor, nomifensine, on voltammetric recordings near microdialysis probes. So, here we report results from other drugs that increase dopamine levels by mechanisms other than uptake inhibition.

PROCEDURES FOR VOLTAMMETRY NEAR MICRODIALYSIS PROBES

Voltammetric Electrodes and Procedures

Microcylinder voltammetric electrodes (7 μm in diameter, 400 μm long) were prepared with individual carbon fibers (T300, Amoco Performance Products, Greenville, SC). Each microelectrode was electrochemically pretreated once before use with a triangular potential waveform (0–2 V vs. Ag/AgCl, 200 V/s, 1 s). Each microelectrode was calibrated in dopamine standard solutions before and after every in vivo experiment. Dopamine detection was accomplished by fast-scan cyclic voltammetry (FSCV) performed with a computer-controlled potentiostat (Ensman Instruments, Bloomington, IN) and a software package developed in-house. Electrode pretreatment and calibration were performed in artificial cerebrospinal fluid, aCSF: 145 mM Na^+, 2.7 mM K^+, 1.0 mM Mg^{2+}, 1.2 mM Ca^{2+}, 152 mM Cl^-, 2.0 mM phosphate, pH 7.4.

For FSCV, the potential started at 0 V and was linearly scanned at 300 V/s to 1 V, then to −.5 V, and back to 0 V vs. Ag/AgCl. Scans were repeated at 200 ms intervals. Dopamine oxidation was monitored between 0.5 and 0.7 V on the initial potential scan. Conversion of in vivo oxidation current to dopamine concentration was based on post-calibration of each electrode. Dopamine was identified during all in vivo experiments by comparing the background-subtracted voltammograms obtained in vivo and during post-calibration of each electrode.

Dialytrode Design

This work involves a device we call a dialytrode. It is a microdialysis probe equipped with a carbon fiber microelectrode mounted onto its outer surface.[2,3] Concentric dialysis probes (200 μm i.d., 3 mm long) were constructed with a hollow fiber membrane (SpectaPor RC, MWCO 13,000, Spectrum, Inc., Houston, TX) and a fused silica outlet line (75 μm i.d., 10 cm long, Polymicro Technology, Phoenix, AZ). A carbon fiber microelectrode (800 μm long) was mounted immediately adjacent to the outer surface of the probe. The integrated carbon fiber had no impact on probe performance, which is as expected since the cross section of the fiber is negligible (~0.3%) compared to the area of the probe itself.[2] For the sake of comparison, dopamine was also monitored by fast scan cyclic voltammetry with a carbon fiber microelectrode (400 μm long) mounted in the probe outlet. The probe was perfused with aCSF at 0.109 μL/min.

Surgical Procedures

The University of Pittsburgh Institutional Animal Care and Use Committee approved the procedures involving animals. Chloral hydrate anesthetized male Sprague–Dawley rats (250–350 g) were wrapped in a homeothermic blanket (EKEG Electronics, Vancouver, BC) and placed in a stereotaxic frame (David Kopf Instruments, Tujunga, CA) with the upper incisor bar raised 5 mm above the interaural line.[12] Holes were drilled through the skull. Electrical contact between brain tissue and a Ag/AgCl reference electrode was established with a salt bridge.

Dopamine was measured at two ipsilateral sites in the striatum. The posterior site was 1.8 mm anterior to bregma and 2.5 mm lateral from midline. The anterior site was 2.8 mm anterior to bregma and 2.5 mm lateral from midline. These sites are placed 1 mm apart because several studies show that tissue within 1 mm of microdialysis probes is disrupted by traumatic injury and the depletion of dialyzable substances.[4–6]

Except where noted, a microcylinder electrode was placed 4.5 mm below dura at the posterior site. Then, either a second microelectrode or a dialytrode was placed at the anterior site. A stainless steel bipolar stimulating electrode (MS303/1; Plastics One, Roanoke, VA) was placed in the ipsilateral MFB. The stimulus waveform (a constant current, biphasic square wave with a pulse width of 2 ms, a pulse height of 280 μA, a frequency of 45 Hz, and a train length of 25 s) was optically isolated from the potentiostat (Neurolog 800; Medical Systems, Greenvale, NY). Experiments were continued only when stable voltammetric responses of at least 30 nA in amplitude were obtained. The microelectrode at the posterior site remained in place throughout all experiments.

Dialytrodes were mounted on a high-resolution stereotaxic manipulator and slowly lowered over a time period of 20–30 min to 7.0 mm below dura. The probes were perfused in-place for a minimum of 2 h before experiments began. Previously, we showed that this implantation procedure produces a stable, impulse dependent, nanomolar-level of dopamine in basal dialysate samples.[13] Unless noted otherwise, stimulation was delivered to the MFB at 20-min intervals.

Dual Electrode Voltammetry

Experiments involving simultaneous voltammetry at both sites were conducted according to two designs. First, microelectrodes were placed 4.0 mm below dura at both the posterior and anterior sites and were subsequently lowered 0.5 mm between consecutive stimuli to a final depth of 6.0 mm below dura. Second, microelectrodes were placed at a fixed depth of 4.5 mm below dura. Nomifensine, L-dihydroxyphenylalanine (L-DOPA), pargyline, or sulpiride were administered after the third stimulus and post-drug stimulus responses were recorded at the intervals indicated.

Voltammetry near Microdialysis Probes

Experiments involving voltammetry in the presence of a dialytrode were conducted according to each of the following designs.

Design 1: Initially, stimulus responses were recorded with microelectrodes placed at both the posterior and anterior sites. The anterior microelectrode was then removed and replaced with a dialytrode, which was perfused in-place for 2 h before MFB stimulation was continued. Responses were recorded at the posterior, adjacent, and outlet microelectrodes. Nomifensine was administered systemically after the third stimulus, and post-drug responses were recorded 20 min later.

Design 2: A microelectrode was placed at the posterior site and a dialytrode was placed at the anterior site. The dialytrode was perfused in-place for 16 h, during which the rat received injections of glucose in saline (0.5 mL, 5 mM, i.p.) every 2–3 h, before MFB stimulation was continued. Responses were recorded at the posterior, adjacent, and outlet microelectrodes. Nomifensine was administered systemically after the third stimulus, and post-drug responses were recorded 20 min later.

Design 3: A microelectrode was placed at the posterior site and a dialytrode was placed at the anterior site. The dialytrode was perfused in-place for 2–5 h, after which MFB stimulation continued. Responses were recorded at the posterior, adjacent, and outlet microelectrodes. Nomifensine, L-DOPA, pargyline, or sulpiride was administered systemically after the third stimulus and post-drug responses were recorded at the intervals indicated. Each rat received a single drug.

All drugs were administered by intraperitoneal injection in phosphate-buffered saline vehicle at the following doses: nomifensine (20 mg/kg); L-DOPA (250 mg/kg administered 30 min after 150 mg/kg carbidopa); pargyline (75 mg/kg); sulpiride (100 mg/kg). The drugs were obtained from Sigma (St Louis, MO).

RESULTS

Dual Electrode Recording

Stimulus responses at the posterior and anterior sites were characterized by simultaneous voltammetry in a total of 15 rats. The depth profiling experiment was performed in three of these rats. Figure 22.1 shows that in one rat a robust stimulus response was observed at all vertical electrode

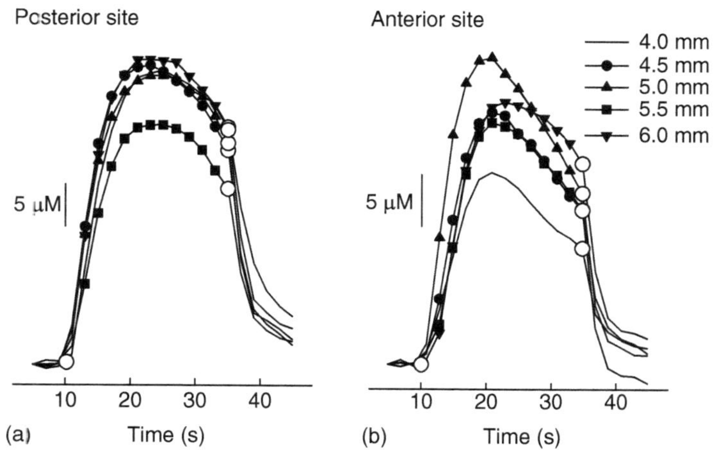

FIGURE 22.1 Stimulus responses recorded with a pair of voltammetric microelectrodes at the posterior (a) and anterior (b) sites in striatum. The open circles mark the beginning and end of the stimulus. Robust stimulus responses were observed at all vertical positions tested at both the posterior and anterior sites.

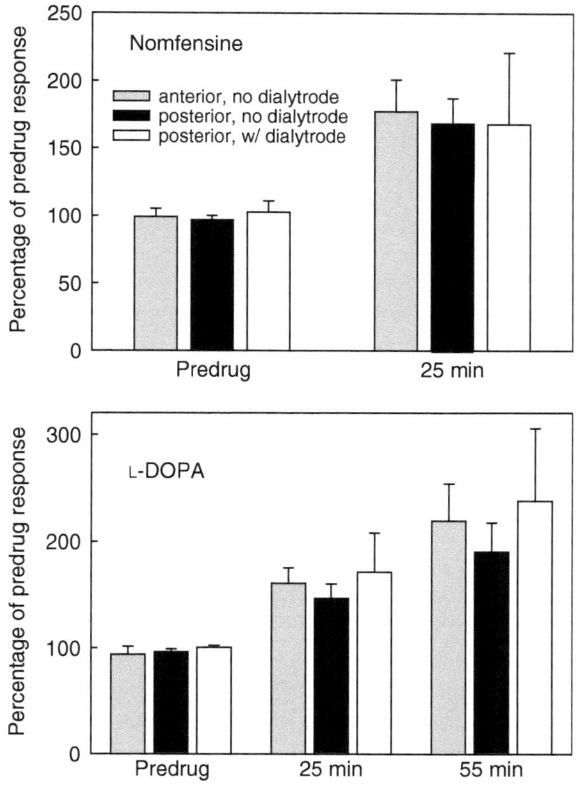

FIGURE 22.2 The effect of nomifensine, L-DOPA, pargyline, and sulpiride on the relative amplitude of the stimulus response obtained at the anterior site in the absence of a dialytrode (gray bars), the posterior site in the absence of a dialytrode (black bars), and the posterior site in the presence of a dialytrode at the anterior site (white bars). The bars report the mean ± s.d. of three observations. Except in the case of sulpiride, there were no significant differences between the results obtained at the three types of recording location (statistical analysis discussed in the text).

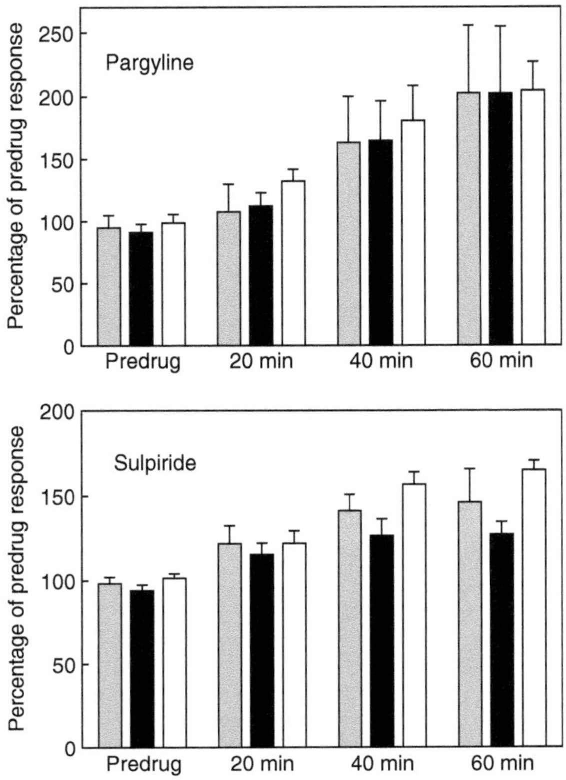

Figure 22.2 *(continued)*

placements examined at both the posterior and anterior sites, which is typical of the results obtained in all three rats. At a fixed depth of 4.5 mm below dura, the response amplitudes observed at the posterior and anterior sites in 12 rats corresponded to 17.9 ± 8.76 μM and 15.9 ± 7.39 μM (mean ± s.d.), respectively, which are not significantly different according to the *t*-test. The black and gray bars in Figure 22.2 show the relative effects of four drugs on the response amplitude at the posterior and anterior sites, respectively.

Probe Implantation Affects the Voltammetric Responses

Experiments were performed according to Designs 1–3 of Voltammetry Near Microdialysis Probes to examine the impact of probe implantation on the general features of the voltammetric stimulus responses at the two sites within the striatum.

Design 1: Figure 22.3 illustrates the effect of probe implantation on evoked responses observed at the posterior and anterior site in a single rat. Robust responses were evoked at both the posterior and anterior sites prior to implantation of a dialytrode. A robust response was still observed at the posterior site 2 h after the anterior microelectrode had been replaced with a dialytrode. Nevertheless, 2 h after implantation of the dialytrode no evoked response was detectable at the anterior site, i.e., at the microelectrode placed immediately adjacent to the microdialysis probes.

The response amplitude at the posterior site 2 h after dialytrode implantation (12.5 ± 7.51 μM, mean ± s.d., $n = 12$) was not significantly different than the amplitudes (reported above) obtained at the posterior and anterior sites without implantation of a dialytrode (one-way ANOVA: $f = 1.09$; d.f. = 2,33).

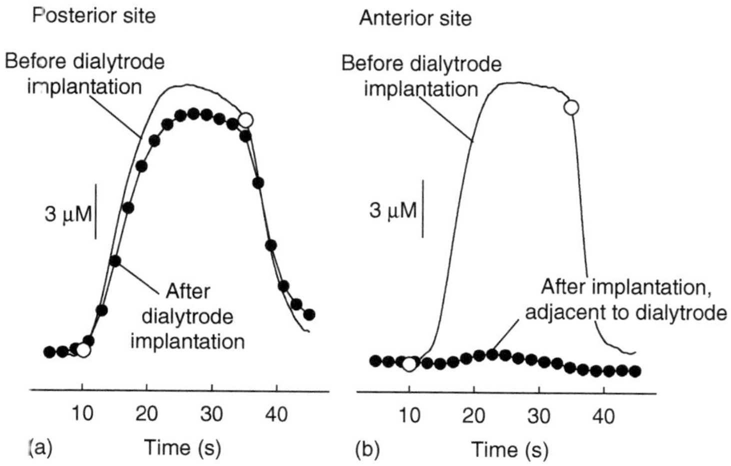

FIGURE 22.3 Stimulus responses recorded with a voltammetric microelectrode at the posterior (a) and anterior (b) sites just before and 2 h after implantation of a dialytrode at the anterior site. At the posterior site, the stimulus response was not significantly affected by dialytrode implantation at the anterior site. At the anterior site, evoked release was readily observed prior to dialytrode implantation, but was below voltammetric detection limits at the microelectrode adjacent to the microdialysis probe 2 h after dialytrode implantation.

In a group of three rats, the experiment of Figure 22.3 was continued after the systemic administration of nomifensine. Figure 22.4 shows that the response amplitude at the posterior site increased after nomifensine administration. Moreover, evoked responses were detectable at the adjacent and outlet microelectrodes after nomifensine administration. Equivalent results were obtained in all three rats in which this protocol was used.

Design 2: Stimulus responses were recorded at the posterior, adjacent, and outlet microelectrodes 16 h after implantation of the dialytrodes (Figure 22.5). Before nomifensine

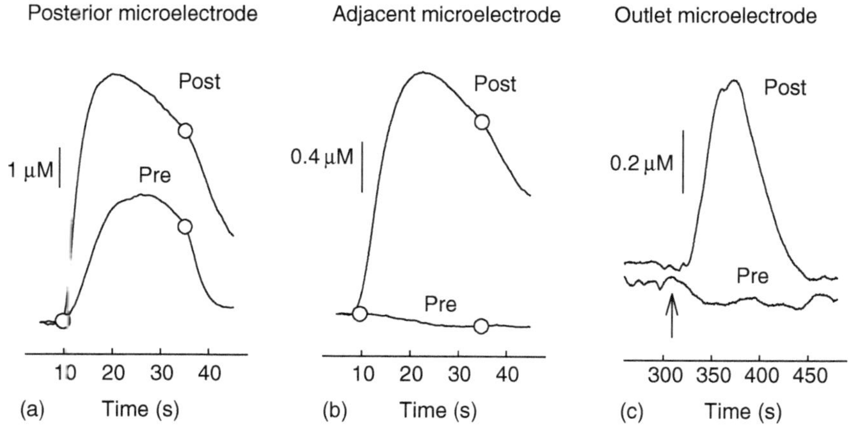

FIGURE 22.4 Pairs of stimulus responses recorded simultaneously with voltammetric microelectrodes at the posterior site (a), adjacent to a microdialysis probe (b), and at the outlet of a microdialysis probe (c). The experiment was initiated 2 h after implantation of the dialytrode at the anterior site. The responses were recorded before (pre) and 25 min after (post) systemic administration of nomifensine (20 mg/kg, i.p.). The open circles show the beginning and end of the stimulus. The vertical arrow in (c) indicates the time after the start of the stimulus the response was expected at the outlet electrode, based on the delay time measured during probe precalibration. The vertical scale bars were determined by post-calibration of each microelectrode.

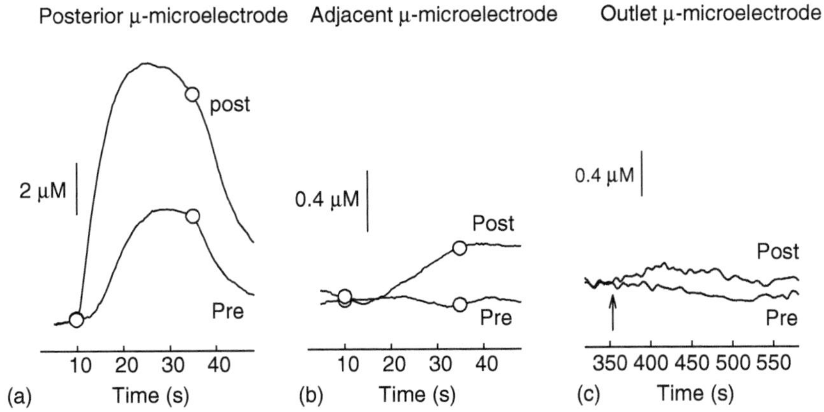

FIGURE 22.5 Results presented in the same format as in Figure 22.5 except that the experiment was initiated 16 h after implantation of the voltammetric electrode at the posterior site and the dialytrode at the anterior site. The responses recorded at the posterior site were not significantly different from those observed 2 h after dialytrode implantation, whereas the responses at the adjacent and outlet microelectrodes were significantly reduced (statistical analysis described in the text).

administration, the response amplitude at the posterior site was slightly but not significantly (t-test) smaller at 16 h after implantation (9.1 ± 3.1 μM, mean ± s.d., $n = 3$) than at 2 h (12.5 ± 7.51 μM, mean ± s.d., $n = 12$). The relative increase in the amplitude induced by nomifensine at the posterior site was not significantly different (t-test) at 16 h (196 ± 40%, mean ± s.d., $n = 3$) and at 2 h (164 ± 50%, mean ± s.d., $n = 3$). Prior to nomifensine, evoked responses were again not detected at the adjacent and outlet microelectrodes at 16 h post implantation. At the adjacent microelectrode, the response observed after nomifensine was significantly smaller at 16 h than at 2 h: 1.09 ± 0.41 μM at 16 h vs. 2.94 ± 0.64 μM at 2 h (mean ± s.d., $n = 3$, $p < .015$, t-test). The response at the outlet electrode at 16 h was too small to reliably quantify.

Design 3: Stimulus responses were recorded simultaneously at the posterior and adjacent microelectrodes before and after the administration L-DOPA, pargyline, or sulpiride (Figure 22.6). These experiments were initiated 2 h after implantation of the dialytrodes. Each drug was examined in a different group of three rats. Even though each drug increased the evoked response at the posterior site, evoked responses at the anterior and outlet microelectrodes remained undetectable (outlet microelectrode data omitted). The white bars in Figure 22.2 show the effect of each drug on the relative amplitude of the stimulus response at the posterior site after dialytrode implantation. The results for each drug were analyzed by two-way ANOVA with a repeated measures design. The main effects were the time with respect to drug administration and the recording site. The effect of time was significant in all cases (nomifensine, $f = 36.2$, d.f. = 1,12, $p < .001$: L-DOPA, $f = 34.0$, d.f. = 2,18, $p < .001$: pargyline, $f = 26.0$, d.f. = 3,24, $p < .001$: sulpiride, $f = 55.9$, d.f. = 3,24, $p < .001$). The effect of recording site was significant only in the case of sulpiride ($f = 15.8$; d.f. = 2,24; $p < .001$). Interactions between time and recording site were not significant.

DISCUSSION AND IMPLICATIONS

Several lines of evidence suggest that the tissue surrounding a microdialysis probe is disrupted from its normal state.[1–6] Hence, it is relevant to ask how the trauma affects the dopaminergic system. Voltammetry in conjunction with carbon fiber microelectrodes presents the opportunity for direct observations, since the electrodes are small enough in their dimensions to be placed with the

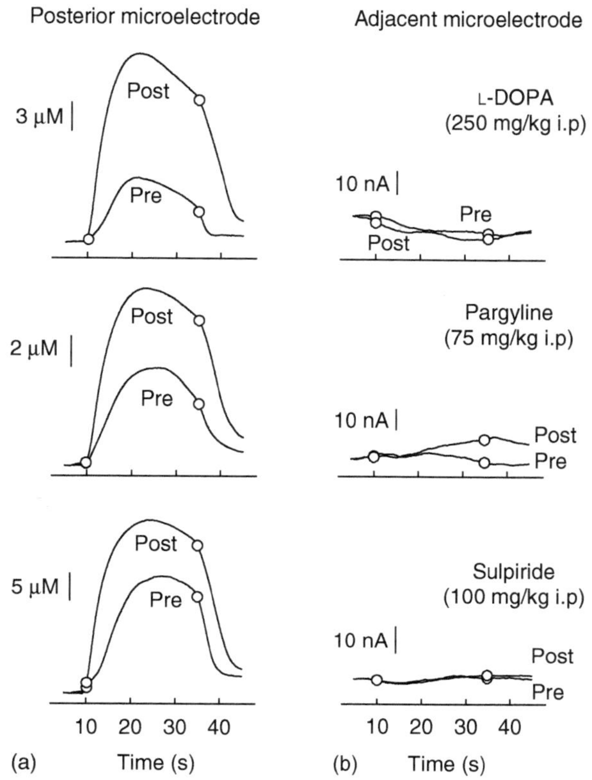

FIGURE 22.6 Pairs of stimulus responses recorded simultaneously with voltammetric microelectrodes at the posterior site (a) and adjacent to a microdialysis probe at the anterior site (b). Responses were recorded before (pre) and after (post) the systemic administration of L-DOPA (top row), pargyline (middle row), or sulpiride (bottom row). During these experiments responses were also recorded at the probe outlet but, since no evoked response was evident, the data are omitted for the sake of clarity.

probe's trauma layer. The results presented here were collected in the striatum of chloral hydrate anesthetized rats, since there is an indication that the anesthetized preparation may minimize implant trauma. The use of anesthesia makes it possible to implant microdialysis probes very slowly[14–16] and makes it possible to keep the animal completely immobilized thereafter. This produces a stable, highly impulse dependent baseline of nanomolar dopamine levels in dialysate obtained 2 h after implantation of the probe.[13] The rapid attainment of a stable, impulse dependent baseline level of dialysate dopamine strongly implies that the procedures followed during this study permit the tissue surrounding the probes to stabilize quickly after probe implantation. Nevertheless, as a consequence of the decision to use anesthetized animals for this study, the outcome of this initial investigation of the feasibility of a voltammetric approach to the in vivo calibration of microdialysis is relevant only to anesthetized conditions.

Measurements of extracellular dopamine in the striatum were performed at two sites placed 1 mm apart, a distance selected on the basis of recent estimates of the physical extent of probe-induced tissue disruption.[4–6] First, voltammetric stimulus responses were recorded with microcylinder carbon fiber electrodes at both sites. At several depths of penetration beneath the brain surface, responses were consistently detected at both sites (Figure 22.1). At a fixed depth of 4.5 mm below dura the magnitude of the pre-drug evoked responses at the two sites were not statistically different. Furthermore, there was no statistical difference between the effects of four dopaminergic drugs on the amplitude of the stimulus responses at the two sites (Figure 22.2).

The similarity between the responses obtained at the two sites, despite the existence of local heterogeneity in striatal tissue,[17–20] can be attributed to the use in this work of microcylinder electrodes, which by virtue of their length, measure the spatial average of extracellular dopamine concentrations.[21] Overall, voltammetric responses observed with these microcylinder electrodes at the posterior and anterior sites are statistically indistinguishable. Local heterogeneities in the evoked dopamine response, therefore, are excluded as a possible explanation for the differences in results between electrodes adjacent to the probes and 1 mm from the probe.

For the most part, stimulus responses at the posterior site 1 mm from the site of probe implantation were unaffected by implantation of a dialytrode at the anterior site. The amplitudes of the responses observed at the posterior site (before and after probe implantation) were not statistically different. Also, the effects of nomifensine, L-DOPA, and pargyline on the response amplitude were not changed by probe implantation (Figure 22.2). Implantation of the dialytrode affected the results obtained after sulpiride, however, which suggests that probe implantation may have induced some disruption at the posterior site. This effect was so slight (Figure 22.2), however, that it appears reasonable to conclude that evoked responses recorded 1 mm away from a microdialysis probes provide an adequate representation of the evoked responses expected in the absence of probe-induced tissue disruption.

Overall, the results in Figure 22.1 and Figure 22.2 confirm that evoked responses at the two sites are inherently similar. Moreover, the evoked responses observed at the posterior site 2 h after implantation of a microdialysis probe at the anterior site are inherently similar to those observed without probe implantation. Hence, the lack of a detectable response at microelectrodes placed immediately adjacent to microdialysis probes cannot be attributed to any inherent variability of evoked responses in the striatum. Rather, it seems more likely that the absence of a detectable response at microelectrodes placed immediately adjacent to microdialysis probes is a consequence of probe-induced tissue disruption. Figure 22.3, which shows that an evoked response was observed at the anterior site with a conventional voltammetric electrode just prior to implantation of a dialytrode at the same location, strongly supports this interpretation.

Consistent with previous related work,[1–3] evoked responses were detectable at microelectrodes placed adjacent to microdialysis probes after the systemic administration of the dopamine uptake inhibitor, nomifensine (Figure 22.4). Moreover, after uptake inhibition evoked responses were also detectable at the microelectrode placed at the outlet of the microdialysis probe (Figure 22.4). It appears from these results that uptake inhibition, which prolongs the lifetime of dopamine in the extracellular space, enables dopamine to diffuse further from sites of evoked release and thereby increases the amount of dopamine that diffuses to the probes. The appearance of an evoked response adjacent to the probe and at the probe outlet after uptake inhibition forms the basis of our suggestion that uptake inhibition increases the relative recovery of dopamine from the striatal extracellular space.

Because many studies report that microdialysis results are affected by the post implantation interval,[22–26] the experiment with nomifensine was also performed with a post implantation interval of 16 h. Without uptake inhibition, evoked responses remained undetectable at the microelectrodes adjacent to the probe and at the probe outlet. After uptake inhibition, the evoked responses at the adjacent microelectrodes observed 16 h after probe implantation were substantially smaller than those observed 2 h after probe implantation. Moreover, at 16 h after probe implantation, no post-nomifensine response was detected at the probe outlet. Together, these observations show that, despite the combined use of slow probe implantation and anesthetized animals, extending the post implantation interval does not eliminate the disparity between evoked responses recorded immediately adjacent to microdialysis probes and those recorded 1 mm from microdialysis probes. Indeed, the disparity after uptake inhibition is greater, leading us to wonder if the 24 h wait after probe implantation really constitutes a recovery period as it is often described.

Also, the systemic administration of drugs that increase the evoked response at conventional microelectrodes (Figure 22.6) was not effective at promoting voltammetric stimulus responses at the microelectrodes adjacent to microdialysis probes. This is consistent with the concept that uptake plays a central role in determining the distance that dopamine can diffuse from sites of evoked release. Although the other drugs studied here increase the amount of dopamine released during stimulation, they do not act to increase the extracellular lifetime of dopamine, and hence, do not increase the likelihood that dopamine will diffuse to the probe.

Overall, the results in Figure 22.1 through Figure 22.6 extend our earlier suggestion that due to the combined effects of tissue trauma and dopamine uptake, the microdialysis recovery of dopamine from the striatal extracellular space is inherently low. Dopamine release is suppressed in the tissue immediately adjacent to the probe, so normal dopamine release amplitude is observed only at several hundred micrometers from the probe.[3] Nevertheless, dopamine uptake appears to remain relatively intact, compared to release, next to the probe. Uptake inhibition has a far greater effect on the amplitude of evoked dopamine release when the microelectrode is in the trauma layer (Figure 22.1 and Ref. [3]). Once dopamine is released, it has to diffuse through the trauma layer in order to be recovered by the probe. It is highly likely that dopamine will be taken up before it reaches the probe unless the transporter is inhibited.[27] So, uptake inhibition increases dopamine recovery.

The conclusion that uptake inhibition increases dopamine recovery is supported by our observations of stimulus responses after the administration of drugs that increase the response amplitude in the absence of the probe. Three drugs that we examined, L-DOPA, pargyline, and sulpiride, all increased the response amplitude in the absence of the probe, and 1 mm from the probe, but none of these drugs caused detectable responses adjacent to the probe or at the probe outlet. These drugs affect the evoked response by enhancing dopamine release, in contrast to nomifensine, which slows dopamine uptake. Enhancing evoked release did not appear to have the ability to boost dopamine recovery, supporting our suggestion that uptake kinetics are a key determinant of recovery.

The results presented here lend further support to the idea that uptake lowers dopamine recovery, which is exactly the opposite of its effect on dopamine extraction.[28] Under conditions where the recovery is smaller than the extraction fraction, the concentration of no-net-flux underestimates extracellular concentrations.[3] The exact extent to which extracellular dopamine concentrations are underestimated remains unknown, but in recent studies we found evidence for a resting striatal dopamine concentration in the micromolar range.[29,30] At present, however, our micromolar estimate of extracellular concentration is difficult to resolve with the previous work of Gonon and Buda,[31] who estimated 25 nM in pargyline-treated rats. A potentially more important finding is that the recovery factor, which is key to the quantitative performance of the microdialysis probe, is sensitive to drugs that inhibit dopamine uptake.

Finally, it is interesting to note that the voltammetric recordings performed during this work at the outlet of the microdialysis probe correlated best with the voltammetric recordings performed immediately adjacent to the probe. The recordings at the outlet did not correlate with the recordings performed 1 mm away or in the absence of the probe. This observation strongly suggests that the dopamine content of the microdialysate samples most often reflects the dopamine content in the extracellular fluid in the tissue immediately adjacent to the probe, i.e., that tissue most profoundly traumatized by the probe. This emphasizes one of the key attributes of the carbon fiber microelectrode for in vivo measurements, which is that their small dimensions inflict minimal trauma at the implantation site.[7]

ACKNOWLEDGMENTS

This work was supported by NIH (Grant NS 31442) and by a grant from the National Parkinson's Foundation Center of Excellence at the University of Pittsburgh. Hua Yang was the recipient of a Graduate Summer Fellowship from the Division of Analytical Chemistry of the American Chemical Society.

REFERENCES

1. Lu, Y., Peters, J. L., and Michael, A. C., Direct comparison of the response of voltammetry and microdialysis to electrically evoked release of striatal dopamine, *J. Neurochem.*, 70, 584–593, 1998.
2. Yang, H., Peters, J. L., and Michael, A. C., Coupled effects of mass transfer and uptake kinetics on *in vivo* microdialysis of dopamine, *J. Neurochem.*, 71, 684–692, 1998.
3. Borland, L. M., Shi, G., Yang, H., and Michael, A. C., Voltammetric study of extracellular dopamine near microdialysis probes acutely implanted in the striatum of the anesthetized rat, *J. Neurosci. Methods*, 146, 149–159, 2005.
4. Benveniste, H. and Hansen, A. J., Practical aspects of using microdialysis for determination of brain interstitial concentrations, In *Techniques in the Behavioral and Neural Sciences*, Robinson, T. E. and Justice, J. B. Jr., Eds., *Microdialysis in the Neuroscience*, Vol. 7, Elsevier, Amsterdam, pp. 82–102, 1991.
5. Dykstra, K. H., Hsiao, J. K., Morrison, P. F., Bungay, P. M., Mefford, I. N., Scully, M. M., and Dedrick, R. L., Quantitative examination of tissue concentration profiles associated with microdialysis, *J. Neurochem.*, 58, 931–940, 1992.
6. Clapp-Lilly, K. L., Roberts, R. C., Duffy, L. K., Irons, K. P., Hu, Y., and Drew, K. L., An ultrastructural analysis of tissue surrounding a microdialysis probe, *J. Neurosci. Methods*, 90, 129–142, 1999.
7. Peters, J. L., Miner, L. H., Michael, A. C., and Sesack, S. R., Ultrastructure at carbon fiber microelectrode implantation sites after acute voltammetric measurements in the striatum of anesthetized rats, *J. Neurosci. Methods*, 137, 9–23, 2004.
8. Lindefors, N., Amberg, G., and Ungerstedt, U., Intracerebral microdialysis: I. Experimental studies of diffusion kinetics, *J. Pharmacol. Methods*, 22, 141–156, 1989.
9. Bungay, P. M., Morrison, P. F., and Dedrick, R. L., Steady-state theory for quantitative microdialysis of solutes and water *in vivo* and *in vitro*, *Life Sci.*, 46, 105–119, 1990.
10. Morrison, P. F., Bungay, P. M., Hsiao, J. K., Mefford, I. N., Dykstra, K. H., and Dedrick, R. L., Quantitative microdialysis, In *Techniques in the Behavioral and Neural Sciences*, Robinson, T. E. and Justice, J. B. Jr., Eds., *Microdialysis in the Neurosciences*, Vol. 7, Elsevier, Amsterdam, pp. 47–80, 1991.
11. Parsons, L. H. and Justice, J. B. Jr., Quantitative approaches to in vivo microdialysis, *Crit. Rev. Neurobiol.*, 8, 189–220, 1994.
12. Pellegrino, L. J., Pellegrino, A. S., and Cushman, A. J., *A Stereotaxic Atlas of the Rat Brain*, 2nd ed., Plenum Press, New York, 1979.
13. Qian, J., Wu, Y., Yang, H., and Michael, A. C., An integrated decoupler for capillary electrophoresis with electrochemical detection: application to analysis of brain microdialysate, *Anal. Chem.*, 71, 4486–4492, 1999.
14. DeBoer, P. and Abercrombie, E. D., Physiological release of striatal acetylcholine in vivo: Modulation by D_1 and D_2 dopamine receptor subtypes, *J. Pharmacol. Exp. Ther.*, 277, 775–783, 1996.
15. Bickerdike, M. J. and Abercrombie, E. D., Striatal acetylcholine release correlates with behavioral sensitization in rats withdrawn from chronic amphetamine, *J. Pharmacol. Exp. Ther.*, 282, 818–826, 1997.
16. Abercrombie, E. D. and DeBoer, P., Substantia nigra D_1 receptors and stimulation of striatal cholinergic interneurons by dopamine: a proposed circuit mechanism, *J. Neurosci.*, 17, 8498–8505, 1997.
17. Jimenez-Castellanos, J. and Graybeil, A. M., Subdivisions of the dopamine-containing A8–A9–A10 complex identified by their differential mesostriatal innervation of striasomes and extrastriasomal matrix, *Neuroscience*, 23, 223–242, 1987.
18. Loopuijt, L. D., Sebens, J. B., and Korf, J., A mosaic-like distribution of dopamine receptors in rat neostriatum and its relationship to striasomes, *Brain Res.*, 405, 405–408, 1987.
19. Marshall, J. F., O'Dell, S. J., Navarrete, R., and Rosenstein, A. J., Dopamine high-affinity transport site topography in rat brain: major differences between dorsal and ventral striatum, *Neuroscience*, 37, 11–21, 1990.
20. Garris, P. A., Ciolkowski, E. L., and Wightman, R. M., Heterogeneity of evoked dopamine overflow within the striatal and striatoamygdaloid regions, *Neuroscience*, 59, 417–427, 1994.

21. Peters, J. L. and Michael, A. C., Changes in the kinetics of dopamine release and uptake have differential effects on the spatial distribution of extracellular dopamine concentration in rat striatum, *J. Neurochem.*, 74, 1563–1573, 2000.
22. Benveniste, H., Drejer, J., Schousboe, A., and Diemer, N. H., Regional cerebral glucose phosphorylation and blood flow after insertion of a microdialysis fiber through the dorsal hippocampus in the rat, *J. Neurochem.*, 49, 729–734, 1987.
23. Westerink, B. H. C. and De Vries, J. B., Characterization of in vivo dopamine release as determined by brain microdialysis after acute and subchronic implantations: Methodological aspects, *J. Neurochem.*, 51, 683–687, 1988.
24. Santiago, M. and Westerink, B. H. C., Characterization of the *in vivo* release of dopamine as recorded by different types of intracerebral microdialysis probes, *Naunyn–Schmiedeberg's Arch. Pharmacol.*, 342, 407–414, 1990.
25. Robinson, T. E. and Camp, D. M., The feasibility of repeated microdialysis for within subjects design experiments: Studies on the mesostriatal dopamine system, In *Techniques in the Behavioral and Neural Sciences*, Robinson, T. E. and Justice, J. B. Jr., Eds., *Microdialysis in the Neurosciences*, Vol. 7, Elsevier, Amsterdam, pp. 189–234, 1991.
26. Moghaddam, B. and Bunney, B. S., Depolarization inactivation of dopamine neurons: Terminal release characteristics, *Synapse*, 14, 195–200, 1993.
27. Yang, H., Peters, J. P., Allen, C., Chern, S.-S., Coalson, R. D., and Michael, A. C., A theoretical description of microdialysis with mass transport coupled to chemical events, *Anal. Chem.*, 72, 2042–2049, 2000.
28. Smith, A. D. and Justice, J. B. Jr., The effect of inhibition of synthesis, release, metabolism and uptake on the microdialysis extraction fraction of dopamine, *J. Neurosci. Methods*, 54, 75–82, 1994.
29. Kulagina, N. V., Zigmond, M. J., and Michael, A. C., Glutamate regulates the spontaneous and evoked release of dopamine in the rat striatum, *Neuroscience*, 102, 121–128, 2001.
30. Borland, L. M. and Michael, A. C., Voltammetric study of the control of striatal dopamine release by glutamate, *J. Neurochem.*, 91, 220–229, 2004.
31. Gonon, F. G. and Buda, M. J., Regulation of dopamine release by impulse flow and by autoreceptors as studied by in vivo voltammetry in the rat striatum, *Neuroscience*, 14, 765–774, 1985.

Index

I

K

L

M

Q

R

S

T

U

V

W

Z